Design Systems for VLSI Circuits

NATO ASI Series

Advanced Science Institutes Series

*A Series presenting the results of activities sponsored by the NATO Science
Committee, which aims at the dissemination of advanced scientific and technological
knowledge, with a view to strengthening links between scientific communities.*

The Series is published by an international board of publishers in conjunction with the
NATO Scientific Affairs Division

A	Life Sciences	Plenum Publishing Corporation
B	Physics	London and New York
C	Mathematical and Physical Sciences	D. Reidel Publishing Company Dordrecht, Boston, Lancaster and Tokyo
D	Behavioural and Social Sciences	Martinus Nijhoff Publishers Dordrecht, Boston and Lancaster
E	Applied Sciences	
F	Computer and Systems Sciences	Springer-Verlag Berlin, Heidelberg, New York
G	Ecological Sciences	London, Paris and Tokyo
H	Cell Biology	

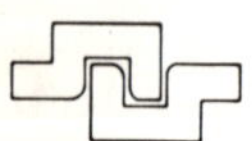

Series E: Applied Sciences – No. 136

Design Systems for VLSI Circuits

Logic Synthesis and Silicon Compilation

edited by

G. De Micheli

Stanford University
Standford, California
USA

A. Sangiovanni-Vincentelli

University of California
Berkeley, California
USA

P. Antognetti

University of Genova
Genova
Italy

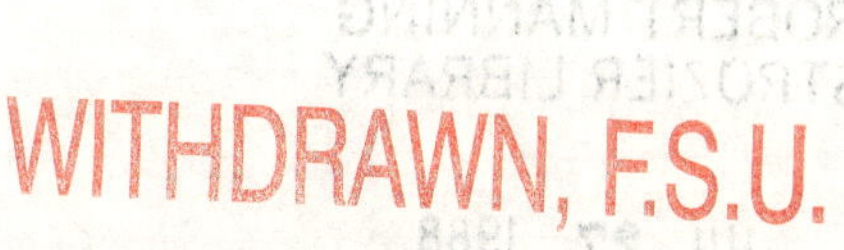

1987 Martinus Nijhoff Publishers
Dordrecht / Boston / Lancaster
Published in cooperation with NATO Scientific Affairs Division

Proceedings of the NATO Advanced Study Institute on "Logic Synthesis and Silicon Compilation for VLSI Design", L'Aquila, Italy, July 7-18, 1986

Library of Congress Cataloging in Publication Data

NATO Advanced Study Institute on "Logic Synthesis and
    Silicon Compilation for VLSI Design" (1986 : L'Aquila,
    Italy)
    Design systems for VLSI circuits.

    (NATO Advanced Science Institutes series. Series E,
Applied sciences ; v. 136)
    "Proceedings of the NATO Advanced Study Institute on
'Logic Synthesis and Silicon Compilation for VLSI
Design,' L'Aquila, Italy, July 7-18, 1986"--T.p. verso.
    Includes index.
    1. Integrated circuits--Very large scale integration--
Congresses. 2. Silicon compilers--Congresses.
I. De Micheli, G. II. Sangiovanni-Vincentelli, Alberto.
III. Antognetti, Paolo. IV. Title. V. Series.
TK7874.N338 1987   621.381'73   87-14188

ISBN 90-247-3561-0 (hardback)
ISBN 90-247-2689-1 (series)

Distributors for the United States and Canada: Kluwer Academic Publishers,
P.O. Box 358, Accord-Station, Hingham, MA 02018-0358, USA

Distributors for the UK and Ireland: Kluwer Academic Publishers, MTP Press Ltd,
Falcon House, Queen Square, Lancaster LA1 1RN, UK

Distributors for all other countries: Kluwer Academic Publishers Group, Distribution
Center, P.O. Box 322, 3300 AH Dordrecht, The Netherlands

Printed in The Netherlands

PREFACE

Synthesis is a crucial component of future CAD systems.
The competitive edge of IC design will most probably come
from the use of effective synthesis tools.

A complete synthesis system should generate layout masks
from a high-level algorithmic, behavioral or functional description of a VLSI system, a description of the target technology and a description of the constraints and cost functions. The design should be completed in reasonable time and
with the quality a human designer could obtain.

Working designs have been produced with silicon compilers,
but the quality of the design has always been a problem.
While for a restricted class of designs, such as Digital
Signal Processors (DSP), the use of a fixed floor-plan has
been successful, its use for less constrained applications
results in inefficient utilization of area and poor performance. In addition, the structure of the control logic is often too rigid and not optimized, thus yielding a slow and
large chip. The present trend is to break the synthesis process into stages, and to use tools that optimize real estate
and/or performance to go from one stage to the next.

This book covers most of the topics in the design of digital VLSI circuits, and focuses on theory, as well as algorithms and computer implementations of design systems.

The book covers in depth the three basic components of a
synthesis system:
Physical synthesis or layout, including floor-planning, par-

titioning, placement, routing and compaction; Logic synthesis, including combinational logic, sequential logic and algorithmic or behavioral synthesis; Procedural design and module generation.

The authors of the different chapters represent a well-balanced blend of academic and industrial backgrounds. In this way, the readers have a unique opportunity to obtain a full description of the present (industrial) and future (academic) concepts in silicon compilation for VLSI design.

The content of this book results from the lecture notes handed out at the NATO Advanced Study Institute (ASI) held at the Scuola Superiore G. Reiss Romoli (SSGRR), L'Aquila (Italy), from July 7 to July 18, 1986.

We would like to thank Dr. C. Sinclair of NATO's Scientific Affairs Division for his helpful assistance in organizing the ASI, and the lecturers for their presentations and for the careful preparation of the text.

P. Antognetti

G. De Micheli

A. Sangiovanni Vincentelli

TABLE OF CONTENTS

VIII

SYNTHESIS OF LSI CIRCUITS

Alberto Sangiovanni-Vincentelli
Department of Electrical Engineering and Computer Sciences
University of California at Berkeley

1. INTRODUCTION

Computer aids have been used for both the design and verification of electronic systems for many years prior to the introduction of commercial Integrated Circuits (ICs) in the early 1960s. Such tools have found their way into virtually every aspect of the design of such systems, from integrated circuit process technology to the design of complex computer architectures. The recent explosion in the complexity of electronic systems that the advent of Very Large Scale Integration (VLSI) has allowed, has made the use of sophisticated computer-aided design tools indispensable. Not only are computer aids necessary for both the design and verification of integrated circuits today but, as the semiconductor processing technologies mature, computer aids will soon also provide key proprietary advantages as semiconductor and system design houses vie for the promising Application-Specific IC (ASIC) market of the next decade. We believe that the pivotal technologies in future IC CAD systems include

1- *tools for IC synthesis*, such as placement and routing, combinational and sequential logic synthesis tools, and architectural design aids,

2- *design system management tools*, including the management of design versions and alternatives in a distributed computing environment, data dependency management, and efficient and flexible interfaces to new tools,

3- *verification tools*, including physical and electrical rules checking, simulation, and formal verification techniques.

2. APPLICATIONS SPECIFIC INTEGRATED CIRCUITS AND CAD

As the competition for designs increases, driven by the increasing number of companies in the ASIC business and by the high capital cost of a modern IC processing facility, there is increasing demand for designers to be able to differentiate their IC product from that of their competitors. Higher performance, lower cost, more features, or a faster time to market are all major factors which differentiate IC products in the ASIC marketplace. In the past, different companies have been able to provide such product differentiation through their IC fabrication technologies: the ability to pack more transistors on a given chip or to provide a higher switching speed per gate drove the designs and their advantage in the marketplace. However, the silicon process technology is maturing rapidly — significant gains in performance and density are becoming increasingly expen-

2

sive and many companies are resorting to "joint ventures," often with former competitors in the United States, Europe, or Asia, to maintain their position in IC process technologies. Because of this decrease in the relative competitive advantage obtained from process technology, semiconductor companies and "silicon foundries" must emphasize other aspects of the design process if they are to compete effectively for the ASIC market. The two avenues available are in system architecture and computer-aided design. Among other tasks, system architecture involves matching the specifications at the system level with the available technology. This process is in general very expensive and requires complex trade-offs involving hundred of variables that only a few human designers can handle. Even these designers have great difficulties with the process. Computer-aided design offers a set of tools for designers that allows the reduction of design time and a better match between system specifications and technology by permitting a better and faster exploration of design alternatives.

The design task involves three major components: CAD programs, support for specific design styles, and support for component libraries. A lack of CAD support in any of these areas may result in a significant reduction in the competitiveness of the designers' final product. On the other hand, a significant proprietary advantage in any or all of these areas will maintain a companies' position as a force in the marketplace.

Among the CAD tools, with a few notable exceptions, design synthesis systems for ASIC designs are far less mature than the tools in the other two categories, and large gains in circuit efficiency and design time are still to be had. In addition, many of the state-of-the-art synthesis techniques involve far more insight into the problem to be solved and command of complex optimization techniques ("inspiration" to use a term introduced by R. Newton) than brute force programming techniques and implementation tricks ("perspiration" according to R. Newton) that often make the difference in layout editors and schematic data capture systems, and, as a result, can form the basis of a proprietary and differentiating technology for ASIC design. Techniques for efficient synthesis (system design, register-level design, logic design, placement, routing, and array compilation) will provide a major focus for both University research and Industrial competition over the coming years.

3. DESIGN METHODS AND DESIGN LIBRARIES

The use of a particular class of circuit structures is referred to as a *design method*, or design style, and while the development of new algorithms and techniques for CAD continues, a significant contribution to the design of VLSI circuits will continue to come from the development of new circuit design methods. However, while the implementation of a design method does not *require* the use of computer aids *per se*, the most successful design methods will be those designed to take maximum advantage of the computer in both the circuit design and verification phases. The design method must provide the *structure* necessary to use both human and computer resources effectively. For VLSI, this structure also provides the reduction in design complexity necessary to reduce design time and to ensure that the circuit function can be verified and the resulting circuit can be tested.

Design methods can be classified in four categories: programmable arrays (e.g. gate-arrays and sea-of-gates), standard-cell, macro-cell, and custom design.

A *programmable array is* a one- or two-dimensional array of repeated cells which can be customized by adding or deleting geometry from specific mask layers. Since a number of processing steps are completed prior to customization, the locations of components on those layers are independent of a particular circuit implementation. Examples of programmable arrays include the Gate-Array, Sea-of-gates [HUI85] and Read-Only Memory (ROM).

The *gate-array* (also referred to as master-slice, or uncommitted logic array) is by far the most common programmable array designed by computer. It is also the case that the computer aids for gate-array design are the most advanced and the most mature. In this approach, a two-dimensional array of replicated transistors is fabricated to a point just prior to the interconnection levels. A particular circuit function is then implemented by customizing the connections within each local group of transistors, to define its characteristics as a basic cell, and by customizing the interconnections between cells in the array to define the overall circuit. Generally a two-level interconnection scheme is used for signals and, in some approaches, a third, more coarsely defined layer of interconnections is provided for power and ground connections. Because one or more interconnection layers are used within a group of transistors to define the function of a cell, these intra-cell interconnections often block the passage of more global inter-cell connections. For that reason, and to simplify the placement and routing problems associated with these arrays, the inter-cell connections are implemented on a rectilinear grid in the *channels* between the cells. In some cases, channels are also provided which run over the cells themselves and in some arrays, wider channels are provided in the center of the array to alleviate the congestion often found in that area if particular routing strategies are employed. Figure 1 shows the architecture of a gate-array chip.

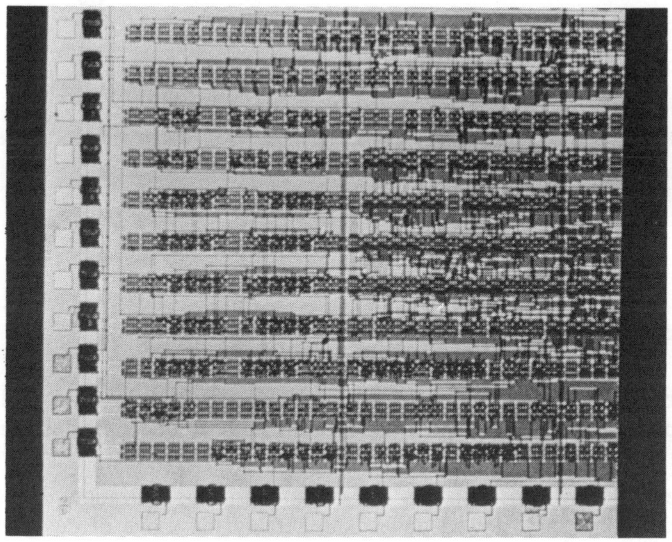

Figure 1. A section of a gate-array chip.

As processing technology advances and additional levels of high-quality interconnect are provided, the need for pre-determined wiring channels is less clear. The gate-array placement and routing problem becomes much closer to that of a printed circuit board, where routing is possible over all cells, and the individual transistors can be packed closely together. This recent variation of the gate-array structure is often referred to as a *sea-of-gates* array[HUI85] and, as well as providing new challenges for the CAD community, promises to replace many of the designs previously undertaken in the conventional gate-array style.

There are two basic architectures that have been proposed for this design style: one often called *sea-of-transistors* and the other called *channelless gate-array*.

Figure 2 shows a channelless gate-array, a channelless gate-array designed by California Devices Inc. Note that the routing area is not organized *a priori* into channels as in the case of standard gate-arrays as the one shown in Figure 1. Note also that the devices occupy the entire area of the chip. In Figure 2, the basic building blocks are similar to the ones used in standard gate-arrays. A similar architecture is used by Hughes Aircraft.

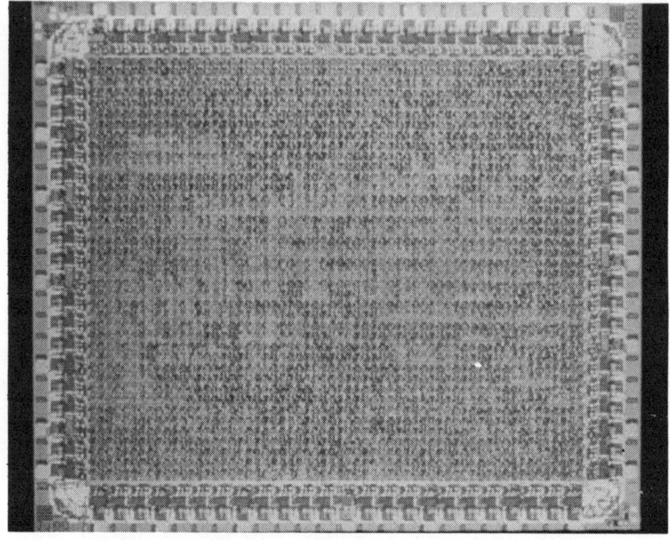

Figure 2. A sea-of-gate chip (courtesy of CDI).

In the sea-of-transistor architecture, transistors are arranged in rows (columns) of complementary pairs. Figure 3 shows a sea-of-transistors architecture designed at LSI Logic.

In both cases, two level metal layers are available for customization. The introduction of these architectures has slowed down considerably the move of designer towards less structured approaches to ASIC such as standard cells or macro-cells. While we believe that the macro-cell design style to be described below still maintains certain advantages over the sea-of-gates approach, the future of the standard-cell design style is

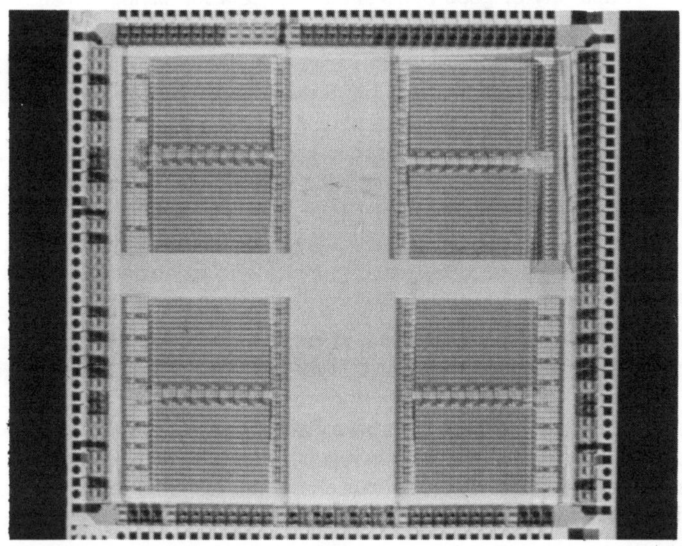

Figure 3. A sea-of-gate chip with a RAM (courtesy of LSI Logic).

more uncertain if we exclude the use of standard-cell blocks in a macro-cell design style floor-plan.

Because of the very fine granularity of the sea-of-transistors architecture, it is very convenient to use macros that implement a given logic function in an optimized collection of transistors. There are two basic approaches followed to cope with the sea-of-gates design problems. The more conventional one is based on pre-designed macros that make use of a certain number of transistors arranged in rows (or columns) where all macros have the same height and different width. In the other approach, macros can have any different aspect ratio and shape and may involve neighboring rows (columns) of transistors. Because this design style is in its infancy, libraries are not reach and effective in terms of area utilization as they could be. Some of the vendors translate existing standard-cell libraries into macros either by hand or using automatic tools[ROW86]. Given the tight arrangement of the transistors it is possible to design large macros effectively with little area penalty. For example, it is possible to design RAM blocks using the transistor array (Figure 3). This capability makes the sea-of-gates approach competitive with the standard-cell and macro-cell design approach, but makes the automatic generation of chips more difficult than in the standard gate-array design style.

If regular macros are used, the interconnections inside the transistor array to implement the macro, are realized with the first metal layer. Then to make the interconnections of different macros possible, routing regions are carved out of the array where active area is traded-off with routing area. The routing regions do not have blockages on the two interconnection layers and can be used as in the case of the standard-cell layout style, except that the width of the channel can take only a limited set of values corresponding to one or more rows of transistors being used for routing region and hence unused as active area. Note that in this case standard-cell placement and routing tools can be adapted to the sea-of-gates design style rather easily. I suspect that the macro libraries presently offered reflect the availability of good place and route tools for standard cells.

On the other hand, I believe that macros could be designed more effectively making use of the full structure of the array. In this case, however, the standard-cell place and route tools could not be used. In my opinion, the presence of large macros not arranged in rows and/or columns make the placement and routing problem of the sea-of-gates approach very similar to the macro-cell layout problem. For this reason, I expect the development of new tools that will solve both the sea-of-gates and the macro-cell layout problem. In addition, I expect the use of three layer of metal interconnections to be introduced soon in this design style. The active area utilization could then be made much higher, but the wiring problem becomes more complex and new routing tools will have to be developed.

The *standard cell* (or polycell) approach refers to a design method where a library of custom-designed cells is used to implement a logic function. These cells are generally of the complexity of simple logic gates or flip-flops and may be restricted to constant height and/or width to aid packing and ease of power distribution. The interconnections among cells are implemented in dedicated routing regions between two rows of cells and are called channels. Figure 4 shows the echo-canceller chip designed at AT&T Bell Laboratories, where the cells are organized in columns and a large macro has been inserted.

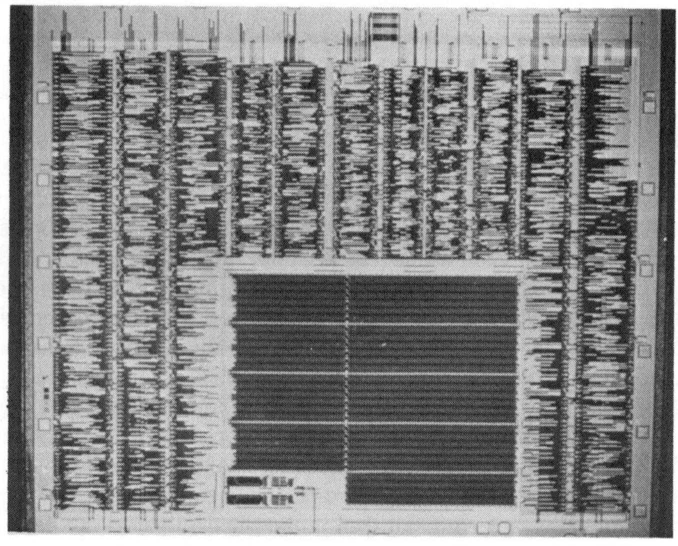

Figure 4. The AT&T echo canceller chip (courtesy of AT&T Bell Labs).

Nowadays state-of-the-art standard cell systems permit cells of different height *and* width to be included in the same design. This results in non-uniform routing channels between adjacent rows and requires a more sophisticated channel routing capability if the silicon area is to be used to its maximum efficiency. Unlike the programmable array approach, standard cell layout involves the customization of all mask layers. This additional freedom permits to tune finely the width of the channels as opposed to the gate-array approach where the channels are of fixed size and to the sea-of-gates approach in use today where the channel width is not fixed, but must vary by a fixed amount determined by the number of tracks that can be accommodated in two adjacent rows of complementary devices.

While most standard cell systems only permit inter-cell wiring in the channels between rows of cells or through cells via pre-determined "feed-through" cells, some systems permit over-cell routing if additional levels of interconnect are available. Standard cell systems are also used extensively in a variety of technologies including bipolar and CMOS.

It is often relatively inefficient to implement all classes of logic functions in a single design approach. For example, a standard cell approach is inefficient for memory circuits such as RAM and stack. In the *macro-cell*, or building block, method, large circuit blocks, customized to a certain type of logic function, are available in a circuit library. These blocks are of irregular size and shape and may allow functional customization via interconnect, such as a PLA or ROM macro, or they can be parameterized with respect to topology as well. With the parameterized cell, the number of inputs and outputs may be parameters of the cell. In some systems macro-cells may also be embedded in gate-array or standard-cell designs. Parameterization is also possible with respect to transistor sizes to optimize performances. In fact, this level of customization is the basic advantage of the macro-cell design style over the emerging sea-of-gates approach. However, the level of sophistication of the tools for the macro-cell design style is higher than the one needed for more regular design styles. Assuming that adequate tools will be soon available, the macro-cell floor-plan style will evolve as the floor-plan of choice for large, ASIC designs.

In the *custom-design* approach, all the mask geometries are designed *ad hoc* by human designers with little help from CAD tools. In general, layout editors such as CALMA systems are used to enter the mask geometries. This design method can be clearly highly optimized since there are no constraints on the positions of the devices nor on the components. For this reason, micro-processors and other high-volume complex chips were designed with this methodology. However, the design time grew more than linearly with the size of the design, as a consequence, design costs were too high to be economically viable and, more importantly, it was almost impossible to predict accurately when a design would have been ready for the market. This last consideration is of paramount importance when time-to-market is a crucial factor in guaranteeing the success of a product. For this reason, more structured techniques were used. For example, regularity was used to decrease the design time. In this way, large regular blocks were assembled by designing the basic cells and replicating the basic cells as many times as needed in rows and/or columns with the help of a layout editor.

More recently, micro-processors were designed using regular arrays for components such as RAM, ROMs, data-paths, and blocks of standard-cells or PLAs for control logic. The DEC mVAX CPU chip (Figure 5) was designed using regular arrays laid out in the full-custom design style. The floor-plan is similar to the one of a macro-cell chip.

The Intel 80386 (Figure 6) is an example of design using a hybrid between the standard-cell and the macro-cell approach. In the case of the Fairchild Clipper (Figure 7), the design included almost entirely blocks of standard cells, yielding the design approach known as *hierarchical standard cell*. I believe that this hybrid is a good approach today when the CAD tools for the macro-cell design style are still in their infancy. However, I

Figure 5. The CPU of the DEC mVAX (courtesy of DEC).

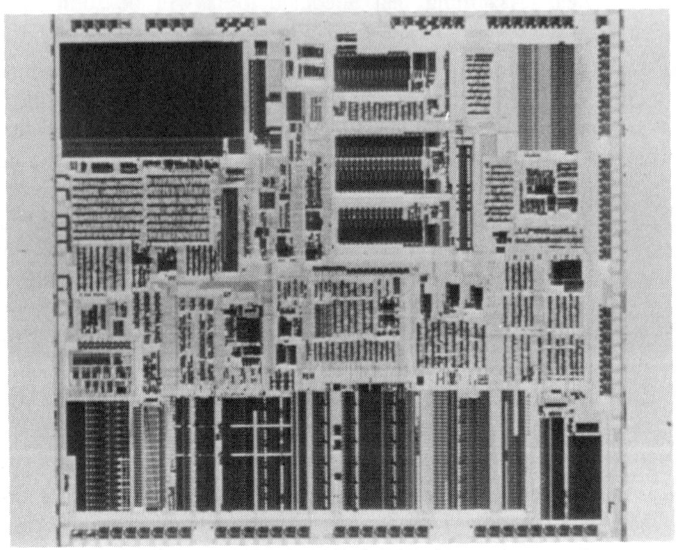

Figure 6. The Fairchild Clipper chip (courtesy of Fairchild).

expect a more pronounced shift towards the macro-cell design style in the near future when better placement and routing tools will become available.

Other hybrid approaches can be defined: for example, several of the blocks of the Intel 80386 were designed by hands, yielding a hybrid between custom and macro-cell

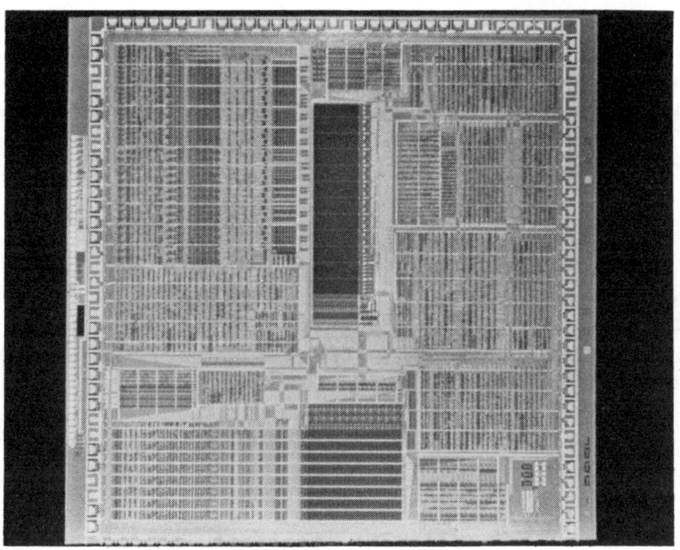

Figure 7. The Intel 80386 (courtesy of Intel).

design style.

All ASIC systems require a library of primitive components, whether they be individual transistors, logic gates, or entire subsystems. These library cells may be invariant designs, such as the traditional standard cell or gate array building blocks, they may be parameterized cells, such as those in the libraries offered by the "silicon compiler companies", or they may be sophisticated, module-generator-based libraries, where different cell topologies are generated on the fly as a function of the user's input description. Recently there has been a growing concern about the level of electrical characterization that is available for cells obtained with module-generators. Tools in this area are rudimental and the designer often resorts to generating a number of cells of predetermined structure, characterizing with great care their electrical behavior and storing these structures in static libraries, thus somewhat defeating the purpose of module-generation. It is therefore necessary to develop tools and design methodology that can solve the electrical characterization problem, if module-generation is to capture that part of the design market that it is addressing.

If a designer is to compete in the ASIC marketplace of tomorrow, he must be able to customize his CAD design environment in all three of these areas.

4. SYNTHESIS SYSTEMS

As we pointed out in the introductions, synthesis is a crucial component of future CAD systems. A complete synthesis system should generate layout masks from a high-level *algorithmic, behavioral* or *functional* description of a VLSI system, a description of the target technology and a description of the constraints and cost functions. The design should be completed in reasonable time and with the quality a human designer could obtain.

Very few design-aids are available to assist the VLSI designer at the algorithmic level. At this level, the designer describes the system by specifying its operations or functions without necessarily giving implementation details such as the "hardware" components needed to implement the system. Design at this level involves the translation of a required algorithmic-level specification into an *architectural* or *register-transfer-level* implementation. The architectural representation of the design includes components such as registers, memories, processors, which specify the high level implementation of the system.

Once the functional partitioning of the design is completed, estimates of the layout size, power-supply requirements, and speed of the high-level circuit blocks used to implement the various sub-functions are required. A chip-plan must also be constructed to determine the relative placement of these building blocks. This chip plan is then further refined as the design proceeds. These tasks are often performed manually, perhaps with the help of the computer to perform book-keeping tasks such as the storage of interconnection data.

Silicon compilers have been proposed to carry out the entire synthesis process. Since the task is so complex, early silicon compilers assumed that a target technology and a floor-plan were chosen by human designers. In this case, the difficult steps of linking the high level synthesis task to the layout problem was resolved by eliminating computer intervention. Among the most important contributions of the early research on silicon compilers is the development of *procedural design languages*. These languages are used to write programs which, when executed, generate in a flexible and possibly technology independent way the layout of entire chips and/or of leaf cells, i.e., the basic low level cells such as NAND gates, NOR gates, inverters, register cells. The work by D. Johannsen at Caltech[JOH79] and the work on DPL[BAT80] at MIT were examples of such important contributions. Procedural design languages can be used effectively to generate the layout of regular structures such as ROMs, RAMs, PLAs and data paths. In particular, the use of these languages eases the construction of parametrized and technology independent *module generators,* i.e., of computer programs that generate the layout of a block such as a ROM, RAM or PLA, given a functional representation, such as a truth table.

Working designs have been produced with silicon compilers, but the quality of the design has always been a problem. While for a restricted class of designs, such as Digital Signal Processors (DSP), the use of a fixed floor-plan has been successful, (e.g. the LAGER silicon compiler developed by R. Brodersen at Berkeley[RAB85]), its use for less constrained applications results in inefficient utilization of area and poor performance. In addition, the structure of the control logic is often too rigid and not optimized, thus yielding a slow and large chip. J. Fox[FOX83] has illustrated the pitfalls of an existing silicon compiler, MacPitts[SOU83], by comparing a design for a telecommunications chip generated by a silicon compiler with a design obtained with the use of a standard-cell place and route system.

The present trend is to break the synthesis process into stages, and to use tools that optimize real estate and/or performance to go from one stage to the next. At first, attention has been paid to the optimal generation of regular arrays such as ROMs, RAMs and PLAs. For example, module generators have been built by VLSI Technologies Inc. and Silicon Compilers Inc. using a procedural design language. Simple routing techniques were also offered to connect the modules generated by these tools. These two companies were the first to introduce the concept of silicon compilation and procedural design languages in industry. While these concepts have now gained considerable attention in the industrial community and several other companies are offering procedural design languages (e.g. Silicon Design Laboratories and SDA) and module generators (e.g. Silicon Compilers Inc., SDA and Seattle Silicon), a few years ago when these companies were founded, traditional designers expressed a great deal of resistance towards these new design techniques.

The SILC silicon compiler under development by J. Fox at GTE and the Yorktown Silicon Compiler (YSC) being developed at the T.J. Watson Research Center by R. Brayton[BRA84], are two examples of systems where layout optimization and efficient logic synthesis are introduced. While SILC addresses the problem of translating an algorithmic description of the system to be designed into an architectural description, YSC starts with an architectural description of the design leaving the task of determining the architecture of the chip to the human designer. The Design Automation Assistant under development at AT&T Bell Laboratories[KOW85] is the most recent entry in the automatic synthesis arena. This system is based on the work done at Carnegie-Mellon University for high-level synthesis and on the work done at Stanford for layout. One of the most interesting aspect of this system is the use of a knowledge-based expert system to carry out the translation of the behavioral level description into a register transfer level description and to generate an optimized floor-plan. Research work on silicon compilation is also carried out at the University of Illinois with the ARSENIC silicon compiler developed by D. Gaijski[GAI84].

The procedural design aspects of early silicon compilers has been neglected for some time. Recently Silicon Design Laboratories (SDL), founded by the developers of the PLEX system[BUR83] for automatic synthesis of micro-processor-based designs, introduced a procedural design language for IC design which is available commercially[BUR85]. This language is an important tool to develop module generators. We believe that procedural design systems are important for the synthesis systems of tomorrow. However, designs generated by writing procedures in textual form are very difficult to debug and there is a need for the development of debugging tools that combine textual and graphic representation of a design, a problem that in my opinion, has not been solved well.

Another approach has been proposed recently for module generation development that adds to a procedural design language the use of diagrams to describe the topology or structure of the module to be generated. In this approach graphics is used to represent the overall structure or floor-plan of the array. Law and Mosby [LAW85] of SDA proposed first this approach consequently generalized to include many levels of hierarchy by R. Mayo[MAY86] of the University of California, Berkeley.

5. ORGANIZATION OF THE BOOK

This book includes contributions by the foremost experts in synthesis systems. The book is organized in two basic parts: the first part is dedicated to tutorials covering broad aspects of the synthesis problem, with some emphasis given to systems or algorithms developed by the contributors, the second part collects specific contributions to synthesis systems and tools.

In the first part of the book we review the three basic components of a synthesis system:

- - Procedural design and module generation; with the papers by J. Allen, who reviews the problem of module generation, H. Shrobe, who reviews the work originated at MIT on procedural design and presents the use of AI techniques for synthesis, and R. Newton, who reviews procedural design and symbolic layout including compaction.

- - Physical synthesis or layout, including floor-planning, partitioning, placement, routing and compaction; with the paper by A. Sangiovanni-Vincentelli, who reviews the state-of-the-art in placement and routing with particular emphasis on the latest results in this very vast field.

- - Logic synthesis, including combinational logic, sequential logic and algorithmic or behavioral synthesis; with the papers by R. Brayton, who presents the basic algorithms for combinational logic synthesis that he devised and the MIS system developed in collaboration with researchers of University of California, Berkeley, G. Hachtel, who reviews the work on multi-level logic minimization and its relationships with the testing problem, D. Brand, who presents a different approach to combinational logic synthesis based on rules, G. DeMicheli, who reviews the state-assignment and the encoding problems encountered in sequential logic design, and D. Gaijski, who reviews the work on expert systems for high-level synthesis.

The second part of the book includes the work by S. Law and M. Buric on module generator tools and procedural design languages as developed by their companies who have a leadership role in the new generation CAD vendors, by R. Otten who reviews the physical design aspects of the Yorktown Silicon Compiler, by A. DeGeus who presents the SOCRATES system, a rule-based logic synthesis system, by Curtois and Baitinger, who present their work on silicon compilation and finally, by B. Brodersen and H. DeMan who developed the first examples of synthesis systems which can map effectively system specs into mask layout for a well defined domain of application: digital signal processing.

6. ACKNOWLEDGEMENTS

The work presented here is the result of many discussions and joint research with Richard Newton, who is also gratefully acknowledged for the many years of exciting collaboration. The authors of the papers included in this book had all an impact on this paper through many (sometimes heated...) discussions. In particular, I wish to thank Giovanni DeMicheli for his careful reading of this manuscript and for the editorial suggestions.

Thanks to CDI, LSI Logic, AT&T Bell Labs, DEC, Fairchild, and Intel for providing and allowing the publication of the photographs of the chips shown in this paper.

This work has been partially supported by SRC, DARPA and NSF. The support of the industrial sponsors, of the CAD/CAM Center of the University of California, Berkeley, and, in particular, of SGS Microelettronica, AT&T Bell Laboratories, DEC, Fairchild, GE, Harris Semiconductors, Hewlett-Packard, Hughes Aircraft, IBM, Intel, Itaucom, LSI Logic, Olivetti, Philips, Silicon Compilers Inc., Silicon Design Laboratories, and Xerox, is gratefully acknowledged.

7. REFERENCES

[BAT80] J. Batali and A. Hartheimer, "The Design Procedure Language Manual" MIT AI Lab, VLSI Memo 80-31, Spt. 1980.

[BRA84] R. Brayton, G. Hachtel, C. McMullen and A. Sangiovanni-Vincentelli, *"Logic Minimization Algorithms for VLSI Synthesis"* Kluwer Academic Publishers, 1984

[BUR83] M. Buric et al. "The Plex Project: VLSI Layouts of Microcomputers Generated by a Computer Program", *Proc. 1983 Int. Conf. on CAD*, pp. 49-50, Oct. 1983.

[BUR85] M. Buric and T. Matheson, "Silicon Compilation Environments" *Proc. 1985 Cust. Int. Circ. Conf.*, May 1985.

[FOX83] J. Fox, "The MacPitts Silicon Compiler: A View from the Telecommunications Industry", *VLSI Design*, May 1983.

[GAI84] D. Gaijski and J. Bozek, "ARSENIC: Methodology and Implementation" *Proc. 1984 Int. Conf. on CAD*, pp. 116-118, Santa Clara, Nov. 1984.

[HUI85] A. Hui et al., "A 4.1K Gates Double Metal HCMOS Sea of Gates Array", *Proc. 1985 Cust. Int. Circ. Conf.*, pp. 15-17, Portland, May 1985.

[JOH79] D. Johannsen, "Bristle Blocks: A Silicon Compiler", *Proc. 16th Design Automation Conf.*, pp. 310-313, 1979.

[KOW85] T.J. Kowalski et al., "The VLSI Design Automation Assistant: From Algorithms to Silicon" *IEEE Design and Test of Computers*, Aug. 1985.

[LAW85] S. Law and J. Mosby, "An Intelligent Composition Tool for Regular and Semi-regular VLSI Structures" *Proc. 1985 Int. Conf. on CAD*, pp. 169-171, Santa Clara, Nov. 1985.

[MAY86] R. Mayo, "Mocha Chip: A System for the Graphical Design of VLSI MOdule Generators" *Proc. 1986 Int. Conf. on CAD*, pp. 74-77, Santa Clara, Nov. 1986.

[RAB85] J. Rabaey et al., "An Integrated Automated Layout Generation System for DSP Circuits" *IEEE Trans. on CAD of ICAS*, Vol. CAD-4, n. 3, pp. 285-296, July 1985.

[ROW86] J. Rowson et al. "Gate-Array Macro Layout Automation" *Proc. 1986 Int. Conf. on CAD*, pp. 448-451, Santa Clara, Nov. 1986.

[SOU83] J.R. Southard, "MacPitts: An Approach to Silicon Compilation", *Computer*, Dec. 1983.

PART I: SYNTHESIS METHODS

MODULE GENERATORS FOR VLSI DESIGN*

by
Jonathan Allen
Research Laboratory of Electronics
and
Department of Electrical Engineering and Computer Science
Massachusetts Institute of Technology
Cambridge, Massachusetts 02139

1. Introduction and Overview

Until the mid-1970s, integrated circuits were designed manually with the benefit of device equations and simulators. The artwork was then laid out by hand, usually by a draftsperson who had memorized the design rules of the process being used. Questions regarding layout structure and organization, together with considerations of packing, placing, and routing, were expert skills acquired by the draftsperson, and were not codified in any readily reproducible manner. The hand-generated layout was then digitized into machine-readable form, and converted into statements in a low-level description language which served only to characterize the geometry of the several mask layers. This style of layout design resulted in many errors, the expenditure of much time and effort, and great difficulties in editing beyond the scope of a few devices. As the design complexity increased, these difficulties became intolerable. There was a clear need to impose <u>discipline</u> on layout strategies so that the benefits of <u>regularity</u> and <u>repetitiveness</u> could be increasingly appreciated.

Once layout design was viewed as a disciplined procedure, subject to process design rules and well-established layout structures, it was natural to conceive of replacing much of the manual layout generation with automated procedures. At the same time, there was great interest within the artificial intelligence field in the idea of <u>procedural coding of knowledge</u>. In other words, knowledge was not viewed so much as a static entity as it was a procedure to accomplish a specific effect. A good example comes from the natural language processing field. When a question is to be "understood" by a computer, that understanding can be considered as a procedural response which is appropriate to the knowledge base's state and the question's content. Integrated circuit layouts, previously regarded as strictly static entities, could then be regarded as the result of executing a procedure, possibly modified by a set of parameters. This notion is particularly useful, since many regular designs vary only in ways that can be parametrically characterized. Thus, there is a great saving in design characterization due to capturing these dependences in a procedural way.

---

* Research sponsored by the Air Force Office of Scientific Research, Air Force Systems Command, USAF, under Grant Number AFOSR 86-0164.

It must also be emphasized that the LISP programming language played an important role in furthering the idea of procedural representation of static entities. In LISP, both procedures and data are represented by a single, common form, namely the S expression. It is natural, in LISP, to treat procedures as data (and vice versa), when it is convenient in a programming context. Thus, in many ways, introducing procedural generation of VLSI layouts was a natural extension of programming methodologies that were already well-developed. This extension was accelerated by the need for these techniques as a result of the increasing complexity of the designs being generated.

In many ways, VLSI module generators are like subroutines with parameters. These parameters include: the design's logic personalization, size and orientation; the designation and names of input/output ports; connectivity between constituent cells; and the nature of replication used within the overall design. From this viewpoint, a trivial module generator could be constructed to generate an N-bit shift register from identical constituent cells, where N is a parameter to the generator. It can be seen from the sequel, however, that the degree of programmatic control over cell generation is far more powerful than that implied by this simple example.

These subroutines can be viewed as one large transformation from an input functional specification to the output layout. Many times, however, the input is not specified at any purely functional level, but has some structural aspects as well. Constructing the overall layout procedure as one large transformation can often lead to a high degree of complexity, so it must be broken up in order to be manageable. This can be done in a natural way since there are many individual <u>constraints</u> (such as those imposed by architecture, logic, circuit form, and device design) which permit overall design factorization into these aspectual views. Each of these constraint domains have well-formedness conditions, and a major focus of current research is the formal codification of these constraints. Furthermore, as previously suggested, the layout itself is often constrained to have predictable structural semantics, so that the final design has a general form characteristic of the particular module generator. Thus, program logic arrays of a particular style, for example, all look the same at a relatively high level of the structural hierarchy. This factorization of the overall procedural transformation into constraint domains is also important, otherwise the cell's nature may be excessively distributed throughout the procedure's code. A major focus in module generation has been to separate the various constraint domains, and to provide the user with a systematic means to rigorously specify each of these constraints in a natural way.

It is best to remember that effective module generators cannot merely compose geometries. All constraint domains must be <u>well-formed</u> at each level. Thus, the circuit representation must perform well electrically (without races and hazards, and with appropriate logic levels), and the layout level must satisfy all process design rules. Furthermore, the circuit's structural coherence must be instantiated in the layout so that the design's important sub-units (such as an individual transistor or a contact) are viewed as a unit comprising all mask levels inherent to their specification.

Another important aspect of procedural design is the use of <u>symbolic representation</u> with constraint propagation. This is an important instance of <u>delayed binding</u>, and provides the designer with a means to specify numerical relationships in terms of symbolic identifiers which are later bound to particular numbers. For example, the width-to-length ratio of a transistor can be characterized symbolically, and later bound to a particular number when the performance requirements on this transistor can be readily assessed. This also illustrates the nature of constraint propagation on a local level. Thus, if the width-to-length ratio is specified (even symbolically), then fixing either the device's width or length automatically constrains the other parameter in terms of the specified width-to-length ratio. Generalizing these constraints to two devices, the width-to-length ratios needed in ratioed NMOS design form another constraint which must be satisfied in order to preserve logic levels within such designs. Clearly, the idea of procedural design is an instance of delayed binding, since it is not necessary to store the final layouts, but only to know which parameters to use in instantiating a given procedure to provide these layouts.

The notion of delayed binding can be extended even further to provide valuable features in the procedural generation of layout. In many cases, the designer wishes to obtain a "<u>performance summary</u>" of the final design without generating a detailed layout of the final cell. In this way, the design's bounding box, time delay through various module ports, power dissipation, and the specification of connection points can be obtained from a module generator very rapidly, particularly when there are highly constrained and predictable structural design semantics (such as in register arrays and program logic arrays). Given this initial summary of the final design, it is often unnecessary to generate the complete, detailed layout until very late in the design process, since the module is adequately characterized for design purposes in terms of its functional summary and the performance specifications obtained through module generator use. The use of these performance summaries leads naturally to the notion of <u>design exploration</u>, since through their use, it is possible to investigate procedurally many different design options in a short time and at reasonable computational cost. In this way, alternate logic forms, circuit design styles, and architectural forms may be readily assessed at the performance level without the need to generate an entire layout. It is important to emphasize that performance can be regarded as having two aspects. There is, of course, circuit performance specified in terms of area, speed, and power dissipation. There is also architectural performance which characterizes the amount of parallelism in the design, which to some extent can be traded off against electrical performance. Unquestionably, the goals of high-performance module generators are directed to the formal constraint specifications at the various levels of representation with a well-coordinated exploration strategy utilizing both electrical and architectural performance summaries.

Once the availability of design exploration through many iterations is available, due to fast procedural techniques with delayed binding, then it is natural to introduce the further notion of <u>design optimization</u>. There are certainly several forms of optimization relevant to the various levels of design description. For example, architectural transformations can be used to optimize a design's throughput, and techniques are available for constructively achieving a design with such performance. At the logic level, techniques are available for logic minimization, both algorithmically and heuristically, in terms of universal two-level logic and multi-level logic forms. Finally, at the layout level, folding and compaction techniques lead to the effective minimization of the layout area with concomitant improvements in electrical performance. In many regular structures, optimum interconnect packing can also be achieved, and indeed, high-performance channel routers can be viewed as circuit interconnect optimizers encapsulated in a procedural form.

An inevitable tension resulting from procedural design techniques is the trade-off between efficiency and generality. On one hand, when the resulting layout structure is highly constrained (as in a RAM, ROM, or PLA), high efficiency can be obtained in terms of minimum layout area at the cost of reduced generality. When highly general forms are desired, as in the case of unrestricted state machines, then the use of predictable structural semantics forces these designs into a particular layout structure which may perform satisfactorily if the layout style is well-suited to· the particular architectural form. Most of the current module generation and silicon compilation techniques utilize these structural constraints. Thus, a major challenge for improving such generators is to provide for increased generality while retaining the efficiency of the more highly specific generators. In this paper, one particular technique described is focused on the notion of "design by example" which provides these benefits. On the other hand, when the structure is not predictable from the input functional specification, then the generation of an efficient layout can be exceedingly difficult due to the lack of appropriate constraints. This problem is compounded further when it is necessary to provide for large variations in device size, often specified in terms of routing requirements, which complicates efficient layout generation.

In order to achieve generality with efficiency, many module generators and compilers seek to provide bus-oriented designs using bit slice layout with standard routing. This leads to a simple composition strategy consisting of adjacent cell abutment, achieved through a simple "tiling" strategy that is easily implemented. It must be emphasized, however, that such straightforward techniques are often inadequate. Obtaining simple geometric connectivity may not be sufficient to guarantee appropriate circuit conformance between the constituent cells, in terms of logic levels and the avoidance of races and hazards. Furthermore, the composition process needed in a sophisticated module generator can be very complex, utilizing affine transformations for generalized rotations, and providing for cell overlap for a variety of purposes including cell encoding and performance options. Thus, many designs are not merely simple replications of regular structures. More often, they are "quasi-regular" structures highly susceptible to procedural representation which require a great deal of personalization, both within the structure and at its periphery, as a result of the need to satisfy a wide variety of boundary constraints.

Procedural techniques have now reached a substantial level of sophistication and power, although there is still room for improvement. In this chapter, a systematic view is given of these accomplishments. In order to see the specifics of procedural design, an inverter specification in a particular design language is first shown. This is generalized to NOR gate specifications, followed by NOR gate arrays in the form of Weinberger arrays which provide ample opportunity for highly sophisticated optimal layout, including interconnection packing and optimal transistor orientation. Building from the Weinberger arrays, PLAs are examined next, both in terms of layout in a specific process, but also in terms of the more general ability to generate many different PLA styles from a minimal and appropriately abstract specification. Logic optimization and folding techniques are also characterized at this point. These techniques naturally lead to a generalization of the "design by example" methodology in the form of a regular structure generator. This is an exceedingly powerful tool, capable of generating a wide variety of highly efficient quasi-regular structures. In addition to strictly regular arrays (such as ROMs and RAMs), it can be used to generate program logic arrays and architecturally optimized structures such as retimed array multipliers. Finally, an extremely ambitious design, namely the generation of a procedure for generating a floating point adder, is described with an elaboration on the techniques used in this program to assure coordinated well-formedness between all levels of design.

2. Procedural Design of an Inverter

In order to provide a specific example of procedural design, the complete specification of a single CMOS inverter is shown here using a readily available language designed for this purpose called L.[1] The code for this inverter specification is shown in Figure 1, and the corresponding layout is shown in Figure 2.

```
L:: TECH scmos
CELL Inv (nw,pw,gl)
(
NUM t_space = 15;
NUM c_g_space = 6;
NUM c_d_space = 7.5;

TN ntran R90 W=nw L=gl AT (0,0);
TP ptran R90 W=pw L=gl AT
         (ntran.X, ntran.Y + nw/2 + pw/2 + t_space);
MNDIFF c1 AT (ntran.X - gl/2 - c_g_space, ntran.Y) ;
MNDIFF c2 AT (ntran.X + gl/2 + c_g_space, ntran.Y) ;
MPDIFF c3 AT (ptran.X - gl/2 - c_g_space, ptran.Y) ;
MPDIFF c4 AT (ptran.X + gl/2 + c_g_space, ptran.Y) ;
MPOLY c5 ;
c5.X = c2.X;
c5.Y = ntran.Y + nw/2 + c_d_space;

IN POLY Input AT (c1.X, c5.Y);
OUT POLY output AT c5;
VDD MET p_vdd AT c3;
GND MET p_gnd AT c1;

WIRE Input HOR VER ntran.gr;
WIRE Input HOR VER ptran.gl;
WIRE c4 HOR VER c5;
WIRE c2 HOR VER c5;
WIRE c1 HOR VER ntran.d;
WIRE c2 HOR VER ntran.s;
WIRE c3 HOR VER ptran.d;
WIRE c4 HOR VER ptran.s;
)
```

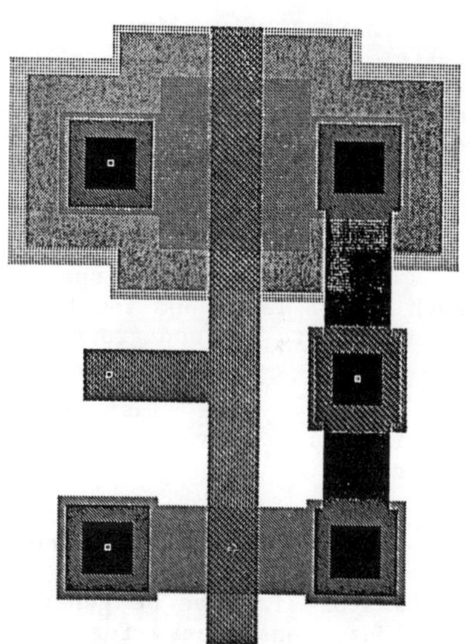

Figure 1.

L Program Inverter Generator

Figure 2.

Inverter layout corresponding to the output of the generator program in Figure 1, with nw=5, pw=10, and gl=3

In the following text, the individual lines of the inverter generator L program are described.

The technology used (a variant of CMOS) is specified in the first line. Next, the calling sequence for the inverter generator is given, which shows that the name of the cell to be generated is "inv." There are three parameters: the width of the n-channel pulldown transistor "nw;" the width of the p-channel pullup "pw;" and the gate length "gl" which is common to both devices. The declaration of three numerical values follows, which are given symbolic names: "t_space" refers to the space between the two transistors; "c_g_space" refers to the space between a gate and a center point of a contact; and "c_d_space" refers to the space between the edge of a piece of diffusion and the center point of a nearby contact from metal to some other layer. These three numbers are important for the proper maintenance of design rule constraints in the resulting layout.

Next, comes the declaration of the n- and p-transistors which includes their orientation, size, and location in the layout. It is important to recognize that although a transistor is specified in terms of a poly rectangle crossing a diffusion rectangle, the transistor is treated as a single symbolic entity. This helps to maintain its coherence in the design. Furthermore, since all transistors are designated explicitly, any inadvertent crossing of poly over diffusion not declared as a transistor can be recognized as a layout error.

The next five lines declare five contacts, c1 through c5. The first letter of all five declarations ("M") refers to the metal layer, and the remaining letters characterize the diffusion or poly layer to which the contact is made. The rest of the specification clearly provides the contacts' location specification. The "period" notation designates either the x or y coordinate of a particular object. So, ptran.Y refers to the y coordinate of the single p-transistor.

The next two lines specify the input and output ports, and designate each port's layer and location. Subsequently, the next two lines show the connection to Vdd and ground.

Finally, the last eight lines declare wires using a layout specification convention which will not be described in detail in this paper.

This code example shows that even a simple circuit requires considerable specification. But, many different inverters that vary in transistor size can be generated from this program. The L language also provides constructs for iteration and arrays, so that substantial structures can be "tiled" from a small number of generator programs. The construction of these programs is not simple, and as demonstrated here, facilitates specification only at the layout level. The designer of such a generator must ensure that the resulting layout follows all design rule specifications. Hence, the designer may choose to insert several spacing constraints, such as the three constraints shown in the inverter example previously described. Module generators have sometimes been described as providing layout that is "correct by construction," but it is evident that the correctness of the resulting layout is only as good as the care taken in writing the generator program. In most situations, users choose to use a design rule checking program to verify layout wellformedness. It is also increasingly common to find layout specification performed at the purely symbolic level, with a later transformation (coupled with compaction) utilized to generate the correct geometric layout. Although this practice is a good example of delayed binding, some spacing constraints may still be needed to meet pitch matching constraints at cell boundaries.

Cell generation programming languages are usually embedded in a common language such as LISP or C, and hence, they build on a powerful set of control constructs. In this way, conditional assembly is readily provided so that a single program may serve the needs of many different applications, depending on particular functional and input/output constraints. For example, it is common practice to provide register cell generators with a variety of ports that may be selectively deleted, depending on the application.

3. NOR gates and NOR gate arrays

In order to build structures larger than the inverter, a way to modularly build random logic gates is needed. In NMOS technology, the NOR gate is particularly well adapted to this requirement, since the size of a given NOR gate can be expanded by simply adding additional pulldown devices in parallel with the existing pulldown structure. It is essential to recognize that when such an extension is instantiated, there is no need to change the size of the depletion mode NMOS pullup, since the gate's ratioing constraints will be continually met as additional pulldowns are added. This modular property of NOR gates is responsible for their frequent use in large NOR gate arrays, including Weinberger arrays and program logic arrays. To illustrate this property, Figure 3 shows a two-input NOR gate and a five-input NOR gate as designed by Ayres.[2] There are many possible ways to specify multiple-input NOR gate layout, but the desired modularity property is easily provided by this example.

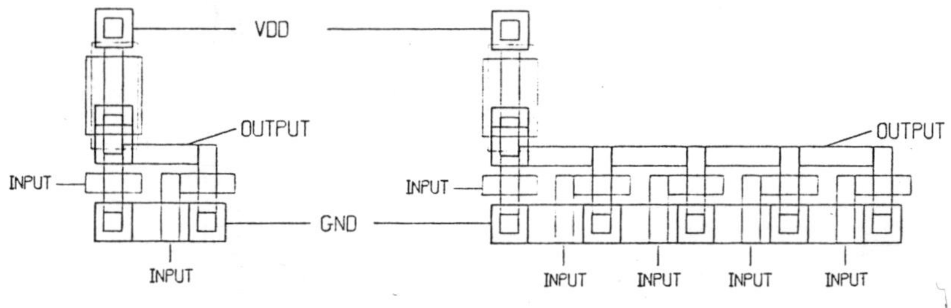

Figure 3. Two-input and five-input NOR gates illustrating the
modular composition of NOR gates

In order to provide further extension of NOR gate arrays, it is possible to introduce the sharing of ground lines as shown in Figure 4. A separate vertical ground bus runs adjacent to the pulldown structure with ground tabs extended to the left in order to facilitate a connection to the pulldown device sources. In Figure 4, it is important to realize that a metal line extends from the butting contact at the pullup device all the way to the bottom output. This avoids undesirable voltage drops as the NOR gate's size increases.

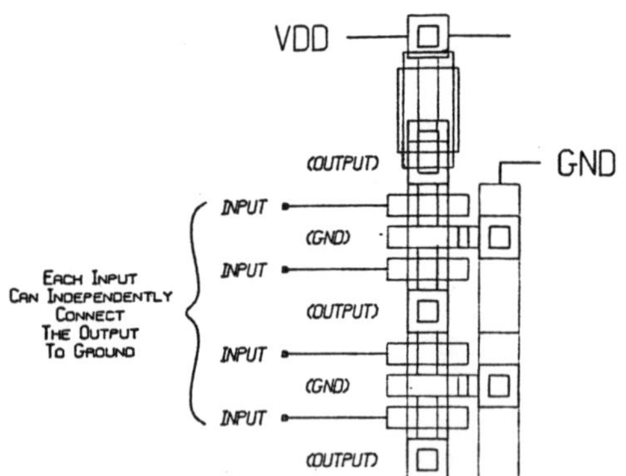

Figure 4. NOR gate layout that allows
the sharing of ground lines

It is now easy to combine multiple NOR gates in a modular way using this layout convention, as shown in Figure 5. Here, three NOR gates share the same set of ground lines and four inputs. In this way, it is natural to build NOR gate arrays systematically so that they can readily be specified by an appropriate layout language. Details of the construction of such generators are given by Ayres.[2]

Figure 5.

NOR gate array with four inputs
bussed across the array and shared
ground lines between pairs of
input lines

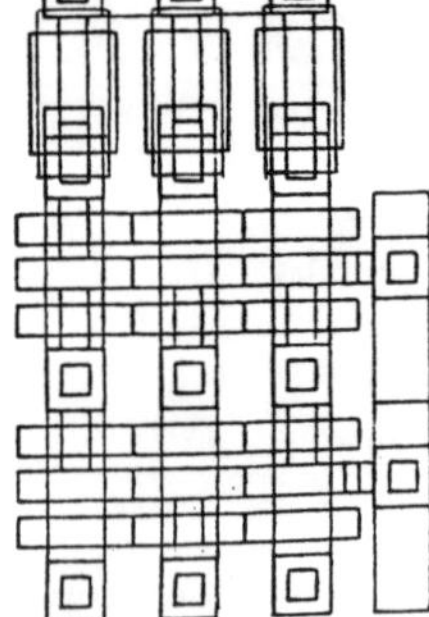

4. Program Logic Arrays (PLAs)

Once a NOR gate array generator is available, it is simple to extend it in order to provide the layout for a complete program logic array. Since these arrays are so important for modern structured design, much effort has been expended to facilitate highly efficient and optimized arrays. These arrays afford two-level logic (commonly referred to as AND-OR logic) implemented in the form of two interconnected NOR gate arrays. Logic optimization for these arrays has been extensively studied, and is comprehensively documented.[3]

Considerable research has also been devoted to the study of folding, where input and output columns and product-term rows are shared by breaking the column or row in appropriate places. Important graph-based heuristic techniques have been devised for folding, and recently, simulated annealing techniques have been applied to give a very high level of performance.[4]

Another aspect of PLA optimization is the use of so-called "two-input encoding." This has the effect of increasing the logical burden of the input lines with very little additional decoding logic that is external to the array proper. Optimal techniques for using this strategy are provided in the context of multi-level logic optimization.[5] In this chapter, further characterization of techniques for logic optimization, folding, or input encoding will not be pursued. But, it should be emphasized that effective programs are available for these purposes which can often provide dramatic reductions in array size and corresponding improvements in circuit performance.

Once these optimizations have been completed at the logic and topology levels, then a layout generator program is needed to construct the detailed PLA geometry. Many programs have been written, and a layout from one particular NMOS PLA generator is illustrated in Figure 6. [6]

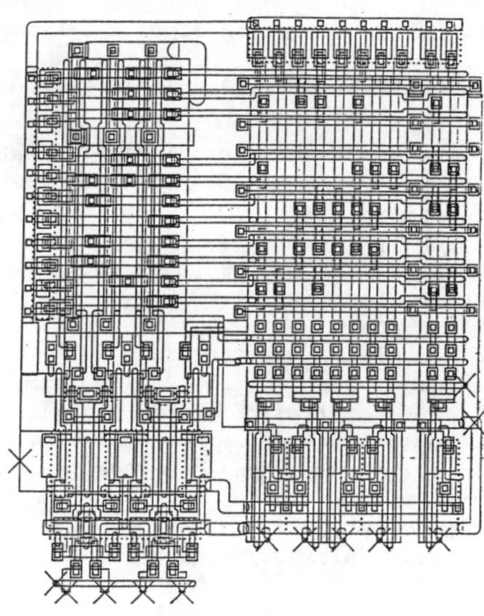

Figure 6.

NMOS PLA generated by a specialized PLA generator. Large Xs are input, output, and ground connection points.

The generator program is capable of highly optimized packing of the pullup load devices in each of the two NOR arrays, and properly sized super buffers are automatically provided to drive the input lines across the AND plane, and to receive the outputs of the OR plane. Within the two NOR planes, the pulldown devices are all minimally sized, but it is necessary to appropriately size the metal busses in order to avoid electromigration difficulties. It should also be noted that in some NOR plane designs, the ground lines or the NOR gate output nodes may be extended substantially in diffusion. This can lead to voltage drops that destroy the array's intended functionality. For this reason, in large PLAs, it is essential that the NOR gate output nodes and ground lines be realized in metal.

In the PLA layout illustrated, the large Xs at the connection points for power and ground (as well as input and output) are used in the performance summary for the resulting PLA. It is relatively easy to compute the bounding box for the overall PLA, and to estimate the propagation delay through the array as well as the power dissipation (which may be logic-dependent). The generator program is capable of specifying the connection points, bounding box, estimated time delay, and power dissipation <u>without</u> generating all of the detailed layout geometry. This is an exceedingly important feature, since it allows the designer to assess the PLA's important performance aspects without the necessity for the computational expenditure required to generate all of the detailed layout. This is a good example of delayed binding, and such summaries are essential design exploration tools. Ideally, all module generators should be provided with rapid performance estimators to facilitate the high-level process of floor planning and clock frequency estimation in a realistic way. It is also important to emphasize that properly constructed module generators must maintain wellformedness at all levels, including logic functionality, circuit behavior (in terms of interconnect connectivity and logic levels), and layout adherence to design rules. Currently, no formal tools are available for maintaining the overall design coherence across all of these levels of representation, and it is regretable that many approaches to module generation focus only on the layout level without explicit consideration of circuit and logical correctness.

5. <u>The Search for Generality: Regular Structures</u>

The previous section demonstrated that it is possible to generate procedurally compact and efficient layouts competitive with manually produced designs for well-defined regular structures. Nevertheless, such efficient procedural generators facilitate early binding on several design parameters which give rise to additional needs. First, the ability to track technological variations is desirable, so that the generator can be easily updated for design rule changes. Symbolic design techniques combined with compaction can address this problem. But it is also possible to minimize this difficulty by exploiting the design's regularity. This will be demonstrated in the sequel. There are also many design variations that represent small perturbations on the basic design. Thus, in program logic arrays, changes in clocking methodology, multiplexing, latching, and other factors can produce a wide variety of distinct designs. It is certainly not desirable to produce a new procedural generator starting from first principles for each variation.

From these considerations, there is not only a need to maintain the efficiency of specialized PLA designs, but also the need to provide flexibility that will allow design alterations without writing an entirely new generator. In a regular structure such as a PLA, there are a small number of distinct building block cells with a few inter-block connections called interfaces. Furthermore, all of the essential information needed to specify a PLA in a given style, but of _any_ size, can be naturally and insightfully represented in a minimally sized example (namely a two-input, two-product term, two-output PLA). In other words, this small PLA utilizes all the needed cells together with all of the required interfaces, so that a program can be written to take advantage of this minimal specification in order to generate _any_ size PLA. Such a strategy is known as "design-by-example" generation. Each PLA style (e. g., NMOS three-phase clocked PLAs with input and output latching) utilizes a different set of distinct cells and inter-cell interfaces which vary over the many possible styles. But, the amount of design information that must be specified by the user has been drastically reduced to: 1) the design's regularity, and 2) the specification, within the generator program, of how an arbitrarily sized complete PLA can be composed from the constituent primitive or basic cells together with their local interfaces. In such an approach, any inter-cell routing must be explicitly represented as a cell itself.

In practice, a design community readily builds a cell library for the PLA styles of interest. Then, using a graphic editor, the minimal sample PLA is composed from these cells. This process, which is straightforward for the designer, _implicitly_ defines the requisite interfaces needed to compose _any_ PLA of the style inherent in the sample. By this simple and expedient approach, a large array of different PLAs in an unlimited variety of sizes can be built. To specify the PLA's size, the designer simply supplies the encoding table for the AND and OR planes, which is an array of ones and zeroes. It should be noted that the several constraints on the complete PLA specification have been conveniently separated, thus avoiding a monolithic PLA generator program with little internal structure. Specifically, the constraints are partitioned as follows:

- The detailed graphical layout is provided by the user or from a cell library, and the PLA generator treats these cells as given.

- The cell interfaces are implicitly derived by the generator program from the relative placement of the cells used in the minimal sample example.

- The size and encoding information needed to specify the number of inputs, outputs, and product terms are specified symbolically (ones and zeroes) by the designer.

In this way, PLAs of a specified style can be readily generated. Delayed binding is provided at several levels. Cells can be used in several PLA styles, and they can be connected at any time to form new styles "by example." Once a new style is established, it can be instantiated in any size and with any encoding, simply by specifying the encoding "cross points" symbolically. Of course, the generator can easily provide performance summaries for any style example and a given encoding specification, since all of the requisite information is available in partitioned form to compute the bounding box, delay estimators, and power dissipation. The partitioning of design constraints allows the generator user to focus naturally on the design's different facets, and permits easy modification of any constraint domain without upsetting any other constraints. For this reason, representation of the qualitatively distinct constraints can be regarded as "orthogonal" to one another. This facilitates modular design and ease of design composition.

6. <u>Broad Generality with High Efficiency</u>

Design-by-example PLA generators have proven to be very useful, and this idea can easily be extended to other types of regular structures such as RAMs, ROMs, register arrays, and bit-slice arithmetic logic units. For each of these structures, there is a basic regular "target" architecture that must be incorporated into a generator program. The resulting circuit layouts are very efficient because of the design's regularity and the generator's specialization to the desired target architecture. On the other hand, a new generator must be written for each of these architectures. It would be highly desirable to derive a new kind of generator that can retain the <u>efficiency</u> associated with specialized target forms, yet increase in <u>generality</u> by allowing the ready specification of new architectures in an appropriate design language without writing an entirely new generator for each case.

The variety of architectures referred to here does not merely consist of arrays of identical, abutting cells. In practice, there is always some degree of complexity along the edges of a regular array, and each design instance must be parametrically personalized with respect to problem size and functionality. This requires the placement of a variety of cell maskings that implement such options as transistor and bus sizing, cell interfacing, clock assignment, and functional encoding -- a task which cannot be accomplished with the simple array generating commands found in graphic editors. Although regularity does permit most regular structures to be personalized in an algorithmic manner, a high degree of flexibility is still required in the placement and orientation of the cells and cell maskings.

The Regular Structure Generator (RSG)[7] was developed to meet the needs just described. The goal is the ability to provide for multiple architectures within one generator framework, yet retain the efficiency of those generators specialized to an architecture. The RSG achieves this goal by introducing new representational abstractions, powerful mechanisms to manipulate these abstractions, and a clean separation of graphical (cell layouts) and procedural (how to connect them) domains along a natural boundary.

The RSG uses three distinct kinds of input files to produce the required circuit layout output file, as shown in Figure 7.

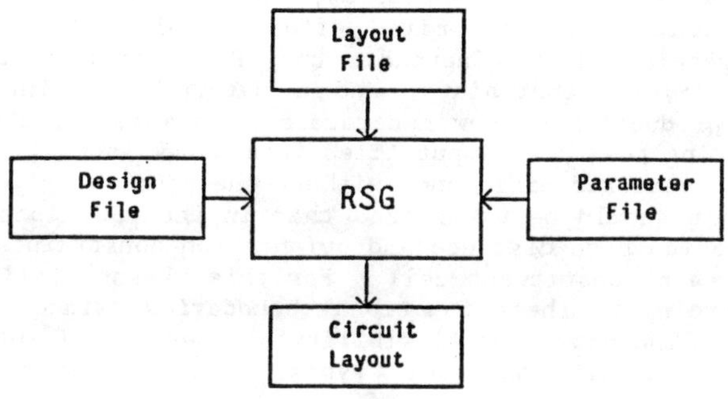

Figure 7. Regular Structure Generator input and output files

In the design file, a parameterized procedural description of the required architecture is provided. This is specified in an input language (a subset of LISP) which supplies abstraction mechanisms that can support a highly functional set of primitives to define the regular structures and evaluate the complex conditionals required for personalization and edge effects. As seen in the design-by-example PLA generator described above, personalization is further supported by the ability to arbitrarily place and orient cells according to _interfaces_ defined by example in the graphical domain. The input layout file provides such a graphical description of the leaf cells and their relative placements. Finally, an input parameter file specifies the size and configuration of a particular structure, for example, by means of a truth table.

It is important to recognize that each input file furnishes qualitatively distinct information. The design file is a procedural representation of architectural structure written in LISP. The layout file graphically specifies the leaf cells and local interfaces. The parameter file is a symbolic coding of size and functional details in the overall design. By completely decoupling the graphical and procedural domains, a level of modularity is obtained that achieves local efficiency in layout generation, and global efficiency in the management of new architectures, layouts, and interfaces to other CAD tools.

The RSG also supports macro abstraction (i. e., the specification of macro cells as interconnections of smaller cells whose binding to actual layouts can be delayed to any desired time). In addition, interface inheritance relations afford a procedural means to define interfaces between any two macro cells: a new interface between two macro cells can be computed from any legal interface between a subcell in the first macro cell and a subcell in the second macro cell. As a result, macro cells can be used to specify even more complex cells in an entirely procedural manner with no need for additional layout.

The RSG's significant contribution is ease of use through appropriate abstraction mechanisms. In other words, the RSG does not produce any circuit layouts which, with unlimited effort, could not be produced by other layout generators. Instead, the RSG permits a convenient partitioning of design constraints, and a natural user interface which facilitates the production of new generators. In a sense, the RSG can be viewed as converting the three input files into a new specialized generator that supplies the required efficiency without the need to write an entirely new generator. It should be emphasized that in the RSG, local interfaces between cells provide both distance and orientation constraints between the local origin of each constituent cell. For this reason, cell layouts can be designed according to their functional boundaries, since cell abutment is not necessary. The power and flexibility of these local interfaces also reduce the number of different cell types required, and encourage cell encoding.

An interesting internal representation used by the RSG is the <u>connectivity graph</u> which is used to build new cells. Once the RSG algorithm reads in the sample layout to define the primitive cells and build up an initial interface table, a connectivity graph is utilized to specify the new cell together with useful delayed binding properties. In the connectivity graph, vertices represent partial instances of cell types, and edges between vertices represent interfaces. Each connectivity graph needs only to be a spanning tree, since cycles in the graph contain redundant information. For a given sample layout, each connectivity graph contributes to a unique layout. Since the local interfaces specify both the separation and orientation of leaf cells, the connectivity graph is a symbolic representation of interconnected leaf cells without the need to establish the absolute geometrical layout position of each constituent cell.

When it is time to generate the complete layout, however, only a root node in the graph must be selected, and arbitrarily placed and oriented to the corresponding instance. The graph is then traversed, and each node in the graph (initially all partial instances) is expanded into a complete instance with a location and orientation. Thus, the RSG uses a two-step process to first determine connectivity, and then to use the connectivity information with cell definition and interface information to build a layout. This provides a clean separation between the graphical and procedural information. The procedural information in the design file is used to build the connectivity graph, and remains constant over the design's different implementations as supplied by the sample layout. The graphical information from the sample layout is used to transform the connectivity graph into a physical layout of a design's particular implementation. Cell spacing parameters related to the graphical information are never accessed or manipulated in the design file. This delayed binding on location and orientation of instances allows for clean macro abstraction in the design file. Since, in the design file, partial instances are connected without assigning actual locations and orientations to them, it is possible to build subgraphs without prior knowledge of where and with which orientation the instances in the subgraph will be used. In practice, the user who writes the design can think readily in terms of cells and interfaces and their utilization in a connectivity graph, and is naturally guided to exploit the abstract regularity of a desired architecture.

Since the design file is specified in a LISP dialect, the user has a powerful programming language available to specify the architecture and additional constraints on the assembly of cells and interfaces. This raises the possibility for <u>optimization</u>, which is an important aspect of design exploration. As an example of such optimization, RSG programs for optimized Weinberger array layouts have been written to allow optimally packed polysilicon gate lines in rows across vertically oriented nodes of corresponding NOR gates. The heuristic optimization code for this purpose can readily be expressed in the RSG's design language, and this capability has been exploited to enhance layout efficiency in many examples.

7. Design Exploration and Optimization

The powerful procedural generation techniques afforded by the Regular Structure Generator and similar tools have been employed in many applications. A good illustration of the ability to generate layouts for nontrivial regular structures arises in the design of pipelined array multipliers.[8] A purely combinational 6x6 signed two's complement multiplier based on the Baugh-Wooley algorithm is shown in Figure 8.

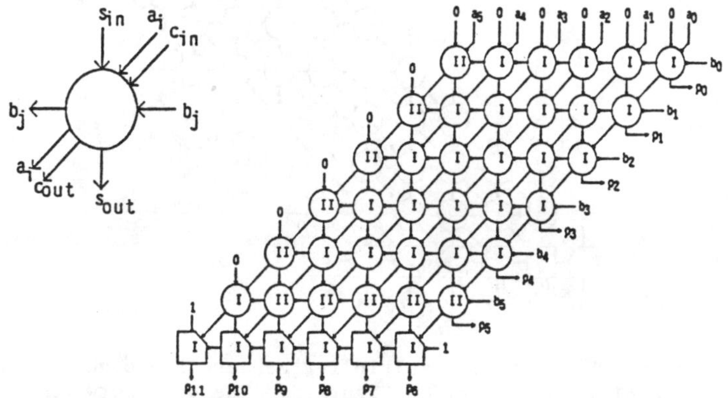

Figure 8. Baugh-Wooley 6x6 combinational multiplier. Inputs
are a_0 . . . a_5 and b_0 . . . b_5. The output product
lines are p_0 . . . p_{11}.

The multiplier consists of an array of two types of carry-save adders that reduce the product to the sum of two words, which are then added in a final row of cells connected as a carry-propagate adder. (The two diagonal connections have been condensed to one for clarity). Each cell type contains an AND gate and a full adder: cell type I adds the bit product $a_i b_i$ to its sum and carry inputs; cell type II adds $a_i b_i$ to its sum and carry inputs. The carry-propagate adder consists of type I cells drawn as polygons to distinguish them from the carry-save cells.

Using retiming transformations,[9] the multiplier can be pipelined to any degree in a manner that preserves the multiplier's functionality and the inner array's regularity. But, this adds irregularity to the array's periphery in the form of input and output register stacks. Retiming is an interesting technique that provides transformations to manipulate register placements within an architecture in order to enhance the design's throughput. In fact, the corresponding theory supplies a constructive optimum for the best throughput in terms of such transformed register placements.

Pipelining can be introduced into any combinational design by first adding delay at the inputs, and then redistributing the delay in the form of registers throughout the design by using retiming transformations guaranteed to preserve the original specification's functionality. Using these transformations, it is easy to design several versions of a multiplier. Thus, a bit-systolic multiplier (Figure 9) with a maximum of one full adder combinational delay between any two registers can easily be furnished.

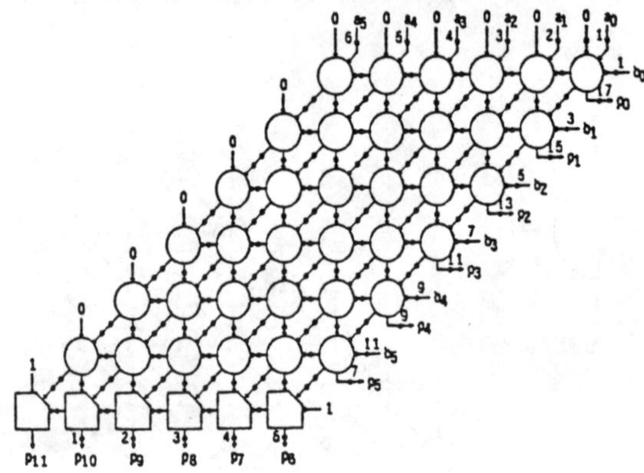

Figure 9. Bit-systolic multiplier achieved through retiming the Baugh-Wooley multiplier in Figure 8. The dots between the cells denote registers, and the integers adjacent to the dots on the periphery indicate a number of cascaded registers. Note that there is at least on register between each pair of adjacent cells, hence the design is systolic.

This design represents the highest possible degree of pipelining, given the choice of the full adder as the largest indivisible or atomic cell.

A second version (Figure 10) implements a lower degree of pipelining, which allows a maximum of two combinational delays between any pair of registers.

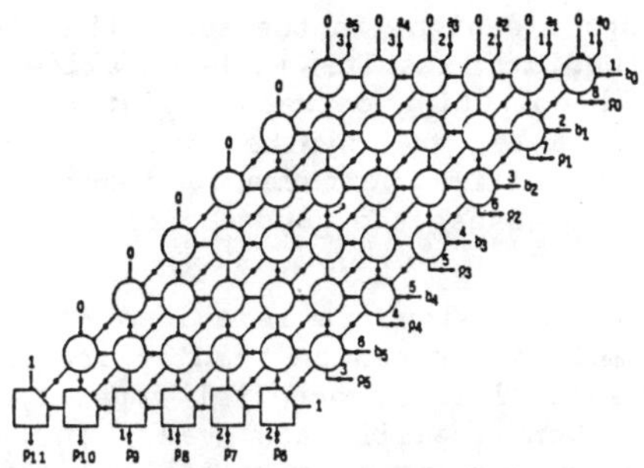

Figure 10. Array multiplier achieved through retiming the multiplier in Figure 8 in such as to provide a combinational path of two cells vertically between some pairs of adjacent cells in the array. Dot conventions are the same as in Figure 9. In this version of the multiplier, a larger amount of combinational delay is provided between register latchings than in the Figure 9 bit-systolic design.

The cell layouts and interfaces are shown in Figure 11, and the layout of a 6x6 bit-systolic multiplier (at the cell level of the design hierarchy) is shown in Figure 12.

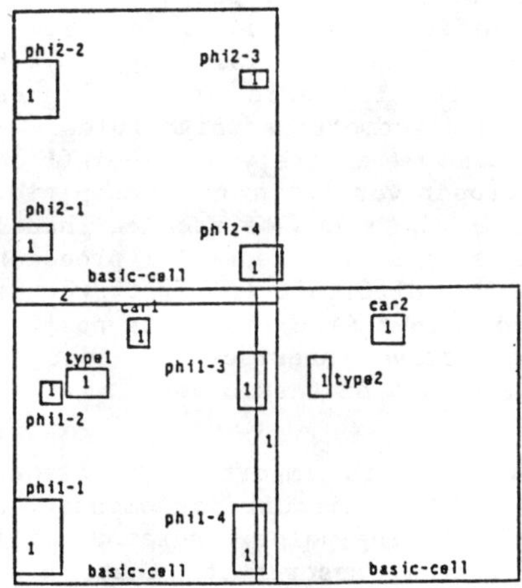

Figure 11.

Graphical realization of a possible layout file (constituent cells and their local interfaces) for a bit-systolic array multiplier.

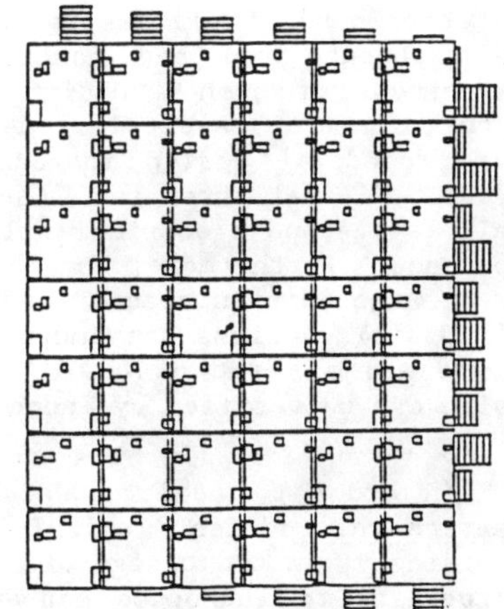

Figure 12.

Layout of a 6x6 bit-systolic array multiplier at the cell bounding box level of the design hierarchy.

The ability to manipulate the amount of combinational delay between registers is exceedingly important in optimized design, because pure architectural manipulation (which ignores the effects of circuit performance and technology constraints) can often result in designs that do not provide an overall optimum. For example, it is possible to introduce too much pipelining into a design, in the sense that too much time is spent in latching up intermediate results without allowing for a sufficient amount of combinational computing between registers. For this reason, it is important to be able to synthesize several different design alternatives, and then utilize performance summaries composed over the entire design in order to assess the goodness of each design alternative.

From the previous discussion, it is evident that the various retimed versions of an array multiplier can be readily specified by a design file in the RSG. Hence, specialized procedural generators for each of these architectural variants can be easily achieved. It is interesting to note that each cell in the array must be personalized with respect to several different options, including cell type, cell interface, register assignment, and clock assignment. There are also edge effects to consider, such as the number of peripheral registers, and the specification of constant inputs along some of the design's edges. The fast generation capability afforded by the RSG, coupled with efficient performance estimation, is the key to modern design exploration, providing the designer with the opportunity to consider many different alternatives at a total system level, where each of the alternatives may be the result of substantial optimization at the architecture, logic, circuit, and layout levels.

8. A Floating-Point Adder Generator

Floating-point adders represent highly complex designs which must be very efficiently laid out in order to provide acceptable performance. On the other hand, it is desirable to parameterize such an adder in terms of the exponent size and mantissa size over a substantial range. The correctness of such a design must be provided in terms of the logical circuit function, electrical design rules, and geometric design rules. In most procedural design instances, the focus is entirely on geometrical layout. Minimal attention is paid to the proper verification of compatible logic levels and electrical wellformedness in the sense of race and hazard avoidance. Furthermore, for these complex designs, the generator procedure has a large "design space," and it may be very difficult to verify whether all possible designs are indeed well-formed. In this case, it is possible to use a small set of cells and their respective interfaces so that any design can be verified by induction on a small set of test cases.

In order to achieve a coherent design, it is important to cleanly separate the modules' logical behavior from their circuit implementation. Accordingly, in the design of a complicated floating-point generator, it is appropriate to introduce conventions for logical description that ensure proper functioning of not only the individual cells, but of the overall system as well.[10] For example, module behavior rules at the logic level must stipulate that each module pin has a "null" state, and one or more "defined" states. Each pin is either an input or an output, but never both. Furthermore, if all inputs are null, then all outputs are null; and if all inputs are defined, then all outputs must be defined. It is exceedingly important in clocked systems that the stipulation of a single-input transition must give rise to a single-output transition. The terminal state of outputs is a function of input terminal state only. The introduction of a null state supplies needed generality in order to properly characterize transitions between defined states, and to permit the use of dual-rail logic. Module composition rules are simple, and specify that outputs cannot be shorted, valid states of connected nodes must conform, and there must be no loops in combinational circuits. For the design of such a floating-point unit, these rules have been implemented with a verification capability that provides for a design's consistent logical specification over all of the constituent leaf cells and their interconnections.

The logic level transition constraints are essential to avoid races and hazards. For example, if a generator permits only geometric composition, then timing delays may be introduced that will lead to logical errors, particularly in clocked circuits such as CMOS domino logic. Although the techniques to avoid these races and hazards are understood, and have been implemented in a few specialized cases, no general purpose procedural layout tools are available at this time. An overall codification of permissible designs at all levels of representation is clearly missing.

The geometric construction of a floating-point adder is facilitated by using fixed-height cells and carefully designed end cells. This approach is very useful for the design of variable mantissa and exponent floating-point units, because each row of arithmetic logic capability in the design can be characterized in terms of a string of internal bit cells together with a left end cell and a right end cell. Because of this structural constraint, it is easy to verify the design's geometric correctness by induction from the smallest versions to those which have an unlimited number of internal bit cells. In fact, it can easily be shown that geometric design rule adherence can be verified for any such design from a simple design that has only two internal bit cells together with the requisite left and right end cells. While such an approach is not amenable to tree-like structures, it is very useful for arithmetic logic units.

Due to the large amount of shifting required in floating-point units, procedural routing capability must also be specified. For variable sized floating-point units, it is not feasible to place all of this routing in predesigned cells. Instead, the routing itself must be generalized in a procedural way, based largely on river routing techniques. This need corresponds to a departure from that capability provided by the Regular Structure Generator, and is usually associated with the capability of a full-chip silicon compiler which must provide for the placement and routing of all constituent cells within the entire chip.

Figure 13 shows an example of a 15-bit floating-point adder designed by this generator. Although the figure cannot be read for detail, it does show the short 4-bit exponent component together with a larger 10-bit mantissa. The shift routing is readily visible. This is an efficient layout which uses dual-rail logic, and it has been completely verified at the logic, electrical circuit, and layout levels. The emerging availability of such coherently constrained procedural generators is an encouraging sign, and indicates the standard to which all generator designs must aspire.

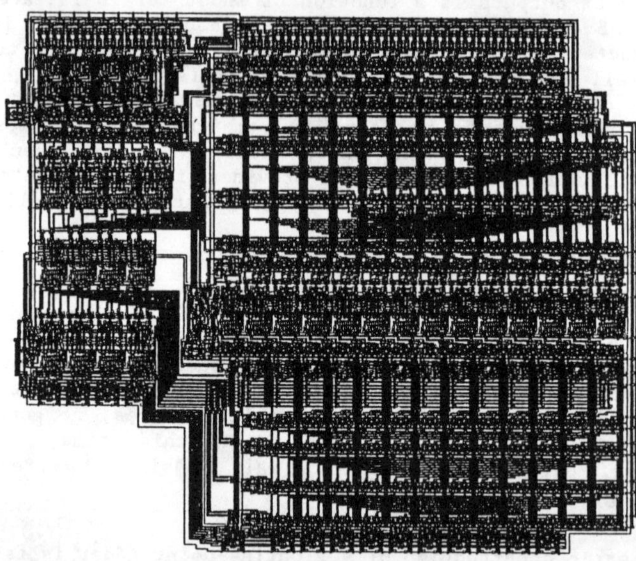

Figure 13. Fifteen-bit floating-point adder layout. Four-bit exponent section is on the left, and ten-bit mantissa section is on the right.

9. The Future

The previous discussion has demonstrated how procedural generators can be constructed for modules that supply correct and optimized designs for a wide variety of functions in an efficient layout that affords adequate performance. The need to consistently separate the various constraint forms into their natural representations has been illustrated, and the way in which these tools can be utilized in design exploration has been discussed.

A unified design discipline is now needed which extends over all of the qualitatively distinct constraint domains including function, architecture, logic, circuit, layout, and device. Currently, no unified representation is available that is capable of simultaneously specifying all of these constraint levels. Yet, this is essential in order to provide a precise representation of design alternatives. From the discussion of multiplier design, the interaction between optimal forms at different levels of representation is apparent. It is clear that designs which are optimized in one area (e. g., architecture), may not be optimal in another area (e. g., electrical circuit performance). For this reason, it is essential to have a unified representation of all constraint levels, and a means to explore their interaction in a systematic and efficient way that utilizes broadly general module generators together with fast and accurate performance estimators. Furthermore, there is a strong need for tools that allow the propagation of changes in one level of representation to all other corresponding representations. This is the so-called "alignment" problem. In its most general form, it is very difficult because the correspondence between the different levels of representation may not be a simple one-on-one projection.

The challenge to provide these new representational and design exploration tools must lead to new design disciplines and formal representations which will serve to place the entire enterprise of digital integrated circuit design on a rigorous basis. It is only through the use of such representations and tools that increasingly complex designs can be generated procedurally. Otherwise, the composition tools are ad hoc, or restricted only to a single level of representation, such as layout. Much of procedural generation has progressed in an intuitive manner, but a complexity barrier is approaching that will demand a consistent and well thought out formal basis. In the future, with this foundation, it will be possible to approach digital integrated circuit design in the context of a well understood theory which affords correctness and optimal performance in a way that is well suited to the perspective of the modern integrated circuit designer.

REFERENCES

1. The L language has been developed by Silicon Design Labs Inc.,
    Liberty Corner, New Jersey, USA.

2. Ayres Ronald F: VLSI Silicon Compilation and the Art of
    Automatic Microchip Design. Prentice-Hall, 1983.

3. Brayton RK, Hachtel GD, McMullen CT, and
    Sangiovanni-Vincentelli AL: Logic Minimization Algorithms for
    VLSI Synthesis. Kluwer Academic Publishers, 1984.

4. DeMicheli G, and Sangiovanni-Vincentelli AL: "Pleasure: A
    Computer Program for Simple/Multiple Constrained/Unconstrained
    Folding of Programmable Logic arrays" Proc. 20th Design
    Automation Conference, (1983), pages 530-537.

    Wong DF, Leong HW, and Liu CL: "Multiple PLA Folding by
    the Method of Simulated Annealing" Proc. 1986 Custom Integrated
    Circuits Conference, pages 351-355.

5. Rudell RL: "Espresso-MV: Algorithms for Multiple-Valued Logic
    Minimization" Proc. 1985 Custom Integrated Circuits Conference,
    pages, 230-234.

6. Glasser LA, and Penfield PL Jr.: "An Interactive PLA Generator
    as an Archetype for a new VLSI Design Methodology." Proc. 1980 Int.
    Conf. on Circuits and Computers, pages 608-611, Rye, New York,
    October 1980.

7. Bamji CS, Hauck CE, and Allen J: "A Design by Example
    Regular Structure Generator" Proc. 22nd Design Automation
    Conference (1985) pages 16-22.

8. Hauck CE, Bamji CS, and Allen J: "The Systematic
    Exploration of Pipelined Array Multiplier Performance" Proc. 1985
    Int. Conf. on Acoustics, Speech, and Signal Processing; pages
    1461-1464.

9. Leiserson CE, Rose FM, and Saxe JB: "Optimizing
    Synchronous Circuitry by Retiming" Third Caltech Conference on
    VLSI, Pasadena, California, March, 1983.

10. Armstrong RC: "Procedural Layout of a High-Speed Floating-Point
    Arithmetic Unit" Research Laboratory of Electronics Technical
    Report 508, Massachusetts Institute of Technology, June 1985.

Abstraction and Layering in Silicon Compilation Tools

Howard E. Shrobe
Symbolics Cambridge Research Center

Abstract

This is a paper about abstractions and layering. Silicon compilers and module generators exist within a broader CAD environment which in turn exists within a broader programming environment. These environments constitute a series of abstraction layers which, if carefully structured, can make the task of building automatic synthesis tools of this type considerably easier.

This paper will describe such a set of layered abstractions. NS [6] is a highly integrated CAD system built on top of the Symbolics 3600 Lisp Machine programming environment [12]. The programming environment provides important abstraction capabilities such as object-oriented programming and the duality of procedures and data. NS provides an additional layer of abstraction through the use of Virtual-Grid Symbolic Layout. This paper will describe two additional layers of abstraction: The first is a set of experimental extensions to NS to support module generation and silicon compilation; this layer provides a duality between graphical and procedural representations as well as an extensible replication protocol. Finally, the paper will describe the Scheme-81 Data Path Generator [10], a complex module generator which was built on top of a predecessor of the current NS environment.

1. Introduction

There is power in specialization. During the last several years many CAD systems have begun to provide a collection of specialized *module generators* for many of the frequently used regular structures such as Rams, Roms, and Plas. These generators provide the end user of the system with a great deal of leverage; they raise the abstraction level of the design process. But the ability to build or extend these generators depends on the power of the underlying environments: the CAD environment within which the generators run, the CAD environment within which the cell library used by the generator is built and the programming environment which supports the coding of the generator.

If the underlying environments do not provide the appropriate abstractions to the developers of a module generator the final product may well suffer. In the worst case, if the underlying environments provide too low of an abstraction level it becomes impossible (in a practical sense) to build higher level tools at all.

This paper is a study in gaining power by appropriately identifying powerful abstractions at a number of different levels. Each level of the system is only a relatively minor extension of the level beneath it. At the base is the Lisp Machine and its software system [12]. This provides powerful programming and graphical facilities, including the Flavors object-oriented programming methodology and the ability to manipulate and store procedures as data.

The next level is the NS integrated CAD environment [6] which is itself internally layered. At the base of this system is a general purpose diagram editor built upon an object-oriented representation of graphical objects. Upon this platform are built a variety of modes for editing the various aspects of a module such as its schematic or its layout. A particularly important abstraction captured at this level is the notion of *Virtual-Grid Symbolic Layout* [1,5,15,16,17]. This allows the designer to ignore design rules and absolute placement and to concentrate rather on the higher level abstraction of relative placement topology.

Built on top of NS is an set of experimental extensions which I will refer to as NS^2C (NS

Silicon Compilation tools). These are intended specifically to support the constrution of higher level module generators. This level introduces a variety of new abstractions such as a duality between the graphical and procedural representations of a leaf-cell design and an extensible protocol for describing nearly regular, array-like structures.

At a higher level still, are module generators such as the Data Path Generator [10]. This was built in a predecessor of NS which provided some, but by no means all, of the facilities of the current NS environment. module generators raise the abstraction level above the layout level altogether, allowing the end-user to think of the module as a functional block described by a set of parameters or perhaps by an application specific description language.

This paper will first consider a relatively simple but illustrative example of a module generator. The purpose of going through this example is to identify features which would make the construction of the module simpler and to motivate the inclusion of these features. The paper will next present the NS environment, concentrating on its basic represenations and on its use of Virtual-Grid Symbolic Layout. It will also briefly survey the other facilities in the NS system. Next it will travel downward through the layers of abstraction to look briefly at the abstractions provided by the Lisp Machine environment which support NS and its higher level tools. Then it will describe the NS^{2}C extensions and how they draw on the underlying facilities. Finally, the Data Path Generator will be presented.

2. An Illustrative Example: An ECC system

An ECC system has the task of determining whether a memory word containing a data field and a *syndrome* field contains an error. The usual goal is to correct all single bit errors and to detect all double-bit errors. However, it is often possible and desirable to detect some classes of multi-bit errors. One of the major components of an ECC system is the *syndrome-calculator* which computes a sequence of bits (typically 7 for a 32 bit word) each of which is the XOR of a specified subset of the bits in the original word. The selection of bits can be specified by a *syndrome-matrix* which shows which data bits enter into each syndrome bit's XOR. ECC systems are used in systems with a variety of word sizes and with a variety of syndrome widths.

Our first observation is that if we are to go to through the effort of making an ECC system at all, then it would pay to *parameterize* the ECC syndrome calculator for an arbitrary sized word, an arbitrary syndrome size and an arbitrary syndrome matrix. Fortunately, the basic design and geometry of the syndrome calculator are consistent with this desideratum.

The basic layout of the syndrome calculator is array-like. The data bits are bussed horizontally while the syndrome signals flow vertically. Each row of the array corresponds to a bit position in the word; each column corresponds to a syndrome bit to be calculated. Each column computes an XOR of those bits of the word specified by the syndrome matrix; thus, the columns are all distinct.

There are a number of different ways to implement the ECC system. I will follow the style used in [7], in which a dual-rail signal is passed down each column; Vdd is fed into the right rail and Gnd is fed into the left. Each basic cell in a column is either a *conditional-switcher* or an *unconditional-passer*. The CONDITIONAL-SWITCHER uses transmission gate logic to conditionally reverse the signals between the two rails. The signals which drives the transmission gates are the value and the complement of the corresponding bit of the data word. If a data bit is high, the signals are reversed between the two rails; otherwise they are passed unchanged. The UNCONDITIONAL-PASSER cell simply passes the signals along the rails unchanged. It follows that the values arriving at the bottom of the rails will be reversed if and only if an odd number of the bits in the word are high.

This system lays out very conveniently if the CONDITIONAL-SWITCHER and UNCONDITIONAL-PASSER cells fall on the same pitch, with their I/O ports lining up. One way to guarantee this is to regard these two cells as two parameterizations of a single CONDITIONAL-XOR cell. A single parameter (called SWITCH?) controls whether the transmission gates for switching the rails are included or not. Figure 1 shows the two configurations of the cell.

The syndrome-matrix can be viewed as a matrix of parameters for the CONDITIONAL-XOR cells in the array. Therefore, the syndrome-calculator as a whole can be regarded as a parameterized cell, where the syndrome-matrix is the value of the PROGRAMMING parameter.

There is still more complexity to this example, but it's useful to draw some lessons before going on:

1. It's important to be able to parameterize cells.
2. Both leaf cells and composition cells can be parameterized.
3. One common form of parameterization of leaf cells is the conditional inclusion or exclusion of components.
4. The (nearly) regular array is one of the most common forms of composition. A common parameterization of this involves chosing the contents of the array slots from a small number of cell types (or from a small number of parameterizations of a single cell type).

Let us return to the design. Unfortunately, there is a very limited word size over which the design as described so far will work. The design in [7] computes 7 ECC syndrome bits for a 64 bit data word. The syndrome matrix includes 33 inputs in each syndrome bit XOR. This requires the signal to pass through 33 transmission gates, which would cause unacceptable delay. Instead, a two-tier approach is used. The first tier computes the XOR of groups of 8 bits; the outputs of the first tier XORs are fed into a second tier XOR which computes the final result. To improve speed the output of each group of eight CONDITIONAL-XORs is sensed by a sense-amp and latched as is shown in Figure 2; the sense-amp outputs are then used as the inputs to the second tier block of CONDITIONAL-XORs which are also sensed by a sense-amp. Obviously, the timing of the precharge and latch signals is critical.

This scheme requires the outputs of the first level sense amps to be *routed* to the final group of CONDITIONAL-XOR cells as is shown in Figure 3. The most compact routing takes the form of a series of nested u-shaped wires routed in a single routing layer. This routing pattern is considerably more efficient than that produced by a general purpose channel router. To use such a router, one merely provides it with the set of points to be connected. These points are, of course, just a set of specific ports of specific cells, namely the output ports of the first level sense amps and the inputs ports of the second level CONDITIONAL-XOR cells. Because of the routing, the columns of the array are no longer contiguous. Space must be left for the routing channels.

It's worth observing at this point that there are further parameterizations of the basic building block cells which would be useful. It is useful to control the size of the transmission gate transistors in the CONDITIONAL-XOR cells, since increasing their channel widths can increase the speed. It is also useful to be able to control the transistor sizes in the sense-amps. Although there are other useful parameterizations, these are enough to illustrate the major points.

We can now add to our list of observations:

1. A common form of parameterization involves controlling the sizes of devices.
2. It is useful to have available simple routers tailored to common situations.
3. It is useful to be able to name components and their ports so that we may easily extract a netlist or other set of points. These are often the input to routers.
4. The layout pitch in a regular array may not always place cells into abutting positions.
5. The design environment must include good simulation facilities to allow speed critical designs to be verified.

In addition, there are two further observations worth making at this point:

1. The layout of leaf cells must be done by graphical editing. Although procedural descriptions of cell designs can be hand coded, they are hard to write and very difficult to extend and maintain.
2. The layout system should be symbolic and design rule independent. There are

two reasons for this. First, this gives technology independence. The cell designs used by a module generator are sufficiently hard to create that we would like to be able to use a single design for a broad range of related processes. Second, it raises the abstraction level of layout and greatly simplifies both graphical editing and the process of hand coding layout procedures in a layout-language.

2.1. Summary of Desiderata:

Reviewing what we have learned from this example, we can see that the environment we need for developing a module generator must include at least the following features:

1. Parameterized leaf and composition cells.
2. Parameterization to control the sizes of devices.
3. Parameterization to control conditional inclusion or exclusion of components.
4. Graphical editing of the layout of Parameterized leaf cells.
5. Symbolic, design-rule independent layout.
6. The accessibility of a location using a sequence of components and port names.
7. Placement of components using such symbolic location descriptions.
8. Programmable variations of the regular array such as chosing the contents of particular array slots from a small number of cell types.
9. Simple routers tailored to common situations.
10. Simulation facilities integrated with design facilities.

These imply that there is need for both a graphical editing environment and for a procedural layout-language and that the two should be closely integrated. In fact, these facilities must be integrated with a yet broader range of tools for simulation and verification.

2.2. A Scenario Illustrating Our Techniques

In this section I will try to convey an impression of how a designer would create an ECC module generator using the extended NS system.

The sense-amp is the electrically most critical component in the design; it also sets the horizontal pitch of each column. Therefore it's the reasonable place to begin. The first step is to enter a schematic for the sense amp and to simulate it. This is done completely through the NS diagram editor interface. After the schematic is entered, the designer enters the SPICE command. A SPICE simulation is begun in the background; when it terminates the designer enters the SPICE PLOT mode in which the screen is divided into a schematic window and two PLOT windows. By pointing at a node in the schematic with the mouse, the designer indicates which nodes he wants displayed in each plot. See Figure 4. Once the simulation gives satisfactory results, the designer can proceed to create a *virtal-grid symbolic layout* of the sense-amp. This may require several trial layouts and compactions to reach a good layout topology. During this process the screen can be split into two windows, one with the virtual-grid layout, the second with the compacted mask diagram. See Figure 5.

The next step is to create the parameterized cell for the CONDITIONAL-XOR. To do this, the designer enters the EDIT GENERATOR command, which brings the NS^{2}C extensions into play. The first step is to create the unparameterized part of the design consisting of the vertical and horizontal bus wires.

The next step is create the conditionalized parts: the transmission gates which are present for the CONDITIONAL-SWITCHER configuration and the wires which replace them for the UNCONDITIONAL-PASSER configuration. First the designer uses the EDIT PROPERTY DESCRIPTOR command to create a new boolean-valued parameter SWITCH? for this cell. Next, the command to create an *overlay* is invoked to create the PASSER overlay. An overlay is a named group of components which are collectively selected by an *overlay condition*, a lisp expression whose value is treated as a boolean switch controlling whether the parts in the overlay should appear in the diagram or not. For the PASSER overlay, the overlay condition is simply (NOT SWITCH?). Only the parts in the *current-overlay* are available for editing, however, the parts in the other overlays are displayed as a background. The PASSER overlay contains a few wires to replace the transmission gates which are omitted in this configuration. Next the designer creates the SWITCHER overlay whose overlay condition is SWITCH?. The designer puts down the transmission gates in the SWITCHER overlay. Figure 6 shows the overlays in this design.

The transistors in the transmission gates need to have parameterized channel-widths. The designer creates a SIZE parameter for the cell. He then uses the SYMBOLIC ATTRIBUTE command to state that these transistors' channel-width parameter can be expressed *symbolically* by the lisp expression SIZE, i.e that the channel-width is equal to the value of the SIZE parameter of the cell. The system keeps track of the fact that the channel-width of these transistors depends on the value of the size parameter of the cell; it has also kept track of which of the cell's parameters enters into each overlay's overlay-condition.

At this point the designer might want to see how the cell will look with various parameter settings. To do this, he uses the SET CELL PARAMETER command to select values for the cell's parameters. When a cell parameter is assigned a new value, it is propagated to those components and overlays which depend on it. The parameters of the dependent components are recalculated and the display is updated accordingly. Those overlays whose overlay-conditions are now false are made invisible.

Having seen how the cell will appear in its various parameterizations, the designer choses to DEFINE the cell. The NS editor creates a *generator*, a Lisp procedure which, when run with a set of parameters, builds a version of the cell corresponding to these parameters. This procedure contains one PART form to build each component in the unconditionalized part of the design. Each of these contiains the name of a cell-type and an alternating series of attribute names and values. In addition, the generator has an OVERLAY form for each overlay, containing the overlay's name, the Lisp expression for its overlay-condition, and PART forms for each component in the overlay. For those component attributes which have a *symbolic attribute value* the lisp expression is used, otherwise the actual value is used. The generator produced is shown schematically below.

```
(defaspect-generator (conditional-xor :virtual-grid) (size switch?)
   (part vg-log :from (pt -4 1) :to (pt 5 1) :layer 'poly :name 'd-strap)
   (OVERLAY SWITCH SWITCH?
      (part n-channel-mosfet :origin (pt 4 1) :orientation 'rot90
            :width (* 2 size))
    ...
   )
   (OVERLAY PASS (NOT SWITCH?)
      (part vg-log :from (pt 4 2) :to (pt 4 0) :layer 'metal)
      (part vg-log :from (pt -3 2) :to (pt -3 0) :layer 'metal)))
```

Now that the building block cells are available, the next step is to create the module generator for the whole ECC. This will be a relatively complex procedure and it will be awkward to design it completely by using the graphical editor. However, much of it is convenient to do this way, so the designer uses the editor to create as much of the procedure as possible. Then he asks the editor to DEFINE the diagram. In response, the graphical editor creates the generator corresponding to the current diagram and also writes the code for this generator into a Lisp Machine text-editor buffer. The designer then proceeds by editing this code and compiling it from the editor buffer. Each time this is done, NS gives the designer visual feedback, updating the display of the ECC systems to correspond to the new generator definition.

What is required for the ECC generator is a regular array which in each slot contains either a sense-amp or one of the two parameterizations of the CONDITIONAL-XOR cell. The system has available *replicator* facilities which can easily be configured to produce the desired behavior. The replicator is passed two numbers indicating how many rows and columns of cells there should be, a *component-cell-list* of the three cells (order is significant), and a *selection-matrix*, an array which has in each of its slots the numerical index of the appropriate cell from the component-cell-list. The *selection-matrix* is almost identical to the syndrome-matrix except that after every eight rows a new row is inserted with each entry in the row containing the index of the sense-amp cell. The array is then automatically built. There are a variety of replication facilities covering many typical ways of building array-like structures; however, since there are always new patterns called for, these facilities are modular and can be combined in new ways or extended easily.

Next the first tier sense amps need to be routed to the second tier CONDITIONAL-XORs. This routing is one of the simple patterns provided in the system, the U-ROUTER. This router is called with a list of points. These are fetched by asking the first-tier sense-amps for the location of their output ports and the second-tier CONDITIONAL-XOR cells for the location of their input ports. The sense-amps are easily found since they are the elements of the array in the rows whose indicies are even multiples of 9 (there are eight CONDITIONAL-XOR and a sense-amp in each block). The *path* (>> OUTPUT-PORT (0 9) ARRAY), for example, finds the location of the output port of the first sense amp. The U-ROUTER returns a routing cell. Since we need one such routing block for each column of the cell array, we make a REPLICATION of the routing blocks. Since the routing columns need to fit between the columns of the cell array, we need to change the HORIZONAL-SPACING parameter of the cell-array to be the width the routing column. We can use the path (>> WIDTH ROUTING-COLUMN) (whose value is the width (in virtual-grids) of the routing column) and then supply this as the value of the horizontal-spacing parameter of the cell-array. This will cause the array to be rebuilt with the correct spacing.

Finally the replication of routing columns needs to be placed so that its columns interleave with the columns of the cell array. This is done by *glue*ing together the first input port of the first routing column to the output-port of the first sense-amp in the first column of the cell array. The GLUE procedure is passed a path specifying each of these points.

At this point, the designer will probably use other NS tools to verify the design. The ISO program can verify that the virtual-grid layout is, in fact, consistent with the schematics. This is a valuable cross-check which often catches simple, but deadly, mistakes. If there is a mismatch this is highlighted on the screen. The switch-level simulator RSIM can be used to simulate a circuit this size quite rapidly. There is a both a graphical interface and a simple procedural interface. Three functions SET-VALUE, VALUE and SIM-STEP allow a program to set a nodes value, to test its value, and to simulate the consequences of changing a node's value. Using these it is simple to build a Lisp test procedure which could simulate how the ECC module would correct data received from a memory system with random errors. Such a test can be coded in a few minutes and can run a quite exhaustive test suite in 20 - 30 minutes.

There are a few more steps needed to finish off the design, but there is little to be gained by pursuing this example further.

3. The NS Environment

3.1. The Major Abstractions of NS

NS [6] is a system within which modules like the ECC can be easily designed. NS is implemented on the Symbolics 3600 Lisp Machine [12] and depends crucially on features provided by that environment. The general structure of NS follows from two observations: First, that a design should be captured by an object-oriented data representation which can be manipulated either by a layout-language or by a graphical editor. The second observation is that layouts can be expressed in a symbolic form using a *virtual grid* [5,17]; such layouts are *compacted* into a representation of the physical mask data.

The layout-language and the graphical editor exist in a common, large, uniform address space; they call on a common set of facilities which manipulate the objects in the data representation. These objects are representations of transistors, wires, contacts and whatever other entities are appropriate. The layout-language is embedded in LISP; it has the full use of the base language, augmented by facilities for manipulating the geometrical data representation underlying NS. Since layouts may be created either graphically or procedurally, NS takes the rather unusual step of using the layout-language as its permanent representation (this technique came from Daedalus [3]). This is facilitated by the Lisp environment which provides a simple representation for procedures that may easily be generated by a program.

3.1.1. The Virtual Grid

Layouts are represented symbolically using a virtual-grid. The objects manipulated in a layout are symbolic representations of the electrical circuit elements: mosfets, wires, contacts, etc. They are not mask rectangles. These symbols, however, are displayed in a form which looks very much like the real mask geometry. Certain of the primitives have parameters; for example, the channel length and width of a mosfet, expressed as multiples of the minimum size allowed in a process.

The grid upon which the objects are drawn is symbolic. The grid represents only relative placement, not physical spacing. If one object is drawn on a grid to the left of another, then this relative placement will be preserved in the final artwork. However, there is no significance to empty grid lines. It is the job of the *compactor* to turn the virtual grid into a physical one by spacing the grid lines far enough apart so that all layout design rules are satisfied. The output of the compactor is a *mask diagram*, i.e. the physical geometry corresponding to the virtual-grid layout. The user manipulates only the virtual-grid layout, remaining unconcerned with design rules. The NS compactor guarantees that objects drawn on the same grid line will remain aligned.

The virtual-grid compaction strategy is two-tiered. Leaf cells are compacted into mask diagrams. However, it is important that the assembly of leaf cells into larger modules can still proceed in the virtual-grid framework. In particular, it is important to guarantee that connectivity expressed in the virtual-grid framework be maintained in the physical mask diagram. Consider a SWITCHER and a PASSER version of the CONDITIONAL-XOR placed to have abutting ports. If the virtual-grid layout of the two cells are compacted separately they will be compressed to different sizes and the ports will not line up in the mask diagram as is shown in Figure 7. If the entire diagram were treated as a single compaction problem, this problem would not arise. However, in that case, aligned virtual grids from different cells which had no port in common would still be treated as a single grid, forcing unrelated objects from different cells to line up. This would result in an inferior compaction. The solution to this problem is the second tier of the compaction system, the *pitch-matcher* [1] whose job it is to stretch the compacted leaf cells so that connection points do align in the mask diagram. Figure 8 shows the results produced by the pitch-matcher.

The advantage of using the virtual-grid is that a designer or a module generator procedure can manipulate the design at a level of abstraction higher than physical geometry. The virtual-grid abstraction lies between physical mask data and a schematic; its entities are the same circuit elements one would see in a schematic. However these are augmented with enough topological information so that they can be displayed in a geometrical form similar to that of the mask. Furthermore, they can automatically be transformed into real mask data by the compaction tools. Most importantly, neither a designer creating a design by hand nor a procedure creating one automatically need ever worry about the intricate variety of physical spacings required by the design-rules. This is a major advance over physical layout-languages such as DPL [3].

3.1.2. Extractors, Networks and Verfication Tools

There is a single representation for electrical networks used in NS. The SPICE and RSIM [14] simulators are driven from this one representation. Also the ISO network comparison program works from this representation; it uses a *graph-coloring* algorithm [8,11,13] to determine whether any two electrical networks are identical. There is also an ERC program which checks that basic electrical rules are being obeyed.

For each electrically meaningful type of diagram (schmatic, virtual-grid layout, mask) there is an *extractor* which produces an electrical network corresponding to the diagram. The network is annotated with pointers to the objects in the diagram and similarly the objects in the diagram are annotated with pointers to nodes or devices in the network. These annotations allow a user to interact with the verification tools in a very natural way. Because the objects in the diagrams know which nodes in the network they correspond to, the user may tell NS to plot a node from a SPICE simulation simply by pointing at a wire in the schematic. Similarly, when using RSIM interactively, the user can set or test a node's value by pointing at an appropriate wire in a virtual-grid layout. Each of these simulators can be driven from any type of diagram. Interaction with RSIM can also be driven by a procedure. The annotations on networks are used by the ISO program. For example, when ISO discovers a node in a schematic which it cannot match to a node in a virtual-grid layout, it can illustrate the problem for the user by highlighting the unmatched node on the window containing the schematic.

3.2. An Overview of The NS Representations

3.2.1. The Hierarchy

A system design consists of a set of descriptions covering many different aspects of the design, such as the logical, electrical, functional, or physical structure. One of the goals of the NS representation is to provide an integrated design system that captures the entire description and maintains links between each of these aspects.

A design is organized as a hierarchy of *modules*. It is useful to think of the design as a single hierarchy, where each module has several *aspects*. Some of the aspects are diagrams, such as a schematic or mask artwork, some are textual, and some are program-generated data. Currently NS supports LAYOUT diagrams (i.e. Virtual-Grid Symbolic Layouts), MASK diagrams (i.e. physical mask layout generated by the compactor), SCHEMATIC diagrams and SCHEMATIC ICONS (the diagram used to represent a module in another module's schematic) as well as an ELECTRICAL NETWORK aspect. The aspects of a particular module are different representations of the same object that should be consistent.

Aspects of modules may be parameterized. The collection of variant aspects is held in an *aspect-selector*, which provides a data index for fast retrieval of that aspect which corresponds to a given a set of parameters. In cases where there is little commonality between the alternate diagrams, they can be entered as separate diagrams through the graphical editor. However, in many other cases, there is enough commonality to merit the use of a procedure to generate the aspects on demand. In this case, the aspect-selector will contain a *generator*.

In addition to diagrams, there are various *primitives* such as *lines* and *points*. A diagram contains a set of parts which may be primitives or *diagram-instances* of other diagrams. A diagram-instance contains a geometrical transform, a pointer to the diagram of which it is an instance as well as the values of its attributes.

An object is constructed by providing the appropriate module with an aspect name and a list of pairs of attribute names and their desired values. The module first provides default values for any attributes that were not provided. If the request is for a primitive object, a new object is always created. Otherwise, the module asks the appropriate aspect-selector either to find or create the requested aspect. The aspect-selector searches its data index to see if it already has the appropriate aspect. If so, this aspect is retrieved and a diagram-instance pointing to it is built. Otherwise, if there is a generator, it is run. The newly built aspect is then indexed and treated as above.

One other important type of object in NS is a buffer. Buffers are the objects which hold a representation of a diagram being edited. In most regards a buffer is very much like a diagram. However, it has associated with it a set of windows (called *Views* in NS) upon which it displays the design. Also to facilitate fast retrieval of objects based upon geometric criteria, buffers store their parts in Quad-trees.

3.2.2. The Graphical Editor

Although there are many types of diagrams in NS, there is only a single diagram editor. This makes it possible for the user to move freely between different aspects of the design hierarchy as well as up and down the hierarchy. It also means that the user must only learn one command interface to edit all types of diagrams. For each different type of diagram there is a corresponding mode that customizes the available commands and primitive objects appropriate to the diagram type.

The diagram editor is used to draw, display, and manipulate the basic graphical primitives such as lines, points, circles, and arcs. These basic primitive objects are specialized into objects specific to the various domains. For example, the graphic primitive BASIC-LINE is an object that has pointers to its end points and knows how to stretch when its end points are moved. Schematic WIRES are BASIC-LINEs that can have signal names. VIRTUAL-GRID LOGs are BASIC-LINEs that have a layer and a signal name and display as a stipple on a black and white screen, or in the layer's color on the color screen. This layered approach to implementing the domain-specific primitives makes it a simple matter to implement new diagram types. Later, we will see how *Flavors* facilitate this layering.

3.2.3. The Basic layout-language

NS stores diagrams as procedures written in an extension of Lisp. The basic layout-language merely adds a few special forms: DEFDIAGRAM, DEFASPECT-GENERATOR and PART. Defdiagram is a form used to define a procedure which can generate a single version of an aspect; it is used to store unparameterized diagrams. The following is a typical use of DEFDIAGRAM:

```
(defdiagram (xor-base :virtual-grid)
   (part vg-log :from (pt -4 1) :to (pt 5 1) :layer 'poly)
   (part vg-terminal
         :x -4 :y 1
         :signal-name "d"
         :connected-layers '(poly))
   ...
   )
```

A DEFDIAGRAM form contains the module's name (XOR-BASE in this case) and the aspect (:VIRTUAL-GRID), followed by a list of forms to be executed. Normally these are simply a sequence of PART statements, but the full generality of Lisp can be used if appropriate. When an NS object is created, it becomes part of some diagram; which diagram can be specified using the :SUPERIOR attribute in the PART form. If no :SUPERIOR is specified the object becomes a part of the *default-superior-diagram* which is always pointed to by the variable *ME*.

A PART form contains a module name (e.g. VG-LOG) followed by an alternating sequence of attribute-value pairs. The result of executing such a form is to create a new object with the specified attribute values.

A DEFASPECT-GENERATOR form is similar to the DEFDIAGRAM form, except that it is used to build parameterized generators:

```
(defaspect-generator (xor :virtual-grid) (switch size)
   (part xor-base :origin (pt 0 0) :name 'xor-base-2)
   (when switch
     (part vg-log :from (pt -1 -3) :to (pt 4 -3) :layer 'metal)
     ...
     )
   (loop for i below size
         do (part vg-log :from (pt 4 i) :to (pt 4 i) :layer 'metal)))
```

A DEFASPECT-GENERATOR form takes a module name and a aspect name followed by a list of *parameters*. These may be referred to within the body of the generator to control device sizes, loop iterations or conditionalization, for example.

4. The Lisp Machine Environment

There are three abstractions of the Lisp Machine programming environment which facilitate the construction of a tightly integrated system as extensive as NS. These are:

1. The *Flavors object-oriented programming* system embedded in Lisp. NS objects are represented as flavor instances; facilities in NS communicate by passing messages and through their common access to the objects in the environment.
2. A large *uniform virtual address* space within which all procedures and data are uniformly accessible. The *Garbage collector* reclaims inaccessbile storage

dynamically, making it available for new allocation. This creates the abstraction of a nearly infinite memory within which NS's objects exist.

3. The *procedure-data duality* in which procedures have a well-defined data representation. Procedures such as an aspect-generator may be stored within data structures such as those which represent NS's aspect-selectors and invoked by virtue of their location in these structures. Furthermore, procedures may be created by other procedures such as by the NS graphical editor. *Dynamic-linking* allows a procedure to be created within the execution environment or to be brought into the environment from permanent storage at any time; when a procedure enters the environment it is executable immediately. Thus NS can create (or load) generators at any time and use these to generate new diagrams on demand.

4.1. Flavors and Protocols

All objects in NS are implemented in the *Flavors* system, an object-oriented extension to the LISP language. A Flavor roughly corresponds to the idea of a Class in Smalltalk or Simula. Each flavor defines a set of *Instance-variables* which any instance of the flavor will possess and a set of *Methods* to which any instance of the flavor will respond. The values of the instance variables are private to each instance of the flavor. A method is simply a named LISP function, compiled to have fast access to the instance variables, which is run whenever an instance is sent a message with the same name as the method.

As in Simula and Smalltalk, the set of messages handled by a particular flavor consists of those messages handled directly by the flavor plus those handled by any of its superiors. In contrast to the hierarchical classes of Simula and Smalltalk, a Flavor may have more than one superior Flavor from which it inherits methods. The existence of multiple-superclasses leads to a different view of inheritance. One doesn't think of a Flavor as *inheriting* behavior from its superior flavors, as much as one thinks of the flavor as *mixing together* the behavior of its various superiors into a larger, more complex behavior. In fact, the natural style of using flavors is to build *mixin* flavors which capture some basic packet of behavior. These mixins are then combined to produce more complex objects which exhibit behaviors gotten from each of its component flavors.

In using Flavors one concentrates on the set of *protocols* that the objects must obey. By a protocol we mean a set of messages that the object must respond to in an appropriate way. As an example consider NS's MODULES, DIAGRAMS and BUFFERS. Modules and diagrams both need to follow the file-saving protocol which means that in response to a SAVE message, they open a file and write a textual representation into it. Buffers (which are temporary workspaces) do not follow this protocol; they cannot be directly saved in the file system. On the other hand, buffers and diagrams both follow the DISPLAY protocol; they each hold a set of parts and display themselves by asking each of their parts to display itself. Modules (which are an aggregation of various aspects) have nothing to do with this. Thus, we get a non-hierarchical pattern of inheritance.

In this example the inherited behaviors from the component flavors are independent. But, this is not always the case. The Flavor system also implements an elegant means for combining nearly (but not totally) independent packets of behavior. This is brought about by the ways in which methods from separate component flavors are combined at the time a new flavor is defined. Often two or more component flavors each want to contribute some part of the overall response to a particular message. Consider the problem of producing a textual descriptions of GRAPHICAL-LINES, WIRES, and VIRTUAL-GRID LOGS. All of these objects are lines which run from one point to another; therefore, they all want their textual description to begin as:

```
(part <type of object> :from <pt1> :to <pt2>)
```

However, VIRTUAL-GRID LOGS and GRAPHICAL-LINES have a width property; VIRTUAL-GRID LOGS and WIRES have signal-name properties. This leads to the inheritance graph shown in Figure 9.

We would like to have the HAS-SIGNAL-NAME mixin provide only the behavior peculiar to it.

The Flavor system facilitates this through the use of *:before* and *:after* methods. When two flavors are combined, the flavor system finds all the :before and :after methods contributed by any of the component flavors. It then builds a *:combined* method for the composed flavor which is a procedure which first calls all of the :before methods, then calls the main method, and then calls each of the :after methods. In our example, the HAS-SIGNAL-NAME and HAS-WIDTH flavors only need to provide :after methods for the :TEXT-FORM message. The :after method associated with HAS-SIGNAL-NAME dumps out the SIGNAL-NAME and the :after method for HAS-WIDTH dumps out the WIDTH. BASIC-LINE provides the code to dump out the two points. It also needs to dump out the type of the object; it can do this by sending itself a :TYPE message which is responded to by the most specific flavors, namely VIRTUAL-GRID-LOG, WIRE, and GRAPHICAL-LINE. Only the following code need be provided:

```
(defmethod (has-signal-name :after :text-form) (stream)
   <code to print the signal name on the stream>)
(defmethod (has-width :after :text-form) (stream)
  <code to print the width on the stream>)
(defmethod (basic-line :text-form) (stream)
  <code to print the word "part",
       the type of the object,
       and the two points>)

(defflavor basic-line (point1 point2) ())
(defflavor virtual-grid-log () (basic-line has-signal-name has-width))
(defflavor wire () (basic-line has-signal-name))
(defflavor graphical-wire () (basic-line has-width))
```

4.1.1. Uniformity, Dynamic Linking and Garbage Collection

In the Lisp Machine environment objects and procedures (procedures are just one distinguished type of object) live within a single virtual address space. One procedure can call another at will, passing references to whatever objects seem appropriate; since all the objects live within the same environment, there is no question of whether the access is possible. Since the procedures live within the same environment, there is no need to transform the objects into a textual form suitable for an external medium. The object, together with all its rich interconnections to the rest of the world, is simply passed onto the callee.

Procedures in our system are recompiled from within the environment. The editor and compiler exist in the same domain as other system and user facilities. Whenever a procedure is recompiled, it is linked into the environment immediately and automatically; running programs will now call the new version of the procedure rather than the old. This dynamic linking means that program development can proceed at a much faster pace. It also means that one part of a program, for example, the NS editor, can create a procedure, such as a generator, to be used by another part.

Procedures and data which are no longer accessible to anyone (the old version of a procedure which has been replaced, for example) are considered garbage. The system is responsible for periodically reclaiming all such garbage and making the reclaimed space available for new allocation. Programmers never reclaim space and never worry about deallocation. There is never the problem of incorrectly freeing space which is still in use.

In summary, our programming environment has several unique features which qualitatively distinguish it from conventional programming systems. Procedures are first class objects which may be combined in novel ways using the Flavors system. All data and procedures live within a uniform virtual memory system which allows open communication between whatever facilities need to communicate. Procedures transmit objects, not character stream representations of objects; there is no loss of the rich set of interconnections when one part of the system communicates with another. Procedures may be freely modified, updated, and extended. Unused virtual storage is reclaimed by the system, creating the illusion of an infinite pool of storage. The collection of all these facilities has the twofold effect of raising the level of abstraction and increasing the degree of integration achievable.

5. The NS²C Extensions

The facilities in NS²C were motivated by experience with the Data Path Generator [10] used in the Scheme-81 chip [2]. The NS²C extensions consist of extensions to the NS editor, such as *overlays* and *symbolic attribute values*, which facilitate the graphical creation of parameterized cells. There are also linguistic extensions, such as the *replication* facility, *paths* for symbolic naming of components and the *glue*ing together of components by using *paths*, which facilitates the writing of module generators. In this section, we will look at how these are implemented and in particular at how they take advantage of the abstractions provided by the NS environment.

5.1. Graphical Extensions

Since there is a rather limited set of types of parameterizations that are useful in leaf cells, graphical facilities can be provided which allow these parameterizations to be expressed conveniently without sacrificing the naturalness of graphical cell design. The ability of a procedural language to iterate, conditionally branch, and treat symbolic expressions as parameters provides a natural means for capturing the kinds of parameterizations that occur in leaf cells. The graphical editor is embedded in the LISP environment which provides a simple means to treat procedures as data and vice versa; this allows the editor to write and install new *generators*. The textual form of a generator procedure can be edited by hand if that proves useful.

GENERATOR-EDITING-BUFFERs are a special kind of BUFFER used for the graphical editing of generators. This flavor is built by mixing the GENERATOR-EDITING-MIXIN flavor with the normal BUFFER flavor. This mixin provides the *overlay* and *symbolic-attribute-value* facilities. GENERATOR-EDITING-BUFFERS maintain a list of attribute names (which will be the formal parameters of the generator) and a list of current values of these attributes. Whenever an attribute's value is changed, the display is automatically updated so that what appears in the window is an accurate representation of what the generated cell would look like if the parameters of the cell were set to these values.

5.1.1. Overlays

Overlays are the means to graphically specify conditional inclusion of a collection of components. An overlay is another buffer attached to the primary buffer. An overlay maintains a small amount of extra information: its name, its *overlay-condition*, and the buffer to which it is attached. A buffer built on the GENERATOR-EDITING-MIXIN flavor maintains a list of all the overlays associated with it.

Each overlay (like any other diagram or buffer) contains a list of parts contained in it. The intent is that all parts in an overlay should be included in the diagram if and only if the OVERLAY-CONDITION evaluates to true in the current environment. Within the editing environment this is accomplished by having a *visibility* property on each overlay which controls whether the parts in the overlay are actually displayed. The *chart* mechanism (to be described in the next section) is used to guarantee that an overlay's visibility is on only when the buffer's attributes have values that would cause the OVERLAY-CONDITION to evaluate to true.

Buffer's normally create the code for a diagram by sending each of the objects in the buffer a :TEXT-FORM message. GENERATOR-EDITING-BUFFERS in addition send a :TEXT-FORM message to each overlay associated with the buffer. The overlay responds by producing an OVERLAY form containing PART forms for the objects contained in the overlay. The generator for the CONDITIONAL-XOR looks like:

```
(defaspect-generator (conditional-xor :virtual-grid) (size switch?)
  (part vg-log :from (pt -4 1) :to (pt 5 1) :layer 'poly :name 'd-strap)
  (OVERLAY SWITCH SWITCH?
    (part vg-terminal :x 4 :y 0 :contact-layers '(metal n-diff))
    (part vg-terminal :x -3 :y 0 :contact-layers '(metal n-diff))
    ...
  )
```

```
(OVERLAY PASS (NOT SWITCH?)
    (part vg-log :from (pt 4 2) :to (pt 4 0) :layer 'metal)
    (part vg-log :from (pt -3 2) :to (pt -3 0) :layer 'metal)))
```

OVERLAY is a Lisp *special-form* or *macro*, meaning that it evaluates in a special way. It treats its first argument as an unevaluated symbol which is the name of the overlay. Its second argument is treated as the OVERLAY-CONDITION. Its remaining arguments are evaluated only if the condition's value is true. It behaves as if it were really the following form:

```
(When (not switch?)
    (part vg-log :from (pt 4 2) :to (pt 4 0) :layer 'metal)
    (part vg-log :from (pt -3 2) :to (pt -3 0) :layer 'metal))
```

However, this is an oversimplification which describes how the OVERLAY form behaves when it is being evaluated within a generator to create a normal diagram. But there is another context for executing the generator code; this is when we want to initialize a GENERATOR-EDITING-BUFFER to edit a generator graphically. In that case, the overlay form cannot be treated as a WHEN form, since we want to have the contents of all the overlays around to edit. When evaluated in this special context, an OVERLAY form simply builds a new overlay with the given name and OVERLAY-CONDITION; while evaluating the rest of the expressions within the OVERLAY form, the overlay, not the buffer, is regarded as the *default-superior-diagram*. Although this sounds complex, the following Lisp code is the definition of the OVERLAY macro.

```
(defmacro overlay (name predicate &body body)
  `(let ((*me* (if (generator-editing-buffer? *me*)
                   (send *me* :add-overlay ',name)
                   *me*)))
     (when (or (generator-editing-buffer? *me*) ,predicate)
       ,@body)))
```

A lisp MACRO is a procedure which produces not a value, but rather another form that should be evaluated instead of the original form. This is one of the standard ways of extending the Lisp syntax. In NS code, *ME* is always bound to the *default-superior-diagram*; when preparing to edit a diagram this is the current buffer. Thus, the following form:

```
(OVERLAY pass (not switch?)
    (part vg-log :from (pt 4 2) :to (pt 4 0) :layer 'metal)
    (part vg-log :from (pt -3 2) :to (pt -3 0) :layer 'metal))
```

is treated as if it were:

```
(let ((*me* (if (generator-editing-buffer? *me*)
                (send *me* :add-overlay 'pass)
                *me*)))
  (when (or (generator-editing-buffer? *me*) (not switch?))
    (part vg-log :from (pt 4 2) :to (pt 4 0) :layer 'metal)
    (part vg-log :from (pt -3 2) :to (pt -3 0) :layer 'metal)))
```

5.1.2. Symbolic Attribute Values and The Chart

The second feature provided by the GENERATOR-EDITING-BUFFER is the ability to relate parameters of objects in the buffer to parameters of the buffer. This relationship is then transferred to the generator itself so that parameters of generated objects are computed by symbolic expressions involving the parameters of the generator. For example, the generator for CONDITIONAL-XOR has a parameter controlling the channel-width of the transistors in the diagram; part of the generator code looks like:

```
(defaspect-generator (xor :virtual-grid) (size switch?)
  (overlay switch switch?
    (part n-channel-mosfet :origin (pt 4 1) :orientation 'rot90
          :WIDTH (* 2 SIZE))
    ...))
```

Which says that the channel-width of the transistor should be twice the value of the SIZE parameter of the generator.

In graphically editing the generator definition we want to be able to specify and illustrate this kind of relationship. As mentioned above, the GENERATOR-EDITING-BUFFER maintains a list of attributes which will play the role of parameters to the generator. In addition, it maintains a special data structure called *the chart* which keeps track of how the symbolic attribute values of a component depend on the attributes of the buffer. The chart associates two pieces of information with each buffer parameter. The first of these is the *current value* of the parameter. The second is a list of *dependency entries*, each of which contains three items: the dependent object, the name of an attribute of this object, and a symbolic expression for the value of this attribute.

The SET SYMBOLIC ATTRIBUTE command is enabled while editing in a GENERATED-EDITING-BUFFER and enables the designer to specify a *symbolic attribute value* for any parameter of any object in the buffer. The mouse may be used to select the object and menus are used to select the attribute name. Once this is done, the user types in a Lisp expression showing how this object's attribute should be calculated.

This leads to one or more new dependency-entries being made in the chart. These are calculated by using a feature of the Lisp evaluator which allows the evaluation of an expression to be monitored by another program (the *evalhook* feature). When a SYMBOLIC ATTRIBUTE VALUE is provided for an object, NS^{2}C evaluates the symbolic attribute expression using the *evalhook* feature. During this monitored evaluation NS^{2}C looks for evaluation steps where a symbol is to be interpreted. If the symbol is one of the buffer's parameters then a dependency-entry is made in the chart linking the buffer-parameter to the object, attribute name, and symbolic expression. Secondly, the CURRENT-VALUE of the buffer parameter is returned as the value of the symbol and evaluation continues. This special evaluation process, therefore, leads both to a value for the expression and to a set of dependency-entries in the chart.

The chart has two uses. The first is to help illustrate how the generator will appear when given different parameter values. There is a command which allows the designer to assign new values to parameters of the GENERATOR-EDITING-BUFFER. When this is done, the CHART is consulted to find objects which have symbolic attribute values which are dependent on the parameter. The value of the dependent object's attribute is recalculated using the EVALHOOK mechanism described above and the object is then sent a message to change its attribute to the new value. This will cause the object to be rebuilt to correspond to the new values and will update the display as well. The VISIBILITY attribute of overlays is also managed using this mechanism, keeping the the conditionalized inclusion of objects up to date as well.

The code in a generator should use a SYMBOLIC-ATTRIBUTE-VALUE if there is one, since this leads to parameterized behavior. However, there is still the problem of graphically editing the generator code at a later date. To do this properly, the symbolic attribute expressions need to be brought back into the editor. This is done by adding to the text form for an object an additional attribute which lists the symbolic attribute expressions:

```
(defaspect-generator (xor :virtual-grid) ( size switch? )
   (overlay switch switch?
     (part n-channel-mosfet :origin (pt 4 1) :orientation 'rot90
          :WIDTH (* 2 SIZE) :name 'nmos-82
          :SYMBOLIC-FORMS '((WIDTH (* 2 SIZE))))
  ;;;
  ))
```

5.2. Linguistic Extensions

The NS^{2}C extensions also include two new additions to the layout-language which make it easier to hand code module generators. The first of these allows objects and their ports to be identified and attached by using symbolic *paths*. The second extension is the replication facility which allows the construction of almost regular arrays. All of these facilities depend greatly on the Flavors system and on the abstractions provided by the NS environment. Cumulatively they amount to less than 2000 lines of code (including comments and blank lines).

5.2.1. Paths and Glueing

In those cases where a generator is created by hand coding, it becomes very important to be able to locate objects and their ports symbolically. We have adapted the *paths* concept from DPL [3]. A PATH is a route through the hierarchy to a specific object or location whose value is expressed in the coordinate system of the asker. For example, we might want to know the location of the left end of the D wire of a particular cell of the cell-array, for example the one in row 5 and column 3. The *path* for this is:

```
(>> cell-array (5 3) D :left-end)
```

Which says: "Start with the cell array, find the object in its fifth row and third column, find its part whose name is D, and then find what appears to me to be its left end". It is important to be clear about the intended coordinate system of a question. For example, D in the above path is inside a cell (the 5 x 3 cell, to be precise) of the cell array and this cell may be rotated or mirrored. Suppose that this enclosing cell has been rotated 90 degrees counter-clockwise, then what appears to be its left end in the top-level coordinate system, would appear to be its bottom in the coordinate system of the enclosing object. The answer returned after following such a path is in the top-level coordinate system.

Path following is implemented by a simple *protocol*. Each type of NS object provides a method for its part of the process; each of these methods are required by the protocol to receive a *partial path* and a geometric transform. If the object is a PRIMITIVE, it expects the partial path to have exactly one item in it which is a meaningful name (such as the name of an end-point or a message name for which the object has a method). It returns this point, transformed by the transform passed in. A DIAGRAM-INSTANCE composes its transform with the passed-in transform and then passes the partial path and the new transform on to its diagram. DIAGRAMS expect the first item in the partial path to be the name of a component, the name of a port, or a message name for which the diagram has a method. If the item is the name of a component or of a port, the diagram locates this object, strips the first item off the path and passes the rest of the path in as the partial path together with the passed-in transform. If the first item is a message name for which the diagram has a method, the diagram sends itself the message; the method which handles the message should return a value. This is then sent the remaining path and transform. The result of composing and passing the transforms along is to build a transform which maps the coordinate system of the ultimate recipient to the coordinate system of the original caller. This is very much like the way in which the DISPLAY protocol works.

The *glue* facility is given two paths each of which mentions one object in the current diagram. The paths should describe points on the boundaries of the two objects. GLUE is responsible for moving the second object mentioned so that the objects touch and so that the points described by the paths abut. It also choses an orientation for the two objects which is most reasonable (i.e. so that the objects overlap as much as possible). This is done by calculating a new transform for the second object and then sending it a message to set its transform to this new value. This causes the object to rotate, mirror and translate itself by the required amount; the display is updated as well if the object is in the current buffer. For convenience, the form used to create an object may specify a locating path:

```
(part foo :source-location (>> A B)
          :destination-location (>> C D E))
```

This will cause the newly created object to be placed so that the B point of its A component is glued to the E point of the the D component of the object named C.

5.2.2. Replications

There are an endless number of ways to construct almost regular arrays. It is very common to mirror alternate rows to allow sharing of power and gnd lines; sometimes it is also desirable to mirror alternate columns as well. Pla structures require a special cell for refreshing the ground line to be repeated at regular intervals. Our ECC syndrome calculator combines some of these features, but it also has a somewhat random pattern for the placement of the SWITCHER and PASSER cells.

Most of these features are almost orthogonal; they interact in minor ways if at all. This suggests the use of Flavors, which requires us to understand the problem in terms of a protocol.

The *replication protocol* is a way of associating with each pair of row and column indices the geometry appropriate to that position of the array. The assumption underlying the protocol is that the array consists of a relatively small number of cell types which may appear in the array, possibly rotated or mirrored. All we need to know is which cell should appear and the transform to apply to the cell. Transforms in NS are specified as a triple of an *orientation* (such as IDENTITY, MIRROR-X, ROTATE-90), an *x-translation* and a *y-translation* (this representation was first used in DPL [3]). The lowest level of the protocol consists of four messages: :INSTANCE-FOR-INDEX, :ORIENTATION-FOR-INDEX, :X-PITCH-FOR-INDEX and :Y-PITCH-FOR-INDEX. The methods which handle each of these messages are passed the row and column indices of the array and are responsible for returning a component of the transform or the instance to be transformed.

Breaking the protocol up into these four components allows the creation of *mixin* flavors which handle certain parts of the protocol independently. For example, there is a mixin when we are using cells of uniform width with a uniform spacing between them without rotation or mirroring. There is a second flavor for when the columns are mirrored. Neither of these knows anything about the y-pitch and so they only provide methods for the :X-PITCH-FOR-INDEX and :ORIENTATION-FOR-INDEX messages. There are similar mixins for the y-dimension. By mixing these together one can get arrays which flip alternate rows, alternate columns, both alternate rows and alternate columns or none of the above. These flavors do not provides methods for the selection of the specific cell to be transformed. There are other mixin flavors for the simplest case of always chosing the same instance, and for the somewhat more complex case of chosing a special instance every N rows or columns. The ECC uses a mixin which choses an instance specified by a *programming matrix*; this style is also used by PLAs.

Replications provide two more features which deserve mention. The first of these is an implicit naming convention for the objects in the array. One of the array elements can be referred to by its row and column numbers using a *path* such as:

```
(>> cell-array (5 3) D :left-end).
```

The second feature provided by replications is storage economy. There is no need for the replication to actually construct all of the objects which appear to be inside it. To display an object we only need a diagram and a transform to apply to it; the replication protocol calculates this information by sending itself messages, without consuming permanent storage. Similarly, to follow a PATH we only need a final diagram and a transform; but we do not need to compose these into a new object requiring permanent storage. A replication is responsible only for maintaining a set of *virtual array contents*. Of course, there is a time penalty in sending messages and running methods, but the replication mixin flavors can choose any place in the spectrum of time-space trade-offs that seem appropriate; it is only during compaction that the virtual array contents need to be made real.

6. A More Complex Example: The Data Path Generator

We have seen how a series of layered abstractions make it possible to build a specialized module generator. However, the ECC module we have been looking at is a relatively simple example of the use of this approach. In this section we will change our focus to a much more complicated example of module generation. The Data Path Generator to be described in this section was developed to build data paths for the Scheme-81 microprocessor [2]. It was tailored to NMOS design. Both the Data Path Generator and the environment it was developed in are now obsolete, but there is a great deal that can be learned from looking at the experience of building and using this system; NS itself drew heavily on this earlier environment.

The Data Path Generator [10] was a tool for automatically constructing a microprocessor data path from a high level description of its contents. It was constructed in the

DPL/Daedalus environment [4], a predecessor to NS which contained features similar to those in NS^{2}C but without the virtual-grid layout of NS. A great deal of difficulty was encountered maintaining parameterized cell designs which needed to be cognizant of the process design rules. In fact, at one point the entire cell library needed to be rebuilt because of a process change. I will describe the Data Path Generator as if it had been built in the NS^{2}C environment.

The Data Path Generator was intended for a highly specialized form of microprocessor, one which directly embeds an algorithm (such as the Lisp interpreter of Scheme-81) on the chip. Microprocessors of this form are radically different from those which implement general purpose machines.

A general purpose processor is intended to provide enough generality so that it may be later tailored to any need; such machines place a high premium on symmetry, typically including a set of interchangeable general purpose registers organized around two busses which connect the registers to a barrel shifter and ALU. The Bristle Blocks system [9], a silicon compiler for this class of chips, inspired the effort to build the Data Path Generator.

In the Data Path Generator each register is assumed to be related in a direct way to a variable of the algorithm being implemented on the chip. It is extremely unlikely that these variables are used identically. Therefore, it is advantageous to implement the data path as a partitioned set, each block of which is associated with a variable of the algorithm. Each such block contains one (or more) registers as well as logic and algorithmic capabilities (operators) which are tailored to realize the functionality of the associated variable. For example, if an examination of the source algorithm reveals that the variable FROBBOZ is often incremented then the data path would include a block including the **frobboz** register and an incrementer. Similarly, other registers may have associated with them a number of operators such as incrementers, constant-sources, and testers.

Since the system's intended use was in designing a microprocessor whose hardware was tailored to a special need, it was necessary for the system to be open-ended. Cells for ad hoc operators (or new registers) may need to be added at any time. Such flexibility was achieved by creating the right platform of abstractions upon which to build the system. This was done by adopting a set of conventions to guide the overall floor plan, by identifying the common patterns of usage that occured given these conventions and then creating a declaration language which referred to these common properties. This made it convenient for a cell designer to tell the Data Path Generator about the properties of a new cell. Next, a specification language was created to allow the end-user to describe a data path in a simple declarative form. Finally, the embedding of the composition program within an environment like the one described above which made it possible to rapidly extend and modify the capabilities of the system as experience revealed weaknesses and shortcomings. Thus, the Data Path Generator continued the process of extending the layers of abstractions.

6.1. Floor Plan Conventions

The overall floor plan of the data path consists of a set of drivers at the top of the data path which drive control signals into the cell array. The cell array is a moderately regular structure containing the registers and operators; each column contains a single register or operator arranged as a bit-sliced structure with the low order bits nearer the top of the array. However, within this overall regular structure a great deal of local irregularity is allowed. There is no requirement that each register or operator occupy each row of its column and obviously there is no requirement for equal horizontal pitching. Each object is as wide as it needs to be.

To keep things simple, there is fixed vertical pitch requirement on all cells within the data path. Each cell must contain two horizontal metal lines for carrying power (i.e. one for Vdd and one for Gnd) and two bus lines called the LOCAL BUS and the GLOAL BUS. As mentioned above, the data path is partitioned into blocks containing a register and the operators with which it communicates. The LOCAL BUS connects the objects in each such block, but is broken at the block boundary. Therefore, the local busses of separate blocks are independent; during any machine cycle, each register may have an operation performed on it, using the local bus as a communication path. The architecture, thereby, achieves a high degree of concurrency. Actually the local bus consists of two wires; one, called VALUE OUT, is driven out from the register to the operators. The second, called LOCAL BUS, is the return path to the register.

Also running through each block is a GLOBAL BUS. The global bus runs through the entire data path, allowing inter-register communication. Figure 10 shows the normal register cell and several associated operators as they would be used in the data path, indicating the position of each of these features. The register uses a master-slave configuration, allowing the register to be both read and altered without danger of race conditions. The INCREMENTER associated with the register in the figure takes advantage of this feature. Since such communication ties up a purely private resource, a data path constructed in this manner can perform many operations during a single machine cycle.

Sense lines in the data path are brought out at the bottom end of the data path. These are typically developed using high fan-in NOR gates; a wire runs vertically through each bit of a column and is selectively grounded by the logic in that bit. The low (or high) order bit is responsible for pulling the line up as well as for its part of the logic. In the figure, there are two examples of this pattern, the EQUALITY TESTER and the CONSTANT TESTER.

Some operators have a natural parameterization into two cases corresponding to the values 0 and 1. The CONSTANT TESTER in the figure is such a cell. A column of such cells can be made to test for any desired value, simply by providing a string of boolean values (i.e. binary digits) which specify which configuration is to appear in each row. Such operators are referred to as *bit-programmed*. Other operators such as the LITERAL in the figure take a numeric parameter which specifies, for example, which of several positions a contact should occur in. Often the value of this parameter is simply the vertical position of the cell in the array, i.e. its row number. Such operators are referred to as *position-programmed*.

A final feature of the data path floor plan is its partitioning into subfields. The Data Path Generator is intended to support a tagged architecture in which each word is partitioned into a datum field and a type field whose bit pattern indicates what class of object is contained in the datum field. The methodology, therefore, allows each register or operator to be broken into subfields which are driven independently. The overall floor plan of the data path includes routing channels as necessary to bring in control lines to the subfields and to route out sense lines produced by the subfields. Vertical channels are allocated next to each partitioned register and horizontal channels are allocated at each subfield boundary. Control lines are then routed from the drivers to the cells in a simple two bend pattern. These channels are present only next to registers which are partitioned into subfields and are only sized large enough to carry the number of control lines needed. Figure 11 shows an example of a subfield boundary in the data path.

6.2. The Composer

The Data Path Generator is a composition program which selects cells from the library and places them in their appropriate places in the replication array. It also selects appropriate drivers and routes the connections between drivers and cells. The composition program supports several cliched patterns of usage. These include:

1. The ability to provide special cells for low order and high order bits of a column.
2. The ability to configure a column by providing a bit-pattern which indicates the setting of a particular parameter for sucessive positions of the column (this is used in constant-sources and constant-tests).
3. The ability to similarly program a column using the row number as the programming parameter.
4. The ability to associate with a register a *shadow register* which can be loaded from the main register using only local busses. This last feature is used to save the contents of all important registers at the beginning of each instruction execution cycle so that the state may be restored in the event of restartable faults. The existence of these local channels of communication between register and

shadow register also allow the register saving to be done in parallel with the operations being initiated.

6.2.1. The Declaration Language

The composition program has access to global information about the data path such as the size of the data path, the parasitic capacitance on the busses and the power consumption of the data path. However, it has little access to the internal structure and properties of the individual cells which it will assemble. On the other hand the generator for each of the cell types embodies a great deal of knowledge about the properties and behaviors of the cells, but has no access to the global context. Thus, there is a communications problem. The solution to this problem is to create a new abstraction level using an *ad hoc* declaration language of cell properties.

The declaration language tells the composer which registers and operators are position or bit-wise programmable and which are specially configured in the low and high-order bits. The declaration language also serves a second purpose. Often a single cell is designed with enough flexibility to serve more than a single role. For example, the equality tester cell can be used to compare a register to the bus or to compare two registers. The first configuration is used when the cell is used as an associated operator of a particular register; the second is used when the tester is a top-level object in its own right. This flexibility is controlled by a single parameter of the cell.

The declaration language is used to say how to set this parameter when the cell is used in either context. The declaration for the equality tester and the normal register cell types are:

```
(defregtype eq-test                      (defregtype normal-register
  (register-mapping                        (control-line-names
   (eq-test (right-source neighbor)))        (load-local load-global
  (associate-mapping                          restore phi-2 read))
   (eq-test (right-source global)))       (driver normal-reg-driver))
  (mirror-next-guy t)
  (connect-to-next-guy t)
  (connect-to-previous-guy t)
  (bottom-bitting? t)
  (sense-line-names (signal)))
```

The REGISTER-MAPPING declaration indicates what actual cell name and what parameters to use to achieve the desired configuration of the cell for top-level use. The ASSOCIATE-MAPPING is used to indicate the configuration desired for use as an associated operator to some register. The bottom-bitting? (and top-bitting?) declarations indicate that the low order bit and the high order bit are to be treated specially, by configuring the cell with its bottom-bit? (or top-bit?) parameter set to true. Declarations such as MIRROR-NEXT-GUY indicate special connectivity properties which this cell takes on when used as a top level object. The CONTROL-LINE-NAMES and SENSE-LINE-NAMES declarations tell the composing program the *names* of the control lines and sense lines used in the cell. The actual locations of these objects is obtained via the *path* facility described earlier. Finally the DRIVER declaration indicate what cell is to be used to develop the control signal for a column built from cells of this type.

6.2.2. The Specification Language

The final layer of abstraction is the one at which an end-user interacts with the system. This is the specification level at which the user states what he wants in the data path, not how to construct it or even what specific cell-types to use.

A fraction of the specification language input specifying the Scheme-81 data path is shown here.

```
(defconst main-regs
  `((env-stack normal-register
      (size ,word-size)
      (starting-position ,word-start)
      (shadowed
       (size ,word-size)
       (starting-position ,word-start))
      (associates
```

```
(rack-count-full? bit-tester
    (size ,count-size)
    (starting-position ,count-start)
    (bit-pattern ,(- (lsh 1 count-size) 2)))
  (increment-count incrementer
    (size ,word-size)
    (starting-position ,word-start)
    (field-size ,count-size)
    (field-start ,count-start))
    ...
    ))
... plus a few pages more ...
))
```

This is a simple, non-procedural and concise description of what is desired. One need state only what objects are desired and how they are to interconnect. The task of the composition program is to select the appropriate cells from the library in order to achieve this task. Conceptually this is a simple task, however certain complexities are faced because of the flexibility supported by the data path floor plan.

6.2.3. The Assembly Process

The greatest complexity faced by the composer is caused by the fact (mentioned above) that the Data Path Generator is intended to support tagged architectures; each register may be broken into subfields (for type and datum) which need to be driven separately. When this occurs, the composer must determine where to leave gaps for control and sense wire routing. The first phase of the composition process is devoted to the simple floor planning needed to locate where these routing gaps must be placed and to determine their size. A second feature computed during this phase is the extent of all local busses so that all cells may be built with appropriately terminated local paths.

The second phase of the process builds token examples of each cell so that a simple estimate of current consumption can be calculated. This is necessary in order to size the Vdd and Gnd line in each cell during the final phase to obey design rules for metal migration and to avoid excessive IR drops.

Having computed this information, the composer program passes over the description again, building appropriately parameterized instances of each cell and placing them into an array. The replication feature makes it unnecessary to create redundant copies of cell instances.

The final phase of the composition process involves the creation and placement of driver cells and the routing of connections between these cells and the cells in the array. This process would be trivial if it were possible to design the drivers to pitch align with the cells they drive. Unfortunately this is not possible Since some of the registers are broken up into subfields which are driven independently and therefore require two drivers. Drivers are placed as close as possible to the cells which they drive so as to allow planar connections requiring only a single layer. The connections are routed using a "greedy river router", another commonly used small routing program.

6.3. Experience

The Data Path Generator was used to construct three separate modules of the Scheme-81 microprocessor. The main data path of Scheme-81 contains 29 blocks of registers and associated operators, 10 of these registers are shadowed. There are 7 incrementers or decrementers, a full adder, several single bit shifters, 2 literal busses for introducing constants as well as several hardwired constant sources in the data path. In addition, the main data path contains several equality tests between a register and the global bus, and several more tests for hardwired constants. Approximately 50 sense lines are produced by the data path and fed to the predicate select module of the chip.

The construction of this module takes about 30 minutes the first time; however, successive iterations occur more quickly since cell structures are saved in the environment and are reused on later runs where possible. This allowed a style of design in which details were committed to as early as possible since it was evident that any design choice could be revoked and reworked with a minimum effort. Within one day of finishing the microcode of Scheme-81 we had a plot of the complete data path ready. As the microcode changed and bugs in the data path were discovered it was simple to maintain an up to date and debugged but tentative version of the system.

7. Summary

Layered abstractions are a natural and powerful way to build a system. The Data Path Generator is one of the more complex module generators built, yet it was only a minor extension of the environment in which it was built. By providing the appropriate abstractions and linguistic features at each level, we make it easy to build very powerful and specialized tools at the next level.

This suggests a set of guidelines for how to proceed when building a specialized tool such as the Data Path Generator. First of all, establish a set of conventions, such as a generic floor plan, which will guide future development. Second, identify the common patterns of usage, i.e. cliche's of the domain. Third, build new abstraction and languages which capture these. This then provides a new platform upon which to build a new layer of abstraction.

References:

[1] Ackland, B. and Weste, N., "An Automatic Assembly Tool for Virtual Grid Symbolic Layout", **VLSI'83**, F. Anceau and E.J. Aas (eds.) Elsevier Science Publishers B.V. (North-Holland), Aug. 1983, Trondheim, pp. 457-466.

[2] Batali, J., Goodhue, E., Hanson, C. Shrobe, H., Stallman, R. & Sussman, G.J. "The Scheme-81 Architecture - System and Chip", **Proceedings of the MIT Conference on Advanced Research in VLSI**, MIT, January, 1982.

[3] Batali, J. & Hartheimer, A., "The Design Procedure Language Manual", Massachusetts Institute of Technology, AI memo 598, September, 1980.

[4] Batali, J, Mayle, N, Shrobe, H., Sussman, G., and Weise, D., "The DPL/Daedalus Design Environment", **VLSI 81**, John Gray (ed), Academic Press, Edinburgh, 1981, pp. 183-192.

[5] Boyer, D. and Weste, N., "Virtual Grid Compaction Using the Most Recent Layers Algorithms", **Proceeding IEEE ICCAD**, Sept, 1983, pp 92-93.

[6] Cherry, J., Shrobe, H., Mayle, N., Baker, C., Minsky, H., Reti, K., & Weste, N., "Ns: An Integrated Design System", **VLSI '85**, E. Horbst (ed), Elsevier Science Publishers, B.V. (North-Holland), 1986, pp. 325-334.

[7] Davis, Harold L., "A 70-ns Word-Wide 1-Mbit Rom With On-Chip Error-Correction Circuits", **IEEE Journal of Solid State Circuits**, Vol. sc-20, No. 5, October, 1985, pp. 958-963.

[8] Ebeling, C. and Zajicek, O., "Validating VLSI Circuit Layout by Wirelist Comparison", **Proceedings of IEEE International Conference on CAD**, Sept. 1983, pp. 172-173.

[9] Johannsen, Dave "Bristle Blocks: A Silicon Compiler", **Proceedings of the 16th Design Automation Conference**, pp. 310-313.

[10] Shrobe, Howard "The Data Path Generator", **Proceedings of the MIT Conference on Advanced Research in VLSI**, MIT, January, 1982.

[11] Spickelmier, R. L. and Newton, A. R., "Wombat: A New Netlist Comparison Program", **Proceedings IEEE International Conference on CAD**, Sept. 1983, pp. 170-171.

[12] Symbolics, Inc. **Lisp Machine Technology Summary**.

[13] Takashima, Makoto, Mitsuhashi, Takashi, Chiba, Toshiaki, and Yoshida, Kenji, "Programs For Verifying Circuit Connectivity Of MOS/LSI Mask Artwork" 1982 , **19th Design Automation Conference**, page 544, paper 32.3.

[14] Terman, Christopher, **Simulation Tools for Digital LSI Design**, MIT/LCS/TR-304, MIT Laboratory for Computer Science, Cambridge Mass, September 1983.

[15] Weste, Neil and Ackland, Bryan, "A Pragmatic Approach to Toplogical Symbolic IC Design", **VLSI 81**, John Gray (ed), Academic Press, 1981.

[16] Weste, Neil, "MULGA - An Interactive Symbolic Layout System for the Design of Integrated Circuits", **The Bell System Technical Journal**, Vol 60, No. 6, July-August 1981, pp. 823-858.

[17] Weste, Neil, "Virtual Grid Symbolic Layout", 18TH DESIGN AUTOMATION CONFERENCE, 1981, pp. 225-233.

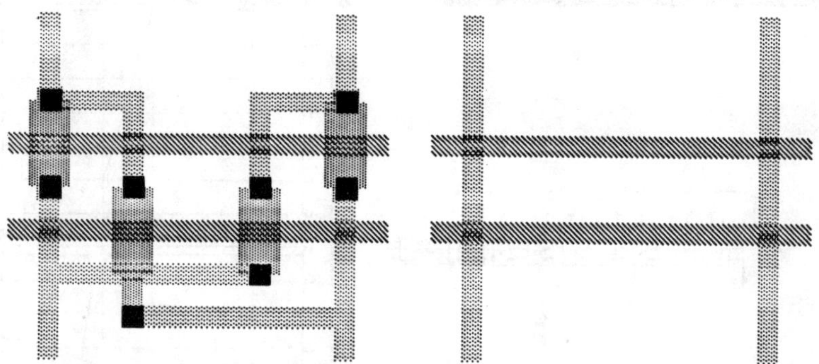

Figure 1: Two Configurations of the CONDITIONAL-XOR cell.

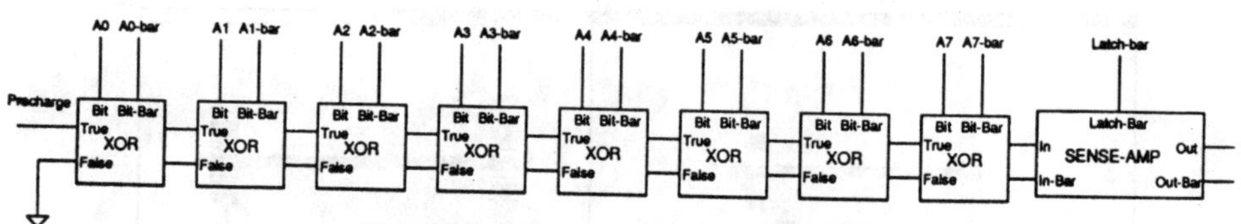

Figure 2: One Tier of the ECC System.

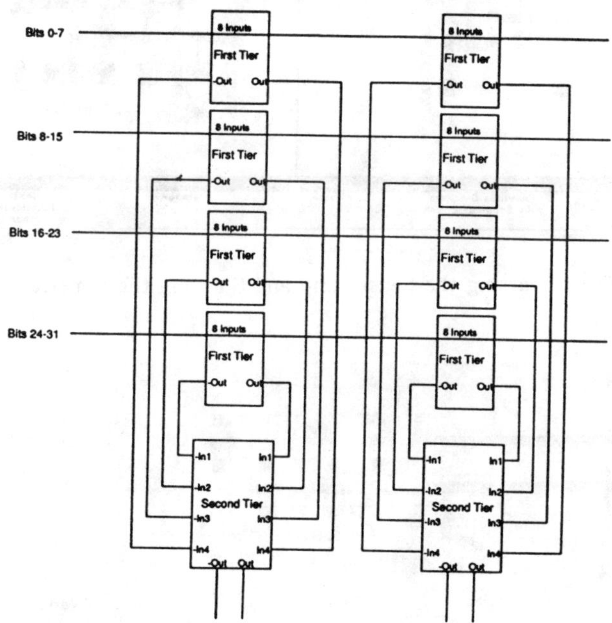

Figure 3: The Routing Pattern Used in the ECC System.

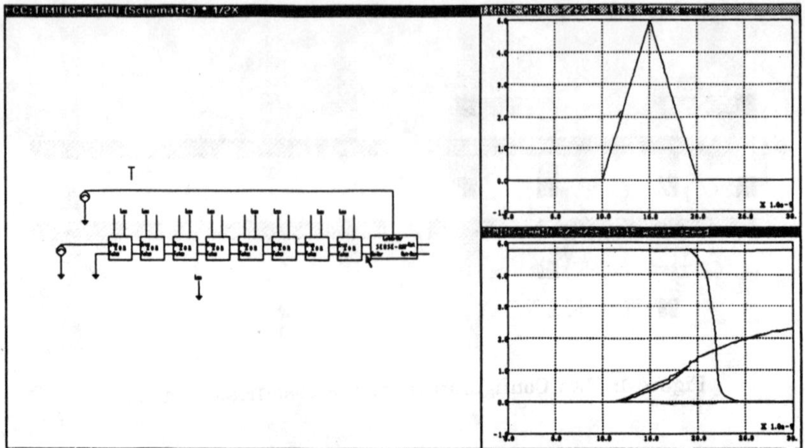

Figure 4: Using NS to Simulate the Sense-Amp and Plot The Results.

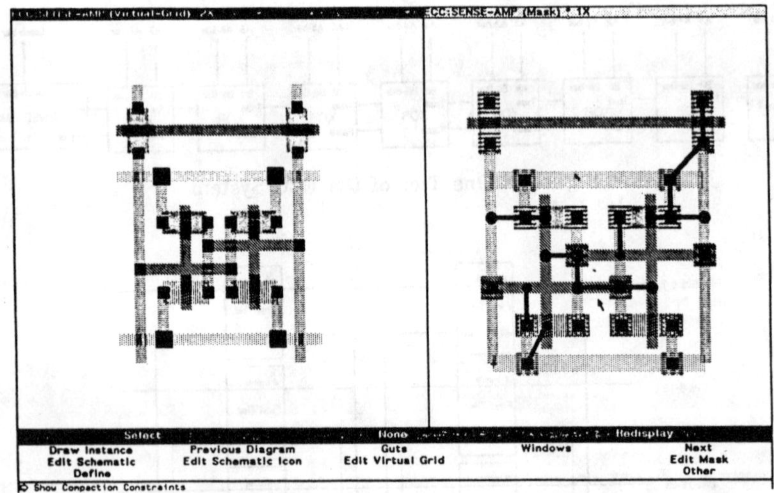

Figure 5: Compacting the Sense Amp and Showing the Constraints.

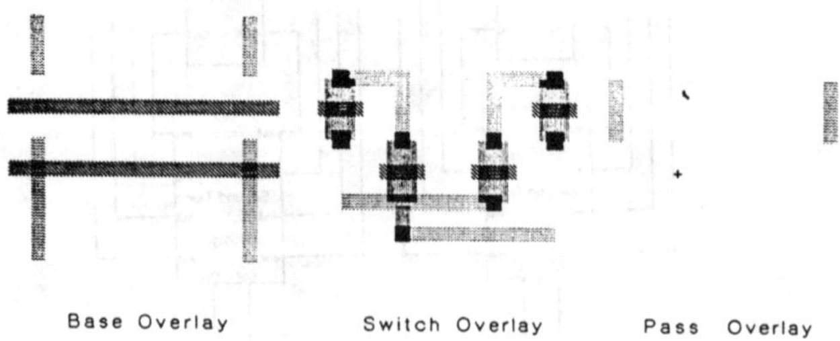

Figure 6: The Overlays in the CONDITIONAL-XOR cell.

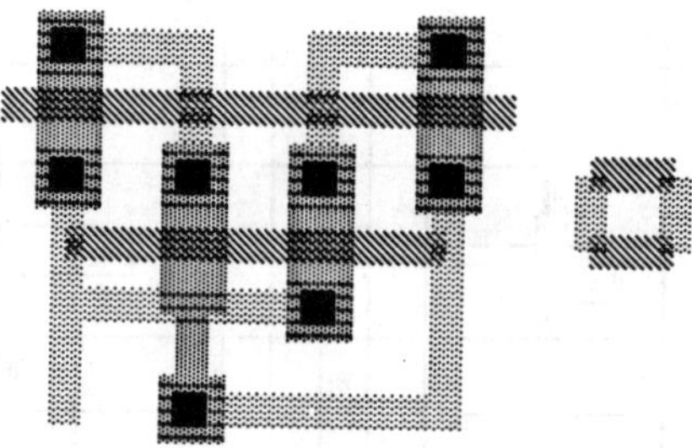

Figure 7: The Mask Diagrams of the two CONDITIONAL-XOR configurations.

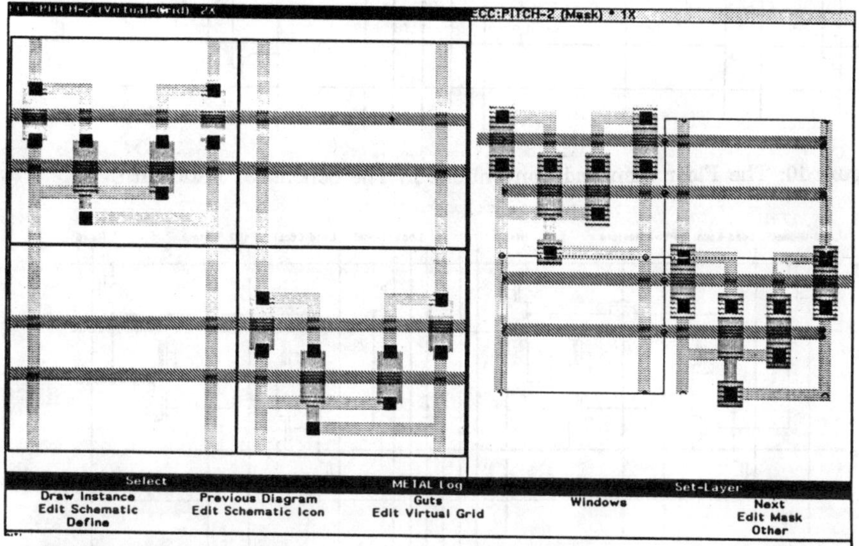

Figure 8: Pitch-matching the CONDITIONAL-XOR cells in the ECC array.

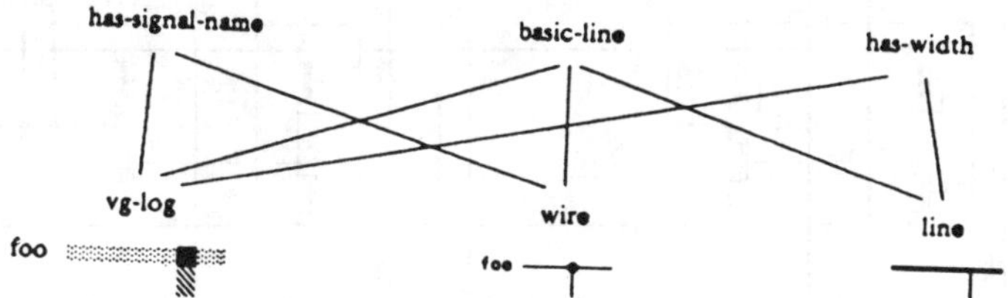

Figure 9: Non-Hierarchical Pattern of Inheritance

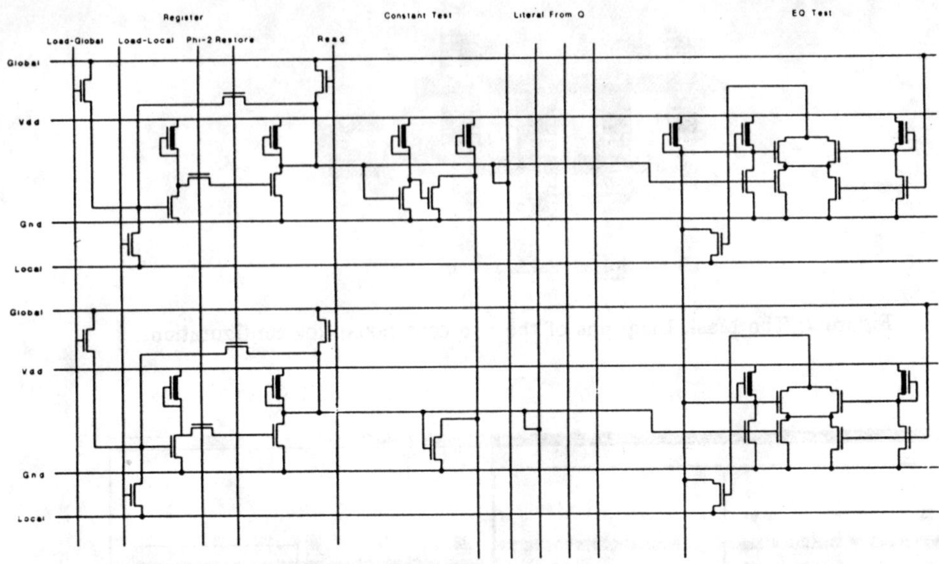

Figure 10: The Floor Plan and Conventions in The Scheme-81 Data Path.

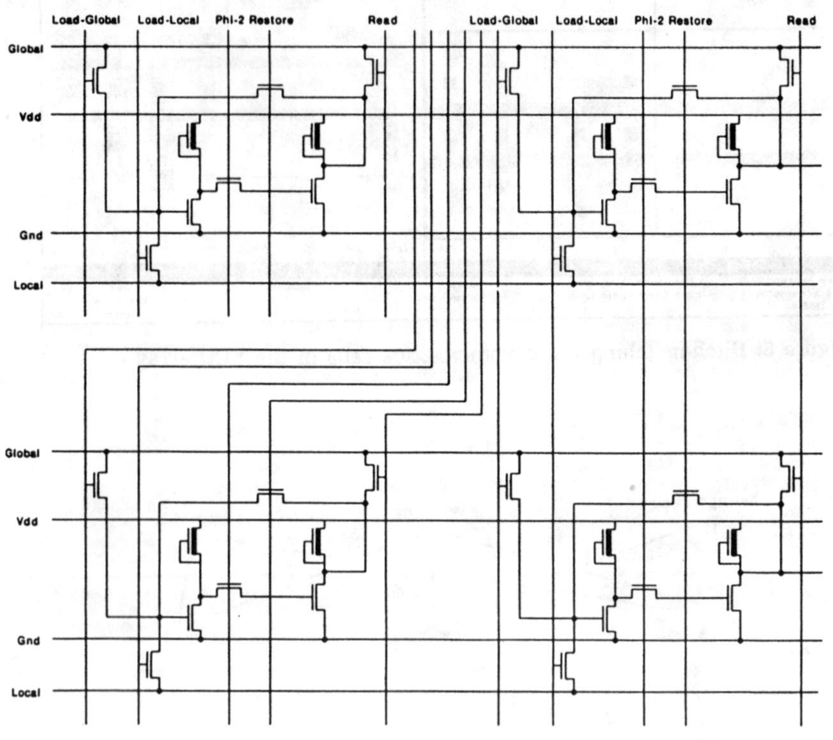

Figure 11: Routing Gaps in The Scheme-81 Data Path.

SYMBOLIC LAYOUT AND PROCEDURAL DESIGN

A. R. Newton
Department of Electrical Engineering
and Computer Sciences
University of California, Berkeley, CA. 94720

1. SYMBOLIC LAYOUT AND COMPACTION

1.1. Introduction

The use of abstractions, or *symbols*, to represent components of a design method (transistors, contacts, cells, *etc.*) can also help reduce the complexity of the design process. In addition, by maintaining the associations among related components (e.g. the rectangles that comprize a contact, or a transistor), non-linear changes in process ground-rules can be accommodated more easily (e.g. "shrink all the metal by 20% in width except around contacts. Maintain all connections.") This ability helps to move designs from a technology which supports one set of ground-rules to another – an ability that is increasing in importance as the cost of fabrication facilities increases and companies rely more heavily on a range of third-party "silicon foundry" services.

In this Chapter, a variety of approaches to representation and manipulation of symbolic layout are reviewed. Following a review of approachs to symbolic layout and compaction, two state-of-the-art programs are described in more detail to illustrate many of the features of modern symbolic design tools and to provide a reference for comparison. These are the *SPARCS* relative-grid spacing program[BUR86] and the *ZORRO* program, which uses a two-dimensional compaction algorithm based on a "zone refining" model[SHI86].

Symbolic IC design has been in use since the early 1970s [LAR71]. The use of symbols to represent IC devices began as a logical mapping between the electrical schematic of circuits and the semiconductor process for creating the devices in silicon. In the circuit schematic, each symbol represents a complete device on a functional level. The electrical properties of each device-type are well modeled and the circuit-level device models are used as a basis for the electrical design of a circuit. In a symbolic design, all geometric mask information is coalesced into a single symbol that represents the circuit-level *device*.

Symbols contain information that is normally lost in present-day rectangle-based IC layout systems. The electrical connectivity of the circuit, as well as the placement of the components themselves, is lost when a design is entered in a rectangle-based artwork system. This information must be extracted from the mask rectangles if, for example, a circuit simulation or testability analysis is to be performed. Layout rule and electrical rule checking have more difficult tasks because they must reconstruct the circuit devices from the mask artwork to identify their context correctly. When symbols are used to represent individual devices, more information is available for use by auxiliary design tools. The symbols and their interconnections are strictly defined and the transformation of the symbols into geometric mask shapes can be an error-free process. Thus, the design and maintenance of IC's with symbols permits the design system to capture more of the design intent than placing individual rectangles on multiple mask layers.

The huge volume of data associated with the integrated circuit (IC) design process can only be managed effectively if the *regularity* – both structural and behavioral – of the circuit is exploited by the designer and the CAD programs. It is now generally accepted that hierarchical data management and abstraction techniques can be used to exploit this structural regularity. For structural descriptions, this regularity is often expressed by the re-use of the same cell more than once in a design. In the behavioral domain, the re-use of parameterized procedural or declarative modules is often used to express aspects of regularity. Structural regularity plays an important part in symbolic design. It can be exploited to reduce the complexity of the layout spacing process in many cases. However, in a system that supports parameterized layout descriptions (see Section 2) or stretchable terminals for pitch-matching (e.g. virtual grid approaches), many of the advantages of the use of hierarchical descriptions may be lost.

In addition to providing a layout capability, symbolic layout systems may be used to guarantee that the spacing between elements in the design satisfies the layout rules for the IC process. Even further, the system can be used to compact the elements together to achieve a minimum size layout that still satisfies all layout rules and maintains circuit connections. The interconnection lines between invariant, primitive symbols (such as transistors, contacts, or circuit cells) are stretched or shrunk to change the size of the IC layout, while the dimensions of the symbols themselves remain unchanged. Thus a symbolic layout system may be used in a number of ways. It may merely provide a layout capability which allows IC mask artwork to be generated from the symbols. It may also provide a *spacing program* which is used to insure that the spacing between elements is correct. In addition, the symbolic design system may provide a *compaction program* which may be used not only to insure that the spacing between elements is correct but also to attempt to minimize the final area of the layout. The FLOSS program[CHO77], developed at RCA for the compaction of circuit cells, paved the way for the development of transistor-level compaction programs for IC's.

Although compaction was first proposed many years ago, it has not been as widely used in IC design as expected. The limitations of early compaction systems that did not allow upper and lower bound constraints, did not take into account important electrical considerations, and where it was difficult to express and edit constraints, are the basic reasons for this situation. To be useful for modern design methods, a layout spacing program should have the following capabilities:

- Full support of upper-bound and user-defined constraints
- *Detection* and *identification* of overconstrained elements
- An efficient constraint-graph solution algorithm
- Adjustable positioning of non-critical-path circuit elements
- Constraints both *within* and *among* circuit elements to provide a uniform approach to technology changes.
- Dependencies *among* constraints to enforcing symmetry during spacing, for analog and high-performance digital layouts

- A hierarchy and technology-independent symbolic layout abstraction

- Ability to handle constraints containing illegal "dead-zone" values.

- Selectable constraint modes, such as virtual grid, relative grid, or a combination of both

Layout compactors have been used in the past to address some of these issues, but to date no spacing program has been able to handle all of these requirements. Recent research has been focussed on developing more general algorithms and systems, as described in the remainder of this Chapter. Once these requirements are met, symbolic layout and spacing programs will become a common features of all modern CAD systems.

1.2. Fixed Grid Systems

Fixed grid symbolic IC design systems have been in use within industry since the early 1970s [LAR71]. The fixed grid represents the allowed locations of geometry for the IC layout. Quantizing the representation of the circuit in this manner allows efficient representation of the IC layout grid with the array constructs found in high-level computer languages. Early programs only allowed layout of highly regular structures, such as programmable logic arrays (PLAs). The *SLIC* program at AMI [GIB76] allows arbitrary layout of IC cells. Each grid point may contain zero or more symbols. If two symbols overlap or are adjacent, simple rules assume the geometries the symbols represent are connected. In SLIC, a circuit is created by drawing the symbols representing the circuit on grid paper (See Fig.1.1.1). This hand-drawn symbolic form is digitized, and the SLIC system generates the actual mask data from the symbols (See Fig. 1.1.2). The *MASKS*

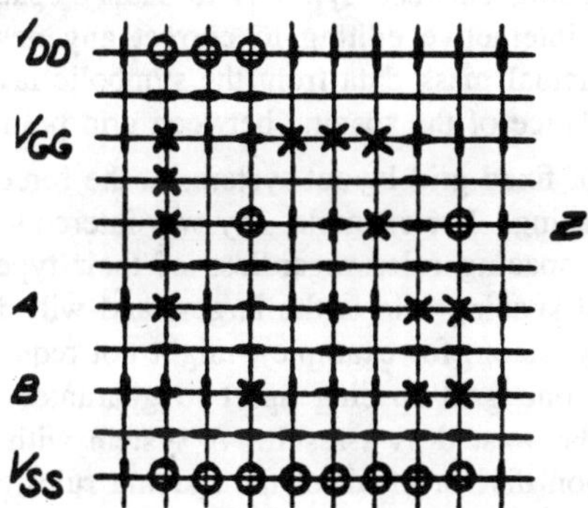

Fig. 1.1.1 Symbolic Input for the SLIC system ([GIB76])

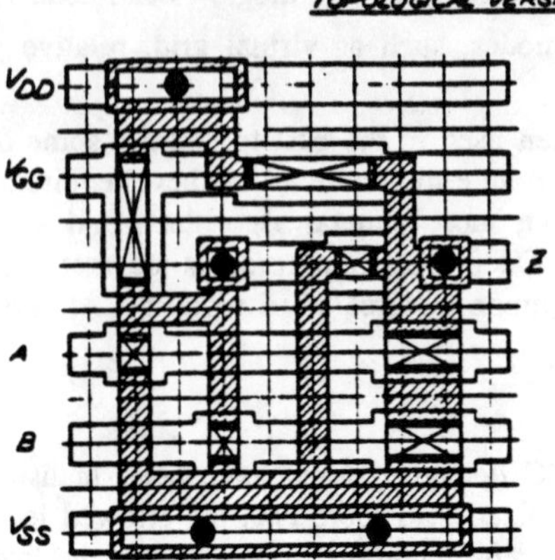

Fig. 1.1.2 Mask Layer Output from the SLIC system ([GIB76])

system from Rockwell International [LAR78] has a similar input format, but the layout symbols are placed in a 'discrete topological schematic', which are entered directly as program data. Figure 1.1.3 shows the symbolic representation of a 2 input NAND gate using this approach.

None of these fixed-grid symbolic layout systems initially provided the capability to dynamically adjust the spacing between symbols to satisfy spacing rules. SLIC provides design rule checking and interactive editing to correct any design errors. Both systems provide for generation of actual mask data from the symbolic layout. Simple spacing rules are enforced through the choice of the spacing between grid points.

The basic limitation of fixed-grid layout systems is the forced choice of the worst-case design rule for the grid spacing. For example, any two interconnection lines one grid spacing apart must satisfy the spacing rules regardless of their type. In a typical NMOS IC process, the metal to metal spacing rule is the largest and will determine the grid spacing. Lines of another type (polysilicon, for example), might not require the largest spacing rule, but they must be at least one grid spacing apart to guarantee that the spacing rules are satisfied. Some area will be wasted as a result. A system with a finer grid, with the grid spacing the greatest common divisor of all of the spacing rules, would overcome this limitation. The symbols used would no longer be simple, since the would occupy many grid locations. Also, the data storage required would increase at least an order of magnitude. Some work along these lines has been ongoing at AMI and Rockwell International.

Simple fixed-grid compaction has been displaced by relative grid and virtual grid techniques, described in detail below.

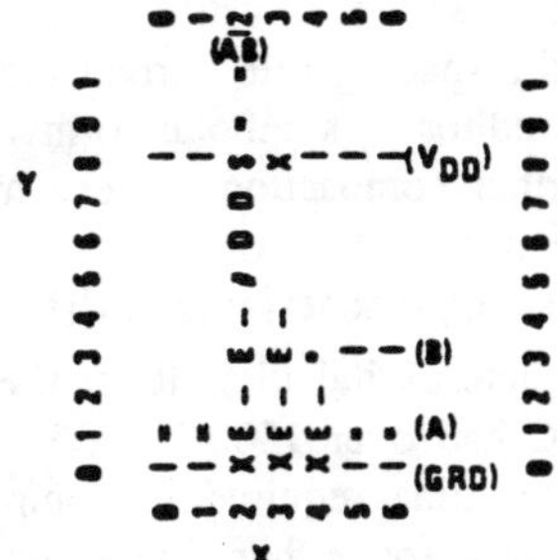

Fig. 1.1.3 MASKS System Input ([LAR78])

1.3. Relative Grid Systems

Relative grid systems only use the grid-based symbolic layout of the IC cells to indicate the relative placement of symbols and to determine the electrical connectivity of the circuit. The locations of symbols are represented as some fraction of the greatest common divisor of the spacing rules. Although the representation of the elements is usually not a grid, the choice of the minimum resolution for symbol location makes the relative-grid approach similar in most respects to the fixed-grid method using a finer grid.

One of the first programs to use the relative-grid method is the *FLOSS* program from RCA [CHO77]. FLOSS is similar to the SLIC program in that it reads a digitized sketch of the layout. The spacing between objects is determined by the construction of the objects and the individual spacing rules between the mask layers of the components that make up each object. Only orthogonal interconnection lines are allowed and the interconnections to cells must be points. While these restrictions are implicit in the nature of fixed-grid layout systems, they are imposed on relative-grid systems only through implementation difficulties.

FLOSS supports a hierarchy for circuit design, and results have been published for the compaction of entire ICs. Compacted results that were 32% smaller than the original symbolic sketch and 19% larger than a hand-drawn layout of the same IC have been reported.

Another symbolic layout aid which used a relative grid and provides compaction is the *STICKS* program, developed at the Hewlett Packard Co [WIL78]. This program compacts the symbolic layout by starting from one side of the layout and sequentially placing elements as far to the side as is possible, given the positions of the previously placed elements and the spacing rules objects on different mask layers. The electrical connectivity of the circuit is also a determining factor in the positioning of elements during the compaction. STICKS reads in a stick diagram [MEA80] as input.

The *CABBAGE* program [HSU79] [HSU79a] is a relative-grid layout compaction tool. It represented the topology of the IC layout with the vertices and edges of a graph, and uses a critical path analysis to determine the spacing rule correct locations of the IC symbols. Designs were entered via a graphics editor in symbolic format. Local modifications to the layout were performed to allow further compaction. These modifications consisted of distortions to interconnect, such as the introduction of "jogs". Other compaction programs have used further refinements of these approaches[WOL85][MAL85].

Cabbage I was used to design a complete digital filter IC at the Katholieke Universiteit Leuven in Belgium [HUR82]. This IC had over 1500 NMOS transistors, included in over 9000 layout symbols, an the computer time required to compact this example was over 4 CPU hours. This demonstrated the need for either simpler compaction strategies or the exploitation of the regularity in the circuit.

The *SLIM* system [DUN80] combines a shear line algorithm used by Akers [AKE70] with the graph representation of the IC symbols. SLIM is a successor to the *SLIP* system [DUN79][DUN78]. In this program, multiple spacing methods are used to optimize both speed and compaction efficiency. The IC symbolic data is partitioned automatically into optimal size groups calculated by the program [DUN79]. A loose initial placement guarantees a layout-rule correct (although not optimal) relative placement of the partitions. Critical path analysis, similar to the method used in *CABBAGE* , is coupled with a local-compaction method. This reduces the total computer time required for solution of the compaction of each partition while maintaining an efficient compaction result. The local compaction procedure clusters together objects on the critical path. Jogs are inserted as zero-length lines perpendicular to the direction of the line into which they are inserted. They allow the objects connected to the top and bottom parts of a line to move independently in the direction parallel to the jog. In SLIM, jogs are only inserted at contact locations. Global rift line compaction [AKE70] removes the excess space between the locally compacted partitions. The worst-case order dependency for the execution time of the composite algorithm is approximately $O(n^{3/2})$, but practical circuits can be compacted in substantially less time than this worst-case.

The *PYTHON* program[BAL82] was developed based on the concepts used in *CABBAGE* and could perform hierarchical compaction of IC designs, through use of a hierarchical database for storage of the symbolic IC layouts and the use of abstractions of cell layouts. To allow technology and IC family independence the compaction was performed on interconnected *protection frames*. Protection frames are an abstraction of a cell and they reduce the amount of data needed at each level in the hierarchy.

By far the most successful commercial application of relative-grid compaction has been by VTI. The Calma STICKS system is also a commercial, relative-grid compaction product. Recently, relative-grid systems have been described by IBM[FIE84]. Another complete design system is presented in [LEN82].

A detailed description of a modern relative-grid compactor is presented in Section 1.2.

1.3.1. Virtual Grid Systems

The *MULGA* system from Bell Telephone Laboratories[WES81][WES81a] was the first to describe this compaction capability which differs from other relative-grid layout systems. It uses a *virtual grid* to perform compaction of the symbolic layout. This virtual grid combines the ideas of both the fixed and relative-grid approaches to IC layout. Symbols are placed at grid locations, as in fixed-grid layout. The spacing between grid rows and columns is adjusted and takes on a real value, dependent on the actual spacing required by objects on each row. Individual rows and columns may have a unique spacing, so the grid spacing is non-uniform. Given the non-uniformity of the grid, there is no area penalty when two adjacent objects have a required spacing smaller than the worst-case layout rule. The resulting compacted layout is more dense than could be attained with fixed-grid layout but generally not as dense as in a compaction with a true relative-grid compaction method. Some comparisons are summarized in[WOL85]. Objects that are originally placed on a single row or column in the grid will remain on that row or column throughout the compaction, whether they are physically connected or not. In a relative-grid system, this restriction is not made and objects that are not physically connected may move relative to one another to achieve a more compact final result.

For reasons of efficiency, the virtual-grid compaction approach does not support upper-bound constraints. The separation of adjacent grid lines is determined using only local operations. Pitch-matching between cells compacted separately is achieved by compacting each cell to its minimum size. Then, since each cell has its own virtual-grid-to-physical-grid map, the mappings are adjusted to ensure a match in the final layout. Virtual grid, like other relative-grid schemes, suffers from problems of preferential compaction directions. In some virtual-grid implementations this problem has been reduced by trying both X- and Y- grid-line compactions at each step and choosing the best one. This still provides only local improvement but it also permits the system to consider corner-corner constraints as well.

Virtual-grid compaction can be implemented as a special-case of relative-grid compaction, as described in Section 1.2. However, by using special-purpose data structures which take advantage of the simplifications possible for virtual grid, speed advantages of from 5 times to 25 times have been reported[BOY83], as shown in Table 1.1. More importantly, the compaction time is approximately linear in the number of grid lines. This may not be the case for more sophisticated approaches.

The VPACK system, described in [BOY83], uses an extended virtual grid approach. As with all virtual grid system, it assumes a matrix of grid lines to which circuit components can be "attached." Such components must be attached to at least one horizontal grid line and at least one vertical grid line. The actual space between each pair of adjacent grid lines, in microns or lamda units, is determined after compaction; the grid is not a "physical" grid, and hence the term virtual grid. Components are attached to the grid graphically, or by specifying a particular (x,y) location in a textual format. An example of such a textual description in the ABCD language[ROS83] is shown in Figure 1.1.4. In the VPACK system, the components available include: *wires,* which are specified as a path list of virtual grid points, including starting and ending points (a pointer to the wire is stored at each of its grid points); *contacts,* which occupy a single virtual grid point; *devices* and *loads.* A device (transistor) takes up at least three virtual grid points, corresponding to legal connection point to the device. At each virtual grid point a maximum of two different wire layers may exist, for a single-metal technology.

The compaction proceeds in two passes: an X-direction pass followed by a Y-direction pass. A "picket fence" approach is used to minimize back-tracking. This approach is identical to "shadowing" used in relative grid compaction schemes[HSU79]. The idea here is that as the compaction moves from left to right in the X-direction, the

Compactor	Computer	Cell	Time (sec)	Time/Dev (sec)
SLIM[2]	not given	5-in nand	45	7.5
		D-flip flop	212	19.27
CABBAGE[1]	HP 1000 E	T-flip flop	21	1.75
	RTE 4	5 T-flops	191	3.18
MULGA	VAX 11/750	T-flop	13.5	1.125
	UNIX C	12 T-flops	336	2.3
STICKS[3]	CALMA	30 - devices	1800	60
Rochester	VAX 11/780	T-flop	11.8	1.0
	UNIX PASCAL	Priority Q	131.4	4.7
VPACK	VAX 11/780	T-flop	3.2	.27
	UNIX C	12 T-flops	22	.15
		48 T-flops	64	.15
		192 T-flops	352	.15

Table 1.1. Compaction times for various compactors. The
last column shows the compaction time per device[BOY83].

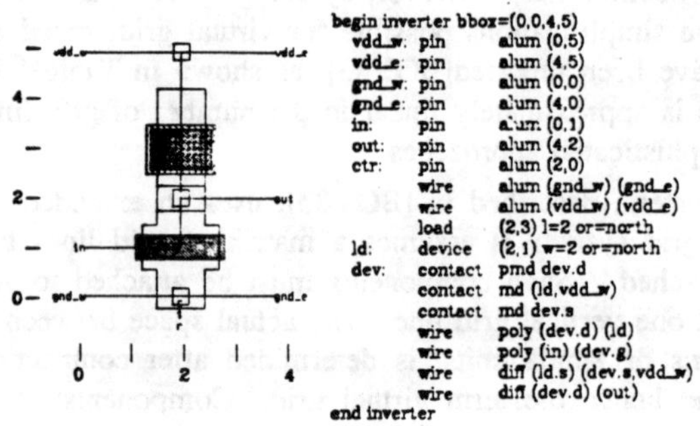

```
begin inverter bbox=(0,0,4,5)
    vdd_w:  pin      alum (0,5)
    vdd_e:  pin      alum (4,5)
    gnd_w:  pin      alum (0,0)
    gnd_e:  pin      alum (4,0)
    in:     pin      alum (0,1)
    out:    pin      alum (4,2)
    ctr:    pin      alum (2,0)
            wire     alum (gnd_w) (gnd_e)
            wire     alum (vdd_w) (vdd_e)
            load     (2,3) l=2 or=north
    ld:     device   (2,1) w=2 or=north
    dev:    contact  pmd dev.d
            contact  md (ld,vdd_w)
            contact  md dev.s
            wire     poly (dev.d) (ld)
            wire     poly (in) (dev.g)
            wire     diff (ld.s) (dev.s,vdd_w)
            wire     diff (dev.d) (out)
end inverter
```

Fig. 1.1.4 The symbolic layout of an inverter
and its corresponding ABCD description[BOY83].

latest contour of checked material on each layer (the picket fence) is maintained. Since all
material on the same layer behind the fence has been checked, the system only needs to
consider the relationship between new material and the fence. Once a grid line has been

processed, its material is added to the fence.

In both relative and virtual grid approaches, it is necessary to perform a number of technology-specific transformations. In particular, most compaction systems developed recently inplement *terminal merging*, where connection points that are on the same net are permitted to overlap if it improves the layout density. In addition, when components do not quite merge (e.g. partial overlap of contacts in the same net), it is necessary to either move them apart or to complete the merge, if the circuit is to be manufacturable. This is generally implemented as a post-processing step. In fact, implementation of efficient algorithms for these general "second order" problems is more difficult than the general compaction strategy itself!

Virtual grid systems have been quite successful. Much of the success can be attributed to the fact that they are fast and that they do not distort the layouts more than perhaps the designers want – the results are predictable by inspection. As a result, users "trust" the system. This has led to the inclusion of a virtual-grid-mode in relative grid compactors, as described below.

1.3.2. Other Approaches

To obtain competitive layouts from automatic compactors, enhanced algorithms have been proposed. W. Wolf [WOL83][WOL84] solved the above problem by finding the critical path, shearing the component pair, and continuing compaction in the preferred direction. This approach can be considered locally-two-dimensional, although there is always a preferred direction of compaction. Similar, local heuristics were employed in [DUN79] and have been used in a variety of programs since that time.

True two-dimensional compaction has been attempted with exponential complexity algorithms. It can be proved that the general two-dimensional compaction problem is NP-complete. Recently there have been efforts to develop algorithms that could perform X- and Y- axis compaction at the same time but the results obtained are not suitable for a practical implementation. In [HAC80], the general placement problem was formulated as a mixed integer programming problem. This formulation subsumes the compaction problem but the author indicates that long run times may result for reasonable problems. However, as machine cycle times decrease and if parallel implementations of the approach are possible, such general solutions may become feasible.

Another approach is to start with a totally collapsed layout and then remove the distance violations one by one [SCH83]. G. Kedem and H. Watanabe [KED84] translated the compaction problem into a special form of a mixed-integer programming problem. While this approach was successful, the problem being solved is closer to the virtual grid approach than to the general, relative-grid layout problem.

At the other extreme, the Simulated Annealing algorithm[KIR83] has been applied to the compaction problem, with good results but with large computer times.

A novel approach to compaction is to treat it as a box-packing problem. The first program to use this approach was the *BEADS* program, developed by W. Crowther[CRO80]. More recently, the *ZORRO* program has been developed which takes a similar approach and is described in detail later in this Chapter.

1.2. THE ANATOMY OF A COMPACTION SYSTEM: SPARCS

SPARCS[BUR86] is part of a large symbolic design system under development at the University of California at Berkeley. Central to the system is an object-oriented database called *OCT*[MOO86]. The graphical interface to the system is called *VEM*[HAR86]. The symbolic layouts to be spaced by *SPARCS* are contained within OCT, and user interaction with *SPARCS* occurs via VEM. Figure 1.2.1 shows these three parts of the symbolic design system.

As is the case with most spacing systems, *SPARCS* consists of two major modules: the *constraint graph builder* and the *constraint graph solver*. The constraint graph is constructed from the relative placement of the symbolic layout elements and a table of spacing rules. The solver performs critical-path analysis, positions the elements that are not on the critical path, and identifies overconstrained problems. In addition to the layout and the spacing rules, explicit constraints are stored in OCT. All constraints may be edited, and new constraints may be added, via the VEM interface.

1.2.1. Symbolic Layout Data Abstraction

The abstraction chosen to represent the circuit has been carefully selected so that the same generic representation can be used on any level of the hierarchy of the design. The abstraction consists of *objects* of two classes: *immutable* objects, which translate during a spacing, and *mutable* objects, which translate and change size. As distinct from most other spacing systems, an object processed by *SPARCS* may be as simple as a point or an edge, or may be as complicated as a collection of Manhattan polygons (an instance).

The immutable objects are used to represent transistors, contacts, standard cells, macro cells, etc. Each mask layer of each immutable object is described by a *protection frame*[KEL84]. The single restriction imposed by the design system on the shapes used to

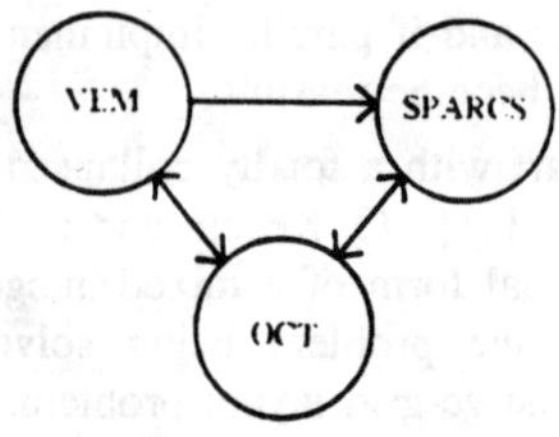

Figure 1.2.1. The SPARCS Symbolic Design System

implement the protection frames is that they be Manhattan polygons.

The *mutable* objects are used to represent elements such as wires, busses, and parameterized cells that are allowed to both change extent and translate. All interconnect is represented by collections of single Manhattan segments with contacts at segment intersections.

1.2.1.1. Protection Frames: Protection frames are used to define the limits of the geometry contained within a cell - to 'protect' all of the geometry contained inside them[KEL81][KEL81a]. The protection frames are a per-layer set of polygons that are required to cover all of the geometry that lies below them. On the simplest level, a single bounding box could be used as the protection frames. Although the derivation of this box is easy and efficient, it does not provide a true representation of the internal geometries it bounds. Quite large unused areas within the cell are possibly 'invisible' outside the cell. The bounding box model of protection frames can be extended by making a separate bounding box for each mask layer in the IC process. Since the limits of each layer are most likely different, less unused area within the cell is 'invisible' outside the cell.

Per-layer sets of bounding rectangular polygons form the basis of protection frames as used in *PYTHON* and *SPARCS*. These allow an arbitrary tradeoff between spacing efficiency and computational efficency. The closer a frame approximates a bounding box, the smaller the amount of information necessary to process to correctly space that frame. This will mean less computer time for spacing the frame. It will also mean the greatest area loss (barring a bounding box bigger than the actual cell geometry). On the other extreme, merging the geometry internal to the cell to form the protection frames will provide the greatest area efficiency, since all of the unused area within the cell may be used at the next level in the hierarchy. This approach reduces the effectiveness of the hierarchy however, since the amount of information contained in these merged protection frames is roughly comparable to examining all of the interior geometries individually. The only gain remaining at this point would be due to multiple instances of the same cell.

1.2.1.2. Terminal Frames: These complement protection frames and define allowable areas of interconnection within the protection frames. One of the key differences between *SPARCS* and most other compactors is this idea of terminal connections that are *areas* instead of *points* (This concept was first introduced in the *PYTHON* program). Rather than requiring a connection at the exact center of a point structure such as a contact, the entire contact area is suitable for termination of the interconnecting line. The connection is limited so that the line always remains within the boundaries of the terminal frame. Thus if a line is the same width as its terminal frame, the effect is the same as the point terminals in other systems. If, however, the terminal frame is *wider* than the interconnect line, the interconnect can move between two constraints. This movement allows a spacing program to take maximum advantage of the instances and interconnections on the most constraining path through the IC geometry to obtain the most area efficient spacing solution.

There are two major restrictions on terminal frames. First, they must have at least one edge coincident with a protection frame edge. With the definition of protection frames presented above, it is not possible for a terminal frame to exist outside a protection frame. To permit this would allow connection to a terminal at a point where there could not possibly exist geometries since frames define 'inviolate' areas on mask layers. Objects such as contacts have protection frames and terminal frames that are entirely coincident. An interconnect line may connect to a contact from any one of four sides.

More than one interconnection line may terminate on the same terminal frame. There is no restriction to the total number of interconnections on each terminal frame or even the number of interconnections per side of the terminal frame. However, each interconnect line must end in a terminal frame. This is a convention used to simplify the treatment of interconnections, and is not directly related to their nature.

1.2.1.3. Hierarchical Spacing Using Protection and Terminal Frames: These two simple concepts of protection and terminal frames together provide the basis for a true hierarchical spacing. As each cell is designed, the elements within the cell are spaced, according to the layout spacing rules, by the compactor. After each cell is properly spaced, protection frames are automatically generated from the geometry internal to the cell[LAI86]. This is done *only once*, and the protection frames are stored with the cell. The terminal frames are defined explicitly by the user, who labels the signals which are to be exported to the next level in the hierarchy. The geometry that implements these electrical nets defines the physical implementation of the terminal frames. Alternatively, the user can define *local terminals* when laying out his cell. These terminals can automatically be promoted to the abstraction of the cell and can be used as terminal frames.

After the terminal frames and protection frames have been established for a cell, the cell can be placed at the next level in the hierarchy. The spacing program only need look at the protection frames of the cell at that level and need never look at the geometries contained within the cell.

The use of protection frames with terminal frames is intended for a bottom-up implementation style. A top-down design style may require additional tools such as interconnect routers. If a cell is placed in its unspaced (original input) form and later compacted, the old terminal locations of the cell may not be at the same locations as the new terminals. The interconnections to the instances of the cell may require patching to physically connect to the now smaller cell.

Figure 1.2.2 shows a transistor represented by its protection frames and terminal frames. Figure 1.2.3 shows a spaced D flip-flop circuit. The spaced flip-flop after framing is

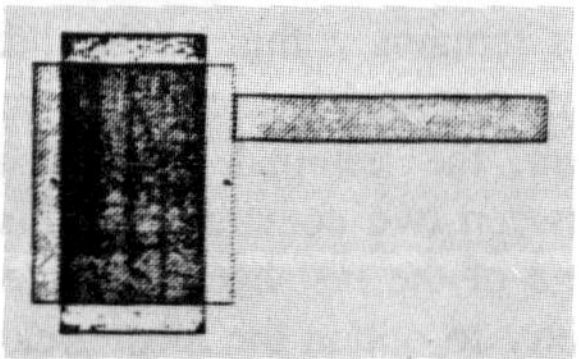

Fig. 1.2.2 Transistor with Protection and Terminal Frames

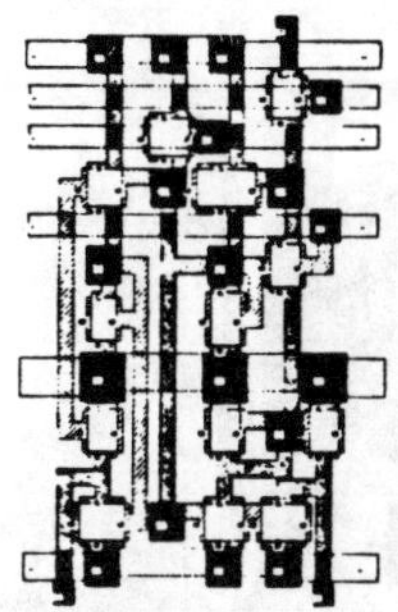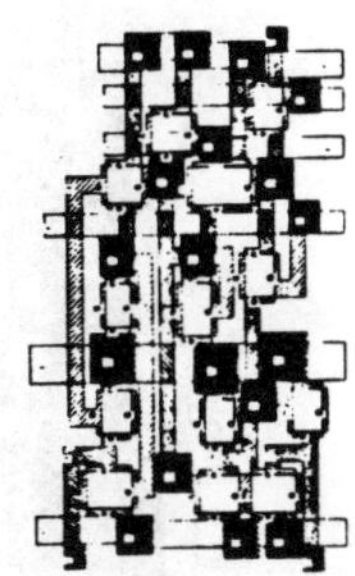

Fig. 1.2.3 Flip-flop spaced by SPARCS using:
(a) Virtual Grid (b) Relative Grid

shown in Figure 1.2.4. A shift register made by connecting and spacing several framed flip-flops appears in Figure 1.2.5. The time required to space the shift register is less than one fourth of the time needed to space a single flip-flop.

Fig. 1.2.4 Framed Flip-Flop from *VULCAN*

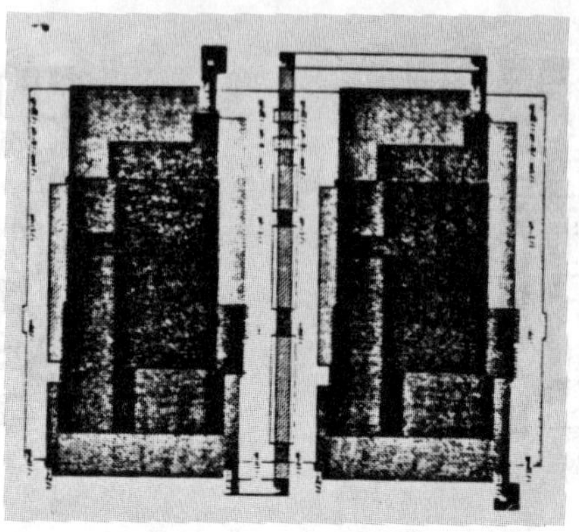

Fig. 1.2.5 Spaced Shift-Register

1.2.2. Parameterized Objects

SPARCS supports *point* and *edge* objects, as well as those composed of rectangles, polygons, and segments. In fact, a rectangle or polygon may be though of as the abstraction (or protection frame) for a set of edges and points. One important application of point and edge objects is in the technology-parameterization of symbolic layouts. The lowest-level layout elements, such as contacts and transistors, are technology-dependent to some degree in all systems. These elements can be stored in the database as points, edges, and constraints. The expressions used to find the values of the constraints, and the structures of the horizontal and vertical constraint graphs, are stored in the database as well. When the technology changes the constraint expressions are evaluated by *SPARCS* using the new design rules. The graphs are then solved using the new constraint values, resulting in updated protection frames and terminal frames for the element. This mechanism is also used to represent higher-level parameterized cells.

Shown in Figure 1.2.6(a) is a MOSFET with an irregular gate geometry. The device is represented by ten point objects. The parameters for this element are given in Table 1.2.1.

Parameter	Meaning
W	device width
L	device length
polyOver	overlap of poly past diffusion
diffOver	overlap of diffusion past poly
cWidth	contact dimension
cInside	min. diffusion width around contact

Table 1.2.1 Technology Parameters for MOSFET

Parameters *W* and *L* are local to the transistor; all other parameters are stored in a technology file along with the spacing rules.

Shown in Figure 1.2.6(b) is the transistor of Figure 1.2.6(a) after increasing *W* by 25%, reducing *L* by 33%, and reducing *cWidth* by 25%. The other constraints are unchanged. Depicted in Figure 1.2.6(c) is the transistor of Figure 1.2.6(a) with the *polyOver*, *cWidth*, and *cInside* dimensions reduced by 50%, and the *W*, *L*, and *diffOver* dimensions reduced by 33%. The *SPARCS* constraint management procedures thus provide a powerful mechanism for scaling layouts either uniformly or non-uniformly as fabrication technologies evolve. The constraint specification is declarative; the constraints among objects are declared and the spacer is then responsible for ensuring that they are met. The importance of this fact will become clear in Section 2. Within leaf cells, most of the constraints are equality constraints while at higher levels of the design the constraints often have unequal lower and upper-bound values.

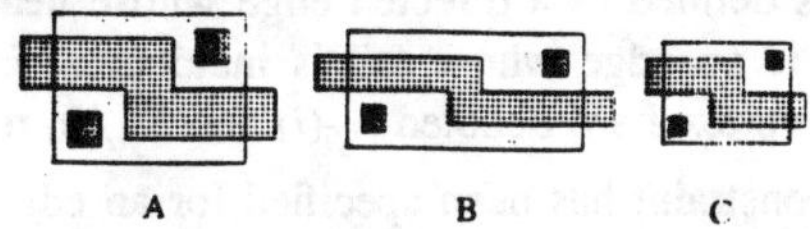

Fig. 1.2.6 MOSFET Re-spaced for Three Technologies

1.3. CONSTRAINT MANAGEMENT

1.3.1. Introduction

All layout spacing programs use lower-bound constraints to specify the minimum spacing between layout elements (transistors, wires, contacts, etc.) In recent years the importance of also including upper-bound constraints has been recognized. For example, in Figure 1.2.2 two constraints, an upper-bound and a lower-bound, are used to describe the terminal area so that the wire segment may slide along it. The concept of *sliding terminals* was first introduced in the *PYTHON* program[BAL82] and can result in substantial increases in layout density, as shown in Figure 1.2.3. In the *SPARCS* program [BUR86], fixed constraints (i.e., an upper-bound and a lower-bound of equal value) are used to allow the user to space by either virtual or relative grid as illustrated in Figure 1.2.3. In virtual grid mode, the fixed constraints that bind elements on a common centerline are added automatically.

1.3.2. Representing the Constraints

Once layout-rule analysis has been completed and user-specified (or module-generator-specified) upper and lower-bound constraints have been added, two constraint graphs $G_x(V, E)$ and $G_y(V, E)$ are constructed. These graphs represent constraints among objects in the horizontal and vertical directions, respectively. While the graphs are formed separately, there are often some constraint relationships between elements of the different graphs. In the following description, the subscripts x and y have been dropped for clarity. Each vertex $v_i \in V$ represents an *object* (point, edge, immutable shape, terminal, or cell instance). Each directed edge $e_{ij} \in E$ represents a constraint between vertices v_i and v_j. Since the layout spacing problem is concerned with the spacing adjustment of a layout that has already been placed topologically, the following conventions are used:

(1) The direction of the edge implies "to the right of" in G_x and "above" in G_y.

(2) A lower-bound constraint (minimum spacing) between vertices v_i and v_j is represented by a *lowerBoundConstraint*, L_{ij}, on edge e_{ij}. Similarly, an upper-bound constraint (maximum spacing) is represented by an *upperBoundConstraint*, U_{ij}, on the same edge. This constraint pair is written (L_{ij}/U_{ij}) [BAL82].

(3) A *fanin* edge at v_i is defined as a directed edge whose head is incident at i. Similarly, a *fanout* edge at v_i is an edge whose tail is incident at i. The total number of fanin and fanout edges at vertex i are denoted $N_{fi}(i)$ and $N_{fo}(i)$, respectively.

(4) If no upper-bound constraint has been specified for an edge, its value is assumed to be infinite (denoted INF). If no lower-bound value has been specified, its value is assumed to be zero.

(5) Without loss of generality, the constraint graph is converted to a directed polar graph with the addition of a *Source* vertex, V_{SO}, and a *Sink* vertex, V_{SI}. V_{SO} is connected to all vertices V_k where $N_{fi}(i)=0$ with an edge $e_{SO, k}$ of weight (0/INF). Similarly, all vertices V_m where $N_{fo}(m)=0$ are connected to V_{SI} with an edge $e_{m, SI}$ of weight (0/INF).

These conventions are equivalent to the more conventional approach of representing lower-bound constraints as forward-pointing edges with positive weights and upper-bound constraints as backward-pointing edges with negative weights. However, the polar representation is more restrictive than the general case in that it does not allow cycles in the graph consisting of only positive-weight edges or only negative-weight edges. Such cycles correspond to a layout that has no topological placement for any values of the edge weights. Cycles of this form correspond to input errors and are easily detected and reported while building the graphs.

The more common occurrence in a spacing program that supports mixed constraints is an overconstraint due to a positive cycle that contains some positive-weight edges and some negative-weight edges (i.e., a conflicting set of lower-bound and upper-bound constraints). We refer to this case as a positive-weight *mixed cycle* in the graph. For such situations, the spacing program should detect the cycle and report all vertices and edges involved in the limiting lower-bound path and in the limiting upper-bound path from the left-most vertex to the right-most vertex of the cycle.

1.3.3. Detecting Overconstraints

Most spacing programs in use today use general algorithms to detect and report overconstraints. The *existence* of positive cycles can be detected in the time required to solve a legal graph [BAL82]. However, this information is of little use since the vertices which comprise the positive cycles are not enumerated.

Some spacers determine the existence of overconstraints by iteration count and give no indication as to which constraints need to be modified. The program reported in [DO85] can detect only those overconstraint loops which have a single upper-bound constraint; another [KIN84] copes with overconstraints by consecutively ignoring constraints until the graph can be solved, resulting in an illegal layout.

The algorithm used in *SPARCS*[BUR86] and described here uses a two-pass process. The first pass is a graph leveling which has time complexity of $O(E)$ [KNU73]. The leveling is performed so that the vertices can be processed in topological order. Leveling also detects cycles composed of all-positive or all-negative-weight edges, which are input errors as described above. Following leveling, the graph is guaranteed to be a polar and acyclic using the formulation defined earlier.

The second pass detects all left-most and right-most vertex pairs in the graph that are involved in an overconstraint. For each pair, the paths that form the overconstraint, i.e., the largest lower-bound path and smallest upper-bound path, are reported. If there are a number of overconstraints between those vertex pairs, the path causing the worst overconstraint is reported.

Since the only overconstraints that may remain are mixed cycles, in the conventional notation, *all potential cycles must contain at least one vertex that has at least one fanout edge with an upper bound constraint and at least one edge that has a lower-bound constraint.* The second phase of the algorithm begins with scheduling these vertices, as described in the following code fragment. A *bubble* is a small data structure that is attached to a node and carries the maximum and minimum constraints to that node from the node where the bubble originated. It is the creating and merging of bubbles as they move through the graph that is used to detect overconstraints.

```
Overcon()
{
    foreach(vertex V_j^ORG with fanout edge e_jk
                        such that U_jk< INF and N_fo(j)>1)   {
        add bubble B_j^j at vertex V_j^ORG;
        V_j^ORG.bubbleCount = 1;
        Lschedule( V_j^ORG);
    }

    while((V_j = LgetNextScheduledNode()) is not NULL and V_j
                        is not marked as source of overconstraint)   {
        foreach(bubble B_k^j at vertex V_j)   {
            if(V_k.count > 1 and B_k^j marked for processing)   {
                processFanouts();
                processFanins();
            }
            if(B_k^j.lower > B_k^j.upper)   {
                B_k^j.overcon = TRUE;
            }
        }
    }

    foreach(node V_j)   {
        foreach(bubble B_k^j where B_k^j.overcon = TRUE)   {
            recursively follow B_k^j.lowerVertex path to V_k;
            recursively follow B_k^j.upperVertex path to V_k;
        }
    }
}
```

---

where:

```
processFanouts()
{
    foreach(fanout edge e_jm of V_j)   {
        if(B_k^m is not present at V_m)   {
            create bubble B_k^m at V_m;
            B_k^m.lower = -INF;   B_k^m.upper = INF;
            V_k.bubbleCount = V_k.bubbleCount + 1;
        }
        if((newLow = B_k^j.lower + L_jm) > B_k^m.lower)   {
            B_k^m.lower = newLow;
            B_k^m.lowerVertex = V_j;
            moved = TRUE;
        }
        if((newUpp = B_k^j.upper + U_jm) < B_k^m.upper)   {
            B_k^m.upper = newUpp;
            B_k^m.upperVertex = V_j;
```

```
        moved = TRUE;
      }
   if(moved and V_m != V_k)   {
      Lschedule(V_m);
      mark B_k^m to be processed;
   }
   delete bubble B_k^j at V_j;
   V_k.bubbleCount = V_k.bubbleCount - 1;
   }
}
```

and:

```
processFanins()
{
   foreach(fanin edge e_mj of V_j)   {
      if(U_mj < INF)   {
         if(B_k^m is not present at V_m)   {
            create bubble B_k^m at V_m;
            B_k^m.lower = -INF;   B_k^m.upper = INF;
            V_k.bubbleCount = V_k.bubbleCount + 1;
         }
         if((newLow = B_k^j.lower - U_jm) > B_k^m.lower)   {
            B_k^m.lower = newLow;
            B_k^m.lowerVertex = V_j;
            moved = TRUE;
         }
         if((newUpp = B_k^j.upper - L_jm) < B_k^m.upper)   {
            B_k^m.upper = newUpp;
            B_k^m.upperVertex = V_j;
            moved = TRUE;
         }
         if(moved and V_m != V_k)   {
            Lschedule(V_m);
            mark B_k^m to be processed;
         }
         delete bubble B_k^j at V_j;
         V_k.bubbleCount = V_k.bubbleCount - 1;
      }
   }
}
```

The most efficient algorithm for solving this problem known until recently $O(V, E)$ worst-case time and storage complexities [TSA85]. In addition, this algorithm detects only a single overconstraint cycle per pass. The algorithm presented recently in [BUR86] has worst-case time and memory complexities of $O(V^{ORG}E)$, where V^{ORG} is the number of vertices with at least one upper-bound-constrained fanout vertex. In most practical situations, this number is small relative to the total number of vertices. In addition, the above algorithm finds all worst-case overconstraints in the graph in a single pass.

Table 1.3.1 contains the execution times of the *SPARCS* overconstraint detection algorithm for three overconstrained graphs. All times given in this chapter are in seconds for a C-language implementation running on a VAX 11/785 under Berkeley Unix.

Nodes	Lower-Bound Constraints	Upper-Bound Constraints	Overconst. Paths	Time
15	217	11	1	.10
40	315	37	7	.41
101	2024	76	4	6.20

Table 1.3.1. Overconstraint Detection Times

1.3.4. Constraint Graph Solution

Once the graph is proven overconstraint-free, a two-phase process for calculating the positions of the circuit elements is usually undertaken[HSU79]]. The first phase involves performing a longest or critical-path analysis of the graph. The elements that lie on the critical path determine the extent of the layout and typically are placed at their minimum legal positions. A substantial number of the layout elements may not lie on the critical path. Such *noncritical* elements may occupy any position in a prescribed interval. Other considerations can be used to locate these elements, such as fabrication yield or delay optimization.

1.3.4.1. Critical-Path Determination:

The solution of the critical-path or longest-path problem for acyclic graphs is well-understood, with worst-case complexity of $O(n^2)$ for a simple implementation[AHO74]. Other implementations with lower complexities have been reported, with the increase in efficiency often due to the data structures used.

The longest-path problem is more difficult when upper-bound constraints are included. The algorithm used in [LIA83] solves such graphs, and requires time of $O(LU)$, where L is number of lower-bound branches in the graph and U is the number of upper-bound branches.

The constraint solver of *SPARCS* uses *event-driven selective-trace* techniques to minimize redundant processing. Loosely stated, event-driven selective-trace algorithms schedule items for processing only when when a neighboring, constraining item changes. In the present context vertices in the constraint graph have their locations updated only when the location of an adjacent vertex changes. The pseudo-code for the algorithm is shown below. Each vertex in the graph is assumed to have three location values – $V_j.lower$, $V_j.upper$, $V_j.placed$ – which correspond to the minimum legal location, the maximum legal location, and the final location of the vertex, respectively. For vertices on the critical path, all three values are the same.

```
Critical()
{
  initialize $V_{SO}$.lower = $V_{SO}$.placed = $V_{SO}$.upper = 0;
  foreach( fanout vertex $V_j^{FO}$ of $V_{SO}$)
    Lschedule($V_j^{FO}$);

  while(($V_j$ = LgetNextScheduledVertex()) is not NULL) {
    lowerBound = 0;
    upperBound = 0;
    foreach( fanin vertex $V_k^{FI}$ of $V_j$)
      lowerBound = MAX(lowerBound, $V_k^{FI}$.lower + $L_{kj}$);
    foreach( fanout vertex $V_m^{FO}$ of $V_j$)
      upperBound = MAX(upperBound, $V_m^{FO}$.lower - $L_{jm}$);

    location = MAX($V_j$.lower, MAX(lowerBound, upperBound));
    if(location and $V_j$.lower are different) {
      $V_j$.lower = location;
      foreach(fanin edge $e_{mj}$ of $V_j$)
        if( $U_{mj}$ < INF)
          Lschedule($V_m$);
      foreach(fanout edge $e_{jn}$ of $V_j$)
        if( $U_{jn}$ < INF)
          Lschedule($V_n$);
    }
  }
}
```

A particular fanin vertex v_{f_i} of vertex v is scheduled if v changes and there is an upper-bound constraint between them. A fanout vertex v_{f_o} of vertex v is scheduled if v changes and there is a lower-bound constraint between them. If there are no upper-bound constraints then there are at most V vertices scheduled. Let the number of upper-bound constraints be K. Each upper-bound constraint branch causes the vertex at its head to to be re-processed at most once (or else the graph has a cycle of net positive weight). In the worst case, each re-processing of a vertex can cause V other vertices to be scheduled. The overall worst-case complexity for this algorithm is therefore $O(VK)$, including the initial graph leveling. In practical cases the worst-case behavior of this algorithm is seldom encountered, because most constraints are local in nature.

Once the lower-limit for each vertex has been established, the algorithm is executed in the reverse direction starting at the sink node to find the upper-limit values. Vertices that are on the critical path have equal upper and lower legal positions at the conclusion of the second pass.

Table 1.3.2 contains run times for several examples with lower and upper-bound constraints for the complete process. Both passes are included in the times.

Vertices	Constraints	Time
15	226	.05
101	2,024	.29
201	40,200	5.15

Table 1.3.2. Execution Times for Critical-Path Algorithm

1.3.4.2. Non-Critical-Path Elements: The critical path solution produces the final positions for a subset of the layout elements. There is considerable freedom available for positioning the other, noncritical elements. One positioning strategy for the noncritical vertices, used in many early compactors, is to locate them at their minimum legal positions (the "lower-left-corner" compactors(e.g. [WIL78],[MOS81])). This is often a poor choice; the layout extent does not change since it is determined by the critical path and fabrication yield may be adversely affected by the close proximity of elements to their neighbors. A better strategy is to position noncritical elements such that the empty space in the layout is uniformly distributed among them. This "average slack" approach was first used in *CABBAGE*[HSU79]. Other early programs allowed the user to choose any one of four corners for the compaction, depending which one worked the best! Another useful strategy is to position highly-connected elements close together at the expense of less connected elements, or to bias the slack with the resistivity of the interconnect to obtain a first-order delay minimization. This approach was used in the *LAVA* program, developed at Stanford[WOL85], but no details are available as to the approach used.

The *SPARCS* program uses a general, force-based heuristic for placing the non-critical components, as described below. Each constraint e_{ij} in *SPARCS* has a nonzero, positive integer *force weight*, denoted F_{ij} associated with it. The weight represents the relative attraction between the vertices incident to the constraint branch. The noncritical vertices are then positioned by a relaxation algorithm. At each step in the algorithm, the current vertex v_j is located such that

$$d_i \sum_{fanins} F_{ij} = d_o \sum_{fanouts} F_{jo}$$

where d_i is the free space to the left (or bottom) and d_o is the free space to the right (or top) of the current vertex. If the current vertex changes position, its fanins and fanouts that are noncritical are scheduled for processing. The pseudo-code for the algorithm is given below.

Figures 1.3.1(a)-(c) show four cells, with the two end cells fixed. If the two middle cells are located at their minimum positions the layout of Figure 1.3.2(a) results. If the free space is equally distributed the layout in Figure 1.3.1(b) is obtained. In the layout shown in Figure 1.3.1(c), the space between cells A and B and cells C and D has been minimized relative to the space between cells B and C. This case results from biasing the spacing to reflect the connectivity between cells; the outer pairs are connected by polysilicon, and the inner pair by metal. Each poly connection contributes a force of "2" to the constraint branch between the cells involved, whereas each metal connection contributes a

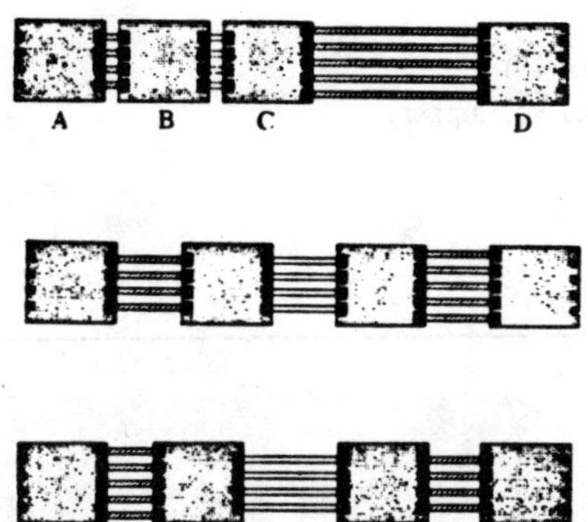

Fig. 1.3.1 Slack distribution using (a) All cells at minimum location
(b) Free space equally distributed (c) Space allocated by connectivity.

force of "1". The algorithm used in this program is described more explicitly below:

```
Slack() {
  foreach(noncritical vertex Vj)
    Schedule(Vj);

  while((Vj = GetNextScheduledVertex()) is not NULL) {
    newLower = 0;              /* new minimum position */
    newUpper = INF;            /* new maximum position */
    leftWeight = rightWeight = 0;
    foreach(fanin vertex Vk^FI of Vj) {
      newLower = MAX(newLower, Vk^FI.placed + Lkj);
      newUpper = MIN(newUpper, Vk^FI.placed + Ukj);
      leftWeight = leftWeight + Wkj;
    }
    foreach(fanout vertex Vm^FO of Vj) {
      newLower= MAX(newLower, Vm^FO - Ujm);
      newUpper= MIN(newUpper, Vm^FO - Ljm);
      rightWeight = leftWeight + Wjm;
    }
```

$$
newloc = \frac{(newUpper - newLower)(rightWeight)}{(leftWeight + rightWeight)} + newLower;
$$

```
    if(newloc and Vj.placed are different) {
```

```
    Vj.placed = newloc;
    foreach(fanin vertex V_k^{FI} of Vj) {
      if(V_k^{FI}.lower is not equal to V_k^{FI}.upper)
        Schedule(V_k^{FI});
    }
    foreach(fanout vertex V_m^{FO} of Vj) {
      if(V_m^{FO}.lower is not equal to V_m^{FO}.upper)
        Schedule(V_k^{FI});
    }
  }
}}
```

1.3.5. Active Constraints

Active constraints are a relatively new concept and are used to force the relative spacing between one pair of nodes to be the same as the relative spacing between other pairs. Active constraints are needed in order to space analog cells, where symmetry relationships must be maintained.

Figure 1.3.2(a) shows four capacitor cells connected to implement two matched capacitors. It is desired to space the layout while maintaining the common-centroid geometry. If two spacings are performed, a horizontal followed by a vertical, the layout shown in Figure 1.3.2(b) results. Symmetry has been lost due to the intervening element in the lower left corner of the layout.

If the spacing steps are carried out with the inclusion of active constraints, the layout of Figure 1.3.2(c) is obtained. The common-centroid geometry has been preserved in the spacing process. In the horizontal step, capacitor cells A and B are actively constrained to the source of the graph and cell pair A and C is actively constrainted with pair B and D. Analogous active constraints are used in the vertical step. This example was produced by the *SPARCS* program which uses a relatively simple iterative approach to solving the active constraint problem at this time.

1.3.6. Constraints with "Dead Zones"

The final class of constraint problem which must be solved to make a spacing program of general utility is that of constraints with non-contiguous ranges of legal values. For example, in many IC technologies a *Reflection Rule* may apply, which states that, for two pieces of geometry $G1$ on layer $L1$ and $G2$ on layer $L2$: "$G1$ may come as close to $G2$ as $RR12$ spacing units, *or* it may lie on directly on top of and be enclosed by $G2$, but neither of its edges may lie within $RR12$ spacing units of $G2$." A similar rule is involved with the merging of contacts, or vias – they may be as far apart at the via-via spacing rule, or they may be collapsed together, but they may not overlap.

While there is active work on-going to find an algorithm that will solve this type of constraint, other than adding "special case" code to hard-wired spacing programs, I know of no general solution to this problem at this time. In particular, I know of no solution for which the requirements for convergence have been described or for which a convergence

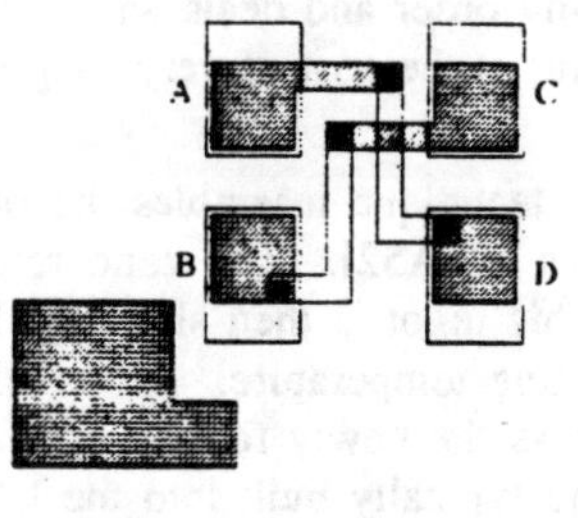

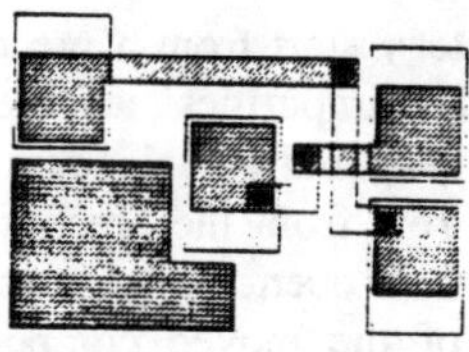

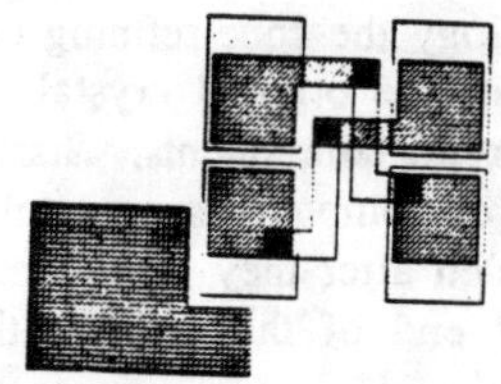

Fig. 1.3.2 Common centroid spacing (a) before spacing
(b) After conventional spacing (c) After spacing with active constraints.

proof exists.

1.4. COMPACTION BY ZONE REFINING

The zone-refining approach [SHI86] for two-dimensional compaction combines combines a generalized one-dimensional compaction procedure with sophisticated lateral movements of the elements to be compacted. This concept is similar to a technique developed by Crowther[CRO80] and implemented in the in the *BEADS* program, however the heuristics used are different in the two approaches.

These lateral movements will "shake" the individual components into more densely packed arrangements. The technique bears a strong similarity to the technique of 'zone-refining' used in the purification of crystal ingots. The change in geometry in zone-refining is not as free as in general 2-dimensional compaction algorithms. On the other hand, the restrictions placed on the movements of individual components have allowed the development of a polynomial algorithm which produces excellent results when compared to traditional methods.

The natural process of crystallization has been used in other areas as a model for obtaining densely-packed layouts. [KIR83]. Since the Simulated Annealing implementations in use today move blocks in any order and deals with the circuit as a whole, globally optimal solutions can be achieved but at the cost of very long run times for systems with a large number of variables.

Th Zone Refining compaction technique resembles the process of zone refining used for the purification of crystal ingots [PFA52]. The zone refining process starts with an already established crystal ingot. This ingot is then slowly pulled through a heating spiral that locally heats the crystal to melting temperature. At the exit end of the heat zone the material recrystallizes. In this process the newly formed crystal has a lower concentration of impurities since the impurities are typically built into the lattice at a lower rate than the proper crystal atoms. Thus most impurity atoms are kept in the molten zone and are eventually swept out of the crystal. If higher purity is required, the process can be repeated.

In close analogy the zone-refining compaction strategy start from a precompacted circuit that represents the original crystal. In this case the 'impurities' are the unnecessary voids between circuit components. Starting from one side, individual circuit components or small clusters of components are peeled off row by row from the precompacted layout and are reassembled after they have been moved across an open zone. In the reassembly step at the other end of the zone, both coordinates of the moved components can be altered.

The implementation of this concept is straight-forward for the simple case of rectangular blocks that need to be placed in a compact manner into a bin of given width. In integrated circuit layouts, the interconnections add a considerable complication. They can severely constrain the freedom with which the individual components can be moved. Thus it is necessary to introduce additional jogs in the wires between them to permit less restricted lateral movements of each component. This additional degree of freedom permits a higher packing density in the newly formed part of the layout. The geometrical design rules are observed by maintaining and using the constraint graphs in both the x- and y-direction.

The advantage of the zone refining approach is that the number of components that must be considered at any one time is dramatically reduced from the general placement case, and the complexity of the algorithm becomes manageable. Just as in the physical zone refining process, the compaction process can be repeated if the results are not yet satisfactory after the first pass.

To illustrate the zone refining process, consider the box-packing problem described in [SHI86]. The assumed task is to pack a collection of rectangular boxes densely into a bin of given width. In this process the individual boxes must maintain their original orientation, and the reordering of the topological arrangement of the boxes should be kept reasonably small.

The starting configuration, analogous to the original crystal, is placed in the top part of Figure 1.4.1. Boxes are removed from the lower end of this *top* configuration and reassembled in a new *bottom* configuration growing upwards. The jagged horizon that delimits the new constellation of blocks against the middle zone is called the *floor*, and the lower boundary of the top configuration is called the *ceiling*.

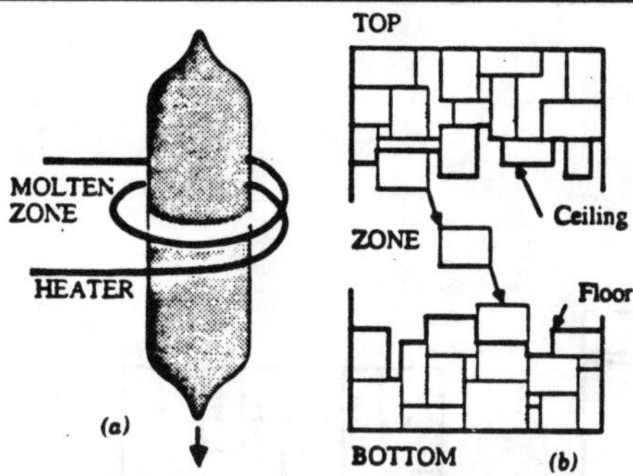

Fig. 1.4.1 Initial Configuration for Zone Refining

1.4.1. Data Structures

In analogy to many scan-line algorithms used in computer graphics, there are two main data structures: one that describes the global state of all components, called the *adjacency graph* (Fig. 1.4.2(b)), and another associated with the moving refinement zone which contains the "currently active components" that must be referred to frequently in each block move.

1.4.2. XY Adjacency Graph

The positions of all blocks during the compaction process are represented in the form of two superposed graphs that represent each block as a node and each horizontal or vertical adjacency as a corresponding arc. These arcs can be labeled with the minimal allowable horizontal or vertical separations between the centers of the block, and this adjacency graph can thus be turned into a constraint graph for properly placing the blocks without overlap. For an actual circuit layout, the constraints attached to these arcs can become more complicated and contain upper as well as lower bounds.

Two additional nodes, *source* (X1 and Y1) and *sink* (X2 and Y2), are used for each graph to keep the boxes within the bounding rectangle. If a node has no predecessors, then an arc is inserted from the source; similarly, if a node has no successors, then an arc is extended to the sink.

1.4.3. Floor and Ceiling

The "currently active" data structure consists of two linked lists of the blocks forming the floor and ceiling respectively. These structures are used for quick reference when blocks are selected and moved across the free zone between top and bottom. They can easily be updated from the information in the adjacency graph. The selected box transferring across the zone moves freely between floor and ceiling (Fig.1.4.2(c)).

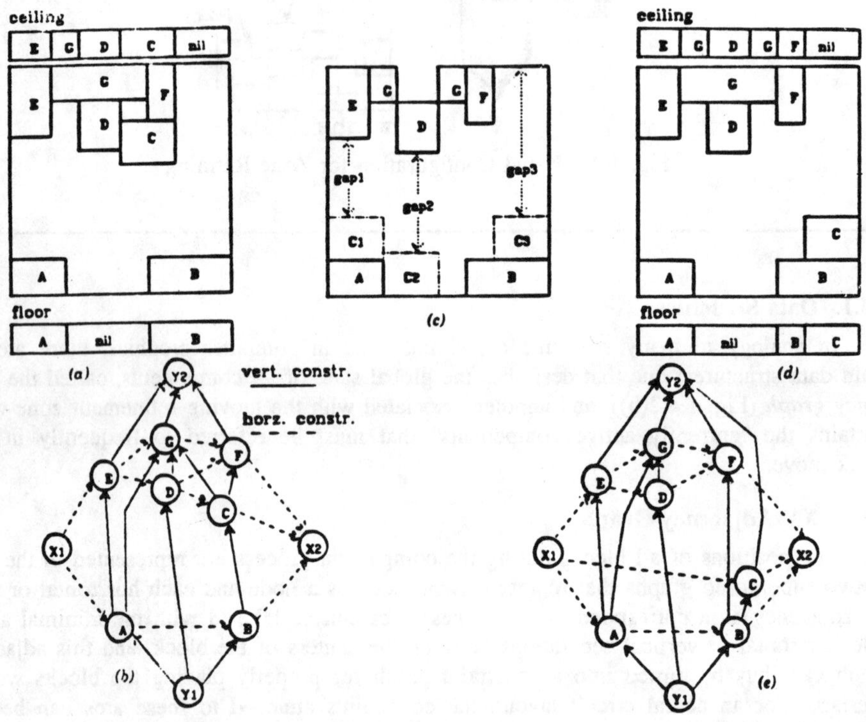

Fig. 1.4.2. Example of box packing in progress. (a) Intermediate constellation of boxes, floor and ceiling data structures.(b) Corresponding adjacency graph. (c) Box C has been selected to be moved; three candidate places C1, C2, C3 are evaluated.(d) Box constellation after box C has been placed and new floor and ceiling structures. (e) Updated adjacency graph.

1.4.4. Zone Refining Algorithm

The main loop of this process then becomes:

```
while( the top is not empty) {
    select individual boxes according to a selection function;
    move them from the top to the bottom;
     place them on the floor according to a placement function.
}
```

A good starting configuration can be achieved with two simple one-dimensional compaction operations. The first compaction is carried out in the horizontal direction and may include an optional "centering" of all components that have some slack in their horizontal positions. This is followed with a vertical compaction from bottom to top. At the start of zone refining, the (empty) floor is thus flat while the ceiling is the profile of the precompacted block constellation.

One of the blocks that is currently part of the ceiling must be selected for transfer across the zone. Several rules can be used to choose a block. One approach that works well is to choose one of the blocks that defines the lowest point in the ceiling; if there is a tie, the left-most block is selected. The block is removed from the top, and the ceiling data structure is updated.

For the selected block, an optimal position must now be found on the floor. For different x-positions the resulting minimum width of the free zone is evaluated. The x-position that maximizes this width is chosen. At this point, there are many possibilities to choose an appropriate cost function. If the original ordering of the blocks should be modified as little as possible, a hard limit on the maximum lateral displacement can be set. Alternatively, a cost can be associated with any displacements in x that is traded off against the gain in zone width.

To find the optimal place for box C in Figure 1.4.2, for example, it is only necessary to compare the three candidate places shown in Figure 1.4.2(c). Since gap 3 is the largest, box C is placed as shown in Figure 1.4.2(d). When evaluating different placements of a box, there may be more than one location that yield the maximal improvement. In our approach, the left-most position is then selected.

Once the optimal position has been determined, the block is placed correspondingly and the floor information is updated (Fig.1.4.2(e)). This figure also shows the modified adjacency graph after placement of box C.

Zone refining of an actual circuit layout follows the same basic algorithm as box-packing. However, its implementation is much more complicated, for several reasons:

(1) There is more than one level of interconnect. Thus proper spacing in all of these levels must be considered. In particular this means a separate pair of ceiling and floor profiles need to be kept for each independent level of interconnect. For an NMOS process we can limit ourselves to just two levels: a metal level and a combined diffusion/polysilicon level. If there are additional metal or polysilicon layers, they need to be represented by separate levels. The profiles are built using the concept of protection frames, described earlier.

(2) The presence of complicated components such as transistors and inter-level contacts that span several levels and the existence of sophisticated geometrical layout rules leads to more complicated constraints for the placement of components. For each arc

there are now more complicated expressions than a simple minimum separation. There can also be upper bounds as in the case when a connection must touch a circuit block within the extent of a contact area of a certain size.

(3) The components in a circuit layout are connected by wires that are neither infinitely thin nor infinitely flexible. This tends to limit the degree to which individual components can be moved and thus the degree of topological rearrangement possible. For a full exploitation of the possible two-dimensional movements, jogs need to be inserted into the wires at appropriate places.

(4) As a consequence of all the points above, components cluster into groups that should be moved as a whole. There are additional difficulties associated with discovering these groups, moving them around, finding an optimal place for them, and dealing with the multitude of wires connected to them.

The *ZORRO* program, which implements the zone-refining compaction algorithm, addresses these issues, as described in [SHI86].

Two-dimensional compaction by zone refining constitutes a novel approach in the broad spectrum of automatic layout techniques between simple one-dimensional, fixed-grid compaction by constraint-graph based methods and simulated annealing techniques for general placement. This intermediate option is seems to give denser layouts than one-dimensional compactors because of the additional degrees of freedom in moving individual components. On the other hand it uses a much more restricted approach to search for a final solution than general simulated annealing techniques, and thus requires much shorter run-times, but cannot be expected to find a global optimum in density if achieving the optimum would require dramatic changes in topology.

Zone refining is a novel compaction method, that, while having a preferred direction of compaction at every stage, must be viewed as a two-dimensional method since, during compaction steps, components are moved in the direction of both coordinate axes to obtain densest local packing. In different versions of the *ZORRO* program and with different kind of layouts, a zone refinement pass takes from 3 to 10 times as long as a complete one-dimensional compaction step of the constraint-graph based compactor Python[BAL82]. The observed reductions in area over one-dimensional compaction range from 4% to 15%, best case. The improvement depend on the initial topology of the circuit as well as the exact sequence of directions in which the compaction steps are applied. The work in this area is very young and significant improvements in both run-time and overall area reduction are likely.

2. LAYOUT LANGUAGES

2.1. INTRODUCTION

In recent years, the notion of procedural circuit design, using a *layout language*, has emerged as a key component in the design process. The use of textual language constructs, rather than just graphics (or data), for describing IC designs was pioneered at Caltech in the ICL language and the Bristleblocks[JOH79] module generator[1], and applied later at MIT in the DPL environment[BAT80] and Daedalus constraint-based design system[BAT81][SHR82]. A number of simple, procedural extensions to existing languages (often called "embedded layout languages") have also been developed. These efforts, and others, have inspired a large number of projects at University and Industrial sites for in-house use. The term *silicon compiler* is often associated with procedural design and a number of new companies are advertising *silicon compilers*. A more detailed overview of the general area of silicon compilation is presented in [GAJ85][GAJ86]. Unfortunately, the majority of systems offered so far do *not* offer the general user a procedural design capability. Rather, they can be characterized as cell-based systems where the circuit building-blocks are parameterized cells. These cells can be assembled in a variety of ways, depending on the *design style* the "compiler" is using.

True procedural design, where the IC designer can write programs which, when executed, produce layout is still of key importance to the productivity of the custom IC designer. In the VTI suite of tools, a procedural design language in the form of an embedded language has been provided. This language allows the user to generate small cells and modules at the mask-layout level, but the designer is responsible for coding layout ground-rule data into the cell descriptions. Silicon Design Labs offers a procedural Design system based on their 'L' language[BUR85], which is an evolution of the 'I' language, developed at ATT Bell Laboratories[BUR83]

Many of the early procedural design systems had limited success for a number of reasons:

- The relationship between graphics and procedure was not exploited sufficiently. Graphics and procedure are generally treated as disjoint descriptions of the design. The graphical mask layout was the *result* of the procedure, rather than an active part of it.

- Verifying the correctness of a procedurally generated design is generally performed at the mask layout level. At that point, no correspondence between the active procedure which created an object and its geometrical layout is maintained. As a result, debugging the design is very difficult.

- Often the designs produced by the early systems did not achieve high density, high performance, or meet power requirements.

Another way of reducing the potential for errors, as well as ensuring the technology-parameterization necessary at the lowest level of design, is to have the procedural design system generate symbolic layout rather than detailed geometry. A spacing program, or constraint-solver, can then be used to guarantee a layout-rule-correct design and to convert

---

[1] The term Silicon Compiler was first coined at Caltech to describe the ICL and Bristleblocks work. At that time, it only referred to the structural aspects of the design process.

the symbolic layout to mask artwork. The ALI system, developed at Princeton[LIP83], was one of the first layout-languages to insist upon constraint-based descriptions.

In the remainder od this section, a number of layout-language-based design systems are reviewed and compared. The DPL system is used to illustrate many of the features found in other layout-language-based design systems.

2.2. A CASE STUDY: THE DESIGN-PROCEDURE LANGUAGE

The Design Procedure Language (DPL) [BAT80] was developed at MIT as an early procedural design language for IC layout. Many of the key concepts of procedural design were first codified in DPL and so it will be used here to illustrate a number of concepts.

DPL is a collection of LISP functions that are used to construct and to manipulate a hierarchically-organized, object-oriented database. Since it is embedded in LISP, DPL inherits the power of a full programming language. In DPL a design procedure constructs a data-structure which hold a description of a design. This description may also contain procedures for further manipulation of the database.

Parameters in DPL use call-by-keyword syntax, and may be assigned default values. Constraints among parameters may be specified by the designer. An overall picture of the DPL system is shown in Figure 2.2.1. The DPL design process involves the following stages:

(1) The designer specifies procedures for constructing pieces of the design.

(2) These pieces are then used to build more complex representations. The procedures may refer to information stored in the structure earlier.

(3) The final "design" is then a complex yet organized hierarchical structure which may be used to produce mask layouts.

2.2.1. Types

A *type* in DPL holds a procedure that builds the data structures; the procedure stored in a *type* is called the *maker function* of the type. The maker functions may take parameters, have default values, and constraints among the parameters may be stored with the type. A type may be thought of as a description of a class of objects that share some common features – these objects are the structures produced when the maker function is run with a particular set of parameter values.

2.2.2. Prototypes

The structure that *is built* by a type is called a *prototype*. A prototype holds the description of a certain actual cell-type in the design. Whereas a prototype could be displayed on a graphics system, a type could not. In DPL, the user never actually touches or edits a prototype, he only specifies how to build one. The importance of this distinction is developed later in the Chapter. So, for example, a type might be used to describe a generalized resistor which takes, as a parameter, its resistance value R. If the type for resistor is invoked and the maker function is executed with R=100, a prototype resistor will be created with resistance value 100 ohms.

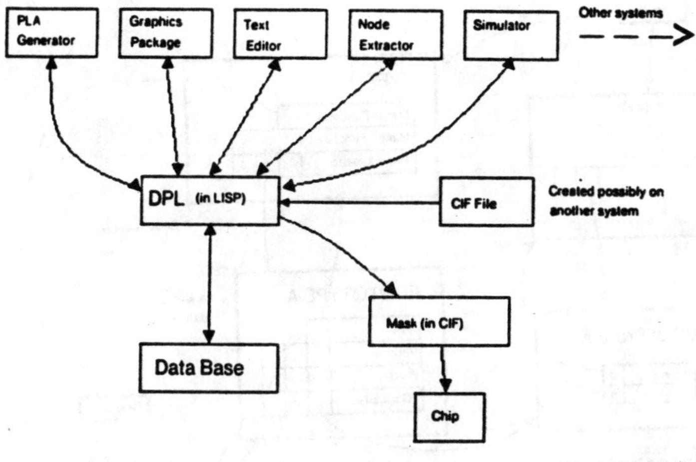

Fig. 2.2.1 The DPL and LISPIC Environment[BAT80]

2.2.3. Virtual Copies

If the maker function of a type "calls" another type, the prototype built by the "called" type will be part of the prototype built by the "calling" type. For example, if a type *inverter* is called with parameter *KRatio* set to 4, in the process of building the actual K=4 inverter, say *inv*4, it may call the type *emos* to construct a pull-down transistor. If *emos* is called with parameters $W=2$ and $L=4$, a prototype emos transistor, say *emos*2x4, will be created. But this prototype transistor is actually part of the *prototype* inverter *inv*4, not part of the *inverter* type itself. A prototype that is part of another prototype is called a *virtualcopy*(VC) in DPL. It is called a virtual copy because the description of the prototype is available from the VC but is not stored there. Rather, it is stored with the actual prototype. That way, a single prototype can be used as a VC in many prototypes. So the VC stores information about who is using it (its parent prototype) and what prototype it actually represents. This is illustrated in Figure 2.2.2. Of course, other information can be stored with the VC than simply its parent and its role-model.

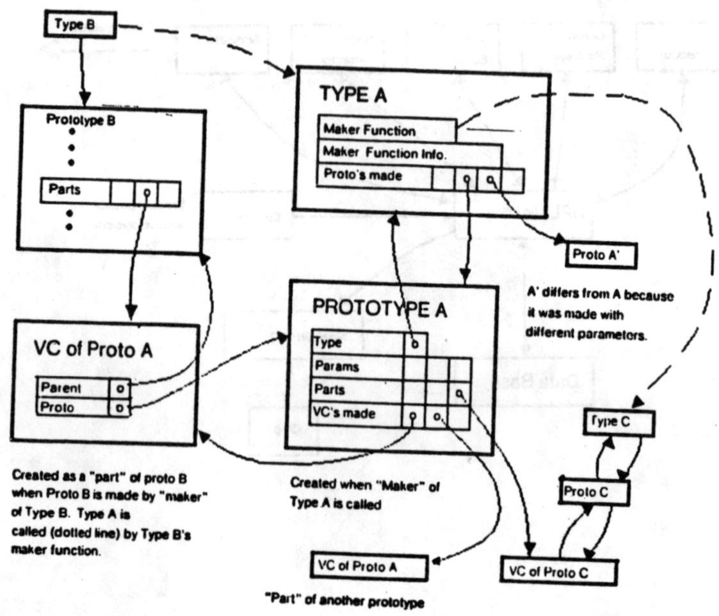

Fig. 2.2.2 The DPL Part Hierarchy[BAT80]

2.2.4. Supertypes and Subtypes

In addition to using a prototype as part of another prototype, it is possible to specify that a certain type includes all of another type. That is, it inherits all of the features of its *supertype*. So a *subtype* has all of the parameters of its *supertype*, plus any others declared when the subtype is defined. In that way, a subtype is really a modified version of its supertype. This relationship is illustrated in Figure 2.2.3.

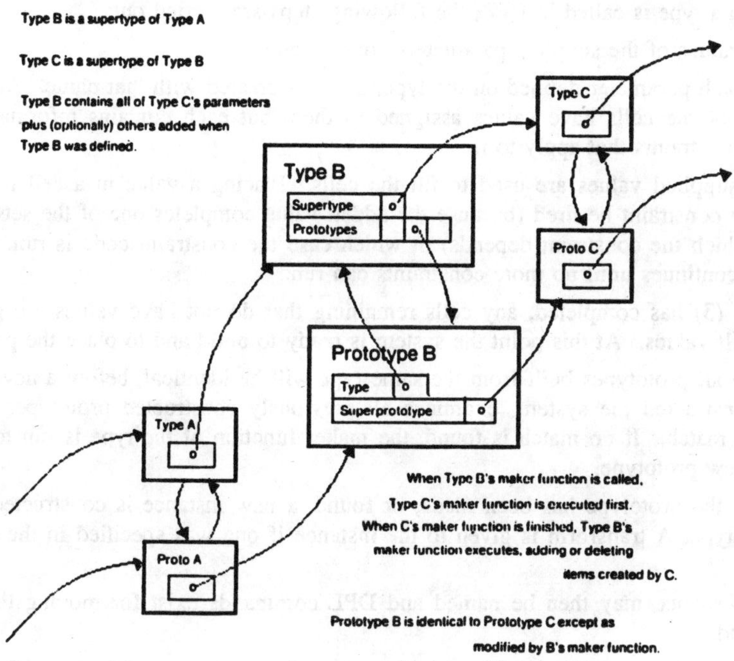

Fig. 2.2.3 The DPL Type Hierarchy[BAT80]

2.2.5. Instances

Some of the information about an object may depend on the context in which the object is used, or "seen"(e.g. the geometric transform associated with a particular use of a VC). The specification of a transform is not part of the VC because the same VC may be viewed in different ways. This is a feature that is unique to DPL. For example, while the pulldown transistor used in an inverter may be viewed in the coordinate system of the inverter as the inverter is being built, it might be viewed later from the coordiante system of which the inverter is a part. For this reason, and others, a VC is contained in an object called an *instance*. The instance also contains other data (like the transformation) that is particular to the context in which the VC is being viewed. In terms of a layout description format, like CIF or EDIF, a prototype can be thought of as a particular CIF cell or EDIF cell view. An instance can be thought of as a CIF call or an EDIF instance.

2.2.6. The Flow of DPL Processing

When a type is called in DPL, the following steps are carried out:

(1) The values of the supplied parameters are evaluated.

(2) For each parameter defined on the type, a cell is created with that name. At this point none of the cells have values assigned to them but each contains information about any constraints that apply to it.

(3) The supplied values are used to fill the cells. Placing a value in a cell may require that a constraint be fired (because the added value completes one of the sets of values on which the constraint depends) in which case the constraint code is run. This process continues until no more constraints can run.

(4) Once (3) has completed, any cells remaining that do not have values are given their default values. At this point the system is ready to build and to place the prototype.

(5) Since all prototypes built from the same type will be identical, before a new prototype is constructed the system examines all previously constructed prototypes and looks for a match. If no match is found, the maker function of the type is run to construct the new prototype.

(6) Once the prototype has been made, or found, a new instance is constructed from the prototype. A transform is given to the instance if one was specified in the call to the type.

(7) The instance may then be named and DPL commands exist for moving the instance around.

For more details concerning DPL and its syntax, the reader is referred to [BAT80]. An example DPL description of a simple pass transistor is included as Figure 2.2.4.

The concepts of *type*, *prototype*, *supertype*, *virtual copy* and *instance* are by no means unique to DPL. Many modern object-oriented programming languages have analogous concepts, perhaps with different names (e.g. LOOPS and SMALLTALK from Xerox PARC, FLAVORS from MIT] and many procedural design systems and languages have either adapted these concepts from DPL or from one of these programming languages, or have discovered the concepts independently.

```
(deflayout pass-transistor                    ;;; PASS-TRANSISTOR
  '((primary-parameters
       ((channel-length 2)
        (channel-width 2))))
  (part 'channel rectangle
        (layer 'channel)
        (length (>> channel-length))
        (width (>> channel-width)))
  (part 'source-diffusion rectangle
        (layer 'diff)
        (length *diff-overhang*)
        (width (>> channel-width))
        (top-center (>> bottom-center channel)))
  (part 'drain-diffusion rectangle
        (layer 'diff)
        (length *diff-overhang*)
        (width (>> channel-width))
        (bottom-center (>> top-center channel)))
  (part 'poly-piece rectangle
        (layer 'poly)
        (length (>> channel-length))
        (width (+ (>> channel-width)
                  (* *poly-overhang* 2))))
  (setq-my p (>> center-right poly-piece)))
```

Fig. 2.2.4 A DPL Description of a Pass Transistor

2.3. PRIMITIVE OPERATIONS

The DPL example above illustrates many of the global concepts associated with procedural design. However, local issues, such as the positioning of objects relative to one another or techniques for handling layout rules, are also critical to the utility of procedural design.

2.3.1. Positioning

Rather than placing components in absolute locations, much of what a designer wants to say is relative positioning: "place this component above that one, so that the top-center of one this component is in line with the bottom-left corner of that one." Most layout languages have constructs for expressing this sort of relationship. The notions of *top–center* and *bottom–left*, etc. are usually defined in terms of the bounding-box of the prototype object. While such definitions are a little crude, more precise reference points are often defined in terms of terminals or reference "markers" placed in the prototype cell.

In DPL, the *align* operator can be used to specify component alignment. It takes the form:

> **(align** *object referencePoint targetPoint* **)**

where *object* is the thing to be moved, *referencePoint* is a point, usually on *object*, and *targetPoint* is the point to which *referencePoint* is to be relocated.

> e.g.
> **(align (>> contact1) (>> top-center contact1)**
> **(>> bottom-center source-diffusion trans1))**

will move *contact*1 so that its *top–center* is a the *bottom–center* of the *source–diffusion* of *tran*1. Note that none of the technology-dependent rules were included in the primitive operation itself. In DPL, it is possible to place objects such that layout rules might be violated, since it is possible to place objects at absolute positions.

In the ALI system, developed at Princeton[LIP83], the user is specifically not permitted to place objects by absolute location. Rather, the separation primitives in ALI specify that their arguments must be separated by in a certain direction but they make no statement about how far they are separated. For example, the ALI fragment:

```
e.g.
boxtype
  htype = array[1..hnumber] of metal;
box
  horizontal:htype
begin
  for i:=1 to hnumber-1 do begin
    above(horizontal[i],horizontal[i+1]);
    xmore(horizontal[i],10)
end.
```

Would generate a stack of *hnumber* boxes on metal, one above the other, separated by the minimum metal-metal spacing (unless some other constraint forced them further apart), and at least 10 units long (the *xmore* statement means "make sure box *horizontal*[i] is at least 10 units in the x-direction." This description is more declarative in nature than some of the positioning constructs available in DPL. When the ALI description is processed, a number of inequality constraints are generated and solved to determine the final layout. On the other hand, an ALI description must be "complete" for it to be realizable – a relationship must be expressed between every pair of rectangles in the design.

It would be possible to emulate an ALI description in DPL, using constraints, but the converse would no be true. On the other hand, the designer is not as concerned with satisfying layout-rule requirements in the ALI system. Just as DPL is built on top of LISP,

ALI is built on top of PASCAL.

2.3.2. Wiring Models

Most of the rectangles required in a large design are actually used to connect cells. At the floor-plan level of the Berkeley SOAR processor[MAR84], over 7,000 rectangles were used to interconnect components while only 4,000 rectangles were needed to describe the protection frames of those components.

As with symbolic layout systems, many layout languages support abstractions for describing interconnections – often called *wires*. Wires are generally specified by indicating the mask layer on which the interconnection is to be implemented, the width of the wire, and the path the wire is to follow. In a constraint-based languages, the path is indicated by constraining the end-point of the wire segments to meet. In other approaches, absolute coordinates for the path corners may be specified. An example of a wire described in DPL is shown below:

```
e.g.
(wire 'wireEx
  (run-layer 'DIFF)
  (from (>> bottom-center source-diffusion trans1))
  (-y 4)
  (to-x (>> center cont1))
  (save-cp 'fork)
  (jog-x (>> bottom-center source-diffusion trans2))
  (restore-cp 'fork)
  (run-layer 'POLY)
  (to-y (>> top-center cont1)))
```

The wire begins as diffusion at the bottom-center of the source diffusion of *trans*1, runs down 4 lambda units (note the explicit distance specification) and connects to the x-coordinate of the contact *cont*1. The current connect-point is then saved and named *fork*. The wire is then run with a *jog–x* command to the bottom-center of the source diffusion of *trans*2. The saved connect-point *fork* is then restored and the layer is changed to polysilicon. The final segment of the wire is then run to the top of *cont*1 (note that the DPL system automatically inserts a butting contact where the layers change for this NMOS technology). The layout created is illustrated in Figure 2.3.1

2.4. IMPERATIVE OR DECLARATIVE?

Most programming languages support unilateral computation. That is, programs perform predetermined operations on their inputs to produce desired outputs. Physical systems, on the other hand, are more often specified as a set of *constraints* among several variables. For example, the constraint $x+y=1$ says no more about how to compute x given y than it says about how to compute y given x. In a programming language, however, the statement $x:=y-1$ is a unilateral expression for the computation of x given y. Most computer programming languages are designed for expressing algorithms – *imperative, procedural* descriptions. In a typical algorithmic language, a compound algorithm is

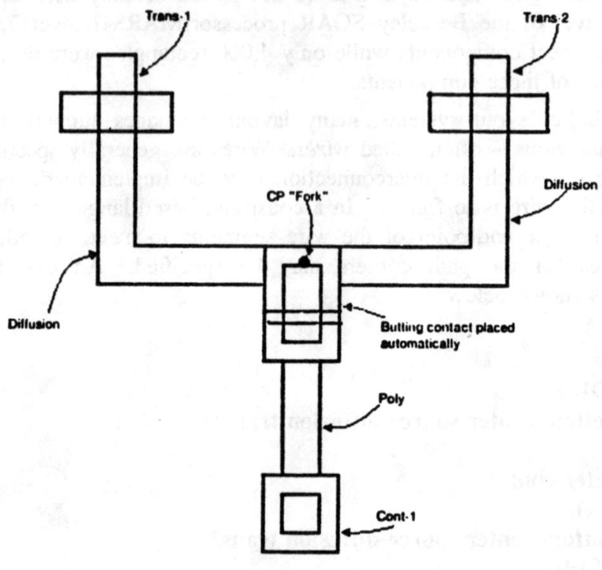

Fig. 2.3.1 A *wire* Example for DPL[BAT80]

constructed from simpler ones by rules of *temporal sequencing* and *data flow*. Algorithms can be temporally concatenated (sequencing), selected among (conditionals), and iterated (loops). Data flow is indicated explicitly by functional composition and implicitly by side-effects on shared variables.

For expressing descriptions of structural relationships and physical situations a *declarative* language is more appropriate[SUS79]. In a declarative, constraint-based language a compound constraint is built from simpler ones by linking some of their parts. The method of concatenation is usually a *equivalence declaration* that the two entities are in fact the same. The difference between these two approaches was introduced in the previous section by example. In this section, the advantages and disadvantages of the two approaches are reviewed.

It might seem that in writing constraints rather than procedures that the algebraic expressions are just in a different, not even convenient, form. This is true to some extent, but there are more important aspects. A constraint-based language can be thought of as a form of "structured algebra"[SUS79] in that the constraints can be composed to form larger systems. The constraint algebra has two concepts of interest: a *notation* for expressing constraints and a *set of transformation rules* which permit the derivation of the consequences of the constraints. It is true that standard algebraic notation is often more concise than a constraint-based description. However, the price we pay for conciseness is asymmetry. Another fundamental reference in this area is [BOR79].

Consider the following simple example. There are three cells, A, B, and C, to be placed in a row (only the x-coordinates will be dealt with, for clarity). The reference coordinate for each cell is its lower-left corner. Each cell is 1 unit square and they must be separated from each other by at least the rule BOXSEP, a technology-dependent constant. The designer wants the cells uniformly spaced but requires that the lower-left corner of C be exactly PITCH units from the lower-left corner of A. This is illustrated in Figure 2.4.1. In an imperative language, the designer might describe such a situation as follows:

```
/* set boxes at minimum legal locations */
A.x = 0;
B.x = A.x + 1 + BOXSEP;
C.x = C.x + 1 + BOXSEP;
/* match pitch */
while( C.x - A.x < PITCH ) {
  B.x = B.x + 1;
  C.x = C.x + 2;
}
/* round down if necessary */
if( C.x - A.x > PITCH ) C.x = C.x -1;
```

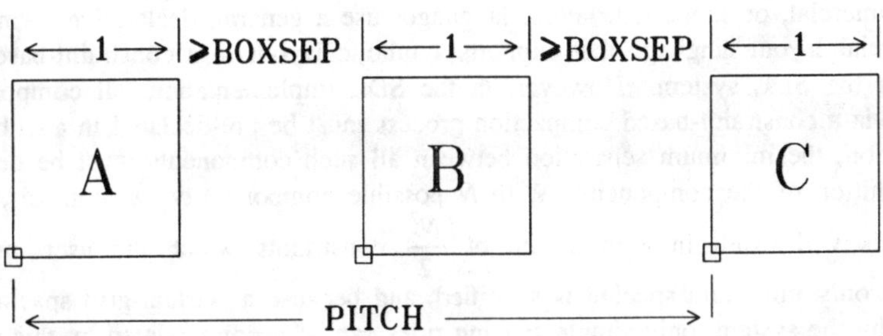

Fig. 2.4.1 Pitch-match Example

of course, for this simple case, the designer could have written:

A.x = 0;
C.x = A.x + PITCH;
B.x = A.x + PITCH/2;

but what happens if PITCH < 2*(1+BOXSEP)! Additional code would be needed to test that condition. If that is the case, and the system is overconstrained for a particular set of design rules, what then? Unless the user is very careful, and considers all possible consequences of the design, possible error configurations will be missed and the system will eventually fail in an unpredictable way. Even if the error condition is detected, how does the writer of the cell description convey to the user of the cell the nature of the of the problem – which cells are involved, what constraints are causing the problem, etc?

In a declarative, constraint-based layout-language, the description might take the form:

minSepX(A, B, 1+BOXSEP);
minSepX(B, C, 1+BOXSEP);
exactSepX(A, C, PITCH);

These functions define the constraints, which will be solved later (similar to delayed-binding in a programming language) using techniques similar to those introduced in Section 1.3.

This approach has three major advantages from a practical point of view:

(1) The designer doesn't have to program for all possible special cases. In general, the code written in such cases is extremely complex anyway.

(2) All designers express their intent in the same way. This is extremely important if multiple designers are working on the same design or if a designer must ever read, and understand, a cell description written by another designer.

(3) The way in which the design interacts with technology-dependent factors is localized. Side-effects in the design caused by changes in technology ground-rules can be identified and corrected.

No commercial, or supported, layout languages use a general, declarative form. The only commercial layout language that supports symbolic layout in a constraint-based approach today is the SDL system. However, in the SDL implementation, all components to be spaced via a constraint-based compaction process must be pre-declared in a technology file. In addition, the minimum separation between all such components must be declared with the definition of the component. With N possible component-types (contacts, transistors, etc.), this will result in a minimum of $\dfrac{N^2}{2}$ constraints which the user must specify.

Because only minimum spacing is specified, and because a virtual-grid spacing approach is used by the system, only simple spacing rules can be accommodated by this system. As with all virtual-grid systems, pitch-matching between cells is handled by stretching. Though this implementation cannot generate dense layouts in the general case, it is a first approach to implementing a declarative style of layout description in a general layout language.

2.5. MIXED APPROACHES

While a textual description of a design is very powerful, it is generally very difficult to debug a design described by a procedure. The designer is continually having to more from the graphical "output" of the system to the textual "input" to the system. Over the past few years, a number of attempts have been made to combine the graphical and textual descriptions more intimately.

The first, and most obvious approach, is to let the designer enter a diagram graphically and translate it into an equivalent textual description. For example, if the user enters the two rectangles shown below using a graphics editor, an equivalent textual representation can be derived.

```
(cell ???
    (rectangle (layer POLY) (point -2 -1) (point 2 1))
    (rectangle (layer NDIF) (point -1 -2) (point 1 2))
)
```

The user can then edit the textual description (e.g add the cell name, replace the absolute coordinates for the points with symbolic values) and create a transistor-generator. Unfortunately, once the generator has been created and is used to generate a layout, the layout cannot be inverted to the generator – as mentioned earlier, some information may be lost in the transformation. For example, the description above might be edited to form:

```
(cell emos (parameter W (default 2)) (parameter L (default 2))
    (rectangle (layer POLY)
        (point (- (+ (/ W 2) GATEOVER )) (- (/ L 2)))
        (point (+ (/ W 2) GATEOVER ) (/ L 2)))
    (rectangle (layer NDIF)
        (point (- (/ W 2)) (- (+ (/ L 2) SDOVER )))
        (point (/ W 2) (+ (/ L 2) SDOVER ) ))
)
```

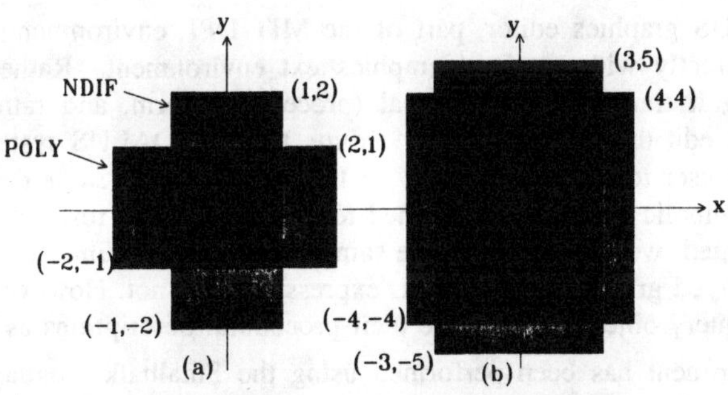

Fig. 2.5.1 (a) Simple Graphically-entered Transistor
(b) Result after parameterization and re-execution.

where GATEOVER and SDOVER are layout-rule parameters obtained from a technology file. If this procedure was called with W=6 L=8, and GATEOVER and SDOVER are assumed to both be 1, the result shown in Figure 2.5.1(b) would be produced. If this layout were converted *back* to a procedural form using the approach described above, it would appear as:

```
(cell emos
    (rectangle (layer POLY) (point -4 -4) (point 4 4))
    (rectangle (layer NDIF) (point -3 -5) (point 3 5))
)
```

and the parameterization of the cell has been lost. A number of approaches have been taken to solving this problem, and some of them are described below. This is an area of active research and the reader should expect a number of new developments over the coming months.

2.5.1.1. Textually-annotated Graphical Cells

A number of graphics editors (e.g. GRED from AT&T Bell Labs.) and module generators (e.g. [MAY83] [WOO86]) have been developed recently which support procedurally-annotated graphical descriptions. In the case of the module generators, the basic cells in the system (often tiles) are described graphically. The composition of the tiles, however, is described by procedures. A significant claim that the developers of such system make is that since much of the work involved with silicon design is structural, a graphical representation should be the heart of the system, rather than a textual one. They claim that such systems are easier to use, and many users agree with them. Such systems often do not provide as many degrees of freedom as the more general text-based systems. This has both advantages (e.g. more expressive power) and disadvantages (e.g. easier to make mistakes).

2.5.1.2. Graphically-entered Procedural Descriptions

The DAEDALUS graphics editor, part of the MIT DPL environment[BAT81], allow the user to work directly with a mixed graphics/text environment. Rather than entering graphics and having it translated to a textual (procedural) form, and rather than simply allowing the used to edit the result of the procedure, the DAEDALUS system was the first attempt to allow the user to work graphically or textually with the *same* description of the design. Both the symbolic expressions attached to objects, and the resulting absolute coordinates they represented, were available at the same time in the system. The absolute coordinates can be displayed graphically; the DPL expressions can not. However, DAEDALUS permits the user to query objects and update their procedural descriptions as well.

A similar experiment has been performed using the Smalltalk language by S. Trimberger. In this case, Smalltalk objects which represented design objects (the Smalltalk Class concept can be used to represent types and super-types) had associated Smalltalk methods for displaying them graphically or procedurally.

Another approach that has been used is to build a prototypical design graphically and then to annotate it with procedures for allowing it to generalize to other prototypes[GLA80][SHR82]. This approach is appealing to the designer and is finding its way into a number of in-house and commercial design systems. Research is also ongoing at MIT[BAM85].

3. ACKNOWLEDGEMENTS

This work is supported in part by the Digital Equipment Corporation, the Semiconductor Research Corporation, and the Defense Advanced Research Projects Agency under contract N00039-86-R-0365. Their support is gratefully acknowledged.

I also thank Jeff Burns and Hyunchul Shin for their help in preparing this manuscript and the many examples used in the course. In addition, Srinivas Devadas, David Harrison, Chuck Kring, Peter Moore and Rick Spickelmier helped with the preparation of computer programs for the course.

4. REFERENCES

[AHO74] A. V. Aho, J. E. Hopcroft, and J. D. Ullman, *The Design and Analysis of Computer Algorithms*, Addison-Wesley, Reading MA. 1974

[AKE70] S. B. Akers, J. M. Geyer, and D. L. Roberts, "IC Mask Layout with a Single Conductor Layer", *Proceedings of the 7th Annual Design Automation Conference*, June 1970, pp 7-16.

[BAL82] M.W. Bales, "Layout Rule Spacing of Symbolic Integrated Circuit Artwork", ERL Memo No. UCB/ERL M82/72, University of California, Berkeley, May 1982.

[BAM85] C. S. Bamji, C. E. Hauck, and A. Jonathan, "A Design by Example Regular Structure Generator," *Proc. 22nd IEEE/ACM Design Automation Conf.*, pp.16-22, June 1985

[BAT80] J. Batali and A. Hartheimer *The Design Procedure Language Manual*, A. I. Memo No. 598, VLSI Memo No.80-31, Massachusetts Institute of Technology, Artifcial Intelligence Laboratory, Sept. 1980.

[BAT81] J. Batali, N. Mayle, H. Shrobe, G. Sussman, and D. Weise, "The DPL/Daedalus Design Environment," *VLSI 81: Very Large Scale Integration*, J. P. Gray, ed., pp.183-192, Academic Press, 1981

[BOR79] A. Borning, "Thinglab - A Constraint-Oriented Simulation Laboratory," SSL-79-3, Xerox PARC, Palo Alto, CA. July 1979.

[BOY83] D. G. Boyer, N. Weste, "Virtual Grid Compaction using the Most Recent Layers Algorithm," *Proc. IEEE Conf. oh Computer-Aided Design*, pp.92-93, Sept. 1983.

[BUR83] M. R. Buric, C. Christensen, T. G. Matheson, "The Plex Project: VLSI Layouts of Microcomputers Generated by a Computer Program," *IEEE Int. Conf. on Computer-Aided Design, Digest of Papers* Sept. 1983.

[BUR85] M. R. Buric and T. G. Matheson, "Silicon Compilation Environments," *Proc. Custom Integrated Circuits Conf.*, pp.208-212, May 1985.

[BUR86] J. Burns, A. R. Newton, "SPARCS: A New Constraint-Based IC Symbolic Layout Spacer," *Proc. IEEE Custom Integrated Circuits Conf.*, pp.534-539, May 1986.

[CHO77] Y. E. Cho, A. J. Korenjak, and D. E. Stockton, "FLOSS: An Approach to Automated Layout for High-Volume Designs", *Proceedings of the 14th Annual Design Automation Conference*, June 1977, pp. 138-141.

[CHO85] Y. E. Cho, "A Subjective Review of Compaction," *Proc.*22nd pp.396-404, June 1985.

[CRO80] W. Crowther, Xerox PARC, Private Communication, 1980.

[DO85] J. Do and W.M. Dawson, "Spacer II: A Well-Behaved IC Layout Compactor", *Proc. VLSI 85 International Conference*, Tokyo, Japan pp. 277-285, August 1985.

[DUN78] A. E. Dunlop, "SLIP: Symbolic Layout of Integrated Circuits with Compaction", *Computer-Aided Design*, Vol. 10, No. 6, Nov 1978, pp. 387-391.

[DUN79] A. E. Dunlop, "Integrated Circuit Mask Compaction", PhD Thesis, Carnegie-Mellon University, Pittsburgh, PA, 17 Oct 1979.

[DUN80] A. E. Dunlop, "SLIM - The Translation of Symbolic Layouts into Mask Data", *Proceedings of the 17th Annual Design Automation Conference*, June 1980, pp. 595-602.

[FIE84] R.D.Fiebrich, Y.Z.Liao, G.Koppelman, and E.Adams, "PSI: A Symbolic Layout System," IBM J. of Research and Development, pp.572-580, Sept. 1984.

[GAJ85] D. D. Gajski, "Silicon Compilation," *VLSI Design*, Nov 1985

[GAJ86] D. D. Gajski, N. D. Dutt and B. M. Pangrle, "Silicon Compilation (Tutorial)," *Proc. IEEE Custom Integrated Circuits Conf*, pp.102-110, May 1986

[GIB76] Dave Gibson and Scott Nance, "SLIC - *Symbolic Layout of Integrated Circuits*", *Proceedings of the 13th Annual Design Automation Conference*, June 1976, pp. 434-440.

[GLA80] L. A. Glasser, "An Interactive PLA Generator as an Archetype for a New VLSI Design Methodology," *Proc. IEEE Int. Conf. of Circ. and Computers*, pp.608-612, October 1980.

[HAC80] G. D. Hachtel, "On the Sparse Tableau Approach to Optimal Layout," *Proc. IEEE Int. Conf. of Circ. and Computers*, pp.1019-1022, October 1980.

[HAR86] D.S. Harrison, "VEM Programmer's Manual", Internal Memorandum, University of California, Berkeley, Jan. 1986.

[HSU79] M. Y. Hsueh and D. O. Pederson, "Computer-Aided Layout of LSI Circuit Building-Blocks", *Proceedings of the 1979 IEEE International Symposium on Circuits and Systems*, pp. 474-477.

[HSU79a] M. Y. Hsueh, "Symbolic Layout and Compaction of Integrated Circuits", PhD Thesis, UCB/ERL M79/80, University of California at Berkeley, Berkeley, CA, 10 Dec 1979.

[HUR82] J. Hurt, Private Communication, Mar 1982.

[JOH79] D. Johannsen, "Bristle Blocks: A Silicon Compiler," *Proceedings of First Caltech Conference on VLSI*, Caltech Computer Science Department, 1979.

[KED84] G. Kedem and H. Watanabe, "Graph Optimization Techniques for IC Layout and Compaction." *IEEE Trans. on CAD of ICs* Vol.3, No.1, January 1984.

[KEL81] K. H. Keller, "KIC: A Graphics Editor for Integrated Circuits", MS Report, University of California at Berkeley, Berkeley, CA, June 1981.

[KEL81a] K. H. Keller, "Squid: A Database System for Integrated Circuits", Preliminary Draft, University of California at Berkeley, Berkeley, CA, December 1981.

[KEL84] K.H. Keller, "An Electronic Circuit CAD Framework", ERL Memo No. UCB/ERL M84/54, University of California, Berkeley, July 1984.

[KIN84] C. Kingsley, "A Hierarchical, Error-Tolerant Compactor", *Proc. 21st. Design Automation Conference*, pp. 126-132, June 1984.

[KIR83] S. Kirkpatrick, C.D.Gelatt, M. P. Vecchi, "Optimization by Simulated Annealing," *Science*, Vol.220, pp.671-680, May 1983.

[KNU73] D.E. Knuth, *The Art of Computer Programming, Vol. 1: Fundamental Algorithms*, 2nd ed., Addison-Wesley, Reading, MA 1973.

[LAI86] T. Laidig, "The VULCAN Protection Frame Generator," Users Manual, Computer-Aided Design Group, Univeristy of California, Berkeley.

[LAR71] R. P. Larsen, "Computer-Aided Preliminary Layout Design of Customized MOS Arrays", *IEEE Transactions on Computers*, Vol. C-20, No. 5, May 1971, pp. 512-523.

[LAR78] R. P. Larsen, "Versatile Mask Generation Techniques for Custom Microelectronics Devices", *Proceedings of the 15th Annual Design Automation Conference*, June 1978, pp. 193-198.

[LEN82] T. Lengauer and K. Melhorn, "The HILL system: A Design Environment for the Hierarchical Specification, Compaction, and Simulation of Integrated Circuits," *Proc. Conf. of Advanced Research in VLSI*, Jan. 1984.

[LIA83] Y.Z. Liao and C.K. Wong, "An Algorithm to Compact a VLSI Symbolic Layout with Mixed Constraints", *IEEE Transactions on Computer-Aided Design of Integrated Circuits and Systems*, vol CAD-2 no. 2, pp. 62-69, April 1983.

[LIP83] R. J. Lipton, et al., "VLSI Layout as Programming," Trans. ACM, Programming Lang. and Systems, Vol. 5, No.3, pp. 405-421,July 1983.

[MAL85] F. M. Maley, "Compaction with Automatic Jog Introduction," *Proc. Chapel Hill Conf. on VLSI*, H. Fuchs, ed. pp.261-283, Computer Science Press 1985.

[MAT82] R. Mathews, et. al., "A Target Language for Silicon Compilers," *Digest of Technical Papers, COMPCON 82*, pp. 349-353, IEEE Computer Society, 1982

[MEA80] C. A. Mead and L. A. Conway, *Introduction to VLSI Systems*, Addison-Wesley, Reading, MA, 1980.

[MOO86] P.P. Moore, "OCT Database Programmer's Manual", Internal Memorandum, University of California, Berkeley, Jan. 1986.

[MOS81] R. Mosteller, "REST: A Leaf Cell Design System," in *VLSI 81: Very Large Scale Integration*, J. P. Gray, ed., pp.163-172, Academic Press, 1981

[PFA52] W. H. Pfann, "Principles of Zone Refining," *Trans. AIME*, 194, 747, 1952.

[ROS83] J. B. Rosenberg, et. al., "A Vertically Integrated VLSI Design Environment," *Proc. 20th ACM/IEEE Design Automation Conf.*, pp.??-??, June 1983.

[SCH83] M. Schlag, Y.Z.Liao, and C.K.Wong, "An Algorithm for Optimal Two-Dimensional Compaction of VLSI Layouts," *Integration*, pp.197-209, 1983.

[SHI86] H. Shin, A. Sangiovanni-Vincentelli, C. Sequin, "Two-Dimensional Compaction by Zone Refining," *Proc. 23rd ACM/IEEE Design Automation Conf.*, pp115-122, June 1986.

[SHR82] Howard E. Shrobe, "The Data Path Generator," *Digest of Technical Papers, COMPCON 82*, pp. 340-344, IEEE Computer Society, 1982

[SUS79] G. J. Sussman and G. L. Steele, "CONSTRAINTS - A Language for Expressing Almost-Hierarchical Descriptions," *Artificial Intelligence*, No. 14, pp.1-39, 1980. North Holland Publishing Co.

[THE78] A. Thesen, *Computer Methods in Operations Research*, Academic Press, 1978, Chapter V.

[TSA85] A.K. Tsakalidis, "Finding a Negative Cycle in a Directed Graph", Memorandum A 85/05, Fachbereich 10, Angewandte Mathematik und Informatik, Universitat des Saarlandes, 6600 Saarbrucken, West Germany, March 1985.

[WES81] N. H. E. Weste, "Virtual Grid Symbolic Layout", *Proc. 18th ACM/IEEE Design Automation Conference*, June 1981, pp. 225-233.

[WES81a] N. H. E. Weste, "MULGA - An Interactive Symbolic Layout System for the Design of Integrated Circuits", *The Bell System Technical Journal*, Vol. 60, No. 6, July - August 1981, pp. 823-857.

[WIL78] J. D. Williams, "STICKS - A graphical complier for high level LSI design", *AFIPS Conference Proceedings*, Vol. 47, June 1978, pp. 289-295.

[WOL83] W. Wolf, R. Mathews, J. Newkirk, and R. Dutton, "Two-Dimensional Compaction Strategies", *IEEE International Conference on Computer-Aided Design*, pp. 90-91, Sept. 1983.

[WOL84] W. Wolf, "Two-Dimensional Compaction Strategies," Ph.D. Thesis, Dept. of Electrical Engineering, Stanford University, 1984.

[WOL85] W. Wolf, "An Experimental Comparison of 1-D Compaction Algorithms," *Proc. Chapel Hill Conf. on VLSI*, pp.165-179, 1985.

[WOO86] G. Wood and H-F S. Law, "SKILL - An Interactive Procedural Design Environment," *Proc. Custom Integrated Circuits Conference*, pp.544-547, May, 1986.

AUTOMATIC LAYOUT OF INTEGRATED CIRCUITS

Alberto Sangiovanni-Vincentelli
Department of Electrical Engineering and Computer Sciences
University of California at Berkeley

1. INTRODUCTION

To build a synthesis system that can rival human designers for the quality of the final design, it is essential to devote attention to the physical aspect of the process. Early "silicon compilers" dealt with the layout phase of the design in a straight-forward fashion: the floor-plan of the chip was fixed and little attention was paid to the quality of the routing of the components [SOU83], [JOH79]. This approach had as its main goal the decrease of design time, while performance considerations were often neglected. Consequently, the area usage, the speed and the power consumption of chips assembled with these tools were poor [FOX83]. This characteristic certainly contributed to the resistance offered by the industrial community to the introduction of these tools as production tools.

Recently, it has become clear that physical design is a crucial component of effective synthesis systems. For example, a great deal of effort has gone into the development of the placement and routing tools of two synthesis systems: the Yorktown Silicon Compiler[BRA85] and the Berkeley Synthesis System[NEW86].

In this chapter, layout techniques are addressed with particular emphasis on the techniques that are relevant in an automatic synthesis environment. Many papers have been written on these topics and it would be impossible to review all of the relevant results in such a wide field. Several review papers and books are available (e.g. [SOU81], [OHT86], [BRE72], [PRE86]), that cover a large part of the published results. In particular, the routing aspects of the layout problem have been nicely reviewed in [OHT86]. For this reason, in this chapter I focus more on floor-planning, partitioning and placement algorithms. In the spectrum of problems in automatic layout of integrated circuits, these three are regarded as the most difficult but as the most relevant to achieve a good final layout.

The newest algorithms will be examined in more details while the historically well established algorithms will be briefly mentioned. In addition, I will draw heavily from our experience in developing a placement and routing system, and the related algorithms, for the Berkeley synthesis environment.

This chapter is organized as follows: in Section 2, the placement and routing problem and some of the existing automatic layout systems will be introduced. In Section 3, partitioning will be examined. In Section 4, floor-planning and placement will be addressed. In Section 5, the routing process will be briefly described and in Section 6, the architecture and some of the algorithms of the Mosaico system developed at Berkeley will be presented.

2. PRELIMINARIES

2.1. Problem Definition

The layout of integrated circuits consists of the placement of the devices (cells) composing the design in a two dimensional finite space and of the interconnection of the pins of these devices according to the schematic of the circuit to be implemented. The goal of this process is to complete the placement and interconnection of the design in the smallest possible area satisfying a set of *design constraints* (such as those based on the positions and sizes of the cells to be placed and routed), a set of *technological constraints* (such as those based on design rules and the number of layers that can be used to route the cells), and a set of *performance constraints* (such as those based on the timing of the logic to be implemented). This optimization problem is a very complex combinatorial optimization problem.

A combinatorial optimization problem for our purposes consists of a finite (often very large) set of configurations, a Boolean function defined for all the configurations that determines whether the configuration satisfies a set of constraints, and a cost function, mapping all the configurations into real numbers. An algorithm for its solution explores a subset of the set of configurations and finds a minimum-cost configuration that satisfies all the constraints.

Even simplified versions of the layout optimization problem are NP-complete or NP-hard.[1] Note that there are NP-complete problems that may be solved exactly and quickly in most of the cases, because the worst case instances that would require a non-polynomially bounded running times, very rarely occur. However, in the optimal layout problem, I believe (supported by practical experience) that most of the problem instances would require an inordinate amount of computer time if we insist on searching for exact solutions. For this reason, most of the algorithms proposed for the optimal layout problem are heuristic.

Heuristic algorithms explore the configuration set quickly but they cannot be guaranteed to produce the optimum solution. In general, heuristic algorithms restrict the size of the configuration space to be examined. Most of them explore the configuration space by generating a new configuration from an old one according to a set of rules. These rules are designed so that they can be computed quickly. The rules for generating new configurations define a topology on the configuration space.

The properties of the solutions generated by a heuristic algorithm depend on the topology induced by the rule set. This topology can be represented by a directed graph. Let $X = \{x_k\}$ be the set of all the configurations under exam. Define a directed graph $G(V,E)$, where V is in one-to-one correspondence with the elements of X and an edge (v_i, v_j) joins vertex v_i to v_j if there is a rule that generates configuration x_j starting from configuration x_i. Then, x is a local minimum of the problem if all the configurations corresponding to the *fanout* vertices, i.e., the vertices $V(\hat{v}) = \{v_i | (\hat{v}, v_i) \in E\}$, have a cost which is larger or equal to the cost of x. The concept of local minimum is used to determine whether a given algorithm produces solutions that cannot be improved by a simple transformation.

Several heuristic algorithms can generate solutions that are local minima. However, heuristic algorithms devised for layout problems are characterized by topologies that yield many local minima, some of which may be very far away from the optimum solutions. For this reason, algorithms that can explore the configuration space in a more sophisticated way such as simulated annealing to be presented in Section 3, are attracting more and more interest.

---

[1] Informally, an NP-complete problem belongs to a class of difficult combinatorial problems such as the traveling salesman problem and the coloring problem for which an algorithm whose complexity is bounded by a polynomial in the size of the input is not know and is unlikely to be found. An NP-hard problem is a combinatorial problem which as at least as hard as the NP-complete problems. For a precise definition and a thorough discussion of NP-complete and NP-hard problems see [GAR79].

Even with the use of heuristic algorithms, the complexity of the layout problem as presented above is such that the problem is traditionally subdivided into several stages. While these stages are often common to the various design styles as presented in the initial chapter of this book, such as gate-arrays, standard cells and macro cells, their complexities vary from one style to another.

2.2. Layout Systems

Automatic layout systems were introduced first for gate-arrays and standard-cells. These design styles use a regular layout for the cells. Hence, automatic layout systems are considerably simpler than the ones for macro-cell where the layout structure is much less regular. Good gate-array systems developed both in house and by vendor companies are now widely available. The Engineering Design System of IBM was among the first complete layout systems to be developed for gate-arrays in the late 1960s, even though the first papers describing the algorithms used in the system were published in the early 1970s. In Japan, the first gate-array place and route system was developed by Oki and reported in 1974. Silicon foundries such as LSI logic offer their customers place and route systems optimized for their gate-array families. A variety of CAD vendors have developed gate-array software; they include Mentor/CADI, Tektronix/VR Systems, Daisy and Silvar-Lisco.

Interestingly, the standard-cell and gate-array design styles were developed at about the same time. However, the first automatic placement and routing tools for standard cells were developed earlier than the ones for gate-arrays. Philco Ford was the first company to do work in the area of automatic place and route of standard cells around 1964. This work was picked up and extended at Fairchild in 1965 with the MOSAIC system. The cells were arranged in rows and pins appeared on both sides of the cells. The routing was accomplished with a precursor of the channel router (which recently became the most popular routing technique). Fairchild also developed an automatic rubylith cutter for this structure. RCA Camden developed in 1966 the PRF (Placement, Routing and Folding) system for placement and routing. In this system cells were placed back-to-back and hence had pins on one side only. This system was then taken to the National Security Agency and improved. In the late 1960s RCA extended this approach to PR2D where a two-dimensional placement algorithm was developed. In the 1970-1972 time-frame the MP2D system was developed at RCA. This system is still widely used. Bell Laboratories developed a poly-cell (a synonym for standard-cell) place and route system called LTX in 1973. The successor of this system, LTX2, was extended to gate-arrays and to macro-cells.

The growing interest in the standard cell design style has prompted a number of CAD vendor companies including Silvar-Lisco, SDA and Tangent, to develop place and route systems for this design methodology.

The recent development of a gate-array technology known as sea-of-gates [HUI85] , [ROW86], where the entire area of the chip is covered by devices densely packed, has created new excitement in the gate-array community. This technology promises excellent area utilization and does not have the routing-area restrictions that standard gate-array architectures have, while allowing the prefabrication benefits of standard gate-array technology. To date there has been little development of automatic place and route systems of sea-of-gates chips, if we exclude obvious extensions of gate-array and/or standard cell software. I believe that in the next few years, many new algorithms and software systems will be developed to exploit fully this architecture.

Macro-cell placement and routing is considerably more difficult. In fact, as described elsewhere in this book [SAN87], cells can occupy any position on the chip and routing areas are not predefined as in the gate-array and standard-cell cases. Japanese companies are the clear leader in this area. NTT, NEC, Hitachi, Sharp and Sony all have good systems. In Europe, Siemens has been among the first companies to develop experimental macro-cell tools. The GAELIC/COMPEDA system was developed by the University of Edinburgh and made available by a commercial company. To the best of my knowledge, the Siemens system was the only system that has been used in the production environment [LAU86]. In the U.S., some companies such as Hughes aircraft have recently developed experimental systems for the VHSIC (Very High Speed Integrated Circuits) program of the Department of Defense. The CICLOPS system first developed by Preas and VanCleemput at Stanford has been improved at SANDIA labs. Because of the importance that macro-cell system are bound to have in the future, I am sure that several offerings by CAD companies as well as by silicon foundries will be soon available.

In the academic world, Japanese and U.S. universities have the lead in the development of algorithms and programs for placement and routing. Recently, universities have placed emphasis on the development of complete place and route systems for a variety of design styles. A few examples are: the PI system [RIV82] for macro cell place and route from MIT, the BBL (Berkeley Building-block Layout) system [CHE83] for macro-cell place and route, the ThunderBird standard cell system [BRA86b], the Mosaico system [BUR87] for floor-planning and macro-cell place and route, all developed at the University of California, Berkeley; and the hierarchical layout system developed at the Osaka University.

Most of the work done on layout has been concentrated on optimizing area. Recently, performance issues, in particular speed, have become a major concern [DUN84], [BUR84]. Coupling timing analysis with placement and routing has been proposed to influence the layout process of gate-arrays with speed considerations [BUR84]. I believe that this is a key issue in the development of future systems.

Most of the systems developed so far decompose the optimal placement and routing problem in a way similar to the flow shown in Figure 2.1. Note however that several systems lack one or more of the stages shown. At first, an estimate of the area of the chip and of its performances is calculated with the floor-planning step. In this step, the components of the chip may or may not be completely characterized in terms of area, power or speed. Next, the components of the layout are partitioned and placed on the area that is assigned to the chip, possibly with some information about the area needed to complete the routing of the components. Then, the area that is not taken by the components, called the routing area, is subdivided into smaller areas (the channel definition and ordering step). A global decision on the paths taken by each of the nets connecting the modules is then made in the global routing step. The issue here is how to use effectively the space resources available for routing. If all the nets cannot be routed in the available routing area, the placement of the components is modified to make more room for routing. Finally, the detailed routing step takes the information provided by the channel definition and ordering step as well as by the global routing step and implements the interconnections.

Note that this part of the flow has been the *de facto* standard for placement and routing of integrated circuits. The reason for this decomposition is essentially computational. Ideally, the placement and routing step should be combined so that the optimization car-

Layout Tools

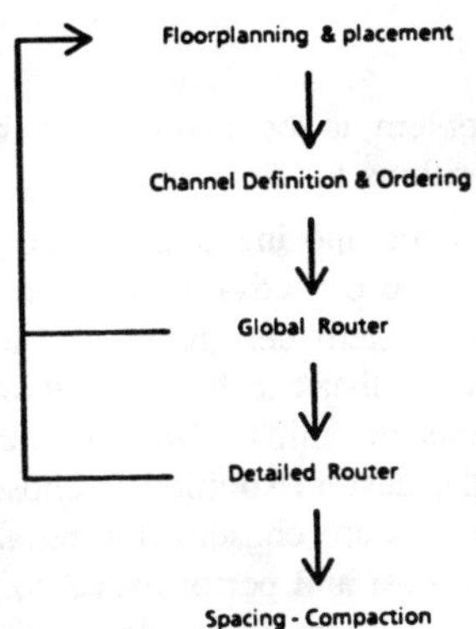

Figure 2.1. The Floor of an Automatic Layout System.

ried out at the placement stage could be done based on an accurate cost function. Given the quality of the detailed routing algorithms developed in recent years, it is fairly simple to estimate accurately the area needed to complete the routing of the chip once global routing has been performed. However, global routing the entire chip at each iteration of the placement algorithm to evaluate the cost function is out of the question at least with the present computing technology and algorithms. Nevertheless, there is a strong trend to incorporate more and more accurate information about the routability of a given configuration when optimizing the floor-plan, placement and partitioning steps. These approaches will be investigated in detail in the section devoted to the newest placement techniques.

In Figure 2.1, the last step of the pipeline is spacing. This step is not as commonly used as the others, but I believe it is crucial for the development of complex placement and routing systems. When carrying out the detailed routing step, many systems use the physical design rules to make sure that these are satisfied by the final layout. However, the detailed routing process may either become quite complex in the attempt to exploit all the intricacies of process technology or inefficient by using worst case design rules that may jeopardize the quality of the overall layout. In addition, any variation in process technology would require a modification of the layout programs with increased cost in terms of human resources and design time. However, if the entire placement and routing process is carried out at a *symbolic* level, a representation level where the technology information is abstracted away, then the algorithms used in the layout process are mostly technology independent and, consequently, simpler to develop and easier to maintain. To guarantee area efficiency and design-rule correctness, a symbolic spacing program such as SPARCS [BUR86] or Zorro [SHI86], reviewed in another chapter of this book, can be used. Spacers can also be used by other tools in an automatic synthesis environment (such as module generators) and are optimized to produce compact layouts that satisfy the design rules. This is important to maintain cleanliness and uniformity across the overall synthesis system, while reducing the cost of its maintenance.

In the next sections, I will review the various stages of the layout process: partitioning, floor-planning, placement, and routing.

3. PARTITIONING

Partitioning is the first problem to be investigated because partitioning algorithms form the backbone of many algorithms for floor-planning and placement.

Partitioning is the task of decomposing a given design into parts so that a given objective function is optimized. The objective function may be quite complex according to the application. For example, system designers have to decompose the circuitry they wish to implement, into components that can be realized with standard parts such as TTL logic or with custom and semi-custom chips. The cost function in this case includes the design time (certainly larger if the custom solution is chosen), performance of the system (in general better if custom solutions are chosen), the reliability of the design, and finally the feasibility of the design given cost and performance constraints. Partitioning is a most important problem with many ramifications in the design process, especially when the process is automated as in the synthesis systems we review in this book. However, at the moment, no good algorithm or heuristic is available for this task and I expect that system partitioning will be a major area of research in the years to come.

At a lower level, partitioning is used to decompose the chip into a set of macros that can be designed with a certain degree of independence. If the partitioning process is done effectively, the design task is simplified without giving up too much in terms of overall performance and area utilization. For example, an architecture designer decomposes the design of a micro-processor into the design of the data-path and of the control logic. The design of these two components is subject to a set of constraints that are determined from the initial specifications as part of the decomposition process.

Partitioning is also used to decrease the complexity of placement. This is certainly useful when thousands of objects have to be placed and the running time of the best algorithms often increases more than linearly (usually quadratically or cubically) with the number of objects. In this case, logic gates or functions are grouped together and assigned to blocks with fixed or variable dimensions. Then, the placement stage determines the positions of the components within the blocks and of the blocks on the chip. Note that partitioning can be and has been applied to all design styles. For example, the Engineering Design System of IBM uses partitioning as a pre-process to a gate-array placement program. In several standard-cell systems, first the cells are partitioned into rows, and then the precise positions of the cells within the rows are determined.

The topics covered in this section are:
    3.1. Cost Function
    3.2. Constructive Methods
    3.3. Iterative Improvement
        3.3.1. Random Interchanges
        3.3.2. The Kernighan-Lin Algorithm
        3.3.3. Simulated Annealing

3.1. Cost Function

Partitioning can be formulated as follows: Let $M = \{m_1, \cdots, m_i, \cdots, m_n\}$, be a set of n objects or *modules,* connected by a set of *nets* $N = \{n_1, \cdots, n_j, \cdots, n_k\}$, each represented as the set of modules it connects, i.e., $n_j = \{m_{j_1}, ..., m_{j_l}\}$. Net n_j has at least a *pin* on each of the modules it connects. We denote by P_j the set of pins of net n_j, by $p(i)$ the set of pins on a module m_i, and by P the total number of pins in the circuit. The problem is to find a partition of M, $\Pi = \{\pi_1, \cdots, \pi_t\}$, $\pi_i \subset M$, $\cup \pi_i = M$, $\pi_i \cap \pi_j = 0, i \neq j$ subject to capacity constraints, i.e., $|\pi_i| \leq K_i$, where $|\pi_i|$ denotes the cardinality (number of elements) of the set and $\sum_{i=1}^{t} K_i \geq n$, which minimizes

$$C(\Pi) = \sum_{m_i \in \pi_k, m_j \in \pi_k, k \neq h} c_{ij}$$

where c_{ij} is the number of nets that connect m_i to m_j, is minimized. Generalizations of this problem include the assignment of sizes, $s(m_j)$, to modules, so that the capacity constraints are now given in terms of the sums of the sizes of the modules in each element of the partition (instead of the number of modules), and assignment of weights, $w(n_i)$, to nets, so that the cost function is in terms of weights of nets (instead of the number of nets).

This problem is NP-hard and hence heuristic partitioning algorithms have been proposed over the years. These algorithms are similar in their architecture to the algorithms used in placement and can be divided into *constructive* and *iterative improvement* methods.

3.2. Constructive Methods

Constructive methods do not need an initial solution to be given. The starting point is in general the un-partitioned set. One of the most commonly used methods in this class follows an *aggregation* strategy, i.e., assigns one module at a time to a partition. Several version of clustering techniques have been proposed over the years. For an extensive review of these methods see [HAN72a], [KUR65].

For the sake of simplicity, assume that the *bi-partition* problem is solved, i.e., that $|\Pi|=2$. Assume further that the weights of all the nets and the sizes of all the modules to be partitioned is the same. Finally, assume that the partition must be balanced, i.e., that the number of modules in the two partitions must be roughly the same. Given a partition $\Pi = \{\pi_1, \pi_2\}$, the cost function is then defined as

$$C(\Pi) = \sum_{m_i \in \pi_1, m_j \in \pi_2} c_{ij}.$$

The aggregation algorithm begins by selecting the first module to be assigned to π_1, called the *seed* of the partition. This module is in general chosen to be the one with the greatest number of nets connected to it. Define the internal cost for a module m_j with respect to the temporary partition to be

$$I(m_j) = \sum_{m_i \in \pi_1} c_{ij}$$

and the external cost

$$E(m_j) = \sum_{m_i \in \pi_2} c_{ij},$$

then the cost function is

$$C(m_j) = E(m_j) - I(m_j). \tag{3.1}$$

The next module to be assigned to π_1 is the module m_j that minimizes Eqn. (3.1). The procedure is repeated until the number of modules assigned to π_1 reaches the limit set for the size of the partition. All the modules not in π_1 are assigned to π_2.

Regardless of the algorithm used, sometimes the capacity constraint may cause the partition to have a large cost. For this reason, the capacity K_1 is in general given with a tolerance ε_1, i.e., a partition is feasible if

$$K_1 - \varepsilon_1 \leq |\pi_1| \leq K_1 + \varepsilon_1$$

Then the cost of all the partitions found by the algorithm that have size satisfying the relation above is recorded and the minimum cost solution is selected.

This constructive algorithm is very fast, but the quality of the results is not good in general. Being "greedy", it often reaches solutions possibly quite far away from the minimum that is sought. Constructive algorithms are in fact mostly used as a starting point for other more elaborate methods such as iterative improvement.

3.3. Iterative Improvement

Iterative improvement algorithms have been used with success in many applications. The common thread among these methods is that an initial solution is given and the algorithms improve the quality of the solution by modifying it incrementally. To illustrate the algorithms, the same assumptions used to present the aggregation algorithm will be used throughout this section.

3.3.1. Random Interchanges

The simplest form of iterative improvement algorithm is the so called *random interchange* algorithm. In this algorithm, a given partition, is modified by first selecting a pair of modules, one in each element of the partition, and then evaluating the cost of the partition by interchanging them. Most of the random interchange algorithms accept an interchange only if the cost function decreases. If the cost increases the interchange is rejected and the modules remain in their previous positions.

Random interchange usually obtains fairly good results, but other iterative improvement algorithms such as the Kernighan-Lin algorithm [KER70] and simulated annealing [KIR83], presented in the next two sections, often outperform this simple heuristic.

3.3.2. The Kernighan-Lin Algorithm

This algorithm not only is rather fast but also it often produces very good results.

The basic idea of the algorithm is again to interchange modules among the two elements of the partition to obtain a better solution. A scoring function is used to evaluate the interchanges. This scoring function measures the difference in cost between the solution before the interchange and the one after the interchange. Given $m_i\,!mo\,\pi_1$ and $m_j\,!mo\,\pi_2$, the scoring function for the interchange of the two modules is

$$D(m_i, m_j) = E(m_i) + E(m_j) - I(m_i) - I(m_j) - 2c_{ij}, \tag{3.2}$$

where the external and internal costs are computed with respect to the two elements of the partitions the modules belong to. D measures the *gain* in the cost of the partition if the two modules are interchanged. In fact, if the two modules are interchanged, the nets that were connecting m_i to modules in π_2, become internal interconnections and do not contribute to the cost function. However, the nets that interconnect m_i to modules in π_1

become external interconnections after the interchange and they now contribute to the cost of the partition. Note that the term $2c_{ij}$ is subtracted from the scoring function since the connections between m_i and m_j are counted twice as external connections in $E(m_i)$ and $E(m_j)$. In Figure 3.3.1, the basic calculations in the algorithm are shown. All the pairs are tried and the pair that maximizes the scoring function is selected for the interchange. Note also that the modules are interchanged even if the scoring function is negative, indicating a net loss in the quality of the solution if the interchange is accepted. To avoid infinite loops, the modules that have been interchanged are now fixed in their locations.

After the first pair is interchanged, the procedure selects another pair according to the same criterion. The procedure stops when all the modules have been interchanged. When this procedure terminates, the net gain is zero, since all the modules which were in π_1 at the beginning are now in π_2 and vice versa. The cost of the partition is monitored by computing

$$G_k = \sum_{i=1}^{k} D_i$$

where k is the number of interchanges attempted so far and D_i is the gain corresponding to the i-th interchange. The actual set of interchanges that are implemented is the one which corresponds to the to the minimum value of the cost, i.e., to the maximum of the gain

$$\max_k G_k.$$

If the minimum value of the cost is attained at the starting partition, then the partition is not modified and the algorithm stops. Otherwise, the operations are repeated on the new partition, until convergence is reached, i.e., until the minimum value of the cost corresponds to the the initial partition.

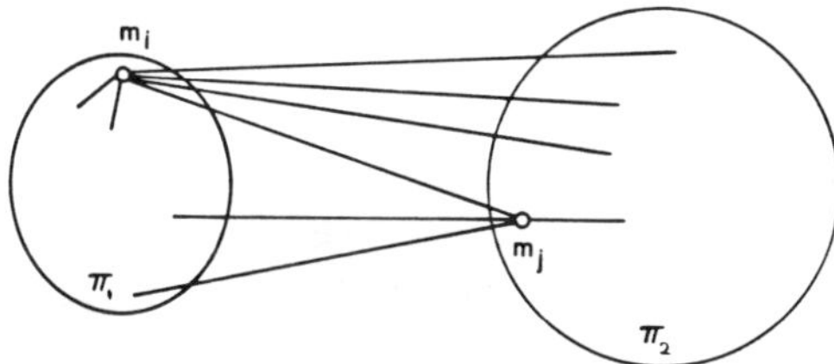

$$E(m_i)=4; E(m_j)=3; I(m_i)=2; I(m_j)=1; c_{ij}=1; D(m_i,m_j)=2.$$

Figure 3.3.1. Cost calculation.

Note that the cost can be decreased after $k+1$ interchanges even though the first k interchanges gave an increase in cost, as in the case of the plot of Figure 3.3.2. If a greedy strategy would have been applied, only interchanges that decrease the cost would have been accepted.

For the random interchange algorithm and the Kernighan-Lin algorithm, it is possible to define the same topology in the configuration space according to the approach presented in Section 2. Let $\Xi=\{\Pi_k\}$ be the set of all the feasible partitions. Define a directed graph

$G_{\Xi}(V,E)$, where V is in one-to-one correspondence with the elements of Ξ and an edge (v_i,v_j) joins vertex v_i to v_j if there is an interchange of two modules that generates the partition Π_j starting from partition Π_i.

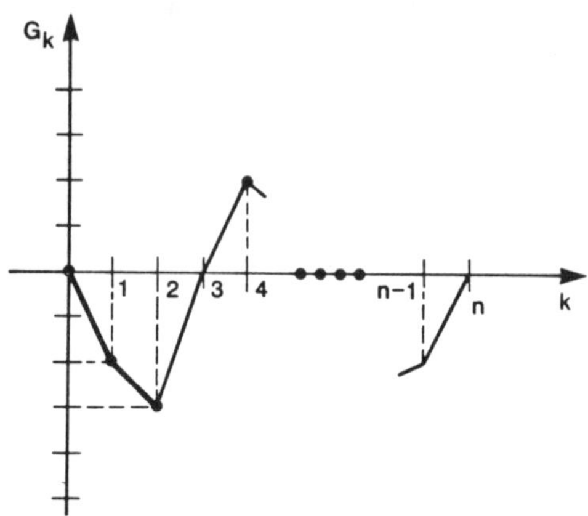

Figure 3.3.2. Plot of the gain as a function of the number of interchanges.

The random interchange algorithm is not guaranteed to find even a local minimum in the topology defined above, while the Kernighan-Lin partitioning algorithm does. However, it could also generate a solution with the same optimality property by selecting at each step the interchange with maximum gain and stopping the search whenever an interchange that would decrease the cost is not found. In practice, it has been found that the Kernighan-Lin algorithm in its present form finds solutions better than those found by the simplified algorithm.

The part of the algorithm that requires the largest amount of work is the gain computation. In the most naive implementation of the algorithm, $O(P^2)$ gain calculations per iteration are necessary, where P is the total number of pins of the nets. Since a gain calculation for a module m_i is $O(p(m_i))$ where $p(m_i)$ is the number of pins of the module, it is clear that the computational complexity of one iteration of the algorithm is more than quadratic in the number of pins. However, a clever implementation due to Fiduccia and Mattheyses [FID82] has complexity linear in the number of pins.

3.3.3. Simulated Annealing

Simulated annealing is a relatively new approach to combinatorial optimization problems. The results that have been obtained on a number of layout problems, from partitioning, to gate-array placement, from floor-planning to global routing, have been so interesting that a fairly detailed explanation is warranted here.

In the partitioning framework, simulated annealing can be seen as a modification of the random interchange algorithm since the interchanges are selected randomly as well. However the acceptance of the interchanges is done in a much more sophisticated manner. Because of the wide applicability of this algorithm, simulated annealing is presented in a general setting.

As we pointed out in Section 2, many combinatorial optimization problems are NP-complete or NP-hard. For these problems, heuristic algorithms are often used. Heuristic algorithms explore the configuration set fast but they cannot be guaranteed to produce the optimum solution. "Greedy" algorithms are a class of heuristic algorithms that explore the configuration set and accept a configuration only if the cost of the new configuration is lower than the old one. In general, greedy algorithms can find local minima in the topology defined by the generation rules.

In this topology, it is possible to "escape" from local minima only if "hill-climbing moves" are accepted, i.e., if configurations are generated and accepted that have a larger cost than the present one.

Simulated annealing as proposed by Kirkpatrick et al. [KIR83], allows "hill climbing" moves. These moves are accepted according to a certain criterion which takes the cost into consideration. The controlling mechanism is based on the observation that combinatorial optimization problems with large configuration spaces exhibit properties similar to physical processes with many degrees of freedom.

In particular, bringing a solid into a low energy state has been considered [KIR83] similar to the process of finding an optimum solution of a combinatorial optimization problem. Annealing consists of melting the solid with a heat bath and then lowering the temperature slowly until a crystal (the minimum energy state) is formed. The rate of decrease of temperature has to be very low around the "freezing temperature" i.e., the value of temperature when the crystal starts to form. The Metropolis Monte Carlo method can be used to simulate the annealing process; this has been proposed as an effective method for finding good solutions of combinatorial optimization problems.

This method when applied to combinatorial optimization generates moves randomly and checks whether the cost of the new configuration satisfies an acceptance criterion based on temperature. If the cost decreases, the move is accepted. If the cost increases, then (by comparison with a random number between zero and one) the move is accepted with probability $f(\Delta c_{ij}, T) = \exp(\frac{-\Delta c_{ij}}{T})$ where Δc_{ij} is the change in cost obtained by moving from configuration i to j and T is temperature, the controlling parameter. (If the random number is larger than f, the move is accepted, otherwise the move is discarded). Note that the higher the temperature is, the more likely it is that a "hill climbing" move is generated and accepted. Note also that "hill climbing" moves are less and less likely to be accepted as the temperature is decreased. A certain number of moves are generated and checked before a decrease in temperature is allowed. The initial temperature, the number of moves generated at each temperature and the rate of decrease of temperature are all important parameters that affect the speed of the algorithm and the quality of the final configuration.

The following pseudo-code gives a precise definition of the operation of the algorithm.

Simulated Annealing Algorithm Structure (j_0, T_0)

```
{

/* Given an initial state j_0 and an
   initial value for the parameter
   T, T_0. */
```

124

```
       T=T₀;
       X=j₀;
     while( "stopping criterion" is not satisfied )
           {
         while( "inner loop criterion" is not satisfied )
              {
            j= generate (X)
            if( accept( c(j), c(X),T )
                  X = j ;
            }
            T=update (T)
           }
     }
```

The acceptance of a new state j is determined by **accept**, whose structure is shown below.

accept($c(j)$, $c(i)$, T)

```
    {

    /* returns 1 if the  cost variation passes a test.
     T is the control parameter. */
```

$$\Delta c_{ij} = c(j) - c(i) ;$$
$$y = f(\Delta c_{ij}, T) ;$$

r=**random** $(0,1)$;

```
    /* random is a function which returns a pseudo
      random number uniformly distributed on the
     interval [ 0 , 1 ] . */
```

if $(r<y)$
     $return(1)$;
else
     $return(0)$;
```
    }
```

The acceptance strategy is represented by the function f which takes values on the interval (0,1]. In the case of the original version of simulated annealing,

$$f(\Delta c_{ij},T) = \min \left[1 , e^{-\frac{c(j)-c(i)}{T}} \right] .$$

The shape of f varies as T is adjusted.

The control parameter T can be updated by means of many different laws: the original law and possibly the most used in practical applications is

$$T_{new} = \alpha\, T_{old} \qquad 0 \le \alpha \le 1$$

A mathematical analysis of the algorithm is very important to understand the essential features which make the algorithm work well and to suggest techniques for controlling its operation. Markov chains can be used as a mathematical model of simulated annealing.

It has been proved that under certain assumptions on the number of moves generated by the algorithm at each temperature and on the annealing schedule, (i.e., the law according to which the temperature is lowered, the time spent at each temperature, and the initial and final temperature), as the temperature goes to zero, simulated annealing produces the optimum solution of combinatorial optimization problems with probability one [GEM84], [ROM85], [MIT85], [LUN86], [HAJ85]. One such set of assumptions is that at each temperature the stochastic process reaches an equilibrium state [ROM85], [LUN86]. A second is that the temperature schedule satisfies the following law:

$$T_{k+1} = \frac{\gamma}{\log(T_k + K)} \tag{3.3}$$

where γ is a parameter depending on the problem [2], K is a constant greater than or equal to 1 and k indexes the infinite sequence of temperatures selected by the algorithm [GEM84], [MIT85], [HAJ85]. The first condition implies that an infinite number of moves has to be taken at each temperature, a sufficient condition to reach equilibrium, but it does not require any particular rule for lowering the temperature. The second condition implies that when $k \to \infty$, $T_k \to 0$ infinitely slowly, i.e., an overall infinite number of iterations have to be taken in this case as well to insure convergence of the algorithm. However, there are no restrictions on the number of moves to be taken at each temperature: even one move per temperature will suffice. Note that there is no bound on the initial temperature value in either approach and that the issue of when to terminate the search is not addressed since the theoretical results are obtained as the number of iterations go to infinity, i.e., the temperature goes to zero.

These results are of interest from theoretical point of view only, but they guarantee that the "heuristic" versions of the algorithm are on firm ground. Two results are available to bound the distance from the solution obtained after a finite number of steps and the optimum solution. This bound is not easy to calculate, but the results point out the role of important parameters of the algorithm such as the number of configurations that can be generated from a given one [MIT85], [GEL85].

In the application of simulated annealing to "real" problems, a number of heuristic rules have been developed to decide how many moves are to be taken at each temperature, how to lower the temperature, the initial temperature and the stopping criterion. Interestingly, all of these are related to the first theoretical approach. The heuristics try to identify when equilibrium is "almost" reached at a given temperature and how to lower the temperature so that the equilibrium at the next temperature is close to the equilibrium at the previous temperature [OTT85], [HUA86].

The initial temperature plays an important role in obtaining good results. Several schemes have been proposed to select a good value for the initial temperature (e.g., [OTT85], [KIR83], [WHI84], [AAR85], [HUA86]), all looking for a value of temperature so that almost all attempted moves are accepted.

---

[2] Hajek [HAJ85] showed that Eqn. (3.3) determines a necessary and sufficient condition for the convergence of simulated annealing if γ is set to a particular value.

The stopping criteria are related to the observed improvement of the cost function. More conservative rules require that no change in the cost function be seen for a number of temperature-update cycles. More aggressive rules terminate the search when the improvements are below a certain threshold.

In the partitioning framework, a move in the simulated annealing algorithm is generally taken to be the interchange of two modules [3] . The cost function is the same as the one used in the other algorithms.

Several experiments have been tried to compare simulated annealing to other techniques for partitioning. For example, in 1983, R. Brayton of IBM selected a difficult example and invited proponents of different techniques to try their algorithms [BRA87]. Clustering techniques, eigenvalue-based algorithms, the Kernighan-Lin algorithm and simulated annealing were tried. Simulated annealing and the Kernighan-Lin algorithm found the optimum solution, while some of the other techniques did not. However, simulated annealing took a much longer time.

D. Johnson of AT&T Bell Laboratories tried a set of experiments on a number of well-known combinatorial problems and found that for some simpler problems such as bipartite matching, simulated annealing did not perform as well as other algorithms, but that for more complex problem such as graph partitioning, simulated annealing seemed to obtain consistently better results on a set of randomly generated examples [ARA84].

A. Dunlop of AT&T Bell Laboratories tried a comparison between the Kernighan-Lin algorithm in the Fiduccia-Mattheyses version and simulated annealing and obtained similar quality results on a set of "real" problems arising from the placement of standard-cells [DUN85].

My own opinion on this issue, based on extensive experience with this algorithm, is that simulated annealing gives consistently good results on difficult problems characterized by complex cost functions, such as placement and floor-planning of gate-arrays, standard-cells or macro-cells. When problems with a very well defined cost function and known mathematical properties, such as graph partitioning, traveling salesman problem, bipartite matching, are attacked, then good *ad hoc* heuristics perform as well if not better than simulated annealing. In addition, it is very difficult to draw general conclusions about the quality of the solutions provided by simulated annealing since they depend very much on implementation issues such as the choice of the cost function (for less well-defined problems) and of the move set. Furthermore, it is necessary to understand deeply the statistical issues related to simulated annealing to run meaningful comparisons. I have often seen runs with few moves per temperature, initial temperature set to ridiculously low values, and convergence criteria based on dubious reasoning. These facts explain why some researchers report excellent results when using simulated annealing when others report disappointing ones.

4. FLOOR-PLANNING AND PLACEMENT

Floor-planning and placement are related problems. In fact, floor-planning can be interpreted as a generalization of the placement problem. Recent results have allowed the generalization of placement algorithms to the floor-planning problem. For this reason, these two problems are examined in the same section.

---

[3] In some implementations a larger number of modules are interchanged to make the search more effective. However, the cost may increase too rapidly to offer a better overall strategy.

Placement is defined to be the task of assigning precise locations on the plane to the components of the design so that area, timing and power consumption of the chip are optimized. Thus, given

- a set of modules with fixed geometries, fixed pins, delay and power consumption;
- constraints on the positions of some of the modules;
- constraints on the total area and on the aspect ratio of the chip;
- a net list specifying which pins have to be interconnected;
- constraints on total power dissipation;
- estimated delay per unit length of interconnection;
- constraints on delays at the chip level;

the problem to be solved is:

determine the locations for all the modules, so that all the constraints are satisfied and so that a weighted sum of total area, net length, delay, and power dissipation is minimized.

The constraints on the positions of the components may take several different forms: the orientation of some of the components may be fixed, the location of the components may be restricted to be in a specified region of the plane (e.g, pads may only be placed at the periphery of the chip), the distance between two components may be bounded above and/or below, two components may have to be placed symmetrically with respect to an axis of the chip.

Placement is a very difficult problem given the variety of constraints and goals. In general, no algorithm is able to cope with the full complexity of placement. Thus, several approximations are used to reduce the problem to one that can be solved in reasonable computer time. The approximations involve the computation of the cost function and of the constraints as well as the restriction of the space where the search for a good solution is carried out.

In contrast with the placement problem which is fairly well-defined, the floor-planning problem is less structured and as such is even more difficult to solve.

Floor-planning is one of the first stages in the design of VLSI circuits. In this step, a designer partitions a large design into macro-modules and selects areas, aspect ratios and I/O pin locations of these modules trying to optimize design quality measured in terms of overall area of the chip, wire density, power dissipated and delays along critical paths.

The aspects of floor-planning that have been most investigated are related to the determination of the relative positions (the precise definition of relative position of a set of components will be given in of the modules on the plane. Timing, power and area estimations are the factors guiding the floor-planning phase of the design. Floor-planning can be used to verify whether it is possible to realize a design in silicon without performing the detailed layout and design of all the blocks and functions.

As pointed out above, the information about the exact shape of the cell may be unknown at the initial phase of floor-planning. For example, in the design of a microprocessor, if the control logic is implemented with blocks of standard-cells, then the number of rows used for the modules is not necessarily fixed. The use of many rows will produce a block that is tall and skinny; using a few rows will produce a block that is short and fat. If the control logic is implemented with PLAs, topological operations such as folding and partitioning can be used to vary aspect ratios, while maintaining logical equivalence. This flexibility can be exploited at the floor-planning stage to obtain better area utilization and performance.

Another degree of freedom at the floor-planning stage is the position of the pins of some of the modules. When standard cells are used for control logic, the signals may leave the block at any point on its periphery. Similarly, when multiple-folding is used in PLA design, the inputs and outputs of the PLA can be positioned on all the sides of the module [DEM83]. Pin positions on the boundaries of the modules can be assigned so that the wiring area and the interconnection length is minimized.

Summarizing, the floor-planning problem can be formulated as an extension of the placement problem as follows:
Given

- a set of modules with variable geometries, pins with variable positions, estimated delay and power consumption;
- constraints on the positions of some of the modules;
- constraints on the total estimated area and on the aspect ratio of the chip;
- constraints on the aspect ratios of the modules;
- constraints on the positions of the pins;
- a net list specifying which pins have to be interconnected;
- constraints on total estimated power dissipation;
- estimated delay per unit length of interconnection;
- constraints on estimated delays at the chip level;

the problem to be solved is:

determine geometries, locations, aspect ratios and pin positions for all the modules, so that all the constraints are satisfied and so that a weighted sum of total estimated area, net length, delay, and power dissipation is minimized.

In the floor-planning step, the interaction among various tools in a synthesis system is particularly important, especially when a macro-cell design style is used. On one side, other tools may be called upon to provide area, power, delay estimation to the floor-planner, on the other side, the floor-planner can pass constraints to other tools while module generators, wiring and placement are finalized. For example, in the Berkeley Synthesis System, the logic synthesis program MIS [BRAY86], reviewed in another chapter of this book [BRAY87], is used to design logic blocks in multi-level form and can be run with constraints on area, pin-position, and timing. In this mode of operation, MIS may take a considerable amount of time and may require user-intervention to trade-off the objectives imposed by the designer. MIS can also be used to generate area estimates for the floor-planner, when run in a completely automatic way with no constraints and with area minimization as unique criterion.

When deciding the relative position of the components of the design, timing data should be provided in addition to the connectivity information to guide the floor-planning algorithm. In YSC [BRA85] and BSS [NEW86], timing information are provided by the following procedure. First the delay through each of the components of the design is estimated, either by running a program or by browsing the data-base for delay information on existing library modules that are already completely characterized. Then, a timing analysis algorithm is run to identify the critical paths assuming zero delay on the interconnections. The critical paths are then compared with the timing requirements of the design and slacks assigned to each of the interconnections. These slacks are reported to the floor-planner which now can either convert this information into weights for the nets or transform it into spatial constraints among blocks, i.e., constraints that require blocks to be within a certain distance from each other [OGA86]. The transformation from the timing constraints to the spatial constraints is based on approximations of delays per unit length of interconnect and may not be very accurate, but the relative importance given to the placement of the components can be very effective in reducing the need of iterating the design process and, with it, the design time.

Because of the many degrees of freedom and of the uncertainties in the estimates of many of the parameters, the optimization problem presented by floor-planning is quite difficult. Some tools are interactive and given a relative placement of the modules, provide estimates for wiring length, channel congestion, area utilization, timing, and power dissipation, thus helping the designer to explore the design space in search of an acceptable solution. However, especially for system designers, it is important to make the floor-planning task as automatic as possible, without eliminating the possibility for the designer to interact with the algorithms. Furthermore, even expert designers may find themselves in a difficult position trying to cope with the many degrees of freedom and the many constraints that floor-planning entails. For this reason, an automated floor-planning tool may be a welcome addition to the expert designer's tool-set. In fact, I believe that good algorithms may obtain better solutions than human designers if a convenient and effective way is provided for the designer to input his design goals, which may dynamically change during the evolution of the search in the design space.

This section has the following organization. The cost function evaluation is a much more complicated problem than in the case of partitioning. Several components of the cost function are not well defined or are very expensive to compute. A nontrivial amount of work has been done in this area and will be presented briefly in Section 4.1.

Assuming that the relative placement of the components of the layout is given, one problem is the determination of a final position of the components such that the total area of the chip is minimized. Polar graphs are used to represent the layout and to provide a fast and simple algorithm to solve the optimization problem. When the components are allowed to change orientation, the algorithm used to optimize area has to take into consideration this additional degree of freedom. Surprisingly, this less constrained problem is NP-complete. Of similar complexity is the problem of determining the aspect ratios of modules such as the ones involved in the floor-planning phase. However, there are some constrained placements that yield a simple and elegant algorithm to determine the optimal area and the corresponding orientations and aspect ratios of the components: the so-called slicing structures. These structures will be investigated in detail and the procedure to determine the optimal area described in Section 4.2.

It is possible to classify algorithms for placement and floor-planning in a way which is similar to the one used in partitioning: constructive and iterative-improvement algorithms. However, another classification can be made according to the structure of the layouts generated by the algorithms: algorithms for slicing-structures and algorithms for unconstrained placement. In my opinion, this classification is more useful to present the recent developments in placement and floor-planning and for this reason I have decided to follow it. The slicing structure algorithms are reviewed first in Section 4.3. In particular, the min-cut algorithm and its variations will be presented, and a simulated-annealing based algorithm will be described.

For unconstrained placements many algorithms have been proposed over the years. Some of the most famous ones, such as the constructive method, the force-directed algorithms, and the eigenvalue approach will be reviewed in Section 4.4. These algorithms cannot handle variable aspect ratio components and hence they have a limited applicability to problems that arise naturally in a synthesis environment. However, a method will be presented in Section 4.4 that can convert any non-slicing structure into one, so that the shape determination algorithms can be applied as post-processing steps to the results obtained with more traditional approaches.

Simulated annealing can be used to take into account all the degrees of freedom of the floor-planning problem at once. The approach used at Berkeley in the TimberWolfMC program will be explained in some detail in Section 4.5.

If a qualitative approach is followed for net-length minimization, it is possible to recast the optimization problem into a graph dualization problem. This approach is mathematically elegant and has spawned some amount of interesting research work even though the practical results have not been particularly good. The rectangular dualization algorithm that follows this approach is described in Section 4.6.

In all the methods previously mentioned, the quality of the placement depends on estimates of the routing area. Recently, there has been an attempt to modify this approach to carry out the placement and routing steps together so that the routing area can be computed with great accuracy. The non placement/routing approach and the top-down, bottom-up approach will be reviewed in Section 4.7.

Summarizing, in this section, the following topics will be covered:

4.1. Cost-Function Evaluation

4.2. Data Structures to Represent Layouts

4.3. Algorithms for Slicing Structures

4.4. Algorithms for Unconstrained Placement

4.5. The TimberWolf Simulated Annealing Approach

4.6. The Rectangular Dualization Method

4.7. Combining Global Routing and Placement

4.1. Cost Function Evaluation

As pointed out in the previous section, the cost function for the floor-planning and placement problems is quite complex, involves several components and varies depending on the nature of the design in question. Area utilization, net-length and performance all depend on the results of the wiring step.

Most of the placement algorithms generate a sequence of placements for the chip. The cost function is evaluated for each of these placements to guide the generation of the next placement in the sequence. Since all the components of the cost function depend on the routing step, the chip should be completely routed for each of the placements in the sequence if the cost function were to be computed exactly. However, it is impractical to route the chip to evaluate the quality of a placement. Hence, approximate calculations have to be used to carry out the floor-planning and the placement steps.

It is important to stress that the cost function calculation is of great importance to yield a viable placement and floor-planning program. If the cost function reflects poorly the final cost of the layout, the best optimization algorithm is bound to give disappointing results. The quality of the approximations affects dramatically the effectiveness of a placement algorithm.

This subsection is dedicated to the approximate computation of the most important components of the cost function: net-length and area. In particular, various ways of carrying out approximate net-length calculations are reviewed in Section 3.2.1.1 and algorithms to compute estimates of area utilization are presented in Section 3.2.1.2.

4.1.1. Net-length Calculation

Net length estimation is difficult especially for multi-terminal nets, i.e., for nets that have more than two pins. Many different, but electrically equivalent, ways of interconnecting a set of pins are possible. In Figure 4.1.1a, three equivalent implementations for a four-pin net are shown. In most VLSI chips, interconnections are implemented exclusively with Manhattan geometries. Thus, in this paper it is assumed that Manhattan geometries are the only configurations used for routing.

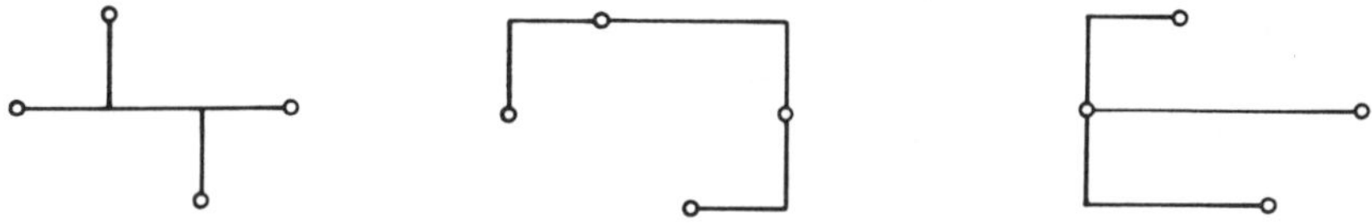

Figure 4.1.1a. Equivalent implementations for a four pin net.

If the problem of implementing the interconnections specified by a net is seen in isolation, then the interconnection of minimum length is preferred. In fact, in general, shorter interconnections imply better electrical performance. The length of an interconnection is measured according to the Manhattan distance: with this metric, two points a and b with coordinates (x_a, y_a) and (x_b, y_b), have distance

$$d(a,b) = |x_a - x_b| + |y_a - y_b|.$$

Finding a rectilinear interconnection of minimum length for a single n-pin net at a first glance seems quite simple, but is in fact NP-hard. It is referred to as the *rectilinear Steiner tree* problem.

Several heuristic algorithms have been proposed for the solution of this problem. To evaluate the quality of a placement, we are interested in a fast-to-compute estimate of the length of an optimal Steiner tree. Among the various estimates that have been proposed, two have had extensive use:

Interconnection Length Estimate 1. The semi-perimeter of the smallest rectangle enclosing all the pins of the net (see Figure 4.1.1b).

This estimate is actually exact for two- and three-pin nets. It is always a lower bound on the length of the shortest rectilinear Steiner tree and is quite accurate for four-

132

and five-pin nets, for which it is possible to demonstrate that the length of the optimal Steiner tree is at most twice as long.

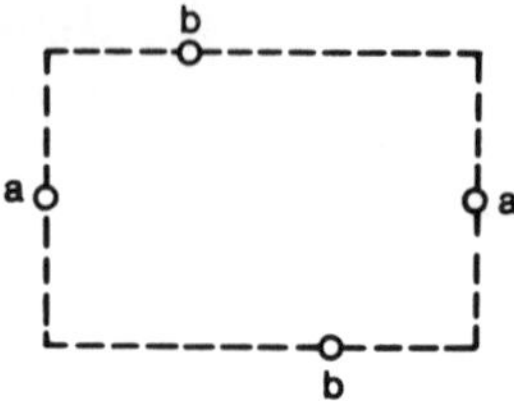

Figure 4.1.1b Semi-perimeter of the smallest rectangle enclosing the pins of the net (a +b).

Interconnection Length Estimate 2. The length of the shortest spanning tree that connects the pins of the net.

The difference between a Steiner tree and a spanning tree is that, while a Steiner tree may have vertices other than the pins themselves, a spanning tree may not. The shortest spanning tree problem can be solved in polynomial time, however its solution is much more expensive than the semi-perimeter calculation. Some placement algorithms, such as simulated annealing, may generate thousands of intermediate results that have to be evaluated. For these algorithms, a net length estimate that can be computed very fast, such as the semi-perimeter heuristic, is essential.

The estimates presented above are approximate in two ways. First, it is assumed that the interconnections are implemented with a minimum rectilinear Steiner tree. This is not necessarily true because routing all the nets in this way can lead to violations of wiring-area constraints. Second, the calculation of the length of a minimum Steiner tree is only approximated since this problem is NP-hard.

An even cruder approximation is often used:

Interconnection Length Estimate 3.

$$\sum_{p_i,p_j \in P_t} d_{ij} \qquad\qquad (4.1.1)$$

where p_i and p_j are two pins of the set of pins P_t of net n_t and d_{ij} is the Manhattan or the Euclidean distance of the two pins.

This approximation may be quite crude for modules of large size since the net length is calculated assuming that all the pins of a net are interconnected separately. Thus the net interconnections are modeled as a complete graph, where the nodes are the pins of the net, instead of the more accurate estimation based on the Steiner tree. This approximation, in spite of its inaccuracy, is widely used.

In some approaches, such as the eigenvalue method and the resistive network optimization method to be described in Section 4.4, the cost function must have certain mathematical properties. The function that has these properties and is close to Estimate 3, is given as: *Interconnection Length Estimate 4.*

$$\sum_{p_i,p_j \in P_t} ((x_i - x_j)^2 + (y_i - y_j)^2),$$

where x_i and y_i are the coordinates of p_i in the plane.

Note that this estimate is the square of the Euclidean distance between the pins of the net.

To cut down the cost of the net length evaluation several algorithms assume that the pins of a module are all located in the center of the module. In this case, Estimate 3 can be computed as

$$\sum_{m_i,m_j,i \neq j} c_{ij} d_{ij} \tag{4.1.2}$$

and Estimate 4 as

$$\sum_{m_i,m_j,i \neq j} c_{ij} d_{ij}^2 \tag{4.1.3}$$

where c_{ij} is the number of nets which connect m_i with m_j and d_{ij} is the Manhattan or the Euclidean distance between the centers of m_i and m_j.. Of course, $c_{ij}=c_{ji}$. Most of the algorithms developed in the past used this version of the net length estimate as cost function. Note that if the modules are large in size, the actual position of the pins may be rather far from the center of the module and the approximation may yield inaccurate results.

4.1.2. Area Utilization

The area used by a particular layout depends on the area of the modules to be placed and on the area needed to complete the interconnections. If the routing area is not considered, then the area used depends on how well the modules can be packed together. However, an optimal packing of the modules may result in an inferior layout after the routing area is added to make the interconnections of all the nets possible. Hence, it is very important, when evaluating a placement, to estimate the amount of area needed to complete the interconnections and to determine where this routing area should be added. Failure to allocate the correct amount invariably results in substantial placement alteration by the global and/or detailed routers. These placement modifications usually result in substantial total interconnect length increases as well as chip area increases because of the introduction of sizable empty areas.

In the case of gate-arrays, the area of the chip is given and, hence, it seems that this component of the cost function should be ignored. This observation would be true if the routing area allocated on the chip in channels around the gates were sufficient to complete the interconnections even if all the cells in the gate-array were used. However, more often than not, this area is not sufficient to implement all the interconnections and some additional routing area must be provided at the expense of an underutilization of the gates of the chip. This is mainly due to two causes: i) the number of interconnections that have to be implemented may be so large that a feasible solution to the routing problem in the fixed area provided by the gate-array architecture does not exist, and ii) the quality of the routing tools may not be adequate to route all the interconnections even though a solution to the routing problem may exist.

It is common to have about 20% of the gates in large gate-arrays not used to provide additional routing area. In this case, estimating how much and where additional area should be allocated is very important.

If many layers of interconnect were available, then there would be no need for a routing area to be reserved on the plane where the modules are placed. However, the most generally used state-of-the-art CMOS technology offers only three layers of interconnect and one of these layers has electrical characteristics that make it undesirable to use.

134

Many early placement programs left the task of inserting routing area where needed to the routing programs, hence obtaining often poor placements. The cost function used was simply the estimated net length since it was experimentally observed that shorter net-length corresponds to a small routing area. While it is true that minimizing net length often optimizes area utilization, it is quite simple to come up with examples where a minimum net-length solution requires more wiring area than other solutions.

For gate-arrays, it is most important to spread evenly the interconnections over the entire area of the chip, to use more effectively the routing area that is in general distributed uniformly across the chip. In general, minimizing net length is not sufficient to obtain a placement that can be completely routed with automatic tools. For gate-arrays, *routing congestion,* i.e., an estimate of the number of interconnections going through each routing region, is used in the cost function [PER76], [KOZ83], [GOT86].

This estimate is difficult to compute because nets can be routed in many different ways and, hence, simplifying assumptions have to be made. These assumptions represent rather accurately the way in which a chip is routed.

Assumption 1. The net is routed entirely in the smallest rectangle containing all the pins of the net.

Assumption 2. One and two-bend, i.e., L-shaped and Z-shaped, interconnections (see Figure 4.1.2a) are the only possible ways to connect two pins of a net.

Assumption 3. When more than one solution exists to interconnect the pins in the net, the solutions are equally likely to be chosen by the routing algorithm.

Figure 4.1.2a. L-shaped and Z-shaped interconnections.

The first two assumptions are used to limit the number of alternatives that need to be considered. However, many alternatives may still be possible as shown in Figure 4.1.2b. The third assumption gives us a tool to compute the estimate of wiring congestion. Each of the possible routing alternatives is characterized by the routing regions traversed. For each of the routing regions, the probabilities associated with each of the interconnections that may traverse it are summed to yield a number which gives an estimate of the congestion of the routing region.

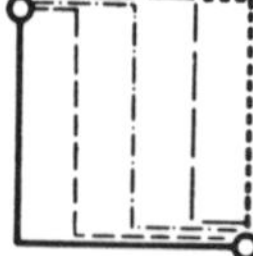

Figure 4.1.2b. Possible interconnections between two points.

For gate-arrays, the maximum number of wires that can traverse a channel is known *a priori* and if the congestion estimated as described above is greater than the capacity of a channel, the routing of the chip would be very difficult and the corresponding placement will be highly penalized. In the standard-cell case, the total sum of the congestions for each of the channels has to be minimized since this number is directly proportional to the area of the channels and, hence, to the area of the chip. However, since there is no limit on the capacity of the channels, the cost function for the standard-cell case is obviously much smoother and hence the optimization problem better conditioned than for the gate-array case.

This calculation, while faster than performing a complete routing step, is still expensive. A cruder but faster estimate for congestion was obtained by counting the number of interconnections that go across a *cut-line* of the chip. A cut-line is a line that traverses the entire chip either vertically or horizontally. A lower bound on the number of interconnections that have to go across this line no matter what routing algorithm is used, is given by the number of nets that interconnect modules laying on different sides of the cut line. This number gives only a rough idea of the congestion of the channels since it does not give detailed information on the number of interconnections that go through a channel, but it gives only global information on the number of interconnections that must go through a set of channels, the ones traversed by the cut-line.

For the macro-cell design style, area estimation is much more complex, since the parts of the chip dedicated to routing do not have a pre-determined shape and location. Most of the approaches proposed thus far ignore the routing area issue and assume that routing areas will be added during global routing. Other approaches involve a mixed top-down bottom-up process that routes clusters of cells while placing them [DAI86].

Another approach is to estimate the routing area needed by adding a halo to (i.e. inflating) each of the modules (Figure 4.1.3). The size of the halo is proportional to the number of pins on the edges of the module. This estimate is of course rather crude, but at least allows the development of algorithms that take into some account routing regions.

Recently, a more complex routing area estimation has been provided still based on the halo idea, but where the size of the halo is determined by a careful analysis of the statistics of routing macro-cell chips [SEC86]. This approach will be described in some detail in Section 4.5.

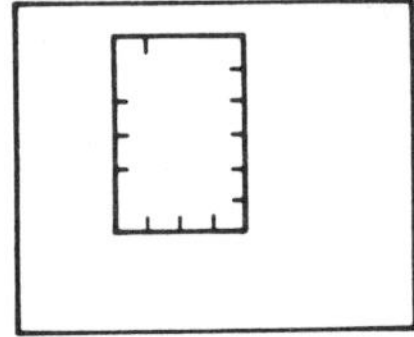

Figure 4.1.3. Halo for routing area estimation.

4.2. Data Structures to Represent Layouts

Several algorithms obtain "relative" positions [4] of the components of the design. For all algorithms, after the placement phase has been carried out, routing is performed in the

---

[4] The precise definition of relative positions of components of the design will be given later in this section.

area that is left available by the components. After routing has been performed, adjustments are needed to accommodate all the nets and to remove unused space. In both these cases, it is important to have a compact data structure that allows the determination of the final positions of the components on the plane. In general the post-processing has as a goal the minimization of the area taken by the layout.

Polar graphs have been widely used as data structure to represent the relative positions of the components. These graphs can be analyzed to find the minimum area occupied by the layout if the relative positions of the components are not to be changed.

All placements can be represented by this data structure. There are special cases of placement that can be represented even more compactly: these structures also offer simple and effective algorithms for the determination of the optimal aspect ratio and orientation of the components of the design if the relative positions of the components have to be maintained. Furthermore, routing on these structures can be performed by the most effective tools available today for this task: channel routers. These appealing special cases are called *slicing structures*.

In this section, both the polar graph and the slicing structure data structures will be presented, and the algorithms for the determination of the final positions of the components as well as of their aspect ratios and orientations will be reviewed.

4.2.1. Polar Graphs and Area Optimization

Given a placement, the area taken by the layout can be defined as the area of the smallest rectangle that contains all the components of the layout. It is possible to subdivide this rectangle into a set of rectangles, each of which contains one and only one of the modules. A placement and the relative rectangle decomposition called *rectangle dissection,* are shown in Figure 4.2.1.1a.

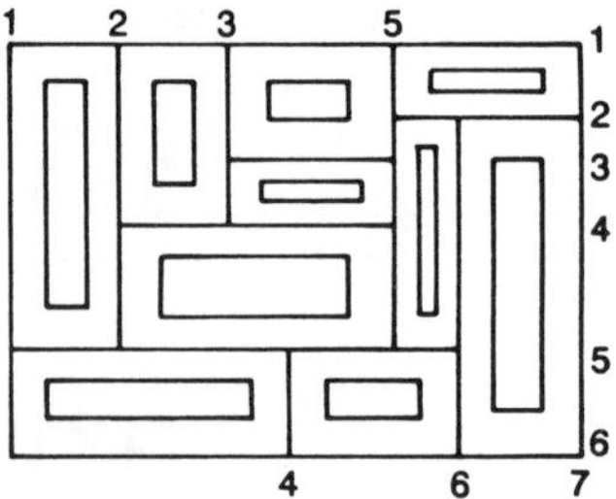

Figure 4.2.1.1a. A placement and the associated rectangle dissection.

A given dissection can be represented by two directed graphs, $G_H(N,E)$ and $G_V(N,E)$, where N is the set of nodes and E is the set of edges of the graphs, called the *polar graphs* of the dissection. The *horizontal polar graph,* $G_H(N,E)$ represents the relative position on the x coordinate of the rectangles containing the modules. The nodes of the horizontal polar graphs are in correspondence with vertical boundaries of the rectangles of the dissection. In particular, a node corresponds to a single vertical line which may bound several rectangles of the dissection that share the same x coordinate (see Figure 4.2.1.1a and b). A source node corresponds to the left-most vertical boundary of the enclosing rectangle. A sink node corresponds to the right-most boundary of the enclosing rectangle. A directed edge between node i and node j represents the fact that there exists a rectangle in the dissection which has its left boundary at the location represented by i and its right boundary at the location represented by j. Then each of the rectangles of the dissection (and hence modules of the layout) is in one-to-one correspondence with a directed edge of the polar graph. Figure 4.2.1.1b shows the horizontal polar graph of the layout of Figure 4.2.1.1a. Note that there may be parallel edges when two rectangles have their vertical boundaries in the same positions.

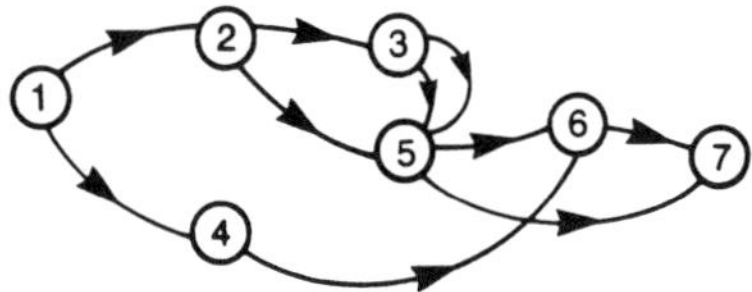

Figure 4.2.1.1b. The horizontal polar graph of the rectangle dissection of Figure 4.2.1.1a.

The vertical polar graph $G_V(N,E)$, is constructed in the same way as the horizontal polar graph, except that now the boundaries being considered are the horizontal ones. Hence, a source node corresponds to the top edge of the enclosing rectangle and a sink node corresponds to the bottom edge of the enclosing rectangle (Figure 4.2.1.1c).

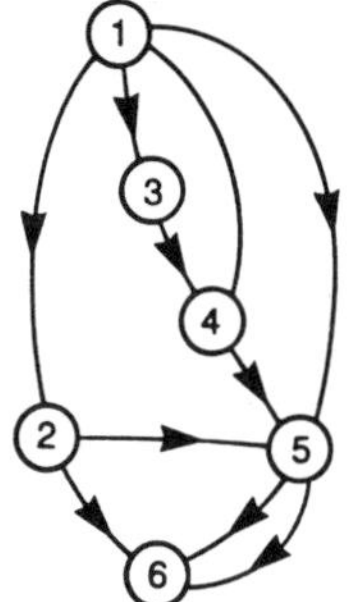

Figure 4.2.1.1c. The vertical polar graph of the rectangle dissection of Figure 4.2.1.1a.

The two polar graphs uniquely define a rectangle dissection and can be used to represent the relative position of the modules in the layout. If we label the arcs of the polar graph with the minimum distance that the boundaries of the enclosing rectangle may have (this distance can be determined by the size of the component and by an estimate of the number of wires that are expected to cross the area near the component), then the *critical path* i.e., the longest path between the sink and the source of the two polar graphs, gives the minimum x and y dimensions of the final layout, if the only change allowed in the layout is the displacement of the components. Several efficient algorithms are available to solve the critical path problem. Note that this is a standard compaction problem and the algorithms described in this book by R. Newton can be applied to solve it.

If we generalize this optimization problem to include the change of orientation of the components of the layout, it is possible to prove [STO83] that the problem is NP-hard. However, there are special cases where this problem (and even its extension to include the change of aspect ratio of the components) can be solved efficiently. This special case is the topic dealt with in the next section.

138

4.2.2. Slicing Structures, Shape Determination and Area Optimization

A special case of rectangle dissection is of particular interest in placement and floor-planning: the *slicing structure* [OTT82]. This dissection has a number of interesting features: (i) it can be represented in a more compact form, (ii) it is ideally suited for hierarchical layout, (iii) the routing problem for the resulting placement is easier to solve, (iv) there are algorithms that can find the optimal shape and the optimal orientation of the components of the design, and (v) there are efficient algorithms to obtain a layout that can be represented by a slicing structure.

4.2.2.1. Slicing Structures and Their Representation

A slicing structure is characterized by particular polar graphs: *series-parallel* graphs. A series-parallel graph is a directed graph that can be reduced to a graph with only two nodes (the sink and the source) connected by a single edge, by applying in sequence two operations: *parallel and series reduction*. Parallel reduction replaces two edges that connect the same two nodes with a single edge. Series reduction removes from the graph a node i with only one incoming and one outgoing edge, (j,i) and (i,k), and adds a new edge (j,k) to connect node j to node k.

Note that parallel reduction can be considered as equivalent to merging two rectangles of the same width in the horizontal polar graph or of the same height in the vertical polar graph. Series reduction corresponds to merging in the horizontal polar graph two rectangles that have the same *height* and in the vertical polar graph two rectangles that have the same *width*.

The layout of Figure 4.2.1.1a does not have series-parallel polar graphs, but the one of Figure 4.2.2.2a does. The corresponding polar graphs are shown in Figure 4.2.2.2b and 4.2.2.2c. In view of the property introduced above, it is possible to show that the two reduction operations are dual for series-parallel polar graphs, so that if two edges are subjected to a parallel reduction operation in the horizontal polar graph, they are subjected to a series reduction in the vertical polar graph. A sequence of dual reductions are performed on the polar graphs of Figure 4.2.2.2b and 4.2.2.2c in Figure 4.2.2.3.

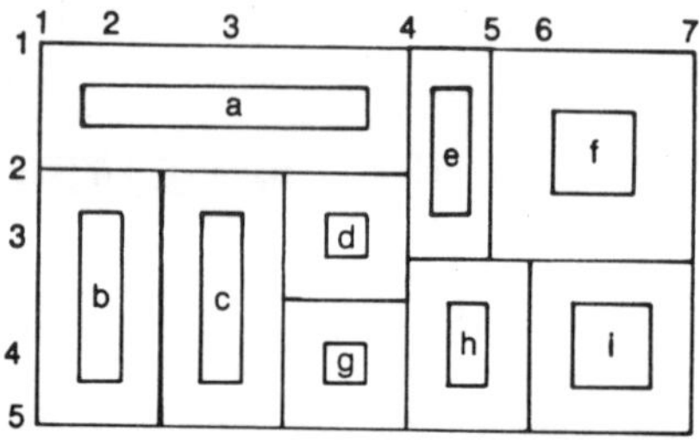

Figure 4.2.2.2a. A placement with a slicing structure.

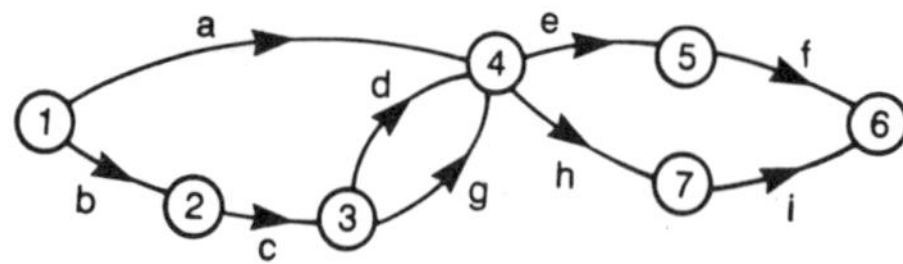

Figure 4.2.2.2b. The horizontal polar graph.

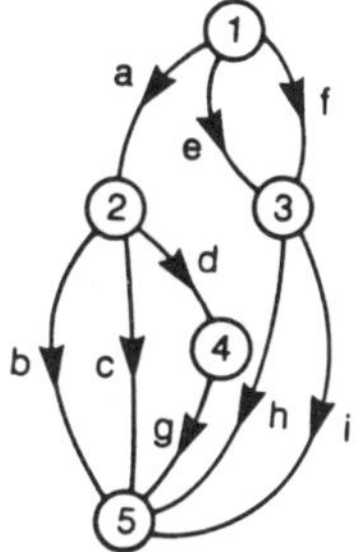

Figure 4.2.2.2c. The vertical polar graph.

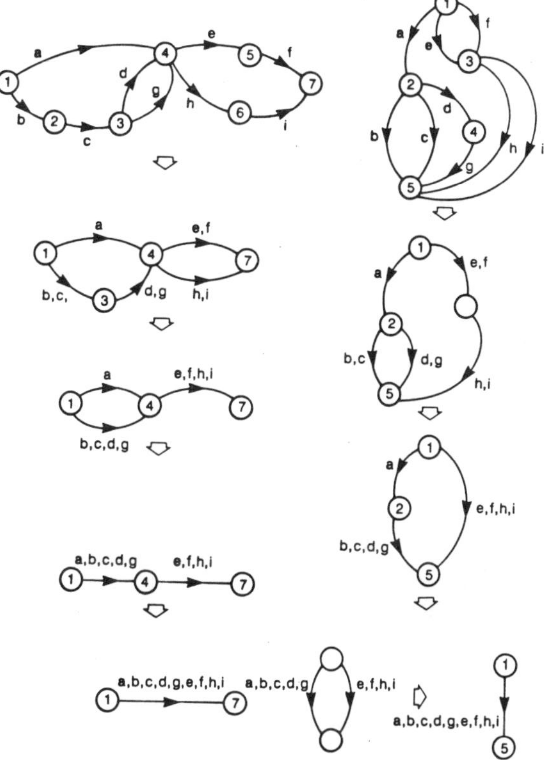

Figure 4.2.2.3. Reduction of series-parallel polar graphs.

The reduction operations are applied iteratively to the polar graphs. The last step of
the iteration must be a parallel reduction in one of the two graphs and a series reduction in
the other as shown in the sequence of Figure 4.2.2.3. The last series reduction implies
that the boundary corresponding to the only internal node (an internal node is defined as a
node which is not the source or the sink of the graph) of the polar graph *bisects* the
enclosing rectangle. For example, for a horizontal polar graph, each of the rectangles of
the dissection lay either to the right or to the left of the boundary corresponding to the

140

internal node. The boundary divides the enclosing rectangle into two rectangles of equal height. If the two resulting rectangles are examined separately, the same considerations can be applied and a boundary can be identified either in the vertical or the horizontal polar graph which bisects each of the rectangles. The boundaries so identified are called *slices*. They recursively divide the area of the rectangle into two parts. The information carried by the two polar graphs can be summarized in a single data structure that identifies the slices of the dissection: a *decomposition tree*. A tree for the dissection of Figure 4.2.2.2a is shown in Figure 4.2.2.4a. Note that each of the nodes of the tree represents a region of the enclosing rectangle.

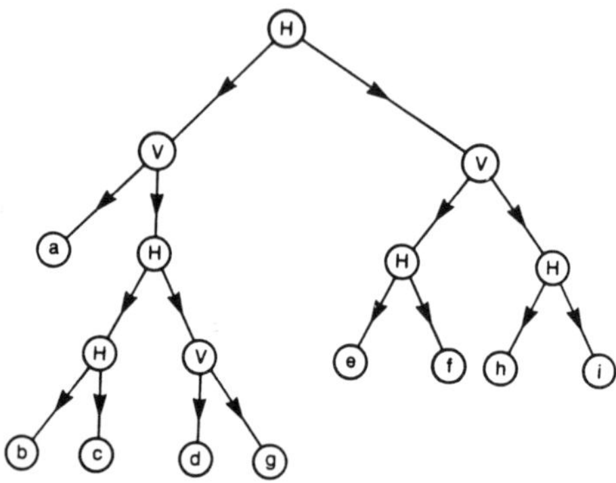

Figure 4.2.2.4a. A decomposition tree.

The decomposition tree represents a hierarchy of the regions of the layout. The root of the tree is the top level of the hierarchy and represents the enclosing rectangle. The children of the root are the two regions of the layout that are created by the first slice. The recursive subdivision ends with nodes which correspond to rectangles that contain a single module of the layout. These leaf nodes are labeled with the name of the module. The edges of the tree represent explicitly the hierarchy. Note that for each non-leaf node of the tree there are two child nodes.

To reconstruct the dissection, the slice has to be identified as vertical or horizontal. This is done by labeling the non-leaf nodes with "H" for horizontal and "V" for vertical bisections. In addition, the relative positions of all the child nodes have to be specified. To simplify the amount of information carried, it is assumed that the left child node is always to the left of the right child node for a horizontal slice and on top of the right child node for a vertical slice.

Even with this additional information, decomposition trees may not be in one-to-one correspondence with a slicing structure. For example, the decomposition tree shown in Figure 4.2.2.4a is non unique. Another decomposition tree which corresponds to the slicing structure of Figure 4.2.2.2a is shown in Figure 4.2.2.4b. This arises from the fact that two nodes in the decomposition tree, say *a* and *b*, where *a* is the father of *b*, corresponds to two slices that are in the same direction (either vertical or horizontal). In this case, the selection of which slice to choose first in the hierarchy expressed by the decomposition tree is arbitrary, but it causes the non unique representation problem.

Two solutions are possible to make the decomposition tree in a one-to-one correspondence with the slicing structure it represents. One is to allow the tree to be non-binary.

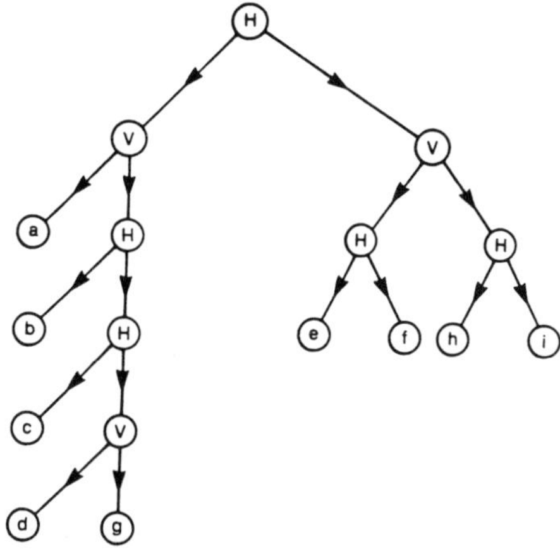

Figure 4.2.2.4b. An equivalent decomposition tree.

In this case, if more than one slice appear in the same direction at a certain level of the hierarchy, they are all represented by a node with as many children as slices. In this case, fathers always correspond to slices that are in the opposite direction to the ones of their children. The tree of Figure 4.2.2.4c is the non-binary decomposition tree representing uniquely the slicing structure of Figure 4.2.2.2a.

Another solution is to fix by convention the first slice to represent in the hierarchy: for example, the right-most and the top-most. In this case, the decomposition tree shown in Figure 4.2.2.4b would be the only correct representation of the slicing structure.

Note that there are algorithms that can only generate binary decomposition trees, e.g., the min-cut algorithm presented in Section 4.3, others that can generate non-binary decomposition trees, e.g., Otten's procedure described in Section 4.4 and in another chapter of this book [OTT87].

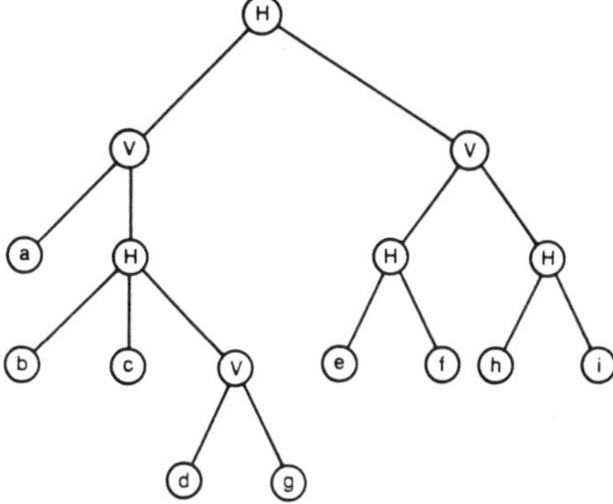

Figure 4.2.2.4c. A unique decomposition tree that is not binary.

4.2.2.2. Determination of Shape and Orientation

The problem to be solved is the determination of the optimal shape and orientation of the modules given their relative positions as a slicing structure. This problem can be solved in polynomial time with an elegant algorithm introduced by Stockmeyer [STO83] for the orientation problem and generalized by Otten for the shape determination problem. Stockmeyer proved that for the orientation problem, the worst case complexity is $O(n^2)$, where n is the number of modules to be considered.

The algorithm is based on the composition of relations in the $x-y$ plane. The basic relation is the *shape constraint relation* as defined by Otten [OTT87].

The Shape-Constraint Relation

Given a rectangular module, the shape constraint relation R for that module is the set of $y-x$ pairs so that a rectangle with width equal to x and height equal to y contains at least a shape/orientation realization of the module. If we assume that the module has fixed area, aspect ratio and orientation, i.e., has width w and height h, then the relation is the set of $y-x$ pairs

$$R=\{(x,y) \mid y \ge h \ , \ x \ge w \}. \tag{4.2.2.1}$$

In Figure 4.2.2.5, the shaded area represents the set of all pairs satisfying Eqn. 4.2.2.1 for the module shown in the same figure. If the module can be rotated by 90 degrees, then the relation is characterized by

$$R=\{(x,y) \mid y \ge h \ , \ x \ge w \ \ or \ \ y \ge w \ , \ x \ge h \},$$

which is represented by the shaded area of Figure 4.2.2.6, obtained by taking the union of the sets of points defined by the first component of the relation and of the set of points defined by the second component. If the module can be rotated by 90 degrees or have n different widths and heights $h_1,.....,h_n$ and $w_1, \cdots ,w_n$, then the relation is

$$R=\{(x,y) \mid y \ge h_1, x \ge w_1 \ \ or \ \ .... \ \ or \ \ y \ge h_n, x \ge w_n \ \ or \ \ y \ge w_1, x \ge h_1 \ \ or \ \ .... \ \ or \ \ y \ge w_n, x \ge h_n$$

obtained by taking the union of the sets of points corresponding to each of the components of the relation. If the module is specified to have a fixed area A and a continuously varying aspect ratio, then the relation is given by

$$R=\{(x,y) \mid yx \ge A \ , \ y \ge h, x \ge w \},$$

where h and w are bounds on the size of the modules (Figure 4.2.2.7).

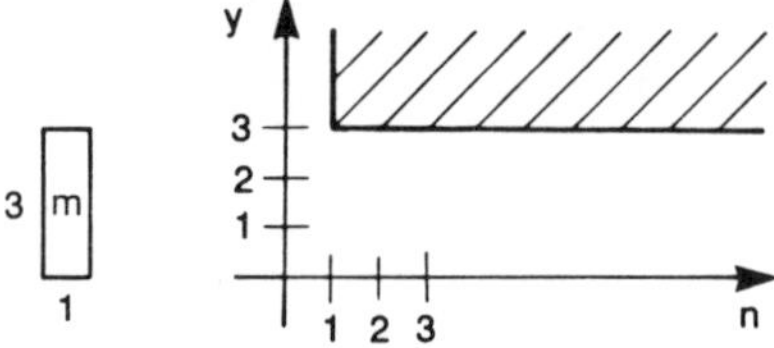

Figure 4.2.2.5. The shape-constraint relation for module m.

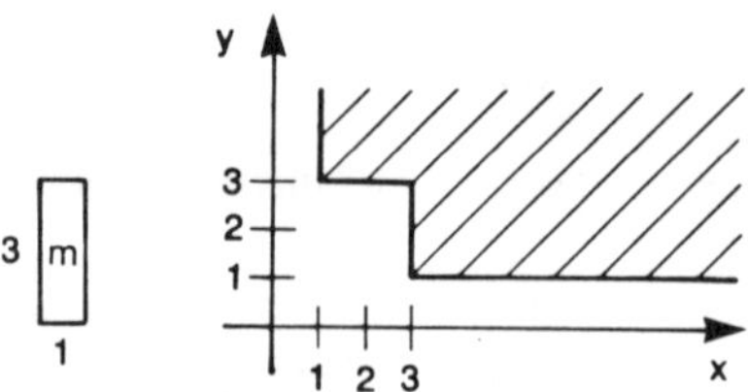

Figure 4.2.2.6. The shape-constraint relation for m when rotation is allowed.

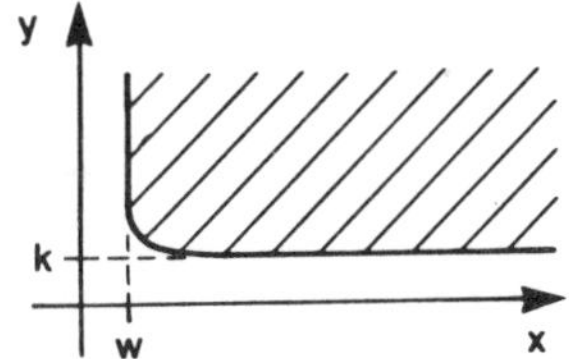

Figure 4.2.2.7. The shape-constraint relation for a flexible module with continuously variable aspect ratio.

If one of the two dimensions, say x, is fixed to a value X, then the other dimension giving the minimum size enclosing rectangle is given by the intersection of the constant line $x=X$ with the shape-constraint relation boundary as shown in Figure 4.2.2.8.

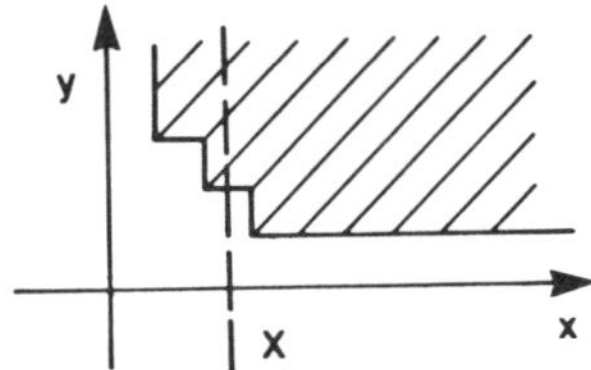

Figure 4.2.2.8. Constraints on x determines the optimum value for y.

The Composition of Shape-Constraint Relations

A slicing structure represented by a decomposition tree describes how rectangles containing the modules of the design are positioned with respect to each other. If, according to the information encoded by the decomposition tree, two rectangles are placed one on top of the other, then the width of the enclosing rectangle is constrained to be equal to the width of the two rectangles, while the height is given by the sum of the heights of the two rectangles. By the same token, if two rectangles are specified to be one to the right of the other, then the enclosing rectangle inherits the height of the two rectangles.

In view of this observation, given the shape constraint relations of two rectangles, a shape constraint relation can be obtained for the enclosing rectangle. Suppose that the two rectangles are to be placed one on top of the other, then the shape constraint relation $R_{1,2}$ for the enclosing rectangle is given by

$$R_{1,2} = \{(y,x) \mid (y_1,x) \in R_1 \text{ and } (y_2,x) \in R_2 \text{ and } y = y_1 + y_2\},$$

where R_1 is the shape constraint relation for the first rectangle and R_2 is the shape constraint for the second rectangle. This set can be obtained by "summing" graphically the shape constraint relations as shown in Figure 4.2.2.9.

If the relative position of the two rectangles is one to the right of the other, then the shape constraint relation of the enclosing rectangle $R_{1,2}$ is given by

$$R_{1,2} = \{(y,x) \mid (y,x_1) \in R_1 \text{ and } (y,x_2) \in R_2 \text{ and } x = x_1 + x_2\}.$$

Here $R_{1,2}$ can be obtained by summing graphically the relations as shown in Figure 4.2.2.10. To make the composition of the shape constraint relations simpler to carry out, continuous boundaries such as the one shown in Figure 4.2.2.6 are usually approximated by piece-wise linear functions. In this case, the resulting shape constraint relation can be easily computed since the boundary is also a piece-wise linear function whose break-points

144

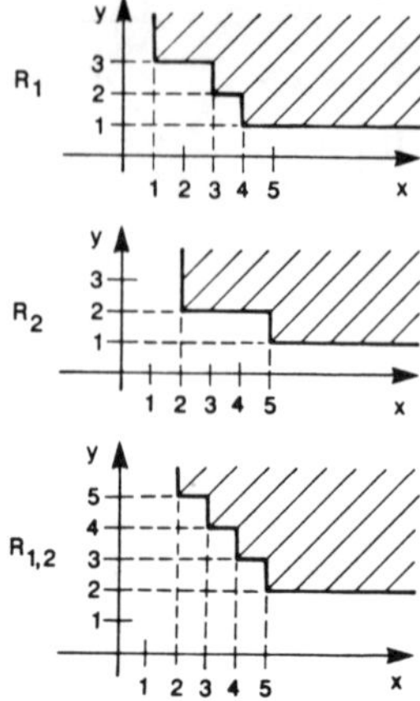

Figure 4.2.2.9. Summing the shape-constraints relation for two flexible blocks to be placed one above the other.

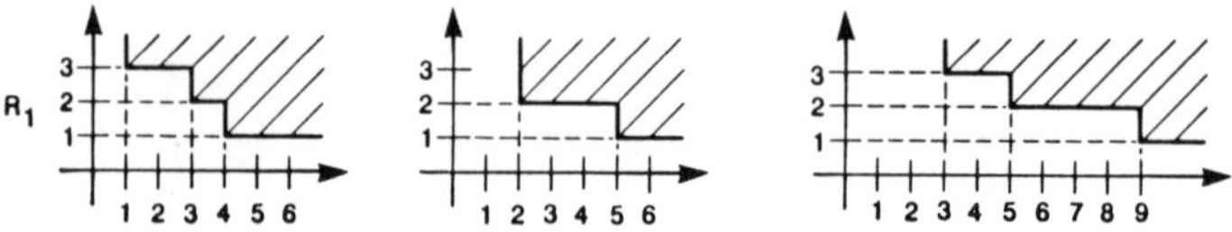

Figure 4.2.2.10. Summing the shape-constraint relation for two flexible blocks to be placed one to the right of the other.

are computed by combining the break-points of the shape constraint relation boundaries of the enclosed rectangles.

If the dimensions of the enclosing rectangle are fixed by external considerations, then the optimal size of the two rectangles enclosed by the constrained rectangle can be easily obtained. Assume that the two enclosed rectangles are placed one on the top of the other. Then the width (the x dimension) of the enclosed rectangles has to be equal to the width of the enclosing rectangle. The height of the enclosed rectangles is obtained by determining the y coordinate of the point of the boundary of the shape constraint relation corresponding to the x coordinate inherited from the parent rectangle. The dual procedure can be applied in the case the two enclosed rectangles are placed one to the right of the other.

To determine the minimum area layout satisfying constraints on aspect ratio width and/or height, the shape constraint relations are composed bottom-up according to the information encoded in the decomposition tree. The last step of this procedure gives the shape constraint of the entire chip. The constraints on aspect ratio, height and width of the chip as specified by the user are imposed on this relation. Then a boundary point of the relation which satisfies all the constraints (if one exists) is selected. This information is then passed top-down to the children of the root node. If the children correspond to a vertical cut, then the x dimension is inherited from the root node. The y dimension of each of the children is determined as follows: on the shape constraint relation of the children, identify the y coordinate that corresponds to the x value inherited from the parent node. This process is repeated until the x and y dimensions of all the rectangles of the decomposition tree are determined.

This procedure can be proven to generate a minimum-area chip that has the relative positions of the modules specified by a given slicing structure. However, the optimization of the aspect ratio is carried out after the relative positions of the modules have been determined and as such may not yield an overall optimal layout.

4.3. Algorithms for Slicing Structures

Slicing structures can be represented in an elegant and compact way. Furthermore, the routing of a chip whose placement can be represented by a slicing structure can be carried out completely by simple and effective tools, channel routers. Hence, it is important to identify algorithms that generate placements represented by slicing structures. On the other hand, since not all placements have slicing structures, these algorithms may produce layouts that are not as good as the ones produced by tools that are not restricted to produce slicing structures.

Several algorithms have been proposed that yield a placement that corresponds to a slicing structure: in this section, the min-cut algorithm and a simulated annealing algorithm are reviewed.

4.3.1. The Min-cut Algorithm

One of the most successful techniques for placement and floor-planning is the min-cut method, proposed by Breuer [BRE77] for printed circuit boards, gate-arrays and standard cells and by Lauther for macro cells. This method produces placements that are represented by slicing structures. In fact, it is interesting to mention at this point that the min-cut algorithm motivated the introduction of slicing structures as placement representations.

4.3.1.1. The Basic Algorithm

The basic procedure is based on the recursive application of a bi-partitioning algorithm. In [LAU80], the Kernighan Lin [KER70] algorithm is used. In [LAP86], in addition to a modified Kernighan-Lin algorithm, an exhaustive search is applied whenever the number of modules to be placed is less than a user-defined limit. This search allows one to obtain optimum bi-partitions for the small subproblems.

The example shown in Figure 4.3.1.1 is used to illustrate the different parts of the algorithm. At first the area of the chip is subdivided into two parts either with a vertical "cut-line" or slice or with a horizontal slice. Using the Kernighan-Lin partitioning algorithm, modules are assigned to the two areas so that the interconnections between the

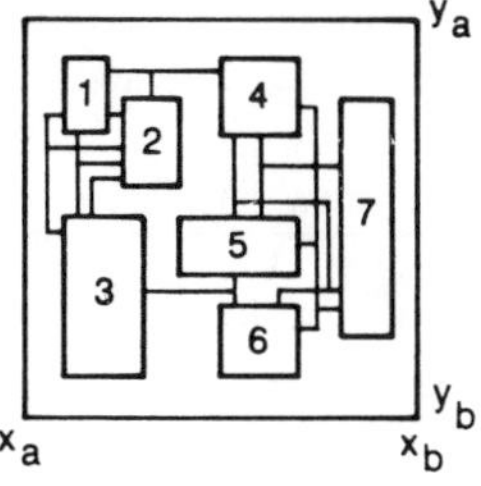

Figure 4.3.1.1. An example of min-cut placement.

146

modules are minimized and the area of the modules assigned to the two parts is roughly equal (Figure 4.3.1.1b). Note that in the macro-cell case, it is very difficult to take into account the aspect ratio of the modules, hence only the area of the modules is used in the partition process.

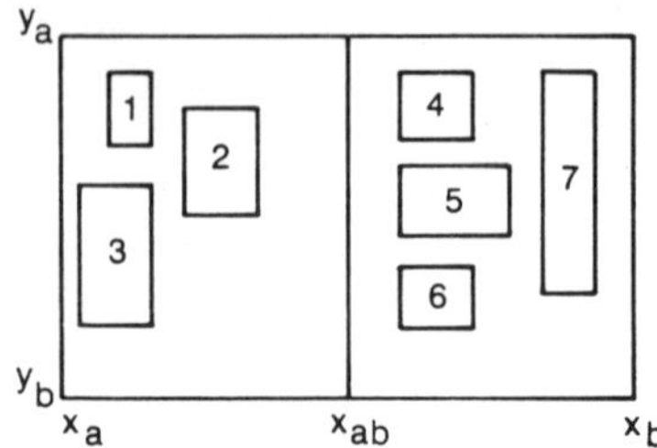

Figure 4.3.1.1b First step of min-cut placement

Once the first partition has been applied, the two areas are subdivided again each into two parts. This subdivision can be obtained with either vertical or horizontal slices (Figure 4.3.1.1c). When an area is occupied only by one module, the area is obviously not subdivided any more. When none of the areas can be subdivided further, the process terminates (Figure 4.3.1.1.d).

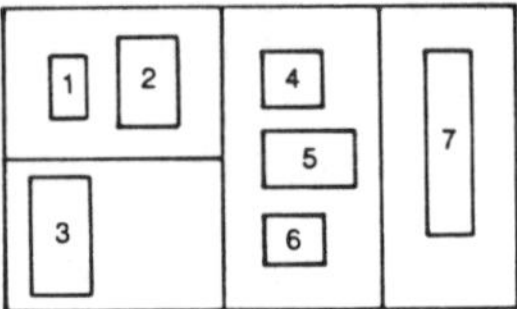

Figure 4.3.1.1c Second step of min-cut placement

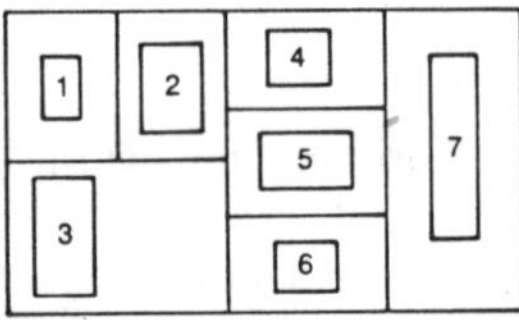

Figure 4.3.1.1d Final min-cut placement

This part of the procedure determines the relative positions, (i.e., the slicing structure), of all the modules but not their absolute positions. A bottom-up traversal of the tree is used to determine the absolute positions of the modules. In this process, the information about aspect ratio and pin positions can be used to optimize the layout. Starting from leaf nodes, the corresponding modules are placed and the smallest rectangle that contains the modules is found. This rectangle now has fixed size and can be composed with other rectangles obtained in this manner until the root node is reached. The last calculation, corresponding to the root of the decomposition tree, gives the overall area of the layout.

For the floor-planning problem, the various alternatives available for the modules are examined to provide the smallest enclosing rectangles. Note however that this process only takes local information into consideration and does not have a global view of the layout. In fact, when the modules are partitioned at a certain level of the decomposition hierarchy represented, for example, by node i of the tree, the only information that is used to determine the location of the modules is the connectivity among the modules contained in the rectangle corresponding to node i. Thus, the positions of the modules as determined by the partitioning algorithm are only locally optimized. For this reason, the final placement can be considerably improved by rotating and mirroring the modules. In fact, the algorithm presented by Lauther includes post-processing steps that improve the original placement. In particular, rotations and mirroring are used to improve both area utilization and net length. Finally, modules can be shifted to eliminate empty areas (Figure 4.3.1.1e and 4.3.1.1f). However, these post-processing steps may destroy the slicing structure of the placement and make the routing problem more difficult.

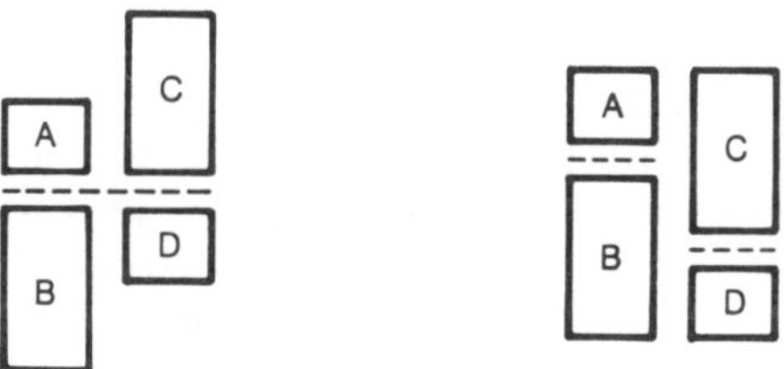

Figure 4.3.1.1e Shift to reduce empty areas

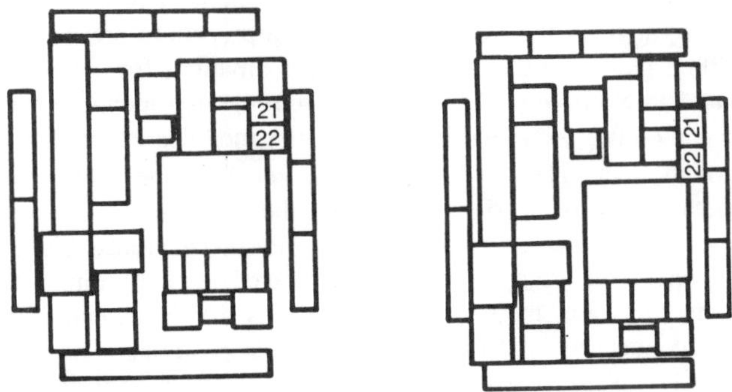

Figure 4.3.1.1f Rotations to reduce empty areas

Rotations, mirroring and shifts are sometimes not enough to overcome the intrinsic limitations of the min-cut algorithm as described above. Several improvements of the basic algorithm have been proposed recently. The algorithms proposed in [DUN85] and in [LAP86] introduce information about the positions of modules partitioned at higher level of the hierarchy. The elegant algorithm proposed by Stockmeyer described above, is used in almost all the floor-planning algorithms, to compute the optimal shape of the flexible modules given a slicing structure.

4.3.1.2. Terminal Propagation

Consider the situation shown in Figure 4.3.1.2. The partitioning process is applied to the rectangle labeled with A in Figure 4.3.1.2. Among the modules in rectangle A, there is a module, m, that is strongly interconnected with modules that have been located by other bi-partitioning steps on the top of rectangle A. Since in the standard min-cut pro-

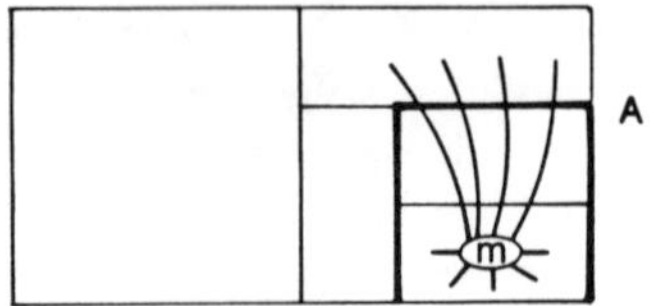

Figure 4.3.1.2. Undesirable solution in min-cut partitioning.

cess there is no information on where the modules of other partitions are located, *m* may end up in the bottom partition. To prevent this from occurring, Dunlop and Kernighan [DUN85] have proposed a modification of the min-cut algorithm, called *terminal propagation,* that takes interconnectivity into account more accurately.

Consider all the modules that are connected to modules contained in the rectangle being partitioned, but that are not themselves in this region. Divide these modules into two sets: the first set contains all the modules that are located on the right (top) of the horizontal (vertical) slice, the second contains all the modules that are located to the left (bottom) of the slice. Replace all the modules of the first set with a dummy module positioned to the right (top) of the slice and all the modules of the second set with a dummy module positioned to the left (bottom) of the slice. All the interconnections are maintained.

In this way, the cost function in the partitioning algorithm takes automatically into consideration the external interconnections. The key point here is to determine whether a module is to the right or to the left of a slice. The procedure presented in [DUN85] carries out a simplified global routing step using approximations for the Steiner tree similar to the ones presented in Section 4.1.1. The global routing step identifies where the interconnections between modules to be partitioned and the modules external to the partition, cut the boundaries of the rectangle which is being dissected. If the intersection with the boundaries of the rectangle is to the right (top), then the corresponding module is assigned to the first set, otherwise to the second set. When the intersection is close to the middle of the region, i.e., where the slice is going to be placed, then no dummy module is introduced. In fact, in this case, there is no need to bias the partitioning algorithm since the difference in interconnection cost for the two possible positions of the modules would not be significant. In [DUN85], it is suggested that any intersection in the middle third of the relevant dimension of the rectangle should be considered too close to the middle to introduce a dummy module.

The procedure presented in [LAP86] is called *in-place partitioning* but it is essentially identical to the terminal propagation algorithm of [DUN85].

This mechanism is simple and effective. Both [DUN85] and [LAP86] report good results with it, when applied to standard-cells and to macro-cells. In particular, about 30% area and net length savings over the standard version of the min-cut algorithm were reported for the standard-cell case[DUN85].

4.3.2. Simulated Annealing for Slicing Structures

Lauther [LAU86], Wong and Liu [WON86] suggested that simulated annealing could be used to generate an optimal slicing structure. The basic idea behind this approach is to start with an initial slicing structure that could be obtained by a min-cut algorithm or by any other means, and then to explore the slicing structure space in search of the slicing structure that would minimize the interconnection cost as well as the total area of the chip.

Given a slicing structure, the moves for the simulated annealing algorithm are transformations on the decomposition tree. The move set has to be chosen so that all the possible relative positions of the modules can be generated and so that the transformation can be computed very fast. In [WON86], a correspondence between decomposition trees and Polish postfix notations for arithmetic expressions is established, where the operands of the arithmetic expressions are the modules and the operators +,* correspond to vertical and horizontal slices respectively. The arithmetic expressions can be thought of as a "bottom-up" representation of the decomposition tree, where the leaf nodes are the operands. In Figure 4.3.2.1, an example of an arithmetic expression representation of the decomposition tree of Figure 4.2.2.4 and its Polish postfix representation are shown.

$$(((b+c)+(d * g)) * a)+((e+f) * (h+i))$$

Arithmetic expression.

$$bc+dg * +a * ef+hi+ * +$$

Polish postfix notation.

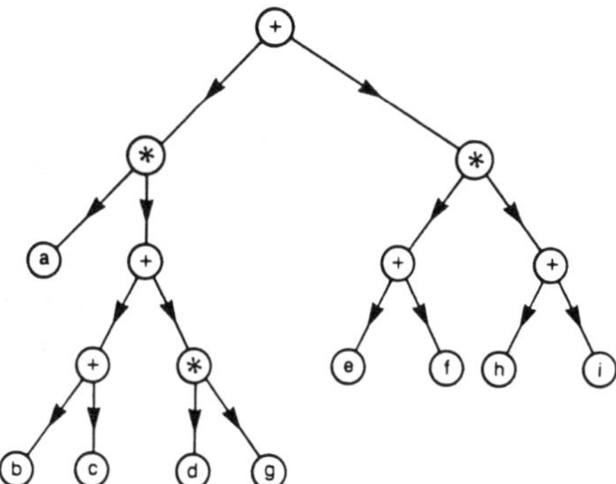

Figure 4.3.2.1. Arithmetic expressions and Polish postfix notation.

Not all Polish postfix expressions correspond to a "legal" slicing structure. For example, an expression that contains more operators than operands, clearly does not correspond to a slicing tree. In fact, a necessary condition for a Polish postfix expression to represent a slicing structure is that given any sequence of operators and operands $\{\alpha_1,\alpha_2, \cdots ,\alpha_n\}$ any subsequence $\{\alpha_1,\alpha_2, \cdots ,\alpha_i\}$ with $1 \leq i \leq n$ is such that the number of operators is less or equal to the number of operands. Another condition is that given any i in the sequence where α_i is an operator, α_{i+1} cannot be another operator. In addition, to make sure that a unique decomposition tree corresponds to a slicing structure, the convention of introducing the right-most or the top-most slice at the highest level in the hierarchy as described in Section 4.2, is used. In this case, the right son of a node cannot be labeled with the same operator as the father. This condition guarantees that there is a one-to-one correspondence between slicing structures and Polish postfix expressions. This correspondence is important because if there were more than one representation of a given slicing structure in terms of Polish postfix expressions, then since the simulated annealing algorithm operates on Polish postfix notations, we would explore a redundant configuration space with implied inefficiency.

150

The set of moves used in [WON86] manipulates a given Polish postfix notation. The move set consists of three moves:

given a sequence of operators and operands corresponding to a slicing structure,

1- *identify a subsequence and "complement" all the operators in the subsequence, where the complementation operation consists of changing a "+" operator in to a "*" operator and vice-versa;*

2- *identify a pair of consecutive operands in the sequence and interchange their position in the sequence;*

3- *identify a operator-operand pair in the sequence and swap them.*

It is rather obvious that the first two moves generate a legal Polish postfix expression, given a legal Polish postfix expression, but the third move may generate a sequence that does not satisfy the conditions identified above. In this case, the move has to be rejected. It can be easily proven that checking whether a move is legal can be done in constant time and hence the checking operation does not overload the simulated annealing calculation.

One of the drawbacks of the min-cut approach to floor-planning was that the optimization of the shapes of the variable modules was carried out only after the relative positions was determined. In this approach, the problem is solved by introducing a cost function that includes two components: the first component is the interconnection cost, computed as the sum of the estimated interconnection lengths between modules. The second component is the total estimated area of the chip.

Since the exact positions of the modules are not known exactly, the total net-length calculation is based on the assumption that all the pins of a module are located in the center of its bounding rectangle. This approximation may be quite crude, but it allows a fast calculation of the interconnection length. The net length estimation is then computed by evaluating the Manhattan distance of the pins of each net, summed over the nets of the layout.

The area calculation is performed by running the Stockmeyer algorithm on the decomposition trees produced by the moves of the algorithm. This part of the cost function takes care of the aspect ratio optimization during the determination of the relative position. Since the complexity of this calculation is not negligible, it is important to identify the necessary computation. Since the move set affects only a part of the decomposition tree, much of the computation needed to estimate the new area can be saved, by recomputing the area corresponding to the subtree which is affected by the change. While the calculation is still rather expensive, it has been observed experimentally that a small part of the tree is actually affected and hence many moves can be applied to obtain a good solution in the simulated annealing algorithm.

While this algorithm is elegant and simple, it does not address the problem of routing area estimation. In addition, the net length estimation may be rather poor since pin positions are only approximated and the interconnection model is quite inaccurate. Finally, only rectangular modules can be handled.

4.4. Algorithms for Unconstrained Placement

In this Section, algorithms that do not produce necessarily a slicing structure are reviewed. In particular, the greedy clustering approach, the force-directed heuristics, the eigenvalue method and the network optimization technique will be briefly described.

Some of these algorithms, in particular force-directed methods, have been used extensively in placement and routing systems. In the last subsection, a method due to Otten to produce a slicing structure from any placement is given.

4.4.1. The Greedy Clustering Approach

Greedy clustering is a constructive placement method similar to the constructive method presented for partitioning. It was one of the first placement methods to be proposed and many papers have described and discussed it. For this reason, it is only briefly reviewed.

Greedy clustering incrementally adjoins modules to the modules already placed. The modules are adjoined according to their size and connectivity. Critical factors for this algorithm are the selection of the seed, (i.e. of the first module to place), the selection of the next module to be placed and the position of the module with respect the modules already placed. A number of papers have been published describing a variety of schemes for each of the factors above. Note that once the position of a module has been assigned, it is not changed during the remaining part of the algorithm. The LILAC system [SUD83] for macro cells developed at Hitachi used clustering techniques for placement. These methods provide fairly good initial placements. However, iterative improvement methods should be applied afterwards to obtain satisfactory results, because constructive algorithms give relatively poor solutions due to their greedy and local nature. For example, the SHARPS system [SUD83] for macro cells and the Oki system [SUD83] for gate arrays have followed this strategy. The running times of these methods are in general short.

4.4.2. Force-directed Methods

The basic idea of this class of methods is to represent the interactions between modules with a set of forces

$$F_{ij} = - c_{ij} \, d_{ij} \qquad (4.4.2.1)$$

where c_{ij} is the possibly weighted sum of the number of nets interconnection module i with module j, d_{ij} is a vector directed from the center of module i to the center of module j and with modulus equal to the distance between module i and j in general measured as the Euclidean ($|d_{ij}| = \sqrt{(x_i-x_j)^2+(y_i-y_j)^2}$) or the Manhattan distance ($|d_{ij}|=|x_i-x_j|+|y_i-y_j|$) between the centers of the modules. Note that Eqn. (4.4.2.1) is the expression for the force between two points connected by a spring with constant c_{ij}. A placement is found by minimizing the sum of the force vectors acting on the modules. This idea is clearly based on a "physical" representation of the problem and the hope is that the cost functions relevant to placement will be minimized by this procedure.

Force-directed heuristics can be used both for constructive placement and for iterative improvement, as described in the next two subsections. They have found wide applications in a variety of gate arrays and macro cell systems such as the PLINT [ANW85] system developed by GE for standard and macro cells, the APLS2 [HSU85] system developed by Hughes for standard cells, the SHARPS system developed by Sharp for macro cells. the BAGEL system developed by the University of California at Berkeley, the MARC system [SUD83] developed by NTT , the MASTER and the LAMBDA systems developed by NEC [SUD83], and the MARS-M3 system developed by Mitsubishi, all for gate-arrays.

4.4.2.1. Constructive Force-directed Algorithms

In constructive force-directed algorithms, an initial placement is constructed by finding the locations of the modules so that the modules are in equilibrium with respect to the forces exercised on them, i.e., the vector sum of the forces exercised on each of the modules is zero.

The positions of the modules subject to the forces described above can be found by solving a system of equations expressing the fact that the sum of the forces acting on each of the modules must be zero. This system of equations is nonlinear in the coordinates of the center of the modules. In general, a Newton-Raphson method is used to solve the system of equations in connection to both metrics. Note that the equations of the system to be solved are not continuously differentiable because of the square root in the Euclidean metric and of the absolute values in the Manhattan metric, and that in this case the Newton-Raphson method may have convergence problem.

If no module has a fixed position then the trivial solution to the problem is given by $x_1=x_2=...=x_n$ and $y_1=y_2=...=y_n$, i.e., all the modules will completely overlap, a clearly non-feasible solution. In fact, the constraint that two modules may not overlap is not captured in the formulation presented above. However, if pads are introduced and their location is maintained fixed, then the force exercised by the connections between the pads and the modules will avoid the complete overlap of all the modules. Nevertheless, some overlaps may remain.

Repulsive forces have been introduced [SOU81] to minimize the overlaps among modules. In particular, a repulsive force is introduced among modules that are not connected. The force can be either modeled as a constant force or as a force inversely proportional to the distance between the modules. In [QUI79] the repulsive force is modeled as a constant force. Note that if a repulsive force is introduced among all the modules that are not interconnected, then the system of equations from sparse becomes full, i.e., each equation depends on the coordinates of all the modules. This aspect makes the solution of the nonlinear system of equations more difficult and expensive.

Even if repulsive forces are introduced, overlaps may still exist at the end of the procedure. Then, a feasible placement is constructed by modifying as little as possible the placement obtained by the force-directed technique.

4.4.2.2. Iterative Improvement Force-directed Algorithms

In general, the constructive force-directed placement is not satisfactory and iterative improvements are needed to obtain a better placement. This is due to the simplifying assumptions about pin locations, to the greedy nature of the algorithms and to the fact that only translations are considered to search the configuration space.

Interestingly, some of the most successful iterative improvement heuristics are force-directed heuristics. The SHARPS, MASTER, LAMBDA systems used variations of force-directed relaxation to improve layouts obtained by other methods such as min-cut.

For example, given a configuration of the modules on the chip, the force vector is computed for all the modules. The selected module is the one subject to the force of maximum modulus. This block is then moved in a position where the resulting force is zero. The location of the center of the module in this ideal position is called *target location* of the module. If the position is not occupied by another module, then another module with the strongest force vector is selected. Otherwise, a variety of different strategies have been proposed. The simplest one is to remove the module which was occupying that position and move it to a position where the total force on it is minimized. To prevent infinite loops, a module cannot be displaced if it has already been considered by the algorithm. This algorithm is often referred to as *force-directed relaxation*.

Force-directed relaxation has been used extensively, but some of the results are not satisfactory because the displaced modules may end up in a poor position.

Pairwise interchange methods are very simple. Two modules are selected for consideration and interchanged if the cost function, is decreased. The key issue here is how to select the pair to be examined. A random selection is often used. An exhaustive examination of all pairs of modules is also possible, although quite expensive. The cost function often used both for selection and evaluation of the quality of the interchange is the total force exercised on the modules before and after the interchange.

A more sophisticated technique, called *force-directed pairwise relaxation,* involves the definition of $\varepsilon-neighborhood$ of a point. An ε-neighborhood of a point is the set of modules m_j whose centers are distance ε apart from it. After a module m_i is selected for the interchange, the modules m_j that are in the ε-neighborhood of the target location for m_i are considered. An interchange is allowed between m_i and m_j, only if m_i is in the ε-neighborhood of the target location of m_j.

Note that pairwise interchange strategies work well if the modules are of similar size, hence are effective in gate-array and standard-cells. If modules have largely different sizes, overlaps may occur and care must be exercised. In [HAN72], [GOT86] a detailed examination of different force-directed heuristics is offered with experimental results ranking the various options. It is interesting to note that a careful evaluation of the quality of force-directed methods in comparison with constructive methods and simple (non forced-directed) pairwise interchange, showed that pairwise force-directed relaxation produces the best results when the cost function being minimized is the interconnection length computed with Estimate 3 described in Section 4.2.

4.4.3. The Eigenvalue Approach

In this approach, the cost function is the square of the total estimated net length (Interconnection Length Estimate 4 presented in Section 4.1).

This approach has been mostly used for gate-array placement where modules have the same dimensions and have to be placed in well defined slots.

To simplify the discussion of the algorithm, the eigenvalue approach is presented for the linear placement problem. The linear placement problem consists of finding a vector $x=[x_1,x_2, \cdots ,x_n]^T$, where x^T denotes the transpose of vector x and x_i is the position on the line of the center of module i, such that

$$\sum_{i,j=1}^{n} c_{ij}(x_i-x_j)^2$$

is minimized, no two modules can occupy the same position, and the values of x are constrained to be the legal positions, i.e., the coordinates of the centers of the available slots.

The cost function can be represented in matrix form as follows:

$$x^T B x \tag{4.4.3.1}$$

where

$$B = D - C$$

is a symmetric matrix, $C=[c_{ij}]$, and D is a diagonal matrix with $d_{ii}=\sum_{j=1}^{n} c_{ij}, i=1, \cdots ,n$. It can be proven easily [HAL70] that B is positive semi-definite, and that if the modules cannot be partitioned into disjoint subsets, the rank of B is $n-1$.

Cheng and Kuh [CHE84] proved that to have a legal placement, the following constraints have to be satisfied:

$$\sum_{i=1}^{n} x_i = \sum_{i=1}^{n} l_i \qquad (4.4.3.2a)$$

$$\sum_{i=1}^{n} x_i^2 = \sum_{i=1}^{n} l_i^2 \qquad (4.4.3.2b)$$

$$\cdots$$

$$\sum_{i=1}^{n} x_i^n = \sum_{i=1}^{n} l_i^n \qquad (4.4.3.2c)$$

where l_i, $i = 1, \cdots, n$ are the legal positions. This is a complex optimization problem, and solving it using standard optimization algorithms would be prohibitively expensive. However, if among all the constraints, a subset is chosen, the problem can be recast into a well-known linear algebra problem: the algebraic eigenvalue problem.

Assume that the only constraint to be considered is Eqn. (4.4.3.2b). Then the problem to be solved is:

$$\min \quad x^T B x$$

subject to

$$x^T x = L$$

where $L = \sum_{i=1}^{n} l_i^2$. Because B is positive semi-definite, the optimization problem is convex and the minimum value of the cost function is unique. The optimality conditions [LUE73] for the constrained problem can be formulated in terms of the Lagrangian

$$\Lambda = x^T B x - \lambda (x^T x - L)$$

as

$$(B - \lambda I) x = 0 \qquad (4.4.3.3)$$

where I is the identity matrix. Eqn. (4.4.3.3) together with the fact that we minimize $x^T B x$, implies that the solution for the constrained optimization problem is given by the eigenvector of the matrix B associated with the smallest eigenvalue. Since B is positive semi-definite and symmetric, all the eigenvalues are real, normal and larger than or equal to zero.

Another way of proving that the eigenvector corresponding to the smallest eigenvalue is the solution to the problem is as follows: Let u_r be a set of eigenvectors of B such that

$$B u_r = \lambda_r u_r$$

where λ_r is the r-th eigenvalue, and

$$u_r^T u_s = 0 , \; r,s = 1,2, \cdots, n, \quad r \ne s$$

$$u_r^T u_r = 1, \; r = 1,2, \cdots, n.$$

Then, since this set is orthonormal, it forms a basis for $\mathbb{R}^n$ and any x has a unique expansion

$$x = \sum_{r=1}^{n} \alpha_r(x)u_r$$

where

$$\alpha_r(x) \triangleq x^T u_r.$$

Then,

$$x^T Bx = \sum_{r=1}^{n} \alpha_r(x)^2 \lambda_r \qquad (4.4.3.4)$$

Hence, the solution of the problem is given by $x=u_1$ where u_1 is the eigenvector associated to the smallest eigenvalue λ_1.

However, since B has rank $n-1$, the smallest eigenvalue is not interesting because it corresponds to an eigenvector with all components equal, clearly a non feasible solution. Thus, the second smallest eigenvalue, λ_2, and the corresponding eigenvalue u_2, are the ones to consider.

When the two-dimensional placement problem is considered, the derivation of the method is very similar, except that now the two smallest eigenvalues have to be computed; the corresponding eigenvectors give the x and y positions of the modules.

Eigenvalues and the corresponding eigenvectors can be computed fairly quickly by means of several algorithms. Since B is symmetric, it is possible to apply effective algorithms, such as the Lanczos algorithm, that exploit the sparsity of B and compute only selected eigenvalues. The cost of this computation is of the same order of the computation needed to solve a linear system of equations with B as coefficient matrix.

Note though that the solution to this mathematical problem does not produce a feasible solution for the original problem since the constraints on the positions of the modules are only partially taken into account. This methods can be considered as an elegant approach to obtain an approximation to an optimal placement. Other methods have to be used to produce a final placement that satisfies all constraints.

The value of the cost function in correspondence to the second and third smallest eigenvalue is a lower bound on the cost function for any feasible solution. Thus it can be used to evaluate the quality of a solution obtained by means of other heuristics. In view of this result, Blanks [BLA85] has demonstrated, using statistical arguments, that exhaustive pairwise interchange applied to a randomly generated problem characterized by the fact that every module is connected to d other modules, yields this lower bound asymptotically, i.e., when the number of modules n goes to infinity. However, Frankle and Karp [FRA86] observed that the exhaustive pairwise interchange yields solutions that are relatively far from the bound when other random problems are considered.

A modification of the eigenvalue approach to yield a feasible solution is given in [FRA86]. Here, the positions x are constrained to have a set of feasible positions. Hence the optimization problem is to select among the legal positions available for the modules, the ones that minimize Eqn. (4.4.3.4).

A clearly non feasible approach to this problem is to evaluate all possible assignments of legal positions to modules and pick the ones that minimize Eqn. (4.4.3.4). In [FRA86], a systematic procedure that identifies good solutions is proposed. In particular, the problem is recast into a maximization problem as follows. For any H, minimizing Eqn. (4.4.3.4) is equivalent to maximizing

$$H - x^T B x = \sum_{r=1}^{n} \alpha_r(x)^2 [H - \lambda_r]. \tag{4.4.3.5}$$

If H is taken to be larger than the largest eigenvalue of B and a matrix V is defined whose columns are the eigenvectors

$$v_r \triangleq u_r \sqrt{H - \lambda_r}$$

scaled so that those with the least cost associated get the greatest length, then the solution to the problem of maximizing Eqn. (4.4.3.5) is given by

$$\| x^T V \|^2$$

so that this maximization problem becomes the problem of finding the point described by $x^T V$ furthest from the origin.

This problem is solved in [FRA86] by searching the Euclidean space with probes d, hyperplanes passing through the origin of dimensions $n-1$. In fact, it is fast to compute the point that has the projection onto d which is furthest from the origin.

Heuristics are proposed to limit the cost of this probing mechanism: probing with hyperplanes of smaller dimensions and increasing the dimensions gradually has given good solutions. This process is faster than the direct probing in $\mathbb{R}^n$ since the directions along which we probe can be focussed towards the points that produced good solutions in the previous probing step.

The method produced good results both in running times and quality of the solutions when compared to the exhaustive pairwise interchange method. It is now being tested on standard cell and gate-array problems.

4.4.4. The Network Optimization Approach

The network optimization method uses the same cost function as the eigenvalue method (Eqns. (4.4.3.1) and (4.4.3.2)). If the constraints specified by Eqn. (4.4.3.2) are ignored, then the problem can be interpreted as an electrical network optimization problem. In fact, B can be considered as the indefinite admittance matrix of a network consisting of linear passive two-terminal components (B is symmetric positive semi-definite, the row sum of the elements of B is zero, and the diagonal elements are the sum of the off-diagonal ones). Then, Eqn. (4.4.3.1) represents the power dissipated in the network and x is the set of node voltages of the network. The sets of current that satisfy the circuit equations minimizes the dissipated power. The node voltages give the optimal positions of the modules.

It is obvious that if no independent source is present in the network, the currents and the voltages must all be zero, a trivial solution to the problem. This corresponds to the obvious fact that if no constraints are imposed on the positions of the modules, the optimal positions are given by all modules placed on top of each other.

In general, pads have fixed positions and they can be represented in the electrical network by independent voltage sources. In this case, the circuit equations corresponding to the optimal placement problem have a non trivial solution. However, this solution may still be non feasible and other techniques have to be used to complete the placement.

This interpretation was proposed by Charney and Plato [CHA68] and used by Cheng and Kuh [CHE84]. It is interesting, but I do not believe it offers any particular insight into the problem. The circuit equations are simply the equations that can be obtained by writing first order necessary conditions of optimality that become sufficient for convex problems. The approach of Cheng and Kuh is valid because of the successive approximations that are used to obtain feasible solutions.

Satisfying the constraints in Eqn. (4.4.3.2a) is easy since they are linear. These constraints can be incorporated in the electrical model with the help of independent current sources. The other constraints are much more difficult to take into consideration with this approach.

In [CHE84], the quadratic constraints described by Eqn. (4.4.3.2b) are satisfied by solving an optimization problem defined on the solution to the simplified problem where only the linear constraints are considered. The optimization problem has as cost function the square of the distance between the solution that satisfies the linear and the quadratic constraints and the one that satisfies the linear constraints only and that was obtained previously. This formulation ensures that the solution to the more constrained problem is as close as possible to the solution of the less constrained problem that we know is optimal.

Finally, a successive relaxation approach is used to obtain a refined solution. First a number of modules on the left are freed and their new position determined by fixing the positions of the other modules. Then, a certain number of modules on the right are relaxed. Finally the modules in central positions are relaxed. This process is iterated three times.

The solutions obtained by the relaxation process may not be feasible but their relative positions have been fairly well optimized, so that the mapping of the modules to the available slots can be done with confidence that the final results are good.

The results obtained by this approach were compared favorably with force-directed methods, both in terms of speed and quality. However, experiments run on standard-cell and gate-array circuits have shown that the quality of these solutions are not as good as the ones obtained by the simulated annealing implementation of the TimberWolf package [SEC85].

4.4.5. Transforming a Placement into a Slicing Structure: Otten's Procedure

The shape determination algorithm can turn any placement algorithm that generates a slicing structure into an effective floor-planning tool. Given a placement, it is of obvious interest, due to the nice properties of slicing structures, to transform it into a slicing structure.

The transformations that can be used are the relative movements of the modules or their deformation. A shrinking procedure has been suggested by Otten [OTT84], [OTT87]. to create a slicing structure. Given a "non-slicing" placement for a set of modules, where the flexible modules are represented by legal instances, i.e., by modules that have aspect ratio and size compatible with the shape constraints, the modules are uniformly shrunk maintaining their aspect ratio until one or more slices appear that bisect the layout. These slices are recorded and the partitioned parts of the layout are considered in isolation. The shrinking procedure is repeated until a complete slicing structure is generated.

The slicing structure gives the relative positions of the modules, while the shape determination procedure gives the optimal aspect ratios and the actual dimension of the chip.

Otten [OTT84], [OTT87] has used this algorithm to solve the floor-planning problem. The details of this procedure are fully explained in another chapter of this book [OTT87] and hence, for the sake of brevity, not described here. In this section, we review the main aspects of the method for the sake of completeness.

In this method, the initial step is carried out by replacing the modules with points in the plane. A simulated annealing algorithm or an eigenvalue approach is used to determine the position of the points in the plane that optimizes interconnection length. After the positions of the points has been determined, the points are replaced by the modules. Flexible modules are replaced by modules that have aspect ratio and area compatible with the shape constraints. Overlaps may easily arise at this step and there is no guarantee that the placement results in a slicing structure. The shrinking procedure is then used to obtain a slicing structure and the optimal shape determination procedure is finally applied.

The simulated annealing algorithm used in Otten's method is fast because of the level of abstraction that is used. Compared to Wong's approach, shape determination and routing area estimation are performed after the point placement problem is solved. Hence, I expect that the optimization will not be as good as the other approach. Note that in both cases, only rectangular modules with fixed pins can be dealt with.

4.5. The TimberWolfMC Simulated Annealing Approach for Floor-planning and Placement

If no assumptions are made on the shape, orientation and pin positions of the modules, the floor-planning and placement problems have many possible solutions and searching for the optimal ones is indeed a very difficult task. As we have seen earlier, assumptions are made so that the space of the possible solutions is restricted. However, there is an algorithm that allows us to consider efficiently the full richness of the solution space and to explore freely different configurations: simulated annealing. In my opinion, this aspect is possibly the most important one in providing excellent solutions often comparable to or even better than the ones obtained by human designers.

Several implementations and modifications of the basic simulated annealing algorithm have been proposed for the gate-array and standard-cell placement problems (e.g., [KIR83], [SEC85], [GRO86]), while only a few results are available for the macro-cell placement problem. For this reason, the application of simulated annealing to the macro-cell problem is emphasized here.

The first application of simulated annealing to the macro-cell placement problem was presented in [JEP83]. This application considers rectangular macro-cells and does not require any special structure for the final layout. The set of moves that is considered allows intermediate non-feasible solutions, i.e., solutions with overlap among macro-cells. The overlap between cells is taken into account by using a penalty term in the cost function. However, experimental results show that to obtain good results and very small overlaps in the final solution, the relative importance of the penalty term in the cost function should be dynamically adjusted. In the first stage of simulated annealing, i.e., when the temperature is high, the cost function emphasizes the estimated net length while at the last stage of the algorithm, i.e., at low temperature, the cost function is heavily biased towards the elimination of overlaps.

The TimberWolfMC program [SEC86] is a more sophisticated implementation and uses simulated annealing to solve both the floor-planning and the placement problem for macro-cells. A list of basic features includes:

- Cells may be represented by any Manhattan polygon
- Cells are allowed to have aspect ratios that vary over a continuous or

discrete range

- Cells can have multiple implementations and the most suitable will be selected
- Cells may have variable pin positions
- Weights can be assigned to each net to bias the placement

The algorithm optimizes net length and area utilization. The net length calculation is based on the exact pin locations (as opposed to assuming that all of the pins associated with a cell are located at the center of the cell). Consequently, all eight possible "in-place" Manhattan transformations (e.g., x-reflection) are considered for each cell. The placement algorithm proceeds in two distinct stages. During the first stage, the interconnect area around the individual cells is determined using a new dynamic interconnect-area estimator. That is, as each cell is moved, its effective area is adjusted. A simulated annealing algorithm is used to minimize the total estimated interconnect length. The second stage of TimberWolfMC consists of several executions of a placement-refinement algorithm. Each execution consists of three steps: (1) A channel definition step, (2) a global routing step, and (3) a placement-refinement step. The information obtained in step 2 is used to compute the density of all of the channels, which then permits accurate interconnect-area determination. The placement of the cells is then refined in step 3 to reflect the required interconnect area. A low-temperature simulated annealing is used to accomplish step 3. The placement refinement step is carried out until convergence is reached. In our experience, only a few iterations (no more than three in the test cases we have considered) are generally needed to achieve convergence.

4.5.1. The Move Set

In this application, the moves are quite complex. They are subdivided into two classes: single module displacement and pairwise interchange. The selection of one of these two options is performed randomly.

If a single module displacement has been selected, a module is randomly chosen. Then, its x and y coordinates are modified at random to bring the module to a new position in the chip. If the move is not accepted, then the same displacement is attempted with the module rotated by 90 degrees. If this new move is also rejected, then the orientation of the module is changed randomly choosing among the eight Manhattan transforms. Whether one of these moves is accepted or not, for a flexible module, the next moves attempted pertain to a change in aspect ratio and pin positions.

If the selected move class is pairwise interchange, two modules are selected randomly and their positions are interchanged. If the move is rejected, then the same interchange is tried with the modules both rotated by 90 degrees.

The ratio of the number of attempted moves that fall in each of the classes is important to obtain good results. Unfortunately there are no theoretical results to back up a particular choice: a number of experiments are attempted with different ratios to identify the "best" value of this parameter.

Some of the moves may create overlap of modules, clearly unacceptable in the final solution to the problem. The moves that create overlaps could be discarded at the onset, however it has been experimentally observed that allowing intermediate stages of the simulated annealing process to accept overlaps, yields better final layouts than discarding such moves. It is therefore important to include in the cost function a component that discourages overlaps from being present at the end of the optimization process.

4.5.2. The Cost Function

As pointed out above, a cost function that reflects accurately the area used and the net length is very important to yield good results. Because of the characteristics of the simulated annealing algorithm, hundreds of thousands or even millions moves have to be tried to obtain reasonable results. For this reason, it is crucial to keep the cost function as simple to calculate as possible.

Unfortunately, it is very difficult to compute the precise amount of interconnect space needed along each edge of a cell without performing a global routing step. Furthermore, it is even more difficult to compute the interconnect space during an intermediate point of a simulated annealing approach where cells may overlap. This makes a channel definition procedure very difficult or even impossible to perform. Thus, while it would be best to maintain a global routing of the placement at any given point in the simulated annealing process, such a procedure is very difficult at best, and surely very cpu-time intensive.

Methods have been introduced [ELG81] [HEL77] which appear to give good results for an accurate interconnect-area estimation for the macro-cell placement problem. However, due to their computational complexity, these methods appear not to be suitable for simulated annealing.

The new approach introduced in [SEC86] estimates the interconnect area needed for each edge of a cell. The estimate is based on three factors:

(i) The *average net traffic*. This is an estimate of the average number of interconnections passing through a channel.

(ii) The relative position of the edge on the chip (channels in the center of the chip tend to be wider than those near the boundary of the chip). Based on an analysis of industrial macro-cell chips, the average channel widths appeared to increase approximately linearly from the chip edges. Furthermore, it appeared that, on average, the center channels were approximately twice as wide as channels located near the chip edge.

(iii) The relative pin density of edge i. The *pin density* of a cell edge i, represented by d_p^i, is defined as the total number of pins belonging to edge i divided by the length of edge i. The *average pin density* for the entire circuit is represented by D_p. This is computed by dividing the total number of pins (for all of the cells) by the sum of the perimeters of all the cells. It is then possible to define the *relative pin density* of a cell edge i as d_{rp}^i, where

$$d_{rp}^i = \frac{d_p^i}{D_p}$$

Note that factor (ii) above must be updated each time a cell is moved during the simulated annealing process. The cpu time required to update this factor is minimal. Factors (i) and (ii) can be determined at the outset and stored.

In the simulated annealing approach used in TimberWolfMC, sufficient interconnect space between the cells is maintained by appending a border around the contours of each cell. Each time the amount of overlap (if any) is updated for two cells, the edges of a cell are expanded outward. The respective outward expansions of the tile edges are updated each time a cell participates in the generation of a new state in the simulated annealing approach. Hence the estimation of the interconnect area attributable to a given edge of a cell is a *dynamic* quantity, depending on the location of the edge at any given time. It follows that if a cell is moved from a corner of the chip area towards the center of the chip area, then the effective area of the cell will increase. Similarly, if a cell is moved from the center towards a corner, then the effective area of the cell will decrease.

Note that this strategy could be used in general by other floor-planning and placement techniques. I believe that this approach is quite powerful and should be further investigated with particular attention devoted to results aimed at bounding the average and worst-case errors.

The cost function for TimberWolfMC consists of three independent terms.

The first term is a function providing an estimate of the total net length. The following definitions are used: (1) N_n is the total number of nets, (2) $X_s(n)$ is the span of net n in the x, or horizontal direction, (3) $Y_s(n)$ is the span of net n in the y, or vertical direction, (4) $H_w(n)$ is the weighting factor for net n for the horizontal direction, and (5) $V_w(n)$ is the weighting factor for net n for the vertical direction. The function $C_1()$ is then given by:

$$C_1() = \sum_{n=1}^{N_n} \left[X_s(n)H_w(n) + Y_s(n)V_w(n) \right]$$

The weighting factors are used to capture different electrical characteristic of interconnection layers as well as the possibility of routing over the cells if two layers of metal interconnections are available.

The second term is the overlap penalty function. The overlaps are measured with respect to the "inflated" area provided by the dynamic routing area estimator presented above.

The third term of the cost function is related to the way in which the pin positions are modified and is of no general interest; hence its description is omitted.

These terms are multiplied by a weighting factor that represents the trade-offs between them. If the factor for overlap is chosen too large, then the simulated annealing algorithm will primarily be concerned with the minimization of the overlap penalty function. That is, relatively little attention will be paid to the quality of the solution, in favor of its feasibility. minimization of the total estimated wire length. On the other hand, if the initial average values of the overlap penalty are substantially smaller than the average values of the estimated net length then the simulated annealing algorithm will pay relatively little attention to the elimination of the cell overlaps, until T is relatively small. At this point, *most* of the attention will then be focused on the removal of the infeasibilities heretofore ignored, with the accompanying increase in the estimated net length.

In this approach, area is minimized by minimizing estimated net length, but a fairly accurate estimate of routing area is provided to make sure that the final placement obtained by the program would not be modified extensively after performing the routing steps.

For the macro-cell design style, it is often important to achieve a particular target area or aspect ratio for the final chip. TimberWolfMC uses the dynamic interconnect-area estimation algorithm to determine the expected interconnect area required for the chip and hence the expected area of the chip. During the course of the simulated annealing algorithm, the cells will be discouraged from extending beyond the expected chip boundaries. If the target chip area is too large, the final aspect ratio generated by the placement algorithm may be quite far from the desired value. On the other hand, if the target chip area is too small, either insufficient interconnection space will be allotted between some of the cells, or cases of cell overlap will be present. For this reason, an accurate area estimate is important.

An alternative to this approach is to add a new term to the cost function proportional to the area of the smallest rectangle that contains all the "inflated" macro-cells for an intermediate placement provided by the simulated annealing algorithm. The smallest rectangle containing all the macro-cells can be computed quickly and gives a very good estimate of the final area occupied by the components of the design. Deviations from desired aspect ratio and/or area can be penalized. This is the cost function used in *Puppy,* the multiprocessor version of the macro-cell placement program presented in [CAS86].

4.5.3. Refinement Steps

Even though the routing area estimate in TimberWolfMC is quite accurate, it does not guarantee that adjustments are not necessary after the routing steps. For this reason, a second stage is provided to produce final results that are satisfactory in terms of the goals set up above.

The second stage of TimberWolfMC consists of several executions of a placement-refinement algorithm. Each execution consists of three steps: (1) a channel definition step, (2) a global routing step, and (3) a placement-refinement step. The quality of the results obtained by stage 1 of the algorithm, is demonstrated by the small changes that have been experimentally observed during the refinement steps.

Simulated annealing with a low starting temperature is used to accomplish the placement refinement step. The placement of the cells is refined to reflect the required interconnect area, as determined by the channel definition and global routing steps. These steps determine the required interconnect area of every channel. Since exactly two cell edges border each TimberWolfMC channel, the spacing requirement between the two cell edges is immediately established. One half of the interconnect space is associated with each of the two cell edges. That is, each edge is expanded outward (from the interior of its cell) by an amount equal to one-half of the required channel width. This estimate is computed only once and is not updated during the refinement steps. The expanded cell edges are used each time the amount of overlap (if any) is updated for two cells.

The same objective function as in stage 1 of TimberWolfMC, described above, is used. The function *generate* () has a much simpler form for stage 2. New states can only be generated by attempting single cell displacements and by alterations of the pin placement. The orientations of the cells are not changed during stage 2. The aspect ratios of the cells also remain fixed during the second stage. This is because these two methods of new-state generation tend to invalidate the calculation of the interconnect area, which would require more placement-refinement steps to obtain convergence in the final chip area and the final total estimated interconnection length.

4.5.4. Experimental Results and Comparisons

The initial temperature, the annealing schedule and the stopping criterion are all determined experimentally. C. Sechen has run a number of examples and has tuned the parameters accordingly. Even though I believe that these parameters are better determined with an adaptive strategy monitoring the statistics of the moves and of the cost function [OTT84], [HUA86], the present strategy has produced good results.

The performance of TimberWolfMC was compared against various industrial, university, and manual placement methods. The two criteria were the net-length and the final chip area. The results for 9 industrial circuits of various sizes are summarized in Table 4.5.1. Note that all the circuits had modules that were completely specified, so that the floor-planning features of TimberWolfMC are not relevant in these tests. All the results obtained satisfied the constraints given to us.

The 26 percent reduction in total estimated net-length and the 14 percent chip area reduction achieved for circuit i1 was in comparison to a placement method based on resistive network optimization described in Section 4.4.4 [CHE84]. Circuits i2 and i3 were in comparison to the CipAR placement and routing package developed by Gould-AMI, based on force-directed heuristics.. The 19 percent total interconnect length reduction and the 50 percent chip area reduction was in comparison to a manual layout performed at Intel Corp. Circuit p1 was in comparison to a manual layout at Hewlett-Packard. Circuit d2 was in comparison to a manual layout at Advanced Micro Devices. Circuits d1 and d3 are also in comparison to manual layouts. The cpu time ranged from 15 minutes for the smallest circuits to 4 hours for the largest circuits on a DEC MicroVax II.

Ckt	#Cells/ #Pads	#Nets	#Pins	Total Inter- connect	Area ($<x>$ by $<y>$)	Percent Inter- connect Reduction	Percent Area Reduction
i1	33/38	121	452	7431	$<236>$ $<223>$	26	14
p1	11/45	83	309	12306	$<293>$ $<294>$	8	18
x1	10/2	267	762	60326	$<875>$ $<744>$	11	15
i2	23/17	127	577	121386	$<2873>$ $<2751>$	49	†
i3	18/10	38	102	7043	$<644>$ $<699>$	46	56
l1	62/0	570	4309	254063	$<1084>$ $<1042>$	19	50
d2	20/139	656	1776	419608	$<1355>$ $<1433>$	13	4
d1	17/0	288	837	37365	$<245>$ $<305>$	23	†
d3	17/5	136	665	325457	$<3398>$ $<3298>$	†	31
avg.						24.4	26.9

† Comparative Result Not Available

Table 4.5.1. TimberWolfMC Performance Comparisons Versus Industrial, University, and Manual Placements.

This simulated annealing approach has great potential for generating excellent solutions because of the many degrees of freedom that are fully taken into consideration by the algorithm. On the other hand, the many degrees of freedom are also responsible for the explosion of the dimensions of the configuration space that is searched by the algorithm, and hence for the long running times.

The simulated annealing approach of Wong and Liu [WON86] restricts the placements to be searched to slicing structures, thus decreasing the number of configurations, and uses manipulations of the decomposition tree to generate new configurations fast. However, slicing structures may not correspond to the best area utilization, as pointed out in [LAU80]. In addition, non rectangular blocks and variable pin-positions cannot be handled, at least in this version of the algorithm.

Otten's approach partitions the floor-planning problem into two phases: relative placement and shape determination [OTT85]. In the relative placement phase, simulated annealing is used to place dimensionless points in the plane, thus eliminating all the configurations that correspond to different shapes and pin positions. By restricting the configuration space, simulated annealing becomes quite fast. However, shape and pin-positions are only considered as a post-processing step and hence the final result may not be as good as the one that can be obtained with TimberWolf.

4.6. The Rectangular Dualization Method

The rectangular dualization method has received great deal of attention in the past five years, e.g., [HEL82], [KOZ84], [LEI84], [TSU86]. This method was proposed first by Heller [HEL82] and is based on a formulation of the floor-planning and placement problem similar to the one used in architectural design [GRA70].

The cost function used in this approach is qualitatively related to the cost functions introduced in Section 4.1. The goal of the rectangular dualization method is to dissect the plane so that all modules connected by a net are adjacent. In some cases, it is possible to find a dissection which satisfies this condition, but in general it cannot be done. For example, the modules shown in the diagram of Figure 4.6.1a cannot be placed so that the adjacency condition is satisfied.

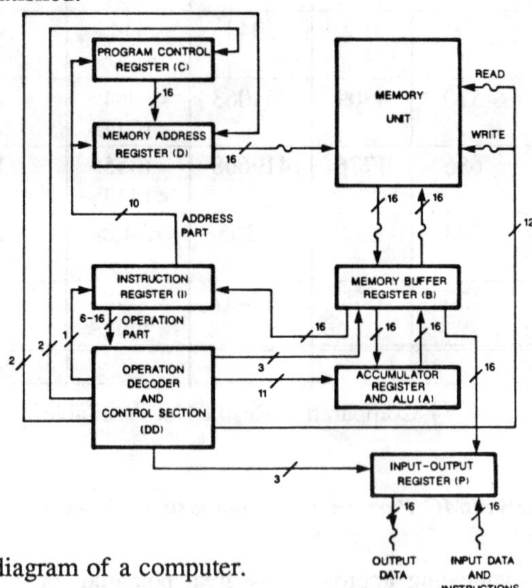

Figure 4.6.1a. Block diagram of a computer.

The rectangular dualization method checks whether it is possible to dissect the layout region so that the shape constraints, as defined in Section 4.2.2, and the adjacency conditions, i.e., modules that are connected should be adjacent on the plane, are satisfied, and, if it is not, finds a dissection that minimizes the number of modules that violate the adjacency condition. Since the pioneering work of Heller and colleagues, several new theoretical results have been found that help fully characterize the steps of the procedure. The rectangular dualization method uses novel representations of a rectangular dissection.

The data of the placement and floor-planning problem are a set of modules, the set of their interconnections and the set of shape constraints. The set of modules and their interconnections can be represented in compact form with an undirected graph, $G_P(V_P,E_P)$, where the nodes are in one to one correspondence with the modules and there is an edge $\{v_i,v_j\}$ in E_P if there is a net which interconnects the modules corresponding to v_i and v_j. Figure 4.6.1b shows the graph for the set of modules of Figure 4.6.1a, where the memory unit is assumed to be outside the boundaries of the chip that is being floor-planned. Note that this gives a complete sub-graph for each net, unless all the nets are reduced to two point nets first.

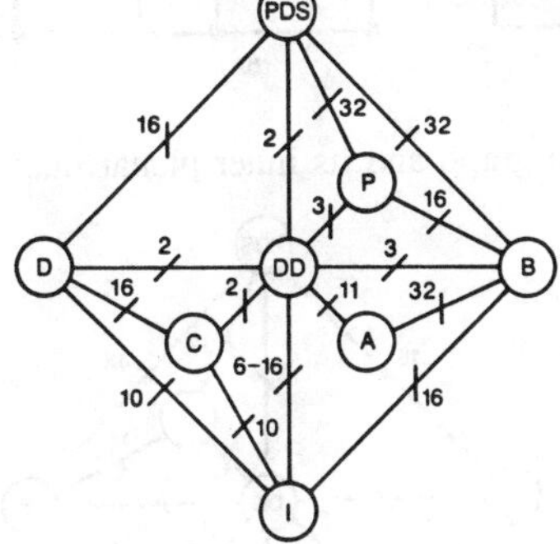

Figure 4.6.1b Graph for the block diagram.

A rectangular dissection can be represented by the vertical and horizontal polar graph described in Section 4.2.1. It can also be represented by the *floor-plan graph, $G(V,E)$*. The floor-plan graph can be obtained trivially from the rectangular dissection as follows. The intersections of the boundaries of the rectangles of the dissection can be divided into just two types, as shown in Figure 4.6.2a, a "T" type and a cross type. Since all the cross type intersections can be represented by two "T" type intersections as shown in Figure 4.6.2b it is assumed that all the intersections are of "T" type. The choice of representation of the cross intersections with "T" type structure has little effect on the steps that follow in the algorithm. Then a node of the floor-plan graph G is in one-to-one correspondence with an intersection. An edge $\{v_i,v_j\}$ belongs to the graph if the corresponding intersections are adjacent. Because of the assumption that all the intersections are of type "T", the degree of each node is three, except for the four corners which have degree two. Note that the floor-plan graph is indeed a very close representation of the rectangular dissection (Figure 4.6.3b is the floor-plan graph of the rectangular dissection of Figure 4.6.3a). In [DAI86], the regions of the dissections are called *rooms*. Rooms are in one-to-one correspondence with the *inner faces*[5] [HAR69] of the floor-plan graph.

---

[5] A planar graph embedded in the plane defines a set of regions of the plane. For the floor-plan graph it is possible to show that the regions of the plane have as boundaries cycles of the graph. These regions are called inner faces if they are bounded. The one region that is unbounded is called exterior face.

166

The *inner dual graph* $\hat{G}(\hat{V},\hat{E})$ represents the adjacency relations of rooms in the floor-plan graph and is shown in Figure 4.6.4. This graph is the planar dual of G except

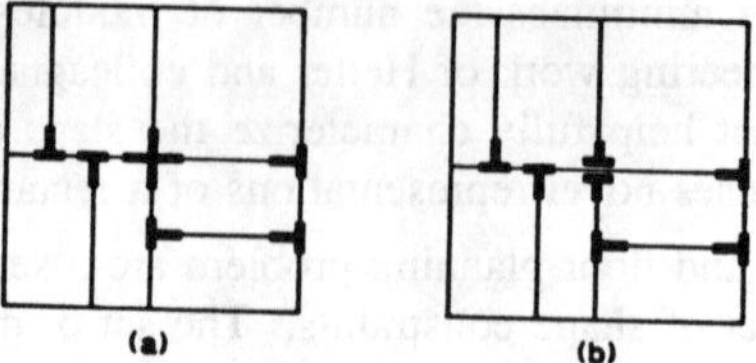

Figure 4.6.2. Converting a cross type intersection.

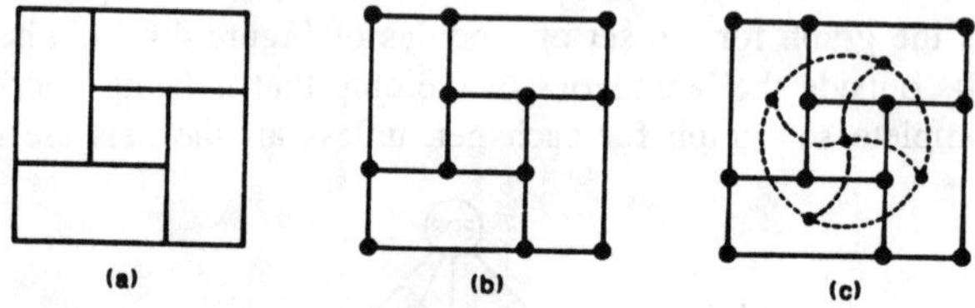

Figure 4.6.3. Floor-plan graph and its inner planar dual.

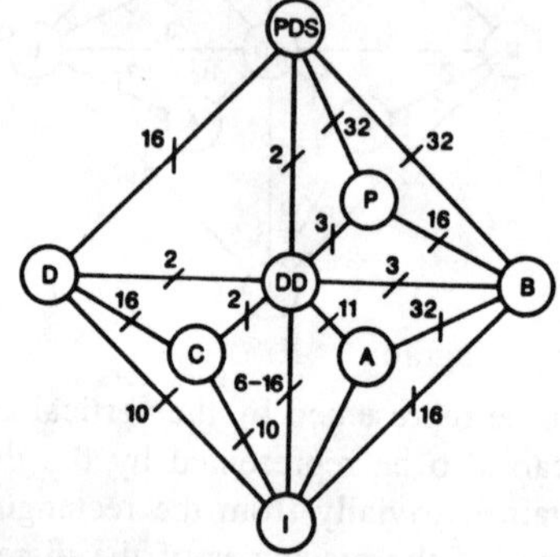

Figure 4.6.4. A planar triangulated graph with complex triangles.

that the node corresponding to the outer face has been eliminated, hence the name inner dual graph. Because of the fact that each node in G has degree three except for the corners, the inner dual graph is a plane triangulated graph, i.e., it can be embedded in the plane and all its faces are triangles.

It is clear that the inner dual graph represents the adjacency relations of modules placed in the rooms described by the rectangular dissection. Now, if the graph of the problem G_P is a subgraph of G, i.e., $E \subseteq E_P$, then all the modules which are interconnected by a net are adjacent.

Thus, given a dissection, it is possible to verify whether it satisfies the adjacency condition for a given problem. However, we wish to obtain a dissection from the graph of the problem. Hence, the key question is whether the given graph of the problem is the inner dual of a floor-plan graph.

As seen above, a necessary condition is that G_P be plane triangulated. However, not all plane triangulated graphs have a floor-plan graph dual. A plane triangulated graph that is the inner dual of a floor-plan graph is said to have a *rectangular dual*. A necessary and sufficient condition for a plane triangulated graph to have a rectangular dual is that no *complex triangles* are present. A complex triangle is a cycle of length three which is not an inner face. The plane triangulated graph shown in Figure 4.6.4, obtained by adding an arc to the graph of Figure 4.6.1b, does not satisfy the condition and indeed there is not way of placing the modules so that the connected modules are adjacent.

Given a problem graph G_P which does not have a rectangular dual, a dissection can be obtained by a sequence of steps. First, G_P must be planarized if it is not planar. Planarization is achieved by introducing new nodes in the graph where two or more edges cross each other. It is important to minimize the number of nodes introduced in the graph, since each of these nodes corresponds to a set of modules that, once placed, will not be adjacent even though they are interconnected. The problem of finding the minimum number of edge crossings in a nonplanar graph has been shown to be NP-complete. Hence, heuristics must be used to carry out this step.

Once a planar graph is obtained, edges have to be added to make it triangulated. This problem has been studied in connection with sparse matrix algorithms [ROS71]. The problem of finding the minimum number of edges so that the graph is triangulated, is NP-hard; however, finding some set of edges that makes the graph triangulated can be done in linear time. In this case, it is important to minimize the number of complex cycles that appear in the graph, a very difficult criterion that can be heuristically approximated by trying to add the minimum number of edges.

Next, complex triangles must be eliminated to yield a graph which satisfies the necessary and sufficient condition stated above. No polynomial time algorithm is known to find the minimum number of edges that need to be removed so that the remaining graph has no complex cycles. However, it is not known whether this problem is NP-hard (even though, given its similarity to a covering problem, I conjecture it is). A heuristic algorithm has been proposed for this task [TSU86].

Once the complex cycles have been eliminated, a floor-plan graph can be obtained via a dualization procedure. Note however, that many different floor-plan graphs correspond to the same graph.

In [MAL82], a three-step algorithm based on linear programming is proposed to determine a floor-plan graph and to select the actual dimension of the layout so that the perimeter and area constraints are satisfied. In the first step, all the floor-plans corresponding to the planar triangulated graph obtained by the steps described above, are enumerated by solving a set of linear integer inequalities. The second step determines all the solutions that satisfy the perimeter constraints and minimizes the area of the overall layout. This mathematical programming problem cannot be solved easily and it is broken into two separate linear programs, whose solutions yields an approximated solution to the problem. Next, the area constraints are introduced by means of a simplex method which determines the amount of deformation and sliding that are necessary to modify the dissection so that all the modules satisfy the shape constraints.

In alternative to this procedure, the shape determination algorithm could be used, once a floor-plan graph is examined, since the floor-plan graph gives the relative positions of the modules.

The idea is elegant mathematically, but the procedure clearly consists of several complex steps. In fact, Heller et al. [HEL82] suggest to cluster modules into larger modules so that this algorithm could be applied in a hierarchical way to minimize its complexity. In addition, many heuristic decisions are taken along the way. A solution that is far from the optimum may be obtained. Equally important, the original model, forcing connected macros to be adjacent, fails to reflect well other factors such as "packing" area important for a good floor-plan.

4.7. Combining Global Routing and Placement

In all the methods presented above, the determination of the routing area was either done *a posteriori* or an estimate was computed to guide the placement. Recently, attention has been focussed on methods that combine the global routing phase and the placement phase to take the routing area into account better.

The floor-planning approach advocated by Otten and followed by Mason can be considered as a three-phase procedure: first, the modules are partitioned top-down either by using the min-cut algorithm or simulated annealing followed by a slicing structure determination step; then, the decomposition tree is traversed bottom-up with the shape-determination phase; finally, the tree is traversed top-down again to transmit the final choice on the shape of the blocks.

Dai and Kuh have proposed a dual approach to floor-planning. In this approach, in a first bottom-up phase, modules are hierarchically clustered according to their size and connectivity. The number of components at each level of the hierarchy is fixed to a maximum of five. At the last step of the clustering procedure, the chip has at most five components, where each of the components contains clusters formed at previous steps. Each of the clusters carries information about shape and aspect ratios, obtained by combining the information available at the lower levels. In the second phase a top-down approach is followed, where the blocks in the cluster are placed and globally routed.

Since the number of components in each cluster is small, many possible relative positions and shapes can be exhaustively examined. When deciding on the number of components that are examined exhaustively, two opposing considerations have to be taken into account. First, the number of components assembled in clusters has to be small enough to allow an exhaustive examination of the possible floor-plans. Second, if the number of components is too small, then the overall floor-plan may not be satisfactory because, during optimization, a large number of possible solutions are not taken into account.

Clusters of five components have been suggested in [DAI86] because the number of different floor-plans of five components is ninety two, a manageable number, and five is the smallest number of components such that at least one of the floor-plans is not a slicing structure. Thus it is possible that the final layout be a non-slicing structure. This is desirable, since there are examples in which limiting the possible floor-plans to slicing structures generates a fair amount of wasted space. (This was one of the motivations in building the simulated annealing approach used in TimberWolf).

Since the top-down phase deals with only five components, not only may all possible floor-plans be exhaustively examined, but also the routing area can be estimated with great accuracy.

In [DAI86], during the clustering step, the net list is modified to carry the information of the connectivity among clusters. When the top-down stage is started at the top level of the hierarchy, all different floor-plans are enumerated and evaluated. The evaluation is carried out by performing a global routing step for each of the floor-plans. The global routing could be done with one of the available heuristics for the rectilinear Steiner tree problem. However, if the floor-plans under considerations are slightly constrained, it can be shown [DAI86] that the Steiner tree problem becomes easy: it can be solved optimally in linear time. Hence, the global routing step is performed exactly for each of the floor-plans considered.

The global routing step at a level of the top-down hierarchy is used to give the terminal positions at the level below. At the end of the traversal of the cluster tree, a complete floor-plan is obtained.

This procedure is quite novel and interesting. Drawbacks are the limitation of the cluster size to five blocks and the clustering algorithm used.

The basic idea here is to simplify the problem enough to be able to run exact algorithms on it. The hierarchical decomposition achieves this goal. More conventional heuristics deal with the full complexity of the problem and find an approximate solution to it. It remains to be seen whether this approach is substantially better than the more conventional heuristics. Preliminary experiments show that the method is very fast and that the results obtained are close in quality to the ones obtained with a simulated annealing approach but not as good. Extensive testing and tuning of the algorithm is needed to explore the full potential of the method.

5. ROUTING

Much work has been done in routing over the past twenty years. Many approaches have been proposed and several have been widely used for the design of integrated circuits. However, many open problems remain to be solved. In particular, new challenges are offered by an ever evolving technology, one outstanding example being multi-layer routing. In this section, I will simply outline the basic problems encountered in routing and mention the most successful techniques. The readers are referred to [OHT86] for several detailed overview papers on the topic.

After the modules have been placed, the interconnections have to be completed in the available space. If routing over the cell is not allowed by the technology or by the design style, the area available for interconnections is that which is not occupied by the modules. In the gate-array and standard cell case, this area is fairly regular, consisting of a collection of rectangular regions alternating with rows or columns of cells. In the case of macro cell design style, the routing area is much less regular and can be "organized" in many different ways.

In general, it is convenient to decompose the routing region into rectangular regions, called channels. The way in which the routing region is subdivided has an effect on the complexity of the problem. This decomposition is called *channel definition* or *routing region definition* stage.

The routing stage where the interconnections are laid out on the chip follows. This stage is in general broken down into two stages: *global* or *loose routing* and *detailed routing*. The global routing stage, sometimes called channel assignment, determines which channels each interconnection will go through. The detailed routing stage determines the actual physical location of the interconnections inside the routing regions. If the region to be routed contains pins on two sides only, then effective detailed routing tools called *channel routers* can be used. Extensions to the basic tools can provide programs that can route channels with pins on three sides [BRA86b] If the routing region has pins on four sides, then a switch box router can be used. In general, channel routers give the best results in terms of the area used to complete the interconnections. Thus routing regions with fixed pins on more than two sides should be avoided. Channel definition and the order in which the channels are routed have a great impact on this issue.

Since global routing and routing region definition depend critically on the routing strategy followed in the detailed routing stage, we review first the work in detailed routing. Then we present channel definition and ordering, and finally global routing.

5.1. Detailed Routing

Given a region with pins on its sides and, possibly, in the middle, detailed routing is the process of implementing the actual geometries of the interconnections among the pins specified by a net list. In the most general case, the regions may be of irregular shape with internal obstructions. However, the most effective algorithms work on regions of regular shape, in general rectangular or close to rectangular, with no obstructions, with fixed pins on two opposite sides and floating pins on the other two sides.

Algorithms for detailed routing can be classified according to the region they can be applied to. The most general routers are called *2-dimensional* or *switch-box routers*. They can route regions with fixed pins on all sides of the region. More sophisticated switch-box routers can handle irregular regions, pins inside the routing region in addition to pins on the boundaries of the region and obstacles in the routing layers. *Channel routers* handle regular regions with fixed pins on two sides of the regions and floating pins on the other two sides.

5.1.1. Switch-box Routing

The *Lee-Moore maze* algorithm also called the grid expansion algorithm [LEE61] has provided the basis for most of the algorithms for switch-box routing.

This algorithm is applied to the interconnection nets, one at a time, on a region where a grid has been superimposed. The grid specifies intermediate locations which can be reached by an interconnection while it is being built. In general, it is assumed that the interconnections have Manhattan geometry. At each point of the grid, the interconnection may change direction.

This router has been applied to gate-array and macro-cell design. Its strengths are its flexibility (it can be applied to irregular regions with pins distributed everywhere and with obstacles and it generates paths with minimum number of bends) and its capability of finding a solution, if one exists; its weakness, besides the running time, is the dependence on net ordering. In fact, the first nets to be routed have a large basically empty region to use, while the last nets to be routed find the region almost full. If the nets are chosen in the wrong order, the last nets may not be routable due to the blockages created by the previously routed nets. Several heuristics are available to speed up this algorithm as well as to choose a good net ordering.

Iterating through the nets, ripping them up and re-routing them, can also result in a significant increase in the power of the method. Recently Ohtsuki et. al [OHT86] reported that Lee algorithm is most useful and powerful when applied to interactive rip-up and reroute. However, very few published results are available, on automatic and general routing modification and re-routing for switch-box routing.

Expert system technology has been applied to the switch-box routing problem to devise an "intelligent" choice of nets to route first and of paths for the nets to avoid early blockages. Weaver, [JOO86] a Knowledge-Based Expert Systems (KBES) router, includes many cooperating experts and several hundred rules to guide the routing. The quality of the routing obtained by this tool was good; however, the running time was quite high (more than 20 minutes for a fairly small problem). Contrary to popular belief about expert systems, according to its author, the tool took along time to develop due to the difficulties in extracting meaningful rules from human experts and in coordinating the work of the various "experts" present in the program.

A recently developed router, Mighty [SHI86b], is based on an incremental routing technique. The novel part of the algorithm is in the modification of connections already implemented to allow blocked or "bad" quality connections to find a better implementation. These modifications are of two kinds: a weak modification step pushes aside existing connections without removing them to make room for the blocked connections, and a strong modification step, invoked when weak modification fails, removes blocking connections.

The basic algorithm consists of four main parts: a path finder that computes the shortest path among two subnets of the same net (a pin is a special case of a subnet), a path conformer that implements a particular path proposed by the path finder, a weak modifier that pushes existing interconnections to make better connections and a strong modifier that removes subnets from the routing region to allow the completion of the routing for a blocked net.

In the path finding phase, the nets are processed in the order they are entered. From each pin of a net, maze routing is used to find the minimum cost path that interconnects two of the pins of the net. The cost used to direct the search is based on the following wiring model. A layer is mainly used for horizontal interconnections and the other for vertical interconnections. (For example, extension of a path in the vertical direction on the horizontal layer, while possible, is penalized). Changing a layer is also penalized to minimize the number of vias.

As soon as a path connecting two pins of the net is found the search is stopped. The path connecting the two pins is recorded on an ordered list organized in increasing cost. Note that the path has not been selected and implemented yet: it has been simply proposed by the path finder.

When all the nets have been processed by the path-finder, the path-conformer takes over. The paths on the ordered list are popped and examined. If the path is feasible, i.e. if no other interconnection has occupied the area needed to implement it, the path is selected. Otherwise the path finder is invoked again and a feasible minimum cost path connecting any two unconnected subnets of the net is sought. Note that when the shortest path is computed, the interconnections laid out already are taken into consideration. If a path is found, then it is examined. If its cost is within a prescribed limit from the ideal minimum cost path that would have been found if no other net was occupying the routing region, then it is inserted in the ordered list. Otherwise, the modification phase is entered.

The weak modifier is first called to push other nets around to make a feasible and "good" connection possible for the net under consideration. If no solution is found, then the strong modification phase is entered. Both modifications imply a modification of the existing interconnections - some interconnections have to be either pushed away or removed. A variety of alternative "good" interconnections for the net under consideration are examined and the minimum cost solution is selected. If such solution is above a preset limit, then the router ends its search with a failure. The process is iterated until either a solution is found or a failure is reported.

This router was tested against existing switch-box routers on a difficult problem described in the literature: the Burstein difficult switch-box (Table 5.1). For this example, Mighty obtained the best results in terms of area and number of vias, while it obtained a routing that was slightly worse than Weaver with respect to total net length. MIGHTY takes 2.7 seconds to complete the connections and the post processing takes additional 1.3 seconds on a DEC VAX 11/785 running 4.3BSD UNIX. Note that WEAVER took 1,390.0 seconds to route the same example.

Router Name	#rows	#columns	#vias	total length
Hamachi	15	23	67	564
Luk	16	23	58	577
Mod Detour	15	23	63	567
Marek	15	23	58	560
WEAVER	15	23	41	531
MIGHTY	15	22	39	541

Table 5.1 : Routing of Burstein's difficult switch-box

Another algorithm has been used quite often for switch-box routing: the Hightower algorithm. This algorithm is gridless in principle. It starts from both pins to be connected and generates an horizontal segment and a vertical one of maximum extension from the pins. Once these four lines are generated, the orthogonal lines of maximum expansion are generated next. (If more than one orthogonal line can be found of the equal maximum extension, the one which is closer to the opposite pin is selected.) This procedure is iterated until two lines expanded from the two pins to be connected intersect. The actual interconnection pattern is then constructed by tracing back the lines at their intersection points.

Note that the algorithm can be quite fast for simple mazes with a small number of barriers and obstructions, while it may be slow for complicated regions, because of the many lines that can be generated before an intersection is found. In addition, it is not guaranteed to produce a solution, even if one exists.

5.1.2. Channel Routing

Hashimoto and Stevens proposed for the first time to use channel routing for IC design. Their contribution is two-fold: i) the abstraction of a routing problem which is simpler to solve than the general problem stated previously; ii) an algorithm to solve the simplified routing problem. Many extensions and improvements have been made to the original algorithms, but the concept of solving the routing problem by "carving" simple routing regions has permeated routing packages for all design styles for many years.

The basic assumptions in the original formulation of the channel routing problem by Hashimoto and Stevens are:

(i)   the routing region is rectangular with no obstructions and with pins on two opposite sides;

(ii)   floating pins on the other two sides of the rectangular region are possible. These floating pins indicate the need to extend some of the nets outside the channel;

(iii)   there are only two layers available for interconnections; all the horizontal segments of the nets are routed on one layer, all the vertical segments on the other;

(iv)   the pins are placed on a regular grid;

(v)   the channel is subdivided into rows or tracks whose spacing is such that interconnections placed on these tracks cannot violate any design rule.

The goal of channel routing is to complete all the interconnections in the minimum number of tracks.

Some of the restrictions have been removed, often exacting a cost in the quality of the solution.

The allowable configuration of each net is either (i) a horizontal segment in one layer, which is connected to the top and bottom pins of the net by vertical segments in the other layer, to minimize the number of changes of layers or (ii) a set of horizontal segments joined by vertical segments (doglegs).

If we define as the *density* of a channel the maximum number of nets crossing any one column of the channel, then the best we can do in accordance to the assumptions above, is to route the channel with a number of tracks equal to the density of the channel. Interestingly, most of the best channel routers available today (Yoshimura and Kuh [YOS82], Rivest[RIV82b], Burstein[BUR83] and YACR[SAN84]) behave well, i.e., in most cases they route channels in a number of tracks which is within one or two of density. In the case of YACR, the wiring model is slightly more general since some of the interconnections may have horizontal segments placed in the "vertical" layer and vice versa. This flexibility makes it easier to obtain a compact routing.

Some of the algorithms described above can be generalized to route a switch-box, for example, Rivest's algorithm has been modified in two different ways to handle switch-box regions [LUK85], [HAM84].

Note that many of the algorithms described here for detailed routing use symbolic data; for example, YACR places horizontal segments with no vertical dimensions and with no information about contact sizes. A post-processor is used to replace the symbolic data with actual geometries. This post-processor is intelligent, i.e., it can change some of the interconnections to maximize the use of the layer with better electrical characteristics and to minimize the number of contact vias.

This approach has the great advantage of keeping the algorithms for routing and their implementation clean, even though it may be possible to improve the quality of the final routing by embedding knowledge about the technology used in the algorithms themselves. I strongly believe that this additional optimization is not worth the effort and may create serious problems for maintenance and extensions to different technologies.

5.2. Routing Region Definition and Ordering

As pointed out above, channel routers are the most effective detailed routers in terms of their efficiency and the compactness of the layout they can generate. Hence, the basic goal of the routing region definition and ordering schemes developed thus far is to decompose the routing area and decide in which order the regions should be routed so as to use channel routers as much as possible.

There are two basic requirements for a routing region to be considered a channel: (i) all the pins on two opposite sides have to be fixed while the pins on the other two sides have to be floating, (ii) once a channel has been routed, only the distance between the two sides with fixed pins may be changed and it may only be increased. This second constraint, also called *rigidity constraint,* is introduced to avoid re-routing a channel which has already been routed. Note that increasing the relative distance of the two opposite sides does not change the difficult part of routing. In fact, updating of the interconnections after such a move amounts simply to extending the vertical segments connecting the horizontal segments to the pins.

Assuming that all the modules are rectilinear (not necessarily rectangular), the routing region can always be decomposed into rectangles. This step can be carried out quite efficiently by using sorting techniques on geometries as commonly done in computational geometry.

The most rigorous approach to channel definition and ordering is followed by Dai et al. in [DAI85]. This approach can also be used as a framework for the work of others. Here, routing regions are represented by *walls*. Walls can meet orthogonally with either a "T" shape connection or a cross connection as shown in Figure 5.2.1a. Cross connections can be represented by pairs of "T" connections in two different ways as shown in Figure 5.2.1b. Assume that one of these representations is chosen; then cross connections can be ignored when describing the algorithms. However, note that the two representations are not equivalent and there are cases where one is better than the other as will be clear later.

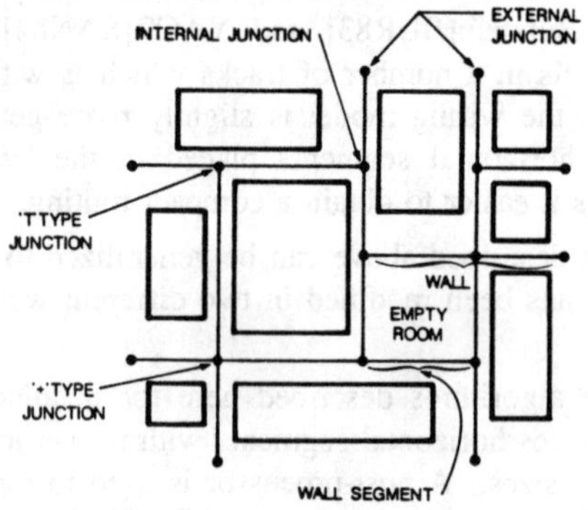

Figure 5.2.1a. A placement and the channel graph with T type and cross (+) type junctions.

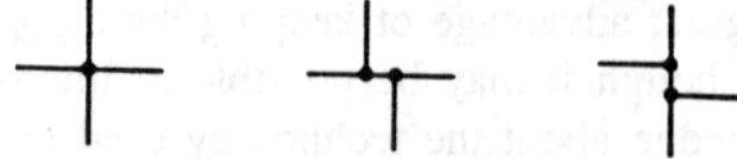

Figure 5.2.1b. Converting a cross junction into T type junctions.

If each routing region represented by a wall is to be considered as a channel, the constraints introduced above induce an ordering relation on the walls. In particular, a wall which is the "vertical" part of a "T" connection, corresponds to a channel that must be routed before the one which corresponds to the "horizontal" part of the junction. This is because if we route the region corresponding to the "horizontal" wall in the "T" connections first, we would have a set of pins whose position is not specified, on one of the sides where fixed pins should be located If the "vertical" part is routed first, then the channel router specifies, when it terminates, the exact locations of the otherwise floating pins.

Thus those "T"s induce a precedence relation among channels. If such a relation is acyclic, i.e., there is no "conflict", then there exists an ordering of the routing regions so that each of the channels considered in the sequence has fixed pins on two opposite sides and floating pins on the other two sides.

Of the two possible representations of a cross connection with a "T" connection, one may yield a cyclic relation, while the other may not. In Figure 5.2.2 such a case is shown. In Figure 5.2.2b a cyclic relation among channels is present, while in Figure

5.2.2c, where a different representation for the cross connection is chosen, the relation is acyclic. Hence, an optimization problem can be defined where the variables are the choices of representations for the cross connections and the objective function is the minimization of the number of cyclic relations among channels. In [DAI85], this problem is not described and the proposed algorithm chooses randomly one of the two possible representations.

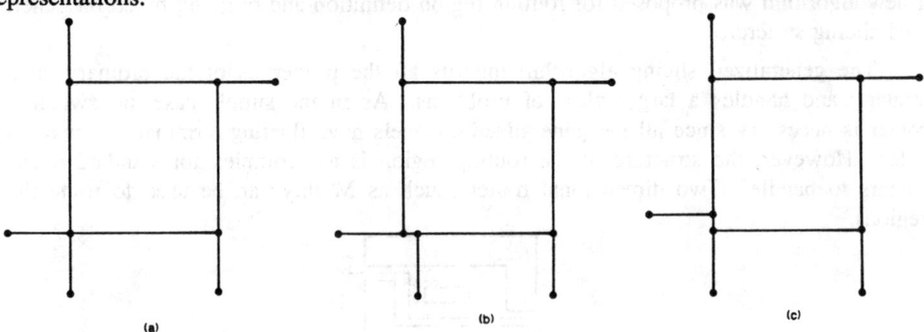

Figure 5.2.2a, b & c

The placement algorithms yielding a slicing structure have the nice property that the relation defined on the walls is always acyclic, and they even provide the routing order One such algorithm is the min-cut algorithm presented in Section 4.3. In this case, each cut corresponds to a routing region and the routing regions corresponding to the cuts producing the leaves of the tree constructed by the algorithm, are the first to be routed since they have pins on fixed positions on the two sides facing the blocks separated by the cut. After the channels corresponding to the leaves of the tree have been routed, the next level cuts identify the channels to be routed next. That is the channels are ordered bottom-up. The routing order is completely specified by the tree.

Unfortunately, as pointed out in Section 4, not all placement algorithms yield a slicing structure; such structures impose a regularity that may not allow the optimum solutions.

Algorithms, such as the simulated annealing algorithm used in TimberWolfMC, may not yield slicing structures, thus creating a cyclic relation among the walls. Several approaches have been tried to solve this problem: some modify an existing placement to create a slicing structure, as described in Section 4.4, others define routing regions which are more complicated than simple channels, (usually switch boxes or L-shaped channels), and route these to break cycles.

The ordering and independence of the routing regions described above are properties of the slicing structure *concept*, regardless of the actual shape of the slices. Typically the term *slicing structure* refers to rectilinear configurations where each slice is a straight cut. However, if the channel definition algorithm is generalized to handle structures where the slices may have bends, no constraints need be imposed to the placement.

The configuration of the slice determines the type of its corresponding detailed routing problem. For example, in the simple case where the slice has no bends, the problem is an ordinary channel-routing problem with two fixed sides (the top and bottom) and two open sides (the left and right). A single-bend slice produces an "L-shaped" channel.

It was proved in [DAI85] that if only rectangular cells are present, an L-shaped slice is the most complex shape that can result from the slicing algorithm. However this result does not hold if the cells are general Manhattan shapes. In this case *k-bend* slices (*k=0,1,2,3,...*) are necessary to guarantee a feasible slicing of the placement for any combination of cells. In Figure 5.2.3 an example of a two-bend slice is presented. In [BUR87], a new algorithm was proposed for routing region definition and ordering based on generalized slicing structure.

The generalized slicing algorithm inherits all the properties of the ordinary slicing strategy and handles a larger class of problems. As in the simple case, no switch-box router is necessary since all the generalized channels have floating terminals on two open sides. However, the structure of the routing region is too complex for standard channel routers to handle. Two dimensional routers such as Mighty can be used to route these regions.

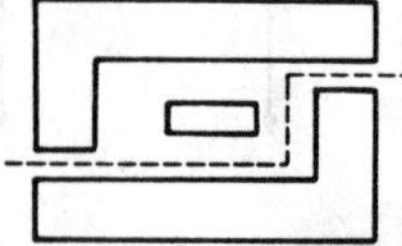

Figure 5.2.3. Two-bend channel.

5.3. Global Routing

We have described ways of subdividing the routing regions into channels. Before applying the channel routing algorithms to the problem, nets have to be assigned to channels. Global routing assigns nets to routing regions, taking into consideration net length, the congestion of the routing regions, priority of signals, and electrical characteristics. It does not specify the route followed by the interconnections inside the channels. Global routing is very important to obtain a good overall layout. In this section, we review briefly the main approaches. The excellent review by Sadowska and Kuh[SAD86] can be consulted for additional information.

Global routing is used in all design styles: gate-arrays, standard-cells and macro-cells. The first mention of global (loose) routing can be found in Nan and Feuer[SOU81].

There are two basic approaches to global routing: one deals with the interconnections one at a time, in this respect similar to the Lee algorithm; the other deals with the interconnections all at once. Of course, the second approach does not suffer from the ordering problem and has a better "global" view of the chip. However, the running time can sometimes be prohibitive and the techniques employed on each individual net must be relatively weak to allow consideration of all nets at once. As we have seen in Section 5.1.1 for the detailed routing problem, iterative, rip-up-and-re-route methods can virtually eliminate the ordering problems of the net-at-a-time methods, while continuing to exercise the power such methods bring to bear on the single-net problems they solve.

The basic problem that a global router has to solve is to distribute the nets in the available channels so that either the density of each of the channel does not exceed a bound (gate-arrays) or the overall size of the chip is minimized. In the latter case, the density of the channels is a variable to be determined by the router. In all the applications of global routing, it is assumed that, the detailed routing can be completed within the density of the channels or just above density; this assumption is justified in practice because of the nature of the problems encountered when routing large chips and the quality of the routers commonly used.

A net-at-a-time approach chooses a path by using shortest-path algorithms that penalize paths for crossing congested channels. This strategy tends to minimize congestion and to optimize wire-length and, correspondingly, the performance of the chip. There are several cost functions that can be used to achieve the goals stated above. A common form of the cost function associated with each "segment" crossing a particular channel is expressed by

$$aL + \frac{b}{c^{T+1}}$$

where a, b and c are parameters to be tuned for the particular application, L is the length of the channel and T is the number of available tracks. Of course, the parameter T for a channel changes as the algorithm proceeds assigning nets to that channel.

If the nets have only two pins, then the approach mentioned above is straightforward. However, if the nets have more than two pins, then many possible interconnection topologies are possible; this is the Steiner tree problem presented in Section 4.1. If the set of modules and channels is modeled as an undirected graph, whose nodes represent the modules and whose edges represent the channels, then the problem becomes the Steiner tree problem on graphs. Unfortunately Steiner tree problems are NP-complete and heuristics are commonly used[GAR77].

The other important approaches deal with the nets all at once. One approach formulates the global routing problem as a mathematical programming problem: a 0-1 linear programming problem. It is well known that this problem is also NP-complete and heuristics have to be used to solve it in a reasonable time. One of the most interesting approaches is to use a linear programming algorithm to find approximate solutions to the 0-1 problem, and then round the solutions to obtain a feasible solution [RAG86].

Simulated annealing has been used by Vecchi and Kirckpatrick[VEC83] to solve the 0-1 programming problem. In this case, the cost function is equal to the sum of the squares of the congestion for each channel, i.e. the number of nets crossing that channel. In this way, the algorithm penalizes congested areas. The "moves" correspond to switching one net from a path to another.

The 0-1 programming methods restrict the solutions to a small subset of the full set allowed by the original problem definition.

Burstein [BUR85] considers the global routing problem hierarchically. The chip is first divided into four quadrants and the nets are routed in these quadrants. Then, each of the quadrants is subdivided into four regions and the implications of the first routing phase are propagated to the next level of hierarchy. The procedure terminates when the real routing regions are considered at the appropriate level of hierarchy. Marek-Sadowska[SAD84] considers a similar approach where the hierarchy is traversed bottom-up instead of top-down. It is difficult to evaluate which approach works better. In my opinion, the two approaches should give very similar results in most cases.

In the gate-array case, if the constraints on density are not met, then a rip-up and reroute phase is added at the end of the main procedure. This process has traditionally been carried out by human designers with the help of a symbolic graphic editor. Rule-based approaches have been used by Marek-Sadowska [SAD85] and by Fujita and Goto [FUJ83] to identify the nets to be ripped-up and re-routed.

6. The MOSAICO System

In the previous sections of this chapter, we have reviewed a variety of algorithms for the floor-planning, placement and routing of large scale integrated circuits. To assemble an automatic layout system, much care has to be exercised in choosing and integrating the tools. Often systems equipped with excellent algorithms have failed when applied to real problems because the tools were not well integrated. A key requirement for automatic layout systems is flexibility in view of the continuous advancements in algorithms and technology. I believe that it is important to review how a complete automatic layout system is put together, describing how the data are stored and how different tools can interact in a complex way. For this reason, in this section, **Mosaico** [BUR87], a complete set of placement and routing tools tightly coupled with logic synthesis, module generation, compaction, verification, and timing analysis, is described. Mosaico has been chosen as an example because of its features and of the experience we acquired developing it.

One of the main requirements for a layout system in a general design environment such as the one being developed at Berkeley [NEW86] is the ability to support a variety of layout styles. The ThunderBird system [BRA86b] had previously been developed to place and route standard-cell designs. The Mosaico system consists of a set of tools for the placement and routing of macro-cells. Mosaico does not impose any constraints on the layout style used to implement the macro-cells. It is tightly coupled to Berkeley's logic synthesis system [.BerkSyn.] through the use of module generators. In the Berkeley Synthesis System, each macro-cell is optimized at the logic level with the characteristics of the target implementation style accounted for. The Mosaico floor-planner is used to produce specifications for the aspect ratio and pin positions of the macro-cell. The combination of the optimized logic equations and the geometric information from the floor-planner comprise the input to the module generator; the generator output is then used in the placement phase of Mosaico.

In Mosaico the floor-planning and the placement tasks are considered in an unified framework; in fact the same program is used for both tasks. The process starts as floor-planning with some or all of the cells not fully specified, and becomes placement as soon as all the macro-cells have fixed implementations. The tool used for floor-planning and placement is TimberWolfMC described in Section 4.5.

The routing methodology used in the system targets the fabrication technologies now emerging that provide multiple interconnect layers. To achieve design-rule independence the entire Mosaico system operates at the symbolic-layout level, rather than the physical-layout level. A spacing step is carried out prior to mask generation to ensure that the layout is design-rule correct.

Mosaico has been configured in a highly modular manner to ease the introduction of new tools into the system. The modularity has been achieved by using one data representation for the design, regardless of the point in the design cycle. No transformation between representation formats is required, and all data is stored using a common data manager.

The remainder of this section is organized as follows. In Section 6.1, the overall structure of the system is presented. In the following sections, the tools that comprise the system are described in the order in which they appear in the design flow. Results obtained on a number of test cases are then presented, followed by some of the directions for future development of Mosaico.

6.1. Mosaico Overview

An overview of the structure of Mosaico, the management of the design data, and the initialization of the system are given in this section.

6.1.1. Pipeline Structure

Mosaico consists of five main steps:

1. Floor-planning and Placement
2. Channel Definition
3. Global Routing
4. Detailed Routing
5. Spacing

The five basic steps are nominally executed in sequence, that is, as a *pipeline*. The pipeline is represented in Figure 6.1 together with the sequence of symbolic views that are stored at each step in the Oct data manager [HAR86].

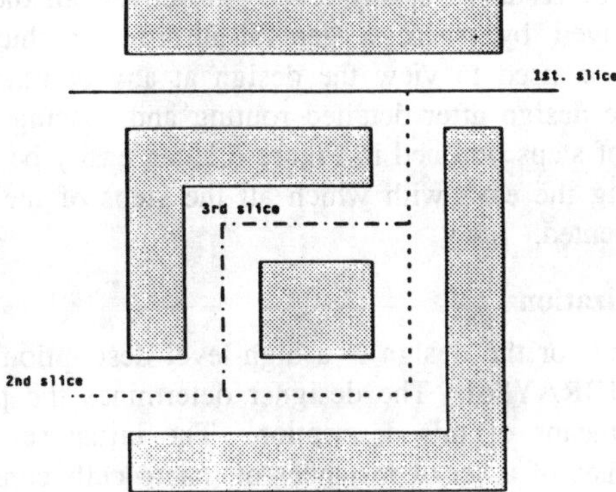

Figure 6.1 Example of zero-bends and one-bend slices.

In an ideal situation each step in the pipeline would be executed once only. In reality iteration of some or all of the steps is usually required. For example, if the routing area estimated in the placement phase is not sufficient, at least one of the macro-cells must be moved. This in turn may require re-processing by the channel definition procedure. A description of the feedback loops in the Mosaico system is included later in this paper.

6.2. Data Management

Figure 6.1 illustrates one of the most important feature of Mosaico system, namely its modularity. Modularity is achieved by enforcing the rule that all the information produced at each stage of the pipeline must be stored in the data manager. Each of the tools in the pipeline reads its input data from a view in the data manager and produces as its output another view which contains the previous data updated with the information added by the tool. This modularity provides the Mosaico system with a great deal of flexibility since every tool can be replaced in the pipeline with another one that requires the same data as input and provide the same data as output.

The key issue in achieving this flexibility is to control the structure of and the amount of data stored at each stage of the pipeline. The set of data has to be rich enough to provide the tool with the information it needs while being general enought to allow interfacing different tools based on different of algorithms. This task is made easy by the characteristics of the design data manager Oct. Oct provides the user with a general way to store the data items and the dependency relations among them. The tools read and write Oct views using a library of procedures. As a result, the tools need not be concerned with the storage format used by Oct, nor are they affected by any changes to the internal structure of Oct.

The particular data items used, and the relations among them, are not part of Oct; rather, they are decided upon by the community of tool-makers. An example of such a decision is the manner used to describe the permutability of terminals. This set of decisions is referred to as the *policy* for the design. Since the policy is not part of Oct it can be changed easily; and since all concerned parties participate in the definition of the policy it provides them with an efficient protocol for storing and exchanging information at each stage of the pipeline.

Another important advantage of storing all intermediate results in the data manager is that the pipeline can be restarted at any level. Moreover, all the intermediate results can be graphically displayed by Vem, a general-purpose graphics editor for Oct views [HAR86]. Vem can be used to view the design at any point, from the floor-plan stage down to the complete design after detailed routing and spacing. Finally, tools that complete more than one of steps outlined in Figure 6.1 can easily be inserted in the pipeline at the expense of loosing the ease with which all the steps of the layout procedure can be unbundled and re-executed.

6.2.1. Pipeline Initialization

The starting point for the design is a high-level description of the chip expressed in the Bdsyn language [BRAY86]. The designer determines the partitioning of the design into macro-cells by means of this description. The initial set of data required for the Mosaico system consists of a set of instances of macro-cells connected by a net-list. The Bdsyn description is used to produce automatically the net-list and the instances, both of which are placed into a symbolic view of the chip named *unplanned*.

The implementations of the macro-cells are generated separately, either by an automatic module generator or by manual design. When the Mosaico pipeline is started not all of the macro-cells need to be completely specified. For some macro-cells the complete layout may be known while for others only estimations of their parameters or a set of constraints may exist. For example, some of the pins may have to be positioned on specific sides of the cell, or a target aspect ratio may be desired even if a range of aspect ratios is feasible.

Since several representations of each cell may be available, Mosaico searches the available views. The views are considered in a specific order so that the view with the most complete information for the floor-planning phase is used. If none of the views is present an artificial view is created in which the macro-cell is represented as a "*soft cell*", i.e., as a cell for which only estimations of its area and of its aspect ratio are available . For such a cell the terminals are left free to move around the boundary of the cell unless ordering constraint are present. Such a description for macro-cells is suitable for the floor-planning step.

6.3. Floor-planning and Placement

As mentioned, in the Mosaico environment, the activity of floor-planning and placement is handled in a unified framework; in fact there is no distinction between floor-planning and placement at the algorithmic level. At the floor-planning stage some of the parameters (aspect ratio, pin positions) of the macro-cells may be varied; at the placement level all the macro-cell parameters are fixed and the only remaining degrees of freedom are the positions and orientations of the cells. The floor-planning and placement maintains a global view of the chip and directs the interactions with module generation and timing verification.

The floor-planning and placement iterations are started by an initial floor-planning operation in which the most detailed description available for each cell is retrieved from the data manager. In this initial representation most of the cells are described only in terms of their input/output signals and estimated area. Some cells may be described at the logical or behavioral level. For such cells an area estimation is performed, based on the complexity of the cell and on the features of the module generator that will be used to produce it.

After each iteration the floor-planner produces as output the position and orientation of each cell and also the shapes and pin positions for those cells that were not completely specified. This information is passed to the module generators, namely Gem, Wolfe, and Topogen. The generators differ in the layout styles they implement. Gem produces macro-cells using a gate matrix approach featuring multiple row and column folding. Wolfe generates macro-cells by assembling standard cells using the TimberWolfSC package [SEC86] combined with the channel routers Yacr[SAN84] and Chameleon[BUR86]. Topogen produces individual cells in the standard-cell style; these cells may be combined into larger cells using Wolfe.

The input to the module generators comes from two sources, one being the floor-planner which specifies geometric constraints as described above. The other data is the logic description produced by the logic optimization tool Mis [BRAY86]. If a module generator is unable to satisfy all the constraints imposed on a particular cell, the actual shape of the generated cell is fed back to the floor-planning and placement and a new iteration is started.

The floor-planning and placement also interacts with the timing analyzer Humming-bird. The job of the timing analyzer is to critique the floor-plan from the point of view of the timing. It performs two tasks; first it checks the arrival times of signals required for the correct functioning of the circuit. Then it marks each net with information about the maximum and minimum delay through that net. If the requirements are not met a new iteration of the floor-planner is necessary. The timing information is used to update the constraints imposed by the floor-planning and placement on the maximum or minimum length for the specific nets. Correction of the violations of the timing requirements may require more than just a new placement. In fact for some of the modules a transistor re-sizing may be necessary or some of the modules may need to be re-synthesized with the new delay targets in mind. In both these cases the intervention of the designer is presently required.

As the floor-planning and placement task is repeatedly executed more and more refined representations of the chip are generated. The macro-cells gain their physical implementation, the information available to the floor-planning and placement tool increases, and the floor-plan task turns more into a placement task.

The program used to carry out the floor-planning and placement task is Tim-berWolfMC described in detail in Section 4.5. When floor-planning and placement is completed the output is stored in Oct as the *placed* view.

6.4. Routing Region Definition and Ordering

Recall from Section 5.3 that the routing *definition* task is to partition the available routing area into smaller regions for the subsequent global and detailed routing phases. Once the regions are determined they have to be *ordered* so that each region can be routed and its width can be adjusted independently of all previously-routed regions.

In Mosaico, the routing regions are particular slicing structure that may have k-bends as defined in Section 5. A divide-and-conquer strategy similar to the one necessary to find a rectilinear slicing structure is used. The differences compared to the standard algorithm are in the method used to determine the next slice. The **Atlas** algorithm consists of the following steps:

- Using a scan-line approach generate a floor-plan graph [DAI85]

- Repeat the slicing procedure until the chip is completely subdivided; i.e., at each step find the minimal-cost slice through the graph.

- The routing order is LIFO (i.e., The last slice found is the first to be routed)

The *cost* of a slice is determined by the number of bends in the slice, the number of orthogonal edges on the slice, and the number of external junctions [DAI85] that are created by the slice. Note that the selection of a path with the smallest number of jogs and the maximum number of orthogonal edges to the path will eliminate as many potential jogs in future slices as possible. This is a greedy approach to find a set of slices with the smallest total number of bends. A more sophisticated algorithm that should produce a nearly "optimal" slicing is under development.

The hierarchy introduced by the slicing procedure corresponds to the binary tree that is used to store the data in Oct. This routing order is used in the router server described below, and the new Oct view is the input to the next step in the pipeline.

6.5. Global Routing

After the channel definition and ordering step described in the previous section, the data must be prepared for global routing. In Mosaico, the data structure used to perform global routing is the *channel graph*. In the channel graph, nodes represent intersections between channels while edges represent channels or sections of them. Each edge in the graph is assigned a weight that represents the maximum number of tracks that can be accommodated in the channel. Once the channel graph is built all the connected terminals of the macro-cell instances are projected to the closest edge in the channel graph. The channel graph with the positions of the terminals on its edges and a net list is the information necessary for global routing. The weights on edges are interpreted as *capacity constraints*. It is important to note that the input to the global router is completely symbolic and therefore totally independent of the layout style.

When the global router finishes, each net consists of a sequence of subnets, each of which is assigned to one of the edges of the channel graph. When a net exits one channel to enter another one a *pseudo-terminal* or *connector* is created.

Presently two different algorithms to perform global routing can be selected in Mosaico. The first algorithm is based on simulated annealing and it gives results that are marginally better at the expenses of a longer computation time. The second algorithm is faster and may be used interactively. The two algorithms are briefly described in what follows.

6.5.1. Simulated-Annealing-Based Global Router

The simulated-annealing global routing algorithm has been developed as part of the TimberWolfMC package. The algorithm has the following features:

- No dependence on the routing order

- Electrically-equivalent pins are utilized to minimize the routing length

The algorithm has two basic stages. During the first stage it attempts to generate the M shortest routes for each net, a task which can readily be accomplished for two-pin nets. For nets consisting of more than two pins, an algorithm has been developed which generalizes the approach in [LAW76].

In the second stage of the algorithm selects a single route from the M alternatives for each net. Let n_i represent a net, where $i \in \{1, \cdots, N\}$ and where N is the number of nets. Furthermore, let n_i^k represent the k-th alternative route for net n_i, where $k \in \{1, \cdots, M\}$. A simulated annealing algorithm is then used to select alternative $n_i^{k_0}$ for each $i \in \{1, \cdots, N\}$ such that the total routing length is minimized subject to the channel-edge capacity constraints. This approach enables the global router to avoid the routing-order dependence problem.

6.5.2. N-Layer Global Router

Nlagr is a global router that can handle n layers of interconnect and over-the-cell routing. Over-the-cell routes are represented by fixed-capacity edges on the graph while routing areas between cells are represented by edges whose capacities can be exceeded with a penalty. Nlagr has the following characteristics:

- The number of bends, and number of channels through which a net passes, are minimized.

- User-specified critical nets are routed first. All other nets are routed based on (estimated) shortest-length first. The estimate used is one-half the perimeter of the minimum enclosing rectangle of the net.

- Nets can be weighted to prefer or avoid specific layers.

There are two different algorithms incorporated in Nlagr. The first uses an extended shortest-path algorithm; the extension accounts for multi-terminal nets.

At each iteration a path is determined from the existing partial path to the nearest unconnected pin on the same net. The search for the new connection is performed by expanding first the nodes of the channel graph whose distance measured in terms of the length of the shortest path from the nodes to the existing path, is within a given limit. The expansion proceeds until an unconnected pin is reached. In the expansion procedure no particular direction is privileged since no unconnected pin has been selected as the target. For this reason the method is referred to as *undirected search*. The results provided by the method depend on the choice of the first pin, and several heuristics are provided to select it.

In the second algorithm, named *directed search,* the pins of the net are connected in a specified order. First a least-cost path is found from the first pin to the second pin. Next the least-cost *incremental* path is found to the third pin (i.e., the first pin among those not yet connected) and so on as in [HAM84]. The order in which unconnected pins are processed is determined by sorting them according to a cost which consists of two terms. The first term is the cost of the path connecting the nodes in the channel graph selected with the criterion described above, to the already-existing path. This cost is the same as that used in the first algorithm. The second term is an estimation of the distance between the unconnected pin and the closest expanded node. The presence of the second cost term makes the search proceed rapidly in the direction of the target. This algorithm is strongly dependent on the pin ordering. Several schemes are provided for ordering the pins prior to routing.

According to experimental results undirected search seems to be better than directed search, although slower. The dependency of directed search on pin ordering can be used to provide interactive optimization. In this mode of operation the global router is typically run first in the undirected-search mode. The order in which the pins were connected is preserved and used to back-annotate the net list. Next the optional interactive mode is entered; the user may then modify the pin order and run the router in directed search mode.

6.6. Router Server

The router server **Spider** performs the detailed routing of the circuit based on the placement, the global router output, the channel order, and the design rules. Spider can handle k-bend routing regions with irregular edges, and wires with different widths, a crucial feature necessary to route special nets like power and ground along with the standard signal nets. Spider also chooses the best layer for floating pins whenever more than one layer is permitted.

The actual routing is performed one channel at a time. The order is determined by Atlas as described in the previous section. Spider retrieves the channel order from the data manager by traversing the binary tree that represents the hierarchy in depth-first order. Then it selects one of the symbolic routers available according to the nature of the channel being processed. Presently the library of routers that can be used by Spider consists of three symbolic routers: the two-layers channel router Yacr2 [SAN84], the multi-layer channel router Chameleon [BUR86], and a general-area router called Mighty [SHI86]. The selection of the particular router is based on a set of rules, such that the simplest router that can successfully route the area is selected. To make it easy to extend the tool library by adding new symbolic routers, the selection rules are kept separate from the core of the program by storing them in a file that can be modified by the user.

Once the symbolic router is selected, Spider prepares the input data in the suitable format. Since all of the routers in the Mosaico library are grid-based, Spider starts by defining a grid. The symbolic grid in the "vertical" direction (the columns) can be built in two different ways. The first choice is to use a uniform grid, where the spacing between grids is determined by the wire widths, the contact widths, and the spacing rules. The second approach is a non-uniform grid. In this case the vertical grid lines are placed to obtain the best possible alignment with the actual locations of the pins. This method is especially useful in dealing with wires that differ in width. Regardless of the type of grid selected, it is not always possible for every pin to be located exactly on a grid line. In this situation an attempt is made to obtain a better alignment of the terminals on the two sides of the routing area by slightly varying the offset between the macro-cells on either side. After the alignment procedure each pin that is still not on a grid line will be connected to the nearest adjacent one by a jog.

In the "horizontal" direction, the grid lines (rows) are initially spaced uniformly. If the general-area router is used, the situation is a little more involved since an estimation of the space necessary to complete the routing must be computed. The spacing is definitively determined only after the routing is completed and the actual width of each wire has been computed. Presently, the non-uniform grid base is the default setting.

After the selected router completes, a post-processing step is performed to try to reduce the number of jogs by shifting wires while maintaining the design rules. A simple optimization step is then performed to minimize the number of vias.

When the routing of a particular region is completed, the compactor is invoked to space the region and the adjacent macro-cells according to the design rules. The combination of the cells and the routing is then considered to be a new, large macro-cell for the following steps in the layout process. A word of caution is in order about the use of the compactor in this channel-by-channel manner. By compacting one channel it is possible to generate a misalignment in the next channel to be routed, possibly increasing its density and possibly increasing the area of the chip. To avoid this problem a more global view of the routing problem is necessary, which is accomplished by looking at the next area to be routed and setting constraints on the compactor to avoid increasing the density in the next channel.

6.7. Power and Ground Routing

The routing of the power and ground nets is more involved than the routing of regular nets because of the following considerations:

- Voltage drops due to the finite conductance of interconnect and contacts have an adverse effect on the noise margins of the cells.

- The maximum current densities tolerated by an interconnect layer cannot be exceeded, otherwise metal migration might result.

The implications of the above are that the power and ground nets might have variable wire segment widths.

In a technology with only one layer of low-resistance interconnect (metal), a planar routing of the power and ground nets might be very desirable. However, in modern technologies two layers of metal are available. Therefore the policy used in Mosaico is to route power and ground using the existing global and channel routers. The exact procedure will be described later.

It is assumed that there are always power and ground rings around the chip. The existence of these rings is important for insuring correct functionality of the chip. If the power pads were not connected through a ring, some of them might be at different potentials due to wiring inductance; this in turn might result in latch-up problems in a CMOS design.

For the sake of the placement task, the part of the ring which passes under a pad is considered to be part of the pad cell itself. Therefore the location of the power and ground rings is automatically adjusted by movements of the pads. After the completion of the placement stage, the unconnected segments of the power and ground rings are tiled together with additional pieces of material of the same layer. Each pad cell has three pins, one for the pad itself and one for each of the power and ground connections. Thus the number of possible connections to each of the power and ground rings is equal to the total number of pads in the chip.

Treating the power and ground nets as regular ones biases the placement algorithm since power and ground are connected to every block of the chip. Furthermore since no assumption is made on the amount of current that can flow across a macro-cell without damaging the internal power (ground) connection, power and/or ground cannot be brought to a module through another one, if a reliable routing has to be performed. This implies that each macro-cell must be connected directly to the rings.

The following steps describe the algorithm used to route power and ground.

1. Before the placement stage, decompose the power (ground) net into smaller nets. Each of these nets contains two sets of equivalent pins, one is the power (ground) pins of a given macro-cell, and the other is the set of all the power (ground) pins on the ring. Then the original power (ground) net is discarded, and the resulting nets are treated as regular ones with high priority in the placement and global-routing stages in order to keep their length as short as possible.

2. After the global routing stage, merge the power (ground) nets in such a manner that every channel contains at most two power (ground) nets. The upper limit of two happens when the two nets enter the channel from its opposite ends and both terminate inside the channel. An implication of this merging is that all the nets which stem from the same power (ground) pad pin are merged into a single net. This step is shown in Figure 6.2.

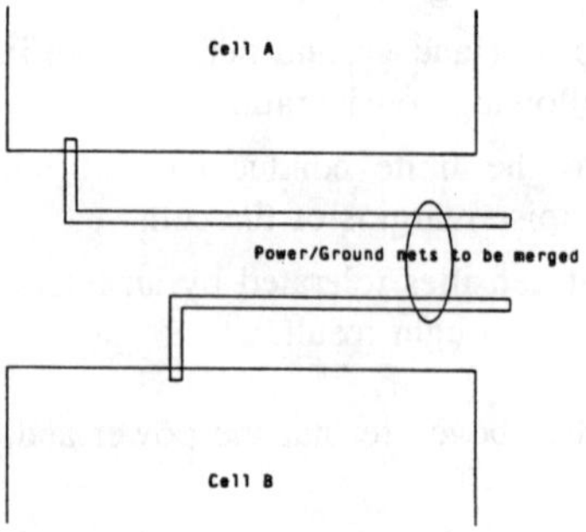

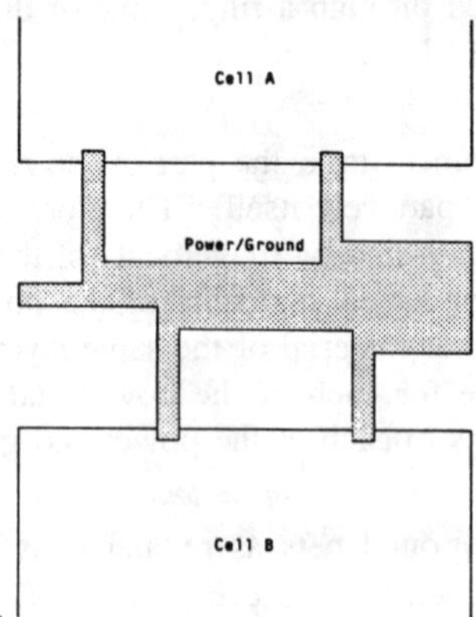

Figure 6.2 Merging of Power/Ground nets in the channel.

3. After the merging step, find the power requirements of the pseudo-pins of the power and ground nets at the boundaries of the channels.

4. After the symbolic detailed routing of each channel, find the subnet widths of power and ground nets in the routed channel. This step is shown in Figure 6.3.

Figure 6.3 Power/Ground sizing.

6.8. Spacing

As noted above, all steps in the Mosaico pipeline are carried out at the symbolic-layout level. In addition, the module generators used in the system generate symbolic layout as well. Symbolic layout spacing (or compaction) is thus an essential part of the system for several reasons. First, all designs must be spaced to ensure that they are design-rule correct; this has the advantageous effect of eliminating design-rule checking, since the layouts are correct-by-construction. Second, the use of generalized symbolic layout provides a mechanism for producing technology-independent designs. The spacing techniques used in Mosaico are capable of updating the symbolic layout primitives themselves (e.g., transistors, contacts), as well as the spacings between them. Also, by the use of a variety of spacing techniques various optimizations of the layout can be performed over a range of area/cpu-time tradeoffs. Presently two spacers are available in Mosaico: SPARCS and Zorro. Both spacers are described in another chapter of this book and are therefore not reviewed here.

6.9. Pipeline Iteration

The two loops shown in Figure 6.1 are provided to account for situations in which the routing area available after placement is insufficient to complete the routing. This situation can occur both during global routing and during detailed routing. In both cases the extra routing area that is required can be provided by spacing apart the macro-cells delimiting the routing region. If the additional area can be provided without drastically changing the order and structure of the routing areas then the pipeline is started again from the point where the exception occurred. In this situation all of the work already completed is preserved.

If the addition of the extra area requires a change such that the organization of the routing area is no longer feasible then two possible solutions exist. In the first the channel definition procedure is re-executed to reorganize the areas not yet routed. The second is performed when the first solution fails; the pipeline is re-started at the placement stage using a more conservative estimate for the routing area requirements.

6.10. Interactive Capabilities

The Mosaico pipeline is operated by means of a script that executes the entire pipeline or subsections of it. The script-based approach provides the user with a great deal of flexibility; the script can be modified both in the sequence of the calls to the tools and in the parameters that are input to the tools, e.g., the parameters used in the simulated annealing procedures.

In addition to the tools described in the previous sections, a set of graphical interactive aids is provided to help the user judge the quality of the layout at any stage of the pipeline. One tool allows the user, for a selected macro-cell, to see a representation of its connections with the other cells drawn from cell center to cell center. Another aid allows the user to see the optimal position of a selected macro-cell as computed by a force-directed algorithm. Both of these programs may be used at any point following the placement step. A third tool uses the information available after the global router completes. It allows the user to select a terminal and highlight the path of the net connected to that terminal as determined by the global router. This aid is used to judge the quality of the placement combined with the channel definition and global routing, prior to the detailed routing.

These tools are closely linked to Vem [HAR86] and are presented to the user as menu options. These aids together with the other Vem editing capabilities allow the user to modify existing views, quickly evaluate the effect of modifications on the quality of the layout, and determine whether or not to run the complete pipeline.

6.11. Applications and Results

The Mosaico system is presently under test on a set of macro-cell examples provided by industry. The examples, whose characteristics are summarized in Table 6.1, consist of macro-cells with fixed aspect ratios and fixed pin-positions. Table 6.2 contains the area statistics after placement, routing, and spacing. The cpu-time required to place and route the examples is reported in Table 6.3 while the results are depicted in Figures 6.4, 6.5 and 6.6.

Circuit	#Macro-Cells	#Pads	#Nets	#Pins	#Channels
ck1	8	5	29	58	17
ck2	12	39	262	691	54
ck3	23	17	129	458	43

Table 6.1. Characteristics of the test circuits.

Circuit	Placed	Routed	Compacted	Area savings (%) after compaction
ck1	1592x1415	1678x1562	1533x1433	16
ck2	17305x15620	17033x14982	15670x14982	1
ck3	2902x3607	4216x4080	n.a.	n.a.

Table 6.2. Area at various stages of the pipeline (in λ).

Circuit	Placement	Routing		Compaction
		Global	Detailed	
ck1	140	5	22	21
ck2	n.a	926	482	960
ck3	1203	74	140	n.a.

Table 6.3. CPU time (seconds, VAX-8650).

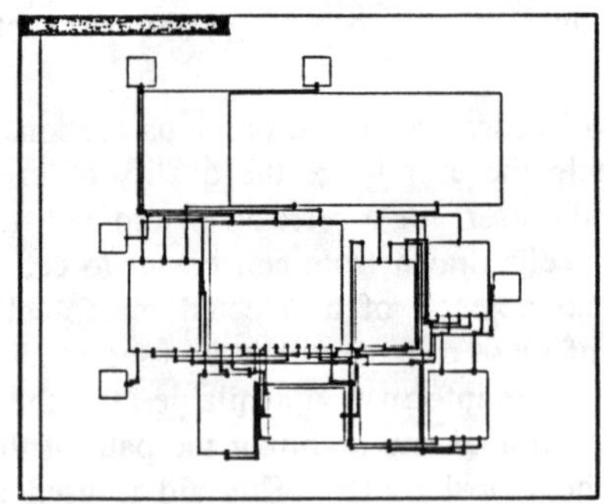

Figure 6.4 Circuit Ck1

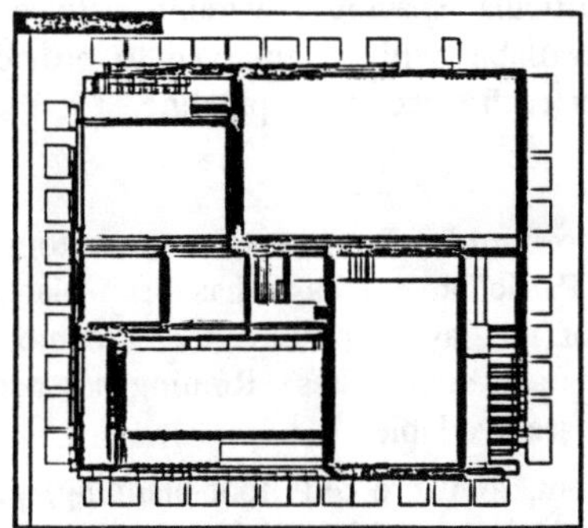

Figure 6.5 Circuit Ck2

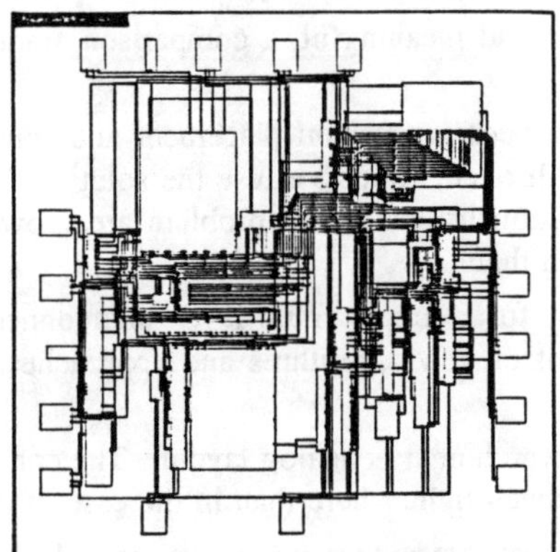

Figure 6.6 Circuit Ck3

The interactions of Mosaico with the module generators and the timing analyzer is still in a preliminary test phase and significant results are not available at the present time.

6.12. Conclusions and Future Work

In this section Mosaico, an integrated set of tools for macro-cell layout, has been presented. Mosaico is designed to interact closely with the other tools in the Berkeley design system. A major point of interaction occurs at the floor-planning stage, where Mosaico exchanges information with module generators and timing analyzers. All the layout procedures are carried out at the symbolic-layout level, which provides the system with technology independence. Furthermore Mosaico is closely coupled with the design manager Oct to achieve an open system with a great deal of modularity and flexibility.

In Mosaico well-tested tools like the detailed routers Yacr and Mighty cohabit with new tools that have been developed explicitly for the system, such as the channel definition and ordering procedure and the power and ground router.

Future development will occur on all steps in the pipeline. In particular the channel definition and ordering algorithm will be enhanced to better take into consideration the influence of the channel definition on the performance of the detailed routers. Also the interactions between the router server Spider and the spacers will be enhanced to provide for layouts in which particular spacing constraints between elements are present. In general the set of tools available will be enlarged to provide the user with the choice of several parallel paths for each stage in the pipeline.

7. CONCLUSIONS

In this chapter, algorithms and systems for the automatic layout of integrated circuits have been reviewed. Particular emphasis has been placed on floor-planning and placement, the two stages of the layout process that are most difficult. Attention has been given to new algorithms and approaches. Routing has been briefly reviewed since many excellent review papers are available.

The Mosaico system, a macro-cell floor-planning, placement and routing tool, has been described with the intention of conveying useful information on how to put together a complex layout system in an automatic synthesis environment.

Whenever possible and meaningful, a comparison among the various approaches has been given.

Many papers have been written on placement and routing of integrated circuits, but several problems are still open, either because the solutions obtained thus far are not satisfactory or because the dimensions of the problem are growing so fast that existing techniques cannot cope with them.

New technologies, for example multi-layer interconnects, pose new challenges and require the development of new algorithms and approaches to the placement and routing problem.

Another "rich" research area is analog layout. The constraints imposed on the performance of the chip are much tighter here than in the case of digital circuits.

In summary, I expect that automatic layout algorithms and systems will continue to be an exciting area of research for many years to come.

8. ACKNOWLEDGEMENTS

This paper would not have been possible without the help of my friends and graduate students who went through the pain of reading this manuscript several times and who suggested modifications that greatly improved its quality. In particular, I thank F. Romeo, G. Sorkin and G. De Micheli.

Some of the results described in the paper are due to the work of the layout group of the Laboratory for Computer-Aided Design of Integrated Circuits, University of California at Berkeley. The contributions of C. Sechen, A. Casotto, M. Igusa, F. Romeo, H. Shin, H. Yaghutiel, D. Braun and F. Marron are gratefully acknowledged.

This work has been partially supported by SRC and DARPA. The support of the industrial sponsors, of the CAD/CAM Center of the University of California, Berkeley, and, in particular, of SGS Microelettronica, AT&T Bell Laboratories, DEC, Fairchild, GE, Harris Semiconductors, Hewlett-Packard, Hughes Aircraft, IBM, Intel, Itaucom, LSI Logic, Olivetti, Philips, Silicon Compilers Inc., Silicon Design Laboratories, Telebras and Xerox, is gratefully acknowledged.

9. REFERENCES

[AAR85]   E.H. L. Aarts and P.J. M. van Laarhoven, "Statistical Cooling: A General Approach To Combinatorial Optimization Problems", *Philips J. Res.*, Vol. 40, pp. 193-226, 1985.

[ANW85] H. Anway, G. Farnham and R. Reid, "Plint Layout System for VLSI Chips" *Proc. 22nd Design Automation Conf.*, pp. 449-452, 1985.

[ARA84] C.R. Aragon, D.S. Johnson, L.A. McGeoch, and C. Schevon, "Simulated Annealing Performance Studies", *Workshop on Statistical Physics in Engineering and Biology*, Yorktown Heights, April 1984.

[BLA85] J. P. Blanks, "Near-Optimal Placement Using a Quadratic Objective Function," *Proc. 22nd Design Automation Conf.*, pp. 609-615, 1985.

[BRA85] R. K. Brayton, N. L. Brenner, C. L. Chen, G. De Micheli, C. T. McMullen, and R. H. J. M. Otten, "The Yorktown Silicon Compiler," *Proc. Int. Symp. Circ. Syst. (ISCAS-85)*, June 1985, Kyoto.

[BRA86a] D. Braun, J. Burns, S. Devadas, H.K. Ma, K. Mayaram, F. Romeo, and A. Sangiovanni-Vincentelli, "Chameleon: A New Multi-Layer Channel Router", *Proc. 1986 Design Autom. Conf.*, Las Vegas, June, 1986.

[BRA86b] D. Braun, C. Sechen, and A. Sangiovanni-Vincentelli, "ThunderBird: A Complete Standard-Cell Layout System," *Proc. 1986 Custom Integrated Circuits Conference*, Rochester, New York, May 12-14, 1986.

[BRA87] R. K. Brayton, private communication, 1987.

[BRAY86] R. Brayton, A. Cagnola, E. Detjens, S. Krishna, P. McGeer, L. Pei, N. Phillips, R. Rudell, R. Segal, T. Villa, A. Wang, R. Yung, and A. Sangiovanni-Vincentelli, "Multiple-Level Logic Optimization System," *Proc. IEEE Inter. Conf. on Computer Aided Design*, Santa Clara, pp. 356-359, Nov 1986.

[BRAY87] R. Brayton, "Multi-level Logic Optimization", this book.

[BRE72] M. A. Breuer, *"Design Automation of Digital Systems,"* Prentice-Hall, 1972.

[BRE77] M. Breuer, "Min-cut Placement," *J. Des. Automat. Fault Tolerant Comput.*, pp. 343-362, Oct. 1977.

[BUR83] M. Burstein and R. Pelavin "Hierarchical Wire Routing," *IEEE Trans. on CAD of ICAS*, pp. 223-234, Oct. 1983.

[BUR84] M. Burstein and M. Youssef, "Timing Influenced Layout Design," IBM Research Report #48563, 11/15/84.

[BUR85] M. Buric and T. Matheson, "Silicon Compilation Environments," *Proc. IEEE 1985 Cust. Int. Circ. Conf.*, pp. 208-212, 1985.

[BUR86] J.L. Burns and A.R. Newton, "SPARCS: A New Constraint-Based IC Symbolic Layout Spacer," *Proc. IEEE 1986 Custom Integrated Circuits Conference*, pp. 534-539, May 1986.

[BUR87] J. Burns et al., "Mosaico: An Integrated Macro-cell Layout System," submitted for publication.

[BURI83] M. Buric, C. Christensen and T. Mateson, "The Plex Project: VLSI Layouts of Microcomputers Generated by a Computer Program", *Proc. IEEE Int. Conf. on CAD*, pp. 49-51, 1983.

[CAS86] A. Casotto, F. Romeo and A. Sangiovanni-Vincentelli, "A Parallel Simulated Annealing Algorithm for the Placement of Macro-Cells", *Proc. 1986 IEEE Int. Conf. on CAD*, Santa Clara, pp. 30-33, November, 1986.

[CHA68] H. Charney and D. PLato, "Efficient Partitioning of Components," *Proc. of the 5th Annual Design Automation Workshop*, pp. 16-1 to 16-21, July 1968.

[CHE83] N. P. Chen, C.P. Hsu, E.S. Kuh, C.C. Chen, and M. Takahashi, "BBL: A Building Block Layout System for Custom Chip Design," *Proc. IEEE Int. Conf. on Computer- Aided Design*, pp. 40-41, 1983.

[CHE84] C.K. Cheng and E. S. Kuh, "Module Placement Based on Resistive Network Optimization", *IEEE Trans. on CAD of ICAS*, vol. CAD-3, no. 3, pp.218-225, July 1984.

[DAI85] W.M. Dai, T. Asano, and E.S. Kuh, "Routing Region Definition and Ordering Scheme for Building-Block Layout," *IEEE Trans. Computer-Aided Design*, vol. CAD-4, pp. 189-197, July 1985.

[DAI86] W. Dai and E. Kuh, "Simultaneous Floor-planning and Global Routing for Hierarchical Building Block Layout", *Proc. 1986 Int. Conf. on CAD*, Santa Clara, Nov. 1986.

[DEM83]   G. De Micheli and A. Sangiovanni-Vincentelli, "Multiple Constrained Folding of Programmable Logic Arrays: Theory and Applications", *IEEE Trans. on CAD of IC and Syst.*, Vol. CAD-2, N. 3, pp. 151-167, July 1983.

[DUN84]   A. Dunlop, V. Agrawal, D. Deutsch, M. Jukl, P. Kozak and M. Wiesel, "Chip Layout Optimization Using Critical Path Weighting," Proceedings of the 21 Design Automation Conference, pp. 133-136, June 1984.

[DUN85]   A. Dunlop and B. Kernighan, "A Procedure for Placement of Standard Cell VLSI Circuits," *IEEE Trans. on CAD*, vol. CAD-4, no.1, pp. 92-98, Jan. 1985.

[ELG81]   A. ElGamal and Z. Syed, "A Stochastic Model for Interconnections in Custom Integrated Circuits," *IEEE Trans. on Circ. and Syst.*, Vol. CAS-28, September 1981.

[FOX83]   J. Fox, "The MacPitts Silicon Compiler: A View from the Telecommunications Industry", *VLSI Design*, May 1983.

[FRA86]   J. Frankle and R. Karp, "Circuit Placements and Cost Bounds by Eigenvector Decomposition," *Proc. 1986 Int. Conf. on CAD*, pp. 414-417, Santa Clara, 1986.

[FUJ83]   T. Fujita and S. Goto, "A Rule-based Routing System," *Proc. 1983 Int. Conf. on Comp. Design*, 1983.

[GAR77]   M. R. Garey and D. S. Johnson, "The Rectilinear Steiner Tree Problem is NP-complete," *SIAM J. Appl. Math.*, Vol. 32, No. 4, pp. 826-834, 1977.

[GAR79]   M. Garey and D. Johnson, "*Computers and Intractability: A Guide to the Theory of NP-Completeness,*" Freeman, 1979.

[GEL85]   S. Gelfand and S. Mitter, "Analysis of Simulated Annealing For Optimization," *Proc. 24th IEEE Conf. on Decision and Control*, pp. 779-784, Ft. Lauderdale, Dec. 1985.

[GEM84]   S. Geman and D. Geman, "Stochastic Relaxation, Gibbs Distributions, and the Bayesian Restoration of Images", *IEEE Transactions on Pattern Analysis and Machinery Intelligence*, Vol. 6, pp. 721-741, 1984.

[GOT86]   S. Goto and T. Matsuda, "Partitioning, Assignment and Placement," in T. Ohtsuki, ed., *Layout Design and Verification*, North Holland, 1986.

[GRA70]   J. Grason, "A Dual Linear Graph Representation for Space-Filling Location Problems of the Floor-planning Type," in G. Moore (ed.), *Emerging Methods of Environmental Design and Planning*, The MIT Press, Cambridge, Mass. 1970.

[GRO86]   L. Grover, "A New Simulated Annealing Algorithm for Standard Cell Placement," *Proc. 1986 Int. Conf. on CAD*, pp. 378-380, Santa Clara, Nov. 1986.

[HAJ85]   B. Hajek, "A Tutorial Survey of Theory and Applications of Simulated Annealing", *Proc. 24th IEEE Conf. on Decision and Control*, pp. 755-761, Ft. Lauderdale, Dec. 1985.

[HAL70]   K. M. Hall, "An r-dimensional Quadratic Placement Algorithm," *Management Science*, vol. 17, no. 3, pp. 219-229, Nov. 1970.

[HAM84]   T.G. Hamachi and J.K. Ousterhout, "A Switch-box Router with Obstacle Avoidance", *Proc. 21nd Design Automation Conf.*, Jun 1984, pp 173 - 179.

[HAN72]   M. Hanan and J. Kurtzberg, "Placement Techniques," in M. Breuer, ed., *Design Automation of Digital System*, Prentice-Hall, 1972.

[HAR69]   F. Harari, "*Graph Theory*", Addison-Wesley, 1969.

[HAR86]   D. Harrison, P. Moore, Rick L. Spickelmier and R. Newton, "Data Management and Graphics Editing in the Berkeley Design Environment", *Proc. IEEE Inter. Conf. on Computer Aided Design*, Santa Clara, Nov 1986.

[HEL77]   W.R. Heller, W.F. Mikhail and W. Donath, "Prediction of Wiring Space Requirements for VLSI", *Proc. 14th Des. Aut. Conference*, pp. 32-42, 1977.

[HEL82]   W. Heller, G. Sorkin and K. Maling, "The Planar Package for System Designers," *Proc. 19th Design Automation Conf.*, pp. 253-260, 1982.

[HSU85]   C.P. Hsu et al., "APLS2: A Standard Cell Layout System for Double-layer Metal Technology", *Proc. 22nd Design Automation Conf.*, pp. 443-448, 1985.

[HUA86] M.D. Huang, F. Romeo and A. Sangiovanni-Vincentelli, "An Efficient General Cooling Schedule for Simulated Annealing," *Proc. IEEE Inter. Conf. on Computer Aided Design,* pp. 381-384, Santa Clara, November 1986.

[HUI85] A. Hui et al., "A 4.1K Gates Double Metal HCMOS Sea of Gates Array", *Proc. 1985 Cust. Int. Circ. Conf.,* pp. 15-17, Portland, May 1985.

[JEP83] D. Jepsen and D. Gelatt, "Macro Placement by Monte Carlo Annealing," *Proc. 1983 IEEE Int. Conf. on Comp. Design,* pp. 495-498, Nov. 1983.

[JOH79] D. Johannsen, "Bristle Blocks: A Silicon Compiler", *Proc. 16th Design Automation Conf.,* pp. 310-313, 1979.

[JOO85] R. Joobbani, *"WEAVER: A Knowledge- Based Routing Expert",* Kluwer, 1986.

[KER70] B. Kernighan and S. Lin, "An Efficient Heuristic Procedure for Partitioning Graphs," *Bell Systems Technical Journ.* vol. 49, no. 2, pp. 291-307, Feb. 1970.

[KIR83] S. Kirkpatrick, C.D. Gelatt and M.P. Vecchi, "Optimization by Simulated Annealing", *Science,* Vol. 220, N. 4598, pp. 671-680, 13 May 1983.

[KOZ83] T. Kozawa et al., "Automatic Placement Algorithms for High Packing Density VLSI", *Proc. 20th Design Automation Conf.,* pp. 175-181, 1983.

[KOZ84] K. Kozminski and E. Kinnen, "An Algorithm for Finding a Rectangular Dual of a Planar Graph for use in Area Planning for VLSI Integrated Circuits," in *Proc. 21st Design Automation Conf.,* pp. 655-656, 1984.

[KUR65] J.M. Kurtzberg, "Algorithms for Backplane Formation", in *Microelectronics in Large Systems,* Spartan Books, pp. 51-76, 1965.

[LAP86] D. LaPotin and S. Director, "Mason: A Global Floor-planning Approach to VLSI Design", *IEEE Trans. on CAD of ICAS,* Vol. CAD-5, pp. 477-489, Oct. 1986.

[LAU80] U. Lauther, "A Min-cut Placement Algorithm for General Cell Assemblies Based on a Graph Representation," *Journal of Digital Systems,* Vol. IV, Issue 1, pp. 21-34, 1980.

[LAU86] U. Lauther, Private Communication, 1986.

[LAU86a] U. Lauther and H. Yaghutiel, "Simulated Annealing for Slicing Structures," EECS 290-C final report, Department of EECS, University of California at Berkeley, Jun. 1986.

[LAW76] E. Lawler, *Combinatorial Optimization: Networks and Matroids,* Holt, Rinehart and Winston pp. 102-104, 1976.

[LEE61] C. Y. Lee, "An Algorithm for Path Connections and its Application", *IRE Trans. Electron. Comp.,* pp. 346-365, Sept. 1961.

[LEI84] S. M. Leinwand and Yen-Tai Lai, "An Algorithm for Building Rectangular Floor-plans," in *Proc. 21st Design Automation Conf.,* pp.663-664, 1984.

[LUE73] D. Luenberger, *"Introduction to Linear and Nonlinear Programming,"* Addison-Wesley, 1973.

[LUK85] W. K. Luk, "A Greedy Switch-box Router", *INTEGRATION, the VLSI journal 3,* 1985, pp 129-149.

[LUN86] M. Lundy and A. Mees, "Convergence of the Annealing Algorithm," *Mathematical Programming,* Vol. 34, pp. 111-124, 1986.

[MAL82] K. Maling, S. Mueller and W. Heller, "On Finding Most Optimal Rectangular Package Plans," *Proc. 19th Design Automation Conf.,* pp. 663-670, 1982.

[MIT85] D. Mitra, F. Romeo and A. L. Sangiovanni-Vincentelli, "Convergence and Finite-Time Behavior of Simulated Annealing," *Proc. 1985 Cont. Dec. Conf.,* Dec. 1985 also *J. of Applied Probability,* to appear.

[NEW86] R. Newton, A. Sangiovanni-Vincentelli and C. Sequin, "The Berkeley Synthesis System," subitted for publication, 1986.

[OGA86] Y. Ogawa et al., "Efficient Placement Algorithms Optimizing Delay for High-speed ECL Master-slice LSIs", *Proc. 23rd Design Automation Conf.,* pp. 404-411, June 1986.

[OHT86] T. Ohtsuki (ed.), *"Layout Design and Verification,"* North Holland, 1986.

[OTT82] R. Otten, "Automatic Floor-plan Design," *Proc. 19th Design Automation Conf.,* pp. 261-267, June 1982.

[OTT84]   R. Otten and L. van Ginneken, "Floor-plan Design using Simulated Annealing," *Proc. Int. Conf. on CAD*, pp. 96-98, 1984.

[OTT87]   R. Otten, "Layout Compilation", this book.

[PER76]   G. Persky, "PRO- An Automatic String Placement Program for Polycell Layout", *Proc. 13th Design Automation Conf.*, pp. 417-424, 1976.

[PRE86]   B. Preas and P. Karger, "Automatic Placement: A Review of Current Techniques," *Proc. 1986 Design Automation Conf.*, pp. 622-629, July 1986.

[QUI79]   N. Quinn and M. Breuer, "A Force-directed Component Placement Procedure for Printed Circuit Boards", *IEEE Trans. on CAS*, vol. CAS-26, pp. 377-388, June 1979.

[RAG86]   P. Raghavan, "Randomized Rounding and Discrete Ham-Sandwich Theorems: Provably Good Algorithms for Routing and Packing Problems," Report No. UCB/CSD 87/312, Computer Science Division, University of California, Berkeley, July 1986.

[REE85]   J. Reed, "YACR: Yet Another Channel Router," Electronics Research Laboratory, Memo. No. UCB/ERL 85/12, University of California, Berkeley, February 1985.

[RIV82a]   R. Rivest, "The 'PI' (Placement and Interconnect) System," *Proc. 19th Design Automation Conference*, pp. 475-481, 1982.

[RIV82b]   R. Rivest and C. Fiduccia, "A Greedy Channel Router," *Proc. 19th Design Automation Conference*, pp. 418-424, 1982.

[ROM85]   F. Romeo and A. Sangiovanni-Vincentelli, "Probabilistic Hill-Climbing Algorithms: Properties and Applications," in H. Fuchs, ed. *1985 Chapel Hill Conference on Very Large Scale Integration*, Computer Science Press, May 1985.

[ROS71]   D. Rose, "A Graph Theoretic Study of the Numerical Solution of Sparse Positive Definite System of Linear Equations," in R.C. Read (ed.), *Graph Theory and Computing*, Academic Press, 1971.

[ROW86]   J. Rowson et al. "Gate-Array Macro Layout Generation," *Proc. 1986 Int. Conf. on CAD*, pp. 448-454, Santa Clara, Nov. 1986.

[SAD84]   M. Marek-Sadowska, "Global Router for Gate Array", *Proc. IEEE Int. Conf. on Comp. Design*, pp. 223-234, 1984.

[SAD85]   M. Marek-Sadowska, "Two Dimensional Router for Double Layer Layout," *Proc. 22nd Design Automation Conf.*, pp. 117-123, 1985.

[SAD86]   M. Marek-Sadowska and E. S. Kuh, "Global Routing," in T. Ohtsuki, ed., *"Layout Design and Verification"*, North Holland, 1986.

[SAN84]   A. Sangiovanni-Vincentelli, M. Santomauro and J. Reed, "A New Gridless Channel Router: Yet Another Channel Router the Second (YACR-II)", *Proc. IEEE Int. Conf. on CAD*, Santa Clara, Nov. 1984.

[SAN87]   A. Sangiovanni-Vincentelli, "Synthesis of LSI Circuits," this book.

[SEC85]   C. Sechen and A. L. Sangiovanni-Vincentelli, "Timber Wolf: A Placement System for Integrated Circuits", *Journal of Solid-State Circuits*, May 1985.

[SEC86]   C. Sechen and A. Sangiovanni-Vincentelli, "TimberWolfMC: Chip-planning, Placement and Global Routing of Macro/Custom Cell Integrated Circuits Using Simulated Annealing," submitted for publication, 1986.

[SHI86a]   H. Shin, A. Sangiovanni-Vincentelli and C. Sequin, "Two-dimensional Compaction by 'Zone Refining'", *Proceedings of 23rd Design Automation Conference*, pp. 115 - 122, June 1986.

[SHI86b]   H. Shin and A. Sangiovanni-Vincentelli, "MIGHTY: A 'Rip-up and Re-route' Detailed Router," *Proc. 1986 Int. Conf. on CAD*, Santa Clara, pp. 2-5, Nov. 1986.

[SOU81]   J. Soukup, "Circuit Layout", *Proc. of the IEEE*, pp. 1281-1305, Oct. 1981.

[SOU83]   J.R. Southard, "MacPitts: An Approach to Silicon Compilation", *Computer*, Dec. 1983.

[STO83]   L. Stockmeyer, "Optimal Orientations of Cells in Slicing Floor-plan Designs," *Information and Control*, Vol. 59, pp. 91-101, 1983.

[SUD83]   T. Sudo, T. Ohtsuki and S. Goto, "CAD Systems for VLSI in Japan," *Proceedings of the IEEE* Vol. 71, N. 1, Jan. 1983.

[TSU86] S. Tsukiyama, K. Koike, and I. Shirakawa, "An Algorithm to Eliminate all Complex Triangles in a Maximal Planar Graph for use in VLSI Floor-plan," in *Proc. 1986 IEEE Int. Symp. on Circuits and Systems*, pp. 321-324, 1986.

[VEC83] M. Vecchi and S. Kirkpatrick, "Global Wiring by Simulated Annealing," *IEEE Trans. on CAD of ICAS*, vol. CAD-2, pp. 215-222, 1983.

[WHI84] S. White, "Concept of Scale in Simulated Annealing", *Proc. IEEE Int. Conf. on Computer Design*, pp. 646-651, Port Chester, New York, 1984.

[WON86] D. Wong and C. Liu, "A New Algorithm for Floor-plan Design", *Proc. 24th Design Automation Conf.*, pp. 101-107, June 1986.

[YOS82] T. Yoshimura and E.S. Kuh, "Efficient Algorithms for Channel Routing", *IEEE Trans. on CAD of ICAS*, pp. 25-36, Jan. 1982.

ALGORITHMS FOR MULTI-LEVEL LOGIC
SYNTHESIS AND OPTIMIZATION

R. K. Brayton
IBM Research
Yorktown Heights
N. Y. 10598

1. Introduction

These notes provide some of the details concerning algorithms, implementation, data structures and general strategy on which several logic synthesis systems have been based. The algorithms were developed initially at IBM Research in Yorktown [Bra82a, Bra82b, Bra84c] with some later development and refinements at Berkeley [Bra86]. Other systems [Bar85, deG86, Gre86, The85] are based partly on these ideas, primarily those which are technology independent. These concepts have been formed over a period of about six years and have been tested on many sets of logic examples of various types from various sources. Although different in many ways, our notions of multi-level logic minimization and synthesis are influenced by our work on PLA (two-level) minimization [Bra84b] which developed in parallel over that same period.

Multi-level logic synthesis plays a key role in a synthesis path for an automatic chip design system in several places. The first place follows extraction of the logic functions from the high level specification. Later, during timing optimization, logic synthesis is used to trade off delay and area to meet timing constraints. At both Yorktown and Berkeley, the logic synthesis systems YLE and MIS respectively are part of such synthesis paths (the Yorktown Silicon Compiler [Bra85a] at Yorktown and the Berkeley Design Environment [New86] at Berkeley). YLE is implemented in APL and MIS is implemented in C, but both have been designed to allow interaction since multi-level logic synthesis is still very much a developing area. However, both are exercised automatically if required by the automatic chip design systems. MIS contains some more recent developments and improved data structures and, being written in C, is easily portable to different computers. It interfaces to common logic interchange format (LIF) files and hence should be easily incorporated into other synthesis systems. MIS should become available soon through the Industrial Liaison Program at Berkeley.

The techniques described in these notes are largely independent of the kind of technology and design style finally used in the implementation. For example, the synthesis systems which use these techniques are targeted at different circuits and different layout styles. The system, YLE, at Yorktown focuses on a dynamic domino-like CMOS logic with the logic macros implemented using a pluricell layout style [Bra84a]. At Boulder [Bar85] and GE [Gre86], the final implementation uses a fixed standard cell library and an expert system [deG86] to handle the technology dependent details. The MIS system was used at Berkeley initially for static CMOS complex cell circuits with the final logic macro being one of either two styles, a generalized gate matrix style [Dev86] or a pluricell style for static CMOS [Ada86, Nai85], and standard cells have been added recently. Although each of these systems requires a final mapping to the specific technology, the relative development effort for this has been modular and small in most cases.

Input to these logic systems is through a common logic interchange format, LIF, a technology independent format for describing multi-level logic networks.

This format has been refined from its original *linked PLA* description, to LIF or BLIF [Rud86d] descriptions which are hierarchical and more general. LIF (or BLIF) files are ASCII character files and are suitable for exchanging logic examples between groups interested in comparing logic synthesis systems.

Also, recently there have been several developments in languages for describing hardware with synthesis instead of simulation in mind, for example YLL, CSIM, and BDSYN [Bre84, Lig85, Rud86]. YLL is embedded in APL, CSIM embedded in C and BDSYN is DECSIM-oriented (DEC mixed mode simulation language). Although these are different in many ways, several common features have emerged. Each starts with a level of specification, in the register-transfer range, which is unambiguous with respect to the logic equations generated. Each preserves a 1-1 correspondence between user-specified variables and the logic functions of the multi-level network generated. Each outputs a variation of a LIF file. Don't-cares can be specified and used immediately (YLL) or passed (BDSYN) to the logic synthesis system. Thus languages and logic synthesis have moved closer to each other, with the languages providing useful information (like don't-care sets) to the synthesis system. Conversely, some of the logic algorithms are used as part of the language processing and logic extraction.

As practical and more optimal logic synthesis becomes a reality and various synthesis systems are made available, better understanding of this area by a broader group of scientists is needed. Logic tools will be put in the hands of chip and system designers, and as part of larger CAD systems, they will interact and be integrated with other tools. Therefore, increased knowledge about multi-level logic synthesis can aid the system designers by broadening their horizons to additional opportunities for more efficient chip and system designs. Similarly, CAD tool developers can benefit by understanding some of the capabilities and algorithms built into a multi-level logic system. Finally, and more generally, the multi-level logic algorithms represent capabilities for manipulating logic functions which are probably more efficient in quality and speed than most people may imagine. It is not difficult to project also that these capabilities should be useful in entirely unrelated subjects since logic is universal.

These notes are incomplete in a number of ways. Multiple function factoring and common sub-expression extraction are not treated as fully and the algorithms for equally optimal results have not been included as for those described for the single-function case. We have chosen to omit proofs of some of the theorems stated. In other cases, observations which should be stated more formally and proved have been left informal. A serious treatment of a Boolean network as an incompletely specified multi-output logic function and its relation to testing is not included. The notes of Hachtel in this same volume should be referenced in this regard. These notes will form a basis for the forthcoming book, "VLSI Multi-Level Logic Synthesis" by R. K. Brayton, G. D. Hachtel, R. Rudell and A. Sangiovanni-Vincentelli to be published by Kluwer Academic Publishers. As such, the opportunity of preparing these notes for the NATO ASI workshop in L'Aquila, Italy has helped to initiate the writing process and to present this material to a larger audience.

2. Basic Notions and Definitions

2.1. Two-Level and Multi-Level Logic

Two level logic minimization is a well developed and well understood area. Techniques, exact and heuristic, for minimum or near minimal answers are known. We can determine if two functions are equivalent and when logic is redundant. These notions have been extended to multi-output functions, and functions of multi-valued variables. Significant progress has been made on the state-assignment problem [DeM85a] and other encoding problems with two-level logic in mind [DeM85b]. In short, two-level minimization is a well developed science.

In contrast, multi-level minimization is less structured, more difficult and relatively new. Our goal is to bring this up to the level of science obtained for two level minimization. As a science, multi-level minimization suffers from the same things that make it attractive for implementing logic, namely, it is very flexible. Hence, the problems are not so well defined. For example, for two-level minimization, we usually have a PLA implementation in mind, and therefore the minimization problem can be abstracted and made largely independent of the technology of the implementation. In these notes, we focus on abstracting the problems and developing procedures to achieve similar objectives for multi-level logic minimization.

Multi-level synthesis has the advantage over PLA synthesis in that it is good for representing and implementing any type of logic. Typically, logic is divided into two groups, control and data-flow logic, with control logic perceived as suitable for PLA implementation while data-flow usually requires multi-level logic. Sometimes, this forces an unnatural decomposition, with the control logic made by PLA generators and the data-flow hand designed or obtained from parameterized libraries. Since multi-level logic is suitable for all types of logic, the ability to automatically and optimally synthesize multi-level logic forces no such dichotomy on the user. In many applications it is more suitable to mix the two types of logic, for example because of layout considerations or for easier specification at the functional level.

In these notes, we formulate, as much as possible, multi-level logic minimization problems independent of technology, and by doing so make some of the problems to be solved better posed. There are two general concepts which aid in this,

the concept of a Boolean network and
the concept of a factored form representation of a logic function.

A Boolean network is a multi-level representation of a set of logic functions just as a PLA is a two-level representation of a set of logic functions. In concept, a Boolean network can be seen as a set of PLA's linked together so that some outputs of one PLA are inputs to another. Each PLA of this set, is not restricted in size or other properties; thus the Boolean network is independent of technology.

The logic function of a PLA is represented in disjunctive form. However, multi-level logic in disjunctive form does not represent very well how the multi-level network might ultimately be implemented, whatever the technology. The second concept, the factored form, is partly a solution for this. One can view a factored form as an additional multi-level representation of a logic function. Roughly, a factored form is a parenthesized logic expression e.g. $(a+b)(c+d)+d'\,e$. We regard factored forms as more closely approximating an implementation of multi-level logic than other methods of representing logic functions. Factored forms can be used to decompose and minimize multi-level logic, and are vital in estimating implementation details such as area and delay during the logic optimization process. While disjunctive forms are the most natural representation for PLA's, we argue that factored forms are most natural for multi-level logic.

Multi-level logic synthesis consists of minimizing logic functions, finding common subexpressions, combining expressions, substituting one expression into a second etc. We use Boolean networks and factored forms as methods of abstracting these processes in order to develop and analyze algorithms which are largely independent of technology.

2.2. Boolean Networks and Data Structures

A Boolean network is a technology independent structure for representing a multi-level logic function. It is a directed acyclic graph where each node i is associated with

a) a variable y_i, and
b) a representation f_i of a logic function.

In the graph, an arc connects from node i to node j if function f_j depends explicitly on the variable y_i. The network has n inputs denoted by $x_1, \ldots, x_n$ and m outputs $z_1, \ldots, z_m$. The inputs and outputs are special nodes (called terminals) of the graph since there is no logic function associated with them. We can view the Boolean network as a subroutine and the input/output nodes as its arguments.

A Boolean network includes the usual PLA representation. In fact, it can represent a standard multiple output PLA in two ways. The first has one internal node for each output. The output of this node is attached to the appropriate output terminal, and the logic function of the node is the disjunctive form for that single output. Thus the Boolean network has one-level only i.e. at most one "gate" separating input from output.

The second method of representing a PLA consists of nodes with only NAND (or NOR) functions, and it can directly represent the output of a PLA minimizer. There is one node for each term of the PLA and one node for each output. As such it is a two-level (or two-stage) Boolean network. In cases where the logic can be implemented as a PLA very well, perhaps because of much term sharing among the outputs, this representation has been found to be a good starting point for the multi-level logic synthesis. However, because it is not limited to NAND's and two levels, multi-level techniques can proceed much further with the optimization. Recently, PLA methods have been been extended into the multi-level logic domain by Sasao [Sas86]. There, two-bit input decoders are used to pair the inputs, creating single four-valued variables. Then a multi-valued PLA minimizer is used to minimize the logic as a PLA and an optimal output phase assignment heuristic is used to select the best phase for each output. At this point, the network is effectively three-level. Finally, common cubes are extracted and the network is further refined to correct for fanout limitations.

A node in a Boolean network can serve also as a repository for information about the node. In our initial implementations, YLE [Bra84c], we stored only the disjunctive form at each node. More recently, MIS [Bra86], we have experimented with storing the logic function in disjunctive form initially, but as the synthesis proceeds, we deposit other information. Typically, a function's complement or factored form may be computed in order to evaluate the network or substitute one node into another. Hence in MIS, we have adopted the strategy of storing the complement, the factored form, the value of a node, delay information, etc. at each node as the information becomes available. We also store a don't-care set, the user's name for the node, if such a name exists, and a short name for display purposes.

Another evolution in data structures has been to store the network in a sparse data format which allows larger circuits to be manipulated. We envision this to be useful when manipulating chip designs for timing and global layout considerations, during which the entire chip description might be resident in memory. The Boolean network can be seen most naturally as a sparse data structure. Using a sparse structure, the disjunctive form at each node is stored compactly as a bit matrix where two columns are used for only those variables which are inputs to the node. The list of input variables (the fanin) is stored at the node (optionally also its fanout). In contrast, the choice used in YLE was a bit map data structure where the disjunctive form is an expanded bit matrix with two columns for each variable in the entire Boolean network. Although this allows for easier detection of common inputs across several functions, ultimately it limits the size of logic. With the sparse data structure, the size of the Boolean network is relatively unlimited, but we pay some run time penalty by having to identify common inputs as we combine or manipulate the various nodes of the network.

As the logic synthesis proceeds, nodes are created and destroyed, and the information at each node becomes more detailed. During the synthesis we need to compute the node's **value** which approximates how much area would be saved if the node were to be implemented. If we have constraints on the delays of the network, we also need to estimate the additional delay caused by that node. Both of these values can be estimated well by using a factored form representation of the logic function of the node. Hence, in the more recent developments at Berkeley, we factor each node during these estimations and deposit the factored form at the node for later use.

2.3. Prime Irredundant Boolean Networks and Equivalence

In general, a Boolean network as an entity, like a PLA, **represents** an incompletely specified multi-output combinational logic function. As we change the network, this representation changes but the underlying logic function is always the same. The several representations are **equivalent** and we seek an optimal one. Although the objective of PLA minimization is primarily to obtain the least number of terms and secondly, the least number of literals, the concepts of prime and irredundant implicants (cubes or terms) of a PLA can be extended to and are useful for Boolean networks.

Let $x_1, \ldots, x_n$ denote the inputs and $y_1, \ldots, y_q$ the intermediate and output variables. Each output is associated with an intermediate variable through a mapping function $o(i)$ in the sense that $y_{o(i)}$ is the ith output function. We assume that there are m outputs. (Since two outputs may be identical, we do not require that o be invertible). Each output may be incompletely specified; hence a don't-care function, d_i, is specified for each output. We view each d_i as a property of the ith output node. The d_i are completely specified functions of the inputs $x_1, \ldots, x_n$; $d_i = d_i(x_1, \ldots, x_n)$.

A **cube** is a subset of literals chosen from the set

$$\{x_1, x_1', \ldots, x_n, x_n', y_1, y_1', \ldots, y_q, y_q'\}.$$

The empty cube is denoted by 0 and the universal cube by 1.

Let $x = (x_1, \ldots, x_n)$ be a point in Boolean n-space, $x \in B^n$. We say that two Boolean networks N_1 and N_2 are **equivalent** if, for each output i, $x \in d_i$, implies that $y^1_{o^1(i)} = y^2_{o^2(i)}$, i.e. when we evaluate the two networks under the same input in the care set, the values for the ith output are equal.

Synthesis transforms an initial Boolean network, N_1, into an equivalent one, N_2, where N_2 is minimal in some sense. Two notions of minimality for PLA's are useful for Boolean networks also.

Suppose each function f_i is represented in disjunctive form as a cover of cubes. Let c be a cube of the cover of the logic function at node i. c is said to be **prime** if no literal can be removed from c without causing the resulting Boolean network, N_2, to be not equivalent to N_1.

A cube c of the cover of the logic function at node i of a Boolean network N_1 is said to be **irredundant** if c cannot be removed from the cover without causing the resulting Boolean network, N_2, to be not equivalent to N_1.

These definitions relate to the logic function at each node being represented in disjunctive form, as a cover of cubes. This is not quite general. We extend these notions to a case where a factored form represents the logic function at each node.

A factored form is formally defined by the following.

A **sum form** is a literal or the sum of two or more product forms.

A **product form** is a literal or the product of two or more sum forms.

A **factored form** is either a sum form or a product form.

Thus a factored form is a sum of products of sums of products , . . . , of arbitrary depth and breadth. It can be represented as a factored algebraic expression, for example

$$(ab' + a'b)(cd(a' + e) + c'(b'e + f)) + bce.$$

or as an alternating series-parallel tree where each node is either a series connection (product) of its inputs or a parallel connection (sum) of its inputs. The alternating series-parallel tree for the above example is

parallel														
series												series		
parallel				parallel								b	c	e
series		series		series					series					
a	b'	a'	b	parallel		c	d	c'	parallel					
				a'	e				series		f			
									b'	e				

Each of the levels of the tree are alternately series or parallel except for the leaves which are literals. The root node is shown as the top node in the diagram above and is arbitrarily chosen as a parallel connection. The children of a node are shown directly below the node. A factored form is "evaluated" starting at the leaves. The value of a series connection is the AND of its children's values and the value of a parallel connection is the OR of its children's values.

Now consider a Boolean network N_1 where the logic function at the nodes are represented by factored forms. A factored form of N_1 is said to be **prime** if no literal of any series connection of the factored form can be removed (shorted) without causing the resulting Boolean network, N_2, to be not equivalent to N_1. A factored form is said to be **irredundant** if no term of a parallel connection can be removed (opened) without causing the resulting Boolean network, N_2 to be not equivalent to N_1.

One goal of multi-level logic synthesis is to develop algorithms for efficiently testing if two Boolean networks are equivalent. A by-product of this would be methods for making the network, whether represented by disjunctive forms or factored forms, prime and irredundant. It has been shown that such networks are fully testable in a sense which is broader that the usual input/output stuck-fault notion [Bar86].

2.4. Factored Forms and Other Representations

A logic function can be represented in a number of ways, as a truth table, in disjunctive or conjunctive form, in Bryant's differential form [Bry86] or in a factored form. The truth table is canonical in that there is only one such representation for each logic function. However, in manipulating logic functions in a technology independent manner, we seek a representation which can be computed quickly and which provides good estimates of such things as the complexity of its implementation or the delay needed by the function. Although a factored form is not unique, with new faster methods for both factoring (see algorithm QF in these notes) and simplifying functions [Bra84b, Bra82b], it is reasonably easy to obtain a near canonical factorization, and it is a representation which provides realistic estimates. For example, the function

$$x = (a + b(c + d))(e + f + g)$$

has 7 literals in its factored form but 24 in its disjunctive form. Clearly, the count of 7 literals is more closely related to the complexity of an efficient implementation than the 24 literals. A factored form also has the property, like Bryant's differential form, of simultaneously representing both the function and its complement (obtained by dualizing the factored form), for example by duality

$$x' = a' \, (b' + c' \, d' \,) + e' \, f' \, g' \; .$$

We show how factored forms can be obtained for representing logic functions, and provide a number of methods for factoring logic functions. These methods range from those which are extremely fast (which are used for estimation purposes) to those which try to provide a near optimal factorization. These later methods are used typically once at the end of the synthesis to obtain a final factorization.

Since factoring and finding common subexpressions are very similar operations, factored forms represent more closely how a function might be implemented with multi-level logic. However, one important difference should be noted. Consider the expression

$$x = (ab + a' \, b' \,)(c + d) + (ab' + a' \, b)(e + f)$$

The expression $ab + a' \, b'$ is a sub-expression that can be extracted and implemented as a separate function;

$$y = ab + a' \, b'$$

$$x = y(c + d) + y' \, (e + f).$$

Note that once the sub-expression is implemented separately, it can be used in both the positive and negative phase. It can also be used to fanout to other logic functions. Even if only used in one function, sub-expression extraction provides more flexibility than factoring. For example in

$$x = (a(b+c)+d)(eg' + g(f+e'))+(b+c)(h+i),$$

$b+c$ is a common subexpression but in the above factoring, which seems to be optimum, $b+c$ is repeated twice in the best factored representation of x. On the other hand, if $b+c$ is extracted;

$$y = b+c$$

$$x = (ay+d)(eg' + g(f+e'))+y(h+i)$$

then only y has to be repeated in an optimal factorization of x. Thus a factor in a factorization of a logic function can be "used" only once in its positive phase, but an extracted subexpression can be re-used many times in both phases.

Nevertheless, factoring and common subexpression extraction are closely related and many of the techniques used for locating good factors and common subexpressions are almost identical. The main difference in these techniques is how a "best" factor is chosen among possible factors. One that is best for factoring may not be the best choice for common subexpression extraction.

2.5. Node Value and Node Elimination

In logic synthesis, we may start with a Boolean network which comes from a high-level specification, or from an implementation in a older technology, or it may be the output of a logic tool such as a PLA minimizer. During the synthesis process, at each stage the network is changed in some way. Thus at any one stage, we may have a network with a mixture of unoptimized parts and other groups of logic decomposed in a near optimal way.

Thus,we require a method for preserving the good part of the network and eliminating the other. For this purpose we define the **value** of a node and we use this to preserve the good structure of a partly decomposed network. The value of a node is a local property, depending on the node and the node's immediate fanout. The value of a node is roughly what would be lost if the node were eliminated. Since what is lost depends on the objective of the synthesis, the definition of value may vary between different stages of the synthesis process. For example, initially our objective may be just area alone, but later in the design process, we may be concerned with trading area for improved performance along the critical paths. We define the value of a node, j, of the network as a linear combination of area and delay objectives,

$$\text{value}(j) = \alpha \; \text{area\_value}(j) + \beta \; \text{delay\_value}(j).$$

We discuss the area value estimate first. Area will be measured in terms of the total number of literals in the factored form representations of the node functions. (Enhanced area estimates may also depend on the number of input literals for estimating wiring area.) Ideally, then the area value is the difference between the number of literals of the resulting network if the node is eliminated (absorbed) and the number of literals of the current network. To compute this exactly would require that the node in question be **eliminated**, i.e. everywhere y_j is found in another expression, it is replaced by f_j and everywhere y'_j is found it is replaced by f'_j. Then each of the modified nodes must be re-factored to obtain the new literal count. It is of course possible but somewhat time consuming to do this, so we seek a method which indirectly provides a good estimate of the result of this elimination.

Once a node's value has been estimated, it can be stored at a node. Since the value only depends on the immediate fanout nodes and the node itself, the value remains valid if none of these nodes change. However, it is important to invalidate the value deposited at the node if any of these nodes change after the value is computed.

To estimate the literal savings, we note that if the node j is eliminated into a fanout node l and then node l is re-factored, we can approximate the resulting number of literals using the current factorization of node l. Let n_l be the number of times either literals y_j or y'_j appears in the factored form for f_l, and let L_j be the number of literals in the factored form for f_j. Then we define

$$\text{area\_value}(j) = \left(\left(\sum_{l \,\epsilon\, \text{FANOUT}(j)} n_l \right) - 1 \right) \times (L_j - 1) - 1.$$

The motivation for this is that we can replace each occurrence of y_j by making a copy of the factored form for f_j, and placing it directly into the factored form for f_l. Similarly, y'_j can be replaced by the dual (see section 4.3.3) of the factored form for f_j. Since the dual has the same number of literals, we treat the occurrences of y_j and y'_j equally. The -1's in the formula account for the number of literals needed to implement node j, and to use y_j and y'_j in other nodes.

To estimate the delay value, we assume that we have a formula or table which for a given factored form would provide an estimate of the delay through a node. We also assume that a slack value, s_j, and signal arrival time, a_j, are given at each node. These are estimates done at the system level. The slack is an estimate of how much more delay a node can introduce before increasing the critical delay through the system. Again in estimating the delay value of a node, we could eliminate node j and recompute the new arrival times of the signals exiting the fanout nodes. Then the delay value is obtained by

$$\text{delay\_value}(j) = \max\left\{ 0, \max_{l \,\epsilon\, \text{FANOUT}(j)} \left\{ a_l^{\text{new}} - a_l^{\text{old}} - \max(0, s_l) \right\} \right\}$$

Note that there is 0 delay value if the new delay is less than the old one, or if the difference between the new delay and the old delay does not exceed the positive slack. Thus for nodes that are not critical (positive slack), there is 0 delay value even if the delay is increased when the node is eliminated providing it does not increase more than the slack. This means, in this case, that the value of the node is given completely by the area value estimate.

The question remains how to estimate the delay value of a node without going to the expense of eliminating it. One answer is similar to the area estimation. We can estimate the new factored forms for the fanout nodes by replacing y_j by the factored form for f_j directly in the factored form for fanout node l. Then the new arrival time a_l^{new} can be computed using the delay estimation for factored forms.

The value of a node is used in several places in logic synthesis. First, during a crude collapsing process which is used to decrease the number of nodes so that local transformations at a node become more effective and more global. Thus we use the value of a node in the following collapsing process, in order to keep the good structure and eliminate the rest.

COLLAPSE(θ):
```
while (any node has changed) {
    for (each node j in the network) {
        if (value(j)≤θ) ELIMINATE(j)
    }
}
```

This procedure iterates until there are no nodes of the network with value $\leq\theta$. Since, as elimination proceeds, the value of a node may change, it is not possible to eliminate all nodes with value $\leq\theta$ in one pass through the network. Also, if $\beta>0$, one should re-compute the arrival times and slacks, since these are affected. However, since the slack computation must be done at the system level, it does not seem feasible to perform this computation during COLLAPSE. Therefore, we assume that the slack values are relatively constant so if $\beta>0$, only the arrival times are updated.

The second place that the node value can be used is during the process of trading area to meet delay constraints. In order to obtain the best trade-off, we would like to identify nodes on the critical paths where this delay-area trade-off is optimal. This is measured by value (j) and this is defined to be the weight of the node. This is used to find a minimum weighted cut-set of nodes (node separation set) which separates the inputs of the critical graph (the part of the network on the critical paths) from the outputs. These nodes are then re-synthesized to effect an area-delay trade-off.

3. Algebraic and Boolean Division

The concept of division applied to logic functions plays a key role in factoring and finding common subexpressions.

We say that a logic function g is a **Boolean divisor** of f if

$$f = gh + r,$$

where h and r are logic functions and h is not null. Similarly, we say that g is a **Boolean factor** of f if $f = gh$. Thus a divisor is a factor of a subset of f.

In the above definitions, any manipulation of f that produces an equivalent logic function is allowed. For example, the function $x = a + bc$ can be expressed as $x = (a+b)(a+c)$, but in order to achieve this knowledge about logic functions is necessary. On the other hand, if $x = ac + ad + bc + bd$, then expressing

$x = (a+b)(c+d)$ requires only algebraic factoring. We have use for both types of manipulations. The true logic manipulations require more time but in principle can achieve superior results. On the other hand, the algebraic manipulations can be made much faster and in many cases give almost as good results. In logic synthesis each kind is used selectively in order to obtain a balance between run time efficiency and quality of results.

In what follows, we are careful to distinguish a logic function from its algebraic representation. We focus partly on disjunctive forms which are minimal with respect to single cube containment. This means that any term (cube) in an algebraic representation is not contained in any other single term. In practice we normally use disjunctive form representations which are an irredundant sum of prime implicants, but these requirements are not strictly necessary. Formally, a disjunctive form which is minimal with respect to single cube containment is called an algebraic or logic **expression** or sometimes just expression. Thus an expression is a particular kind of representation of a logic function. Algebraic, not logic, operations on these expressions transform them into algebraic factorizations.

In order for the manipulation of algebraic expressions to remain in the domain of algebraic expressions, **multiplication** of two expressions g and h is defined only when the sets of variables, on which g and h depend, are disjoint sets. Thus the multiplication of $h = a + b$ with $g = a' + c$ is not defined in the algebraic domain since we would have to know that $aa' = 0$ to obtain another algebraic expression. However, $g = a + b$ multiplied by $h = c + d'$ is defined and the resulting expression is obtained by simple polynomial expansion;

$$(a + b)(c + d') = ac + ad' + bc + bd'$$

The disjunctive form associated with a product of two expressions is an algebraic expression and is unique. Hence the process of multiplication is invertible and we can extend the notion of division.

We say that a logic expression g is an **algebraic divisor** of f if

$$f = gh + r$$

where h and r are logic expressions and h is not null. Here equality is meant in the sense that when we expand the expressions of the right hand side, we obtain exactly the same terms as in f. Similarly, we say that g is an **algebraic factor** of f if $f = gh$.

Some useful relations between Boolean divisors and algebraic divisors of the function and its complement are stated below.

Lemma 3.1: A logic function g is a Boolean factor of a logic function f if and only if $fg' = 0$, i.e. $g' \subseteq f'$ $(f \subseteq g)$.

Proof: If $f = gh$, then $g'f = 0$. Conversely, $g'f = 0$ implies that $f = fg + fg' = fg = g(f + r)$ where $r \subseteq g'$, i.e. $f = gh$.

Lemma 3.2: g is a Boolean divisor of f if and only if $fg \neq 0$

Proof: Since $f = fg + fg' = (f + k)g + r$ where $k \subseteq g'$, then if $fg \neq 0$, $f = hg + r$ where $gh \neq 0$. Conversely, $f = gh + r$ implies that $fg = gh + gr$. Since $gh \neq 0$, then $fg \neq 0$.

Lemma 3.3: If f is a sum of prime implicants, and g is an algebraic divisor of f, then g is a Boolean divisor of f'.

Proof: $f = gh + r$ implies that $f' = g' r' + h' r'$. If $g' r' = 0$, then $f' = h' r'$ or $f = h + r$ implying that f was not a sum of primes. Thus $g' r' \neq 0$ and hence g is a Boolean divisor of f'.

Lemma 3.4: If g is an algebraic divisor (algebraic factor) of f, then g is a Boolean divisor (Boolean factor) of f.

Proof: Trivial.

The first two Lemmas show that for any logic function f there are many Boolean divisors and factors, in fact, **any** function containing f is a Boolean factor of f and **any** function not orthogonal to f is a Boolean divisor of f. This poses a problem in choosing a best factor since there are so many. Lemma 3.4 states that the algebraic divisors are a subset of all the Boolean divisors. Experimentally we have found that the algebraic divisors provide a reasonably good subset from which to choose. The fact that these can be determined quite quickly, as we shall see, makes them good candidate divisors in a logic synthesis system, for either factoring the logic functions or finding common divisors among many functions.

This leads to our first paradigm for multi-level logic synthesis.

a. Minimize each logic function to obtain an algebraic expression.

b. Perform algebraic operations on these expressions.

In general, we face two tasks in using either notion of division. First, to find a good candidate divisor, and the second to effect the division, i.e. to determine, given g and f, the coefficient h and remainder r so that $f = gh + r$.

For notation later, we define the **quotient** of f and g (denoted f / g) as the largest algebraic expression h such that $f = gh + r$. Thus $f = (f / g)g + r$, and the **remainder** r has the property that g is not one of its algebraic divisors.

The following is a sketch of the algorithm for carrying out algebraic division with complexity of $O(n \log n)$ where n is the total number of terms in f and g.

ALG_DIV(f ,g):
$U =$ restriction of f to the literals in g .
$V =$ restriction of f to the literals not in g .
/* note that $u_j v_j$ is the jth term of f */
$V_i = \{ v_j \epsilon V : u_j = g_i \}$.
$h = \cap V_i$.
$r = f - gh$
return (h ,r)

Care should be taken to make this algorithm as fast as possible since it is a key subroutine of many of the algorithms used in a logic synthesis system. One way to accomplish this with complexity of $O(n \log n)$ where n is the total number of terms in f and g , is to numerically encode the cubes of U, V and g . By sorting these numbers, the comparisons required to compute V_i and h can be made efficient, and by keeping track of the indices during the sorting process, the remainder r is an easy consequence. Recently, it has been shown that algebraic division can be done in linear time with the resulting quotient and remainder obtained in the proper order [McG86].

One of the transformations used on Boolean networks, which makes direct use of algebraic division, is called **algebraic re-substitution** and consists of algebraically dividing each function f_i at node i in the network by each function f_j or by f_j' at node j . During re-substitution, if f_j is an algebraic divisor of f_i , then f_i is transformed into

$$f_i = h y_j + r ;$$

similarly for f_j' . In practice, we attempt this for each pair f_i , f_j in the Boolean network, implying as many as $2q^2$ algebraic divisions.

The following observations are trivial but important in circumventing most of these divisions.

The function f_j is **not** an algebraic divisor of f_i if

1. f_j contains a literal not in f_i .
2. f_j has more terms than f_i
3. for any literal, the count in f_j exceeds that in f_i .
4. y_i is in the transitive fanin of f_j .

In some cases, we are not interested in the result of division if the quotient (f_i / f_j) is only a single cube. This can be detected sometimes by another useful filter,

if for any literal, the count for f_j equals the count for f_i , then (f_i / f_j) is either null or a single cube.

4. Algebraic Procedures

4.1. Kernels

The notion of a kernel [Bra82a] of a logic expression was introduced to provide a means for efficiently finding common subexpressions. Kernels are our bridge between algebraic expressions and factored forms. A **kernel** of an expression, f, is defined by the following two rules.

1. A kernel k of an expression f is the quotient of f and a cube c; $k = f / c$.

2. A kernel k is "cube free". (No cube is an algebraic factor of k, i.e. k cannot be rewritten as $k = dg$ where d is a nontrivial cube and g is an expression.)

For example, suppose that

$$f = abc + abde.$$

Then $f / a = bc + bde$ is the quotient of f and the cube a, but it is not cube free since the cube b is a factor of f / a,

$$f / a = b(c + de).$$

However, $f / ab = c + de$ is cube free and hence a kernel.

Since no single cube is cube free, a kernel must consist of at least two cubes. Also, since the universal cube, 1, is a cube and $f / 1 = f$, then if f is cube free, f is considered one of its own kernels.

Associated with each kernel is a cube, called its **co-kernel,** which is simply the cube divisor used in obtaining the kernel. Since the same kernel may be obtained in several different ways by dividing f by different cubes, the co-kernel of a kernel is not unique.

A kernel is said to be of **level 0** if it has no kernels except itself. Similarly, a kernel is of **level** n if it has at least one level $n - 1$ kernel but no kernels (except itself) of level n or greater.

For example, (in most examples, we show the equivalent factored form for clarity and brevity)

$$x = (a(b + c) + d)(eg' + g(f + e')) + (b + c)(h + i).$$

has, among others, the kernels $b + c$ and $a(b + c) + d$, which are level 0 and level 1, respectively, while x itself is a kernel of level 2 since it has level 1 kernels but no level 2 kernels other than itself. Note that

$$y = j(a(b + c) + d)(eg' + g(f + e')) + (b + c)(h + i),$$

is a kernel of level 3 since it contains the level 2 kernel

$$(a(b + c) + d)(eg' + g(f + e'))$$

which has co-kernel j.

The following is an algorithm for computing all the kernels of an expression f.

KERNELS(f):
$c =$ largest cube factor of f
$K = $ KERNEL1($0, f / c$)
if (f is cube free) {
    return $f \cup K$
    }
return K

KERNEL1(j, g):
$R = g$
for ($i = j + 1$; $i \leq n$; $i++$) {
    if (l_i appears in only one term) continue
    $c =$ largest cube dividing g / l_i evenly
    if (l_k not in c for all $k \leq i$)
        $R = R \cup$ KERNEL1($i, g / (l_i \cap c)$)
    }
return R

The algorithm works as follows. The argument j in KERNEL1 is a pointer to the literals already factored out (all literals $\leq j$ have been processed). KERNEL1 is designed to find all kernels associated with any cube divisor **not** containing any of the literals l_i for $i \leq j$. The recursive call to KERNEL1 restricts the terms to those containing literal i. These are then kernels which have as co-kernel a cube whose literals include l_i and the literals of c, the largest cube factor of g / l_i. The recursion is done only if the cube c has no literals $k \leq i$, since all co-kernels associated with this recursion will involve the literals of c, and if one of these has been factored already, we would just reproduce a kernel and co-kernel already found. This makes the algorithm such that we only process unique co-kernels. Also it gives the algorithm a very effective tree trimming strategy for searching for kernels.

A few of the important properties of kernels are discussed below. We use the notation $K(f)$ to refer to the set of all kernels of f.

Theorem 4.1.1: If two expressions f and g have the property that $k_f \in K(f)$ and $k_g \in K(g)$ imply that k_g and k_f have at most one term in common, then f and g have no common algebraic divisors with more than one term.

This theorem is used for detecting if two or more expressions have any common algebraic divisors other than single cubes. This can be done by computing the set of kernels for each logic expression, and forming nontrivial (more than one term) intersections among kernels from different functions. If this intersection set is empty, then we need only look for divisors consisting of single cubes (which is an easier task). Thus, we do not need to compute the set of all algebraic divisors for each expression to determine if there are common nontrivial algebraic divisors. This leads to greatly improved run time efficiency since the set of all kernels is much smaller that the set of all algebraic divisors, and, secondly, in the algorithm for computing kernels, the cube-free property of kernels allows us very effectively to trim the search tree for the kernels. This theorem is useful in finding nontrivial common divisors of two or more expressions. We find common cube divisors later by a different algorithm.

The next theorem leads to very fast methods for factoring and finding the prime factorization of a single expression, and for finding the greatest common divisor of several expressions.

Theorem 4.1.2: If p is a kernel of f and is prime (cannot be factored algebraically), then p is a kernel of exactly one of the prime factors of f .

Proof: Suppose $f = \prod f_i$. Let d be a co-kernel of p. We can uniquely write $d = \prod c^i$, where support(c^i)$\subseteq$support(f_i). Since $p = f/d = \prod(f_i/c^i)$, then p can be prime only if c^i is a cube of f_i for all but one i. Thus $p = f_i/c^i$ and since p is cube free, then p is a kernel of f_i.

Corollary 3.2: Let k be a level 0 kernel of f . Then k is a kernel of exactly one of the algebraic prime factors of f .

Proof: A level 0 kernel is prime.

To explain the meaning of this result consider the expression

$$(a(b+c)+d)(eg' + g(f+e')).$$

This factors into two prime factors $a(b+c)+d$ and $eg' + g(f+e')$. Note that the level 0 kernel $b+c$ occurs in only one of the prime factors, while the level 1 kernel $(b+c)(f+e')$ spans across several of the prime factors. We see later how this can be used to obtain a very fast and effective method for factoring logic expressions.

4.2. Rectangles and Kernels

We now give a useful alternate way of looking at kernels. Among other things, it leads to a fast method for finding a single level 0 kernel. Consider the expression $x = abd' + acd + bcd$ but represented as a Boolean matrix, B, where there is one row for each term in the disjunctive form and one column for each different literal. For example, the expression x is represented as

	a	b	c	d	d'
abd'	1	1	0	0	1
acd	1	0	1	1	0
bcd	0	1	1	1	0

A **rectangle** of B is defined as a subset of rows, R, where $|R| \geq 2$, and a subset of columns, C, such that for each $i \epsilon R$ and each $j \epsilon C$, $B_{ij} = 1$. For example $R = \{2,3\}$, $C = \{4\}$ is a rectangle in the example above, but $(\{2,3\},\{2,3\})$ is not a rectangle. A **prime rectangle** is one which is not contained in another rectangle, i.e. both the row set and column set are not contained in those of another rectangle. The rectangle $(\{2,3\},\{3,4\})$ contains the one above but is prime.

The correspondence between rectangles of the Boolean matrix for f and kernels for f is given by the following discussion, and was suggested by an observation of A. Wang [Wan85] that intersections of kernels can be obtained by the kerneling algorithm.

A **co-rectangle** of a rectangle R,C is the pair R,C' where C' is the set of columns not in C. Thus the co-rectangle of $(\{2,3\},\{3,4\})$ is $R = \{2,3\}$, $C = \{1,2,5\}$. The expression corresponding to a co-rectangle is determined by the entries in the Boolean matrix restricted to the rows and columns of this co-rectangle. In the above example this is $a + b$

Theorem 4.2.1: k is a kernel of f if and only if it is an expression corresponding to the co-rectangle of a prime rectangle of f.

The rectangle itself corresponds to the co-kernel, i.e. the cube divisor used to obtain the kernel. The cube divisor is obtained by "anding" together the literals corresponding to the columns C; in the above example, we obtain the cube equal cd.

Thus the set of all kernels of a logic expression f is in 1-1 correspondence with the set of all prime rectangles of the corresponding Boolean matrix. A level 0 kernel is the co-rectangle of a prime rectangle which is also maximal in the sense that no other rectangle contains its column set. In other words it corresponds to an prime rectangle of maximal width. A prime rectangle of maximal height corresponds to a kernel of maximal level, i.e. one whose rows are not contained in another rectangle.

Rectangles and rectangular coverings (see section 4.4) of Boolean matrices seem to be fundamental concepts in which several logic synthesis problems can be commonly formulated.

This alternate point of view suggests several possibilities concerning kernels. One is that it leads to a polynomial method for computing a set of kernels which contain all level 0 kernels without computing all kernels. The technique would be to

1. Compute and record all pair-wise intersections of the Boolean matrix. Each intersection is a co-kernel.

2. Eliminate all duplicate intersections.

3. For each remaining intersection, (set of columns) find the set of rows of the corresponding prime rectangle. The co-rectangle provides the kernel.

This was coded as part of the MIS development, but it was found that the standard method for computing kernels was superior. Apparently, although the above method has complexity at most $n^2 \log n^2$, where n is the number of terms in the function, it generates many duplicate intersections, so that too much time is spent in computing and then eliminating them. In contrast, the kerneling procedure KERNELS never generates a duplicate co-kernel (intersection) although it must iterate through all the kernels to find the level 0 kernels.

There are several algorithms, such as quick factor QF and greatest common divisor GCD, where we need to find just one level 0 kernel. Here the above method can be effective, but an alternate method for finding a level 0 kernel is obtained by a trivial modification to algorithm KERNEL1 to make it return the first level 0 kernel found.

ONE_LEVEL_0_KERNEL(g):
```
if ( | g | ≤1) return 0
if ((L = LITERAL_COUNT(g )) ≥ 1) return g
for (i =1 ; i ≤ n ; i ++) {
        if (L (i)≤1) continue
        c = largest cube dividing g /lᵢ evenly
        return ONE_LEVEL_0_KERNEL(g /(lᵢ ∩ c ))
        }
```

Here LITERAL_COUNT returns a vector of literal counts for each literal. If all counts are ≤ 1, then g is a level 0 kernel. The first literal with a count greater that one is chosen. It and all literal factors are divided out and the algorithm recurs until no literal appears more than once.

4.3. Factoring and Decomposition

4.3.1. Generic Factoring

We now discuss three different algebraic methods for factoring logic expressions and two which use Boolean methods. The generic factoring method is described by the following recursive procedure.

GFACTOR(f):
If (the number of terms in f is one) return f
$k =$ CHOOSE_DIVISOR(f)
(h ,r)$=$ DIVIDE(f ,k)
return (GFACTOR(k) GFACTOR(h) + GFACTOR(r))

The method first chooses a divisor of f and performs the division to obtain $f = kh + r$. At this level in the recursion, k ,h , and r are expressions which each must be recursively factored. Hence in the last line, GFACTOR is called for each of these expressions. The factored product plus factored remainder is then formed and returned. Internally, the factored forms can be represented by either series-parallel trees or parenthesized expressions.

We can build several variations of factoring by choosing different algorithms for CHOOSE_DIVISOR and DIVIDE. Thus if CHOOSE_DIVISOR merely selects the best literal divisor (CHOOSE_LITERAL) and DIVIDE performs single literal division, we have a very fast but non optimal method for factorization. We call this method LF for "literal factorization".

As an example of literal factorization, consider the expression,

$$x = ac + ad + ae + ag + bc + bd + be + bf + ce + cf + df + dg.$$

CHOOSE_LITERAL might choose $k = a$ and ALG_DIV would return $h = c + d + e + g$ and $r = bc + bd + be + bf + ce + cf + df + dg$. Continuing in this way, LF would obtain, depending on which literals are chosen in sequence, as a factorization of x

$$LF(x) = a (c + d + e + g) + b (c + d + e + f) + c (e + f) + d (f + g).$$

Replacing CHOOSE_DIVISOR with a procedure CHOOSE_KERNEL which computes all kernels and chooses one of greatest "value" would lead to a slower but better quality factorization which we call XF. Continuing with the example, we obtain using XF the factorization,

$$XF(x) = (c + d + e)(a + b) + f (b + c + d) + g (a + d) + ce.$$

Although XF is slower because it takes longer to compute the kernels and to choose a best one, it always gives better factorizations than LF (14 versus 16 literals).

Replacing CHOOSE_DIVISOR with CHOOSE_KERNEL and DIVIDE with BOOL_DIV leads to a procedure BF (for Boolean Factorization). (BOOL_DIV is an algorithm which implements Boolean division and is defined in section 5.4.) This applied to the above example gives the same results as XF. However, for the example

$$x = ab' + ac' + ad' + a' b + bc' + bd' + a' c + b' c + cd' + a' d + b' d + c' d$$

Boolean factorization gives

$$BF(x) = (a + b + c + d)(a' + b' + c' + d')$$

whereas

$$LF(x) = a'(b + c + d) + b'(a + c + d) + c'(a + b + d) + d'(a + b + c),$$

$$XF(x) = a'(b + c + d) + (a + b)(c' + d') + c(b' + d') + c'd$$

Note that the result of BF is not an algebraic factorization.

These three methods, LF, XF, BF, use a generic recursive factoring scheme GFACTOR and two different methods for selecting a divisor,

CHOOSE_LITERAL - pick the best literal (very fast)
CHOOSE_KERNEL - compute all kernels and choose best one (slow)

and the two variations of DIVIDE,

ALG_DIV - algebraic (weak) division (fast)
BOOL_DIV - Boolean (strong) division (slow)

These provide us a spectrum of speed and quality of results for factoring but more recently, we have developed a new method which is almost as fast as LF and produces almost as good results as XF.

4.3.2. Quick Factoring

A variation on the generic factoring schema above is

GFACTOR2(f):
If (the number of terms in f is one) return f.
$k = $ CHOOSE_DIVISOR(f).
$(h,r) = $ DIVIDE(f,k)
If (h is not a cube) { $k = $ CUBE_FREE(h)}
     else { $k = $ ONE_LITERAL_OF(h)}
$(h,r) = $ DIVIDE(f,k)
return GFACTOR2(k) GFACTOR2(h)+ GFACTOR2(r).

The heuristic used here is that having chosen the divisor k, we obtain h and are in the process of writing $f = kh + r$. Given that we are going to use factors k and h, we might as well collect everything that can be multiplied by h. Before this is done we eliminate any literal factors of h (CUBE_FREE) in order to obtain the largest k possible. Then we perform the second DIVIDE to obtain this new k, which must include at least the old k. This can be quite effective if we did not choose the best divisor k perhaps because of run time considerations.

A fast variation of factoring is obtained then by replacing CHOOSE_DIVISOR with ONE_LEVEL_0_KERNEL which determines a single level 0 kernel using one of the algorithms discussed. This leads to the procedure QF (for quick factor).

In the example

$$x = ac + ad + ae + ag + bc + bd + be + bf + ce + cf + df + dg ,$$

QF give and XF give different but similar in quality factorings,

$$\mathrm{XF}(x) = (c + d + e)(a + b) + f(b + c + d) + g(a + d) + ce.$$

$$\mathrm{QF}(x) = g(a + d) + (a + b)(c + d + e) + c(e + f) + f(b + d)$$

but QF is much faster because it only needs to determine one level 0 kernel for each factor. However, for

$$x = (a(b + c) + d)(eg' + g(f + e')) + (b + c)(h + i),$$

we obtain

$$\mathrm{QF}(x) = (a(g(e' + f) + eg') + i + h)(b + c) + d(g(e' + f) + eg')$$

but XF obtains the better result (13 literals versus 16 literals)

$$\mathrm{XF}(x) = (a(b + c) + d)(eg' + g(f + e')) + (b + c)(h + i).$$

by working harder to find the best kernel to be used at the top level of the recursion.

In our more recent implementations of logic synthesis algorithms, [Bra86]. QF has become an important tool since it is so fast and effective. Currently we are using QF continuously during synthesis to estimate area and delay.

4.3.3. Factoring Using Duality

The final factoring method simply recognizes that a factored form can be obtained by factoring the complement function by any of the methods discussed and using duality or DeMorgan's law to obtain a factoring of the function. Consider the example

$$x' = (ab' + a'b)(cd(a' + e) + c'(b'e + f)) + bce.$$

and suppose we want to obtain a factoring for x. The series-parallel tree for x' is

parallel														
series												series		
parallel				parallel								b	c	e
series		series		series				series						
a	b'	d	b	parallel		c	d	c'	parallel					
				a'	e				series		f			
									b'	e				

and using DeMorgan's law a factored form for x is obtained as

$$x = ((a' + b)(a + b') + (c' + d' + ae')(c + (b + e')f'))(b' + c' + e')$$

or

<table>
<tr><td colspan="15">series</td></tr>
<tr><td colspan="12">parallel</td><td colspan="3">parallel</td></tr>
<tr><td colspan="4">series</td><td colspan="8">series</td><td rowspan="5">b'</td><td rowspan="5">c'</td><td rowspan="5">e'</td></tr>
<tr><td colspan="2">parallel</td><td colspan="2">parallel</td><td colspan="4">parallel</td><td colspan="4">parallel</td></tr>
<tr><td rowspan="3">d'</td><td rowspan="3">b</td><td rowspan="3">a</td><td rowspan="3">b'</td><td colspan="2">series</td><td rowspan="3">c'</td><td rowspan="3">d'</td><td rowspan="3">c</td><td colspan="3">series</td></tr>
<tr><td>a</td><td>e'</td><td colspan="2">parallel</td><td rowspan="2">f'</td></tr>
<tr><td></td><td></td><td>b</td><td>e'</td></tr>
</table>

The procedure is outlined below and is denoted by DF for DeMorgan Factoring.

$$\mathbf{DF}(f\,):$$
$$f' \;=\; \mathrm{COMPLEMENT}(f\,)$$
$$F' \;=\; \mathrm{FACTOR}(f'\,) \quad (\text{using LF, QF, XF, or BF})$$
$$\text{return } \mathrm{DUAL}(F'\,)$$

4.3.4. Decomposition

The operation **decomposition** is similar to factoring except that each divisor is formed as a new separate node in the Boolean network and the associated variable is substituted into the function being decomposed. Decomposition is one of the transformations used to break down functions considered too large into smaller pieces.

For each method of factoring, we have the associated method for decomposition. In particular, the method **quick decomposition** based of quick factoring provides a very fast method for quickly breaking down a Boolean network. This can be combined with algebraic re-substitution to provide a means of finding common subexpressions. The method is

1. Apply quick decomposition to each node of the network.

2. Perform algebraic re-substitution of each node into every other node where possible.

3. Eliminate all single literal functions and all functions with small value.

At the end of quick decomposition, each node of the network cannot be factored; each literal appears only once. Resubstitution identifies identical nodes and one is substituted into the other leaving a node whose logic function is a cube with a single literal. These are eliminated along with the nodes that have small value, typically those which do not fanout.

The motivation behind this is that QF is very fast but still identifies good kernels for factoring each single function well. The kernels used for this become nodes of the Boolean network and re-substitution identifies common ones. Thus the common divisors identified in this way are ones that are also near best for factoring. Of course, this is not always the best choice, and not all common divisors are found, but the method is very fast and the results quite good.

4.4. The Rectangle Covering Problem

A **rectangle** of a Boolean matrix B, $B_{ij} \in \{0,1\}$, was defined as a subset of rows and columns (R,C) such that

$$B_{ij} = 1 \quad \text{for all } i \in R, j \in C.$$

This concept has been useful so far in giving an alternate interpretation of kernels. In fact, rectangles and rectangle coverings of Boolean matrices seem to have extensive applications in logic synthesis. We will discuss several of these here (namely optimal factoring and common cube extraction) where the Boolean matrix B_{ij} has different interpretations depending on the problem posed.

In general we will be interested in the weighted rectangle covering problem where we assign each rectangle (R,C) a weight given by a weight function, $w(R,C)$, depending on the rows R and the columns C. We say that B is **covered** by the set of rectangles $\{(R^k, C^k)\}$ if

$$B_{ij} = 1 \quad \text{implies} \quad i \in R^k, j \in C^k \quad \text{for some rectangle } k.$$

A covering need not be disjoint; a 1 in B may be covered by several rectangles. The weight of a cover $\{(R^k, C^k)\}$ is defined as

$$\sum_k w(R^k, C^k).$$

The rectangle covering problem is to find a cover of B with minimum weight.

One straight-forward way to approach this, is to formulate it as a row covering problem which is the second part of the Quine-McCluskey procedure for minimizing logic functions. Recall from section 4.2, that a **prime rectangle** $P = (R,C)$ of B is one not strictly contained in another rectangle of B. We first would find all prime rectangles of B. (Note that because of the relation with the procedure KERNEL for finding kernels, Theorem 4.2.1, we can use an almost identical procedure to find all prime rectangles of B). Now construct another Boolean matrix M which has a row for each 1 in B and a column for each prime rectangle. We put a 1 in M_{ij} if the prime P^j covers the 1 in B associated with row i of M.

A standard row covering problem for M can be stated as

$$\text{minimize: } |x|$$
$$\text{such that: } Mx \geq 1.$$

This would yield a rectangle covering with the least number of rectangles. In general, we want to find a covering of least weight, hence each column of M is given a weight W_i equal to the weight of the associated prime rectangle P^i; $W_i = w(P^i)$. Solving

$$\text{Minimize: } W^T x$$
$$\text{such that: } Mx \geq 1,$$

yields a minimum weighted prime rectangle covering.

As a side remark, if M is a prime implicant table for a logic function and if we associate with each prime implicant P^i, a weight W_i equal to the number of literals in P^i, then we would obtain from the above weighted minimization problem, a minimum literal cover for a logic function. This is because the weight of a prime implicant is minimal, i.e. any implicant contained in a prime has more literals.

Unfortunately, the weight of a rectangle in the other applications to be considered below is not always an increasing function in the size of the rectangle, so that the optimal solution may involve non prime rectangles. Thus in some applications we may be required to build an even larger matrix M which has one column for each rectangle of B, and then find the minimum cover. Later, we discuss heuristic methods, similar to those used in ESPRESSO-II which allow us to focus on only prime rectangles.

4.4.1. Common Cube divisors

The problem of finding common cube divisors is an application of rectangle covering. Consider a matrix B which is derived from a cover of cubes $C = \{c^i\}$ of a logic function f,

$$B_{ij} = \begin{array}{ll} 1 & \text{if } x'_j \in c^i,\ j \leq n \\ 1 & \text{if } x_{j-n} \in c^i,\ j > n \\ 0 & \text{otherwise.} \end{array}$$

Recall that a supercube of a set of cubes is the smallest cube (most literals) which contains each of the cubes in the set. It is a common sub-expression (sub-cube) of the set. A rectangle of B_{ij} corresponds to the supercube of a set of cubes of C i.e. the rectangle (R^k, C^k) is the supercube of the set of cubes $\{c^i\ ;\ i \in R^k\}$. It represents a common cube divisor which can be extracted or used in factoring the function f. If the number of literals in the supercube is the cost of implementing it, then

$$w(C^k) = |C^k|$$

By substituting a single variable to replace the supercube in each of the cubes $c^i,\ i \in R^k$, we remove $(|C^k| - 1)|R^k|$ literals from the cover. Thus the savings in literals (or value) associated with the supercube is

$$-w(R^k, C^k) = |R^k|(|C^k| - 1) - |C^k| = (|R^k| - 1)(|C^k| - 1) - 1.$$

Note that this weight function is monotone decreasing function of the size of the rectangle. Hence, a prime rectangle has minimal weight, so in this case we seek a solution among primes only. Unfortunately, because of overlap of these rectangles, this still may not give us the solution which saves the most literals.

4.4.2. Heuristic Covering Methods

Because of the obvious complexity of finding an optimum rectangle cover, we resort as usual to good heuristics. The following is a greedy procedure for determining a near optimal rectangle cover. It proceeds in two phases. In the first, we attempt to find an optimal prime rectangle cover. If the weight function is not a monotone decreasing function of the size, we proceed to a second phase. In the second phase, we attempt to decrease each of these rectangles in an optimal way to obtain a cover with smaller rectangles. Since the weight of a rectangle decreases with decreasing size, then the reduced covering has less weight. During the iteration, as primes are chosen, an increasing number of 1's in B will be marked as "don't-care" points indicating that the 1 is covered but that it is permissible for another rectangle to cover the same point. However, the weight function should be such that there is neither an advantage nor disadvantage in covering a don't-care point.

RECTANGLE_COV(B):
$P = 0$
while (there are "cares" in B) {
    P_i = prime rectangle of greatest value.
    Mark as don't-care any 1 in B *associated* with P_i
    $P = P \cup P_i$,
    }
P = IRREDUNDANT(P,B)
return P = REDUCE_R(B,P)

Like in ESPRESSO-II, IRREDUNDANT selects an optimal subset of P which covers B. In REDUCE_R (similar to REDUCE in ESPRESSO-II), we try to maximally reduce each of the rectangles in the cover by deleting rows or columns of the rectangle to decrease its weight. This can be done if the other rectangles of the cover jointly cover a row or column of the rectangle.

REDUCE_R(B,P):
for ($i = |P| - 1$ to 1) {
    $\hat{P} = P - \{P_i\}$.
    $\hat{P}_i$ = rectangle of least weight covering $P_i - \hat{P}$.
    }
return $\{\hat{P}_i\}$

Note that in RECTANGE_COV, 1's that are "associated" with the 1's of B covered, are marked as don't-care. Later we have a general notion of a 1 of M being associated with a rectangle. In most applications this means simply that it is covered by the rectangle, but it is important to be able to have 1's indirectly associated with other 1's. The procedure RECTANGLE_COV is analogous to the EXPAND, IRREDUNDANT, and REDUCE sequence of ESPRESSO-II [Bra84b] with REDUCE being necessary because larger rectangles have increased weight. These operations can be iterated, as done in ESPRESSO, by restricting the first part simply to selecting and expanding to prime each rectangle of a given rectangle-covering. This is then made irredundant and reduced, with the reduced rectangles becoming the input to the first part for re-expansion. Iteration would continue until no decrease in weight is obtained.

We will see how the RECTANGLE_COV procedure can be applied to the problem of extracting an optimal set of cube divisors from a cover, and how optimal algebraic factoring can be approached by formulating a related rectangle covering problem. This type of factorization will be extended to multiple functions in section 4.4.6.

4.4.3. Optimal Algebraic Factorization Using Rectangle Covering and Overlap

Sometimes an improved factorization can be obtained by allowing one or more terms to be repeated i.e. a term may be produced by more than one product. As an example consider the function

$$x = a(c+d+e+g)+b(c+d+e+f)+c(e+f)+d(e+f+g)$$

which can be factored optimally as

$$x = (a+b+e+f)(c+d)+(a+d)(e+g)+b(e+f).$$

Note that if this is multiplied out, the term de is produced twice. However, each of the products combine expressions with disjoint variables and hence are algebraic multiplications.

The idea of allowing repeated terms (**overlap**) leads to improved algebraic factorizations and is a natural consequence of viewing the factoring problem as a rectangle covering problem.

Given an expression f, consider its optimum factorization allowing overlap. An optimum factorization is one which has the least number of literals. It is a sum of products of sums of products etc. At the top level, the first sum is a set of products, each of which represents a subset of the terms of f which can be algebraically factored. Our goal in optimal factoring is first to choose these subsets correctly and then to recursively optimally factor their prime factors.

We say that a subset of terms of f is a **product** of f if it can be factored algebraically into two or more prime factors. A **maximal product** of f is a product not contained in another product of f. For example, for

$$x = ac+ad+bc+bd+ec+ed+ef,$$

the subset $ac+bd+ad+bc$ is a product of x but it is not maximal since

$$(a+b)(c+d)\subseteq(a+b+e)(c+d)\subseteq x$$

However, both $e(c+d+f)$ and $(a+b+e)(c+d)$ are maximal products of x.

An optimum algebraic factorization of f is a sum of products, but in general each product need not be a maximal product. However, each product is contained in some maximal product. Thus an appropriate set of maximal products could in theory be reduced in an optimum way thereby obtaining an optimum factorization. We will relate these ideas to a rectangle covering problem using Theorem 5.3.1 below, which relates maximal products to kernels. In turn, kernels are related to rectangles of an appropriate Boolean matrix.

Definition: A subkernel of f is any expression with at least two terms, obtained as the maximal set of terms common to each kernel of any set of kernels of f.

For notation, if $\{e_i\}$ is a set of expressions, then $S(\{e_i\})$ is the maximal set of terms common to all the expressions;

$$S(\{e_i\}) = \{t^j \; ; \; t^j \in e_i \text{ for all } i\}.$$

Thus a subkernel is $S(\{k_i\})$ if it has at least two terms.

Theorem 5.3.1: Let s be a maximal product of f. Then s can be written as

$$s = c \prod_i q_i$$

where c is a cube and each q_i is a cube-free subkernel.

Proof: Since s is a product, it can be written as

$$s = cgh$$

where c is a cube and g and h are either cube-free expressions or 1. Let $h = \{h^i\}$. If $cg \neq S(\{f/h^i\})$, since $cg \subseteq S(\{f/h^i\})$, then $S(\{f/h^i\})h$ is a product of f containing $s = cgh$ contradicting that s is maximal. Thus $cg = S(\{f/h^i\})$ and

$$s = cS(\{f/ch^i\})h.$$

Similarly for h, so

$$s = cS(\{f/ch^i\})S(\{f/cg^i\}).$$

Hence, s can be written as a cube times the product of cube free subkernels of f.

Given a Boolean matrix B, recall the KERNEL procedure of section 4.1. During this procedure, we record a kernel defined as a cube free quotient f/c for some cube (co-kernel) c. Each of the co-kernels is unique by the nature of the KERNEL procedure. Further, since f is minimal with respect to single cube containment, then any set of co-kernels which multiply the same cube must also have this property. Therefore, any such subset of co-kernels is an algebraic expression.

We build a Boolean matrix, M, for the rectangle covering problem as follows. Each row corresponds to a unique co-kernel of a kernel, and each column corresponds to a unique cube, d^j, of any kernel found. $M_{ij}=1$ if kernel k^i contains cube d^j; otherwise $M_{ij}=0$. Since the cubes d^j are unique, then a prime rectangle (R,C) of M corresponds to the intersection of the kernels associated with the set of rows R.

For example, let

$$x = (a+bc)(de+f)+g$$

then M is

	a	bc	de	f
a	0	0	1	1
bc	0	0	1	1
de	1	1	0	0
f	1	1	0	0

Here we see that the second kernel (row 2) was obtained by dividing x by bc (the second cube on the left). This kernel is $de+f$.

Next we order the terms of $f=\{t^l\}$ and **associate** with each $M_{ij}=1$ a term number $m_{ij}=k$ if $c^i d^j=t^k$, i.e. if the product of cubes c^i and d^j is the kth term of f, we make $m_{ij}=1$. Thus two 1's in M are associated if they correspond to the same term.

In the example, if the terms of x are ordered

term number	1	2	3	4	5
terms	ade	af	$bcde$	bcf	g

then the terms associated with the Boolean matrix are

	a	bc	de	f
a	0	0	1	2
bc	0	0	3	4
de	1	3	0	0
f	2	4	0	0

i.e. $m_{13}=1$ since $(a)(de)=ade=t^1$, $m_{14}=3$ since $(de)(bc)=bcde=t^3$ etc.

A rectangle (R,C) of M is then said to **cover the terms** associated with the 1's covered by the rectangle;

$$T(R,C)=\{t^l : t^l=m_{ij}, i \epsilon R, j \epsilon C\}$$

Note that no term is covered more than once by a single rectangle, i.e. if $m_{ij}=m_{kl}$ and $i,k \epsilon R$, $j,l \epsilon C$ for some rectangle (R,C), then $i=k$, $j=l$. An entry $M_{ij}=1$ in M is said to be **associated** with a rectangle (R,C) if $m_{ij} \epsilon T(R,C)$. Note that this is different than a rectangle simply covering that entry. For example entry $M_{32}=1$ is associated with rectangle $(\{1,2\},\{3,4\})$ since $m_{32}=3 \epsilon \{1,2,3,4\}$, the terms covered by the rectangle.

In applying the procedure RECTANGLE_COV to M and m, it is necessary to give a weight to a rectangle (R,C). In the greedy procedure in the first part, we would like to choose a prime rectangle P_i with two objectives. First, we would like to cover as many care points as possible with the next choice of P_i. On the other hand, we would like to choose a rectangle with least weight.

This motivates the following weight function for a rectangle. First, let the "cost" of a cube, $c(c^i)$, be equal to the number of literals in it. Suppose (R,C) is a prime rectangle. Let $\hat{R} \subseteq R$ be the unique set of rows of R each of which contains a care point in one of the columns of C, and similarly for $\hat{C} \subseteq C$. The cost of (R,C) is then defined as

$$\hat{c}(R,C) = \sum_{i \in \hat{R}} c(c^i) + \sum_{j \in \hat{C}} c(d^j).$$

Here, $\{c^i\}$ and $\{d^i\}$ are the cubes of the co-kernels and kernels respectively. We sum over only those rows and columns which contain "care" points, since we don't want to penalize prime rectangles for covering don't-care rows or columns.

In the example, if all points are care points, the cost of $(\{1,2\},\{3,4\})$ is the number of literals in the factorization represented by the rectangle, $(a+bc)(de+f)=6$. However, if points $(1,4)$ and $(2,4)$ are don't-care, then the cost is 5 since $(a+bc)(de+f)$ can be implemented as $(a+bc)de$.

Finally, the weight of (R,C) is defined as

$$w(R,C) = \frac{\hat{c}(R,C)}{\# \text{ care points covered by } (R,C)}$$

In other words, the weight of (R,C) is the average cost of each care term covered by (R,C). In this way we balance our desire to cover many care points against the low cost of small rectangles.

With these notions of the weight of a rectangle, and of which of the 1's in M are associated with a rectangle, we can apply RECTANGLE_COV to obtain a near optimal factorization of f. This will be discussed in section 4.4.5.

4.4.4. Choosing a Rectangle of Least Weight

We next discuss a strategy for finding the rectangle of least weight. It is composed of three algorithms GREEDR, GREEDC and PING_PONG. Each receives as input a matrix B, $B_{ij} \in \{0,1,2\}$ where 2 indicates a don't-care point, and a "cost" for each row and column given by vectors w^R, w^C. GREEDR and GREEDC also have an index, k, as an argument, indicating a given row or column respectively.

GREEDR(B, k, w^R, w^C):
$R = \{k\}$
$C = \{j \; ; \; B_{kj} \neq 0\}$
$P = (R, C)$
while $(\; |C| > 1) \;\{$

$$l = \underset{l \notin R}{\text{argmin}} \; \frac{w_l^R}{\displaystyle\sum_{B_{lj}=1, j \,\epsilon\, C} (w_j^C)^{-1}}$$

$R = R \cup \{l\}$
$C = \{j \; ; \; B_{ij} \neq 0, \text{ all } i \,\epsilon\, R\}$
if $(w(R,C) \text{ is best yet}) \; P = (R, C)$
$\}$
return P

In the above procedure, we start the rectangle with the given row k, and the columns are simply the nonzero (1's or 2's) columns of row k. Next we choose a row l not in R which minimizes the quotient shown. The sum in the denominator is maximized by choosing a row with many care columns ($B_{lj}=1$) in common with C, each of small weight. Then l is added to R, and C is updated in a way which allows don't care points in the rectangle. If the present rectangle is the best seen so far it is recorded. This is iterated until C has only one column. In this algorithm, we see a sequence of prime rectangles starting with one of maximal width and ending with one of maximal height.

The procedure GREEDC is the same as GREEDR except that the roles of rows and columns are interchanged.

The procedure PING_PONG below alternates passes through GREEDR then GREEDC until a rectangle is repeated.

PING_PONG(B, w^R, w^C):
$k = \text{argmin} \; (w_k^R)/ \displaystyle\sum_{B_{kj}=1} (w_j^C)^{-1}$

$P_1 = P_2 = 0$
while $(\text{ either } P_1 \text{ or } P_2 \text{ has not repeated }) \;\{$
$\quad P_1 = (R_1, C_1) = \text{GREEDR}(B, k, w^R, w^C)$
$\quad j \,\epsilon\, C_1 = \text{argmin} \; (w_j^C)/ \displaystyle\sum_{B_{ij}=1} (w_i^R)^{-1}$
$\quad P_2 = (R_2, C_2) = \text{GREEDC}(B, j, w^R, w^C)$
$\quad k \,\epsilon\, R_2 = \text{argmin} \; (w_k^R)/ \displaystyle\sum_{B_{kj}=1} (w_j^C)^{-1}$
$\}$
return the repeated rectangle P_i.

4.4.5. Optimal Factoring, OF

Finally, a near optimum factoring procedure is the following.

QF(f):
Build Boolean matrix M and term map m using kerneling procedure on f.
Compute w^R, w^C from co-kernels $\{c^i\}$ and kernel cubes $\{d^i\}$.
$B = M$
$P = 0$
while ($B_{ij} = 1$ for some ij) {
      $P_1 = (R_1, C_1) = \text{PING\_PONG}(B, w^R, w^C)$
      $P_1 = \text{EXTEND}(P_1, B, w^R, w^C)$
      $P = P \cup P_1$
      for (all ij) {
            if (i, j associated with P_1) $B_{ij} = 2$
            }
      }
$P = \text{IRREDUNDANT}(P, M, m)$
$P = \text{REDUCE\_R}(M, m, P)$
for ($i = 1$; $i \leq |P|$) {
      $f_i = \{c^j ; j \in R^i\}$
      $g_i = \{d^j ; j \in C^j\}$
      }
return ($\sum \text{OF}(f_i) \times \text{OF}(g_i)$)

OF uses PING_PONG to determine a near optimal selection of the first level of factorization, i.e. the first set of products in the sum of products representation. After each rectangle, P_1, has been selected, we recognize that PING_PONG is not perfect in finding the best rectangle. Hence, similar to what is done in GFAC-TOR2, we try to use this rectangle to find a better one by the procedure EXTEND. This examines the expression determined by the columns of P_1 to see if it factors. Note that, if there are factors of P_1, then each factor will be represented by a subset of columns since the kerneling process used in forming the Boolean matrix, M, found all kernels. The associated rectangle of each factor is examined to see if it is better that P_1, and if so, it replaces P_1. EXTEND returns with the best such factor found.

An irredundant subset of P is selected and each of these are reduced. These are then recursively optimally factored by applying OF to the expressions f_i and g_i formed by the co-kernels of the rows and cube kernels of the columns of the reduced rectangles.

4.4.6. Optimal Decomposition: Multiple Functions

We now discuss the problem of finding an optimal set of subexpressions for simultaneously decomposing many functions.

4.4.6.1. Cube Divisors

In section 4.4.1, the problem of finding an optimal set of common cube divisors was transformed into a related rectangle covering problem. This could be solved by computing all the prime rectangles and solving a related row covering problem for an incidence matrix as discussed in section 4.4. The prime rectangles can be computed efficiently by using a minor modification to the KERNELS procedure. Here we discuss the heuristic of applying the greedy PING_PONG procedure to the problem of finding an optimal set of common cubes for decomposing a set of functions. In section 4.4.6.2, we discuss a similar strategy for finding an optimal set of common kernels for decomposing a set of logic expressions. In practice, these procedures are applied to an entire Boolean network.

First, we build the cube-literal incidence matrix as before where there is one row for each cube c^i in the network and one column for each literal,

$$B_{ij} = \begin{array}{ll} 1 & \text{if } x'_j \in c^i, \ j \leq n \\ 1 & \text{if } x_{j-n} \in c^i, \ j > n \\ 0 & \text{otherwise.} \end{array}$$

The weight w_i^C associated with a column is 1 since each column is a single literal. Also, $w_i^R = 1$. In this way, the next row l chosen in GREEDR is the one with the most cares in the columns of C. The corresponding rectangle is judged using the weight function

$$w(R,C) = - \left(\left(|R|-1 \right)\left(|C|-1 \right) - 1 \right)$$

as discussed in section 4.4.1.

Next, the "optimal" set of rectangles is selected as in OF and made irredundant and reduced. However, instead of factoring, our objective here is to build a node for each of the cubes determined, and to substitute these everywhere. The only problem is that we cannot use algebraic division algorithm here in a straight-forward way since in general the rectangles selected may have overlap. Thus if we were to substitute one cube into all functions and then the next cube, in some sequence, the resulting substitution may block the algebraic division algorithm from obtaining the desired result. Thus we must substitute all the cubes simultaneously. We will face a similar problem in the next section in having to simultaneously substitute kernels into many functions because of the overlap of the rectangles.

However, in the common cube case, the problem of simultaneous substitution is easy. Since we have all the rectangles in the cover, we know in which terms each cube can be used. This is just the row set R^i of the rectangle. Hence, for each term (row), we form a set of new literals which include one for each new cube node that covers a literal in that term, and one for each original literal of the term not covered by a cube (it is possible that a cover of rectangles may not cover all 1's in the Boolean matrix if we terminate the covering problem because the rectangles chosen have too much weight). Then each term is replaced by the corresponding new term. This is summarized in the following procedure.

OPT_CUBE_EXTRACTION:
Build cube-literal incidence matrix M for the entire Boolean network
Set the association relation m to the identity.
$w^R = w^C = 1$
$B = M$
$P = 0$
while ($B_{ij} = 1$ for some ij) {
$\qquad P = P \cup P_1 = (R_1, C_1) = \text{PING\_PONG}(B, w^R, w^C)$
$\qquad$ for (all ij) {
$\qquad\qquad$ if (i, j associated with P_1) $B_{ij} = 2$
$\qquad\qquad$ }
$\qquad$ }
$P = \text{IRREDUNDANT}(P, M, m)$
$P = \text{REDUCE\_R}(P, M, m)$
for ($i = 1$; $i \leq |P|$) {
$\qquad$ form new node $q + i$ with function $f_{q+i} = \{ l_j \; ; \; \text{if } j \in C^i \}$
$\qquad$ }
for (each term t^i in network) {
$\qquad L = $ literals of t^i not covered by P
$\qquad$ replace t^i by $\{ y_{q+j} \; ; \text{if } i \in R^j, 1 \leq j \leq |P| \; \} \cup L$
$\qquad$ }

4.4.6.2. Common Kernels

With multiple functions, we would like to determine a set of kernels or kernel intersections, which are common to more than one function. Then the new node created for this kernel would fan out to at least two other nodes. We would like to find an optimal set of such common kernels.

One greedy method for this is to compute the set of kernels for each function of the network. Next, we identify common kernels or common kernel intersections. Among these, the one with greatest value is selected and substituted everywhere. The next kernel of greatest value is then chosen and substituted. The value of a kernel can be determined during the kerneling and intersection process.

However, after one kernel is selected and substituted, it is possible that the values of the remaining kernels change. In fact, after one substitution, a kernel may cease to be a kernel of the resulting network. Of course, only the most valuable kernel could be selected and substituted, and then the entire kerneling process repeated on the resulting network to obtain the next kernel. However, this is quite expensive.

Several variations on this have been tried. One is to try to substitute the leading p kernels where p is some parameter. After p substitutions, the network is re-kerneled and the procedure repeated until no common kernels are found. Then common cube extraction is performed on the network.

Another variation is to compute only level 0 kernels during the kerneling process, and only "level 0" intersections. This limits the number of kernels generated and the number of intersections required, and hence is quite a bit faster. A heuristic for using only the level 0 kernels is that by iteratively kerneling at level 0, effectively higher level kernels are found implicitly if one level 0 kernel is used by another one.

The computation to find the intersections of all kernels is not trivial. In some systems, the intersections are restricted to only intersections of pairs of kernels. Because of the relation between kernels and rectangles, we see that the problem of finding kernel intersections is the same as kerneling, and limiting intersections to pair-wise intersections is the same as the procedure for finding level 0 kernels described in section 4.2.

In this section, we take a more global approach. It is an approach similar to OF, except that at the end, we must simultaneously substitute everywhere, the kernel or kernel intersections which are associated with the rectangle cover. Note that no kernel intersections need be determined, since this is done implicitly during REDUCE_R.

The first step is to build a Boolean incidence matrix for the kernels and co-kernels as was done in section 4.4.3. This is done for each node of the network, and a merged Boolean matrix is formed, composed of each of the individual matrices. In the merging process, we identify common kernel cubes so there is one column for each unique cube of any kernel. The co-kernels are not merged, but instead each is marked indicating which function (node) it came from. Thus the Boolean matrix M has one row for each co-kernel, c^i, and one column for each unique cube, d^j, of any kernel. An entry $M_{ij}=1$ if kernel i contains cube d^j.

Now consider the interpretation of a rectangle of M. It is associated with a set of columns C which corresponds to a set of cubes $k=\{d^j \; ; \; j\in C\}$. k is an algebraic expression representing either a kernel or an intersection of kernels. On the left-hand side, the set of cubes associated with the rows, $\{c^i \; ; \; i\in R\}$ determines in which functions k can be substituted.

To determine the association mapping m, we give each of the terms of the functions a unique integer and as before associate a 1 in M with a term t^l if $c^i d^j = t^l$ and $M_{ij}=1$. The weights associated with the rows and columns, w^R, w^C are the same as in OF i.e. equal to the number of literals in c^i, d^j respectively.

After PING_PONG, REDUCE_R, and IRREDUNDANT have selected an "optimal" rectangle covering P, we create a new node for each expression associated with the columns C^i and simultaneously substitute these everywhere. Because of the possible overlap of the rectangles, it is more difficult to effect an optimal substitution of these expressions using algebraic methods.

This problem is illustrated by the following example. Suppose we have as one of our expressions in the network

$$v = ((a+e)(b+c)+(a+c)(b+d))(f+g),$$

and suppose that during the optimal decomposition, we have selected, possibly

because of other functions in the network, to be created the following new nodes
of the network,

$$x = a + e$$
$$y = b + c$$
$$z = a + c$$
$$w = b + d$$
$$u = f + g.$$

These can be substituted into v in an optimal way to obtain $v = (xy + zw)u$, but
this is not easy. For instance if we do the substitutions in the sequence,
x, y, z, w, u using ALG_DIV, we get $(xy + zd + bc)u$. One answer is to use
BOOL_DIV and simultaneously divide v by x, y, z, w, u by forming the don't-care
set

$$x'(a+e) + xa'e' + y'(b+c) + yb'c' + z'(a+c) + za'c' + w'(b+d) + wb'd' + u'(f+g) + uf'g'$$

However, this can be expensive, and is probably an overkill, since the result we
seek can be obtained by algebraic operations on the expressions.

Since, currently, we do not have a clean solution to the simultaneous algebraic substitution of kernels, we will omit further discussion here, and in the
algorithm below, simply refer to this process as SIM_ALG_SUB for simultaneous
algebraic substitution.

OP_DECOMP:

Build Boolean matrix M and term map m using the
kerneling procedure on each function in the network.

Mark each row k with $FM(k) = $ network node co-kernel k came from.

Set w^R , w^C to the literal counts of the
co-kernels $\{c^i\}$ and kernel cubes $\{d^i\}$ respectively.

```
B = M
P = 0
while (B_ij = 1 for some ij) {
        P_1 = (R_1, C_1) = PING_PONG(B, w^R, w^C)
        P_1 = EXTEND(P_1, B, w^R, W^C)
        P = P ∪ P_1
        for (all ij) {
                if (i,j associated with P_1) B_ij = 2
                }
        }
P = IRREDUNDANT(P, M, m)
P = REDUCE_R(P, M, m)
for (i = 1 ; i ≤ | P | ) {
        S_i = {k ; for k = FM(j), j ∈ R^i)}
        y_{q+i} = f_{q+i} = {d^j ; j ∈ C^j}
        }
for (i = 1 ; i ≤ q) {
        f_i = SIM_ALG_SUB(f_i, {f_j ; i ∈ S_j})
        }
```

In this algorithm, after REDUCE_R, we create new nodes associated with the
column sets of the covering rectangles $\{P^i\}$. These become nodes
$q+1, \ldots, q+|P|$ of the network. Next we effect the simultaneous algebraic substitution of these into the original nodes $1, \ldots, q$ where we use the information
given by the row sets R^i and the row marking function FM to determine which
of the new nodes should be used for simultaneous substitution into the original
nodes.

It should be noted that these ideas have not been fully developed and that
the final implementation may differ in some details from the one described here.

5. Boolean Procedures

We discussed extensively some manipulations of algebraic expressions, which
preserve their algebraic nature. The algebraic procedures manipulate monomials;
polynomials where no power is greater than 1. Algebraic multiplication and division are defined only under the condition that two factors have disjoint variable
sets. A general paradigm was developed for synthesis of Boolean networks by

minimizing each logic function to an irredundant sum of primes, and then applying various algebraic procedures.

By restricting attention to algebraic expressions of the Boolean functions, we are able to develop very fast methods for transformation of the Boolean network. However, while we gained in speed, some of power of fully Boolean function manipulation was lost.

For instance, the Boolean function

$$f = a'\,(b+c+d)+b'\,(a+c+d)+c'\,(a+b+d)+d'\,(a+b+c)$$

can be factored algebraically as

$$f = (d+c)(a'+b')+(a+b)(c'+d')+c'\,d+a'\,b+cd'+ab'$$

However, using a Boolean procedure, BF, f can be expressed more simply as

$$\mathrm{BF}(f) = (a+b+c+d)(a'+b'+c'+d').$$

Since the two factors in the result above have a some variables in common, this could not have been achieved by the algebraic methods. The general factored form, such as the one above, also points out that the notion of each node being represented by an algebraic expression is too restrictive. For instance, what is the algebraic expression for

$$(a+b+c+d)(a'+b'+c'+d')?$$

Is it $a'\,b+ab'+a'\,c+ac'+b'\,c+bc'$ or $ac'+b'\,c+a'\,b$? The conclusion seems to be that, in general, each node should be treated as a logic function with the possibility of a variety of different representations.

Boolean procedures are those that use the identities of Boolean algebra such as $aa'=0$, $aa=a$ and $a+a'=1$. An example of a Boolean procedure is any of the two level logic minimization algorithms, such as ESPRESSO-II [Bra84b] and all of its sub-procedures for manipulating logic such as COMPLEMENT, TAUTOLOGY, REDUCE etc. The Boolean procedures are typically slower than the algebraic procedures and hence must be used discriminantly.

In this section we derive some of the Boolean procedures that have proved useful in manipulating logic functions. The focus will be single functions possibly with don't-care sets. Later in section 5.6.2, we apply some of these procedures to the case where there are many functions interacting in a multi-level Boolean network.

5.1. Minimizing Logic Functions with Different Objectives

Let f be a logic function with a don't-care function d. If we minimize f with respect to d, using a heuristic logic minimizer, such as ESPRESSO-II, the final representation is somewhat dependent on the heuristics and procedures used. Even using exact minimizers, such as McBOOLE [Dag84] or ESPRESSO-EXACT [Rud86a], the result is only guaranteed to have a minimum number of terms. There may be several such minimum covers, some more desirable, such as the one with the least number of literals. Indeed, another irredundant prime cover, not even minimum in the number of terms, may have the minimum number of literals. In general, there are various objectives desired in minimizing a logic function. Typically, two-level logic minimizers have as their primary objective, the minimum number of terms, because the output of the minimizer is usually implemented as a PLA where the area is proportional to the number of terms. However, with multi-level logic as the method of implementation, the number of terms is one of the least important objectives. Rather minimization objectives such as the minimum number of

> literals
> variables that the function depends on (variable support)
> literals that the function depends on (literal support)
> literals in a factored form

are much more important. The last objective reflects an implementation not represented by a logic function in disjunctive form, so even the notion of finding a minimum (in some sense) disjunctive form is not sacrosanct. The middle two objectives reflect heuristics which relate the least number of variables or literals possibly leading to good factoring or easier wiring problems. These are called minimum support objectives. In practice, the user may prefer a result which reflects a combination of these minimization objectives.

5.2. Don't-Care Sets

We know it is important, in minimizing logic, to take advantage of don't-care sets. In multi-level logic, this becomes even more important because of additional don't-cares implied by the nature of multi-level logic. Don't-care sets can arise in any kind of logic. The most frequent is in control logic where a set of control signals may have been encoded using $b \geq \log_2 n$ bits where n is the number of controls to be encoded. If $n < 2^b$ then there are unused code points, and since these should never occur, they can be used (except in an error function) as don't-cares to help simplify the logic. Don't-cares are also implied in parallel datapaths where one path does something different than the other. If there are instructions which do not exercise a part of the datapath, then we can let that part compute nonsense under that instruction, knowing that the result will never be used. These are two examples where the don't-care set can be inferred by the user because of some more global knowledge he has about the system design. However, because of the nature of multi-level logic,there are many more such don't-care points.

In multi-level logic, the first kind of don't-care set are external ones specified by the user. These are denoted by $d_1, \ldots, d_m$ one for each of the m outputs. We refer to these as the **output** don't-care set. In some cases, they are all the same don't-care set. Each of the d_i is a function of the inputs only; $d_i = d_i(x_1, \ldots, x_n)$.

Another kind of don't-care arises because of the new variables introduced at the intermediate nodes. Since $y_i = f_i$, where y_i is the variable at node i and f_i is the logic function at node i, then we don't-care if $y_i \neq f_i$. Thus another kind of don't-care is the function

$$\sum_{i=1}^{q} (y'_i f_i + y_i f'_i)$$

which is the union of all the don't-cares from all the intermediate nodes. We call this the **global** don't-care set since it applies to all functions.

Finally, there is the don't-care set due to the way a node's output is used. It is local to the node. This is called the **fan-out** don't-care set. As an illustration of the fan-out don't-care set of a node, consider the network

$$x = a' b + ab' c'$$
$$z_1 = ax + c$$
$$z_2 = a' x' + c'$$

A don't-care set for x can derived by looking at its fan-out and seeing how specifically x is used. In this case we can derive that for z_1, if $c = 1$ or $a = 0$, then it does not matter what x is since it does not effect the value of z_1. This is written as $c + a'$. Similarly, for z_2, x has no effect if $c' + a = 1$. Thus we don't care what x is under the product of these conditions;

$$(a' + c)(a + c') = ac + a' c' .$$

Since this is a don't-care condition for x, we can use it to simplify the logic function for x and obtain, for example,

$$x = bc + b' c' .$$

If we replace the logic function at node x by this, we obtain an equivalent Boolean network but with one less literal. Thus the original network was not prime in the sense discussed in section 2.3.

In general, the individual don't-care set for each fan-out is obtained by computing the logic function $f_x = f_{x'}$, where f_x is the cofactor of function f with respect to x. The immediate fan-out don't-care set for x is then the product of each of these functions;

$$\prod_{i \,\epsilon\, \text{fanout}(x)} (f_{ix} f_{ix'} + f'_{ix} f'_{ix'})$$

In the example, $z_{1x} = a + c$ and $z_{1x'} = c$, so that $(z_{1x} = z_{1x})' = a' + c$. Similarly, $(z_{2x} = z_{2x'}) = a + c'$ and the product of these is $ac + a' c'$, as above.

Continuing in this same vain, if z_1 has a don't-care set, d_1, and z_2 has a don't-care set, d_2, then the fanout don't-care set for x is

$$(a' + c + d_1)(a + c' + d_2)$$

In general, if the fanout of a node j is only to outputs, then the fan-out don't-care set is

$$\prod_{i \,\epsilon\, \text{fanout}(j)} (f_{ix} f_{ix'} + f'_{ix} f'_{ix'} + d_i)$$

where d_i is the output don't-care set associated with node i, which is an output.

It would appear then that it is possible to extend this to the case where the fanout may be to another internal node, say node k. The fanout don't-care set used node k would be the one derived for it by considering where it fans out. This could be done recursively starting at the output nodes and working back to the inputs. At each node we would have computed already, since the graph is acyclic, the fanout don't-care set for each of the node's fanouts, and these would be used in the expression above in place of the d_i's. Unfortunately, this does not always work correctly and can cause errors. The trouble is caused by possible reconvergent fanout.

These are familiar notions for people who work in testing, and it seems clear that the use of these don't-care sets is another approach to some testing issues. These questions are fully developed in [Bar86, Bar87].

The don't-care set generated for a node can include the global and fanout don't-care sets. Since these can be extremely large, in practice the don't-care set used is usually severely limited, for example to the immediate fanin and fanout of a node. Better methods for limiting the don't-care set to a "useful" subset are required and need further investigation.

5.3. Minimum Support Algorithms for Incompletely Specified Functions

We have discussed how different objectives are appropriate when minimizing multi-level logic. Two of these concern an objective which was motivated by decreasing the connectivity of the network. Here we discuss some such algorithms for minimizing with these objectives in mind, making use of any don't-care set available.

Let $ff = \{f, d, r\}$ be an incompletely specified logic function, and suppose that F, D, and R are given cube covers for f, d, and r respectively. Given a cube $c^i \in F$, we define the **variable blocking matrix**

$$(B^i)_{jk} = \begin{array}{ll} 1 & \text{if } c_k^i = 1 \text{ and } r_k^j = 0 \text{ for } r^j \in R \\ 1 & \text{if } c_k^i = 0 \text{ and } r_k^j = 1 \text{ for } r^j \in R \\ 0 & \text{otherwise.} \end{array}$$

This is the same as the blocking matrix used in the EXPAND step of ESPRESSO-II. The **super variable blocking matrix** B is composed of all the rows of all the B^i. Note therefore that B has $|F| \times |R|$ rows.

A **row cover** of B is a binary vector v, $v_i \in \{0,1\}$, such that $Bv \geq l$, where l is the vector of all 1's.

Theorem 5.3.1: If v is a row cover of B, and $v_i = 0$, then there exists a cover for ff which is independent of the variable x_i.

Corollary 5.3.1: If v is a minimum row cover of B ($|v|$ is minimum), then a minimum variable support for ff is the set of variables $\{x_i ; v_i = 1\}$.

Of course, there may be several minimum row covers for a binary matrix, hence there are several associated sets of variables which are minimum support sets for ff.

In the above formulation, the resulting prime irredundant cover for ff may depend on both x_l and x'_l, but this only counts once in determining the minimum variable support. Since each unique literal appearing in a cover may represent a wire which must be routed to the function in an implementation, a minimum literal support may be more desirable. This is solved in a similar way.

For each $c^i \in F$, define the **literal blocking matrix** $\hat{B}^i$ as

$$(\hat{B}^i)_{jk} = \begin{array}{ll} 1 & \text{if } c_k^i = 1 \text{ and } r_k^j = 0, \ r^j \in R \\ 0 & \text{otherwise} \end{array}$$

$$(\hat{B}^i)_{j,n+k} = \begin{array}{ll} 1 & \text{if } c_k^i = 0 \text{ and } r_k^j = 1, \ r^j \in R \\ 0 & \text{otherwise} \end{array}$$

Let $\hat{B}$ be the **super literal blocking matrix** composed of all the rows of all the $\hat{B}^i$.

Theorem 5.3.2: If v is a row cover of $\hat{B}$ and $v_l = 0$, $l \leq n$, then there exists a cover of ff independent of x_l; if $v_{n+l} = 0$, $l \leq n$, then there exists a cover of ff independent of x'_l.

Corollary 5.3.2: If v is a minimum row cover of $\hat{B}$, then a minimum literal support for ff is the set

$$\{x_l \ ; \ v_l = 1\} \cup \{x'_l \ ; \ v_{n+l} = 1\}.$$

These various row covering problems can be solved heuristically. One of the best such techniques is the one used in ESPRESSO_EXACT [Rud86a].

5.4. Boolean Division for Incompletely Specified Functions

As with algebraic division, Boolean division is defined in terms of Boolean multiplication, i.e. intersection or ANDing. Thus given logic functions f, g, h, e, we say that g is a Boolean divisor of f if

$$f = gh + e \quad \text{and} \quad h \neq 0,$$

i.e. the Boolean function formed by intersecting g and h and ORing this with e is a cover of f. Extending this to an incompletely specified function $ff = \{f, d, r\}$, we say that g is a **Boolean divisor** of ff if

$$f = gh + e \quad \mod d \quad \text{and} \quad gh \neq 0 \mod d$$

meaning that the equality is not required on the don't-care set d, in other words, $gh + e$ need only be some cover of ff. We require either that g, h and e be completely specified functions, or are functions which all have d as their don't-care set. Note if $\hat{f}$ is any cover of ff, then any factorization of $\hat{f}$ is also a cover of

ff .

If there exists a cover of ff that can be expressed as gh mod d, then we say that g is a **Boolean factor** of ff, or that g divides ff evenly.

Theorem 5.4.1: Suppose $ff = \{f, d, r\}$ is an incompletely specified function and g is contained in $d + r$. Then g' is a Boolean factor of ff (divides ff evenly).

Theorem 5.4.2: If g is a Boolean divisor of $ff = \{f, d, r\}$, then g' is a Boolean divisor of $ff' = \{r, d, f\}$. In applying Boolean division to factoring, there two distinct problems to be solved are:

1) given a logic function g, and an incompletely specified function ff, compute logic functions h and e such that $gh + e$ is a cover of ff and such that h and e are minimal in some sense.

2) given ff, find a function g such that $gh + e$ is a cover of ff and such that g, h and e are minimal in some sense.

The minimal conditions on g, h, e make these problems interesting. It is not hard to find divisors of ff, but it is difficult to find divisors which lead to simple factorizations.

Theorem 5.4.3: Given g and $ff = (f, d, r)$, define an extended don't-care set $\bar{d} = d + xg' + x'g$, where x is a new variable. Suppose $\hat{f} = xh + e$ is a cover of $\overline{ff}$ (with the extended don't-care set $\bar{d}$), where h and e are independent of x and x'. Then $gh + e$ is a cover of ff.

Theorem 5.4.3 provides a solution to the first problem, in that $\hat{f}/x$ is the minimized cofactor h, given a candidate divisor g. This method of division has two basic steps,

1. Form a larger don't-care set $d + xg' + x'g$ expressing that $x \neq g$ is a don't-care condition, where x is a new variable.

2. Minimize f using this new don't-care set with any two-level logic minimizer that uses don't-cares.

3. Return $(\hat{f}/x, e)$, where e, the remainder, is the terms of $\hat{f}$ which do not include x.

In general, f may be divided by several functions simultaneously. For example, given g, h, we form the don't-care set

$$\hat{d} = d + x'g + xg' + y'h + yh' .$$

A minimal cover $\hat{f}$ of ff using $\hat{d}$ provides us with a minimal factorization

$$f = g(\hat{f}_{xy}h + \hat{f}_x) + \hat{f}_y h + e.$$

As mentioned earlier, in minimizing multi-level logic, we may have several objectives. With division defined as discussed, we focus on minimal factored representations. We see that a minimal cofactor h and remainder e are obtained by extending the don't-care set and using a two-level minimizer. We next take one more step to reduce the support of the final representation.

The procedure BOOL_DIV below uses the heuristic for obtaining the minimum literal support as discussed in section 2.2.2. This is done by a call to procedure MINLIT passing as arguments the care onset f of ff and the care offset $\bar{r}$ of $\overline{ff}$. MINLIT solves the relevant row covering problem and returns a slightly expanded cover, where the literals not in the minimum support have been eliminated.

```
BOOL_DIV(f ,g ,d ):
n = number of variables
g' = COMPLEMENT(g )
        /* add the implied don't-care set to d  using */
        /* the n +1 variable as a new one */
        /* DC =d +gv'_{n+1}+g' v_{n+1} */
DC = APPEND((d ,2),APPEND((g ,0),(g'  ,1)))
        /* form the care offset */
r = COMPLEMENT(APPEND(f ,DC ))
        /* form the care onset */
f = COMPLEMENT(APPEND(r ,DC ))
        /* expand f  to its minimum literal support removing v'_{n+1} */
f = REMOVE(f ,v'_{n+1})
f = MINLIT(f ,r )
        /* expand f  into primes and make an irredundant cover */
f = EXPAND(f ,r )
f = IRREDUNDANT(f ,DC )
        /* compute f /v_{n+1} */
h = COFACTOR(f ,v_{n+1}))
        /* REMAINDER returns terms not multiplying v_{n+1} */
e = REMAINDER(f ,v_{n+1})
return(h ,e )
```

We assume here that f is a given cover of ff, d is the don't-care set for ff, and g is a given candidate divisor. The first step forms a new don't-care set and the corresponding onset and offset, thus giving a new incompletely specified function (f ,DC ,r) in the larger Boolean space. Next v'_{n+1} is removed from the onset and the minimum literal heuristic is applied. At this point, any two-level logic minimizer could be applied, but here only the EXPAND and IRREDUNDANT steps are used. Finally, the coefficient of v_{n+1} is returned as h, and the remaining terms as e.

5.5. Candidate Boolean Divisors and Boolean Re-substitution

We have discussed how functions which are nodes in the Boolean network may be reused if they are algebraic divisors of other functions. Also, in using Boolean division, we have the problem of finding good Boolean divisors. We discuss one way of solving this by extending re-substitution to the Boolean domain, but adding filters so that unnecessary Boolean divisions are not performed.

Recall Lemma 3.1;

Lemma 3.1: A logic function g is a Boolean factor of a logic function f if and only if $fg' = 0$, i.e. $g' \subseteq f'$ $(f \subseteq g)$.

We extend this to a result which is useful in finding candidate Boolean divisors.

Theorem 5.5.1: Let f be a cover of cubes $C = \{c^i\}$ and g be a given logic function. Partition C into C_1 and C_2, where $c^i \in C_2$ if $c^i g' = 0$; otherwise $c^i \in C_1$. Let f_1 and f_2 be the logic functions corresponding to the covers C_1 and C_2. Then

$$f = f_1 + g f_2.$$

This theorem can be used to test if a candidate logic function g may be useful as a divisor of f. The procedure is to compute g' first. (If g is a node of the Boolean network, then g' is usually available at the node, so no complementation needs to be done.) Next, we select the cubes of f orthogonal to g'. These are the set C_2, and if there are sufficiently many, there is some hope that g could be used to simplify f. Note that with f written as $f = f_1 + g f_2$, there is no gain in simplicity yet, since originally $f = f_1 + f_2$. However, with f written as $f = f_1 + g f_2$, we may be able to simplify f_1 and f_2. This is done by trying to re-express f_2 using the don't-care set $d = f_1 + g'$. Let $\hat{f}_2$ be the result of

$$\begin{aligned}
&\text{Minimize} &&f_2 \\
&\text{with} &&\text{don't-care} = f_1 + g'
\end{aligned}$$

Similarly, f_1 potentially can be simplified using the don't-care set $d = \hat{f}_2 g$.

However, before we do either of these minimizations, we build another filter. Note that when examining a potential divisor g, we can quickly access if f_2 can be simplified as follows. In expanding f_2, we use the don't-care set $f_1 + g$, so the offset against which we expand is $(f_2 + f_1 + g)' = f' \cap g'$. Since, typically we already have both f' and g' at each node, the test whether any cube of f_2 can expand is easy. Further, the expansion can be terminated if we determine that any one of the cubes of f_2 can be expanded, since then the factorization $f_1 + g f_2$ now has no more literals than f. Of course, since we are not expanding f_1 in this quick expansion test, we may miss some potential simplifications, but we are using this as a quick test to see if the candidate divisor is good.

If the two filters, a) the count on the cubes orthogonal to g' and b) the quick expansion test, are passed, then we call Boolean division BOOL_DIV to determine the final result of the Boolean substitution.

This leads to the following algorithm, for **Boolean re-substitution** where we try to substitute one of the nodes f_j of the Boolean network into another f_i.

> **BOOL_RESUB(f_i, f_j):**
> f_{i2}= cubes of f_i orthogonal to f'_j
> if $(|f_{i2}|/|f_i| < \theta)$ return (f_i)
> if $($not EXPANDTEST$(f_{i2}, f'_i \cap f'_j))$ return(f_i)
> $(h, e) =$ BOOL_DIV(f_i, f_j)
> return $f_i = y_j h + e$

Here we assume that functions f_i and f_j are logic functions at nodes i and j of the Boolean network. θ is a parameter used to determine if there are enough cubes of f_{i2} to continue. Next we use EXPANDTEST to see if any of the cubes of f_{i2} expands. Since the complements of f_i and f_j are usually parts of the nodes i and j, expansion is quite simple; see [Bra84b]. Finally, if these two filters are passed, we invest the time to try the full Boolean substitution using the algorithm BOOL_DIV already described.

5.6. Node Minimization Methods

In logic synthesis, we require a spectrum of different methods for simplifying the logic function at a node of the network. Ideally, we should be able to simplify a node completely in the sense that the node is prime and irredundant for the entire network, as discussed in section 2.3. This is usually not feasible and certainly not always necessary during each phase of a logic synthesis process. For the most part, minimization methods of ESPRESSO-II are used with different don't-care sets generated according to how thorough and fast we want to be. The first method is an extension of the method called SIMPLIFY in the ESPRESSO-II reference [Bra84b]. The main idea of this extension is to compute the complement simultaneously as we simplify. Then, during the merging step, we can perform an EXPAND operation against this complement.

5.6.1. SIMPCOMP

The SIMPCOMP procedure is a form of the unate recursive paradigm [Bra82b], and is a slight modification to the SIMPLIFY and COMPLEMENT procedures. SIMPCOMP computes both a simplified representation of the function and a simplified representation of the complement simultaneously. It requires about the same amount of time as COMPLEMENT, but gives better results. Also, whereas both SIMPLIFY and COMPLEMENT produced simplified covers, in general, there is no guarantee that the result is either prime or irredundant. In contrast, the result of SIMPCOMP is guaranteed to be prime, but not necessarily irredundant. If desired, at the expense of more time, it can be made irredundant by applying IRREDUNDANT afterwards. In practice, SIMPCOMP is about 5 times faster than ESPRESSO-II. On the other hand, it is limited by not allowing don't-cares.

```
SIMPCOMP(f ):
if (f is unate) {
        F = UNATE_SIMPLIFY(f )
        R = UNATE_COMPLEMENT(f )
        return (F,R )
        }
j = BINATE_SELECT(f )
```

$$(F_0,R_0)= \text{SIMPCOMP}(f_{x'_j})$$
$$(F_1,R_1)= \text{SIMPCOMP}(f_{x_j})$$
$$F_0 = \text{EXPANDJ}(j,x'_j,F_0,R_1)$$
$$R_0 = \text{EXPANDJ}(j,x'_j,R_0,F_1)$$
$$F_1 = \text{EXPANDJ}(j,x_j,F_1,R_0)$$
$$R_1 = \text{EXPANDJ}(j,x_j,R_1,F_0)$$
$$F = F_0 \cup F_1$$
$$R = R_0 \cup R_1$$

```
return (F,R )
```

Here, if the function is not unate, a variable j is selected from the binate variables in a way to balance the splitting of terms evenly between f_{x_j} and $f_{x'_j}$. Each of these is then recursively processed by calling SIMPCOMP which returns both the function and complement F_0,R_0, and F_1,R_1. The next step is to merge these results which is done with a special expand operation. Since, as Theorem 5.6.1.1 below states, the result of SIMPCOMP is prime, it is only possible to expand in the x_j coordinate. Therefore EXPANDJ need only expand against the appropriate complement; in the case where we are expanding x'_j,F_0, this is R_1. The EXPANDJ operation can be done by a fast distance-1 test [Rud86c];

a cube $x_j c$ can be expanded to c if and only if c is distance ≥ 1 from each cube of the offset.

The following theorem can be proved by induction, starting from the fact that the algorithms UNATE_SIMPLIFY and UNATE_COMPLEMENT return the minimum cover of primes.

Theorem 5.6.1.1: Each of the cubes in F and R returned by SIMPCOMP is prime.

This theorem is used inside SIMPCOMP in that the results returned by simplifying f_{x_j} etc. are prime, and hence none of the coordinates, other than possibly x_j, can be expanded. Hence only x_j needs to be tested to obtain a prime at each stage.

5.6.2. ESPRESSO-based Minimization Procedures

Several minimization procedures can be built from ESPRESSO. These vary only in the don't-care sets constructed. In our application of ESPRESSO-II, we use the latest version ESPRESSO-MV [Rud85] which has been greatly optimized. ESPRESSO-MV has more function and capability than required by multi-level optimization. For example, it allows multi-valued inputs, multiple outputs, and performs an iteration which may work too hard for the quality of results desired at a particular stage in the multi-level synthesis. Also, it would be more appropriate in multi-level synthesis to minimize a node with the objective of minimizing the number of literals in the resulting expression. However, because ESPRESSO-MV is extensively optimized, we use it directly and unaltered as a subroutine.

SIMPLIFY1(f_j):
$DC = 0$
$\hat{f}_j = \text{ESPRESSO-MV}(f_j, DC)$
$\hat{F} = \text{QF}(\hat{f}_j)$
if $(\text{LIT\_IN}(\hat{F}) \leq \text{LIT\_IN}(\text{factored form of } f_j))$ $f_j = \hat{f}_j$

Here LIT_IN counts the number of literals in a given factored form. A node is replaced with the result if is simpler than the existing representation as measured by the resulting factored form computed by QF. SIMPLIFY1 is the most direct application of ESPRESSO applied to a single node f_j of the network. No don't-cares are used.

SIMPLIFY2(f_j):
$DC = d_j$
for (k in FANIN(j) and k not an input node) {
$\quad DC = DC + y_k f'_k + y'_k f_k$
$\quad$ }
$\hat{f}_j = \text{ESPRESSO-MV}(f_j, DC)$
$\hat{F} = \text{QF}(\hat{f}_j)$
if $(\text{LIT\_FAC}(\hat{F}) \leq \text{LIT\_IN}(\text{factored form of } f_j))$ $f_j = \hat{f}_j$

Here the immediate fanin don't-care set is generated and used in ESPRESSO-MV, and if there is an output don't-care set specified for node j, d_j, then it is used also.

The final method uses additionally the immediate fanout don't-care set, constructed according to the discussion in section 5.2.

SIMPLIFY3(f_j):

$$DC = d_j$$

for (k in FANIN(j) and k not an input node) {

$$DC = DC + y_k f'_k + y'_k f_k$$

}

for (k in FANOUT(j)) {

$$DC = DC + d_k + f_{ky_j} f_{ky'_j} + f'_{ky_j} f'_{ky'_j}$$

}

$$\hat{f}_j = \text{ESPRESSO-MV}(f_j, DC)$$
$$\hat{F} = \text{QF}(\hat{f}_j)$$

if (LIT_FAC($\hat{F}$)$\leq$ LIT_IN(factored form of f_j)) $f_j = \hat{f}_j$

In general, there are two strategies for using any of the simplification methods in multi-level minimization. The first is that we require an algebraic expression to be minimal with respect to single cube containment in order to apply the algebraic methods. In practice, it is better to minimize each expression to make it prime and irredundant. Hence either SIMPCOMP followed by IRREDUNDANT or SIMPLIFY1 would be appropriate.

The second strategy is that simplification can be seen as a local transformation. If we collect the don't-cares implied by the node's immediate environment, then this transformation is made more global. In addition, during synthesis we eliminate nodes if they do not have sufficient "value" (see COLLAPSE in section 2.5). Hence, the network gets smaller, the nodes get larger, and simplification becomes, in effect, more global.

More experimentation remains to be done on different ways to use simplification. For example, we would like to use more of the transitive fanin and fanout don't-care sets. However, the don't-care sets can become quite large, and considering the fact that these are complemented during ESPRESSO-MV, we can quickly reach the limits of time and space allowed. Thus we seek methods which can strip down the don't-care set to a sub-set which is estimated to be useful in the simplification process. In addition, the correct use of the fanout don't-care set has just been determined and this needs to be incorporated into the procedures.

Acknowledgements

These notes will become part of a book, and the ideas presented are due to many intense interactions with my co-authors, Gary Hachtel, Curt McMullen, Rick Rudell, and Alberto Sangiovanni-Vincentelli. Secondly, Albert Wang has contributed in many ways through the implementation of MIS and through close interactions during 1986 at Berkeley. The logic "team" in Course 290 at Berkeley, Alessandro Cagnola, Ewald Detjens, Suresh Krishna, Tony Ma, Patrick McGeer, Li-Fan Pei, Nathan Phillips, Russel Segal, Nicholas Weiner, Robert Yung, and Tiziano Villa helped form some of these ideas. Logic synthesis was greatly aided by experience with its use in the Yorktown Silicon Compiler [Bra85a] and in the design of a 32-bit microprocessor [Bra85b] where N. Brenner, C. L. Chen, G. DeMicheli, J. Jess, M. Mack, R. Otten, L. van Ginneken, J. Katzenelson and Y.

Yamour have contributed. Earlier interactions with various people in IBM such as
G. Marinen, G. Machol, W. Griffin, R. Kilmoyer, J. Davis, and N. Portuoando
were also of great value.

References

[Ada86] G. Adams, S. Devadas, C Kring, F. Obermeier, P. Tzeng, A. R. Newton, A. Sangiovanni-Vincentelli, and C. Sequin, "Module Generation Systems," *ICCAD86* (November 1986).

[Bar85] K. A. Bartlett and G. D. Hachtel, "Library Specific Optimization of Multi-Level Combinational Logic," *ICCAD85*, (October 1985).

[Bar86] K. Bartlett, R. K. Brayton, G. D. Hachtel, R. Jacobi, R. Rudell, A. Sangiovanni-Vincentelli, and A. Wang, "Algorithms for Multi-Level Logic Minimization Using Implicit Don't-Cares," *ICCD86* (October 1986).

[Bar87] K. A. Bartlett, R. K. Brayton, G. D. Hachtel, R. M. Jacobi, C. R. Morrison, R. Rudell, A. Sangiovanni-Vincentelli, and A. Wang, "Multi-Level Logic Minimization Using Implicit Don't-Cares," submitted *IEEE Trans. on CAD*,

[Bra82a] R. K. Brayton and C. T. McMullen, "The Decomposition and Factorization of Boolean Expressions," *ISCAS Proceedings*, (April 1982).

[Bra82b] R. K. Brayton, J. Cohen, G. D. Hachtel, B. Trager, D. Y. Y. Yun, "Fast Recursive Boolean Function Manipulation," *ISCAS Proceedings*, (April 1982).

[Bra84a] R. K. Brayton, C. L. Chen, C. T. McMullen, R. H. J. M. Otten, and Y. Yamour, "Automated Implementation of Switching Functions as Dynamic CMOS Arrays," *CICC84* (May 1984).

[Bra84b] R. K. Brayton, G. D. Hachtel, C. T. McMullen, and A. Sangiovanni-Vincentelli, *Logic Minimization Algorithms for VLSI Synthesis*, Kluwer Academic Publishers, Boston (September 1984).

[Bra84c] R. K. Brayton and C. T. McMullen, "Synthesis and Optimization of Multi-Stage Logic," *ICCD84*, (October 1984).

[Bra85a] R. K. Brayton, N. L. Brenner, C. L. Chen, G. DeMicheli, C. T. McMullen, and R. H. J. M. Otten, "The YORKTOWN Silicon Compiler", *ISCAS'85*, (June 1985).

[Bra85b] R. K. Brayton, N. L. Brenner, C. L. Chen, G. DeMicheli, J. Katzenelson, C. T. McMullen, R. H. J. M. Otten and R. L. Rudell, "A Microprocessor Design Using the Yorktown Silicon Compiler," *ICCD85*, (October 1985).

[Bra86] R. K. Brayton, A. Cagnola, E. Detjens, S. Krishna, A. Ma, P. Mcgeer, Li-Fan Pei, N. Phillips, R. Rudell, R. Segal, A. Wang, N. Weiner, R. Yung, T. Villa, A. R. Newton, A. Sangiovanni-Vincentelli, C. H. Sequin, "Multiple-Level Logic Optimization System," *ICCAD86* (November 1986).

[Bre84] N. Brenner, "The Yorktown Logic Language: An APL-like Design Language for VLSI Specification," *ICCD84*, (October 1984).

[Bry86] R. Bryant, "Graph-Based Algorithms for Boolean Function Manipulation," *IEEE Trans.*

on Computers, (1986).

[Dag84] M. R. Dagenais, V. K. Agarwal, and N. C. Rumin, "The McBoole Logic Minimizer," draft (1984).

[DeM85a] G. DeMicheli, R. K. Brayton, and A. Sangiovanni-Vincentelli, "Optimal State Assignment for Finite State Machines" *IEEE Trans on CAD/ICAS,* (July 1985).

[DeM85b] G. DeMicheli, "Symbolic Minimization of Logic Functions," *ICCAD85* (November 1985).

[deG86] A. de Geus and W. Cohen, "Optimization of Combinational Logic Using a Rule-Based Expert System," *J. Design and Test,* (to appear).

[Dev86] S. Devadas and A. R. Newton, "Genie: A Generalized Array Optimizer for VLSI Synthesis," *Design Automation Conference,* (June 1986).

[Gre86] D. Gregory, K. Bartlett, A. de Geus, and G. Hachtel, "SOCRATES: A System for Automatically Synthesizing and Optimizing Combinational Logic," *Design Automation Conference* (June 1986).

[Hac86] G. D. Hachtel, "Verfication Algorithms for VLSI Synthesis," *NATO Conference on Logic Synthesis and Silicon Compilation for VLSI Design,* L'Aquila, Italy, (July 1986).

[Lig85] M. R. Lightner, H. Vellandi, B. Vellandi, P. Moceyunas and H. Mueller, "CSIM: The Evolution of a Behavioral Simulator from a Functional Simulator; Implementation Issues and Results," *ICCAD85,* (November 1985).

[McG86] P. McGeer, R. Brayton, "Efficient Stable Algebraic Operations for Boolean Expressions", EECS Technical Report, UC-Berkeley. (Nov. 1986).

[Nai85] R. Nair, A. Bruss, and J. Reif, "Linear Time Algorithms for Optimal CMOS Layout," *VLSI: Algorithms and Architecture,* (1986).

[New86] A. R. Newton, A. L. Sangiovanni-Vincentelli, and C. H. Sequin, "The Berkeley Synthesis Project," *ICCAD86* (November 1986).

[Rud85] R. L. Rudell, "ESPRESSO-MV: Algorithms for Multi-Valued Logic Minimization," *CICC85,* (May 1985).

[Rud86a] R. L. Rudell, "Exact Minimization of Multiple-Valued Functions for PLA Optimization:," *ICCAD86,* (November 1986).

[Rud86b] R. L. Rudell and R. Segal, "BDSYN Users Manual," Berkeley CAD group tools documentation, (April 1986).

[Rud86c] R. L. Rudell, private communication.

[Rud86d] R. L. Rudell, "BLIF Reference Manual," Berkeley CAD group tools documentation, (May 1986).

[Sas86] T. Sasao, "MACDAS: Multi-Level AND-OR Circuit Synthesis using Two-Variable Function Generators," *Design Automation Conference* (June 1986).

[The85] J. F. M. Theeuwen, P. T. H. M. van Paassen, "Automatic Generation of Boolean

Expressions in NMOS Technology," *ICCAD85*, (November 1985).

VERIFICATION ALGORITHMS FOR VLSI SYNTHESIS

GARY D. HACHTEL and REILY M. JACOBY

Department of Electrical and Computer Engineering
University of Colorado, Boulder, CO 80309

1. Introduction

This paper presents novel and efficient methods of testing whether a given multilevel logic function is the tautology. Given two multilevel logic functions, F and G, and the Boolean function D representing inputs which can't occur, then the precisely stated condition,

$$D(v) + F(v) \cdot G(v) + \overline{F}(v) \cdot \overline{G}(v) \equiv 1, \ \forall v \in B^{|v|} \tag{1.1}$$

where $B^{|v|}$ is the set of all possible inputs, amounts to testing the equivalence of F and G. The treatment will be quasi-tutorial. In order to control the size of this chapter, and to achieve readability for non-expert readers, emphasis will be on a qualitative treatment. Thus examples will be emphasized, and when theoretical results are stated, no proofs are given. Proofs, and a more rigorous (and terse) presentation, will be presented in [HaJ86]. Similarly, although we give algorithms expressed as structured code to guide other implementers of multilevel tautology checking, details such as data structures are also deferred to [HaJ86]. We borrow definitions, theory and our perspective on multilevel logic minimization and its relation to test generation from [BBH86] and [BBH]. Our view toward future research in logic verification is expressed in [HaJ86] and [HJM].

A major aspect of VLSI silicon compilation is the verification of the Boolean equivalence of two sets of logic functions - one derived, say, from a hardware description language, and one from a proposed hardware implementation. Thus our methods have application to the area of silicon compilation. The increasing complexity of the VLSI design task has generated a growing interest in automatic logic synthesis and compilation techniques. This has spawned a demand for new algorithms for logic minimization, synthesis, and decomposition which are better suited to the size and complexity of VLSI circuits. It is in this framework that this technique for logic comparison was developed.

There are numerous other uses for a multilevel tautology checking tool in a silicon compiler system, since this tool forms the basis of 1) correctness verification for logic transformations in the synthesis process, e.g., minimization, decomposition or technology remapping, [BHM84] [DJB80] [DJB81] [DBG84] [BCD86] [deC85] 2) redundancy detection and elimination algorithms, [Bra83], 3) automatic test pattern generation algorithms, [Fuj85a], and 4) inference testing for Logic Programming.

The enormous design and fabrication costs of VLSI circuits do not allow automatic compilation systems the luxury of a trial and error approach to digital system synthesis. Hence verification tools are essential to any logic synthesis or logic transformation procedure to insure that the input and output representations are functionally equivalent (i.e. logically consistent). Moreover, since silicon compilers are generally large and complex systems, efficient verification tools can also serve as useful program development aids.

As new fabrication technologies emerge, there is increasing interest in mapping existing designs into these new technologies. Here again there must be some way to verify that this mapping process has produced a circuit that is functionally equivalent to the original design.

Besides providing the foundation for verification tools, functional equivalence also forms the core of redundancy detection and automatic test pattern generation algorithms. A node of a circuit is termed redundant if it can be replaced by a constant logic value without altering the output behavior of the circuit. To determine whether or not a node is redundant, two copies of the circuit description are created. One is left unaltered and one is altered so that the node in question reflects a constant logic value. If these two circuits are functionally equivalent, the node is redundant. If they are not functionally equivalent, the node is not redundant. Moreover the process of determining non-equivalence between the two circuit descriptions generates a test pattern that tests the node for stuck-at faults.

Silicon compilation is generally concerned with mapping a functional description of circuit behavior into the primitives of some technology (e.g. gate array or standard cell libraries). Unless the target technology is a PLA, the circuit descriptions used in this mapping will not be in a sum-of-products form. Rather, multilevel circuit descriptions will be employed. At present the only algorithms available for comparing two functions operate on two-level representations. A partial exception to this is a sketch of an unavailable algorithm, given in [SBH82]. However, no detailed algorithm or experimental results are given. Thus, when the need arises to compare two multilevel functions in the silicon compiler environment, the only currently available means of doing so is to flatten each multilevel representation to a two-level representation and then apply some existing two-level comparison technique.

This has some disadvantages. Flattening can be prohibitively expensive in terms of time, since it requires repeated application of the complementation and intersection operations. Moreover, the two-level descriptions resulting from such iterated cube intersections tends to produce minterms, usually leading to large flattened covers. This can cause the ensuing two-level comparison to be quite slow. In fact, it may be impossible to flatten certain networks, such as 32 bit ALU's or multipliers. Thus, a preferred strategy for multilevel comparison is to operate directly on the multilevel representations and avoid the inefficiencies of having to transform the circuit descriptions to a two-level representation before comparing them. On the other hand, as we show below, multilevel logic produced by the "WEAK DIVISION" [BrM82] [BCD86] logic synthesis method seem to flatten quite efficiently.

The concept of a Boolean Network, which is our basic mathematical model for multilevel logic and is central to all the concepts developed in this paper, is illustrated in Figure 1.1. As defined in Section 2, the structure of a Boolean network is defined by a set of single output, 2-level functions, F, and a set of fan in (Adjacency) relations, FI, which describe how these functions are interconnected and a set, PO identifying which of these function outputs are primary outputs, i.e., are available to the external environment.

Before discussing how to answer the multilevel equivalence and tautology questions for Boolean Networks, we briefly and informally summarize some essential aspects of answering the two-level tautology question for an arbitrary Boolean function $H(x):B^{|x|} \to B$. The **Shannon Expansion** of H is given by

$$H = x^j H_{x^j} + \overline{x}^j H_{\overline{x}^j} , \tag{1.2}$$

where x^j is the positive half of $B^{|x|}$ in the j-th coordinate direction, and H_{x^j} is the projection of the Boolean function of H onto this half space, and similarly for $H_{\overline{x}^j}$. Then the cofactor H_{x^j} [BHM84] is formed from the set $\{H_i\}$ of cubes of H in which H_j appears positively. Once this set is formed, H_{x^j} is obtained by deleting literal x_j for all cubes in the set. A more rigorous definition of cofactor is given in section 2.2 below.

We summarize here some results proved in [BHM84]. If c is a cube, e.g., $c = x^1 \overline{x}^2 x^3 \subseteq B^{|x|}$, then

$$H_c = ((H_{x^1})_{\overline{x}^2})_{x^3} . \tag{1.3}$$

$$c \subseteq H \text{ iff } H_c = 1 , \tag{1.4}$$

$$\overline{(H_c)} \equiv (\overline{H})_c \ . \tag{1.5}$$

If F and G are two level functions, then

$$(FG)_c = F_c G_c \ . \tag{1.6}$$

The critical identity for our purposes is

$$H(z) \equiv 1 \ \textit{iff} \ H_c \equiv 1, \ \forall c \in B^{|z|} \ , \tag{1.7}$$

since (1.7) forms the basis for the tautology algorithm in [BHM84], which uses recursive Shannon Expansion to obtain a set of disjoint cubes c^j (half spaces) such that $\bigcup_j c^j = B^{|z|}$.

Note that in this case, $H_{c^j} \equiv 1, \ \forall j \Leftrightarrow H \equiv 1$. If any cube $c \in B^{|z|}$ exists such that $H_c \not\equiv 1$, then $H \not\equiv 1$. Our multilevel tautology algorithm is based on a strategy similar to that of [BHM84].

The plan of the paper is as follows. In Section 2, definitions are given for Boolean Networks, multilevel equivalence, and cofactoring in the presence of don't care conditions. In Section 3 we establish the synthesis, logic minimization, and testing context, in which multilevel tautology and equivalence checking has primary importance. In Section 4 we present theory for 4 methods for multilevel tautology and equivalence checking. Simple examples are given of the application of these methods. In Section 5 we present an algorithm frame for multilevel tautology, which forms a basis for all our implemented tautology equivalence, test generation and redundancy checking tools. The "flattening" method for multilevel tautology is discussed briefly in this section as well. Section 6 discusses experimental results on multilevel tautology. Section 7 gives some results on the application of multilevel tautology to the problem of test generation and redundancy checking. We conclude in Section 8 with a summary of contributions and an assessment of suggested future work. We will refer frequently to various algorithmic procedures which we have developed, and structured code for these procedures is given in Section 9. Section 9 is followed by data tables, followed in turn by the figures referred to in the text.

2. Background and Basic Definitions

This section begins by defining the concept of a Boolean Network, and then discusses how the imprecise notion of equivalence used in (1.1) can be made precise for Boolean Networks. Then the idea of cofactored Boolean Network is presented. Primality and irredundancy are then defined for the multilevel case, and related to the idea of a don't care set. The section concludes with discussion of some examples of equivalent Boolean Networks and their don't care set.

2.1 Boolean Networks

The primary object in our approach to multilevel logic tautology and equivalence checking is a Boolean Network, illustrated in Fig. 1.1, and defined formally below, which is a technology independent multilevel structure for representing an **incompletely specified logic function** [BHM84]. The Boolean Network may be regarded as an abstraction of an interconnected set of logic gates, as might be specified by a netlist of standard cells. Considered in isolation, each gate in this network realizes a completely specified logic function, but in the context of the network, it realizes an incompletely specified subfunction. Each interconnection represents a signal net associated with the output of one of the gates. Before formally defining a Boolean Network, we briefly introduce the concepts of a) completely and incompletely specified Boolean Functions, and b) their representations.

An **incompletely specified Boolean function** (f,d,r) is a set of 3 **completely specified functions** $f:B^t \rightarrow B$ (the on set), $d:B^t \rightarrow B$ (the don't care set) and $r:B^t \rightarrow B$ (the off set). The minterms of f, d, and r completely **partition** the vertices of the Boolean t-cube B^t. Here f may be thought of as a function, $f(v)$, of a t-dimensional vector

252

$v = (v_1, v_2, \cdots v_t)$. A simple incompletely specified logic function is illustrated in Figure 2.1. In this example $v = (v_1, v_2, v_3)$, $t = 3$, and $f(v) = 1$ for $v \in \{000, 101, 010, 111\}$ else $f(v) = 0$; $d(v) = 1$ for $v \in \{100, 110\}$, else 0; and $r(v) = 1$ for $v \in \{001, 011\}$, else 0. An incompletely specified logic function reduces to a completely specified function when $d = \varnothing$, i.e., there is no don't care set.

Note that a completely specified function, $f(v)$, may be **independent** of certain of the v_i, and this fact is usually reflected in the selection of a representation, F, of $f(v)$. The variables **explicitly** represented in F are called the **support** of F. One representation of f is the sum of products form, e.g.,

$$F = v_1 \overline{v}_3 + \overline{v}_1 v_3 + \overline{v}_2 \, v_2 \, ,$$

which is also called the disjunctive normal form. Note that here the **support** of F is $\{v_1, v_2, v_3\}$, and that a variable which a function does **not** depend on, like v_2 in the above example, may appear explicitly in the support of the function. Other representations are possible and significant, e.g., conjunctive form, factored form [Law64], etc. However, in this paper we use disjunctive form since we rely heavily on 2-level, sum of products based, Boolean minimization procedures such as the subprocedures of ESPRESSO_II, [BHM84].

Product terms like $v_1 \overline{v}_3$ and $\overline{v}_1 v_3$ will be called **cubes** in the sequel. Each cube consists of a set of literals, and each literal appears in one of the two forms v_i or $\overline{v}_i$. If "v_i" appears it stands for the predicate "$v_i = 1$", and if $\overline{v}_i$ appears it stands for the predicate "$v_i = 0$". Thus the cube $v_1 \overline{v}_3$ stands for the conjunction of predicates $v_1 = 1$ and $v_3 = 0$. In this sense we can view the cube $v_1 \overline{v}_3$ as the intersection of the subcubes (half spaces of the Boolean t-cube) $v_1 = 1$ and $v_3 = 0$ (cf. Figure 2.1 in which the dimension of the Boolean t-cube is $t = 3$). Cubes are the basic Boolean data object, as defined below.

Definition 1 (Boolean Networks)

As illustrated in Figure 1.1, a **Boolean Network**, η, is a pair $(\mathbf{F}, PO\,)$, where $\mathbf{F} = \{F_j, j = 1, 2, \ldots m\}$ is a set of m given **representations** of the on sets f_j of incompletely specified functions $(f_j, \ d_j, \ r_j)$, $j = 1, 2, \ldots m$. Each such representation is a set of cubes. With each F_j is associated a "local output" logic variable, y_j, in the set $IV = \{y_1, y_2, \cdots y_m\}$, which we call the **Intermediate Variable** set. The specified primary output set, $PO \subseteq \{1, 2, \ldots m\}$, identifies the subset $\{y_i \,|\, i \in PO\} \subseteq IV$ of the outputs of the F_i as observable **Primary Outputs** of the Boolean Network. It is convenient to refer to this subset as the primary output vector z, defined so that $z_k = y_{PO(k)}$, $k = 1, 2, \ldots, p$, $p = |PO|$.

Since η is completely determined by the pair $(\mathbf{F}, PO)$, we write $\eta = (\mathbf{F}, PO)$. However, η has further structure, determined by the **support sets** $SUPP(F_j)$, of the representations, F_j. These sets determine the structure of a **directed graph,** $G = (N, E)$, with nodes $N = \{1, 2, \ldots m, m + 1, \ldots t\}$, $t = |\bigcup_{j=1}^{m} SUPP(F_j)| > m$. With each node $i \in N$, we associate a logic variable v_i. With the first m nodes of N we associate the representation F_i and its corresponding variable y_i, so that $v_i = y_i$, $i = 1, 2, \ldots m$. (This duplication of notation is quite useful in the sequel). However, **no** F_i is associated with the last $n = t - m$ logic variables in the vector v. Instead, these nodes are identified with **Primary Inputs** of η, and are associated with duplicate logic variables $PI = \{x_1, x_2, \cdots x_n\}$. Thus $v_{m+i} = x_i$ $i = 1, 2, \ldots n$. Note that the vector v can be viewed as the catenation $v = (y, x)$.

For each node $j \in N$, we define the **Fan In Set** (for short **Fan In**) FI_j, as follows. If $j \leq m$ (Intermediate Variables), $FI_j = \{i \,|\, v_i \in SUPP(F_j)\}$, but if $j > m$ (Primary Inputs) $FI_j = \varnothing$. Thus the Primary Input nodes are terminal nodes of the directed graph G. The edge set E of G contains directed edge (i, j) if node i is in the Fan In of node j, i.e., $i \in FI_j$. Then we may define the **Fan Out,** FO_j, of node j to be the set of all nodes $i \in N$ for which there is an edge $(j,i) \in E$. Similarly, we define the **Transitive Fan Out**, TFO_j, to be the set of all nodes $i \in N$ such that there exists a (directed) path from j to i in G, and the **Transitive Fan In**, TFI_j, to be the set of all nodes $i \in N$ for which there exists a path from i to j in G. By convention, $j \in TFO_j$, but $j \notin TFI_j$. $\square$

It is important to note that although F_j depends explicitly only on the variables $v_k \in SUPP(F_j)$, we may formally view each F_j as a function of the entire vector $v = (y,x)$ (recognizing, of course, that f_j may be functionally independent of many of the v_k). Thus we may write

$$v_j = y_j = F_j(v) = F_j(y,x) = F_j(y(x),x), \quad j = 1,2,...m \tag{2.1}$$

as the basic constitutive relations of η. Note that as indicated in the last identity, if equations (2.1) are satisfied for $j = 1,2,...,m$, then the solution vector $v(x) = (y(x),x)$ is the vector of values appearing at the nodes of the Boolean Network in response to the Primary Input vector x. In particular, $z(x)$ represents the values at the Primary Outputs of η in response to x, i.e., the same values that would have been obtained by logic simulation of the vector x. Thus given that (2.1) is satisfied, $z(x)$ represents the IO (input/output) map of η.

Thus a Boolean Network η is a **representation** of a set of incompletely specified functions, $(f(i), d(i), r(i))$, one for each primary output $z_i(x)$. Thus $z_i(x)$ can be regarded as the "IO map" from PI to PO_i of the Boolean Network η. A representation, DX_i, of the completely specified "don't care" function $d(i)$ must come from the system designer. We refer to DX_i as the **External** don't care set, which arises from two phenomena. First, for a particular design the designer may decree that a particular primary input vector $x \in B^n$ will **never occur**. The vector x constitutes a don't care minterm, and such minterms are don't care for **all** Primary Outputs. The set of all such minterms is labeled DXP. Second, the designer may state that for any of the outputs z_i, $i \in PO$ the value of z_i will **not be used** for a set of Primary input vectors (minterms) in the set DXO_i. Thus for each primary output the total external don't care set can be written

$$DX_i = DXP + DXO_i, \quad i = 1,2,...p = |PO| \, . \tag{2.2}$$

Equation (2.2) gives a representation of the completely specified functions $d(i)$, $i = 1,2,...p$ (don't care sets) associated with the primary outputs of a Boolean Network. A principal objective of the sequel is to identify representations of the analogous don't care sets for each of the incompletely specified functions associated with the intermediate variables of a given Boolean Network (and their corresponding internal nodes).

Note that a Boolean Network can be thought of as an acyclic directed graph (V,E), in which there is a vertex $v \in V$ corresponding to each function F_i and each primary input. The edges correspond to the fan in relationships between the functions F_i: $E = \{ (j,i) \mid 1 \le i \le m \text{ and } j \in FI_i \}$. The acyclic directed graph corresponding to some Boolean Network, $\eta = (F,PO)$, is termed the **function graph** of η. The **reverse function graph** of η is derived from the function graph of F by reversing the direction of all edges in the function graph.

For ease of reference we collect some basic definitions and notation to be used throughout the sequel. A **cover matrix** is a matrix representation of an algebraic sum-of-products description of a Boolean Function. A **cube** is a single row of a cover matrix. Each cube corresponds to a single product term in the corresponding algebraic sum-of-products description. A cube is to be viewed as a row vector. A 1 in the j-th position (component) of a cube indicates that the corresponding algebraic product term contains literal x_j. A 0 in the j-th position (component) of a cube indicates that the corresponding algebraic product term contains literal $\overline{x}_j$. A 2 in the j-th position (component) of a cube indicates that the corresponding algebraic product term does not depend on x_j in either its positive or negative form. Note that x_j may denote either the variable itself or the positive value of the variable and $\overline{x}_j$ always denotes the negative value of the variable. Generally the context will make clear how to resolve the ambiguity of notation for x_j. If c is any cube, then c_j refers to the j-th component of c (i.e. $c_j = 0$, 1 or 2). Two cubes c and d are **disjoint** if there exists an $i \leq m + n$ such that 1)$c_i = 1$ and $d_i = 0$ or 2)$c_i = 0$ and $d_i = 1$. The **intersection** of two cubes c and d is defined to be $\varnothing$ if c and d are disjoint. Otherwise the intersection of c and d generates a third cube e. The i-th component of e is given by the following: if $d_i = 2$ then $e_i = c_i$, otherwise $e_i = d_i$. If c is a cube for which $c_i \neq 2$, $\forall i$ then c is called a **minterm**.

The following summarizes the notation used in the sequel. Note we shall have cause to refer to the **fan out** FO_j of node j as well as the fan in FI_j. If FO_j is altered, it is assumed that FI_j is correspondingly altered, and vice versa.

Notation:

$B = \{0,1\}$.

$B^{|x|} =$ the Cartesian product of B with itself $|x|$ times.

$\eta = (F, FI, PO) =$ Boolean network with functions F_i and fan in relations FI_i.

$x =$ a vector consisting of all primary input variables in η.

$y =$ a vector consisting of all intermediate variables in η.

$y(x) =$ a vector of intermediate values resulting from primary inputs p.

$v = (y,x) =$ a vector consisting of all variables in η.

$v_j =$ refers to a variable in η or the positive literal of the variable.

$\overline{v}_j =$ refers to the negative literal of the variable x_j.

$v^j =$ a special cube with the single literal v_j.

$\overline{v}^j =$ a special cube with the single literal v_j.

$c^i =$ refers to the i-th cube in a collection of cubes.

$c_j =$ refers to the j-th component of cube c.

$c_j^i =$ refers to the j-th component of the i-th cube in a set of cubes.

Let $H =$ the cover of any completely specified two-level function.

$H_k = k$-th cube (row, product term) of H.

$|H| =$ the number of cubes in H.

Let $F_i =$ the cover of the i-th function in a Boolean network.

$F_{ik} =$ the k-th cube of F_i.

$(F_i)_{x^j} =$ cofactor of F_i with respect to x^j. (see Definition 3 below).

$(F_i)_c =$ cofactor of F_i with respect to cube c.

$|F_i| =$ the number of cubes in F_i.

$cd = c \bigcap d =$ intersection of two cubes.

$cH = c \bigcap H =$ intersection of the cube c with every cube in H.

$FG = F \cdot G = F \bigcap G =$ intersection of each cube in F with every cube in G.

$F + G = F \bigcup G = F,G$.

$FI_j =$ fan in of $F_j = \{i \,|\, F_i$ depends directly on the variable $x_i\}$.

$FO_j =$ fan out of $F_j = \{i \,|\, F_i$ depends directly on the variable $x_j\}$.

2.2 Equivalence of Boolean Networks

Boolean Equivalence is a key concept in Logic Optimization. In the multilevel context, we wish to establish when a given Boolean Network, η, can be replaced by another one, η', with an equivalent IO map, $z'(x) \equiv z(x)$. That is, the relation between Primary Inputs and Primary Outputs is preserved. Thus, η' represents the same set of incompletely specified functions $(f(i), d(i), r(i))$, $\forall i \in PO$.

Definition 2 (Equivalence)

Boolean Networks $\eta = (F, PO)$ and $\eta' = (F', PO')$ are said to be **equivalent** (written $\eta \equiv \eta'$) if there exists a permutation q of $\{1,2,...p\}$ such that for each Primary Output $z_i'(z)$, $i \in PO'$, $z_i'(z) = z_{q(i)}(z)$, $q(i) \in PO$, for all $z \in DX_{q(i)}$. If all the outputs of Boolean Network η are **individually** tautologous, i.e., if $z_i(z) = 1$, $\forall z \in B^{|z|}$, $\forall i \in PO$, then we say η is tautologous (written $\eta \equiv 1$). $\square$

The permutation, q, in Definition 2 is needed to identify the proper correspondence between the primary outputs of the two Boolean Networks which may be very different structurally. For simplicity, we assume, without loss of generality, that q is the identity permutation.

Note that Definition 2 requires only that the **primary outputs** of two Boolean Networks match for each **care** primary input vector. In particular, it is **not** necessary to have identity or even correspondence between the intermediate variables of the two networks. For example, as shown in Figure 2.2, a multilevel Boolean network (Figure 2.2a) could be equivalent to a 1-level network (Figure 2.2b). The "flat" 1-level network of Figure 2.2b is specified by the following equations.

$$F_1 = z_1 z_2 + \overline{z}_1 \overline{z}_2 + z_3 z_4 , \tag{2.3a}$$

$$F_2 = z_1 \overline{z}_2 + \overline{z}_1 z_2 . \tag{2.3b}$$

We shall discuss the equivalence of the Boolean Networks of Figure 2.3 in greater detail below. For now we complete this section by defining the **cofactor** operation on both representations of functions and on Boolean networks.

Definition 3 (Cofactor Operation)

The **cofactor** of a sum of products representation, $F = \{c_i\}$, of a Boolean Function with respect to a literal v_j is defined to be

$$(F)_{v_j} = \bigcup_i (c_i)_{v_j} .$$

Here if literal v_j is contained in c_i, $(c_i)_{v_j}$ is just c_i with literal v_j deleted, else if literal $\overline{v}_j$, appears in c_i, $(c_i)_{v_j} = \varnothing$. If neither v_j or $\overline{v}_j$ appears in c_i, then $(c_i)_{v_j} = c_i$.

The **cofactor** of a Boolean Network $\eta = (F, PO)$ with respect to a literal v_j is a Boolean Network, $\eta_{v_j} = (F_{v_j}, PO)$, where $F_{v_j} = \{(F_i)_{v_j}\}$ is the set of cofactors of the representations $\{F_i\}$ of the original Boolean Network. Note the literal v_j may be **either** a primary input (i.e., $j \in PI$) or an **intermediate variable** (i.e., $j \in IV$). We denote the vector of logic variables in this cofactored network to be $(v)_{v_j}$, with components $(v_1)_{v_j}, (v_2)_{v_j}, \cdots (v_t)_{v_j}$.

Similar definitions apply when the cofactor is with respect to the literal $\overline{v}_j$. $\square$

This definition is computationally crucial to the task of answering the tautology and equivalence questions for Boolean Networks. Note in particular that the edges of the cofactored Boolean Network η_{v_j} are defined by the support of the $(F_i)_{v_j}$, from which the variable v_j is now totally missing. Thus each node i in the fanout of node j in η is disconnected from node j in η_{v_j}.

2.3 Implicit Don't Care Sets of Boolean Networks

Considered in isolation, the F_j are representations of **completely specified** functions, but embedded in the network, they are representations of **incompletely specified** functions (f_j, d_j, r_j), i.e., node j has an associated a don't care set d_j. Thus a network of individually prime and irredundant functions, such as that shown in Figure 2.3a, may be neither prime nor irredundant. Such redundancies arise from the don't care sets generated by the structure of a Boolean Network. We illustrate this by identifying a representation, D_1, of the don't care set d_1 of node 1 of the Boolean Network of Figure 2.3a. As shown in [BBH86], the 5-cube set

$$D_1 = \overline{y}_3(x_1 x_2 \overline{y}_2 + \overline{x}_1 \overline{x}_2) + y_3 \overline{(x_1 x_2 \overline{y}_2 + \overline{x}_1 \overline{x}_2)} + \overline{y}_2(x_1 \overline{x}_2 + \overline{x}_1 x_2) + y_2 \overline{(x_1 \overline{x}_2 + \overline{x}_1 x_2)}$$

is a valid representation of d_1, assuming $DX_i = \varnothing$, $i \in PO = \{1,2\}$, i.e., that the Boolean Network η has no external don't care set. For primary outputs, the derivation of such implicit don't care sets may be generalized as follows.

Definition 4

The "Overall" Intermediate Variable don't care set, DIV, is defined by

$$DIV = \sum_{j=1}^{m} DIV_j , \tag{2.4a}$$

where

$$DIV_j = y_j \overline{F}_j + \overline{y}_j F_j = y_j \oplus F_j . \tag{2.4b}$$

Note by DeMorgan's Law, we have

$$\overline{DIV} = \prod_{j=1}^{m} (y_j \equiv F_j) = \prod_{j=1}^{m} (\overline{y}_j \overline{F}_j + y_j F_j) . \; \Box \tag{2.4c}$$

We shall show below that for any vertex in $v \in B^t$ (represented by the overall vector v of η) which satisfies (2.1) for $j = 1,2,\ldots m$, it follows that $v \in \overline{DIV}$. Conversely if any of the equations $y_j = F_j(v)$ are not satisfied, we have $v \in DIV$. Since these observations will be used repeatedly in the sequel, we summarize them formally as follows.

Lemma 1. Suppose DIV is computed for Boolean Network $\eta = (F,PO)$ according to (2.4). Let v be any vertex of B^{m+n}. Then v satisfies (2.1) if and only if $v \notin DIV \subseteq B^{m+n}$.

3.0 The Context of Equivalence Checking: VLSI Synthesis, Logic Minimization, and Testing

When used in its strictest sense, **verification** means equivalence checking and is to be distinguished from simulation, which usually refers to **partial** verification. Equivalence checking is typically carried out whenever a representation is optimized, i.e., changed, at a given level of abstraction, or when the representation is mapped into a new level of abstraction, e.g., from the Behavioral Language level to the Boolean Network level. In this paper we focus on equivalence checking at the Boolean Network level, and in this section we discuss how equivalence checking is used in the context of 1) logic minimization, and 2) testing. Not surprisingly, these two contexts are fundamentally related.

3.1 Primality, Irredundancy and Logic Minimization of Boolean Networks

The task of minimizing a Boolean Network η consists of iteratively transforming η into an equivalent network η' where η' is smaller than η in some sense. Two properties of minimality, similar to those for the classical 2-level case, are especially relevant to equivalence and tautology checking in the multilevel case, since Boolean Networks having these properties are shown below to be 100% testable for stuck faults.

Definition 5 (Prime and Irredundant Boolean Networks)

Given a Boolean Network $\eta = (F,PO)$, a cube c of the 2-level representation of F_i is prime if no literal of c can be removed without causing the resulting network η' to be not equivalent to η. In more formal terms, $\eta'=(F',PO)$ is a Boolean Network for which $F_j'=F_j$, $\forall j\neq i$ and $F_i'=(F_i-\{c\})\bigcup c'$, where c' is c with one of its literals removed. Similarly, a cube c of F_i is **irredundant** if c cannot be removed from the representation of F_i without causing the resulting network η' to be **not** equivalent to η. A **Boolean Network** $\eta=(F,PO)$ is said to be **prime** if all the cubes in each of the representations F_i of η are prime, and **Irredundant** if all of these cubes are irredundant. $\square$

Note that these two concepts are associated with local minima of a cost function which is nondecreasing in the total number of cubes and literals required to represent the incompletely specified logic functions, realized by the given Boolean Network. Note that the properties of primality and irredundancy arise from **both** the representation F_j and the Boolean Network, η, in which it is embedded.

Given a Boolean Network η, it is of obvious interest to obtain an equivalent prime and irredundant network η'. One possible procedure to obtain this simplification is to examine each cube as well as each literal in first encounter order, and for each such cube or literal construct a simplified network η', identical to η except for the removal of the selected cube or literal. Then Definition 2 may be embodied in a computer program such as [HaJ85] to check if $\eta=\eta'$. If so, the cube or literal is redundant, and can be removed from η. If no cube or literal can be so removed, then the resulting Boolean Network is Prime and Irredundant (Definition 5). This elementary minimization procedure is the one used to obtain the equivalent prime and irredundant Boolean Network of Figure 2.3b from that of Figure 2.3a. This procedure is intimately related to the way in which the basic D-algorithm (and its variants), [Fuj85b] is used in test generation algorithms. A somewhat subtler, but closely related procedure [BBH86] leads to the second and third more highly optimized but equivalent Boolean Networks of Figures 2.3c and 2.3d.

3.2. Test Generation

We can use the idea of multilevel cofactor in the generation of tests for stuck-at faults in a multilevel network. We can also use the basic notion of multilevel equivalence for the more general fault models of transistor level testing as well. In this section we explore these basic testing categories in turn.

Given some Boolean network $\eta=(F,PO)$, we can view the cofactor network $(\eta)_{v_j}$ given by Definition 3 as a copy of η in which v_j is stuck at 1. Note if $j\in PI$, v_j is a primary input, and if $j\in IV$, v_j is an intermediate variable. A **test** for v_j stuck at 1 exists if and only if there exists a cube c, containing only primary input literals, such that a) $v_i(c)\neq(v_i(c))_{v_j}$, and b) $c\subseteq\overline{DX_i}$ some $i\in PO$. We state this as

Theorem 1.

There exists a test for the output fault v_j stuck at 1(0) if and only if

$$\eta\neq(\eta)_{v_j(\overline{v_j})}\,.$$

Note that we use parenthesis here (and in the sequel) to denote a "dual" statement, i.e., "v_j" corresponds to "1", and "$\overline{v_j}$" to "0".

The Algorithms given in Section 5 for checking multilevel equivalence are constructive in the test generation sense. That is in discovering non-equivalence, they all produce one or more "contradiction cubes" $c\subseteq\overline{D}$. The cube c has only primary input literals, and any of its minterms is a test for the output fault stuck at 1. It is possible, simply by running these algorithms to exhaustion, to discover **all** such contradiction cubes, i.e., all such tests.

Corollary 1

Let $c \in B^{m+n}$ be any cube whose literals all correspond to intermediate variables of Boolean Network η. Then there exists a test for the multiple output stuck fault corresponding to c if and only if

$$\eta \neq (\eta)_c .$$

If $I = \{i \,|\, c_i = 1\}$ and $J = \{j \,|\, c_j = 0\}$ then Corollary 1 implies there exists a test for the fault condition in which all z_i, $i \in I$, are stuck at 1 and all z_j, $j \in J$, are stuck at 0.

Input stuck faults may be similarly categorized. Suppose we are interested in the input stuck fault where variable $z_k \in FI_j$ is stuck at 1.

Definition 6 (Partial Cofactor)

The **partial cofactor of a Boolean Network** $\eta = (\mathbf{F}, PO)$ with respect to a literal v_j is a Boolean Network, $\eta_{v_j}^J = (\mathbf{F}_{v_j}^J, PO)$, where if $i \in J$, $(F_i)_{v_j}^J$ is the set of cofactors of the representations F_i of the original Boolean Network, else, $(F_i)_{v_j}^J = F_i$. Note the literal v_j may be **either** a primary input (i.e., $j \in PI$) or an **intermediate variable** (i.e., $j \in IV$). We denote the vector of logic variables in this cofactored network to be $(v)_{v_j}^J$, with components $(v_1)_{v_j}^J$, $(v_2)_{v_j}^J$, $\cdots (v_t)_{v_j}^J$.

Similar definitions apply when the cofactor is with respect to the literal $\overline{v}_j$. □

Corollary 2

Let Boolean Network $\eta_{v_j}^J$ be defined according to Definition 6, with $J = \{k\}$. Then there exists a test for input v_j of function F_k stuck at 1(0) if and only if

$$\eta \neq \eta_{v_j}^J .$$

If $J = FO_j$, then Corollary 2 is indistinguishable from Corollary 1. Tests for multiple input stuck faults may be similarly defined.

The above theorem and corollaries show one simple way to define, and even set up the computation of, test generation problems, **assuming** that a general case equivalence checker for Boolean Networks is available. Of course, it must be kept in mind that the minimization and/or testing contexts for equivalence checking offer additional opportunities for algorithm efficiency ([Fuj85b]) which are not available in the general use.

To complete this section, we consider the case of **combinational transistor level** testing (sequential aspects are being investigated by Jacob Abraham at the University of Illinois.) Here we again exploit the universality of the general case equivalence checker. The key to transistor level testing is generation of the Boolean function, F_j', corresponding to an actual transistor level with some fault asserted.

Definition 7

Let $\eta' = (F', FI)$ be a Boolean Network where

$F_i = F_i'$, $\forall i \neq j$;

$F_j' = \mathrm{MAP}\ (\mathrm{FAULT}\ (\mathrm{EXTRACT}\ (\mathrm{CIF}\ (F_j)))) .$ □

Here CIF $(\cdot)$ is a function whose input is a Boolean function (e.g., one produced by the Boulder Logic Synthesis System), and whose output is a CIF file. Thus the **function** CIF is essentially a generation, place, and route tool. EXTRACT $(\cdot)$ is a function whose input is a CIF file and whose output is a transistor net list of the type used as input to switch level simulators (such functions are available as part of the tool set on most modern IC workstations). FAULT $(\cdot)$ is a function whose input is a "good" transistor net list and whose output is a related netlist with some fault asserted, e.g., short, open, or "stuck-at". This simply is the effect of the user editing the net list and replacing, say, differing net names with the same net name. Finally MAP $(\cdot)$ is a function whose input is the (switch-level) transistor net list and whose output is the corresponding **combinational** Boolean Function (we do not consider the case where the faulted net list exhibits sequential or oscillatory behavior). Bryant at CMU, Hajj at the University of Illinois, and Lightner and Heydemann at the University of Colorado are working on implementations of such functions.

Corollary 3

Let $\{F_0'\}$ be the output function of the Boolean Network given by Definition 7. Then there exists a test for the transistor-level fault asserted by function FAULT $(\cdot)$ if and only if

$$\eta \neq \eta' \, .$$

4. Multilevel Equivalence Checking: Methods and Examples

In this section we give qualitative descriptions of 4 methods for multilevel equivalence checking: a) the "Flattening" method, b) the "Don't Care" method, c) the "Simulation" method, and d) "Algebraic String Comparison" method. Each of these methods may be regarded as a variant of the basic method of recursive Shannon Expansion described in Section 1 for 1-level Boolean Networks, i.e., PLA's, [BHM84]. Examples are given of the application of each of these methods to the simple Boolean Networks of Figure 2.2. Supporting theory is developed as needed. A key concept in this theory is the idea of a "Comparison Network".

Definition 8. (Comparison Network). Given Boolean Networks $\eta = (\mathbf{F}, PO)$ and $\eta' = (\mathbf{F}', PO')$, the **Comparison Network** $\eta^C(\eta, \eta')$ is a Boolean Network which is constructed as follows:

a) For each pair of corresponding outputs $j \in PO$ and $j' \in PO'$, connect y_j and $y_{j'}$ as inputs to an equivalence gate (XNOR) whose defining equation is $y_k^C = F_k^C(y^C, x) = (y_j \equiv y_{j'})$;

b) Connect each equivalence gate output y_k as input to an AND gate, which thus has $|PO| = |PO'|$ inputs in all. The output of this AND gate is the single primary output of η^C. $\square$

The role played by a Comparison Network is to convert the **multilevel equivalence** question into an equivalent multilevel **tautology question**. That is, $\eta^C \equiv 1$ if and only if $\eta \equiv \eta'$, i.e., if the constituent networks are equivalent. Figure 4.1a gives an example of the Comparison Network formed from Boolean Networks η^a and η^b of Figures 2.2a and 2.2b. Note that nodes 1 and 2 of Figure 2.2a are relabeled 4 and 5, and nodes 1 and 2 of Figure 2.2b are relabeled 6 and 7. The binary cofactoring tree for the comparison network η^C of Figure 4.1a is given in Figure 4.1b. The defining equations for the cofactors of η^C (cf., Definition 3) for some nodes of this tree are given in Figures 4.1c, 4.1d and 4.1e. Note that the equations for a given node of the cofactoring tree of Figure 4.1b can be derived by cofactoring η^C by the conjunction of the literals associated with each edge of the tree which is in the unique path from the given node to the root. Note the "left son" edge of a node labeled "x_j" is associated with the positive literal, x_j, and the right son edge with the corresponding negative literal, $\overline{x}_j$. Thus the cube $c = x_1 x_2 x_3 x_4$ is associated with the path from the root to the lower left most node of Figure 4.1b. Consequently the defining equations of $(\eta^C)_c$ are $(F_7^C)_c = (x_1 \oplus x_2)_c = 0$, $(F_6^C)_c = (x_1 \equiv x_2)_c = 1$, $(F_5^C)_c = (x_1 \oplus x_2)_c = 1$,

$$(F_4^C)_c = (\overline{y_5^C} + x_3 x_4)_c = 1 \qquad (F_3^C)_c = (y_4^C \equiv y_6^C)_c \to 1, \qquad (F_2^C)_c = (y_5^C \equiv y_7^C)_c \to 1, \qquad \text{and}$$
$$(y_1^C)_c = (F_1^C)_c = (y_2^C y_3^C)_2 \to 1.$$

The reader will note that the symbol "$=$" is used to denote **direct** Boolean evaluation, whereas the symbol "$\to$" is used to denote **transitive** evaluation. Thus, according to (1.2)-(1.7) $(y_7^C)_c = (y_5^C)_c = (x_1 \overline{x}_2 + \overline{x}_1 x_2)_{x_1 x_2 x_3 x_4}$ evaluates directly to 0, and $(y_6^C)_c = (y_4^C)_c = (x_1 x_2 + \overline{x}_1 \overline{x}_2)_{x_1 x_2 x_3 x_4}$ directly to 1. On the other hand $(y_3^C)_c = (y_5^C \equiv y_7^C)_c$ evaluates to 1 only after substituting the **previously evaluated** logic values for $(y_5^C)_c$ and $(y_7^C)_c$. Thus a very elementary form of **logic simulation** is going on in the multilevel cofactoring step. We shall give a detailed description of multilevel cofactoring in Section 4 below.

4.1 The Flattening Method for Multilevel Tautology

In this section we give a conceptual procedure for obtaining a flattened, 1-level representation, η^1, of a given Boolean Network η. We then show that the flattened representation is equivalent to the original. Thus the question $\eta \overset{?}{=} 1$ can be answered by checking if $\eta^1 \equiv 1$ by conventional PLA methods (cf., [BHM84]), which are based on the 1-level cofactoring techniques of (1.2)-(1.7). Note, as shown below, η^b is the flattened version of η^a in Figure 2.3, and $\eta^b \equiv \eta^a$.

From $\eta = (F, PO)$ and Procedure 1 below we derive a second Boolean network $\eta^1 = (F^1, PO)$. η^1 is a 1-level network consisting of only primary output functions which have only primary inputs as fan ins.

Procedure 1.

The 1-level Boolean network η^1 is conceptually derived from η by the following procedure:

1) for all $x \in B^n$

    1a) Let $v = (y(x), x)$ be the self-consistent solution of η for each primary input x.

    1b) If $y_j(x) = 1, \forall j \in PO$, add minterm x to the representation of F_j^1.

2) Minimize the representations obtained in 1), and call the resulting prime and irredundant Boolean Network η^1.

Note that for each $x \in B^n$, there is a corresponding value for each intermediate variable $y_1(x)$, $y_2(x)$, ... $y_m(x)$. We emphasize that this is only a **conceptual** procedure. Practical computation of η^1 is discussed in Section 5.

Lemma 2.

Let η^1 be the "flattened" Boolean Network derived from $\eta = (F, PO)$ by Procedure 1. Then, according to Definition 2,

$$\eta \equiv \eta^1 .$$

We illustrate Lemma 2 by presenting a truth table (Table 4.1) for checking the equivalence of η^a and η^b. The 16 possible primary input vectors for η^a and η^b are given in the first 4 columns of Table 4.1. By performing mental logic simulation, the reader may verify that columns 5 and 6 give the corresponding logic values for the primary outputs y_1^a and y_2^a of η^a. Direct logic evaluation of the 1-level Boolean Network η^b gives the corresponding values for y_1^b and y_2^b listed in columns 7 and 8. The last column shows the result of testing if $y_1^a(x) = y_1^b(x)$ and $y_2^a(x) = y_2^b(x)$, for each primary input vector x. Since both are true for each x, we have verified, according to Definition 2, that $\eta^a \equiv \eta^b$. Note the last column gives the values which would obtain for the unique primary output of the comparison network η^C of Figure 4.1a, so we have also verified that $\eta^C \equiv 1$.

As discussed in Section 5 below, the flattening method for testing equivalence and tautology is conceptually simple, but can be extremely expensive in practice for certain types of Boolean Networks, e.g., multipliers, parity trees, ALU's, etc. However, the method should not be dismissed, since it is useful for a significant class of verification problems. In this class, Boolean Networks obtained from a given 1-level Boolean Network (PLA) by the process of **algebraic decomposition** [BCD86], [BrM82], have been found to flatten with ultimate efficiency, so verification via flattening is a clear win for problems in this class (cf., the experimental results of Section 6 below).

4.2 The Don't Care Method for Multilevel Tautology

In the "don't care" method for multilevel tautology checking, the basic idea is to exploit the don't care set DIV of (2.4) by obtaining from a given Boolean Network η a second, equivalent, 1-level network, η^D, which has only the primary output nodes of η. As shown conclusively in Section 6 below, the don't care method, per sec, is comparatively inefficient, often grossly so. However, it remains attractive because that it requires the least software development expenditure, since neither flattening nor multilevel cofactoring and logic simulation are required, and because (cf. Lemma 3 below) it provides a basis for emulating the important "single path backtrace" heuristic of advanced test generation algorithms [Fuj85b].

The don't care method derives from the viewpoint that the defining equations (2.1) represent **constraints** on the validation of vertices $v \in B^{m+n}$ as vectors of logic variables of η which occur as a response to primary inputs v_{PI}. This constraining concept is the basis for Lemma 1 above. We (and others, cf [WeS85]) have discovered that it is possible to relax these constraints, and replace them with suitably defined don't care sets. We now show that this leads to a new multilevel tautology method. In this method, **primary output functions** (only!) of a given Boolean Network are used to define a 1-level Boolean Network η'' (i.e., a PLA). With the construction given below of a don't care set D'' for use in Definition 2, we can show that $\eta \equiv \eta''$. Consequently, $\eta \equiv 1$ can be tested by answering the equivalent PLA tautology question $\eta'' \equiv 1$.

We originally thought this method would be quite efficient, since it would be exploiting an existing 1-level tautology algorithm, [BHM84], which had been quite successful in practice. However, as shown in Section 6 below, this was not actually the case. There were unfortunate inefficiencies, due both to the size and the structure of the don't care set D, and to the fact that $\eta \equiv 1$ can be tested by exploring B^n (cf., Table 4.1) whereas $\eta'' \equiv 1$ requires exploration of B^{m+n}. That is, 2^m additional vertices must somehow be tested, where $m = |IV|$ is the number of intermediate variables in the original Boolean network, η.

Before describing the Don't Care method, we give a Definition and Lemma which provides a firm theoretical foundation for it.

Definition 9

Given a Boolean Network $\eta = (F, PO)$, with intermediate variable set IV, and associated don't care sets **DX**, define a set $J \subseteq IV$ and a corresponding "relaxed" Boolean Network, η^J, as follows.

0. Initialize η^J to η;

1. For each $j \in J$

   a) Create a **new, primary input** variable $y_{j,D}$.

   b) For all $i \in FI_j$, replace all occurrences of literals y_j and $\overline{y}_j$ in F_i with the corresponding new literals $y_{j,D}$ and $\overline{y}_{j,D}$;

   c) Augment the don't care set, i.e.,

$$D^j = y_{j_D}\overline{F}_j + \overline{y}_{j_D}F_j \ ,$$

$$\mathbf{DX}\leftarrow\mathbf{DX}\bigcup D^j \ . \ \square$$

Note that the "relaxed" Boolean Network defined by Definition 9 has $|J|$ more primary inputs that the original boolean Network η. Note also that nodes $j \in J$ are still present in η^J, and the defining equations

$$y_j = F_j(y,x), \ \ \forall j \in J$$

are still satisfied by the logic variables of η^J. However, all nodes $j \in J$ have been, in effect, **disconnected** from their original fan out, with dependencies on y_j in the original network replaced by a dependency on the new primary input y_{j_D} in η^J. According to Lemma 1, any vertex, $v \in B^{m+n+|J|}$ will be a self-consistent solution of (2.1) for η^J if and only if $v \notin D^j, \forall j \in J$. Thus we may state the following result.

Lemma 3

Let the "relaxed" network η^J and its associated don't care set be derived from Boolean Network η and its external don't care sets $\mathbf{DX}$, according to Definition 9. Then, according to Definition 2,

$$\eta \equiv \eta^J \ .$$

This may seem confusing at first since η was defined in the space B^{m+n}, whereas η^J is defined in $B^{m+n+|J|}$. However, as illustrated in the example of Figure 2.2b and Table 4.2, we may simply regard the y_{j_D} as primary input variables which are present in η, but are not in the support of any of the Boolean functions of η.

Note that the Boolean Network of Fig. 2.2c may be derived from that of Fig. 2.2a by the procedure of Definition 9. By Lemma 3, with $\mathbf{DX}$ originally empty and $J=\{2\}$, $\eta^J \equiv \eta^a$. Hence we may test $\eta^a \equiv \eta^b$ by asking if $\eta^J \equiv \eta^b$. This latter question is answered in the affirmative by the data of Table 4.2, which is compilation of the logic variables of η^a and η^J (here $J=\{2\}$) which result from the 32 unique primary input combinations of x_1, x_2, x_3, x_4, and y_{2D}, which are given as the first 5 columns of Table 4.2. The next pair of columns gives the result of simulating η^b for these inputs, and the pair after that is the result of simulating η^J. The 10th column gives the comparison of corresponding outputs for each primary input. Observe that corresponding outputs differ in 12 of the 32 cases (see asterisks). However, as shown in the last column, each of these "difference" inputs is in the don't care set D^J, where

$$D^J = y_{2D}\overline{F}_2^{\{2\}} + \overline{y}_{2D}F_2^{\{2\}} = y_{2D}\overline{y}_2^{\{2\}} + \overline{y}_{2D}y_2^{\{2\}} \ .$$

Consequently we have $\eta^a \equiv \eta^J \equiv \eta^b$.

Since the binary relation "$\equiv$" is transitive, Lemmas 2 and 3 may be subsumed by

Theorem 2

The Boolean Networks η^1 obtained from Boolean Network η by Procedure 1 and η^J from Definition 9 are equivalent to each other, as well as to η, i.e.,

$$\eta \equiv \eta^1 \equiv \eta^J \ .$$

4.3 The Simulation Method for Multilevel Tautology

In this section we give a brief, qualitative treatment of the "Simulation" method of multilevel tautology checking. An example of the application of this method is given, but a more complete treatment is deferred until Section 5. Here we give as Procedure 2 (below) an informal, English language procedure for tautology checking. Procedure 2 is recursive and can be thought of as being called at every node of a recursive binary cofactoring tree like that of Figure 4.1b. Thus Procedure 2 accepts as input a Boolean Network, η, which, presumably, has just been cofactored by some primary input variables according to Definition 3. Its first action is logic simulation, carried out in steps 1) and 2). Note that if no functions of η were simplified, i.e., evaluated to 1 or 0 after the cofactoring, then steps 2)-4) are vacuous. In this case Procedure 2 selects a new splitting variable, and creates 2 new nodes on the recursive binary cofactoring tree (cf. Figure 4.1b). On the other hand, if any functions do simplify, then steps 1) and 2) are carried out iteratively until the network stabilizes, i.e., until the transitive logical implications of all simplifications have been identified. Observe that steps 1) and 2) carry out the process of logic simulation.

Procedure 2 (η)

Input: A Boolean Network η upon which some cofactoring
             operation has been performed (cf. Definition 3)

Output: A logic variable T, equal to 1 if $\eta \equiv 1$, else 0.

Method: Binary recursion, cf., Figure 4.1b.

Actions

1) Directly evaluate the logic functions of η which were affected by the preceding cofactor operations;

2) If there exists a function F_j which has evaluated to 1(0), replace η by $(\eta)_{v_j(\overline{v}_j)}$ and go to 1;

3) If a primary output has evaluated to 0, return $T = 0$;

4) If all primary outputs have evaluated to 1, return $T = 1$;

5) Select a new variable, v_j, for recursive Shannon Expansion;

6) Recur in the positive halfspace, i.e., call Procedure 2 with $(\eta)_{v_j}$ as input, storing the returned value as $T1$;

7) Recur in the negative halfspace, i.e., call Procedure 2 with $(\eta)_{\overline{v}_j}$ as input, storing the returned value as $T0$;

8) Return $T = (T1 \text{ AND } T0)$

Note that in step 1), **evaluate** means evaluation of sum of products representation of a Boolean function. For example, if $F_1 = (x_1 \equiv x_2) + x_3 x_4 = x_1 x_2 + \overline{x}_1 \overline{x}_2 + x_3 x_4$, we have $(F_1)_{x_1 \overline{x}_2} = (F_1)_{\overline{x}_1 x_2} = x_3 x_4$, $(F_1)_{x_1 \overline{x}_2 \overline{x}_3} = 0$, and $(F_z)_{x_1 \overline{x}_2 x_3 x_4} = 1$. The following two simplification rules (called **special cases** in Section 5) are sufficient to define evaluation.

    R0:   A sum of products representation of a Boolean
            Functions evaluates to 0 if it is empty, i.e., if it has no cubes (product terms),
    R1   A sum of products representation evaluates to 1 if it
            contains a cube with no literals.

The results of applying Procedure 2 to the Comparison Network of Figure 4.1a is given in Figures 4.1c-f. Figure 4.1c gives the original defining equations (2.1) for η_C. First consider the result of cofactoring positively on both x_1 and x_2, which takes us to the first node of the third row of Figure 4.1b. Procedure 2 applies to this node as follows. Since x_1 and x_2 have both been evaluated to 1 by the cofactoring, we see that both F_5 and F_7 evaluate directly to 0 in Step 1. Setting y_5 and y_7 to 0 by the further cofactoring of

Step 2 and repeating step 1 as necessary causes F_2, F_3, and F_1 to evaluate **transitively** to 1, i.e., $y_2 \to 1$, $y_3 \to 1$, and $y_1 \to 1$. At this point Procedure 2 falls through to step 3, whose condition is not satisfied, and then to step 4. The condition of step 4 **is** satisfied, so the value $T = 1$ is returned immediately. Note there is no further recursion from this node, i.e., the binary tree is **pruned**, and the node is a **leaf** of the recursive binary cofactoring tree.

The result of recursively applying Procedure 2 to η^C of Figure 4.1a is given in Figure 4.1f. Note this recursion tree has 8 leaves, instead of 16, which would have been the case if none of the nodes had experienced "early" simplification in step 1. This implies that Procedure 2 offers a factor of 2 speedup over exhaustive logic simulation of all $2^4 = 16$ distinct primary input vectors. Note also that the longest paths in the recursion tree are $x_1 \bar{x}_2 x_3 x_4$ and $\bar{x}_1 x_2 x_3 x_4$ which have length $|PI| = 4$. Early simplification does not occur on these paths since after cofactoring with respect to $x_1 \bar{x}_2 x_3$ (or $\bar{x}_1 x_2 x_3$) we have $F_7 = F_5 = F_3 = 1$, but $F_6 = x_4$, $F_4 = x_4$, $F_2 = (y_4 \equiv y_y)$, and $F_1 = y_2$. Even though it may be clear to the reader already that $y_1 \to 1$, the method of direct logic evaluation embodied in Procedure 2 does not perceive this. Consequently, the 4th cofactoring step, on x_4, is necessary. In Section 4.4 below we will discuss a string grammar method which is capable of such perceptions [SBH82].

We state the proof of the correctness of Procedure 2 as follows.

Theorem 2

Suppose Procedure 2 is called with a given Boolean Network, η, as input. Then Procedure 2 returns $T = 1$ if and only if

$$\eta \equiv 1 .$$

4.4 The Algebraic String Comparison Method

The fourth method for tautology checking can be regarded as a variation of the simulation method described in Procedure 2 above. The variation adds the following "algebraic string comparison" mechanism to those described above for pruning the binary recursion tree. The idea is to view the Boolean Comparison Network as an algebraic factored form, [Bra86] of a single Boolean Function, represented by a continuous algebraic string comprised solely of primary input literals and various Boolean operators ($\cdot$ (AND), $+$ (OR) $\equiv$ (equivalence), $\oplus$ (exclusive OR), $\sim$ (not),...). If it is discovered that both sons of a given node have had early simplification of the primary outputs, thus identifying $F_1 = 1$ on the output of the comparison network, then the recursion backs up to the parent of the current node. In this case the algebraic string for the current node of the recursion tree may be computed, mapped into a canonical form according to a specified grammar (e.g., reverse polish notation, [SBH82]), and stored away for future reference. Then, at any other node of the recursion tree, the canonical algebraic string may be generated and checked for identity (i.e., string comparison) with previously stored strings known to be tautological. If a match is obtained, the tree is pruned. The string comparison is made efficient by 1) judicious choice of the canonical form (i.e., the string grammar), 2) judicious choice of the operator primitives, and 3) length check. Apparently, it is unlikely that many non-identical strings will have the same length.

In various forms, this basic mechanism seems to have been discovered, rediscovered, and/or developed by Lee [Lee59], Akers [Ake78b], [Ake78a], Bahnsen [Bahnsen82.], and Bryant [Bry86]. Actually, Akers and Bryant reported identity checking by graph isomorphism, rather than algebraic string comparison, but this distinction appears to be more related to computational data structure than to concept.

At the root of the recursive binary cofactoring tree for the primary output of the Comparison Network of Fig. 4.1a, the algebraic string can be generated as follows.

$$y_1 = y_2 \cdot y_3 \,,$$

$$= (y_u \equiv y_y) \cdot (y_5 \equiv y_7) \,,$$

$$= ((\overline{y}_5 + x_3 x_4) \equiv ((x_1 \equiv x_2) + x_3 x_4)) \cdot$$

$$((x_1 \oplus x_2) \equiv (x_1 \oplus x_2)) \,,$$

$$= (((\,\neg(x_1 \oplus x_2)) + x_3 x_4) \equiv ((x_1 \equiv x_2) + x_3 x_4)) \cdot$$

$$((x_1 \oplus x_2) \equiv (x_1 \oplus x_2))$$

If this string is "cofactored" with respect to x_1, it reduces to

$$(y_1)_{x_1} = (((\,\neg(\overline{x}_2)) + x_3 x_4) \equiv ((x_2) + x_3 x_4)) \cdot (\overline{x}_2 \equiv \overline{x}_2) \,,$$

and when cofactored with respect to $\overline{x}_2$ it becomes

$$(y_1)_{x_1 \overline{x}_2} = (((\,\neg(1)) + x_3 x_4) \equiv ((0) + x_3 x_4)) \cdot (1 \equiv 1)$$

$$= ((x_3 x_4) \equiv (x_3 x_4)) \,.$$

This string would be discovered to be tautological after cofactoring on x_3 and x_4. Note this **same** string would also be generated at the $\overline{x}_1 x_2$ node of the recursion tree for Fig. 4.1a. Thus the string comparison method would prune the recursion tree at $\overline{x}_1 x_2$ without further cofactoring, **so that the pruned tree would have only 6 nodes, versus 8 for the simulation method.**

The efficiency of the string comparison method also depends on how the string is cofactored and simplified. Complex operators like $\oplus$ and $\equiv$ require more complex computations for these purposes, but may produce earlier simplification. If the operator set is restricted to $\cdot$, $+$, and $\neg$, then string processing is simplified, but some early simplification may be sacrificed.

5. An Algorithm Frame for Multi-Level Tautology.

We discuss in this section a set of procedures and subprocedures, presented as structured code, which constitutes an algorithm frame for multilevel tautology and equivalence checking. The procedures are all collected "en masse" as Section 9 below, and all are subordinate to the "main" algorithm frame, Procedure ML_TAUT. ML_TAUT is a recursive procedure for answering the multilevel tautology question. We think of this procedure as an "algorithm frame," in the sense that by specifying certain of its heuristic ordering and selection subprocedures, it can be "tailored" to perform according to various known algorithms discussed in Section 4, cf [WeS85], [SBH82], [HaJ85], [Ake78a], [Bry86]. Before discussing ML_TAUT per se, we need to introduce its principal subprocedures, ML_COFACTOR and FLATTEN.

5.1. Multilevel Cofactoring and Flattening

Procedure ML_COFACTOR (cf. procedures of Section 9 below), implements Definition 3. Given a Boolean Network, η, and cofactoring cube, c, ML_COFACTOR forms a cofactored Boolean Network described by the equations

$$(y_i)_c = F_c((y)_c, x) \,, i = 1, 2, \ldots m \,.$$

as specified in Definition 3. Each of the two-level functions in the original multilevel Boolean network are individually cofactored, one variable at a time, by a subprocedure call to ML_COF1. Note that the variables in the cube c are then deleted from the corresponding fan in sets of the cofactored functions. An interesting and unique aspect of

266

multilevel cofactoring that has no analog in the two-level cofactor operation is the augmentation of the don't care set by the expression

$$(D_i)_{v_i} \equiv (v_i \overline{F_i} + \overline{v_i} F_i)_{v_i} \equiv \overline{F_i}$$

whenever i is an intermediate variable, i.e., if $1 \leq i \leq m$. According to Lemma 1, this augmentation essentially insures that for any vertex $v \in B^{|z|}$, $v \notin D$ implies that the above Boolean equations for the cofactored Boolean network are consistent with the corresponding solution of the original Boolean network.

Procedure FLATTEN, also given in Section 9, takes a directed edge $(j,i) = e$ of the function graph of the Boolean Network and "flattens" the two-level cover F_j into its fan out cover F_i, $i \in FI_j$. This procedure first identifies the "algebraic cofactor" of F_i with respect to z_j, (i.e., the cubes of F_i which are positive in z_j) denotes it F_i/z_j, and intersects it with F_j. Similarly, $F_i/\overline{z_j}$ is intersected with $\overline{F_j}$. The subset of cubes in F_i (call this subset R) which do not depend on z_j are unaffected by the flattening process. The union of these three sets is taken to form the flattened version of F_i (with respect to F_j) in line 4; and in line 5, the variable z_j is deleted from the fan in of F_i. Note that flattening a fan in, F_j, of F_i can greatly increase the number of cubes in the resulting cover for F_i. In fact, if the support sets of F_j and F_i are disjoint,

$$|F_i'| = |F_i||F_j| \ ,$$

where F_i' stands for the representation of F_i after flattening edge (j,i). Note also that F_i' is not necessarily prime and irredundant, even if F_i and F_j are. In fact the flattened-cubes may tend toward minterms after many flattenings of disjoint or near disjoint support.

Of special interest is the flattening of what we call "simplified" functions, for which $F_j \equiv 1$ or $\overline{F_j} \equiv 1$ ($F_j \equiv 0$), i.e., the cases in which functions (or their complements) consist of a single (tautological) cube. For such functions we can state

Lemma 4

Suppose $F_j \equiv 1$. Then FLATTEN $(\mathbf{F},(j,i))$ returns $\mathbf{F}'$ for which $F'_i = (F_i)_{z_j}$, $\forall$ $i \in FI_j$ and $F'_i = F_i$ otherwise. Suppose $F_j \equiv 0$. Then FLATTEN $(F,(j,i))$ returns F' for which $F'_i = (F_i)_{\overline{z_j}}$, $\forall i \in FI_j$ and $F'_i = F_i$ otherwise.

Note that if $F_j \equiv 1$, $\overline{F_j} = \varnothing$, so $F_i' \leftarrow (F_i/z_j)1 + (F_i/\overline{z_j})\varnothing + R_j$ in line 4. But $(F_i/z_j) + R_j \equiv (F_i)_{z_j}$ by the construction of lines 1 and Definition 3. The case for $\overline{F_j} \equiv 1$ is similar.

5.2 Procedure ML_TAUT

There are 3 basic processes which occur in ML_TAUT, corresponding to the calls to T_FLATTEN, FLATTEN, and ML_COFACTOR. Subprocedure T_FLATTEN examines the specified multilevel functions and checks for trivial cases like $F_i \equiv 1$ (Function F_i has a cube in its representation which has no literals) or $F_i = \varnothing$ (F_i has no cubes). In the first case z_i is set to 1 and in the second case to 0. Such trivial functions are then flattened up into their fan outs, and by Proposition 1, this amounts to simple two-level cofactoring of these fan outs. Note this may simplify the fan out functions to the point where they, too, become trivial. The **while** loop in T_FLATTEN continues until no trivial functions are left in the multilevel cover. In effect, this process is one of logic simulation as discussed in Section 4.

The second basic process in Procedure ML_TAUT is that of subprocedure FLATTEN, which is driven by the edge selection subprocedure E_SELECT. According to a specified heuristic, E_SELECT selects a directed edge, e, from the graph $G = (V,E)$ of the Boolean Network. For each edge (j,i), function F_j is flattened into its fan out F_i. E_SELECT will never return a $\varnothing$ indication if there exists trivial functions, so that on exiting the **while** loop, of line 17, the multilevel cover F has no trivial functions. In fact, a user might specify a selection heuristic which **completely** flattens the multilevel cover into a two-level cover. If this is the case the remainder of ML_TAUT will behave like a two-level tautology checker [Sas85], [BHM84].

The third process in ML_TAUT is procedure ML_COFACTOR (previously discussed in Section 5.1), which is driven by the heuristic cube selection subprocedure CUBE_SELECT. In CUBE_SELECT a set of cubes C is generated, for which,

$$\bigcup_{c^i \in C} c^i \equiv 1 \ \text{ and } \ c^i \bigcap c^j = \varnothing \ , \ \ c^i, c^j \in C$$

i.e., the cubes in C are disjoint, and span the entire expanded space $B^{|z|}$ of the Boolean Network. The multilevel cover is tautologous if and only if it is tautologous on each of the cubes $c^i \in C$, i.e., given the don't care set, D, $(\eta)_{c^i} \neq 1$ (according to Definition 2), $\forall$ $c^i \in C$. If the cubes c^i are half-spaces (i.e. $C = \{\ x^j, \overline{x^j}\ \}$), then the recursive portion of ML_TAUT reduces to the basic recursive Shannon expansion, [BHM84], discussed in relation to (1.2) to (1.7).

The remainder of this paper discusses four specific methods of answering the multilevel tautology question that arises from the general algorithm frame just presented. We present detailed algorithms and/or runtime data for the "flattening" method, the "don't care" method, the "simulation" methods, and the "string comparison" method. The first three methods use the procedures ML_TAUT, ML_COFACTOR, T_FLATTEN, and FLATTEN without modification: examination of the procedures E_SELECT and CUBE_SELECT is sufficient to characterize the differences in the three methods. In fact, the only differences between **any** two of the four methods would be contained in one of the three procedures SPECIAL_CASES, E_SELECT and CUBE_SELECT. For each of the methods mentioned above we present a broad outline of the method, the algorithms used in E_SELECT and CUBE_SELECT, and the results of running the algorithm on a variety of multilevel Boolean Networks.

5.3 The Flattening Method

The "flattening" method has been discussed previously, [HaJ85] and [HaJ86] and in Sections 4.1 and 5.1 above. Three specific flattening methods were investigated, which are derived from the algorithm frame by replacing E_SELECT by any of the three procedures E_SELECT1, E-SELECT2, and E_SELECT3 of Section 9 below. Experimental results are given in Section 6.

5.4 The Simulation Method

In the simulation method the tautology question is answered by operating directly on the multilevel representation. No flattening is performed other than in the routine T_FLATTEN; that is, E_SELECT always returns $\varnothing$. The tautology question is answered by recursively applying the cofactoring operations of ML_COFACTOR combined with the detection and propagation of any resulting simplifications in T_FLATTEN. CUBE_SELECT, through a set of heuristics, chooses a variable of F, z_j (z_j could be a primary input or an intermediate variable), and returns a set of two cubes $\{\ x^j\ ,\ \overline{x^j}\ \}$ to ML_TAUT. Note that with this form of CUBE SELECT, ML_COFACTOR implements the Shannon expansion of F with respect to z_j (it is clear that more sophisticated strategies exist and are worth considering). We call z_j a splitting variable. The particular type of ML_TAUT implemented thus depends on which version of CUBE_SELECT (CUBE_SELECT1, CUBE_SELECT2,...CUBE SELECT6) is employed.

Two variations on the multilevel method suggested by our algorithm frame are discussed here. The difference in the methods is in the range over which the selection of splitting variables is made. In the "restricted" case, only variables belonging to the primary input set are candidate splitting variables. In the "unrestricted" case any variable (primary input or intermediate variable) is a possible splitting variable.

In both the restricted and unrestricted cases two different sets of heuristics are used in selecting the splitting variable. However, there is an idea common to all of our heuristics - the binate select heuristic of ESPRESSO, [BHM84]. The procedure BINATE_SELECT as used here will first form a set containing all columns of the cover matrix that contain both 1's and 0's. From this set, the variable corresponding to the column containing the fewest number of 2's is chosen as the splitting variable. In the event that no column contains both 1's and 0's, the largest cube in the matrix is found (i.e. the cube containing the fewest non-two entries) and a set containing all the non-2 columns of this cube is formed. From this set the variable corresponding to the column (of the entire two-level matrix) containing the fewest number of 2's is selected as the splitting variable.

5.4.1 The Simulation Method - Restricted Splitting Case

As just outlined, this method answers the tautology question by recursive application of the Shannon Expansion on a multilevel representation of a function. The splitting variable used in the Shannon Expansion is selected from the set of primary inputs. Two different heuristics were used in selecting the splitting variable. Let's designate the first heuristic as "voting". "voting" applies BINATE_SELECT to each of the F_i of F. The result of each call to BINATE_SELECT is tabulated and "voting" then selects the primary input variable that was returned most often as the next splitting variable. Designate the second heuristic as "global". "global" creates an artificial two-level function by forming the union of all the F_i of F and then applying BINATE_SELECT to this two-level function. The variable returned from BINATE_SELECT is used as the next splitting variable.

5.4.2 The Simulation Method - Unrestricted Splitting Case

Again this method also answers the multilevel tautology question by recursive application of the Shannon Expansion on a multilevel representation of a function. However, in contrast to the restricted case, the splitting variable used in the Shannon Expansion can be either a primary input or an intermediate variable. The two heuristics described in Section 5.4.1 are also used in this case unchanged except for the fact that here intermediate variables are now candidate splitting variables. Note that if an intermediate variable is selected the don't care set is augmented, as prescribed by procedure ML_COFACTOR.

5.5 The Don't Care Method

In the unrestricted case of the multilevel tautology method it was possible to choose an intermediate variable, say x_j, as a splitting variable. In this event, the procedure ML_COF1 augments the don't care terms associated with the variable x_j (see Equations (2.4a) and (2.4b). In this way only the portions of the don't care set relevant to the current level of recursion need be considered. However, it is also possible to answer the tautology question by initially constructing the entire don't care set and then recursively applying the Shannon Expansion to η with respect to the complete don't care set (see the discussion of Equation (2.4) and Lemma 1, Definition 9, and Lemma 3).

In terms of our algorithm frame, the don't care set is formed prior to the call to ML_TAUT. ML_TAUT is called with the arguments $\mathbf{F}_0$ and D, where $\mathbf{F}_{PO}$ is the set of primary output functions of the given Boolean Network, and D is the entire don't care set as defined by (2.4). In this setting, E_SELECT will always return $\emptyset$. Also CUBE_SELECT applies BINATE_SELECT to the union of $\mathbf{F}_{PO}$ and D.

5.6 The Algebraic String Comparison Method

The string comparison method is implemented by: 1) calling subprocedure STORE_STRING (line 28 of ML_TAUT) to store the string whenever the recursion backs up to the point of the currently active node of the recursion tree; 2) modifying subprocedure T_FLATTEN to check for string identity (this modification is not shown). Data structure play a key role in the efficiency of this method, cf., [HaJ86].

Our implementation of this method show it to be extremely fast in some examples, but rather slow in others. It is storage space intensive, since many strings must be stored in order to capitalize on all possible tree prunings. The experimental results we have to date are not given in Section 6 below, because they are too preliminary and are still varying widely as we refine the basic strategy. However, in comparison to the simulation method, the results consistently show large speedup factors, on the order of 10X on about half the examples, and moderate slowdowns on the order of 2X, on the other half. The method appears to work best on Boolean Networks with low average fan out, but many levels of logic.

6. Experimental Results for Multilevel Tautology Checking

Some points about the nature of the run time data must be made before discussing experimental results for specific instantiations of the algorithm frame.

First, there is a procedure ML_COMPARE that has as arguments the function representations **F** and **G** of two Boolean Networks, along with a common don't care set D. ML_COMPARE compares F and G for equality by forming a third function **H** which defines the comparison network of Definition 8, and invoking ML_TAUT with arguments H and D. (Note that D stands for the primary input don't care set **DX** and may be empty.) All the run time figures indicate how long it took to compare the two multilevel functions listed in the first two columns of each row of the table using procedure ML_COMPARE.

The second point concerns the origin of the Boolean Network pairs to be checked for equivalence. Since one of the primary uses of these algorithms is as a verification tool in a silicon compiler environment, we used Boolean Networks that had been processed through the logic synthesis stage of the Boulder Silicon Compiler. Almost all the Boolean Networks compared represent the input to and the output from a logic transformation stage of a silicon compiler. Table 6.1 provides data about each of the Boolean Networks. Information about size, depth, etc., of each function is tabulated.

All of the algorithms described in this paper were implemented in C. The programs were executed on a Pyramid computer running Berkeley Unix. A number of useful comparisons can be made from studies of the examples described in Table 6.1.

First examine the unrestricted multilevel tautology case of Table 6.2. The results presented here, obtained by applying the heuristics of Section 5.4 to a number of example problems, contain two figures of merit. The first is simply the number of CPU-seconds required to answer the tautology question. The second is the total number of leaves in the binary recursion tree. If we compare the voting method with the global method (using the number of leaves in the binary recursion tree as a figure of merit), we find that the voting method provides the best solution in 10 cases, the global method is best in 6 cases and there are 11 ties. This indicates a slight advantage to the voting method in the unrestricted case.

Next examine the restricted multilevel tautology case of Table 6.3. If we compare the voting method with the global method (again using the number of leaves in the binary recursion tree as a figure of merit), we find that the voting method provides the best solution in 10 cases, the global method is best in 11 cases and there are 7 ties. Neither method can be considered favored here.

Now compare the best result (either voting or global) from the unrestricted case with the best result (either voting or global) from the restricted case. Again we use the number of leaves as our basis for comparison. Here we find that the best results were obtained from one of the restricted methods in 9 cases and from the unrestricted methods in 9 cases with 8 ties between them. Again neither method can be considered favored. As a side note, in 5 of the 9 cases in which an unrestricted method wins, no splitting on intermediate variables was performed. We unfortunately lack this information about the other 4 cases at the time of this writing.

Three flattening methods were presented in [HaJ85]. The results for the top down and the bottom up methods are represented here in Table 6.4. In comparing the bottom up and top down methods, clearly the bottom up method is superior. The bottom up method wins in 23 cases, the top down wins in one case and there are 5 ties. Preliminary results for the top down stepwise method indicates it is roughly comparable to the top down method in efficiency.

The don't care method proves to be very inefficient, as shown by the results of Table 6.5. We attribute this result to the fact that in this method the two-level tautology algorithm is having to work on a considerably expanded variable space (i.e. an input space of $m+n$). Moreover, the construction of the don't care set results in all variables being binate.

Lastly we wish to compare the best multilevel method with the best flattening method with the don't care method. In 21 cases the flattening method provides the best result and in 8 cases there are ties between a multilevel method and the bottom up flattening method.

One final note about the flattening methods are in order. The flattening methods that operate in the algorithm frame presented here all follow a similar pattern. The two functions to be compared are "xnor'd" to create a third function. This function is then flattened with some method and a two-level tautology algorithm is applied to this resulting two-level function. There is another way to achieve this same end however, which we call the "flatten and verify" approach. Each of the two functions to be compared can be flattened separately and then the two-level comparison program of ESPRESSO-IIC [BHM84] run on the two flattened functions. In practice, this proves to be computationally more feasible, as indicated in Table 6.6. In all but five cases (which were specifically constructed as counter-examples), the "flatten and verify" approach proved superior using the computation time as a measure of efficiency.

7. Results on Multilevel Minimization and Test Generation

We describe in this section some results of applying practical implementations of the algorithm frame discussed above to example problems in the areas of multilevel minimization and test generation.

7.1 Multilevel Logic Minimization

The test generation ideas of Section 3.2 were preceded by and are directly related to the concepts of Multilevel Logic Minimization, which was discussed in a paper of that title at the ICCD conference in October of 1986 [BBH86]. That paper established the ideas of primality and irredundancy for the cubes of the functions of a Boolean Network and showed how to compute the **complete** don't care set for each function. Also an algorithm, called ESPRESSO_ML, was presented which exploits the ESPRESSO 2-level logic minimizer, along with the complete Don't Care set $D(j)$ defined for function F_j of Boolean Network η. ESPRESSO_ML was quite successful in reducing the literal counts of example Boolean Networks, and was able to render all the cubes of F_j **prime** and **irredundant**, $\forall$ $j \in IV$.

An implemented version of procedure ML_TAUT was incorporated into a second multilevel logic minimizer, called ESPRESSO_MLT, [HJM], also based on ESPRESSO_II. ESPRESSO_MLT was run on a subset of the examples of Table 6.1. The results are given and compared to the ESPRESSO_ML, [BBH86] in Table 7.1. It is seen that in addition to transforming the given Boolean Networks into prime and irredundant form, substantial reduction in literal counts were achieved in affordable CPU times. It appears that because of the size and logical structure of the don't care sets $D(j)$, the tautology based minimizer has a speed advantage which increases with problem size, indicating a fundamental complexity advantage. For this reason, ESPRESSO_ML used only a subset of $D(j)$ in minimizing F_j. Consequently, the resulting Boolean Networks were not guaranteed to be prime and irredundant.

It was further shown in [BBH86] that a prime and irredundant Boolean Network was 100% testable and that the tests for all input and output stuck faults could be obtained as a byproduct of the minimization. The results of Section 7.2 specifically identify how to generate these and all other conceivable (combinational) faults. Thus a profound relationship, which was known to exist between testing and logic minimization [Rot80], is further established. In fact, it follows that once all input and output stuck faults have been identified and any discovered logical redundancies removed, the Boolean Network is prime and irredundant. In brief, it may be factually stated that prime and irredundant networks are 100% testable, and conversely.

7.2 Results on Test Generation and Redundancy Checking

We conclude our experimental investigation by applying Theorem 1 and procedure ML_TAUT to test for output stuck faults for a subset of the Boolean network examples discussed above. The results are in the last data table presented in Table 7.2 below. The test count refers to the number of output stuck at faults tested, this is equal to (the number of intermediate variables + the number of primary outputs)*2. The time refers to the amount of time required to perform all tests. No fault collapsing was performed. Note that because the tautology checking stops as soon as a counter-example is found (i.e. a test is found), the average time to find a test is much less than the time required to perform a single functional equivalence computation.

8. Summary and Conclusions

We have presented a theory for multilevel tautology and equivalence checking. This theory was used in the formulation of an algorithm from which four specific approaches to multilevel tautology were derived: 1) the flattening method, 2) the don't care method 3) the simulation method and 4) the algebraic string comparison method. The experimental results suggest that the expanded variable space and the increased binate variable count render the don't care method computationally inferior. The flattening method wins in all examples based on the output of a logic decomposition and factorization technique known as WEAK DIVISION. We conjecture that this may be an artifact of the factorization infrastructure of Boolean networks created by WEAK DIVISION and hence is computationally much faster than it might be in a more general setting. The third method wins on Substantial Boolean Networks (ALUs, Multipliers, FSM's, etc.) extracted directly from Behavioral Descriptions [RaL86] and also wins in a set of contrived examples of Boolean networks constructed by interconnecting PLA's. Thus the third method has been used to date for general purpose applications such as multilevel logic minimization and test generation.

Our experimentation with the string comparison method is still underway. But it appears that it will have very substantial advantages for problems with low fan out but many levels of logic. We have tentatively concluded that a general purpose multilevel logic minimization or test generation tool based on Tautology Checking should do some experimentation and/or gather statistics on a given Boolean Network before committing to one single method.

Our theory has shown how to use a multilevel tautology and equivalence checker as a general test generation and redundancy checking tool. A set of experimental results has been used to demonstrate the theory and evaluate the corresponding computational expense. The data indicates that 100% test generation and redundancy checking is generally available and affordable for examples based on WEAK DIVISION. We note that unlike fast ATPG programs like PODEM and FAN, our approach can be used even when the individual functions, F_j, are an arbitrary sum of products description. However, our initial choice of data structures limits the overall size of circuit that can be tested. In fact we have already experimented [HaJ86] [HJM] with a suitable sparse implementation, giving factors of 5-10 savings in CPU time and storage compared to the results of Section 6. This will permit the study of larger problems, especially problems not generated by WEAK DIVISION.

As a final point we have noted in Sections 3 and 7 that tautology is basic to the operations required to make two-level functions prime and irredundant, and that a multilevel minimizer has already been reported based on don't care methods [BBH86]. But the theory of Section 3 forms a fundamental basis for a new class of multilevel logic minimization programs, based on multilevel equivalence checking, [HJM]. For example, for any cube c in any function F_j of a Boolean network η, a conjugate cube c^+, distance 1 in variable l can be added to form a second Boolean network η'. Literal c_l is prime iff η and η' are equivalent in the multilevel sense. Because of the demonstrated superiority of the simulation and string comparison methods we expect this new class of multilevel minimizer to exhibit superior speed performance, as discussed in Section 7.

Some preliminary results in this regard (with the sparse data structure) are given in Table 8.1. Here some of the multilevel versions of the test PLAs are compared to themselves in two different modes. First, a straight comparison of the Boolean Network to itself is made, using the Comparison Network concept. The time taken for determining tautology is given in the second column of the table. Second, statistics were gathered from running the same tautology checker inside the context of the multilevel logic minimizer ESPRESSO_MLT ([HJM]. Average CPU times for true and false tautology cases are given in the third and fourth columns of the table. It is seen that 10X to 40X speedups are observed. The same type of results would be obtained in a test generation context. The speedup is due to 1) trimming away irrelevant portions of the Comparison Network, e.g., all but the transitive fan in of the transitive fan out primary outputs of the function, F_j, being minimized; 2) limiting the BINATE_SELECT splitting heuristics to consider only primary inputs in the transitive fan in of F_j and 3) exploiting the fact that the value assumed by y_j in the non-tautology is **known in advance** in the minimization or test generation context. [HJM].

9. Structured Code for Equivalence Procedures

To avoid duplicate labelling we have simply collected all procedures referred to in the preceding text as a separate section.

Procedure ML_COMPARE (F,G,D)

  /*

This procedure compares two functions for equality.

  /*

Begin
$H \leftarrow F \cdot G + \overline{F} \cdot \overline{G} + D$
If(ML_TAUT(H,D) = 1 **) Return**(1)
Else Return(0)
End ML_COMPARE

Line		
	Procedure ML_TAUT (F,D)	
1	/*	
2	Simplified functions are flattened (line 12) until no further simplification	
3	is possible and return value is checked to see if the tautology question	
4	is answered on B^n where $n = |x|$. Then, selects appropriate further	
5	flattening of Boolean Network (line 16). Then an appropriate support	
6	set J is chosen and a minimal set C of disjoint cubes is defined on	
7	$B^{|J|}$ which cover $B^{|x|}$. Then ML_TAUT is recursively	
8	called on the cofactors of F with respect to each $c \in C$.	
9	/*	
10		
11	**Begin**	
12	$(Done, F) \leftarrow$ T_FLATTEN(F,D)	Flattens "simplified" functions and
13	**If (** $Done$ = 1 **) Return**(x_0)	returns their values in vector x.
14	**Else**	
15	**Begin**	
16	$e \leftarrow$ E_SELECT(F)	Selects one edge to flatten.
17	**While(** $e \neq \varnothing$ **)**	
18	**Begin**	Flatten while appropriate.
19	FLATTEN(F,e)	
20	$e \leftarrow$ E_SELECT(F)	
21	**End While**	
22	$C \leftarrow$ CUBE_SELECT(F,D)	Selects expansion cover.
23	**For(** $c \in C$ **)**	
24	**Begin**	
25	$(F_c, D_c) \leftarrow$ ML_COFACTOR(F,D,c)	
26	**If (** ML_TAUT(F_c, D_c) = 0 **) Return**(0)	
27	**End For**	
28	STORE_STRING (F_c)	
29	**Return**(1)	
30	**End Else**	
31	**End** ML_TAUT	

Procedure T_FLATTEN(F,D)

   /*

Flattens "simplified functions, i.e. functions F_i such that
$F_i \equiv 1$ or $F_i \equiv 0$ into their fan out, returning the
modified Boolean network F, and a condition code indicating whether
or not the primary output F_0 has simplified.

   /*

```
Begin                              First:identify simplified F_i.
  S ← SIMPLIFIED(F,D)              S = {i | F_i ≡ 0 or F_i ≡ 1}.
  P ← Ø
  While( S ≠ Ø )
    Begin
    If( 0 ∈ S )  Return(1,F)       "1" indicates F_0 has simplified.
      Begin
        j ← S(1)                   Select an element of S.
        S ← S − {j}
        For( k ∈ FO_j )            Flatten F_j into its fan out.
          Begin
          F ← FLATTEN(F,(j,k))
          End For
      End If
    S ← SIMPLIFIED(F,D) − P        Update simplified list.
    P ← P ∪ S
    End While
  Return (0,F)                     "0" indicates F_0 not yet simplified.
End T_FLATTEN
```

Procedure SIMPLIFIED(F,D)

```
  Begin
  S ← Ø
  For( i = 1,2, .... ,m )
    Begin
    If ( F_i ≡ 0 OR F_i ≡ 1 )  S ← S ∪ {i}
    End For
  If ( F_0 ∪ D ≡ 0 OR F_0 ∪ D ≡ 1)  S ← S ∪ {0}
  Return(S)
  End SIMPLIFIED
```

Procedure ML_COFACTOR($\mathbf{F}$,D,c)
```
/*
```
Input: Boolean function set $\mathbf{F}$ of Boolean Network $\eta = (\mathbf{F},PO)$, Don't Care Set D,
      and cube c. $\mathbf{F}:B^{m+n} \to B$.
Returned: Multilevel Cofactor $F_c(D_c)$ of $F(D)$ with respect to c.
Inner Loop: Cubes disjoint from c are dropped and variables
      present in c are raised to 2.
```
/*
```

Begin
$D \leftarrow (D)_c$ Cofactor initial Don't Care set.
For($j = 1,2, \ldots, m+n$)
   Begin
   If($z_j \in c$ or $\overline{x}_j \in c$)
      Begin
      $(F_c,D_c) \leftarrow$ ML_COF1($\mathbf{F}$,D,j,c_j)   Cofactor multilevel cover
      End If   with respect to c_j.
   End For j
Return($\mathbf{F}_c$,D_c)
End ML_COFACTOR

Procedure ML_COF1($\mathbf{F}$,D,j,σ)
```
/*
```
Cofactors Multilevel Boolean Function set $\mathbf{F}$ and updates cofactored Don't Care
Set D with respect to x_j, $1 \le j \le m+n$.
```
/*
```

Begin
$FI_j \leftarrow \varnothing$
For($i = 1,2, \ldots, m$)
   Begin
   If($j \in FI_i$) $FI_i \leftarrow FI_i - \{j\}$
   If($i \neq j$)
      Begin
      If($\sigma = 1$) $F_i \leftarrow (F_i)_{x^j}$
      Else $F_i \leftarrow (F_i)_{\overline{x}^j}$
      End If
   Else If($\sigma = 1$) $D \leftarrow D \bigcup \overline{F}_i$ If $x_j \in v$, update the cofactored
   Else $D \leftarrow D \bigcup F_i$ Don't Care set.
   End For
Return(F,D)
End ML_COF1

Line **Procedure** FLATTEN($\mathbf{F}, e$)

/*

$\mathbf{F}$ is the set of Boolean Functions of a Boolean Network ($\mathbf{F}, PO$).
$e = (j, i)$ $j \in FI_i$ (F_j is a fan in of F_i).
FLATTEN flattens function fan in F_j into
Function F_i, returning the correspondingly
altered Boolean Function $\mathbf{F}$.

/*

Begin

1	$(F_i/x_j) \leftarrow \{c \in F_i \mid \dot{x}_j \in c\}$	Cubes positive in x_j.
2	$(F_i/\overline{x}_j) \leftarrow \{c \in F_i \mid x_j \notin c \text{ and } \overline{x}_j \notin c\}$	Cubes negative in x_j.
3	$R \leftarrow \{c \in F_i \mid x_j \notin c \text{ and } \overline{x}_j \notin c\}$	Other cubes of F_i.
4	$F_i \leftarrow (F_i/x_j)F_j + (F_i/\overline{x}_j)\overline{F}_j + R$	
5	$FI_i \leftarrow FI_i \bigcup FI_j - \{j\}$	Delete x_j from F_i, and merge its fan in with that of F_j.
6	**Return(F)**	
	End FLATTEN	

Procedure BINATE_SELECT (F,start,stop)

```
/*
    F is a two-level function. "start" and "stop" define the range of possible
    splitting variables.  1 ≤ start < stop ≤ m + n.
    This procedure returns a variable best_var, where best_var is in the range:
    start ≤ best_var ≤ stop. First, all columns in which there exists both
    1's and 0's are found and from that set one column is selected which
    maximizes the total number of non-2 entries. If there are no columns
    that contain both 1's and 0's, then the largest cube(i.e. the cube containing
    the most number of 2's) of the cover matrix is found. From the set of non-2
    columns in the largest cube, a column is selected that maximizes the
    number of non-2 entries in that column in the cover matrix.
/*
```

Begin
max ← 0
best_var ← −1
For(i = start, start+1, start+2, ... ,stop)
   Begin
   $J \leftarrow \{ j \mid c^j \in F$ and $c_i^j = 1 \}$
   $K \leftarrow \{ k \mid c^k \in F$ and $c_i^k = 0 \}$
   If($|J| > 0$ and $|K| > 0$ and $|J| + |K| > $ max)
      Begin
      max ← $|J| + |K|$
      best-var ← i
      End
   End For
If(best_var = −1)
   Begin
   min ← ∞
   For($c^i \in F$)
      Begin
      $J \leftarrow \{ j \mid$ start$\leq i \leq$stop and $c_j^i \neq 2\}$
      If($|J| < $ min and $|J| \neq 0$)
         Begin
         min ← $|J|$
         largest_cube_index ← i
         End
      End For
   $J \leftarrow \{ j \mid c_j^i \neq 2$ and $i = $ largest_cube_index$\}$
   For($j \in J$)
      Begin
      $K \leftarrow \{ k \mid c^k \in F$ and $c_j^k \neq 2 \}$
      If($|K| > $ max)
         Begin
         max ← $|K|$
         best_var ← j
         End
      End For
   End If
Return(best_var)
End BINATE_SELECT

Procedure E_SELECT1(F)
```
/*
Top Down Flattening Method.

Selects and returns an edge from F to be flattened.
Returns Ø when F is a two-level function.
Note E is initially (1st call) assumed Ø.
/*
```

Begin
If($E=\varnothing$ **)**
   Begin
     $F \leftarrow \{(i,0) \mid i \in FI_0^-$ and $i \leq m\}$ All incoming edges not from PI's
     If($E = \varnothing$ **) Return($\varnothing$)**   Function is two level.
   End If
$e \leftarrow E(1)$ Choose a single edge from E
$E \leftarrow E - \{e\}$
Return(e)
End E_SELECT

Procedure E_SELECT2 (F)
```
/*
Top Down (Stepwise) Flattening Method.

Selects and returns an edge from F to be flattened.
Return Ø when F is a two-level function.
Note E is initially (1st call) assumed Ø.
/*
```

Begin
If($E=\varnothing$ **)**
   Begin
     $E \leftarrow \{(i,0) \mid i \in FI_0^-$ and $LL(F_i) = LL(F_0) - 1$ and $i \leq m\}$
     If($E=\varnothing$ **) Return($\varnothing$)**
   End If
$e \leftarrow E(1)$ Choose any element of E.
$E \leftarrow E - \{e\}$
Return(e)
End E_SELECT

Procedure E_SELECT3 (F)

```
    /*
     Bottom Up Flattening Method.
    /*

    Begin
    Return( DFS(0,F) )
    End E_SELECT
```

Procedure DFS(i,F)

```
    /*
     Performs depth first search of the reverse function graph of F.
    /*
```

$$
\begin{aligned}
&\textbf{Begin} \\
&J \leftarrow \{\, j \mid j \in FO_j \text{ and } j \leq N_{IV} \,\} \\
&\textbf{If}(\ J \neq \varnothing\ ) \\
&\quad l \leftarrow \underset{j \in J}{\mathrm{Min}}\{j\} \\
&\quad K \leftarrow \{\, k \mid k \in FI_l \text{ and } k \leq m \,\} \\
&\quad \textbf{If}(\ K \neq \varnothing\ ) \\
&\quad\quad \textbf{Begin} \\
&\quad\quad m \leftarrow \underset{k \in K}{\mathrm{Min}}\{k\} \\
&\quad\quad \textbf{Return}\ (\ DFS(m,F)\ ) \\
&\quad\quad \textbf{End If} \\
&\quad \textbf{Else Return}(\ (i,l)\ ) \\
&\textbf{Else\ Return}(\varnothing) \\
&\textbf{End DFS}
\end{aligned}
$$

Procedure CUBE_SELECT1 (F,D)

```
    /*
    For all flattening methods.
    /*
```

$$
\begin{aligned}
&\textbf{Begin} \\
&j \leftarrow BINATE\_SELECT(F_0, m+1, m+n) \\
&C \leftarrow \{x^j,\ \overline{x}^j\} \\
&\textbf{Return}(C) \\
&\textbf{End CUBE\_SELECT}
\end{aligned}
$$

```
Procedure CUBE_SELECT2 (F,D)
  /*
  CUBE_SELECT: multilevel tautology - restricted case -  voting.
  /*

  Begin
  For( i = 1,2,...m )
     Begin
     j ← BINATE_SELECT(Fᵢ,m+1,m+n)
     COUNT[j] ← COUNT[j] + 1
     EndFor
  max ← 0
  For( i = m+1,m+2,....,m+n )
     Begin
     If( COUNT[i] > max )
        Begin
        j ← i
        max ← COUNT[i]
        EndIf
     EndFor
  C ← {xʲ , x̄ʲ}
  Return( C )
  End CUBE_SELECT


Procedure CUBE_SELECT3 (F,D)
  /*
  CUBE_SELECT: multilevel tautology - restricted case -  global.
  /*

  Begin
  G ← ∅
  For( i = 1,2,...m )
     Begin
     G ← G ∪ Fᵢ
     EndFor
  j ← BINATE_SELECT(G,N_IV+1,m+n)
  C ← {xʲ , x̄ʲ}
  Return( C )
  End CUBE_SELECT
```

Procedure CUBE_SELECT4 (F,D)

```
/*
CUBE_SELECT for multilevel tautology - unrestricted case - voting.
/*

Begin
For( i = 1,2,...m )
   Begin
   j ← BINATE_SELECT(Fᵢ,1,m+n)
   COUNT[j] ← COUNT[j] + 1
   EndFor
max ← 0
For( i = 1,2,....,m+n )
   Begin
   If( COUNT[i] > max )
      Begin
      j ← i
      max ← COUNT[i]
      EndIf
   EndFor
C ← {xʲ , x̄ʲ}
Return( C )
End CUBE_SELECT
```

Procedure CUBE_SELECT5 (F,D)

```
/*
CUBE_SELECT: multilevel tautology - unrestricted case - global.
/*

Begin
G ← ∅
For( i = 1,2,...m )
   Begin
   G ← G ∪ Fᵢ
   EndFor
j ← BINATE_SELECT( G,1,m+n )
C ← {xʲ , x̄ʲ}
Return( C )
End CUBE_SELECT
```

Procedure CUBE_SELECT6 (F,D)
```
/*
CUBE_SELECT: don't care method.
/*

Begin
```
$j \leftarrow$ BINATE_SELECT$(F \bigcup D,1,m+n)$
$C \leftarrow \{x^j , \overline{x^j}\}$
Return(C)
End CUBE_SELECT

x_1	x_2	x_3	x_4	y_1^a	y_2^a	y_1^b	y_2^b	$(y_1^1 \equiv y_1^b)\cdot(y_2^a \equiv y_2^b)$
0	0	0	0	1	0	1	0	1·1
0	0	0	1	1	0	1	0	1·1
0	0	1	0	1	0	1	0	1·1
0	0	1	1	1	0	1	0	1·1
0	1	0	0	0	1	0	1	1·1
0	1	0	1	0	1	0	1	1·1
0	1	1	0	0	1	0	1	1·1
0	1	1	1	1	1	1	1	1·1
1	0	0	0	0	1	0	1	1·1
1	0	0	1	0	1	0	1	1·1
1	0	1	0	0	1	0	1	1·1
1	0	1	1	1	1	1	1	1·1
1	1	0	0	1	0	1	0	1·1
1	1	0	1	1	0	1	0	1·1
1	1	1	0	1	0	1	0	1·1
1	1	1	1	1	0	1	0	1·1

Table 4.1 Truth Table for $\eta^a \equiv \eta^b$

x_1	x_2	x_3	x_4	y_{2D}	y_1^b	y_2^b	$y_1^{\{2\}}$	$y_2^{\{2\}}$	$(y_1^b \equiv (y_1^{\{2\}})) \cdot (y_2^b \equiv y_2^{\{2\}})$	D^2
0	0	0	0	0	1	0	1	0	1·1	0
0	0	0	1	0	1	0	1	0	1·1	0
0	0	1	0	0	1	0	1	0	1·1	0
0	0	1	1	0	1	0	1	0	1·1	0
0	1	0	0	0	0	1	1	1	0·1*	1
0	1	0	1	0	0	1	1	1	0·1*	1
0	1	1	0	0	0	1	1	1	0·1*	1
0	1	1	1	0	1	1	1	1	1·1*	1
1	0	0	0	0	0	1	1	1	0·1*	1
1	0	0	1	0	0	1	1	1	0·1*	1
1	0	1	0	0	0	1	1	1	0·1	1
1	0	1	1	0	1	1	1	1	1·1	1
1	1	0	0	0	1	0	1	0	1·1	0
1	1	0	1	0	1	0	1	0	1·1	0
1	1	1	0	0	1	0	1	0	1·1	0
1	1	1	1	0	1	0	1	0	1·1	0
0	0	0	0	1	1	0	0	0	0·1*	1
0	0	0	1	1	1	0	0	0	0·1*	1
0	0	1	0	1	1	0	0	0	0·1*	1
0	0	1	1	1	1	0	1	0	1·1	1
0	1	0	0	1	0	1	0	1	1·1	0
0	1	0	1	1	0	1	0	1	1·1	0
0	1	1	0	1	0	1	0	1	1·1	0
0	1	1	1	1	1	1	1	1	1·1	0
1	0	0	0	1	0	1	0	1	1·1	0
1	0	0	1	1	0	1	0	1	1·1	0
1	0	1	0	1	0	1	0	1	1·1	0
1	0	1	1	1	1	1	1	1	1·1	0
1	1	0	0	1	1	0	0	0	0·1*	1
1	1	0	1	1	1	0	0	0	0·1*	1
1	1	1	0	1	1	0	0	0	0·1*	1
1	1	1	1	1	1	0	1	0	1·1	1

Table 4.2 Truth Table for $\eta^{\{2\}} \equiv \eta^b$.
Note $D^2(x) = 1$, $\forall x \in B^5$ such that $y_2^{\{2\}}(x) \neq y_{2D}^{\{2\}}(x)$

PLA #1	PLA #2	pi	po	iv1	iv2	ivc	LL1	LL2	LLc	exit
5xpl.ge	5xpl.wres	7	10	46	28	104	7	4	9	P
9sym.ge	9sym.wres	9	1	51	40	94	8	5	10	P
alul.ge	alul.wres	12	8	3	1	28	2	1	4	P
alupla2-0.ge	alupla2-0.wres	19	1	37	37	77	12	12	14	P
alupla2-1.ge	alupla2-1.wres	22	1	51	51	105	12	12	14	P
alupla2-2.ge	alupla2-2.wres	25	1	67	67	137	14	14	16	P
alupla2-3.ge	alupla2-3.wres	25	1	70	70	143	16	16	18	P
alupla2-4.ge	alupla2-4.wres	20	1	60	60	123	14	14	16	P
clpl.ge	clpl.wres	11	5	4	4	23	5	5	7	P
co14.ge	co14.wres	14	1	12	12	27	4	4	6	P
dc1.ge	dc1.wres	4	7	17	13	51	5	4	7	P
dc2.ge	dc2.wres	8	7	46	26	93	10	5	12	P
dk17.ge	dk17.wres	10	11	28	16	77	6	4	8	P
dk27.ge	dk27.wres	9	9	22	18	67	6	5	8	P
f51m.ge	f51m.wres	8	8	51	34	109	10	5	12	P
link1.ge	link1.wres	8	8	24	24	72	4	4	6	P
link2.ge	link2.wres	8	1	32	32	67	4	4	6	P
link3.ge	link3.wres	12	9	22	22	71	3	3	5	P
link4.ge	link4.wres	15	15	22	22	89	3	3	5	P
link5.ge	link5.wres	13	9	23	23	73	3	3	5	P
misg.ge	misg.wres	56	23	17	12	98	7	4	9	P
rd53.ge	rd53.wres	5	3	14	11	34	7	4	9	P
rd73.ge	rd73.wres	7	3	56	38	103	11	5	13	P
risc.ge	risc.wres	8	31	22	22	137	4	4	6	P
root.ge	root.wres	8	5	51	32	98	9	4	11	P
sqn.ge	sqn.wres	7	3	35	23	67	8	4	10	P
sqr6.ge	sqr6.wres	6	12	33	24	93	6	4	8	P
vg2.ge	vg2.wres	25	8	28	16	68	8	5	10	P
x1dn.ge	x1dn.wres	27	6	29	17	64	9	6	11	P
x6dn.ge	x6dn.wres	39	5	132	70	217	8	5	10	P
z4.ge	z4.wres	7	4	10	10	32	4	4	6	P

pi	number of primary inputs.
po	number of primary outputs.
iv1	number of intermediate variables in function.ge.
iv2	number of intermediate variables in function.wres.
ivc	number of intermediate variables in the xnor of the two functions.
LL1	number of levels of logic in function.ge.
LL2	number of levels of logic in function.wres.
LLc	number of levels of logic in the xnor of the two functions.

Table 6.1

Multi-Level Tautology -- unrestricted splitting					
		voting		global	
PLA #1	PLA #2	time (sec)	leaves	time (sec)	leaves
5xpl.ge	5xpl.wres	22.5	128	10.2	128
9sym.ge	9sym.wres	69.4	345	30.6	349
alul.ge	alul.wres	33.0	1808	26.2	1974
alupla2-0.ge	alupla2-0.wres	1530.9	22692	798.3	23392
alupla2-1.ge	alupla2-1.wres	16898.7	202228	12309.1	253192
alupla2-2.ge	alupla2-2.wres	>43200.0	416575	>43200.0	616918
alupla2-3.ge	alupla2-3.wres	>43200.0	387250	>43200.0	604064
alupla2-4.ge	alupla2-4.wres	34561.4	191751	>43200.0	301692
clpl.ge	clpl.wres	2.9	178	2.6	233
co14.ge	co14.wres	4.3	106	2.5	106
dc1.ge	dc1.wres	1.2	23	0.9	14
dc2.ge	dc2.wres	20.5	298	10.6	212
dk17.ge	dk17.wres	12.5	142	6.9	148
dk27.ge	dk27.wres	4.7	71	2.5	72
f51m.ge	f51m.wres	27.0	256	16.6	256
link1.ge	link1.wres	76.7	256	50.2	256
link2.ge	link2.wres	33.7	178	53.8	136
link3.ge	link3.wres	124.2	3038	85.6	936
link4.ge	link4.wres	-	-	905.4	9720
link5.ge	link5.wres	405.5	12672	237.0	1584
misg.ge	misg.wres	>43200.0	854176	>43200.0	1701090
rd53.ge	rd53.wres	1.7	32	0.9	32
rd73.ge	rd73.wres	29.7	152	12.9	128
risc.ge	risc.wres	5.9	33	4.8	44
root.ge	root.wres	13.8	87	8.5	101
sqn.ge	sqn.wres	10.4	127	5.6	114
sqr6.ge	sqr6.wres	12.4	192	5.3	64
vg2.ge	vg2.wres	4094.3	154588	1013.1	28708
x1dn.ge	x1dn.wres	2741.5	51928	1370.0	46348
x6dn.ge	x6dn.wres	-	-	-	-
z4.ge	z4.wres	5.5	216	2.5	128

Table 6.2

Multi-Level Tautology -- primary input splitting only					
		voting		global	
PLA #1	PLA #2	time (sec)	leaves	time (sec)	leaves
5xp1.ge	5xp1.wres	6.3	128	5.4	128
9sym.ge	9sym.wres	23.1	340	16.3	349
alu1.ge	alu1.wres	17.0	1678	15.5	1873
alupla2-0.ge	alupla2-0.wres	1571.7	38343	665.8	15578
alupla2-1.ge	alupla2-1.wres	8297.4	152758	4200.4	78784
alupla2-2.ge	alupla2-2.wres	-	-	-	-
alupla2-3.ge	alupla2-3.wres	-	-	-	-
alupla2-4.ge	alupla2-4.wres	1649.9	30420	1235.9	25918
clpl.ge	clpl.wres	3.9	384	2.2	223
co14.ge	co14.wres	2.6	106	2.0	106
dc1.ge	dc1.wres	0.5	14	0.5	14
dc2.ge	dc2.wres	6.2	184	5.8	212
dk17.ge	dk17.wres	4.6	148	3.9	148
dk27.ge	dk27.wres	1.7	72	1.5	72
f51m.ge	f51m.wres	8.8	256	8.1	256
link1.ge	link1.wres	8.3	256	15.6	256
link2.ge	link2.wres	14.0	140	15.6	135
link3.ge	link3.wres	84.4	1728	77.8	1314
link4.ge	link4.wres	144.7	4251	516.2	10192
link5.ge	link5.wres	97.5	2428	181.1	2539
misg.ge	misg.wres	-	-	-	-
rd53.ge	rd53.wres	0.7	32	0.6	32
rd73.ge	rd73.wres	6.6	128	6.4	128
risc.ge	risc.wres	2.2	38	2.3	44
root.ge	root.wres	4.7	96	4.5	.103
sqn.ge	sqn.wres	3.5	112	3.1	114
sqr6.ge	sqr6.wres	3.3	64	2.9	64
vg2.ge	vg2.wres	2055.7	79180	2598.9	134986
x1dn.ge	x1dn.wres	1287.7	52062	2354.8	133365
x6dn.ge	x6dn.wres	-	-	-	-
z4.ge	z4.wres	1.9	128	1.7	128

Table 6.3

Multi-Level Tautology -- Flattening					
		bottom up		top down	
PLA #1	PLA #2	time (sec)	leaves	time (sec)	leaves
5xp1.ge	5xp1.wres	5.7	128	>43200.0	0
9sym.ge	9sym.wres	10.5	221	77.5	235
alu1.ge	alu1.wres	125.7	1083	2097.0	1080
alupla2-0.ge	alupla2-0.wres	52.0	46	>43200.0	0
alupla2-1.ge	alupla2-1.wres	222.7	50	>43200.0	0
alupla2-2.ge	alupla2-2.wres	779.0	104	>43200.0	0
alupla2-3.ge	alupla2-3.wres	1214.8	682	>43200.0	0
alupla2-4.ge	alupla2-4.wres	477.2	337	>43200.0	0
clpl.ge	clpl.wres	2.3	16	62.0	16
co14.ge	co14.wres	5.6	106	5.1	106
dc1.ge	dc1.wres	0.6	11	25341.9	12
dc2.ge	dc2.wres	5.1	144	>43200.0	0
dk17.ge	dk17.wres	7.0	61	>43200.0	0
dk27.ge	dk27.wres	2.4	57	>43200.0	0
f51m.ge	f51m.wres	6.6	256	>43200.0	0
link1.ge	link1.wres	29.2	256	16441.3	256
link2.ge	link2.wres	1300.0	104	>43200.0	0
link3.ge	link3.wres	177.4	655	>43200.0	0
link4.ge	link4.wres	2089.2	415	>43200.0	0
link5.ge	link5.wres	350.0	1097	>43200.0	0
misg.ge	misg.wres	>43200.0	0	>43200.0	0
rd53.ge	rd53.wres	0.9	32	1316.5	32
rd73.ge	rd73.wres	10.8	128	>43200.0	0
risc.ge	risc.wres	3.9	31	>43200.0	0
root.ge	root.wres	5.1	74	>43200.0	0
sqn.ge	sqn.wres	2.5	76	>43200.0	0
sqr6.ge	sqr6.wres	2.9	64	>43200.0	0
vg2.ge	vg2.wres	1346.0	376	>43200.0	0
x1dn.ge	x1dn.wres	262.4	360	>43200.0	0
x6dn.ge	x6dn.wres	-	-	-	-
z4.ge	z4.wres	2.6	128	11932.0	128

Table 6.4

Don't Care Method		
PLA #1	PLA #2	don't care time (sec)
5xp1.ge	5xp1.wres	>20640.0
9sym.ge	9sym.wres	>19980.0
alu1.ge	alu1.wres	1000.7
alupla2-0.ge	alupla2-0.wres	-
alupla2-1.ge	alupla2-1.wres	-
alupla2-2.ge	alupla2-2.wres	-
alupla2-3.ge	alupla2-3.wres	-
alupla2-4.ge	alupla2-4.wres	-
clpl.ge	clpl.wres	115.0
col4.ge	col4.wres	1305.7
dc1.ge	dc1.wres	3279.3
dc2.ge	dc2.wres	>25884.0
dk17.ge	dk17.wres	>8640.0
dk27.ge	dk27.wres	>24360.0
f51m.ge	f51m.wres	>25620.0
link1.ge	link1.wres	>18840.0
link2.ge	link2.wres	>19620.0
link3.ge	link3.wres	>23520.0
link4.ge	link4.wres	>26040.0
link5.ge	link5.wres	>20760.0
misg.ge	misg.wres	>73440.0
rd53.ge	rd53.wres	385.4
rd73.ge	rd73.wres	>47700.0
risc.ge	risc.wres	>25590.0
root.ge	root.wres	>55200.0
sqn.ge	sqn.wres	38311.0
sqr6.ge	sqr6.wres	>14520.0
vg2.ge	vg2.wres	>18300.0
x1dn.ge	x1dn.wres	>60000.0
x6dn.ge	x6dn.wres	-
z4.ge	z4.wres	1293.9

Table 6.5

Multi-Level Tautology -- individual flatten and verify time in seconds						
PLA #1	PLA #2	bu	td	tds	sotd	orig
5xpl.ge	5xpl.wres	1.3	1.9	1.7	1.2	1.4
9sym.ge	9sym.wres	2.5	2.0	2.4	2.1	1.8
alul.ge	alul.wres	0.2	0.2	0.3	0.3	0.2
alupla2-0.ge	alupla2-0.wres	25.3	128.5	24.2	127.7	2811.5
alupla2-1.ge	alupla2-1.wres	96.5	2207.5	360.5	2226.3	29399.2
alupla2-2.ge	alupla2-2.wres	329.5	4694.2	7321.3	4746.6	>43200.0
alupla2-3.ge	alupla2-3.wres	552.2	6006.4	14477.2	5968.6	>43200.0
alupla2-4.ge	alupla2-4.wres	240.9	121.1	208.3	118.9	153.2
clpl.ge	clpl.wres	0.2	0.2	0.3	0.3	0.2
col4.ge	col4.wres	0.4	0.4	0.4	0.4	0.4
dc1.ge	dc1.wres	0.3	0.4	0.4	0.3	0.4
dc2.ge	dc2.wres	0.8	1.1	1.1	0.8	0.9
dk17.ge	dk17.wres	0.8	1.1	1.1	0.9	1.0
dk27.ge	dk27.wres	0.5	0.5	0.5	0.4	0.5
f51m.ge	f51m.wres	1.5	2.1	1.7	1.3	1.4
link1.ge	link1.wres	17.7	3297.3	39.9	1589.1	81.9
link2.ge	link2.wres	1446.9	>43200.0	41601.0	>43200.0	5520.6
link3.ge	link3.wres	104.4	>43200.0	37185.1	9930.0	3270.6
link4.ge	link4.wres	71.3	126.7	188.7	34.8	41.9
link5.ge	link5.wres	152.5	>43200.0	>43200.0	>43200.0	38064.6
misg.ge	misg.wres	1.8	2.1	2.0	1.7	1.7
rd53.ge	rd53.wres	0.4	0.6	0.4	0.4	0.4
rd73.ge	rd73.wres	3.5	5.9	4.0	3.5	4.0
risc.ge	risc.wres	1.0	1.6	1.3	1.1	1.3
root.ge	root.wres	1.4	2.3	1.7	1.4	1.6
sqn.ge	sqn.wres	0.8	1.3	0.9	0.8	0.9
sqr6.ge	sqr6.wres	0.9	1.3	1.1	1.0	1.0
vg2.ge	vg2.wres	3.7	4.6	4.2	3.7	4.2
x1dn.ge	x1dn.wres	3.7	4.7	4.1	3.7	4.0
x6dn.ge	x6dn.wres	7.4	14.2	10.7	8.1	10.7
z4.ge	z4.wres	0.7	1.0	0.9	0.8	0.8

bu   bottom up method
td   top down method
tds   top down stepwise method
sotd   single output top down
orig   original method

Table 6.6

Table 7.1
Comparison of ESPRESSO_MLT to ESPRESSO_ML [BBH]

PLA	START CUBES	START LIT	START IV	DELTA LIT	CPU (SEC)	SPEEDUP t_{ML}/t_{MLT}
dec2	74	149	28	4	2416	6.7
plac	53	191	22	8	1450	5.3
adder	22	44	14	0	20	4.3
z4	39	58	16	0	62	3.8
plab	69	119	24	9	1051	3.6
f4	37	75	11	1	110	2.4
f3	31	73	8	0	62	1.9
f5	43	75	14	1	232	1.0
mark	6	8	2	0	1	1.0
f2	18	28	8	0	10	.93
8fun	48	83	15	-1	382	.80
exam2	43	73	13	2	186	.79
clpl	14	19	4	0	8	.67
f1	9	14	2	0	2	.67
dec1	34	52	9	1	52	.66
rd53	30	62	11	2	161	.64
gerf	12	17	4	0	9	.64
insdec	51	79	11	0	179	.64
fadd2	15	29	5	-1	10	.60
f0	7	16	2	0	3	.40

Test Generation Using Multi-Level Tautology			
PLA	time (sec)	test count	redundant nodes
alu1.ge	1.1	22	0
clpl.ge	8.2	18	0
co14.ge	23.2	26	0
dc1.ge	7.3	48	0
dc2.ge	71.7	106	0
dk17.ge	38.8	78	0
dk27.ge	16.8	62	0
f51m.ge	100.2	118	0
rd53.ge	8.2	34	0
rd73.ge	168.1	118	0
risc.ge	20.0	106	0
root.ge	116.4	112	0
sqn.ge	68.2	76	0
sqr6.ge	40.6	90	0
wim.ge	5.8	40	0
z4.ge	7.4	28	0

Table 7.2

PLA	general tautology	context of MLMIN true	context of MLMIN false
mark	0.1	0.0104	0.0104
f1	0.1	0.0056	0.0106
f0	0.1	0.0100	0.0183
f2	0.1	0.0198	0.0178
fadd2	0.2	0.0181	0.0178
gerf	0.0	0.0136	0.0215
clpl	0.8	0.0200	0.0202
dec1	0.1	0.277	0.0273
f3	0.7	-	-
f4	0.8	0.0275	0.0288
f5	0.3	0.0335	0.0352
adder	1.9	0.0513	0.0345
rd53	0.2	0.0376	0.0352
z4	0.8	0.0360	0.0304
insdec	0.3	0.0268	0.0265
exam2	0.3	0.0279	0.0319
8fun	0.3	0.0422	0.0454
plab	0.9	0.0663	0.0692
plac	1.3	0.1347	0.1115
dec2	1.5	0.0793	0.0842
5xp1	1.9	0.0665	0.0609

Table 8.1 Comparison of general tautology (PLA $\stackrel{?}{=}$ PLA) time with average time in context of multilevel minimizer [HaJ86]

BOOLEAN NETWORK
(ACYCLIC)

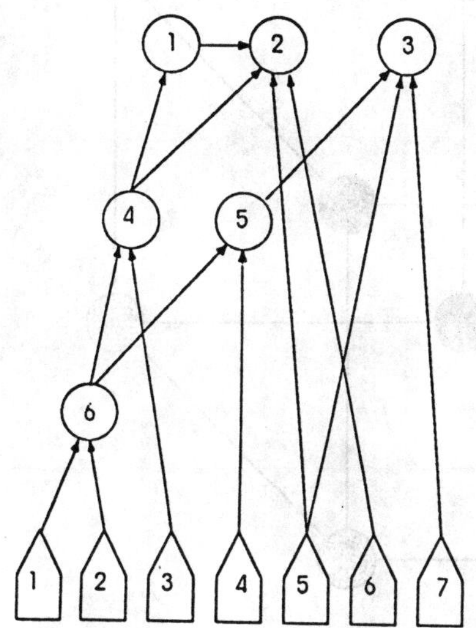

Figure 1.1 A Multilevel Boolean Network

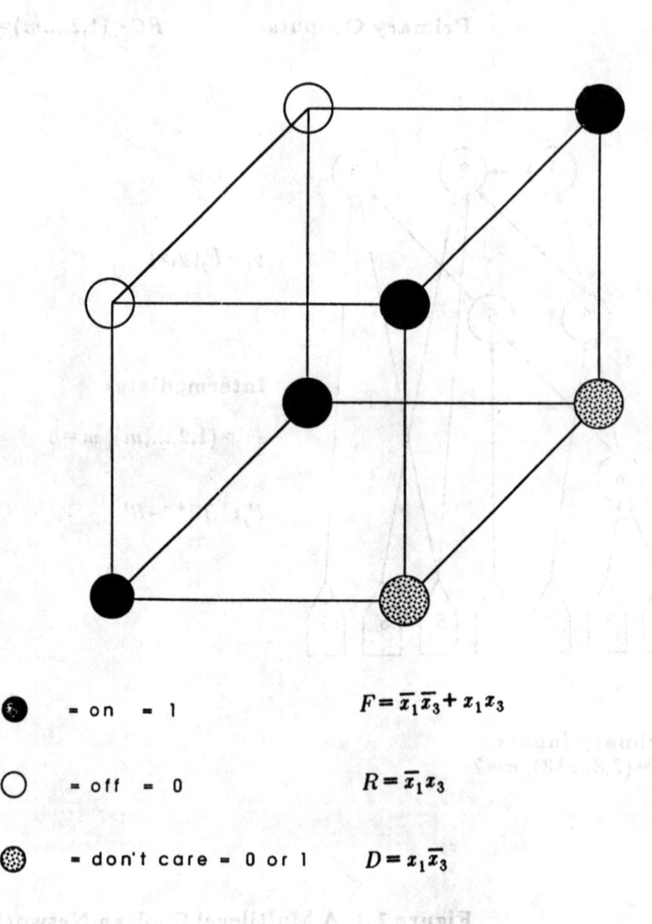

Figure 2.1 Incompletely Specified Boolean Function.

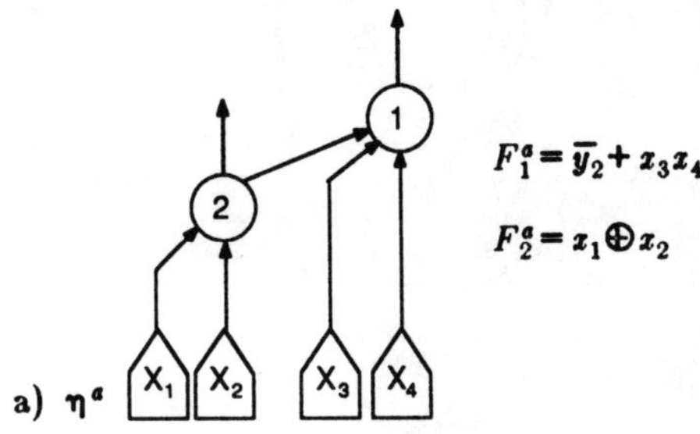

$$F_1^a = \overline{y}_2 + x_3 x_4$$
$$F_2^a = x_1 \oplus x_2$$

a) η^a

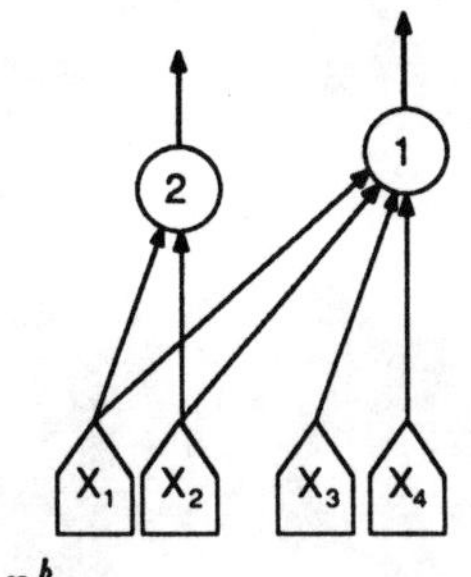

$$F_1^b = (x_1 \equiv x_2) + x_3 x_4$$
$$F_2^b = x_1 \oplus x_2$$

b) η^b

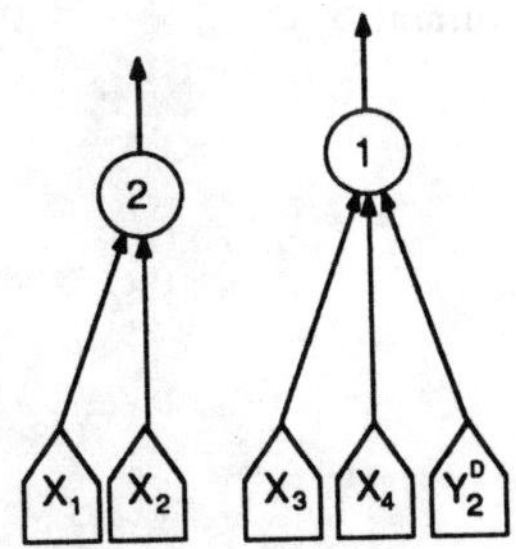

$$F_1 = \overline{y}_{2D} + x_3 x_4$$
$$F_2 = x_1 \oplus x_2$$
$$D^2 = y_{2D} \overline{F}_2 + \overline{y}_{2D} F_2$$

c) η^J, $J = \{2\}$ as derived, according to Definition 6, from η^b.

Figure 2.2 Three Equivalent Boolean Networks

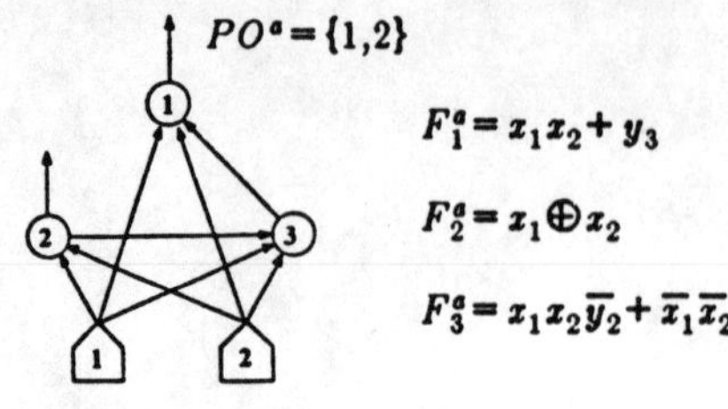

$$PO^a = \{1,2\}$$
$$F_1^a = x_1 x_2 + y_3$$
$$F_2^a = x_1 \oplus x_2$$
$$F_3^a = x_1 x_2 \overline{y}_2 + \overline{x}_1 \overline{x}_2$$

a) $DX_1 = DX_2 = \emptyset$.

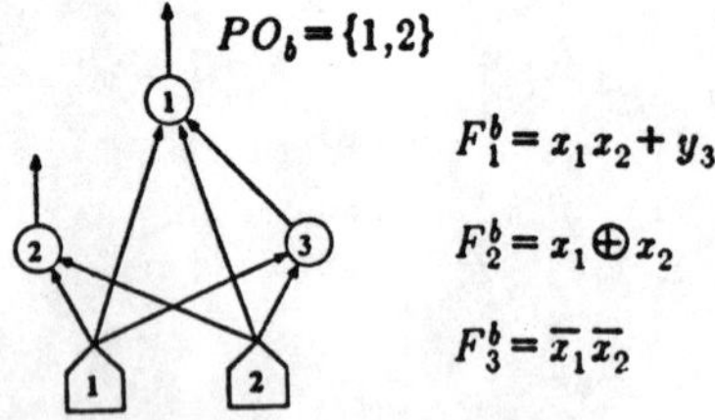

$$PO_b = \{1,2\}$$
$$F_1^b = x_1 x_2 + y_3$$
$$F_2^b = x_1 \oplus x_2$$
$$F_3^b = \overline{x}_1 \overline{x}_2$$

b) $DX_1 = DX_2 = \emptyset$. Prime, Irredundant and 100% Testable.

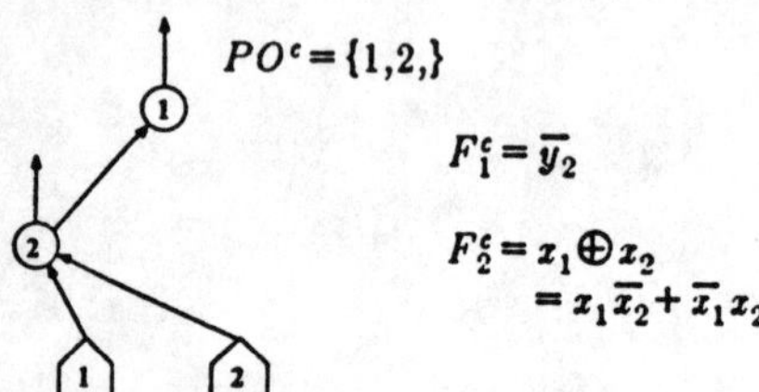

$$PO^c = \{1,2,\}$$
$$F_1^c = \overline{y}_2$$
$$F_2^c = x_1 \oplus x_2$$
$$= x_1 \overline{x}_2 + \overline{x}_1 x_2$$

c) $DX_1 = DX_2 = \emptyset$. Prime, Irredundant, 100% Testable, and R-Minimal.

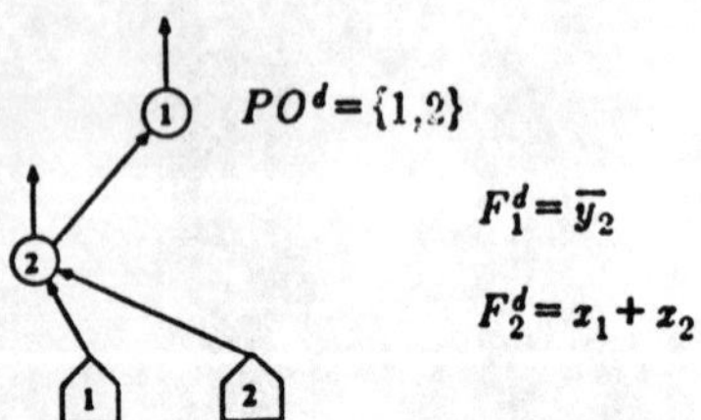

$$PO^d = \{1,2\}$$
$$F_1^d = \overline{y}_2$$
$$F_2^d = x_1 + x_2$$

d) $DX_1 = x, DX_2 = x_2$. Prime, Irredundant, 100% Testable, R-Minimal, Lower Cost.

Figure 2.3
Equivalent, but Progressively Optimized Boolean Networks.

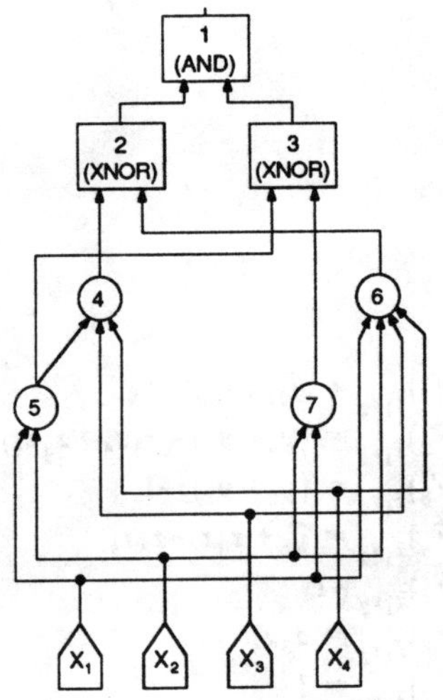

a) Comparison Network: $\eta^C \equiv 1$ if and only if $\eta^a \equiv \eta^b$.

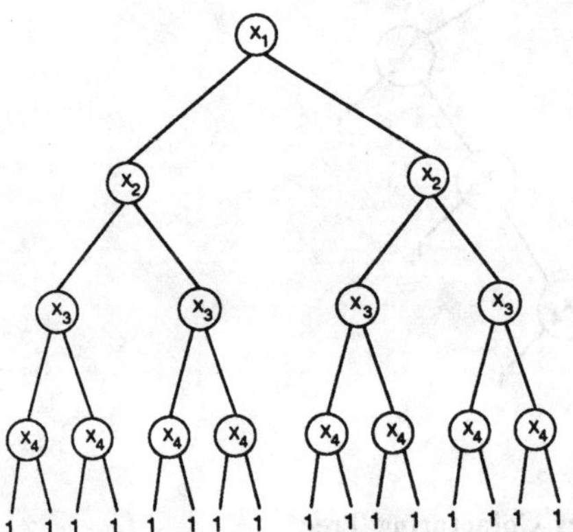

b) Recursive Binary Cofactoring Tree for Testing $\eta^C = 1$

Figure 4.1 Comparison Network and Cofactoring Tree for Testing $\eta^a = \eta^b$.

298

$$(F_1)_{x_1 x_2} = y_2 y_3 \to 1$$
$$(F_2)_{x_1 x_2} = (y_4 \equiv y_6) \to 1$$
$$(F_3)_{x_1} = (y_5 \equiv y_7) \to 1$$
$$(F_4)_{x_1 x_2} = \overline{y}_5 + x_3 x_4 \to 1$$
$$(F_5)_{x_1 x_2} = 0$$
$$(F_6)_{x_1 x_2} = 1 + x_3 x_4 = 1$$
$$(F_7)_{x_1 x_2} = 0$$

c) Boolean Equations of $\eta^C_{x_1 x_2}$

$$(F_1)_{x_1 \overline{x}_2} = y_2 y_3 \to y_2 \qquad (F_1)_{\overline{x}_1 x_2} = y_2 y_3 \to y_2$$
$$(F_2)_{x_1 \overline{x}_2} = (y_4 \equiv y_6) \to (x_3 x_4 \equiv x_3 x_4) \qquad (F_2)_{\overline{x}_1 x_2} = (y_4 \equiv y_6) \to (x_3 x_4 \equiv x_3 x_4)$$
$$(F_3)_{x_4 \overline{x}_2} = (y_5 \equiv y_7) \to 1 \qquad (F_3)_{\overline{x}_1 x_2} = (y_5 \equiv y_7) j \to 1$$
$$(F_4)_{x_1 \overline{x}_2} = \overline{y}_5 + x_3 x_4 \to x_3 x_4 \qquad (F_4)_{\overline{x}_1 x_2} = \overline{y}_5 + x_3 x_4 \to x_3 x_4$$
$$(F_5)_{x_1 \overline{x}_2} = 1 \qquad (F_5)_{\overline{x}_1 x_2} = 1$$
$$(F_6)_{x_1 \overline{x}_2} = x_3 x_4 \qquad (F_6)_{\overline{x}_1 x_2} = x_3 x_4$$
$$(F_7)_{x_1 \overline{x}_2} = 1 \qquad (F_7)_{\overline{x}_1 x_2} = 1$$

d) Boolean Equations of $\eta^C_{x_1 \overline{x}_2}$
e) Boolean Equations of $\eta^C_{\overline{x}_1 x_2}$

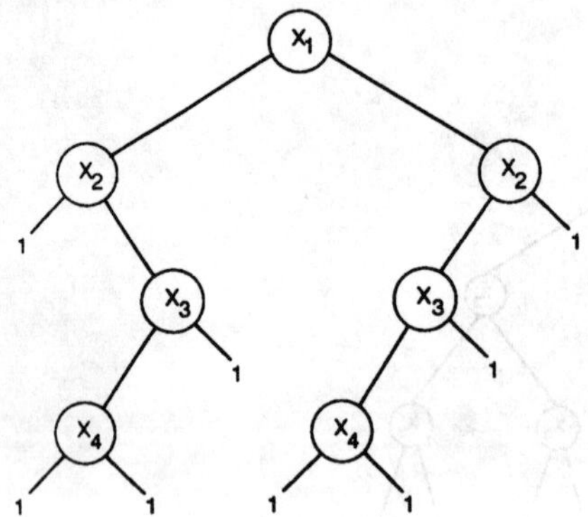

f) Pruned Recursive Binary Cofactoring Tree

Figure 4.1 (Continued)

References

[Ake78a] S. B. Akers, Functional Testing with Binary Decision Diagrams, *Proc. 8th Ann. IEEE Conf. Fault-Tolerant Comput,* , 1978, 75-82.

[Ake78b] S. B. Akers, Binary Decision Diagrams, *IEEE Trans. Comput. C-27,* (June 1978), 509-516.

[BBH86] K. A. Bartlett, R. K. Brayton, G. D. Hachtel, R. M. Jacoby, R. Rudell, A. Sangiovanni-Vincentelli and A. Wang, Multilevel Logic Minimization Using Implicit Don't Cares, *ICCD '86,* Port Chester, New York, October 1986, 552-557.

[BCD86] K. A. Bartlett, W. Cohen, S. DeGeus and G. D. Hachtel, Synthesis and Optimization of Multilevel Logic Under Timing Constraints, *IEEE Transactions on CAD CAD-5,* 4 (October 1986), 582-596.

[BBH] K. A. Bartlett, R. K. Brayton, G. D. Hachtel, R. M. Jacoby, R. Rudell, A. Sangiovanni-Vincentelli and A. Wang, Multilevel Logic Minimization Using Implicit Don't Cares, *IEEE Trans. on CAD (submitted),* .

[Bra83] D. Brand, Redundancy and Don't Care in Logic Synthesis, *IEEE Transactions on Computers C-32,* 10 (October 1983), 947-952.

[BrM82] R. K. Brayton and C. T. McMullen, The Decomposition and Factorization of Boolean Expressions, *Proceedings of the International Symposium on Circuits and Systems,* Rome, 1982, 49-54.

[BHM84] R. K. Brayton, G. D. Hachtel, C. McMullen and A. Sangiovanni-Vincentelli, *Logic Minimization Algorithms for VLSI Synthesis,* Kluwer Academic Publishers, Boston, 1984.

[Bra86] R. K. Brayton, Algorithms for Multilevel Logic Synthesis and Optimization, *Proceedings, NATO ASI on Logic Synthesis and Silicon Compilation for VLSI,* The Netherlands, 1986.

[Bry86] R. E. Bryant, Graph-Based Algorithms for Boolean Function Manipulation, *IEEE Transactions on Computers C-35,* 8 (August 1986), .

[DJB80] J. A. Darringer, W. Joyner, L. Berman and L. Trevillyan, Experiments in Logic Synthesis, *Proceedings of the IEEE Internationial Conference on Circuits and Computers,* , 1980, 234-237A.

[DJB81] J. A. Darringer, W. H. J. Jr., C. L. Berman and L. Trevillyan, Logic Synthesis Through Local Transformations, *IBM J. Res. Develop. 25,* 4 (July 1981), 279-280.

[DBG84] J. Darringer, D. Brand, J. Gerbi, W. Joyner and L. Trevillyan, LSS: A System for Production Logic Synthesis , *IBM Journel of Research and Development,* , September 1984.

[deC85] A. J. deGeus and W. Cohen, Optimization of Combinational Logic Using a Rule-Based Expert System, *Journal of Design and Test,* , 1985.

[Fuj85a] H. Fujiwara, FAN: A Fanout-Oriented Test Pattern Generation Algorithm, *Proceedings, IEEE ISCAS,* , July 1985, 671-674.

[Fuj85b] H. Fujiwara, *Logic Testing and Design for Testability,* The MIT Press, 1985.

[HaJ85] G. D. Hachtel and R. M. Jacoby, Algorithms for Multi-Level Tautology and Equivalence, *Proceedings of the IEEE International Symposium on Circuits and Systems,* Kyoto, Japan, June 1985.

[HaJ86] G. D. Hachtel and R. M. Jacoby, Algorithm for Multilevel Tautology Checking, *IEEE Trans. on CAD (submitted),* , January 1986.

[HJM] G. D. Hachtel, R. Jacoby and C. R. Morrison, Algorithms for Multilevel Logic Minimization Based on Multilevel Tautology Checking, *to be submitted,* .

[Law64] E. L. Lawler, An Approach to Multilevel Boolean Minimization, *Journal of ACM*, , 1964.

[Lee59] C. Y. Lee, Representation of Switching Circuits by Binar-Decision Programs, *Bell Syst. Tech. J. 38*, (July 1959), 985-999.

[RaL86] D. L. Ravenscroft and M. R. Lightner, Functional Language Extractor and Boolean Cover Generator, *ICCAD 86*, , 1986, 120-123.

[Rot80] J. P. Roth, *Computer Logic, Testing and Verification*, Computer Science Press, 1980.

[Sas85] T. Sasao, HART: A Hardware for Logic Minimization and Verification, *Proceedings IEEE ICCD*, Rye, New York, to be presented October 1985.

[SBH82] G. L. Smith, R. J. Bahnsen and H. Halliwell, Boolean Comparison of Hardware and Flowcharts, *IBM Journal Research Development 26*, 1 (January 1982), .

[WeS85] R. Wei and A. Sangiovanni-Vincentelli, VICTOR-II: Global Redundancy Identification, Test Generation, and Testability Analysis for VLSI Combinational Circuits, *Proc. VLSI Conference*, Taiwan, May 1985.

LOGIC SYNTHESIS

DANIEL BRAND

IBM T.J.WATSON RESEARCH CENTER
YORKTOWN HEIGHTS, NEW YORK
USA

1. INTRODUCTION

This is an overview of a logic synthesis system [1] developed at the IBM Thomas J. Watson Research Center and other IBM locations over a period of seven years. Major contributors to the system are listed in the acknowledgements.

The input into the system is a register transfer description of the desired function and its output is a netlist specification of the implementation. The output is produced from the input by a large number of local and global transformations. Most transformations are independent of the target technology; only transformations used towards the end depend on the technology in the sense that they may be given different parameters for different technologies, or that they may be applicable only for one specific target technology. First we will describe the sequence of steps we go through and then concentrate on specific problems and methods.

2. SEQUENCE OF STEPS

2.1. Input

The logic synthesis system accepts several languages as input; these languages are also input for functional simulators, which determine their syntax and semantics. For our system we require that the input be on the register transfer level. This implies that the designer already decided how many registers he wants and how his states are to be encoded. He has to declare all the registers, primary inputs and primary outputs. The function of each primary output and each register is a boolean combination of primary inputs and register outputs. This function can be expressed using the usual programming language constructs such as assignments and conditional statements. There are no restrictions on the operators; how they are handled is described under Step 1 and Step 2. Only register variables are allowed to be self referencing, that is, every loop in the logic must be cut by a register. Figure 1 is an example of an input language.

Besides the function, the designer also needs to supply other information used primarily for timing analysis and correction. Examples of these are: cycle time, expected arrival times of primary inputs, required departure times and capacitances of primary outputs. In some applications the designer may want to specify different clock lines to go to different registers or other information pertinent to the target technology.

The input description is first mapped in a very straightforward way into our internal representation. The representation is a graph of boxes connected by signals; Figure 2 shows the representation of the output R(0) from Figure 1. Examples of box types are AND gates, registers, ALUs, transistors. If the original description is hierarchical then the function of a box would be specified by another graph of the same form. The transformations described next then operate on these graphs.

2.2. Step 1

Step 1 consists of a sequence of transformations that operate on the usual boolean gates (AND, OR, etc.) as well as on the large boxes introduced by the designer: adders, comparators, decoders. These transformations combine adders, propagate boxes through multiplexors, take advantage of the relationship between decoder outputs. Figure 3 shows the result of a local transformation based on distributivity of AND and OR. Figure 4 results from another local transformation, which knows about the relationship among the outputs of a decoder.

Even if the designer did not use such higher level operators we will try to recognize them in his logic. For example, instead of a decoder the designer might have used an if-statement for each decoder output; we will replace that logic by one decode box so that the above mentioned transformations can operate on it.

2.3. Step 2

Step 2 starts with the expansion of each large box into a network of AND, OR, and NOT gates, see Figure 5. How a particular box should be expanded is specified by a file that is read at this point. Thus the user can try a look-ahead adder or a ripple-carry adder by supplying the corresponding file. Alternatively he can use transformations that do the expansion according to user's parameters, e.g., amount of carry look-ahead.

The expansion of large boxes is followed by some simple local transformations to reduce the logic. For example, constants are propagated, some gates resulting from the expansion may be combined with others.

2.4. Step 3

The objective of step 3 is boolean minimization. For convenience the logic is first changed so that it contains only NOR gates or only NAND gates. The choice depends on what is easier to implement in the target technology. The transformations in step 3 range from a simple local transformation eliminating double inverters to expensive global transformations improving testability. Figure 6 results from converting our example to NANDs followed by some simplifications. The first input of the three input NAND in Figure 6 is not testable for stuck at 1, see reference [7], and therefore can be disconnected, see Figure 7. After that some simple transformations result in Figure 8.

2.5. Step 4

The previous steps are essentially the same for all technologies. The transformations in step 4 depend on the target technology in two ways: There are some transformations that are applied only for one technology and no other; however, the majority of transformations in step 4 are common to all technologies, but the tables that drive them may be different. They correct fanin and fanout, they take advantage of special technology features as wired ANDs or AOIs. Finally they generate the legal gates of the target technology. What would happen to our example in step 4 would depend on the target technology. It might be implemented just the way it is in

Figure 8 if NANDs were the only primitives allowed. Alternatively it could be implemented using an AO or using a wired OR if those were available.

2.6. Step 5

Step 5 does not transform the logic at all, but rather presents it for further processing. It generates files for physical design and generates a human readable representation so that engineers can evaluate the design.

2.7. Timing correction

All the above transformations are generally concerned with minimizing area. Some are careful not to increase paths lengths; however they will not increase area in order to meet timing requirements. There are specific timing transformations that try to meet timing requirements even at the expense of area. All these timing transformations have the difficulty of estimating which areas will have timing problems in the final logic. Therefore timing correction is done mostly after step 4, where timing analysis is more accurate. Timing correction is accomplished by rearranging fanin and fanout trees, by using the distributive law, and by adjusting power levels. For example, if the path from S(0) to R(0) in Figure 8 were critical and an OR were allowed in the target technology then the graph could be rearranged to that of Figure 9.

Timing correction after step 4 has the disadvantage of not being able to change the overall structure of the logic. Therefore there are transformations that do timing correction in earlier steps based only on rough estimates. These transformations may decide to do a computation in parallel rather than serially, or to move a signal forward if the signal is late arriving. An example of such a transformation will be given in Section 3.8.

3. PROBLEMS AND METHODS

In Figure 10 each row corresponds to a problem and each column to a method. An X in an entry indicates that we use that method for the corresponding problem. This section will explain the heading of each column and row, and will describe the solutions corresponding to the entries marked X.

As mentioned before, we process logic by a long sequence of transformations. Each transformation consists of three parts:
(i) "Move" is the starting and resulting pattern of the transformation. For example, in a transformation generating AO boxes out of NANDs, X = NAND(NAND(A,B), NAND(C,D)) is a starting pattern and X = AO(A,B,C,D) is a resulting pattern.
(ii) "Acceptance" is the criterion for performing a move. A move may be unacceptable because the target technology does not allow the resulting pattern, or because it would not be advantageous. For example, an AO box may be smaller but slower than three NANDs; in that case we may not want to use it on a critical path.
(iii) "Strategy" determines at which place in the logic a move should be applied. Good strategy is important because performing a move at one location may prevent the same or other moves from applying somewhere else.

304

3.1. Local Transformations

A transformation is local if its starting pattern and acceptance involve only a fixed number of levels. Applications of DeMorgan laws are typical local transformations. The main advantage of local transformations is simplicity and linear running time [6]. Many local transformations have a strategy that scans the entire logic to find the best places to apply; this still keeps their running time linear.

3.2. Global transformations

A transformation is global if its starting pattern or acceptance criterion involve an unbounded number of levels. We have only very few transformations whose staring pattern involves an unbounded number of level, e.g. converting logic into two level representation. Most global transformations have acceptance criteria that compute various relationships among signals; in doing so they may examine an unbounded number of levels. That does not necessarily mean that they have non-linear running time; in fact those that use one-way data flow do have linear running time. This is in contrast to multi-way data flow that is more expensive.

3.2.1. <u>One-way data flow</u> [2] refers to transformations that pass over the logic only once, either from inputs to outputs or from outputs to inputs. During the pass they collect relationships among signals. One type of such relationships are so called "control" sets -- $C_{\alpha\beta}(s)$ is the set of signals t that control the signal s. The quantities α and β take on boolean values and define how t controls s, namely whenever $t = \alpha$ then $s = \beta$. That is, for each signal s we define the four sets

$$C_{00}(s) = \{t \mid t = 0 \Rightarrow s = 0\}$$
$$C_{01}(s) = \{t \mid t = 0 \Rightarrow s = 1\}$$
$$C_{10}(s) = \{t \mid t = 1 \Rightarrow s = 0\}$$
$$C_{11}(s) = \{t \mid t = 1 \Rightarrow s = 1\}$$

For example, for the graph in Figure 11

$$
\begin{array}{llll}
C_{00}(A) = \{A\}, & C_{01}(A) = \{C\}, & C_{10}(A) = \{C, D, E\}, & C_{11}(A) = \{A\}, \\
C_{00}(B) = \{B\}, & C_{01}(B) = \phi, & C_{10}(B) = \{D, E\}, & C_{11}(B) = \{B\}, \\
C_{00}(C) = \{C\}, & C_{01}(C) = \{A\}, & C_{10}(C) = \{A\}, & C_{11}(C) = \{C, D, E\}, \\
C_{00}(D) = \{C, D, E\}, & C_{01}(D) = \phi, & C_{10}(D) = \{A, B\}, & C_{11}(D) = \{D, E\}, \\
C_{00}(E) = \{C, D, E\}, & C_{01}(E) = \phi, & C_{10}(E) = \{A, B\}, & C_{11}(E) = \{D, E\}.
\end{array}
$$

It is not necessary to compute the sets C exactly, in fact it would be prohibitively expensive to do so. It is sufficient to compute only their subsets. This will not cause any of the transformations below to give incorrect results, merely to miss opportunities for simplification. On the other hand computing supersets would result in incorrect logic.

We compute the sets C, or more exactly their approximation, by passing over the logic from inputs to outputs as follows. If s is a primary input or register output then

$$C_{00}(s) = \{s\}$$
$$C_{01}(s) = \phi$$
$$C_{10}(s) = \phi$$
$$C_{11}(s) = \{s\}$$

If s = NAND($s_1, ..., s_n$) then

$$C_{00}(s) = \{s\} \cup \bigcap_{i=1}^{n} C_{01}(s_i)$$

$$C_{01}(s) = \bigcup_{i=1}^{n} C_{00}(s_i)$$

$$C_{10}(s) = \bigcap_{i=1}^{n} C_{11}(s_i)$$

$$C_{11}(s) = \{s\} \cup \bigcup_{i=1}^{n} C_{10}(s_i)$$

For boxes of other types than NAND similar equations apply. For the example of Figure 11 we get these approximations

$$
\begin{array}{llll}
C_{00}(A) = \{A\}, & C_{01}(A) = \phi, & C_{10}(A) = \phi, & C_{11}(A) = \{A\}, \\
C_{00}(B) = \{B\}, & C_{01}(B) = \phi, & C_{10}(B) = \phi, & C_{11}(B) = \{B\}, \\
C_{00}(C) = \{C\}, & C_{01}(C) = \{A\}, & C_{10}(C) = \{A\}, & C_{11}(C) = \{C\}, \\
C_{00}(D) = \{D\}, & C_{01}(D) = \phi, & C_{10}(D) = \{A, B\}, & C_{11}(D) = \{D\}, \\
C_{00}(E) = \{C, D, E\}, & C_{01}(E) = \phi, & C_{10}(E) = \{A, B\}, & C_{11}(E) = \{E\}.
\end{array}
$$

Besides the control sets C, another type of information collected is blockage, which is calculated by processing the logic from outputs to inputs. For illustration assume that we have only NAND gates.

$B(s)$ = set of signals t so that t is connected along every path from s to any output.
This set is important because whenever t = 0, then the value of s is irrelevant. The sets B are calculated recursively as follows. If s is an input into any box other than NAND then

$$B(s) = \{s\}.$$

Otherwise let $s_i = NAND(s_{i,1}, ..., s_{i,k_i})$, for i = 1...n, be all the boxes with input s. Then

$$B(s) = \bigcap_{i=1}^{n} [\, B(s_i) \cup \{s_{i,1}, ..., s_{i,k_i}\}\,]$$

The sets B and C can be represented using bit vectors and then calculated efficiently by bit operations. Having calculated them, they can be used by several transformations as we will describe later.

3.2.2. <u>Multi-way data flow</u> [3] also calculates the sets C, but does that by passing over the logic back and forth until no more information can be added. This takes more CPU time, but collects more implications.

For example, consider Figure 11. If D = 1 then A = 0 and B = 0. If A = 0 then C = 1. If C = 1 and D = 1 then E = 1. These are inferences that multi-way data flow would do starting from D = 1, concluding that $D \in C_{10}(A)$, $D \in C_{10}(B)$, $D \in C_{11}(C)$, $D \in C_{11}(E)$.

The reasoning illustrated above starts with a value assignment to one signal and finds values that other signals must have. In general, the analysis starts with a value assignment to several signals and searches for implications. The analysis stops when no more implications are possible; in particular we never split into cases. For example, in Figure 11 we will derive no conclusions from E = 0, because from E = 0 we could derive only that C = 0 or D = 0. We do not split into cases for performance reasons, and this causes multi-way data flow to compute only an approximation to the sets C, albeit better than one-way data flow.

3.3. <u>Greedy algorithms</u>

The strategy of many transforms is to first scan the logic to decide the best place to apply, and then perform the move. Even if that decision should later prove to be poor, no backtracking takes place; instead we rely on other transformations to correct the problem. In that sense we consider those algorithms "greedy".

3.4. <u>Two level minimization</u>

The traditional approach to logic synthesis has been two level minimization followed by factoring. Therefore we compared it with our approach based on transformations. Even in examples sufficiently small to apply two level minimization the results were usually worse. We believe that this is because two level minimization destroyed the original structure of the logic as created by the designer, and our factoring programs were unable to reconstruct it.

Two level minimization is useful on some small examples where our approach does not find the optimal implementation. However, to be applicable in general, one has to partition a large piece of logic into small pieces where two level minimization is useful. Unfortunately, we have not found a good partitioning algorithm and with our heuristic algorithm we often get worse results.

We use two level minimization only in "open book set" synthesis. There is no library of primitives, but only general criteria for acceptability of a gate. For example, any function implementable in a 6 by 4 CMOS gate might be acceptable. Two level minimization is applied to the function of each gate and the algorithm we selected is particularly tailored to this application. It is exponential in its asymptotic behavior, but it is fast for small examples; in particular it is fast if we restrict ourselves to 32 variables, because then all the bit operations on which the algorithm is based can be done by one machine instruction.

The input into the algorithm is a one output function whose on-set is represented by a list of cubes to minimize; in addition there could be a list of don't care cubes. A cube is a conjunction of literals; a literal is a variable or its negation. The output is a list of cubes that is equivalent to the original on the care domain. The algorithm simplifies the given cubes by applying the following two rules:

Rule 1: A cube C can be deleted if no value assignment to variables makes C evaluate to 1 and all other cubes to 0.

Rule 2: A literal L in a cube C can be deleted if no assignment makes $L = 0$, all other literal of C equal 1, and all other cubes 0.

The procedure terminates when neither rule is applicable, or when applying no rule would make the set of cubes acceptable.

3.5. <u>Area analysis</u>

Both the acceptability criterion and the strategy of a transformation need to know how the move will affect the result in the target technology. For example, consider whether it would be advantageous to transform the logic of Figure 12 into the logic of Figure 13. If the target technology allows wired OR then the logic of Figure 12 represents smaller area.

One move may have different effects in different technologies and such differences are resolved by technology estimators. These are external procedures, different for each technology, which a transform can call to evaluate the impact of a move. Technology estimators solve only partially the problem of area analysis because they cannot predict very well how other transformations will change the logic later on.

3.6. <u>Area Optimization</u>

We have already discussed how greedy algorithms and two level minimization are used in area optimization. There are many local transformations used and some were demonstrated in processing the example of Figure 1.

We will illustrate the use of one-way data flow by following the example of Figure 14. One transform that uses the one-way data flow information will connect a signal S to any NAND gate which has an input T with $S \in C_{00}(T)$. It is easy to see that in Figure 14, $C_{00}(K) = \{K, S\}$; therefore we can connect S as the input of the last NAND gate resulting in Figure 15. This transformation can be justified as follows. For any input pattern which gives the value 1 to S the value of M will be the same whether S is connected there or not. For any input pattern which gives the value 0 to S the value of K must also be 0, since $S \in C_{00}(K)$. With K = 0 the value of M = 1 whether S is connected to the source gate of M or not.

Then we apply a transformation that disconnects a signals S from any gate with output T where $S \in B(T)$. In Figure 15
$B(M) = \{M\}$, $B(K) = \{M, K, S, L\}$, $B(L) = \{M, K, S, L\}$, $B(V) = \{V\}$, $B(G) = \{G\}$,
$B(H) = \{M, K, S, L, H, G\}$, $B(I) = \{M, K, S, L, I, J\}$, $B(J) = \{M, K, S, L, I, J\}$.
Thus we can disconnect S from the gates with outputs H and I, resulting in Figure 16. This transformation can be justified as follows. For any input pattern which gives the value 1 to S the value of any gate is the same whether S is connected there or not. For any input pattern which gives the value 0 to S the value of a gate may be different depending on whether S is connected there; however, before this different value can propagate to any primary output it must pass another gate with input S (in Figure 15 the gate with output M), which gives this other gate the value 1 regardless of anything else.

The combined affect of these two transformations is Figure 16. It was necessary to use global transformations for this simplification because in general there could be many levels between the signals S and M.

3.7. <u>Timing analysis</u>

We assume that we are given arrival times of primary inputs, required departure times of primary outputs, and times when registers are clocked. In addition we need to know delays through all the gates and capacitances on the signals; this information can be supplied by technology estimators as described under area analysis. Likewise these estimators have the weakness of making only estimates and of not being able to anticipate the impact of later transformations. But in timing analysis these weaknesses are more serious than in area analysis. In order to determine whether a particular transformation will reduce area it is sufficient to predict what future transforms will do at the point of application. However, to determine whether a particular transformation will reduce the delay it is necessary to predict what future transforms will do everywhere, because that will determine the critical path.

Given all the above data we calculate the arrival time of each signal by passing over the logic from inputs to outputs in the usual way [4]. Then by passing from outputs to inputs we calculate the required departure time for each signal. The difference between the two is the slack of a signal; negative slack means a timing problem, i.e. the signal lies on a path that is too long or too short.

One-way data flow approach to timing analysis, as described in the previous paragraph, can be improved by calculating the information described under multi-way data flow analysis. This allows us to detect that a critical path may be unexecutable and therefore can be ignored for timing analysis [5]. In order to detect that, we form the conditions for executing the path, namely that all the signals coming into the path must have non-controlling values. If it is impossible then the path is not executable.

3.8. Timing optimization

There are several local transformations for timing optimization. For example, rearranging fanin and fanout trees, using the distributive law, inserting delay blocks into short paths, changing power levels.

The information collected by one-way data flow can be used to determine that a late signal can be moved forward on a path. For example, assume that in Figure 16 the path S G K M is critical. Recall that we could not disconnect S from the gate G because $S \notin B(G)$, and this is because G fans out to V. There is a transformation that will replicate the gate G, to get two signals -- G1 going to V only and G2 going to K only. Then $B(G1) = \{G1, U, V\}$, $B(G2) = \{G2, H, M, K, S, L\}$. Thus we can now disconnect S from the gate G2, resulting in Figure 17, where the critical path has been eliminated.

3.9. Testability

In order to ensure sufficiently high testability coverage one needs to find and eliminate redundant connections. Those are connections that can be replaced by a constant without changing the function of the chip. Therefore they are eliminated by replacing the connection by the corresponding constant and simplifying the logic.

There are several local transformations eliminating redundancy; an example is in Figure 18. Global transformations for removing redundancy are based on the following fact.
>R: A connection of a signal s is not testable for stuck at a value v if every path from the connection to an observable point has the following property:
>It is not possible to set all signals into the path, which are not affected by the connection of s, to a non-controlling value and at the same time set s to $\neg$v.

There is a transformation [3] using multi-way data flow based on exactly the above condition. Suppose that in Figure 19 we wish to determine whether the input A into the gate D is testable for stuck at 0. Any test pattern that tests that connection for stuck at 0 must set A to 1. Then there must be a path from the connection of A to some output along which the fault can propagate. In order to propagate along the path A D G, we would have to have B = 1 and A = 0 (non-controlling value of an OR gate), but this contradicts our previous assumption that A = 1. In order to propagate along the path A D F I, we would have to have B = 1, C = 0, and E = 1; but B = 1 and E = 1 cannot happen at the same time. In order to propagate along the path A D H, we would have to have B = 1; we cannot require that F = 0 because F is affected by the connection we are testing. Since there is no contradiction we find that the connection of A is testable for stuck at 0 through the output H.

Even though the above transformation does not guarantee to find all redundancies, it still takes a lot of CPU time. Therefore there are faster transformations based on one-way data flow. These transformations implement the condition R with a restriction on the length of paths. A connection is declared redundant if it can be determined to be redundant after following paths of length 0 or 1.

An example of a transformation that considers paths of length 0 is the following:
A signal S is the constant 0 if $C_{00}(S) \cap C_{10}(S) \neq \phi$.
A signal S is the constant 1 if $C_{01}(S) \cap C_{11}(S) \neq \phi$.
The reason is simple. Suppose that there is T so that $T \in C_{00}(S)$ and $T \in C_{10}(S)$. Any input that makes T = 0 makes S = 0 because $T \in C_{00}(S)$, any input that makes T = 1 makes S = 0 because $T \in C_{10}(S)$. Therefore any input pattern makes S = 0.

An example of a transformation that considers paths of length 1 is illustrated in Figure 20. The connection X is not testable for stuck at 1 if for each of the signals S either $X \in C_{01}(S)$, or one of the signals $T \in C_{11}(S)$.
The reasoning is the following. To test X for stuck at 1 there would have to be an input pattern for which X = 0 and each T = 1. Then the above condition implies that every S = 1, which makes the signal U = 0 and hence the fault cannot be propagated.

3.10. Partitioning

Partitioning arises in various contexts. Sometimes we have to partition logic into smaller chips, or partition it for human readability, or partition it into gates implementable in target technology. In all these contexts we use greedy algorithms; only the measure for determining which pieces of logic should belong together are different for different applications.

As an example consider the problem of forming partitions that could be implemented as some large CMOS gates. This is done by merging small gates into large ones as long as the resulting gate is legal in the target technology, e.g. does not exceed stacking of N or P transistors. Which two small gates get merged into a larger one depends on the following criteria:
- What is their slack and would the path be speeded up by the merge?
- How many late paths do they lie on?
- How much area will be saved by the merge?
- Are they attached to primary input, primary output or some gate that is not likely to be merged with another?
Determining the above criteria and their relative importance is very heuristic and is done empirically.

3.11. <u>Hierarchy</u>

The importance of hierarchical synthesis increases with the increase in the size of chips. Ideally designers should design a chip in several partitions, synthesize each partition in isolation, and then just put the synthesized partitions together. We have no good solution for hierarchical synthesis because the ideal is unattainable for the following reasons.
- In order to do fanout correction between partitions it is necessary to give up I/O compatibility between the designer's function and the synthesized partition; this compatibility is important for checking purposes.
- Area may be improved by transformations operating on a piece of logic shared by two partitions.
- Outputs out of a partitions are not necessarily observable for testing purposes. Therefore testability must suffer if synthesis can see only one partition at a time.
- Critical paths may extend over several partitions. This makes it very difficult to decide in which partition to increase area in order to meet timing requirements.

Since there is no perfect solution to the problem of hierarchy designers must be careful about partitioning logic. For example, they have to decide which polarity of a signal to send between partitions, they may have to provide two copies of the same signals that goes to too many partitions. Different partitions cannot be synthesized in isolation; when synthesizing one, neighboring partitions must be available to the synthesis system.

CONCLUSIONS

This paper outlined some of the problems experienced and some of the solutions adopted in our logic synthesis system. There are several observations based on the experience of having the system used in production.

The goal of our logic synthesis system is not to produce optimal logic, it is to produce acceptable logic. Acceptability includes fitting into a given area, satisfying timing requirements, and satisfying testability requirements. Logic that is optimal with respect to area may be unacceptable with respect to delay and vice versa. Logic that is 100% testable and has minimal area for given timing requirements would require too many computer resources to be worth generating.

The major benefit of a logic synthesis tool is the faster feedback loop between design and implementation. It allows a designer to evaluate many more alternatives that if he had to implement each alternative by hand.

ACKNOWLEDGEMENTS

The logic synthesis system was developed and contributed to by L. Berman, V. Berstis, R. Brayton, J. Darringer, K. Day, J. Gerbi, J. Gilkinson, J. Goetz, S. Gundersen, D. Hathaway, P. Hauge, V. Iyengar, W. Joyner, T. Nix, G. Sutherlin, L. Trevillyan, and a number of others.

REFERENCES

[1] J.A. Darringer, D. Brand, J.V. Gerbi, W.H. Joyner, Jr., L. Trevillyan, "LSS: A system for Production Logic Synthesis", *IBM Journal of research and Development*, Vol. 28, No. 5, September 1984.

[2] L. Trevillyan, W. Joyner, L. Berman, "Global Flow Analysis in Automatic Logic Synthesis", *IEEE Transactions on Computers*, Vol. C-35, No. 1, January 1986.

[3] D. Brand, "Redundancy and Don't cares in Logic Synthesis", *IEEE Transactions on Computers*, Vol. C-32, No. 10, October 1983.

[4] R.B. Hitchcock, Sr., G.L. Smith, D.D. Cheng, "Timing Analysis of Computer Hardware" *IBM Journal of Research and Development* January 1982, 100-105.

[5] D. Brand, V.S. Iyengar "Timing Analysis using Functional Analysis", RC 11786, IBM T.J. Watson Research Center, March 1986.

[6] J.A. Darringer, W.H. Joyner, Jr., C.L. Berman, L. Trevillyan "Logic Synthesis through Local Transformations", *IBM Journal of Research and Development* Vol. 25, July 1981.

[7] M.A. Breuer and A.D. Friedman, *Diagnosis and Reliable Design of Digital Systems*, Computer Science Press, Inc., 1976.

```
OTH: PROC TYPE(SLICE);
  DCL S FAC(2) TYPE(SIG, IN);
  DCL R FAC(4) TYPE(SIG, OUT);
  DCL Y FAC(4) TYPE(SIG, IN);
  DCL X FAC    TYPE(SIG, IN);
  DCL Z FAC(4) TYPE(SIG, IN);

  Y = Z & REPEAT(X, 3);

  CGOTO (L0, L1, L2, L3), S;
    L0: R =          X || Y(1:3);   DIE;
    L1: R = Y(0:0) || X || Y(2:3);   DIE;
    L2: R = Y(0:1) || X || Y(3:3);   DIE;
    L1: R = Y(0:2) || X;             DIE;

  END OTH;
```

Figure 1.   Input description

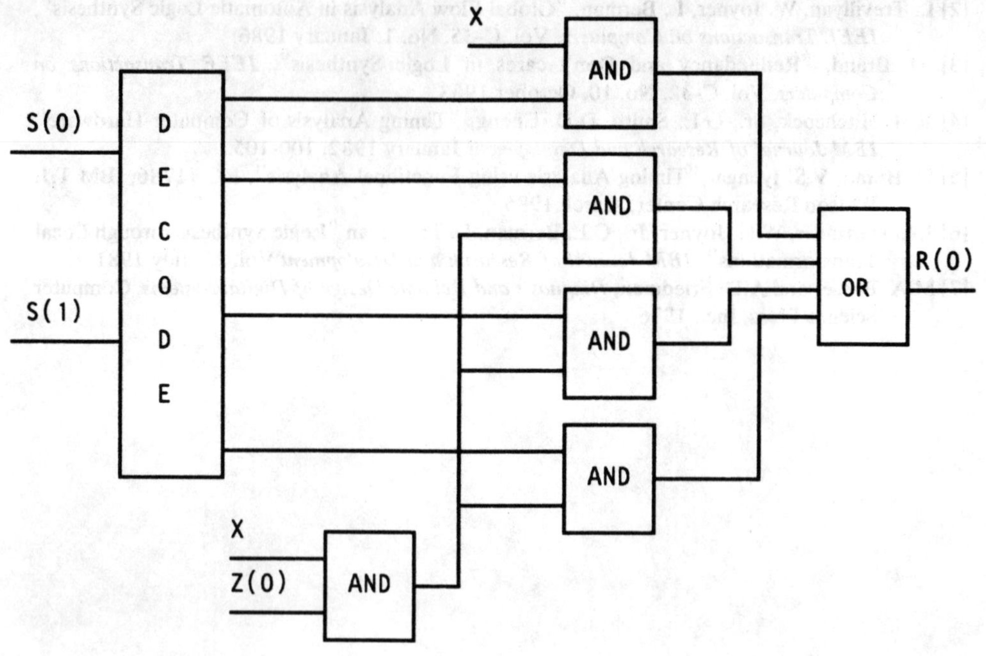

Figure 2. Initial representation

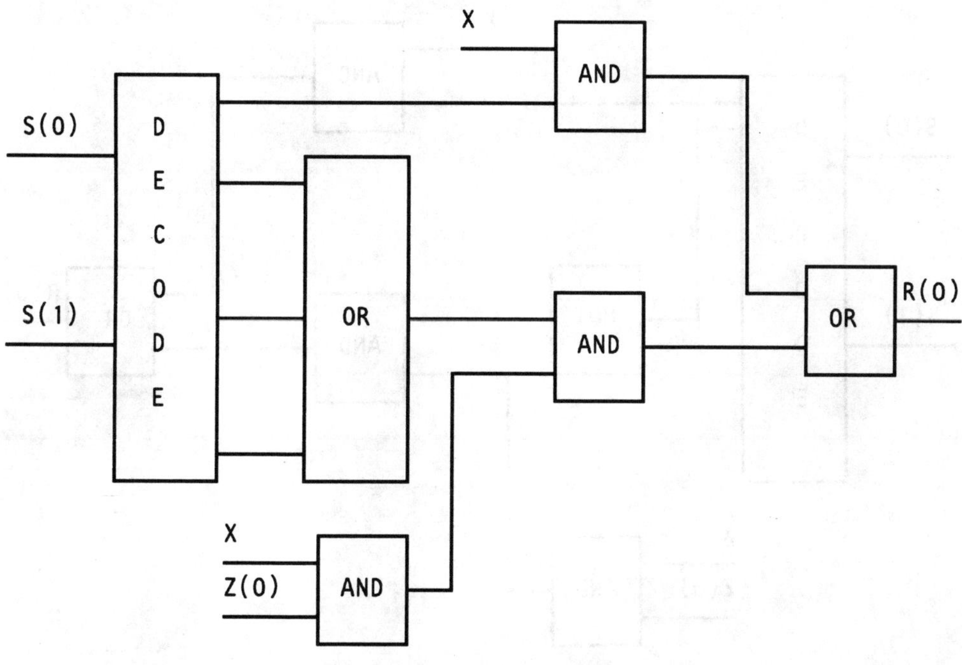

Figure 3.   After propagating OR through three ANDs with
common input

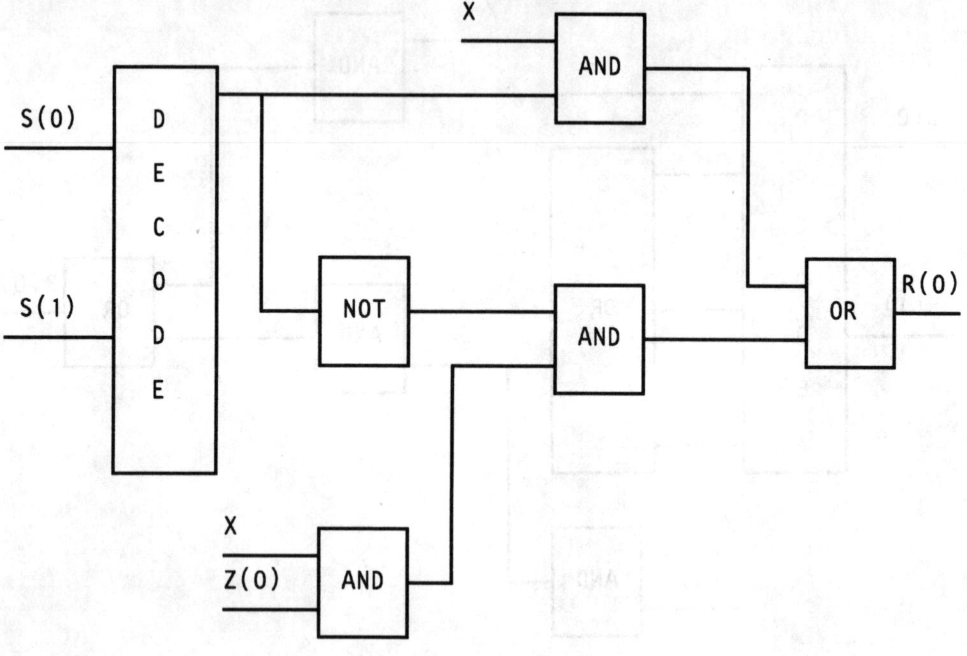

Figure 4. Transformation knowing about decoders

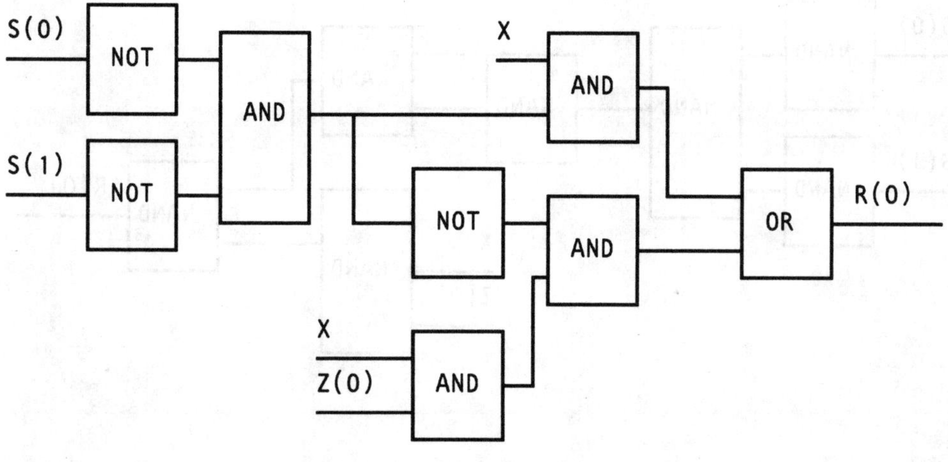

Figure 5.   After expanding decoder

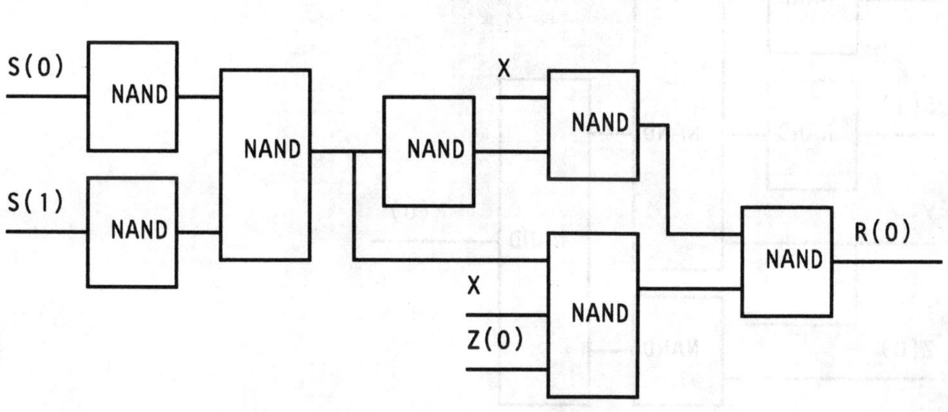

Figure 6.   Conversion to NANDs and simplification

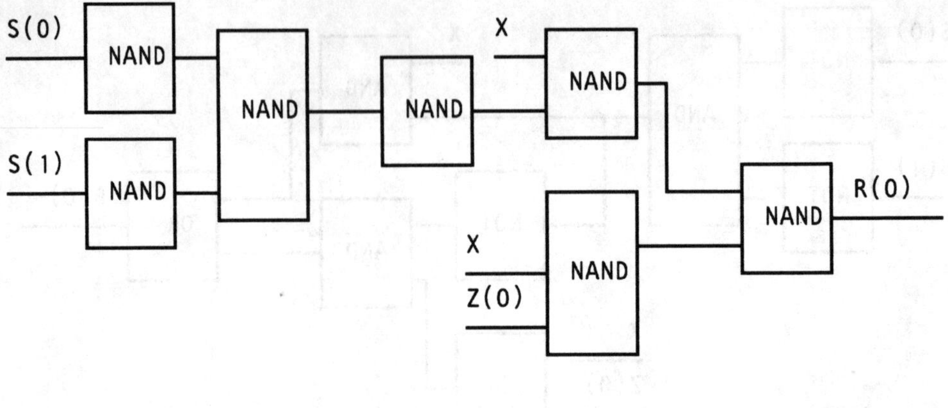

Figure 7.   After removing non—testable connection

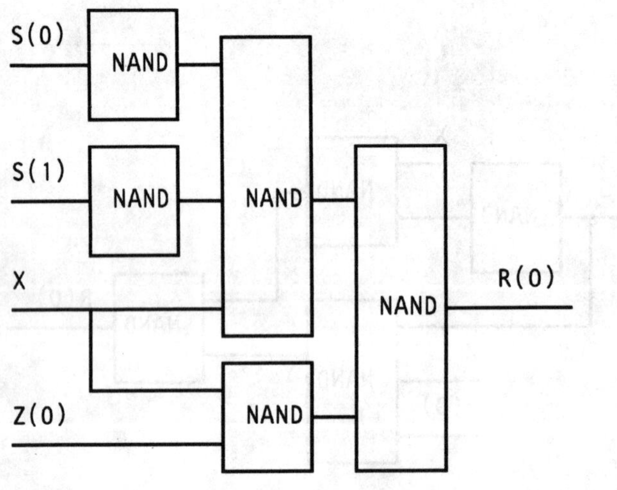

Figure 8.   Final implementation in terms of NANDs

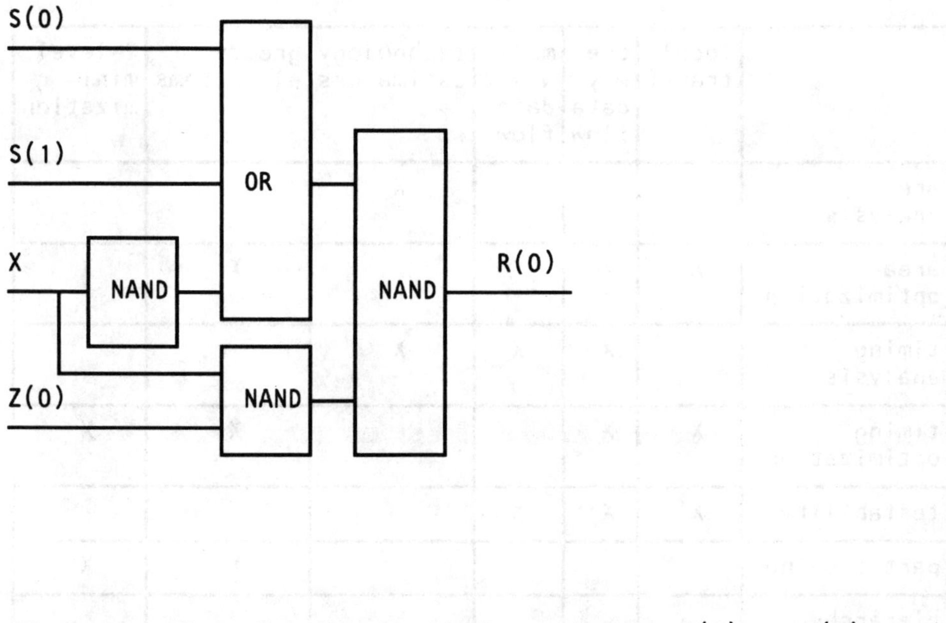

Figure 9.   Reducing levels on the path from S(0) to R(0)

	local transf	one way data flow	multi way data flow	technology estimators	greedy algorithms	2-level mini-mization
area analysis				X		
area optimization	X	X			X	X
timing analysis		X	X	X		
timing optimization	X	X			X	X
testability	X	X	X			
partitioning					X	X
hierarchy						

Figure 10. Table of problems and solutions

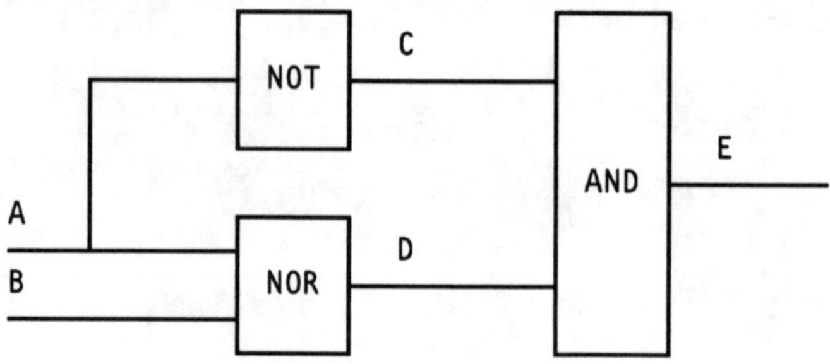

Figure 11. For illustration of the sets C

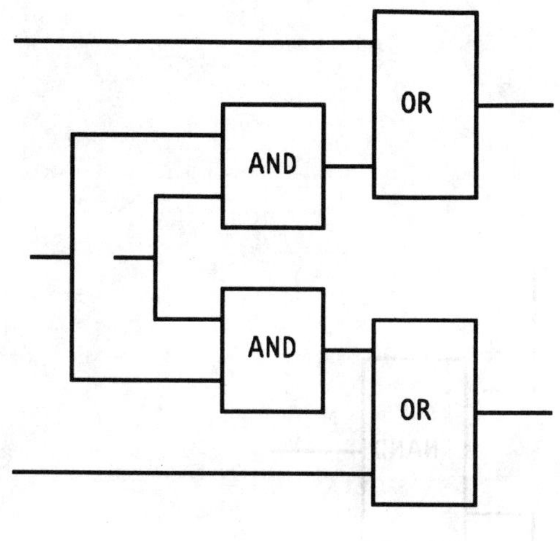

Figure 12.   Implementable with two ANDs and two wired ORs

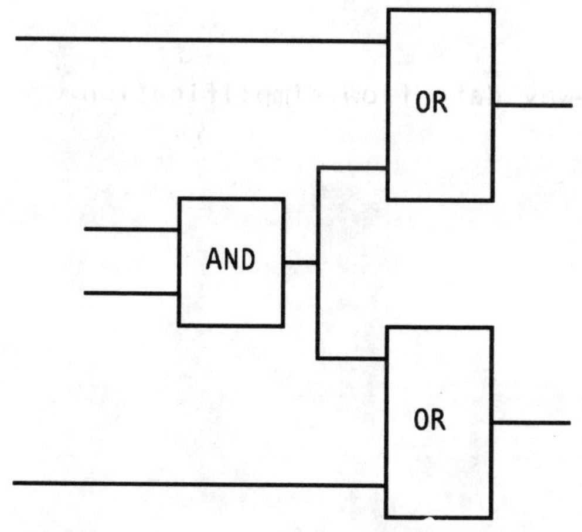

Figure 13.   Not implementable using wired ORs

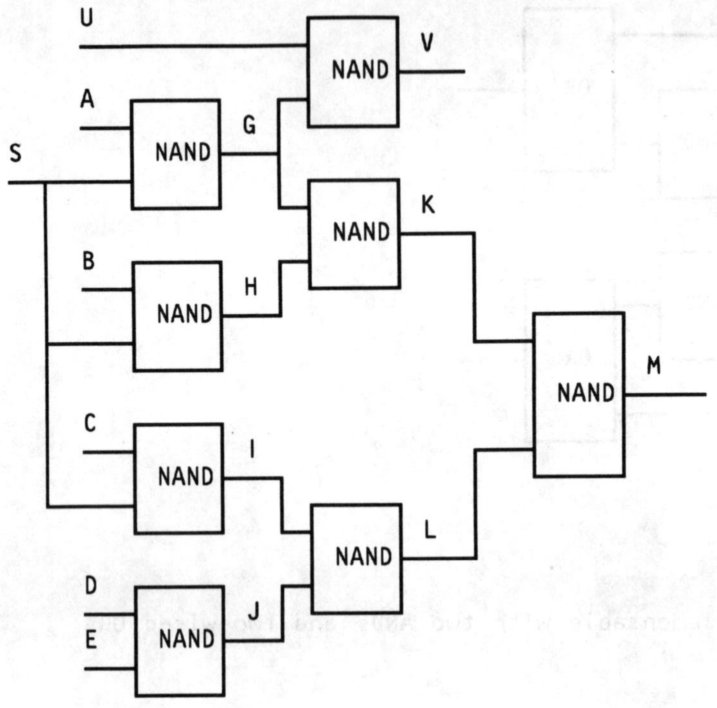

Figure 14.   Logic before one—way data flow simplification

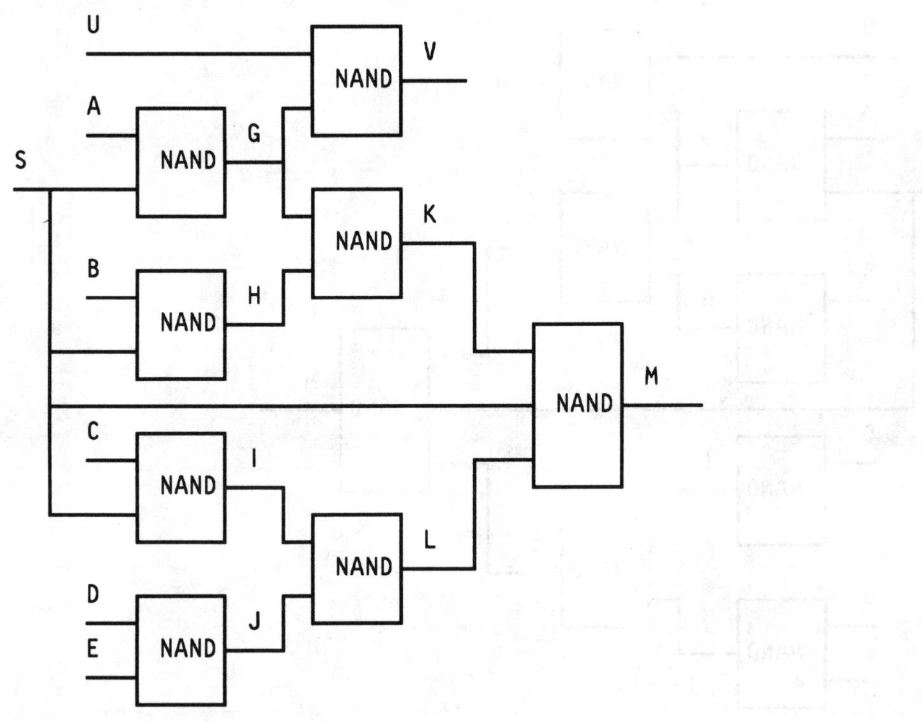

Figure 15.   Logic after propagating S forward wherever possible

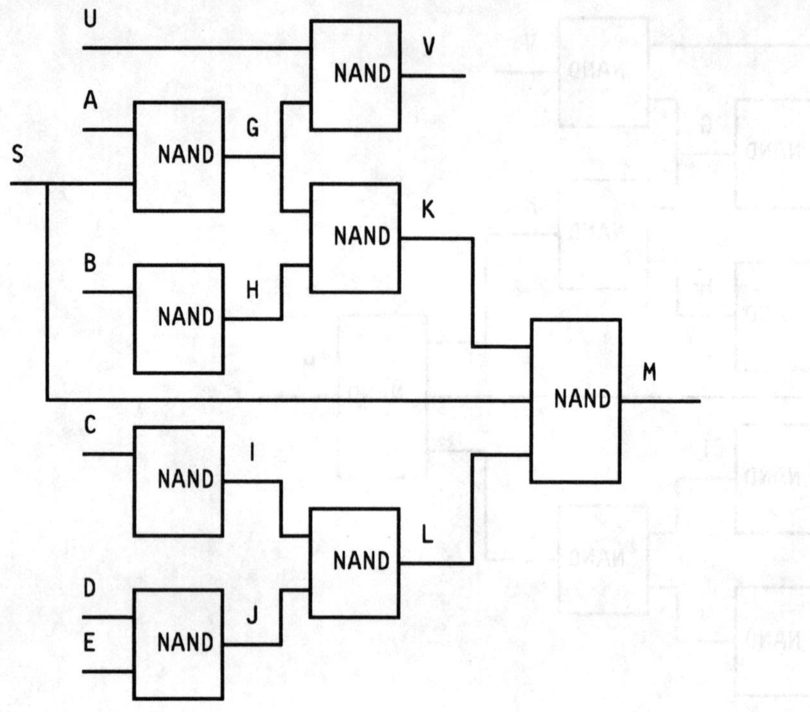

Figure 16.   Logic after disconnecting S wherever possible

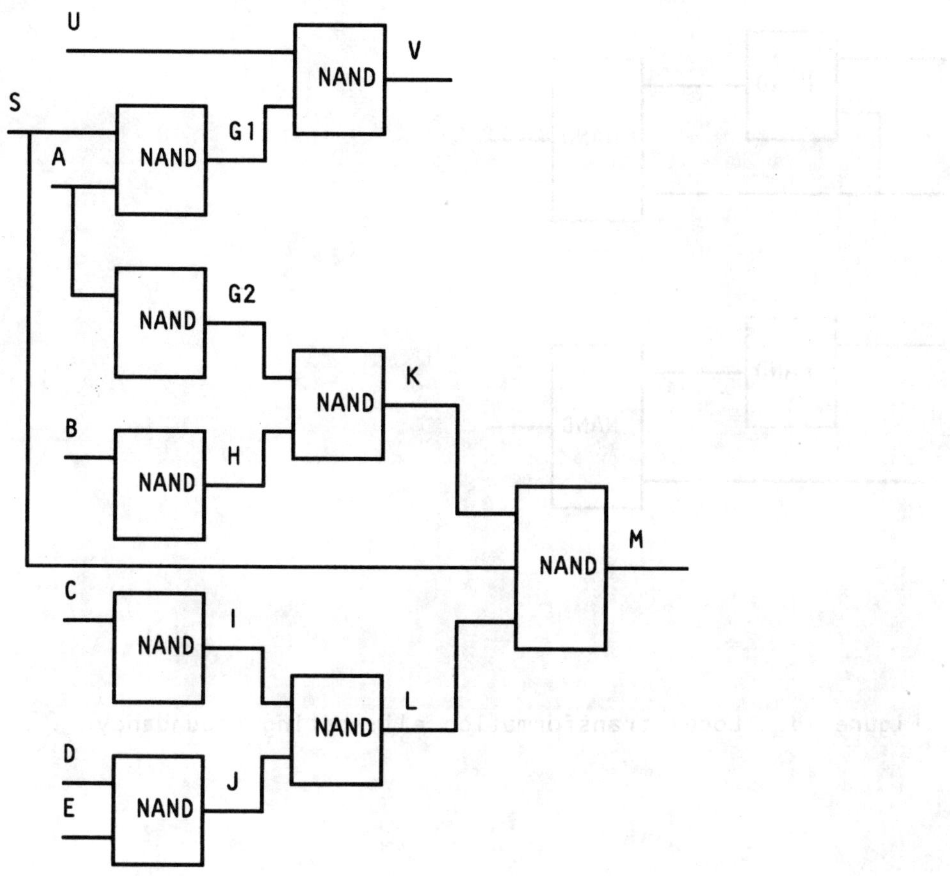

Figure 17.   Logic after replicating G and disconnecting S

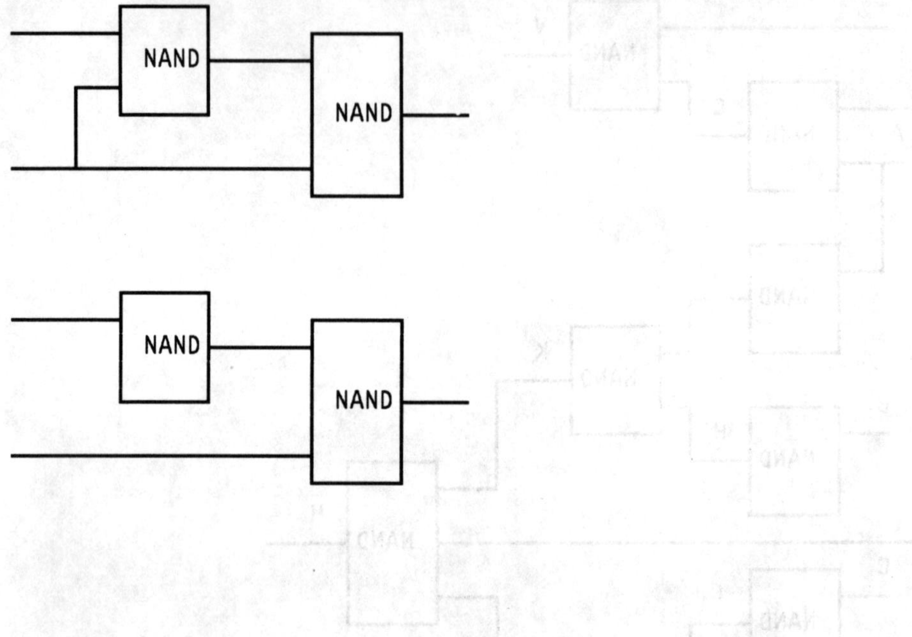

Figure 18.   Local transformation eliminating redundancy

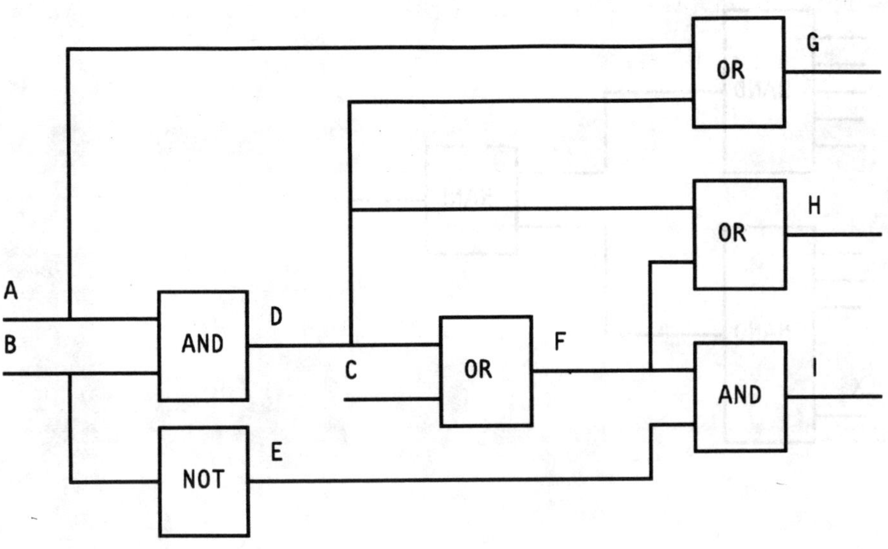

Figure 19.   Is the connection of signal A at the gate
testable for stuck at 0?

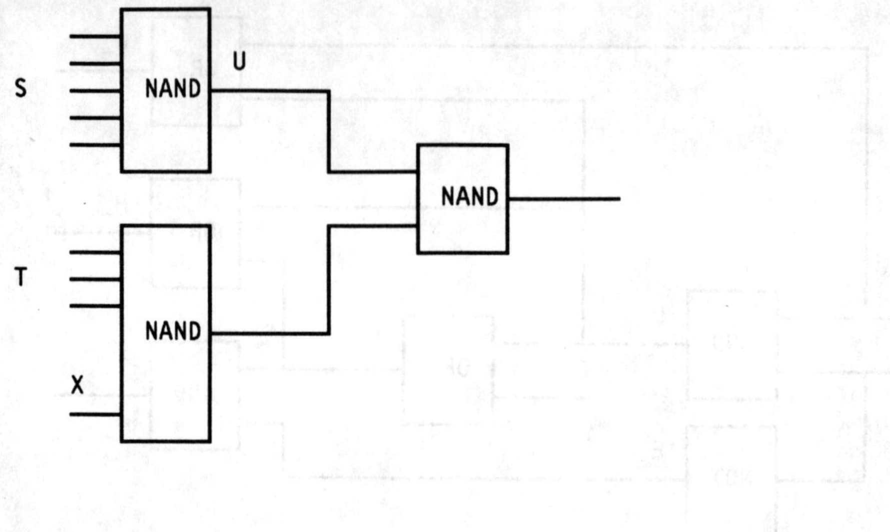

Figure 20.   Is X testable for stuck at 1?

SYNTHESIS OF CONTROL SYSTEMS

GIOVANNI DE MICHELI

IBM T.J.WATSON RESEARCH CENTER
YORKTOWN HEIGHTS, NEW YORK
USA

1. INTRODUCTION

Computer-Aided Design (CAD) tools have made the design of Very Large Scale Integration (VLSI) circuits possible. Research emphasis is now directed towards design methods for automated synthesis and verification of high-performance circuits from behavioral-level specifications. In particular, the capability of designing rapidly high-performance processors is a key to commercial competitiveness. Fast turn-around designs allow to achieve system-level trade-off comparisons, by trying several architectures in the search for the best match between system structure and implementation technology. In this perspective, integrated synthesis systems are necessary for the research and development of integrated circuits and systems in the future years.

Several approaches to automate the synthesis process have been proposed and are referenced in this book. In the beginning it was expected that automated synthesis systems could solve most of the designers' needs: in particular that the "push-button" conversion of high level specifications into the mask geometries could yield circuits acceptable for industrial standards. Unfortunately the synthesis system of the first generation, called silicon compilers, did not meet the expectations: the "compiled" circuits did not have an efficient utilization of the silicon area and the timing performances were not comparable with those obtained by designing the circuit with other methodologies. For this reason the synthesis systems of the subsequent generation were designed with circuit performance optimization in mind. Because of the numerous possibilities in exploiting trade-offs at the design level, the silicon compilers evolved into design environments, i.e. complex CAD systems supporting synthesis, verification, testing, ...

We present here structures, methods and algorithms for circuit synthesis and optimization of control units for VLSI processors. We consider first the design methodologies used for control units and how they evolved with the advent of VLSI circuit technology and computer-aided design synthesis tools. Then we review the steps in the synthesis of control structures and how they are related to optimization techniques. We show how control functions can be described by tables of symbols and how tables can be processed. In particular we present the symbolic design methodology that can be applied to to the tabular description of interconnected combinational and sequential circuits. The goal of this methodology is to achieve an optimal synthesis of the circuits implementing control units in terms of silicon area. The computer implementation of the algorithms and experimental results are then presented.

2. CONTROL STRUCTURES

With the design of single chip micro-processors, the design of data-path and control-units has evolved in the direction of a unified implementation style. In particular, high performance processor designs benefit from the physical proximity of data-path and control and the possibility of tuning the control-unit to the data-path design.

Different design styles have been used for control-unit design. In the past, there was a clear distinction between custom and structured control-unit implementations. Custom implementations tend to exploit particular choices to enhance performance. For example, the control part of the Z8000 micro-processor was implemented by a custom interconnection of (random logic) gates. Structured implementations, as in the case of the M68000, take advantage of circuit design regularity to support complex function designs, as well as engineering changes. In a structured implementation of the control part the sequencing and control functions are implemented by a structured array that is referred to as sequencing and control-store or more simply control-store. The control-store is implemented by a Read Only Memory (ROM), a Programmable Logic Array (PLA) or a more elaborate structure [ANDR80].

Control units with structured implementations have been called micro-programmed, when the control mechanism can be programmed by the personality of the control store, which is a binary representation of a sequence of operations defined by the micro-program. In the seventies, with the advent of VLSI circuits, micro-programming became a way to increase the regularity of the chip structure [LEWI81]. Davis [DAVI72] recognized this new view of micro-programming and gave the following definition: "One particular class of control mechanism uses regularly organized storage arrays to contain a large part of the control information. Machines employing such control mechanism are said to be micro-programmed."

There are several advantages in using micro-programming. The design of the control unit is flexible, can be defined at a later stage of the processor design and can be easily modified. Moreover micro-programming allows a processor to emulate the instruction set of other machines. Different implementation strategies of micro-programmed processors have been used: the control-store can be located on-chip or off-chip and can be writable or not. There is a trade-off in these choices. High performance processors take advantage of the proximity of on-chip control store. However, only processors with writable or off-chip control store can take full advantage of the micro-programming feature of altering the micro-program when the processor design is complete. Only in this case it is correct to talk about micro-programmable processors. For practical reasons, processor have mainly read-only on-chip control-store and therefore should be called micro-programmed. We will consider this class of processors in the sequel.

The advent of automated synthesis systems, or silicon compilers, has changed the perspective of micro-programming. In reality, the entire processor is described as a program to the synthesis system which generates the layout of the masks needed to fabricate the chip. This process is automatic and takes a time which is negligible with regard to that needed to write the program describing the processor. The program can be altered quickly to perform a change in the processor design. Different approaches to synthesis exist as well as different families of high-level language representations [GOLD85]. Some synthesis systems are based on structural language descriptions, which specify the circuit structure and in particular that of the control unit. In this case the micro-program can be seen as the routine defining the operation of the control unit. Therefore, the concept of micro-programmable processor is now related to the existence of a representation of the control structure as a structured sequence of elementary steps. Moreover, the distinction between custom and structured implementation of the control-store is fading as powerful synthesis systems [DARR84] [BRAY85a] have been able to generate implementations of multiple-level logic functions as interconnection of gates in a quasi-structured fashion. (For example as stacks of arrays of custom cells and channels providing the interconnection [OTTE84].) Some automated synthesis systems [SISK82] [BLACK85] [CAMP85] use behavioral-level circuit descriptions. In this case, the control structure is not even specified as an input datum; the structure is derived automatically from the system behavioral specifications and the resources (e.g. ALU, shifter, ...) used to synthesize the data-path.

The computer-aided synthesis of control units is based on the assumption of a specific model for the control structure. In principle, a general-purpose micro-sequencer structure could be used. However, such a choice would not exploit the particular features of each architecture and would therefore be inefficient. For this reason, we look for the common features in different control structures and we propose a general methodology for the optimal synthesis of the corresponding circuits. Before describing the circuit model, it is interesting to show how control structures have been implemented in two micro-processor designs.

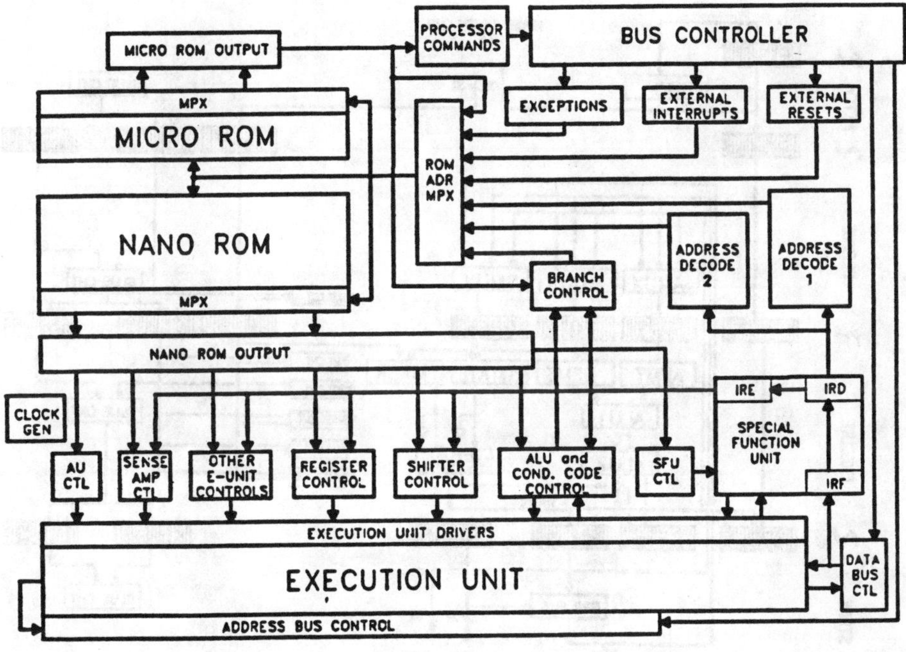

Fig 2.1 The u370 processor.

330

We consider first the design of a VLSI implementation of a Complex Instruction Set Computer (CISC): the micro-370 [HADS85]. The micro-370 is a 32-bit, single-chip, System/370 micro-processor. It executes 102 System/370 instructions. The control-unit is sketched in Fig. 2.1. The sequencing and control stores are kept in two ROMs which are called micro and nano ROM respectively. The sequencing may be altered by commands from the bus controller, the branch control and the special function unit. In essence, the control-unit consists of a Moore-type finite automaton (micro ROM) and a control decoding unit (nano ROM).

Let us consider now the design of the control part of a 801 processor, which has a streamlined architecture. This architecture has been the archetype of a family of processors, called Reduced Instruction Set Computers (RISCs), because the instruction set was simplified by deleting those instruction which are seldom executed and penalize the processor performance. We consider a particular implementation of the 801 micro-processor, done using the Yorktown Silicon Compiler [BRAY85c]. The 801 processor was architected and designed so that each instruction could be executed in a single cycle. The processor was implemented as a 4-stage pipeline. The execution control and interrupt circuit has the structure of a decoding network, that maps the representation of the instruction corresponding to each stage of the pipe to the controls of the related resources (Fig. 2.2). In essence, the control structure is a combinational decoding network.

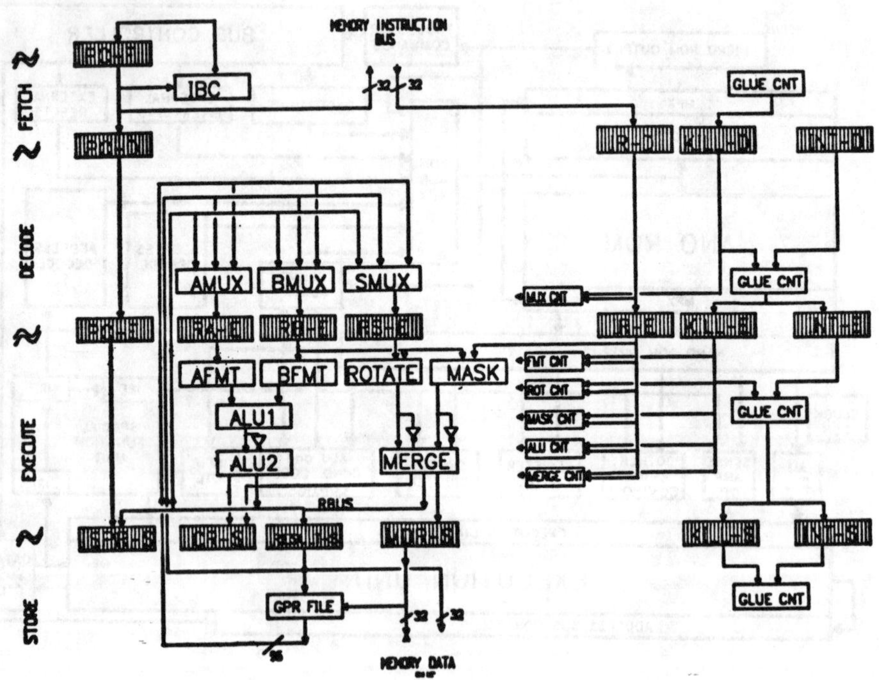

Fig 2.2 The 801 processor, as designed by the YSC.

Though the two example use radically different choices for the control implementation, there is a common feature in the representation of the two control structures. The two control units are personalized by one (or more) blocks of combinational logic. In the first case, the combinational function was implemented by a ROM, in the second by an interconnection of gates. The programmability feature of the control unit is retained in both cases by the personality of the combinational logic. In the first case, this personality is represented by the 0-1 patterns of the ROMs. In the second, the personality is represented by an interconnection of gates, which is a direct translation of the high-level tabular representation of the control operation. In the chip design described in [BRAY85c], the translation was performed automatically by the Yorktown Silicon Compiler.

In general, a control unit can be modelled as a **deterministic automaton or deterministic Finite State Machine** (FSM). Since sequential functions can be represented, using the Huffman model [HILL81], as the combination of a combinational circuit and a memory, then a generic control unit can be represented as an interconnection of a combinational circuit and memory. In particular, if the control function is combinational in nature, as in the case of the 801 design [BRAY85c], the memory component is not implemented and the automaton has only one state. If the control function is not combinational, we assume that memory is synchronized to the system clock (synchronous FSM) to avoid race conditions, as done in most designs. Memory is then implemented by clocked registers, which store the control state of the machine. We assume here the use of Delay-type registers.

Though this model is fairly general, a control unit implementation using an interconnection of one combinational block and a set of registers can be effective only for control functions of small to medium-sized complexity. When the control function is very elaborate, as in the case of some micro-processor designs, it is convenient to break the combinational block into an interconnection of circuits. For example, multiplexers may be useful to implement the next control-state function following a conditional branch. Auxiliary circuits, such as counters and stacks may be used. The structure of the control unit can be therefore generalized to an interconnection of combinational circuits, whose personality can be programmed, and fixed-structure circuits, such as registers and auxiliary circuits. We consider this generalization as a canonical structure for control units and we concentrate on the design of the combinational blocks that personalize the control unit and their optimal implementation.

3. COMPUTER-AIDED DESIGN OF CONTROL STRUCTURES

The synthesis of a control unit can be partitioned in the following tasks: **functional design, logic design** and **physical design**. Functional design consists of mapping the high-level specification into a control structure. Logic design consists of transforming this structure into Boolean equations (or an equivalent representation) and to optimize this representation. Physical design addresses the generation of the mask geometries from the logic specifications. Several computer-aided tools have been developed for the last two tasks, because the underlying problems are well understood. The automatic synthesis from behavioral-level specifications is still an exploratory area for design automation tools.

Functional design consists of defining the structure of a control unit from the high-level specification of the entire circuit. Its starting point is the description of the processor in a behavioral language. Its goal is to define the sequencing of the micro-operations and the timing of the control signals to the data-path as well as the partitioning of the control unit into functional blocks. In other words, functional design provides a control structure as an interconnection of logic blocks and timing blocks (registers). The information needed to perform the functional design is embedded in the behavioral language description. In particular, this description contains the sequencing of the operations independently of the resources used as operators.

Because of its complexity, in the past this task has been performed directly by the system designer, who transformed the description of the architecture in a natural language into a flow-chart or equivalent representation. For example, Tredennick reported in [TRED81] on the functional design of the M68000 control unit, which was achieved by the designer himself without the use of computer aids. Similarly, Brayton *et al.* reported in [BRAY85c] on the process of transforming the description of a 801 processor in a natural language into the YLL structural language [BREN84]: the control of the processor was specified using YLL routines and tables. In other cases, the specifications of the control unit are given directly by means of a micro-program. Micro-compilers are used to to generate the tables that represent the sequencing and the control signals. This tables are then translated into the 0-1 patterns of a ROM (if ROM are used for the control store) or into the specifications of a PLA [PAPA79] or other hardware circuit. In the case of ROM implementation, it is important to minimize the number of the words representing the micro-program. This kind of optimization at the functional level can be done by using compiler optimization techniques [AGER76].

Computer-aided tools for functional design, that make automated synthesis possible starting from the processor behavioral specifications, have been developed only recently [SISK82] [BLACK85] [CAMP85]. The most straight-forward approach to behavioral synthesis is to determine the structure of the data-path first and then to infer the timing and sequencing of its control signals. In this case, the states of the processor and the timing of the control signals to the data-path are derived from the sequencing information in the behavioral description and the timing specifications of the resources used in the data-path as operators. The specification of the control unit can then be generated in a form equivalent to a table that specifies the control states and the micro-operations. Note that the synthesis of the control unit does not necessarily have to follow the data-path synthesis. Indeed, they could be synthesized at the same time. Optimization techniques can then be used to exploit the trade-off between data-path cycle time and resource allocation by introducing/reducing cycles to perform each instruction. As a result, the number of machine states and/or state transitions is modified accordingly. There are some difficult problems associated with optimal functional design which are still under investigation. One is the determination of the optimal pipe-lining of a processor. Another one is to determine a partition of the control unit into resources (e.g. a hierarchy of finite automata) which fits the architecture. Knowledge-based systems, such as the Design Automation Assistant [KOWA85], could be used in this perspective.

The output of the functional design phase is the interconnection of the control structure and the specification of its blocks. Since combinational and sequential functions can be expressed in tabular format, we use here tables of symbols (mnemonics) to represent them. In general, control functions can be represented by interconnection of sequential and combinational functions as a set of linked tables. The formalism of the tabular representation will be presented in the next section.

Logic design consists of mapping the structural description into a representation in terms of logic variables. The straightforward mapping of functional tables into Boolean equation is a trivial procedure but yields inefficient implementations in terms of silicon area and switching-time performances. Logic design includes a set of optimization procedures aiming at simplifying the circuit implementing the function. This simplification correlates with silicon area reduction and switching-time improvement. Logic optimization includes a vast set of techniques, that apply to sequential and combinational functions with different flavors according to the target implementation technology. We consider in this section methods that apply to ROM and two-level logic implementations of combinational functions (including the combinational component of sequential functions). If multiple-level logic implementations are used, the techniques for optimal two-level logic design described here are used to obtain a "good" starting point for multiple-level synthesis. Circuit optimality is related to an implementation choice. For ROM implementations, the optimality is measured by the the number of words and the number of bits. For two-level logic implementations, the optimality is usually measured by the cardinality of the corresponding Boolean cover. For multiple-level logic implementations, the optimality may be measured as a function of the total number of gates and literals. These last two measures correlate to the silicon area penalty of the implementation.

An important task in optimal design of sequential circuits is **state minimization**, which is the search for a representation of the control function using a minimal number of states. Reducing the number of states corresponds to reducing the number of transitions in the sequencing function and eventually to the reduction of the number of words in a ROM implementation and of the number of implicants in a two-level logic implementation. Moreover, a reduction of the number of states corresponds to a reduction of the number of bits needed to represent the states and therefore the number of bits in a ROM and to a simplified transition function in a two-level logic implementation. State minimization has been the object of extensive research. We refer the reader to [HART66] for the basic formalism and to [REUS86] for some recent results and an updated set of references. It was shown in [HOPC71] that the reduction of completely specified finite automata can be achieved in $O(n \log n)$ steps. The minimization of incompletely specified automata is a NP-complete problem [PLEE73]. Unfortunately, most automata derived from control structures are uncompletely specified. Therefore heuristic techniques are used.

The following steps in optimal logic design described in this section depend on the nature of the circuit implementation. If ROMs are used to implement the sequencing and control store, then logic optimization aims at reducing the number of words and/or the number of bits. The number of words can be reduced by several techniques. In particular the detection of possible concurrency of micro-operations as well as data and resource dependencies are used in this perspective. We refer the interested reader to [AGER76] for a set of references. The number of bits in the ROM can be reduced by encoding techniques. For example, consider a ROM implementing the control store only. (Assume the sequencing function is stored elsewhere.) If n resources have to be controlled, then n bits are needed by a fully horizontal microcode implementation. Note however that only as many bits as the ceiling of $\log_2 n$ are necessary in the case of a vertical microcode organization. While vertical microcode allows to reduce maximally the bit dimension of the ROM, it requires a decoding network at its outputs (Fig 3.1). A complete decoding network may be expensive in terms of silicon area and offset the advantage of the ROM bit reduction. Intermediate solution are often used. The basic idea is to group the bits controlling the resources into fields and use one decoder for each field (Fig 3.2). In this case, no word in the ROM can activate two resources whose control is encoded in the same field. This restriction is rather mild, because the possible combinations of concurrent resource controls is often much smaller than 2^n. Schwartz [SCHW68] proposed a an encoding method that minimizes the number of decoders.

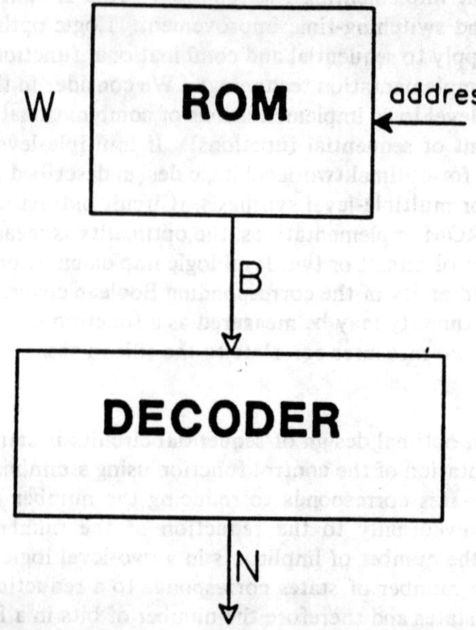

Fig 3.1 ROM and decoder.

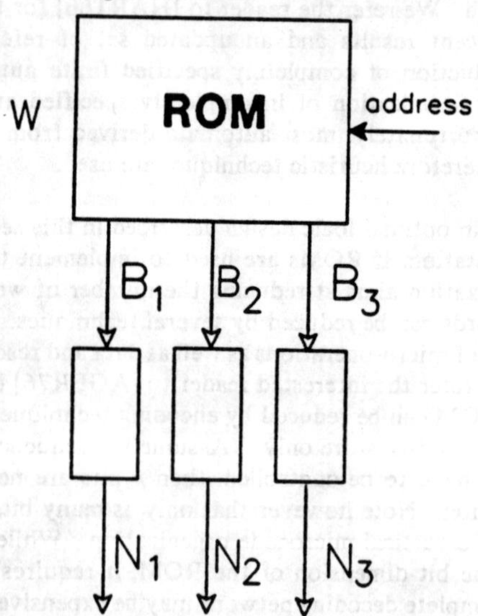

Fig 3.2 ROM with multiple decoders.

However Schwartz's method does not minimize the number of bits under the assumption that resources controlled simultaneously by at least one word are assigned to different decoders. Grasselli *et al.* [GRAS70] reformulated the method in terms of switching theory and presented an algorithm for computing an encoding that minimizes the number of bits in the ROM. When a multiple output two-level logic circuit is used to implement the sequencing and control store, then the logic optimization problem corresponds to minimizing the number of implicants and the number of bits used to represent the control state. The control store may be implemented by one logic function or by the interconnection of two (or more) functions. We consider in this section the case in which the sequencing and control store in implemented by one function, i.e. a two-level function implements the combinational component of the control FSM. The extension of logic optimization to the interconnection of logic circuits will be shown in the next section.

The problem of assigning a binary representation to the control states is a classical problem of switching theory and is referred to as **state assignment or state encoding** [HART66]. State assignment affects the number of implicants in a minimal Boolean cover of a two-level implementation of the control store. If a PLA is used to implement the Boolean function,[1] we may assume that each PLA row implements an implicant and each column is related to an I/O signal (primary input/outputs and the encoding of the present/next states.) The PLA area is proportional to the product of the number of rows times the number of columns. Both row and column cardinality depend on state encoding. The (minimum) number of rows is the cardinality of the (minimum) cover of the FSM combinational component according to a given assignment. The encoding-length is related to the number of PLA columns. Therefore the PLA area has a complex functional dependence on state assignment. For this reason two simpler optimal state assignment problems are considered: i) find the assignment of minimum code length among the assignments that minimize the number of implicants; ii) Find the assignment that minimizes the number of implicants among the assignments of given code length. The optimum solution to the state assignment problem which minimizes the PLA area can be seen as a trade-off between the solutions to problem i) and ii). Note that the above problems are still computationally difficult and to date no method (other than exhaustive search) is known that solves them exactly. Therefore heuristic strategies are used to approximate their solution. We refer the reader to [DEMI85a] for an extended set of references and a critical survey of most of the previous techniques for the state assignment problem. The selection of the type of registers used to store the machine state affects also the size of the implementation. This problem was addressed by Curtis [CURT69] [CURT70] in connection with the state assignment problem.

If the control function is combinational, then logic design consists of encoding the controls into Boolean variables and determining an optimal combinational function. For two level-logic implementation, this is the classic logic minimization problem [BRAY84b].

Physical design is the synthesis of the layout of the circuit implementing the control unit layout from the logic specification. It depends on the layout style and the processing technology. If PLA based implementations are chosen, the circuit can be further optimized by using folding or partitioning techniques [DEMI84a] [RUDE85b]. If custom cells implementations are used, as in the case of the Yorktown Silicon Compiler, then the methods described in [BRAY85a] apply.

---

[1]   It is assumed that the PLA is not folded nor partitioned for the sake of simplicity.

4. SYMBOLIC DESIGN

We consider here an innovative approach to the state encoding problem and the minimization of the combinational part of a control unit. This approach is called **symbolic design methodology**.

In the standard approach to synthesis [HILL81], Boolean representations of switching functions are obtained from the structural description by representing each each mnemonic entry in a table by Boolean variables. The optimization of logic functions, and in particular two-level logic minimization, is performed on the Boolean representation. The result of logic optimization is heavily dependent on the representation of the variables. As an example, the complexity (in particular the minimal cardinality of a two-level implementation) of the combinational component of a finite-state machine depends on the assignment of Boolean variables to the internal states [HART66].

The **symbolic design** methodology presented here avoids the dependence on the variable representation in the optimization process and consists of two steps: i) determine an optimal representation of a switching function independently of the encoding of its inputs and outputs; ii) encode the inputs and outputs so that they are compatible with the optimal representation. This technique can be applied to solve the following problems of logic design:

P1) Find an encoding of the inputs (or some inputs) of a combinational circuit that minimizes its cost.

P2) Find an encoding of the outputs (or some outputs) of a combinational circuit that minimizes its cost.

P3) Find an encoding of both the inputs and the outputs (or some inputs and some outputs) of a combinational circuit that minimizes its cost.

P4) Find an encoding of both the inputs and the outputs (or some inputs and some outputs) of a combinational circuit that minimizes its cost and such that the encoding of the inputs is the same as the encoding of the outputs (or the encoding of some inputs is the same as the encoding of some outputs).

Finding an optimal state assignment of a sequential circuit is equivalent to solving problem P4, when the sequential circuit is implemented by feeding back (possibly through registers) some outputs of a combinational circuit to its inputs. Similarly, finding the encodings of the signals connecting two (or more) combinational circuits, that minimize the total cost, can be reduced to problem P4. The author presented in [DEMI84c] [DEMI85a] an approximation to the solution of the state assignment problem, in which the cost was minimized with regard only to the encoding of the inputs. In particular, the technique presented in [DEMI84c][DEMI85a] solved only problem P1. Problem P2 was attacked by Nichols [NICH65], but the algorithm he presented could deal only with small-scale circuits. A solution to problems P1-P4 was presented for the first time in [DEMI86a].

Though the symbolic design methodology is fairly general, we restrict our attention here to two-level *sum of products* implementations. Since the area of the physical implementation has a complex functional dependence on the function representation (even by using PLA implementations [DEMI85a]), we consider a simplified optimization technique that leads to quasi-minimal areas. In particular we attempt to find first a *sum of products* representation that is minimal in the number of products, and then a representation of the input/outputs that is minimal in the number of Boolean variables.

The difficulty in solving problems P2-P4 is related to finding a minimal two-level representation of a switching function independently of the encoding of both inputs and outputs. We use here a technique called symbolic minimization. Symbolic minimization consists of determining a minimal encoding-independent two-level *sum-of-products* representation of a switching function. It is minimal in number of product-terms and independent of the encoding of all (or part of) the inputs and outputs [DEMI85c]. The minimal symbolic representation is an intermediate step towards the determination of a corresponding Boolean representation. For this reason, three encoding problems are introduced in Section 6 to transform the minimal symbolic cover into an equivalent Boolean representation.

We present first an informal overview of symbolic design. The methodology is introduced by elaborating on an example. We consider first a combinational circuit (Fig. 4.1) and we use symbolic design to find a representation of its inputs and outputs and a corresponding two-level implementation that minimize its cost. This is equivalent to solving problem P3. Note that problem P1 or P2 can be derived from problem P3 by considering only the circuit inputs or outputs.

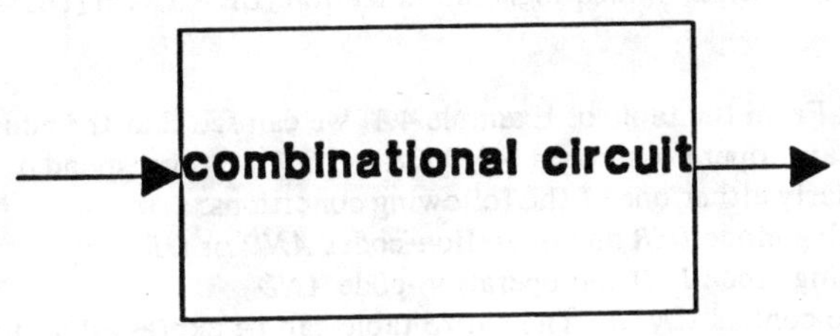

Fig 4.1 Combinational circuit.

Example 4.1: The following truth table specifies a combinational circuit; in particular an instruction decoder. There are three fields: the first is related to the addressing-mode, the second to the operation-code and the third one to the corresponding control signal. The circuit has two 2 inputs and 1 output. Each row specifies a symbolic output for any given combination of symbolic inputs.

INDEX	*AND*	*CNTA*
INDEX	*OR*	*CNTA*
INDEX	*JMP*	*CNTA*
INDEX	*ADD*	*CNTA*
DIR	*AND*	*CNTB*
DIR	*OR*	*CNTB*
DIR	*JMP*	*CNTC*
DIR	*ADD*	*CNTC*
IND	*AND*	*CNTB*
IND	*OR*	*CNTD*
IND	*JMP*	*CNTD*
IND	*ADD*	*CNTC*

In the standard approach to synthesis, each word (mnemonic string) in the table would be encoded by a string of binary symbols (i.e. 1's and 0's). Then, the encoded table would be minimized by a logic minimizer. In symbolic logic design, the table is minimized first (i.e. a table consisting of a minimal number of rows is computed) and then the table is encoded.

A first approach to symbolic minimization can be achieved by grouping the set of inputs that correspond to each output symbol. This process of reducing the size of the table is called here *disjoint minimization*, because the table is considered as set of independent sub-tables corresponding to each output symbol. Remarkably, disjoint minimization can be achieved by using techniques of multiple-valued logic minimization [BRAY84b] [DEMI85a], as shown is Section 5.

Example 4.2: From the table of Example 4.1, we can see that the addressing mode *INDEX* and any operation-codes *AND OR ADD JMP* correspond to the control *CNTA*. Similarly either one of the following conditions:

    addressing mode *DIR* and operation-codes *AND* or *OR*

    addressing mode *IND* and operation-code *AND*

correspond to control *CNTB*. The entire table can be expressed as a set of conditions:

INDEX	*AND OR ADD JMP*	*CNTA*
DIR	*AND OR*	*CNTB*
IND	*AND*	*CNTB*
IND	*OR JMP*	*CNTD*
DIR IND	*ADD*	*CNTC*
DIR	*JMP*	*CNTC*

Note that this table is more compact than the previous one, because it requires only 6 rows instead of 12.

The problem now is to find a Boolean representation of the symbols, corresponding to a Boolean cover representation of the function, with as many rows as the compacted table. While the encoding problem will be presented in detail in Section 6, we show here by an example the consequence of the choice of a particular encoding.

Example 4.3: Consider this particular Boolean encoding of the words.

INDEX	= 00	*AND*	= 00	*CNTA*	= 11
DIR	= 01	*OR*	= 01	*CNTB*	= 01
IND	= 11	*ADD*	= 10	*CNTC*	= 10
		JMP	= 11	*CNTD*	= 00

Then the function can be represented by a Boolean cover as:

$$
\begin{array}{ccc}
00 & ** & 11 \\
01 & 0* & 01 \\
11 & 00 & 01 \\
11 & *1 & 00 \\
*1 & 10 & 10 \\
01 & 11 & 10
\end{array}
$$

where a *don't care* condition on a binary input variable is represented by *. Note that the fourth Boolean implicant can be deleted, because its output part is 00. Moreover note that by deleting this implicant this cover is minimum, i.e. there exist no Boolean covers corresponding to this encoding with fewer than five implicants. (This can be proven experimentally by running an exact minimization algorithm [MCCL56], like that implemented by program ESPRESSO-II with the"exact" flag [RUDE85a].)

We question now to what extent symbolic design guarantees the minimality[2] of the Boolean cover, obtained by replacing the words by their corresponding binary encoding.

Example 4.4: Consider this other Boolean encoding of the words, in which we changed only the encoding of the output symbols:

$$
\begin{array}{lll}
INDEX = 00 & AND = 00 & CNTA = 00 \\
DIR \quad = 01 & OR \ \ = 01 & CNTB = 01 \\
IND \quad = 11 & ADD = 10 & CNTC = 10 \\
& JMP = 11 & CNTD = 11
\end{array}
$$

Then the function can be represented by a Boolean cover as:

$$
\begin{array}{ccc}
00 & ** & 00 \\
01 & 0* & 01 \\
11 & 00 & 01 \\
11 & *1 & 11 \\
*1 & 10 & 10 \\
01 & 11 & 10
\end{array}
$$

Note that the first Boolean implicant can be deleted, because its output part is 00. However note that this cover is not minimum: there exist now a minimum Boolean cover corresponding to this encoding with three implicants and that can be computed from the above one by a standard minimization technique. Namely:

$$
\begin{array}{ccc}
1 & 0 & 01 \\
1 & 1 & 10 \\
11 & *1 & 11
\end{array}
$$

Note that the first cover has two pairs of Boolean implicants with 01 and 10 in the third field and that are merged into two single implicants in the minimum cover. This is possible because the third implicant of the minimum cover has 11 in the third field, and 11 covers 01 and 10.

---

[2] The optimality of a Boolean cover is measured by its cardinality, i.e. by the number of its implicants. A Boolean cover of a function is minimum, if there exist no cover of that function having a smaller cardinality. A Boolean cover of a function is minimal, if its cardinality is minimum with regard to some local criterion. Usually a Boolean cover of a function is said to be minimal, if no proper subset is a cover of the same function [BRAY84b].

The reason for this additional reduction is in the covering relations among the encoded output symbols. Note that when we optimized the symbolic table by disjoint minimization, our goal was only to group the input symbols corresponding to each output symbol independently. The relations among the output symbols were neglected. For this reason it is important to exploit the relations among the output symbols at the symbolic level. **Symbolic minimization**, formally defined in Section 5, is a technique that determines an optimal ordering of the output symbols. This ordering is related to the covering relations among the binary encodings of the output symbols, and is responsible for the additional reduction of the table size, as shown by Example 4.4.

Example 4.5: Consider the following table:

INDEX		AND	OR	ADD	JMP		CNTA
DIR	IND	AND	OR				CNTB
DIR	IND	ADD	JMP				CNTC
IND		JMP	OR				CNTD

Here, an ordering relation is assumed that allows control *CNTD* to override control *CNTB* and *CNTC* when both are specified. The table, together with this ordering relation, is an equivalent representation of the function specified by Example 4.1. It is an example of the result of a symbolic minimization. Note that this table can be transformed into the Boolean cover of Example 4.4 by replacing each symbol by its encoding. Moreover, the encoding of *CNTD* covers bit-wise the encoding of *CNTB* and *CNTC* and allows *CNTD* to override *CNTB* and *CNTC*. Note that the first implicant can be deleted, by assuming that *CNTA* is the default output and that the encoding of *CNTA* is 00, which is covered bit-wise by the encoding of all other outputs.

Once a minimal table has been found, the encoding of the words into Boolean variables is driven by the grouping of the input symbols (group of symbols appear on the same row of a minimal table) and the ordering of the output symbols generated by symbolic minimization. Note that disjoint minimization deals with each output symbol independently, and therefore does not provide information for an encoding of the output symbols that optimize the table size. Therefore disjoint minimization can be used only to solve problem P1 of symbolic design.

We describe in detail in Section 6 how to compute an encoding of the input and output symbols that is compatible with a minimal table. Such an encoding allows to transform the minimal symbolic cover into a Boolean representation with as many product-terms as the minimal symbolic cover. Even though this mapping is not sufficient to imply the minimality of the Boolean cover, the encoded cover can be considered a good solution to the problem. It is important to remark that the length of the encoding (i.e. the number of binary variables) needed to encode each symbol may have to be larger than the minimum length required to distinguish all the symbols in each field (i.e. the ceiling of the logarithm in base 2 of the number of elements in each field). Therefore it is interesting to compute minimal length encodings compatible with a minimal symbolic representation and to trade off possibly the minimality of the number of rows in a table for the number of bits required to encode each field.

Let us now consider the design problem P4. A finite state machine, implementing a sequential circuit, can be implemented by feeding back the (some of the) outputs of a combinational circuit to its inputs, possibly through a register (Fig. 4.2). Finite state machines are generally represented by state tables. State tables consist of four fields related to the primary inputs/outputs and to the present/next state representation. Each field may be partitioned into sub-fields. Optimal state assignment can be solved by symbolic design by minimizing the state table using symbolic minimization and by computing a state encoding compatible with the minimal table [DEMI86a]. The feedback path makes this problem different from designing combinational units. In particular, the state symbols appear in both an input and an output column of the state table and must be encoded consistently: the set of state symbols must be encoded while satisfying the group and the ordering constraints simultaneously. The limitations and the encoding procedure is described in Section 6.

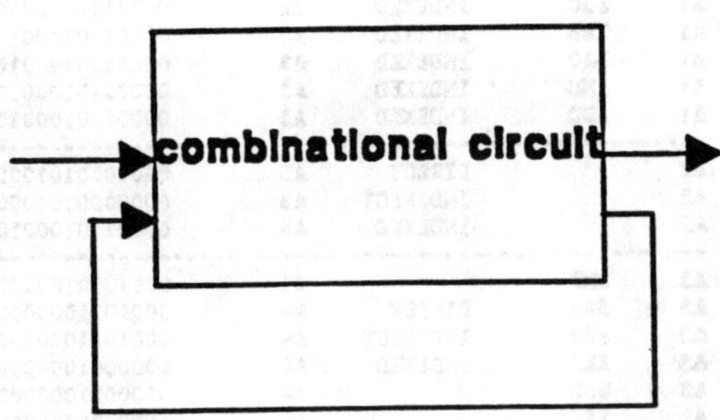

Fig 4.2 Sequential circuit.

STATE	OP-CODE	MODE	NEXT-STATE	CONTROL-SIGNALS
I1			I2	00000000000000111
I2			A1	00000000111011000
A1	JMP	DIRECT	I1	00000100000100100
A1	SRJ	DIRECT	A3	00000000001000001
A1	SAC	DIRECT	A4	00000010000000000
A1	ISZ	DIRECT	A4	00000000000000000
A1	LAC	DIRECT	A4	00000000000000000
A1	AND	DIRECT	A4	00000000000000000
A1	ADD	DIRECT	A4	00000000000000000
A1	JMP	INDIRECT	A2	00000000000000000
A1	SRJ	INDIRECT	A2	00000000001000001
A1	SAC	INDIRECT	A2	00000000000000000
A1	ISZ	INDIRECT	A2	00000000000000000
A1	LAC	INDIRECT	A2	00000000000000000
A1	AND	INDIRECT	A2	00000000000000000
A1	ADD	INDIRECT	A2	00000000000000000
A1	JMP	INDEXED	I1	00001101000100100
A1	SRJ	INDEXED	A2	00000000001000001
A1	SAC	INDEXED	A3	00001101000100000
A1	ISZ	INDEXED	A3	00001101000100000
A1	LAC	INDEXED	A3	00001101000100000
A1	AND	INDEXED	A3	00001101000100000
A1	ADD	INDEXED	A3	00001101000100000
A2		DIRECT	A3	00000001010001000
A2		INDIRECT	A3	00000001010001000
A2		INDEXED	A3	00001101000100100
A3	JMP		I1	00010101000000100
A3	SRJ	DIRECT	A4	00010110000000110
A3	SRJ	INDIRECT	A4	00010110000000110
A3	SRJ	INDEXED	A4	00000010000000111
A3	SAC		A4	00000010000000000
A3	ISZ		A4	00000000000000000
A3	LAC		A4	00000000000000000
A3	AND		A4	00000000000000000
A3	ADD		A4	00000000000000000
A4	JMP		E1	00000001010000001
A4	SRJ		I1	00000001000000001
A4	SAC		I1	00000001000000001
A4	ISZ		E1	00000000010000000
A4	LAC		E1	00000001010000001
A4	AND		E1	00000001010000001
A4	ADD		E1	00000001010000001
E1	LAC		I1	00110100000000000
E1	AND		I1	01000100000000000
E1	ADD		I1	00100100000000000
E1	ISZ		E2	00010100001000010
E2	ISZ		E3	10010110000000010
E3			I1	00000001000000101

Fig 4.3 State table of control unit.

Example 4.6: We consider here the finite state machine implementing the control unit of the microprocessor design described by Langdon in Chapter 5 of [LANG82]. The state table of Fig. 4.3 describes the memory-reference instructions only. The control-unit has nine states, corresponding to instruction-fetch, operand-address evaluation and instruction execution. The states are labeled by mnemonic strings, namely: I1, I2, A1, A2, A3, A4, E1, E2, E3. Seven operations are considered, namely: JMP (jump), SRJ (subroutine jump), SAC (store accumulator), ISZ (increase and skip on 0), LAC (load accumulator), AND (and), ADD (add). Three modes of memory addressing are considered: DIRECT, INDIRECT and INDEXED. The operation and addressing mode are specified by two instruction fields. Each row of the table shows an action as a consequence of particular conditions. There are five fields in each row. Two fields correspond to the present and next control states. Two fields correspond to the primary inputs, i.e. the operation code and addressing mode. The last field corresponds to the control signals. For the sake of clarity in Fig. 4.3, an unspecified field in the table corresponds to a *don't care* condition, i.e. to the specification of all possible words in the field. For example, the first row shows that when the control-unit is in state I1, the state of the control-unit at the next cycle is I2 and control signal 00000000000000111 (incrementing the program-counter) is issued. The third row shows that when the control unit is in state A1, the op-code is JMP and the addressing mode is DIRECT, then the next control-state is I1 and the control signal is 00000100000100100.

Eventually symbolic design can be applied to interconnected logic circuits. Consider two units, to be implemented by two-level logic macros, that communicate through a bus (Fig.4.4). If the representation of the information is transmitted across the bus is irrelevant to the design, symbolic optimization can be used as follows. The transmitting unit can be represented by a table with a symbolic output field and the receiving unit by another table with a symbolic input field. The tables corresponding to both units are optimized by symbolic minimization and the set of symbols, representing the communication signals, can be encoded as in the previous cases. Needless to say, this method can be extended to the interconnection among any number of modules, implementing combinational or sequential functions.

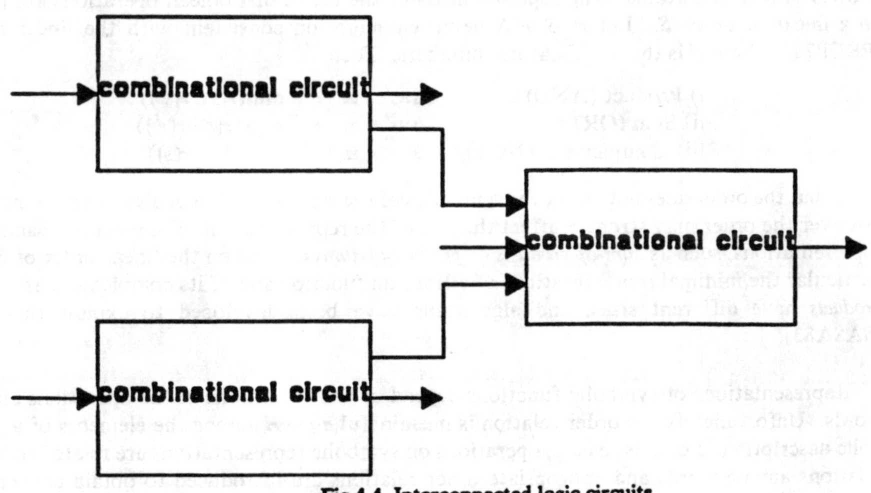

Fig 4.4 Interconnected logic circuits.

5. SYMBOLIC MINIMIZATION

5.1 Definitions

Symbolic functions are switching functions whose variables can take a finite set of values. Each value is represented by a **word** (or mnemonic), i.e. by a string of characters. A symbolic variable s has a set of admissible values S. The symbol ϕ is reserved to denote that variable s does not take any value of S. For functions of n input variables and m output variables, let S_i^I, $i = 1, 2, \ldots, n$ and S_i^O, $i = 1, 2, \ldots, m$ be the set of admissible values for the corresponding variables s_i^I and s_i^O. Then the domain of the symbolic function is the Cartesian product $S^I \equiv S_1^I \times S_2^I \times \cdots \times S_n^I$ and the range is the Cartesian product $S^O \equiv S_1^O \times S_2^O \times \cdots \times S_m^O$. A generic element of the domain is denoted by s^I and one of the range by s^O.

A completely specified symbolic function of n input variables and m output variables is a function $f\colon S^I \to S^O$, that maps each element of the domain to an element of the range. An incompletely specified symbolic function is a function having the property that, for some inputs, some output variables can take any value in the corresponding range. The collection of these points of the domain is called the *don't care* set of that particular output variable.

> **Example 5.1:** The truth table of Example 4.1 is a representation of a completely specified symbolic function with $n = 2$ inputs and $m = 1$ outputs. Here,
> $S_1^I = \{DIR, IND, INDEX\};$ $\qquad\qquad S_2^I = \{AND, OR, ADD, JMP\};$
> $S^O = \{CNTA, CNTB, CNTC, CNTD\}$.

Boolean or **binary-valued** functions are symbolic functions, whose variables can take the values $S = \{0, 1\}$. The domain is $\{0, 1\}^n$. The range of a completely specified function is $\{0, 1\}^m$. For incompletely specified functions, let the symbol * represent the *don't care* condition. Then the range is $\{0, 1, *\}^m$. Similarly, the variables of **multiple-valued** functions can take the values $S = \{0, 1, \ldots, p - 1\}$, where p is the radix of the representation [RINE77] [HURS84]. Algebras have been developed for both the Boolean and the multiple-valued [POST21] representations. The representation of the result of Boolean operations are based on a linear order of S. Let $r\colon S \to N$ be an enumeration consistent with the linear order [PREP73], where N is the set of natural numbers. Then:

$$
\begin{aligned}
&\text{i) Product (AND):} &s \wedge s' &\equiv r^{-1} \min(r(s), r(s')) \\
&\text{ii) Sum (OR):} &s \vee s' &\equiv r^{-1} \max(r(s), r(s')) \\
&\text{iii) Complement (NOT):} &\bar{s} &\equiv r^{-1} (p - 1 - r(s))
\end{aligned}
$$

Note that the order does not affect the semantics of the representation of a switching function; however the order may strongly affect the size of the representation. For example, canonical representations, such as *sum-of-products* or *product-of-sums* depend on the linear order of S : in particular the minimal representations of a Boolean function and of its complement as *sum of products* have different sizes, and algorithms have been developed to exploit this fact [SASA83].

Representations of symbolic functions depend on the definitions of the operations among words. Unfortunately, no order relation is meaningful *a priori* among the elements of a symbolic description. For this reason, operations on symbolic representations are related to order relations among words and appropriate order relations are introduced to obtain convenient representations of symbolic functions. In the following presentation, single-output functions are considered first, i.e. $m = 1$, to simplify the notations. The extension to multiple-output functions is then shown.

Symbolic functions are represented here in a particular canonical notation: *sum of products* or more exactly *sum of products of symbolic literal functions*. Let S be the set of admissible values for a variable s. A **symbolic literal** is a non-empty subset $\sigma \subseteq S$. For any variable $s \in S$, the **symbolic literal function** is defined as follows:

$$l(s, \sigma) \equiv \begin{matrix} \text{TRUE if } s \in \sigma \\ \text{FALSE else} \end{matrix}$$

Example 5.2: Consider the set $S_1^I = \{DIR, IND, INDEX\}$ corresponding to the set of admissible values of the first input variable in Example 5.1. An example of a symbolic literals is $DIR\ IND$. The corresponding literal function is TRUE for either $s = DIR$ or $s = IND$.

By using a *sum of product of symbolic literal functions* representation, only the order in S^o affects the representation, because the literal function maps words into the pair of values (TRUE,FALSE) independently of the order in S^I. Note that if a linear order relation is applied on the range, symbolic function representations in this canonical form are equivalent to multiple-valued logic function representations [RINE77] in the same form. In particular, a multiple-valued representation can be obtained from a symbolic representation by interchanging each symbol s with $r(s)$, where $r(\bullet)$ is the appropriate enumeration.

Since an order in S^o is not necessarily given, the definitions of the representation of a symbolic function are compatible with a set, possibly empty, of **partial order relations** among the elements of the range. Let $R = \{(s, s');\ s, s' \in S^o\}$ be a partial order on S^o. We say that s covers s' if either $s = s'$ or $s' = \phi$ or $(s, s') \in TR$, where TR is the transitive closure of R [AHO74]. The **symbolic sum** of two words s and s' is well-defined only if a covering relation exists among the elements involved. In particular:

$$s \vee s' = \begin{matrix} s & \text{if } s \text{ covers } s' \\ s' & \text{if } s' \text{ covers } s \end{matrix}$$

Else the symbolic sum is ambiguous or ill-defined.

A **symbolic product-term** (or **symbolic product**) of literals is the $n + 1$-tuple $(\sigma_1, \ldots, \sigma_n, \tau)$, where $\sigma_i \subseteq S_i^I, i = 1, 2, \ldots, n;\ \tau \in S^o$. The word τ is called the **output-part** of the literal. A **symbolic product** $p(s^I, \tau)$ of literal functions $l_i(s_i^I, \sigma_i),\ i = 1, 2, \ldots, n$ is a function:

$$p(s^I, \tau) = \begin{matrix} \tau & \text{if } l_i(s_i^I, \sigma_i) = \text{TRUE}, \ \forall i = 1, 2, \ldots, n \\ \phi & \text{else} \end{matrix}$$

Example 5.3: A symbolic product of the function in Example 2.1 is:

$$DIR\ \ AND\ \ CNTB$$

The symbolic product function takes the value $CNTB$ when the two inputs take the values DIR and AND respectively.

Two products p_1, p_2 intersect ($p_1 \cap p_2 \neq \phi$) if $\exists s^I \in S^I$ such that $p_1(s^I, \tau_1) \neq \phi$ and $p_2(s^I, \tau_2) \neq \phi$. Two sets of products, P_1 and P_2 intersect ($P_1 \cap P_2 \neq \phi$) if $\exists p_1 \in P_1$ and $\exists p_2 \in P_2$ such that p_1 and

p_2 intersect. Two products are **output-disjoint** if either they do not intersect or they have the same output-part, i.e. $p_1(s^J, \tau_1)$ intersects $p_2(s^J, \tau_2)$ implies $\tau_1 = \tau_2$. A set of products is output-disjoint if the products are pair-wise output-disjoint.

Example 5.4: Consider the two symbolic products:

$$
\begin{array}{llll}
DIR\ IND & ADD & & CNTC \\
DIR & ADD & JMP & CNTC
\end{array}
$$

The two symbolic product intersect because, for the symbolic input $s^J = DIR\ ADD$, the corresponding symbolic product functions specify a symbolic value. Since both symbolic product functions take value $CNTC$, the product terms are output-disjoint.

A symbolic function can be represented in a *sum of product* form, if $\forall s^J \in S^J$ for which the function is specified, the operation of symbolic sum among products is well-defined. In particular, such a representation always exists in the following two cases: i) for any linear order on S^o; ii) if the representation is a sum of pair-wise output-disjoint products. In the former case symbolic sum is always well-defined because a covering relation is defined between each pair of symbols in S^o. In the latter, only symbolic sum of identical values is required.

Sum of product representations are conveniently represented in tabular forms, as a stack of product-terms. A **symbolic implicant** is a symbolic product $p(s^J, \tau)$ such that $\forall s^J \in S^J$ for which the symbolic function is specified, $f(s^J)$ covers $p(s^J, \tau)$. A **symbolic cover** of a symbolic function is a set of implicants $P = \{p_1, p_2, \ldots, p_{|P|}\}$ whose sum is $f(s^J)$, $\forall s^J \in S^J$ for which the symbolic function is specified. Since symbolic sum depends on the order R on S^o, we denote a symbolic cover by the pair $C(P, R)$. The **cardinality** of a symbolic cover is $|P|$ and depends on R. A minimum symbolic cover of a symbolic function is a cover of minimum cardinality. A **minimal** (local minimum) symbolic cover of a symbolic function is a cover such that no proper subset is a cover of the same function.

Example 5.5: The following table is a symbolic cover of the function specified in Example 5.1.

$$
\begin{array}{lll}
INDEX & AND\ OR\ ADD\ JMP & CNTA \\
DIR & AND\ OR & CNTB \\
IND & AND & CNTB \\
IND & OR\ JMP & CNTD \\
DIR\ IND & ADD & CNTC \\
DIR & JMP & CNTC
\end{array}
$$

Note that the product-terms are pair-wise output-disjoint. Therefore this representation is output-disjoint and is compatible with any set R of partial order relations on S^o, and in particular the empty set. The following table is another symbolic cover of the function specified in Example 5.1.

$$
\begin{array}{lll}
INDEX & AND\ OR\ ADD\ JMP & CNTA \\
DIR\ IND & AND\ OR & CNTB \\
DIR\ IND & ADD\ JMP & CNTC \\
IND & JMP\ OR & CNTD
\end{array}
$$

Here, $R = \{(CNTD, CNTB); (CNTD, CNTC)\}$. Note that the fourth product-term has an intersection with the second and third one and these products are not output-disjoint. By choosing this particular partial order, the cover cardinality is reduced by two. Moreover, note that the first implicant can be removed by assuming that $CNTA$ is the default output, as pointed out in Example 5.5.

5.2 Symbolic minimization

Symbolic minimization is a procedure that attempts to determine a symbolic cover of a symbolic function in a minimum number of product-terms. Finding a minimum symbolic cover is a difficult task. An analysis of the computational complexity of the problem has not been done yet. However we conjecture that any method to find a minimum cover should involve the solution of a covering problem, which is a NP-complete problem [GARE78]. Therefore heuristic algorithms are used to determine a minimal (local minimum) solution. It is important to remark that recent progress in heuristic logic minimization has led to techniques which very often yield minimum solutions in the binary [HONG74] [BRAY84b] and multiple-valued [RUDE85a] case. We assume that the reader is familiar with heuristic logic minimization [HONG74], [BRAY84b]. Since most of the routines in the symbolic minimization algorithm are based on logic minimizer ESPRESSO-II, we refer the reader to [BRAY84b] for details.

Symbolic minimization is achieved by an iterative improvement of the initial cover $C^o(P^o, R^o)$. The symbolic function is described as input by the set of products P^o, while R^o is an empty set of ordering relations, because no order relation is meaningful *a priori* among the elements of a symbolic description. Therefore P^o is always a set of output-disjoint products. The main idea of symbolic minimization is to generate the order R during the minimization process. The symbolic minimizer detects partial order relations that are necessary to define sums of product-terms which would decrease the symbolic cover cardinality. As a result the order relations are determined *a posteriori* by the minimizer. The output of the minimizer is a minimal cover $C(P, R)$.

Symbolic minimization is a very complex technique and is fully documented in [DEMI86a]. To give a flavor of this technique, we present here a simplified version of the main loop of the symbolic minimization algorithm, which applies to symbolic functions with one symbolic output only. In this perspective, symbolic minimization is achieved by an iterative loop, that uses a multiple-valued-input, binary-valued output minimization procedure. This procedure can be regarded as a black box, that takes as input the representation of a multiple-valued-input function and its *don't care* set and returns a minimal representation. Computer programs ESPRESSO-II [BRAY84b], ESPRESSO-MV [RUDE85a] and MINI [HONG74] can be used in this regard.

It is convenient to represent the partial order R by a directed acyclic graph $G(V, A)$, where the vertex set V is in one-to-one correspondence with S^o. The edge set A is initialized empty and is constructed during the minimization process. An edge between two vertices defines a order relation between the corresponding elements of S^o. Therefore the sum of two distinct elements of S^o is well-defined if there is a directed path between the corresponding vertices.

Let us arbitrarily label the elements of the range: $S^o = \{s_i^o; i = 1, 2, \ldots, q\}$. Let $ON_i, i = 1, 2, \ldots, q$ be the subset of the initial set of product-terms P^o consisting of the product-terms whose output part $\tau = s_i^o$. Note that the set ON_i does not intersect the set ON_j if $i \neq j$, because P^o is output-disjoint. Each set ON_i specifies a symbolic single-output single-valued function. The original symbolic function can be seen as a collection of q multiple-valued-input, binary-valued-output functions whose *on* set corresponds to the points of the domain mapped into s_i^o, whose *off* set corresponds to those points mapped into s_j^o , $j{\neq}i$, and whose *don't care* set corresponds to the unspecified points [BRAY84b]. A representation of each set ON_i by a minimal number of product-terms, denoted here by M_i , can be obtained $\forall i = 1, 2, \ldots, q$ by using a multiple-valued-input, binary-valued-output minimization technique. In principle, by performing q minimizations in this way, a minimal cover of the original function can be computed as $P = \cup_{i=1}^{q} M_i$, i.e. as a collection of the q minimal covers M_i. It is shown in [BRAY84b] how to perform the q minimizations simultaneously. This procedure is called here **disjoint minimization**, because the minimal cover P of any completely specified function is output-disjoint.[3] Disjoint minimization does not exploit the benefit of choosing an order R to minimize the cover, and therefore is a weak optimization technique.

The main idea of symbolic minimization is that the cover P is not constrained to be output-disjoint, by introducing appropriate order relations among the elements of S^o. For example, suppose that $(s_j^o, s_i^o) \in R$. Then any point of the domain represented by ON_j can be used to reduce the cardinality of ON_i in the minimization process. In the minimal symbolic representation, such point is still mapped into s_j^o because $(s_j^o, s_i^o) \in R$. In other words, the subset of the domain represented by ON_j is a part of the *don't care* set while minimizing ON_i. We represent the *don't care* set by the set of product-terms DC_i In this case, $ON_j {\subseteq} DC_i$.

Example 5.6: Consider the first cover of Example 5.5. Let $s_i^o = CNTB$ and $s_j^o = CNTD$. Then, ON_i is:

DIR	AND OR	CNTB
IND	AND	CNTB

Suppose $(CNTD, CNTB) \in R$. Then DC_i includes:

IND	OR JMP	CNTD

Therefore the point of the domain $s^I = IND$ OR can be used to reduce the cardinality of ON_i, that can be represented as:

DIR IND	AND OR	CNTB

To minimize ON_i, an explicit representation of the corresponding *don't care* set is needed. Equivalently, the *off* set can be specified and the *don't care* set obtained by complementation of the *on* and *off* sets. To take advantage of the order relations, we use a definition of the *off* set different from that used in [BRAY84b] and mentioned before. For our purposes, the *off* set corresponding to ON_i is the subset of S^I that is mapped by the function f to a value different than s_i^o and covered by s_i^o, because $\forall s^I \in S^I$ s.t. $f(s^I){\neq}s_i^o$ and $f(s^I)$ is covered by s_i^o, $p(s^I, s_i^o)$ is not an implicant of the function. If $G(V, A)$ represents the partial order, then the *off* sets can be defined as a set of product-terms: $OFF_i = \cup_J ON_j$; $J = \{ j \text{ s.t.} \exists \text{ a path from } v_i \text{ to } v_j \text{ in } G(V, A)\}$.

At each iteration of the symbolic minimization loop, M_i is obtained by minimizing ON_i, using a routine that performs multiple-valued-input binary-valued output minimization. We invoke the minimization routine with the pair (ON_i, OFF_i), so that the corresponding *don't care* set DC_i , computed by the minimizer by complementation, includes by construction all the sets ON_j for which no path exists in $G(V,A)$ from v_i to v_j. As a result, minimization may be very efficient in reducing the cardinality of ON_i because of the particularly advantageous *don't care*

---

[3]   If the original function is incompletely specified, the minimal cover P is still output-disjoint if we restrict the definition of intersection among implicants by considering S^I as the *care* set of the function.

set. If M intersects ON_j, the relation (s_j^o, s_i^o) is recorded, by adding (v_j, v_i) to the edge set of the graph.

SYMBOLIC MINIMIZATION LOOP

```
Data ON_i, i = 1, 2, ... , q;
Data G(V, A);
A = φ; P = φ ;
for (k = 1 to q) {
    i = select (k) ;
    OFF_i = U_j ON_j;     J = {j | ∃ a path from v_i to v_j in G(V, A) };
    M_i = minimize (ON_i, OFF_i) ;
    A = A ∪ {(v_j, v_i) s.t. M_i ∩ ON_j ≠ φ} ;
    P = P ∪ M_i ;
};
```

Procedure **select** sorts the sets ON_i according to a heuristic criterion. Procedure **minimize** is a call to a multiple-valued-input binary-valued-output minimizer. The algorithm generates a set of symbolic products $P = \cup_{i=1}^q M_i$ and the directed graph $G(V, A)$. It is proven in [DEMI86a] that the graph $G(V, A)$ generated by the symbolic minimization loop is acyclic (i.e. R represents a partial order relation on S^o) and that $C(P, R)$ is a minimal cover of the original symbolic function represented by $C^o(P^o, R^o)$.

The order R depends on the heuristic sorting of the sets ON_i , done by procedure **select**. As a result, the cardinality of the cover $C(P, R)$ generated by the algorithm strongly depends on this routine.

The simplified algorithm described above invokes q times the minimization procedure. Its computational complexity grows linearly with the number of elements in the range of the function. The minimization procedure is a heuristic procedure: no theoretical bounds on the computational complexity have been proven for heuristic minimization; however exper- imental results have shown that it is practical to minimize a wide range of logic functions [HONG74] [BRAY85] and therefore the algorithm can be used in this perspective.

Several heuristics have been tried. Note that as more edges are added to the graph, it is more likely that the *off* sets become large and the *don't care* sets small. An effective heuristic is to sort the sets ON_i in descending order of cardinality, so that the largest sets will benefit from large *don't care* sets. A key ingredient for an effective reduction of the cover minimality is that the graph should be kept as sparse as possible by introducing only the ordering relations needed to reduce the cover cardinality. Keeping the graph sparse corresponds to keep many degrees of freedom to order S^o in the later iterations of the algorithm. Note that if **minimize** is a "standard" minimization algorithm, it aims at reducing both the product and literal cardinality in each ON_i. Therefore, a local optimization of the number of literals in the mini- mize procedure may introduce new edges in the graph and reduce the likelihood of reducing the symbolic cover cardinality at a later step. For these reason, the symbolic minimization loop has to be modified for efficiency, by tuning the **minimize** routine to the symbolic min- imization problem. To obtain an efficient algorithm for symbolic minimization it is necessary to "open the black box" and examine more carefully the operations that procedure **minimize** performs. We refer the interested reader to [DEMI86a] for details.

6. ENCODING PROBLEMS AND ALGORITHMS

6.1. Encoding problems

Symbolic minimization is used as an intermediate step in solving problems P1-P4 of Section 1. Since the final result must be a binary-valued logic circuit implementation, the symbolic representation has to be translated into a binary-valued (Boolean) one. If a multiple-valued circuit implementation technology were available (including logic gates implementing the literal function [RINE77]), then the (minimal) symbolic representation could be mapped into a multiple-valued representation with the same cardinality by interchanging the words s with $r(s)$, where $r(\bullet)$ is an appropriate enumeration. If a partial order relation exists on a set of words, then the enumeration must be consistent with it.

Let us consider first problems P1-P3. The goal of the following encoding technique is to find a binary-valued *sum of products* representation of the switching function with as many product-terms as the (minimal) symbolic representation. To construct such a Boolean cover, it is sufficient to determine: i) an encoding of the words related to each symbolic input variable such that each symbolic implicant can be represented by one Boolean implicant; ii) an encoding of the words related to each symbolic output variable that preserves the covering relations, i.e. such that the encoding of the sum of any subset of symbolic products is the sum of the corresponding Boolean products. For this reason we consider two encoding problems: the former is related to the encoding of the symbols representing the input variables; the latter to the encoding of the output variables.

Let us consider first the encoding of the input variables. If a variable can take at most two symbolic values, it has a trivial Boolean encoding. In the general case, a variable that can take more than two symbolic values is represented by more than one binary-valued variables. Then, to achieve the goal of encoding each symbolic implicant by one Boolean implicant, we must represent each symbolic literal by one product of Boolean literals. A product of Boolean literals is called **cube** or **face**, because it is a subspace of the Boolean hyperspace that can be represented by a hypercube. Therefore the encoding of the words must be such that each symbolic literal can be represented by a face (Boolean cube) that is a subspace of the Boolean space that contains the encoding of all and only the symbols in the literal [DEMI85a]. If *don't care* words are specified in that literal, than it is indifferent whether the face representing the literal contains the encoding of the *don't care* words or not.

The problem of encoding the words related to the output variables is different, because the output-part of the symbolic implicants corresponding to an output variable consists of one word only (and not of a symbolic literal with possibly more than one word, as in the case of the input variables). However, the encoding of the words related to the output variables must be such that the covering relations are preserved while transforming the symbolic cover into a Boolean cover. Therefore the encoding must be such that, for any two words joined by an order relation, the corresponding encoding are linked by a covering relation, i.e. the first word covers bit-wise the second word.

The encoding problem derived from problem P4 has an additional constraint. In this case, there is one set (ore more sets) of words corresponding to both input and output variables. The encoding of this set of words must satisfy the requirements for the encoding of the input variables and the output variables simultaneously.

We present now formally the encoding problems. Let S be a set of words to be encoded and let $n_s = |S|$. Let n_p be the cardinality of the (minimal) symbolic cover. Let n_b the encoding length, i.e. the number of Boolean variables used to represent S. The encoding problem is studied using matrix notation. Some matrices we consider have pseudo-Boolean entries from the set: $\{0, 1, *, \phi\}$ where $*$ represents the *don't care* condition (i.e. either 1 or 0) and ϕ represents the empty value (i.e. neither 1 nor 0). Logical product and sum on pseudo-Boolean variables is defined as follows:

$$
\begin{array}{c|cccc}
\Lambda & 0 & 1 & * & \phi \\
\hline
0 & 0 & \phi & 0 & \phi \\
1 & \phi & 1 & 1 & \phi \\
* & 0 & 1 & * & \phi \\
\phi & \phi & \phi & \phi & \phi
\end{array}
\qquad
\begin{array}{c|cccc}
V & 0 & 1 & * & \phi \\
\hline
0 & 0 & * & * & 0 \\
1 & * & 1 & * & 1 \\
* & * & * & * & * \\
\phi & 0 & 1 & * & \phi
\end{array}
$$

Let S be the set of values taken by a symbolic input variable. Let us consider the set of literals in the (minimal) symbolic cover related to that variable. The **word–literal incidence matrix** A (or incidence matrix in short) is a matrix: $A \in \{0, 1, *\}^{n_p \times n_s}$

$$
A = \begin{bmatrix} a_{1\cdot} \\ a_{2\cdot} \\ \dots \\ a_{n_p\cdot} \end{bmatrix} = \begin{bmatrix} a_{\cdot 1} | a_{\cdot 2} | \dots | a_{\cdot n_s} \end{bmatrix} = \{a_{ij}\} \quad \text{where: } a_{ij} = \begin{cases} 1 & \text{if word } j \text{ belongs to literal } i \\ * & \text{if word } j \text{ is a } don\,t\,care \text{ word in literal} \\ 0 & \text{else} \end{cases}
$$

where a *don't care* word in a literal is a word that may appear or not in a literal without affecting the function representation [DEMI86a].

> **Example 6.1:** Let S be the set of operation-codes in the symbolic function specified in Examples 4.1 and 5.5, i.e. $S = \{AND, OR, ADD, JMP\}$. Consider the minimal symbolic cover of Example 5.5. Then:
>
> $$
> A = \begin{bmatrix} 1111 \\ 1100 \\ 0011 \\ 0101 \end{bmatrix}
> $$

Let now S be the set of values taken by a symbolic output variable. The **partial order adjacency matrix** $B \in \{0, 1\}^{n_s \times n_s}$ (or adjacency matrix in short) is the adjacency matrix of the graph representing the transitive closure of the partial order R. If word i covers word j, then $b_{ij} = 1$. If word i covers word j and word j covers word k, then $b_{ij} = 1$, $b_{jk} = 1$ and $b_{ik} = 1$. Since covering is a transitive relation, we represent directly all the implied covering relations by matrix B. Moreover, since it is trivial that each word covers itself,we choose not to represent it by convention, i.e. $b_{ii} = 0$, $i = 1, 2, \dots, n_s$.

> **Example 6.2:** Let S be the set of controls in the symbolic function specified in Examples 4.1 and 5.5. Consider the minimal symbolic cover of Example 5.5. Then:
>
> $$
> B = \begin{bmatrix} 0000 \\ 0000 \\ 0000 \\ 0110 \end{bmatrix}
> $$

The **encoding matrix** E is a matrix $E \in \{0, 1\}^{n_s \times n_b}$

$$E = \begin{bmatrix} e_{1\bullet} \\ e_{2\bullet} \\ \cdots \\ e_{n_s\bullet} \end{bmatrix} = \begin{bmatrix} e_{\bullet 1} | e_{\bullet 2} | \cdots | e_{\bullet n_b} \end{bmatrix}$$

whose rows are the encodings of the words.

Definition 6.1: Let $a \in \{0, 1, *, \phi\}$ and $x \in \{0, 1, *, \phi\}$. The **selection** of x according to a is:

$$a \bullet x = \begin{matrix} x \text{ if } a = 1 \\ \phi \text{ else} \end{matrix}$$

Selection can be extended to two dimensional arrays and is similar to matrix multiplication.

Definition 6.2: Let $A \in \{0, 1, *, \phi\}^{p \times q}$ and $X \in \{0, 1, *, \phi\}^{q \times r}$. The **matrix pseudo-Boolean selection** is:

$$A \bullet X = C = \{c_{ij}\}^{p \times r}$$

where: $c_{ij} \equiv V_{k=1}^{q} a_{ik} \bullet x_{kj}$ or equivalently $c_{ij} \equiv a_{i1} \bullet x_{1j} V a_{i2} \bullet x_{2j} V \ldots V a_{iq} \bullet x_{qj}$.

Let us consider the problem of encoding the symbolic input variables first. This problem was presented in [DEMI85a] for the first time. We report here the most relevant results in a more general formulation. We represent the encoding of the symbolic literals by the **face matrix** $F \in \{0, 1, *, \phi\}^{n_p \times n_b}$

$$F = \begin{bmatrix} f_{1\bullet} \\ f_{2\bullet} \\ \cdots \\ f_{n_l\bullet} \end{bmatrix}$$

Each row of F is a face of the n_b-dimensional Boolean hypercube and corresponds to the face that encodes the symbolic literal. The face matrix can be obtained by performing the matrix pseudo-Boolean selection of E according to an incidence matrix A:

$$F = A \bullet E$$

Example 6.3: Consider the incidence matrix of Example 6.1 and the encoding of Example 4.4. Then:

$$E = \begin{bmatrix} 00 \\ 01 \\ 10 \\ 11 \end{bmatrix} \qquad F = A \bullet E = \begin{bmatrix} ** \\ 0* \\ 1* \\ *1 \end{bmatrix}$$

Let now $\bar{A} = \{\bar{a}_{ij}\}$ where $\bar{a}_{ij} = 1$ if $a_{ij} = 0$; else $\bar{a}_{ij} = 0$. Then $\bar{F}^i \equiv \bar{a}_{\bullet i} \bullet e_{i\bullet}$ is a matrix whose rows are: i) the encoding of word i, if word i neither belongs to the symbolic literal j nor is a *don't care* word; ii) empty values. An encoding matrix E is said to satisfy the **input constraint relation** for a given incidence matrix A if:

$$\bar{F}^i \; \Lambda \; F \;\; \equiv \;\; \begin{bmatrix} \bar{f}^i_{1.} \Lambda f_{1.} \\ \bar{f}^i_{2.} \Lambda f_{2.} \\ \cdots \\ \bar{f}^i_{n_l} \Lambda f_{n_l} \end{bmatrix} \;=\; \Phi \quad \forall i = 1, 2, \ldots, n_s$$

where Φ is the empty matrix, i.e. a matrix whose rows have at least one ϕ entry and therefore representing no point in the Boolean space.

Example 6.4: The encoding matrix of Example 6.3 satisfies the input constraint relation. However if we swap the first two rows of E, the input constraint relation is no longer satisfied, because the encoding of the third word, ADD, intersects the fourth face, or equivalently:

$$\bar{F}^3 \; \Lambda \; F \;=\; (\bar{a}_{.3} \bullet e_{3.}) \; \Lambda \; (A \bullet E) \;=\; (\begin{bmatrix} 0 \\ 1 \\ 0 \\ 1 \end{bmatrix} \bullet [\,10\,]) \; \Lambda \; (\begin{bmatrix} 1111 \\ 1100 \\ 0011 \\ 0101 \end{bmatrix} \bullet \begin{bmatrix} 01 \\ 00 \\ 10 \\ 11 \end{bmatrix}) \;=$$

$$=\; \begin{bmatrix} \phi\phi \\ 10 \\ \phi\phi \\ 10 \end{bmatrix} \Lambda \begin{bmatrix} ** \\ 0* \\ 1* \\ ** \end{bmatrix} \;=\; \begin{bmatrix} \phi\phi \\ \phi 0 \\ \phi\phi \\ 10 \end{bmatrix} \neq \Phi$$

The problem of encoding the values taken by a symbolic input variable is equivalent to finding an encoding matrix satisfying the input constraint relation. An optimal solution is one of minimal encoding length. Therefore we can state:

Encoding problem E1: Given an incidence matrix A, find an encoding matrix E with minimal number of columns that satisfies the input constraint relation.

Let us consider now the problem of encoding the output variables.

Definition 6.3: Let $A \in \{0, 1\}^{p \times q}$ and $X \in \{0, 1\}^{q \times r}$. The **matrix Boolean selection** is:

$$A \circ X = C = \{c_{ij}\}^{p \times r}$$

where: $c_{ij} \equiv V^q_{k=1} \; a_{ik} \; \Lambda \; x_{kj}$ and the sum (V) and product (Λ) operators on Boolean variables have the usual meaning:

V	0	1		Λ	0	1
0	0	1		0	0	0
1	1	1		1	0	1

Let now $G = B \circ E$. Row i, $i = 1, 2, \ldots, n$, of matrix G is the logical sum of the encoding of the words that must be covered by the encoding of word i. Therefore we say that a matrix E satisfies the **output constraint relation** for a given adjacency matrix B, if E covers G or equivalently:

$$\bar{E} \; \Lambda \; G \;=\; O$$

where $\overline{E}$ is the Boolean complement of E and O is the matrix of 0 entries.

In this case, the problem of encoding the values taken by a symbolic output variable is equivalent to finding an encoding matrix satisfying the output constraint relation. An optimal solution is one of minimal encoding length. Therefore we can state:

> **Encoding problem E2:** Given an adjacency matrix B representing to a partial order, find an encoding matrix E with minimal number of columns that satisfies the output constraint relation.

The solution of problem P1 (P2 or P3) requires the solution of encoding problem E1 (E2 or both) for each symbolic variable, after symbolic minimization. The solution of problem P4 requires the encoding of one set (or more sets) of words corresponding to both input and output variables, after symbolic minimization.

> **Encoding problem E3:** Given an incidence matrix A and an adjacency matrix B representing a partial order, find an encoding matrix E with minimal number of columns that satisfies both the input and the output constraint relation.

The existence of solutions to the encoding problems E1, E2 and E3 was shown in [DEMI85a] and [DEMI86a]. In particular, there are always encoding that satisfy either the input or the output constraint relation and that can be derived from matrices A and B. However, in general, these encodings are not solutions to problems E1 and E2, because E does not have a minimal number of colums n_b. Problem E3 admits a solution under a restrictive assumption on matrices A and B [DEMI86a]. Therefore, there exist cases in which some constraints have to be relaxed to be able to find a solution [DEMI86A].

6.2. Encoding algorithms

A solution to the encoding problems E1, E2 or E3 is an encoding of minimal length the satisfies the input, output or both constraint relations. Unfortunately, these are computationally complex problems of combinatorial optimization and it is not known whether an optimal solution can be computed by non-enumerative procedures. Since the growth of computation-time as the size of the problem increases is a practical limitation to computer-aided design systems, we consider here heuristic algorithms for the solution of the above problems. Experimental results show that the encodings constructed by these algorithms have reasonably short length, and often equal to the minimum length solution when this is known.

The heuristic techniques presented below solves the encoding problems using greedy strategies. An encoding matrix E is grown from an initial seed matrix by concatenating rows and/or columns. At each step, the best local concatenation is computed, while keeping the previously computed encoding matrix as such. As a result the encoding matrix grows in size, until all the rows are encodings for the words in S that satisfy the constraint relations. The heuristic selections that drive the algorithm attempt to minimize the steps that increase the number of columns to obtain a solution, and therefore guarantee a weak optimality.

There are two major approaches to constructing matrix E: a **row-based** method and a **column-based** one. In the former method, the encoding matrix is constructed row by row, i.e. by computing the encoding of the words one at a time [DEMI85a]. In the latter approach, the encoding matrix is constructed column by column, i.e. by computing one bit of the encoding of all the words at each pass. This idea was introduced first by Dolotta [DOLO64], to solve the optimal state assignment problem, though the encoding problem was stated differently.

A row-based encoding method was presented in [DEMI85a], to solve problem E1. This technique is also effective for solving problem E2. The algorithm can be sketched as follows:

STEP 1: Select a word not yet encoded.
STEP 2: Determine the encodings for that word satisfying the constraint relation restricted to that word and to the previously encoded words.
STEP 3: If no encoding is found, increase the code dimension by adding a column to E and go to STEP 2.
STEP 4: Assign an encoding to the selected word by adding a row to E.
STEP 5: If all words have been encoded, stop. Else go to STEP 1.

We refer the reader to [DEMI85a] for the details of the algorithm when applied to problem E1 and to [DEMI86a] for problem E2. The row-based encoding algorithms generate short encodings (see [DEMI85a] for experimental results) but fail to be effective for large examples. In particular, the candidate encodings generated at STEP 2 may increase exponentially with the encoding length n_b. Therefore, when the encoding length increases, the computer implementation of the algorithm slows down considerably. Moreover, it is complex to handle problem E3 with this approach, because the effectiveness of the algorithm depends heavily on the heuristic ordering of words at STEP 1, and for both input and output constraints there may exist conflicting orderings.

The need to handle both constraints simultaneously, as well as a desired computational-time complexity linear in n_b, has lead to the development of a column-based algorithm. In a column-based algorithm a single-bit encoding of all the words is introduced at each step. While there always exist single-bit encodings of all the words that satisfy the output constraint relation, it is unlikely that such an encoding satisfies the input constraint relation. Therefore we say that, given an incidence matrix A, an encoding matrix E partially satisfies the input constraint relation, if E satisfies the input constraint relation for A', where A' is a subset of the rows of A. The satisfaction ratio is $n_{p'}/n_p$, where $n_{p'}$ is the row cardinality of A'. Then, the column-based encoding algorithm can be sketched as follows:

STEP 1: Select a column vector in $\{0, 1\}^{n_s}$ that satisfies the output constraint relation and corresponding to a maximal satisfaction ratio.
STEP 2: If at the first pass, let E be the selected vector. Else, append the selected column vector to E.
STEP 3: If E satisfies the input constraint relation, stop. Else go to STEP 1.

The algorithm is described in detail in [DEMI86a]. We mention some properties of the encoding problems that are relevant to understand the strategy of this method. If each column of E satisfies the output constraint relation, so does the entire matrix E and *vice versa*. This is not true for the input constraint relation. In particular, if E satisfies the input constraint relation, then a subset of columns of E may not satisfy it. However if E satisfies the input constraint relation for A', then by appending a column to E, the augmented matrix satisfies the input constraint relation for A' as well. In other words, the satisfaction ratio cannot decrease.

Therefore the strategy of the algorithm is to select columns that increase the satisfaction ratio until it reaches unity and the algorithm terminates. The selection of the column to be appended to E is the crucial part of the algorithm, because the goal is to construct a matrix E with minimal number of columns. It is shown in [DEMI86a] that when the algorithm is used to solve problem E1, by selecting as columns the transpose of the rows of A, we construct as a solution $E = A^T$, which is a valid solution if we replace the * entries by 0's or 1's. In other words, there exist a trivial column selection that increases the satisfaction ratio by at least $1/n_p$, i.e. one column is needed to satisfy the input constraint set by each row of A. On the other hand, if the algorithm is used to solve problem E3, there exists also a column selection that increases the satisfaction ratio by at least $1/2 \times n_p$, i.e. at most two columns will be needed to satisfy the input constraint set by each row of A, while satisfying any admissible output constraint. However, since the optimality of the solution is measured by the number of columns of E, we need column selections that maximize the increase of the satisfaction ratio. For this reason, the column selection routines looks for maximal subsets of input constraints (rows of A) that can be satisfied by one column selection. This corresponds to finding a column that maximizes the satisfaction ratio. If an upper bound n_{b_max} on the encoding length is specified, then it is imperative that the rows of E are different from each other after n_{b_max} column assignments. Suppose there are groups of identical rows in E with at most cardinality $2^{(n_{b\_max} - n_b)}$, after having assigned n_b columns. The column selection is done such that, if there are groups of identical rows in $[E \,|\, e]$, their cardinality is most $2^{(n_{b\_max} - n_b - 1)}$, where e is the column being appended. This requirement restricts the selection of the column vector e. We refer the interested reader to [DEMI86a] for the details.

The column-based encoding algorithm can be summarized as follows in a qualitative way. The algorithm constructs an encoding matrix E. If problem E2 or E3 have to be solved, each column is chosen as to satisfy the output constraint relation. Therefore so does the entire matrix E. If problem E1 or E3 have to be solved, each column is chosen as to satisfy partially the input constraint relation, or equivalently to satisfy the input constraint relation for a subset of the rows of A. These rows are then discarded from further consideration. After a finite number of steps, matrix E satisfies the input constraint relation. The heuristic selection of the column attempts to increase maximally the satisfaction ratio at each step, and therefore guarantees a weak minimality of the column cardinality of E. If an upper bound on the encoding length is specified, then the encoding may satisfy only partially the constraint relations, but attempts to satisfy most of the constraints.

The construction of matrix E with bounded column cardinality allows to trade off the minimality of the cover cardinality for that of the encoding length. In particular, given an encoding of length n_b, it is possible to determine the cover cardinality (or a bound on the cover cardinality n_p) on the basis of the satisfied constraints. Therefore, for a particular switching function, it is possible to determine a set of pairs of parameters (n_b, n_p) that relate to the size of the implementation.

As a final remark, we would like to compare the column-based encoding algorithm with the one proposed by Dolotta [DOLO64] and later perfected by Weiner [WEIN67b], Torng [TORN68] and Story [STOR72]. These algorithms addressed the state assignment problem for finite state machines and the mechanism of the encoding algorithm is similar to the one presented here. However the selection of the columns was based on a heuristic criterion: in particular a scoring function was used to select a column on the basis of the likelihood that logic minimization, which would have followed the encoding, could reduce the two-level cover cardinality. (Story's method optimized the number of and-or inputs for each column choice.)

In our approach we relate each column assignment to the satisfaction of some input constraints and therefore to a known reduction of the cover cardinality. Therefore column selection is related to cover minimality in a deterministic way. However our encoding technique is still a heuristic one, because the greedy strategy considers and assigns only one column at a time.

6.3. Implementation of the symbolic method and experimental results

The symbolic minimization algorithm has been implemented in a computer program called CAPPUCCINO, because it is based on the logic minimizer ESPRESSO-II. CAPPUCCINO is written in APL and incorporates a modified version of the ESPRESSO-II original program [BRAY84b]. CAPPUCCINO implements the symbolic minimization algorithm described in [DEMI86a] and sketched in Section 5.2. Table 6.1 summarizes the results:

RESULTS OF SYMBOLIC MINIMIZATION			
Example	Original cover cardinality	Minimal cover cardinality	Minimal disjoint-cover cardinality
EX1	24	9	16
EX2	91	49	57
EX3	170	78	78
EX4	11	5	8
EX5	25	10	11
EX6	24	17	24
EX7	115	89	94
EX8	107	57	92
EX9	184	106	115
EX10	16	14	15
EX11	166	102	111
EX12	49	11	12
EX13	25	10	11
EX14	20	8	13
EX15	56	23	24
EX16	32	15	16
EX17	108	46	55
EX18	32	17	18
EX19	14	9	10
EX20	30	22	23

Table 6.1

The first two numeric columns show the original and final cover cardinality. The last column shows the final cardinality obtained by disjoint-minimization, by using program ESPRESSO-II. In some cases (EX1, EX2, EX8, EX17, ...) CAPPUCCINO does significantly better than ESPRESSO-II in reducing the cover cardinality. In some others, the advantage of introducing covering relations among the output symbols is not a major factor in reducing the cover cardinality. These comparisons have to be understood with caution, because we are comparing different minimization techniques and not program performances.

Computing time ranges from few seconds to about 20 minutes for the largest example on an IBM 3081 computer. Note that APL is interpreted and therefore the execution is much slower with comparison to compiled code programs. Today, CAPPUCCINO is limited to covers of about 2000 symbolic product-terms, due to memory limitations of the APL workspace and computing time. A compiled code implementation of the algorithm based on the data structure and the routines of program ESPRESSO-MV [RUDE85a] (in place of the APL version of ESPRESSO-II) would definitely increase the capability and the performance of the program.

The column-based encoding algorithm has been implemented in program CREAM. CREAM is written in APL and consists of about 25 functions. CREAM is designed to be used with CAPPUCCINO: it takes the representation of a minimal symbolic cover and generates an encoding that can replace the symbolic entries. The final result is a Boolean cover that can be implemented as a PLA (after having been folded or partitioned, if desired), or used as a starting point for multiple-level synthesis, as done by the YLE program in the Yorktown Silicon Compiler [BRAY85a].

CREAM solves the encoding problem E1, E2 or E3 at user's request. It accepts an upper bound on the number of columns to be used in the encoding. For a given bound and the corresponding encoding of all the symbolic fields, it is possible to estimate the area taken by a PLA implementation. Therefore, it is possible to estimate the area as well as the aspect-ratio. This computation can be done for different bounds on the encoding length and therefore a designer can choose among several implementations with different areas and aspect ratios.

CAPPUCCINO and CREAM have been tested on several examples. Some results are reported in the Table 6 .2.

OVERALL RESULTS					
Example	n_i	n_s	n_o	n_p	$n_{p'}$	n_b
EX1	2	6	2	24	11	4
EX2	7	16	7	91	49	8
EX3	1	9	2	170	78	12
EX4	2	4	1	11	6	2
EX5	2	9	1	25	10	5
EX6	1	12	1	24	17	7
EX7	7	48	19	115	89	10
EX8	8	20	6	107	68	7
EX9	11	32	9	184	107	9
EX10	1	8	1	16	14	4
EX11	9	30	10	166	103	12
EX12	4	4	4	49	11	3
EX13	2	11	1	25	10	6
EX14	4	5	1	20	8	4
EX15	8	7	5	56	23	5
EX16	8	4	5	32	15	4
EX17	4	27	3	108	49	11
EX18	4	8	3	32	17	4
EX19	2	7	2	14	9	3
EX20	4	15	3	30	22	7

Table 6.2

The examples are sequential circuits, i.e. we are solving problem P4 by symbolic minimization first and by solving encoding problem E3 after. The symbolic representations have four fields corresponding to the primary inputs/outputs (which are represented by Boolean variables) and the present/next states (which are represented by symbolic variables). The first four numeric columns represent the parameters of the function: n_i is the number of primary inputs, n_s is the number of states, n_o is the number of primary outputs and n_p is the cardinality of the initial symbolic cover. The last two columns show the final cardinality of the Boolean cover and the state encoding length. Note that for a few examples, (as EX1 , EX4 , ...) the Boolean cover cardinality is larger than the minimal symbolic cardinality (reported in the third column of Table 6.1), because it was not possible to satisfy all constraints in the encoding. Note also that a further reduction in cardinality may be obtained by minimizing again the Boolean covers that are not minimal. The computing time is in the order of few seconds for all these examples, on an IBM 3081 computer.

In Table 6.3 we try to evaluate the optimality of the encoding, in terms of encoding length.

ENCODING-LENGTH COMPARISONS					
Example	$n_,$	$\log_2 n_,$	n_b	$n_{b'}$	$n_{b''}$
EX1	6	3	4	3	3
EX2	16	4	8	7	7
EX3	9	4	12	12	12
EX4	4	2	2	2	2
EX5	9	4	5	4	4
EX6	12	4	7	4	6
EX7	48	6	10	8	8
EX8	20	5	7	6	7
EX9	32	5	9	7	7
EX10	8	3	4	4	3
EX11	30	5	12	9	9
EX12	4	2	3	3	3
EX13	11	4	6	4	4
EX14	5	3	4	3	3
EX15	7	3	5	5	5
EX16	4	2	4	4	4
EX17	27	5	11	9	7
EX18	8	3	4	4	4
EX19	7	3	3	3	3
EX20	15	4	7	7	5

Table 6.3

The second column represents the number of words to be encoded $n_,$; the third the minimal number of bits needed to encode the words regardless of any constraint on their encoding (i.e. $ceiling(\log_2 n_,)$); the fourth column the encoding length as constructed by CREAM as a solution to problem E3. These three columns show that the encoding length computed by CREAM is close enough to the minimum length: for most of the examples $ceiling(\log_2 n_,) \leq n_b \leq 2 \times ceiling(\log_2 n_,)$. For some examples (as EX14, EX16, EX18, EX19, ...) it is possible to prove that no encoding of length inferior to n_b can be a solution of the encoding problem. The results of CREAM can be compared with those obtained by program KISS [DEMI85a], which uses a row-based encoding algorithm. The length of a solution to problem E1 computed by CREAM is given in the fifth column of the table and the length of an encoding constructed by KISS in the last column. (The algorithm of KISS can solve only problem E1.) Note that for some examples, the row-based algorithm gives a shorter encoding. However, the corresponding implementation requires more product-terms, as shown by the fourth column of Table 6.1.

7. CONCLUSIONS

We have reviewed some relevant problems in the area of computer-aided synthesis of circuits for control units of VLSI processors. Several algorithms and CAD programs have been developed for the logic and physical design task. However, circuit synthesis from behavioral-level specifications is still an exploratory area for CAD algorithms. The problems arising from optimizing the logic representation are mostly computationally intractable. Heuristic methods have been developed, but they still lack the capability of handling efficiently large designs.

The symbolic design methodology has been presented. It may be applied to the synthesis of interconnected combinational and/or sequential circuits. In particular, it can be used to find good solutions to the optimal state assignment problem. Symbolic minimization allows encoding-independent optimization of switching functions represented by tables of symbols and the encoding algorithms construct a binary representation of the symbols that translate the minimal symbolic cover into a compatible Boolean cover.

Exploratory work in the field of logic design of control circuits needs to address many unresolved problems. First, today's optimization methods must be extended to multiple-level logic representations and a theoretical framework for multiple-level synthesis must be defined. In particular, logic and symbolic minimization must be extended to multiple-level circuits. Then, the relations among different tasks of logic design must be understood and exploited, with particular reference to: i) the partitioning problem of a control unit into functional blocks; ii) the state minimization problem; iii) the selection of the type of registers to store the control-state information; iv) the optimal encoding and minimization of two-level and multiple-level switching functions. Eventually, an efficient CAD system must allow the designer to explore the trade-off among circuit structures, to find the best match between circuit architecture and implementation technology.

The task that CAD scientists are facing for improving computer design methods is very challenging: up to now we have only scratched the surface of a massive body of difficult problems.

8. REFERENCES

[AGER76] T. Agerwala "Microprogram Optimization: a Survey" IEEE Trans on Comput., vol. C-25, pp. 962-973, oct 1976.

[AHO74] A.V.Aho J.E.Hopcroft and J.D.Ullman , "The Design and Analysis of Computer Algorithms", Addison Wesley, 1974.

[ANDR80] M. Andrews, Principle of Firmware Engineering in Microprogram Control, Computer Science Press, 1980.

[BLAC85] R. Blackburn and D.Thomas, "Linking the Behavioral and Structural Domains of Representations in a Synthesis System", Proc. Des. Autom. Conf., Las Vegas, pp. 374-380, Jun 1985.

[BREN84] N.Brenner "The Yorktown Logic Language: an APL-like Design Language for VLSI Specifications" Int. Conf. on Circ. and Comp. Des., Rye, NY, pp.11-15, Sep 1984.

[BRAY84b] R.Brayton, G.D.Hachtel, C.McMullen and A.L.Sangiovanni- Vincentelli, "Logic Minimization Algorithms for VLSI Synthesis", Kluwer Academic Publishers, 1984.

[BRAY84d] R.Brayton, C.L.Chen, C.McMullen R.Otten and Y.Yamour "Automated Implementation of Switching Functions as Dynamic CMOS Circuits", Proc. Cust. Integr. Circ. Conf., Rochester, NY, pp. 346-350, may 1984.

[BRAY85a] R.Brayton, N.Brenner, C.Chen, G.De Micheli, C.McMullen and R. Otten "The Yorktown Silicon Compiler" Proc. Int. Symp. on Circuit and Systems, Kioto, Japan, pp. 391-394, Jun 1985.

[BRAY85c] R.Brayton, C.Chen, G.De Micheli, J.Katzenelson C.McMullen R. Otten and R.Rudell "A Microprocessor Design Using the Yorktown Silicon Compiler" Proc. Int. Conf. on Circuit and Comput. Des., Rye, N.Y., pp. 225-230, Oct 1985.

[CAMP85] R.Camposano, "Synthesis Techniques for Digital System Design", Proc. Des. Autom. Conf., Las Vegas, pp. 475-480, Jun 1985.

[CURT69] H.A. Curtis, "Systematic Procedures for Realizing Synchronous Sequential Machines Using Flip-Flop Memory: Part 1", IEEE Trans on Comput., vol. C-18, pp. 1121-1127, dec. 1969.

[CURT70] H.A. Curtis, "Systematic Procedures for Realizing Synchronous Sequential Machines Using Flip-Flop Memory: Part 2", IEEE Trans on Comput., vol. C-19, pp. 66-73, jan. 1970.

[DARR84] J.Darringer, D. Brand, J.Gerbi, W. Joyner and L.Trevillan, "LSS: A System for Production Logic Synthesis", IBM Journal of Research and Development, Vol. 28, No.5 pp.537-545, Sept. 1984.

[DAVI72] P. Davies "Readings in Microprogramming", IBM Jour. of Res. and Dev., vol. 11, No. 1, pp. 16-40, jan 1972.

[DEMI84a] G.De Micheli, M.Hofmann, A.Newton and A.L.Sangiovanni Vincentelli, "A Design System for PLA-based Digital Circuits" in "Advances in Computer-Aided Engineering Design", Jay Press, 1985.

[DEMI84c] G.De Micheli, R.Brayton and A.L.Sangiovanni Vincentelli, "KISS: a Program for Optimal State Assignment of Finite State Machines", Proc. ICCAD Santa Clara, Nov 1984.

[DEMI85a] G.De Micheli, R.Brayton and A.L.Sangiovanni Vincentelli, "Optimal State Assignment for Finite State Machines", IEEE Transactions on CAD/ICAS, Vol CAD-4, No. 3, pp.269-284, July 1985.

[DEMI85c] G. De Micheli, "Symbolic Minimization of Logic Functions", Int. Conf. on Comp. Aid. Des., Santa Clara, pp. 293-295, Nov. 1985.

[DEMI86a] G.De Micheli, "Symbolic Design of Combinational and Sequential Logic Circuits Implemented by Two-level Logic Macros", IEEE Transactions on CAD/ICAS, Oct 1986 (in print) and IBM Research Report RC 11672.

[DOLO64] T.A.Dolotta and E.J.McCluskey, " The coding of internal states of sequential machines", IEEE Trans. Elect. Comp., vol EC-13, pp. 549-562, Oct. 1964.

[GARE78] M.R.Garey and D.S.Johnson, "Computers and Intractability", W.H.Freeman and Company, San Francisco, 1978.

[GOLD85] A.Goldberg, S.Hirshhorn and K.Lieberherr, "Approaches toward Silicon Compilation" IEEE Circuits and Devices, Vol.1, No.3 , pp. 29-39, May 1985.

[GRAS70] A.Grasselli and U.Montanari, "On the minimization of read-only memories in microprogrammed digital computers" IEEE Trans. on Comput. pp. 1111-1114 Nov. 1970.

[HADS85] R.Hadsell, "Micro/370" in "Microarchitectures of VLSI Computers", Proceedings NATO ASI Series E, No. 96, Martinus Nijhoff, The Netherlands, 1985.

[HART66] J.Hartmanis and R.E.Stearns, "Algebraic Structure Theory of Sequential Machines", Prentice Hall, 1966.

[HILL81] F.Hill and G.Peterson, "Introduction to Switching Theory and Logical Design", Wiley, 1981.

[HONG74] S.J.Hong,R.G.Cain and D.L.Ostapko, "MINI: a Heuristic Approach for Logic Minimization", IBM Jour. of Res. and Dev., vol. 18, pp. 443-458, Sep. 1974.

[HOPC71] J.Hopcroft, "An n log n Algorithm for Minimizing States in a Finite Automaton", in "Theory of Machines and Computation", (Kohavi editor), pp.189-196, Academic Press, 1971.

[HURS84] S.K.Hurst, "Multiple-Valed Logic- Its Status and its Future" IEEE Trans on Comput., vol. C-33, No 12, pp. 1160-1179, Dec 1984.

[KOWA85] T. Kowalski and D.Thomas, "The VLSI Design Automation Assistant: What's in a Knowledge Base" Proc. Des. Autom. Conf., Las Vegas, pp. 252-258, Jun 1985.

[LEWI81] T. Lewis and B. Shriver, "Introduction to Special Issue on Microprogramming", IEEE Trans on Comput., vol. C-30, pp. 457-459, jul 1981.

[LANG82] G.Langdon, "Computer Design", Computech Press, 1982.

[MCCL56] E.J.McCluskey, "Minimization of Boolean Functions", Bell Laborat. Techn. Jour., vol. 35, pp. 1417-1444, Apr. 1956.

[NICH65] A.Nichols and A.Bernstein, "State Assignemnt in Combinational Networks", IEEE Trans on Elect. Comp., vol. EC-14, pp. 343-349, Jun. 1965.

[PAPA79] C. Papachristou, "A Scheme for Implementing Microprogram Addressing with Programmable Logic Arrays", Digital Processes No. 5 pp. 235-256 may 1979.

[PLEE73] C.Pleeger, "State Reduction of Incompletely Specified Finite State Machines", IEEE Trans. on Comput. pp. 1099-1102, 1973.

[POST21] E.L.Post, "Introduction to a General Theory of Elementary Propositions" Amer. J. Math., vol. 43, pp. 163-185, 1921.

[PREP73] F.Preparata and R.Yeh, "Introduction to Discrete Structures", Addison Weseley, 1973.

[REUS86] B.Reusch and W.Merzenich, "Minimal Coverings for Incompletely Specified Sequential Machines", Acta Informatica, No.22, pp. 663-678, 1986.

[RINE77] D.Rine, "Computer Science and Multiple-Valued Logic", North Holland, 1977.

[RUDE85a] R. Rudell and A. Sangiovanni-Vincentelli, "ESPRESSO-MV: Algorithms for Multivlued Logic Minimization" Proc. Custom Int. Circ. Conf., Portland, Oregon, May 85

[RUDE85b] R. Rudell, A. Sangiovanni-Vincentelli and G.De Micheli, "A FSM Synthesis System", Proc. Int. Symp. on Circ. and Sys., pp.647-650, Kyoto, Japan, Jun 85.

[SASA83] T.Sasao "Input Variable Assignment and Output Phase Assignment of PLAs", IBM Research Report, No. 1003, June 1983.

[SCHW68] S.Schwartz, "An algorithm for minimizing read-only memories for machine control" Proc. IEEE Symp. on Switch. and Autom. Theory, pp. 28-33, 1968.

[STOR72] J.R.Story H.J.Harrison and E.A.Reinhard, "Optimum State Assignment for Synchronous Sequential Circuits", IEEE Trans. on Comp., vol. C-21 pp. 1365-1373, Dec. 1972.

[SISK82] J.Siskind, J.Southard and K.Crouch, "Generating High Performance VLSI Designs from Succint Algorithmic Descritions", Proc Conf. on Adv. Res. on VLSI, pp. 28-40, 1982.v. 1970.

[SU72] S.Y.H.Su and P.T.Cheung, "Computer Minimization of Multi-Valued Switching Functions", IEEE Trans on Comput., vol 21, pp. 995-1003, 1972.

[TORN68] H.C.Torng, "An Algorithm for Finding Secondary Assignments of Synchronous Sequential Circuits", IEEE Trans on Comput., vol. C-17 pp. 416-469, May 1968.

[TRED81] N.Tredennick, "How to Flowchart for Hardware", IEEE Computer, pp. 87-102, Dec 1981.

[WEIN67b] P.Weiner and E.J.Smith, "Optimization of Reduced Dependencies for Synchronous Sequential Machines", IEEE Trans on Elect. Comp., vol. EC-16, pp. 835-847, Dec. 1967.

Towards Intelligent Silicon Compilation

Daniel D. Gajski
Forrest D. Brewer

Dept. of Computer Science, University of Illinois at Urbana-Champaign
1304 West Springfield Ave., Urbana, Illinois 61801

1. Introduction

Historically, the design process was carried out completely manually. As each level of the design was completed, it became a specification for the lower levels of the design. These lower levels were designed to fit the downwardly imposed design constraints and the upward physical constraints from their own design components. Failure in one or another level of design resulted in either re-design of that level or relaxation of the imposed constraints by redesign of a higher level of the design. The advent of VLSI technology put severe strain on this manual system by allowing tremendous growth in the design complexity. Thus the manual cycle time for design went from days to months to years, necessitating computer aided design systems.

Present day design systems use Silicon Compilers [JoMa84] [BuMa85] [GaDP86] to generate layout for device fabrication, as well as timing, power, and simulation models for frequently used components (PLA's, RAM's, ROM's, Data-Path's, ALU's, Multiplexers, Counters). In addition, such design systems provide simulation and time verification functions to evaluate the generated design's performance. In these systems, transformations between levels of the design hierarchy are performed by fixed algorithms which optimize the design at each level. Failure analysis and design evaluation must still be done manually, and corrections to the design specification must be iterated through the entire compiler for evaluation.

Rule-Based design systems have begun to appear on the CAD horizon [Kowa84] [JaSi85]. These 'expert designers' replace the procedural algorithms of earlier CAD systems with rule-based expert systems between the design levels. These systems have great potential in CAD as they can encapsulate expert design knowledge as well as the rapidly changing domain knowledge. Since they can be easily extended and modified, rule-based systems allow automated design before general algorithmic models and techniques appear. At present the experts are limited by using a single strategy and conforming to the design methodology of the algorithm they replace. These systems, however, fall short of the goal of completely automated design since they still require manual performance evaluation and redesign.

We therefore propose a new design model that simplifies implementation by a set of communicating expert systems. This model implements both upward and downward constraint propagation, and allows iterative refinement of designs. The iterative refinement procedure requires a goal directed design strategy and closed-loop evaluation of the design performance. To close the loop we choose a simple 'Knob and Gauge' approach with 'Knobs' being downward propagated design constraints, and 'Gauges' being performance evaluations of the proposed design.

This work supported in part by a grant from AT&T Bell Laboratories.

2. A New Model of Design

The design of any complex system is necessarily split into a hierarchy of design abstractions, with more abstract representations at the higher levels and less abstract ones at the lower levels. To accommodate the changing rules and design granularity, each level of abstraction requires its own dedicated designer or design expert. The purpose of the expert on a given level is to create a structure out of the design components predefined for that level and to partition the design constraints into component constraints. This structure implements the behavioral specification produced by the previous higher level designer.

Conventional models of Silicon Compilation pursue a straight top-to-bottom transit of the design hierarchy. Thus, each level of the design is completed, optimized and (sometimes) evaluated before the subsequent design proceeds. The evaluation is performed by a human designer who first determines the relative priority of the design goals, and selects appropriate performance measures. He then analyzes the design to produce corrections in the design specification. Sometimes the changes can be made directly to the present level of design. This results in the specification at the previous level not necessarily matching the final design. In this case the designer is 'on his own' to ensure that the design change does not violate any other constraints. Usually, however, the designer must change the high level specification to effect changes in the final design. Apart from the relatively long turnaround of design changes, the designer has little direct control of the design process at lower levels. Worse, if he does change the specification at a lower level, there is no automatic way for the system to correct the higher-level design.

Design in the new model proceeds from top to bottom as each level is completed, with the provision that each level may fail in its attempt to achieve its goals. When this occurs, control passes back to the parent in the form of a failure report. The higher level task may decide to re-allocate constraints, or change styles, or indeed fail itself. This procedure allows backtracking of earlier design decisions between levels of the design hierarchy, forcing iterative refinement of the design. In addition, this model supports a constraint handling procedure which manages both upward and downward propagation of design styles and parameters. Constraint propagation and failure reporting augment the completed design specification to provide communication between the different design levels. This additional information is used by the expert designer to formulate strategies for the completion of the design. In this way a decision made by the expert is based on much the same design context as that implicitly used by a human designer.

Closed-loop iteration of the design requires that the refinement of the design structure be guided by a functional specification at *each level*. This requirement stems from the need to select components with the required functional abilities in some particular design style. By explicitly saving the desired function along with the detailed implementation, the process of evaluating how well a design performs is simplified. Also, this allows the lower level design procedures greater freedom to select the particular functional equivalent implied by the specification. As an example one can partition any digital design into a set of communicating finite state machines (FSM). The functional decomposition allows the FSM designers to use knowledge about the intended use and required performance of the sub-systems to drive the selection of the implementation strategy. Lastly, the structural design produced by a more abstract level designer is the detailed functional specification needed by a lower level.

Within a design abstraction there are usually several possible alternative structures for a particular desired behavior. Each of these may exhibit differing cost performance characteristics and require different refinement and optimization techniques. These structures can be grouped into sets of similar characteristics called *design styles*. Styles reflect various design approaches forced by different design constraints to achieve the same behavior. A simple example is the choice of ripple-carry addition versus carry-look-ahead. The ripple-carry adder is appropriate if space is at a higher premium than delay time.

As each level is designed, constraints are produced which must be propagated to the designers at the lower levels. These constraints reflect design style decisions, or structural partitions of higher-level design constraints. Style decisions constrain the design styles and strategies of sub-section designers. An example is the decision to use pre-charged carry addition, forcing the use of appropriate implementation components. Structural partitioning refers to the dividing of global constraints such as time, power, or area into local constraints on these values. A requirement of 175nS as maximum cycle time makes demands on the critical path of operations in each cycle. As the design is implemented, this puts a partitioning constraint on the design of each functional component. Figure 1 shows the possible allocation of timing constraints in two stages of the design process. In the first case the allocation has simply divided the cycle time among the function units. In the second case (later in the design) a failure report from the multiply design task has forced a different allocation of time between the functional units.

Iterative refinement of a design requires continuous performance monitoring relative to the design goals. This model assumes a simple approach similar to 'Knobs' and 'Gauges'. A human operator monitoring a process closes the loop manually by reading the appropriate gauges and making adjustments to the knobs (parameters) controlling the process execution. We apply this same simple approach to controlling the design process. Each step of the design process results in subtle changes in the design. These changes are keyed to the desired behavior of the structure and to the process constraints. An evaluation of the partial design is performed periodically to determine if the present style and refinement techniques are driving the performance closer to that desired. The action taken on an evaluation depends on the options available to the designer. It may try to isolate a problem area and optimize, or change styles, or fail and try a reallocation of the constraints.

Figure 2 details the flow of the design paradigm. The proposed paradigm organizes the design experts at each level into several smaller tasks. These tasks are: *Planning*, *Refinement*, *Optimization*, *Constraint Propagation*, and *Evaluation*.

Planning refers to the section of the expert system which performs the control function for the designer. It uses the functional specification and the design constraints to 'envision' possible design strategies and then selects the most promising of these and begins the design process. As the process continues, the planner evaluates the progress of the design to ensure that the plan is feasible. Problems are handled by first determining the cause and then envisioning modifications in either the constraints or the partitioning or if necessary, the style. The planning section defines the present goals and selects the design styles for the

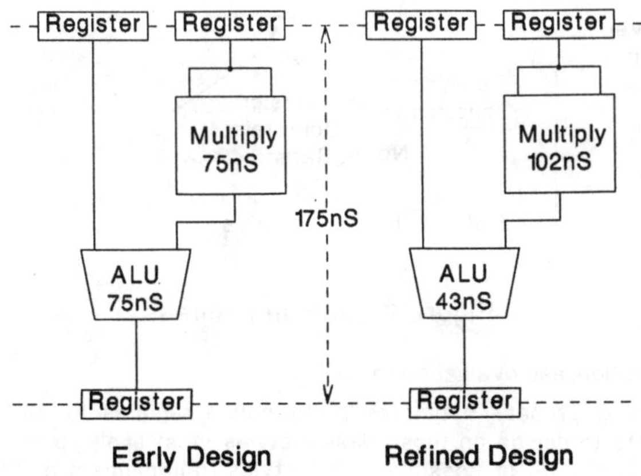

Figure 1. Reallocation of Design Constraints

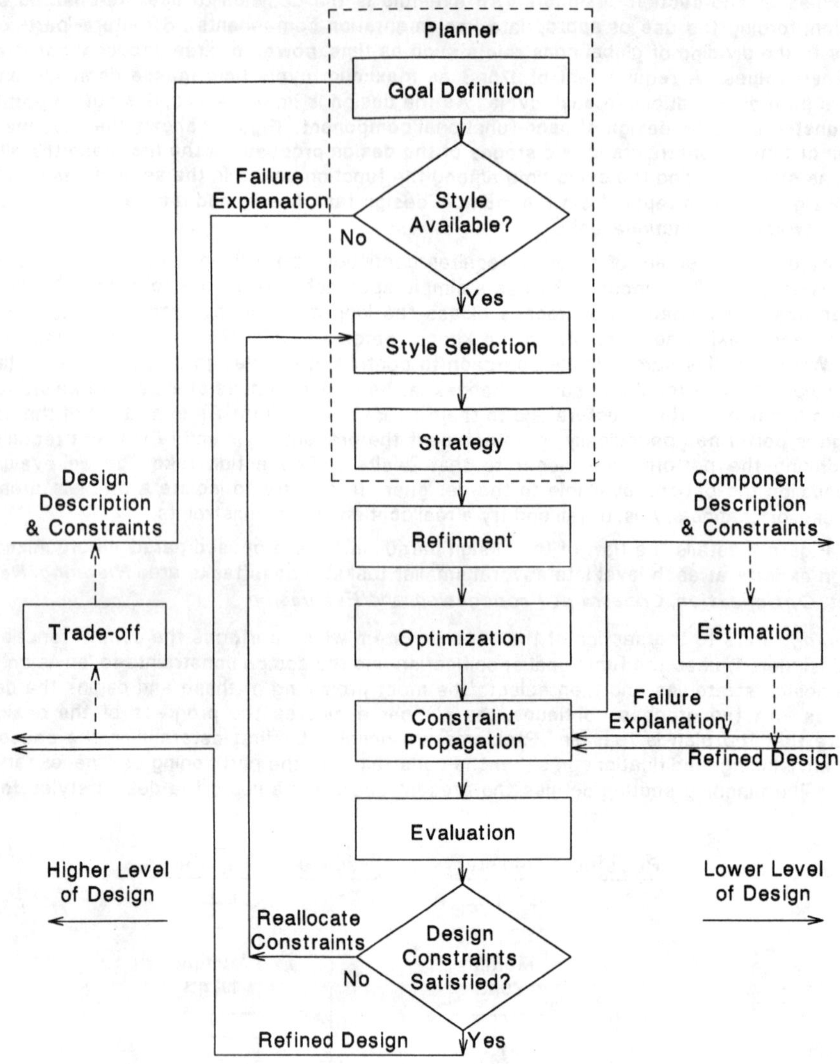

Figure 2. Design Process

refinement, optimization, and evaluation tasks.

The 'Envisioning' process estimates performance regimes for the design styles and uses the constraints to decide on those design styles most likely to provide workable solutions. Figure 3 shows a set of possible design styles to implement a multiplier. Each style has particular advantages and problems in addition to the particular trade-off shown. Downward constraints partition the space of possible designs into regions of acceptability. Goals act to drive the design in particular directions within these regions subject to the design strategy. In the figure, a cell array or iterative multiply style is chosen, depending on the inherited style, constraints and goals. Thus, the envisioning process in the simplest case is simply forward tree pruning of obviously unacceptable branches. It seeks to apply simple rules to determine feasibility of different design styles. ore complex envisioners build a tree of possible design routes, which is then pruned by estimation techniques. The purpose of these complex routes is to let the design expert consider designs which may seem unacceptable in a cursory evaluation, but which may lead to a better overall design. In this way the envisioner can use expert knowledge to avoid the pitfalls of locally minimal designs. This is the same problem addressed by the Simulated Annealing [Kirk83]. An example of this problem would be choice of the cell-array multiplier to appease a speed goal when the design requires a wide ALU which might enhance the iterative multiply approach. Determining which of these designs to implement and exactly how the refinement will proceed is purpose of the design strategy.

A strategy is an ordered set of Refinement, Optimization, and Evaluation steps carried out for a particular design style. The order and character of steps in the strategy may be quite complex when good strategies are known, or very simple when they are not. The strategy may also define explicit backtracking procedures when the design constraints or strategy measures are not met. Strategic measures refer to special evaluations which don't correspond to design constraints but give clues of possible design improvements. A possible example is Function Unit usage statistics over several micro-instructions. These statistics could be used to determine seldom-used or redundant Function Units. A refinement step may impose constraints that cannot be satisfied. The strategy can use this information to perform pruning of the possible designs. This increases the efficiency of the designer by eliminating poor branches early in the design.

The design strategy for each style may be quite different since the style imposes an order on the possible design performance measures. If, for example a carry-save multiplier style was chosen, one may assume that speed is of particular importance and complexity is only of secondary importance. This ordering and the constraints imposed by the decision of a particular design style aid the system in developing a strategy for this level of design. Thus for each design style decision there is a goal-directed reason which can be propagated

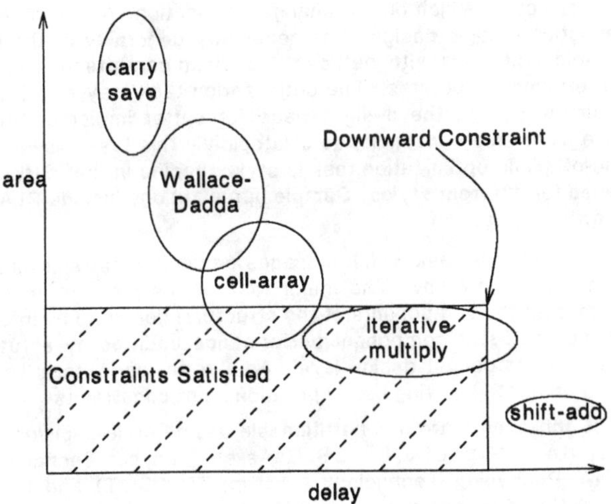

Figure 3. Multiplier Styles

upwards to higher level designers in the event of a design failure. This information can be used by the higher-level designer to reallocate design constraints and allow the design to continue.

Refinement is the task of translating the desired behavioral function into a structure of predefined components from the next lower level of the design abstraction. Thus, for a selected style, the refinement consists of building a structure within the constraints of the style which performs the given function. An example would be a logic implementation of boolean equations, subject to a style requirement of 2-level implementation. This implementation would not necessarily be minimal but would always be functionally correct. Refinement is separated from Optimization in that refinement builds a structure out of the design components that corresponds to the behavioral specification, while Optimization modifies the structural design without changing the function and trys to improve the quality. There are four reasons for this separation. The first is that it is very inefficient to carefully craft and optimize a design which has no possibility of success. The second is that by analyzing the performance of the partial design, the expert designer can concentrate its time optimizing the critical parts of the system. Third, we wish to ensure that the design refinement runs to completion. If the task includes complex optimizations (~g. layout compaction) there is no guarantee that the refinement will succeed. Instead, we refine to an unoptimized (but functionally complete) design and then try possible optimizations. Lastly, the planner may not be extensive enough to provide strategies which match an arbitrary set of constraints. In this case we can use the refiner to explore possible designs by building functionally correct structures which can be evaluated, optimized and iteratively modified until the constraints are met. This iterative approach allows the design paradigm to produce viable designs in cases where one-pass (non-iterative) systems lack strategic knowledge and hence produce poor quality designs. In summary, to provide a time efficient structure generator we simplify the refinement task by: 1. Structuring the design rules around a given style; 2. Removing time-consuming optimization rules from this task.

Optimizations are those rules which search for methods to improve the implementation without making performance/cost tradeoffs. That is, the task seeks to improve performance or reduce cost by intelligent modifications of the design. Optimization routines perform local modifications of the structure which do not change its function. An example is term minimization or cover reduction in logic design. The necessary generality of the refinement functions may lead to implementations with inefficiencies which can often be removed by pattern matching to known efficient structures. The optimization task may also include well-known algorithms which simply search the design space for better implementations. Commonly these algorithms are quite time consuming so strategically this task allows a design time vs. design quality trade-off. The optimization task is style specific in that different patterns and algorithms are needed for different styles. Sample optimizations include PLA folding, register merging, and geometric compaction.

Constraint Propagation is the task which manages passing constraint information both up and down the design level hierarchy. The major task of the propagator is to partition the high-level constraints onto the components of the structural design. The information used by the propagator are estimates of component performance updated by actual design performance figures from the component design level. As the design is iteratively refined, these figures become more accurate allowing better partitioning of constraints.

Another kind of constraints are non-partitionable design style decisions. Examples are: Process Technology (i.e. CMOS, ECL, BiMOS, GaAs etc.), Layout Technology (i.e. Std. Cell, Gate-Array, Custom), Packaging Technology (i.e. Dip, Flat, SOT) and Clocking schemes. These decisions limit the style choices at lower levels of the design by forcing compatible component selection. Such decisions are passed by inheritance to the lower levels. This means that all of the physical partitions of a structure inherit the style decisions made for that structure.

Evaluation is the task of determining how well the design strategy has worked so far. The evaluation task consists of rules for evaluating various performance measures such as power consumption, area, delay time, critical path delay, etc. These rules provide the 'Gauges' used by the strategy to determine how the design will proceed, or whether to backtrack and change a style decision, or finally to admit defeat and produce a failure analysis for the next higher level of design. Because of the simple control approach, the evaluator must be able interpret the operation of the structure relative to the desired functional behavior. This interpretation problem is simplified by requiring that the evaluations be simple numeric parameters, usually related to a performance measure.

There are certain structural evaluations (strategic measures) which are not related to performance figures but which are used by the envisioner to determine possible design improvements. These are usage statistics, layout density, redundancy and testability figures and other parameters relating various local efficiency measures of designs. These measures are used to determine areas where favorable design trade-offs may be made by locating infrequently used or redundant components, or unused space.

The evaluations commonly depend on accurate estimations of the performance of the components at this level of design. The iterative approach of this paradigm insures that as the lower levels of the design are completed, these better performance figures are propagated to the higher level evaluators. Thus the evaluation at a particular point in the design reflects the best knowledge of the design performance available to the system.

Trade-off and Estimation provide facilities at the highest and lowest levels of the design. The estimation task provides design and constraint estimation for some (arbitrary) low-level of the design. In the simplest case the estimator merely queries a data-base for component information of already designed components. Examples would be Standard Cells or Gate Array modules. A more complex case could be calls to Silicon Compiler layout generators. The Trade-off task performs a similar service at the highest level of the design. Its purpose is to determine what to do when the design task fails; Possibilities are to simply query the user, or to provide a method for constraint relaxation.

In summary, the paradigm divides the expert system into 5 smaller tasks: Planning, Refinement, Optimization, Constraint Propagation, and Evaluation. The planner uses the imposed constraints and goals to select appropriate design styles and strategies for the other tasks. Refinement creates a structural design out of the indicated components which corresponds to the behavioral specification. This structure is conditionally optimized and evaluated to determine if the constraints have been met. The planning strategy then takes appropriate action depending on the outcome of the evaluation. The Knob and Gauges approach reduces the complexity of the planning to that of the global design parameters and constraints. Thus the local design decisions are made with respect to their effect on the global design parameters. The style-directed refinement ensures a functionally correct structure that corresponds to the selected style. The evaluation provides the expert system with a means of determining focus to the relevant design problem(s). Finally constraint propagation ensures upward and downward compliance with the imposed constraints and style selections.

3. Advantages of the Design Model

The proposed model has several advantages over present implementations of Silicon Compilers. Most notably it specifically encourages iterative refinement of the design, and removes the need for a human to close the design loop. This allows the computer to complete the entire design process quickly without tedious and error prone intervention of the human operator. If the operator does decide to modify the design, his controls are the same 'knobs' that are used by the design experts. He therefore has control local to each level of the design and assurance that the internal constraint propagation will enforce a correct final design.

Since the design model is completely closed-loop a proposed design can be implemented in several different styles and the best chosen. In present systems, long optimizations are required to perform the same refinement. For these optimizations, unless the design space is very well explored, most of the time spent in optimizing is spent exploring bad designs. Worse, if the final design is simply not good enough the optimization time is wasted. In contrast the proposed design model gets an approximate design quickly by refinement and then seeks to improve on it. Since the design behavior is kept throughout the system, localized evaluations can point out areas where special design effort is needed. In these areas, the optimizations performed enhance the entire design. A pictorial example of this is shown in Figure 4. Although a global optimizing algorithm will in general produce a slightly better design, the space it has to explore may be very large. The design model makes use of local evaluations to determine design choices and by comparison searches a smaller space.

Since the design system at each level is an organized expert, the system is amenable to changes in strategy or to changes in technology without massive changes in the structure. In fact, the partitioning of design knowledge into separate styles reduces the required generality of the rules, and should make system design and update easier. Finally, the control structure of the expert system is quite clear, allowing rule evaluations to become algorithmic when good algorithms are known. This allows the expert designer to make use of rapidly executing programs for certain optimization and evaluation tasks where rule execution would be either inefficient or unnecessarily complicated.

4. Applications of the Design Paradigm

Walker and Thomas define a standard set of design levels of abstraction for Silicon Compilation.[WaTh85] In this section we will describe how the design paradigm can be used to implement a rule-based design system for the Micro-Architecture, Layout, and Logic-Design levels. For another approach to the micro-architecture level see [KnPa86].

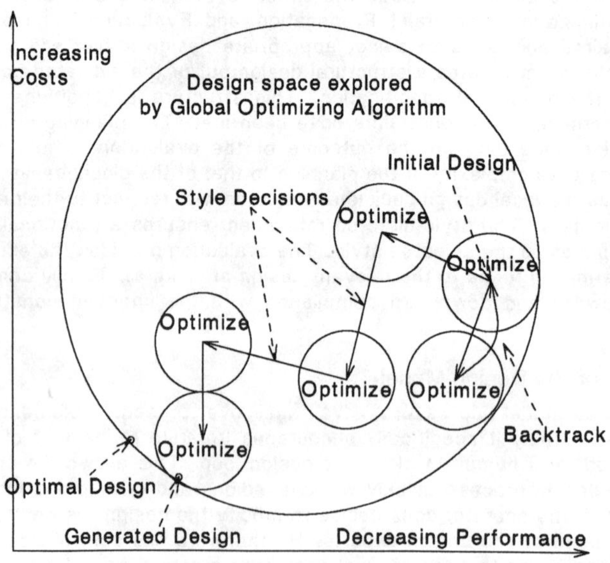

Figure 4. Design Process Comparison

4.1. Micro-Architecture

Design at the micro-architecture level entails creating a register-transfer level design from an algorithmic behavioral specification. The components at this level of design are registers, memories, PLA's, and functional units such as ALU's. Usually a direct correspondence exists between the operators of the algorithm and the component functional specifications. Thus, the design task amounts to creating a network of components interconnected by busses or muxes that conforms to the behavioral specification. Furthermore, the design must satisfy all of the imposed constraints such as speed, power consumption, and area. Common design practice decomposes this problem into the coupled designs of a Control Unit, and a Data Path.

Style Selection and Strategies–– At the micro-architecture level the mapping of the register-transfer level design to quality and performance measures is highly complex because performance estimation is strongly dependent on the behavioral algorithm. Because of this difficulty, we propose a strategy of local exploration of the design space by iterative refinement of the design. Specifically, we can simply carry out a design given some style and component constraints and then evaluate that design to determine the performance and cost. The resultant figures (the gauges) are compared to the constraints to determine a better selection of styles and component limits. This process is shown pictorially in Figure 5. An initial design is carried out using a simple parallel style, assuming infinite resources in chip area for the data path. The control is assumed to be simple and relatively fast. This

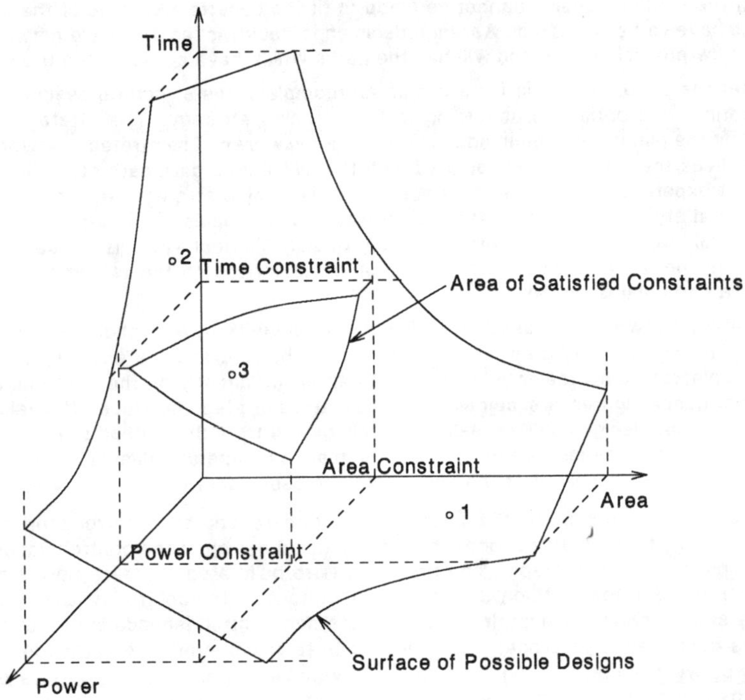

Figure 5. Iterative Refinement of Performance

design is evaluated by using component delays and areas to determine the time/area/power of the design (*point 1*). It is noted that this design satisfies the time and power constraints but exceeds the area limit. A new design is allocated more limited resources in area, but since since power is already satisfied, these resources could be faster. The evaluation of this design is shown as *point 2*. As more designs are evaluated, the resource allocations can be interpolated from those of previous designs. If none of the designs come close to the desired trade-off, a design style change is in order. Finally, if there are no available styles, the planner should generate a failure report to try to relax the constraints at the previous higher-level design. In the example the second design violates the time constraint but does satisfy the area. This design is further modified to increase its performance with only small increases in the area. The resulting design evaluation is plotted at *point 3*.

To achieve the iterative refinement described above, there must be a mapping of design refinement knowledge into the desired performance changes. For this purpose we introduce *Style Networks*, and *Strategy Networks*. A style network is an acyclic directed graph with style decisions at the nodes and desired performance trade-offs labeling the edges. Similarly, a strategy graph is a directed tree with refinements at the nodes and trade-offs marking the edges. A simplified control style network and data path strategy graph are shown in Figure 6. In use the style network indicates the possible styles available from a particular design state. The styles are selected by evaluating the trade-offs which label the arcs. Each decision is inherited in the design so that the design state is the entire path through the graph, not just the last node. Terminal nodes indicate that no further refinement is possible within the imposed constraints of earlier decisions (i.e. the design style selection is complete.) If the resulting design cannot be made to fit the constraints, some of the design decisions will have to be retracted. As each decision is backtracked, the style network will indicate the new possible styles and will flag the paths which have already been tried.

After the style selections for a design are complete, the structural design is generated by the refiner and optimized according to the planning strategy. The strategic knowledge required by the planner is organized into *strategy networks*. The strategy network shown in Figure 6 gives the strategies associated with the distributed data-path style. In general the refinement expert is driven either by a specific style or by a strategy network selection for a more general style. This is necessary since the design models for data-path design are far more general than those for design of the control unit. To deal with this increased generality the strategy network provides a selection of modifications to the refiner which drive the design in a particular direction.

Strategy Networks are used much like style networks but the nodes are not exclusive, at each planning step the graph is consulted from the root and the appropriate refinement action is selected. Once selected, the refinement is run directly on the potential design and new performance figures are calculated. Thus at each step the trade-off evaluations are based on genuine design quality measures. This differs from the method proposed by Knapp [KnPa86] who chooses instead to evaluate the plan in a separate planning space and iterate the design constraints only after the entire plan has been generated.

Refinement-- The tasks for the micro-architecture refiner are: control step partitioning, register, bus and function unit allocation, and functional design of the control module(s). This is commonly done in two steps. The first step is to build a control/data flow graph for the algorithm [OrGa86] and partition the graph into 'states' for the control synthesiser [Park84]. This is a simple means of ensuring that the data and control dependency requirements are met. The next step is to allocate functional units to the 'operation' graph nodes, busses to the arcs between units, and registers to those variables in arcs crossing state boundaries. Finally, the symbolic microcode derived from the last step must be encoded into the control unit.

Optimization-- Optimizations of the register-transfer level structure re-arrange operations subject to their dependencies to minimize either the hardware (by making a particular unit unnecessary) or the cycle time (by moving an operation to an available unit in a previous cycle). This optimization is especially important if a pipeline style is chosen to prevent many

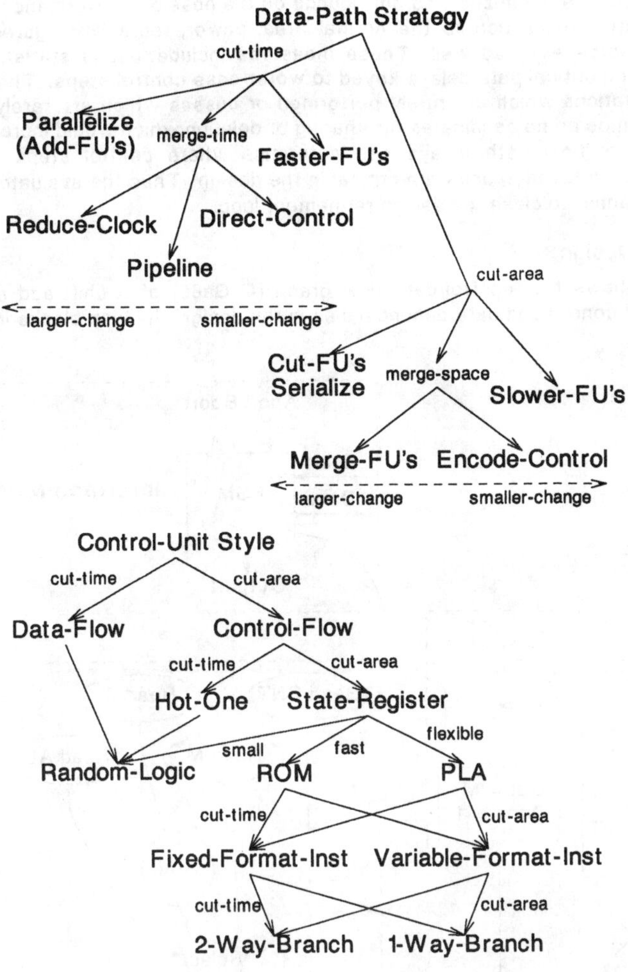

Figure 6. Knowledge Networks

wasted cycles on branching. The optimization may also involve recognizing conditions under which registers and function units can be shared. These conditions arise when the registers, busses, or function units are used exclusively on each time step. In the function unit case the sharing requires that the shared unit is capable of performing both functions. Sharing of registers requires that the variables stored in them are not simultaneously alive. This can be determined by keeping a table of register bindings while the algorithm executes. Busses can also be shared. As an example several registers used exclusively on a bus can be incorporated into a register file, saving significant area. In the control unit design, the optimization usually tries to minimize the size of the ROM or PLA by state encoding [DeMi84], control encoding, PLA and Decoder merging.

Evaluation— The proposed design is evaluated to allow comparisons with the design constraints and to make measurements which aid the expert system in determining how to

376

improve the design. The design evaluations can help to point out critical function units or paths which should be optimized, and put bounds on the possible performance available from a given structure. In addition to the normal area, power, testability figures other useful evaluation measures exist as well. These measures include: usage statistics for various function units, and critical path delays keyed to worst-case control steps. The usage figures can detect operations which are rarely performed or busses which are rarely used. These items would then be prime candidates for sharing or deletion which would increase the design efficiency. The critical path delays can determine where control steps should be re-partitioned or which function units are critical in the design. Thus the evaluator supports the efforts of the planner to close the design refinement loop.

4.2. Example Design

Figure 7 shows the control/data flow graph [OrGa86] of a shift-add multiplier. The graph models all control and data dependencies in the design, the loop blocks indicate control

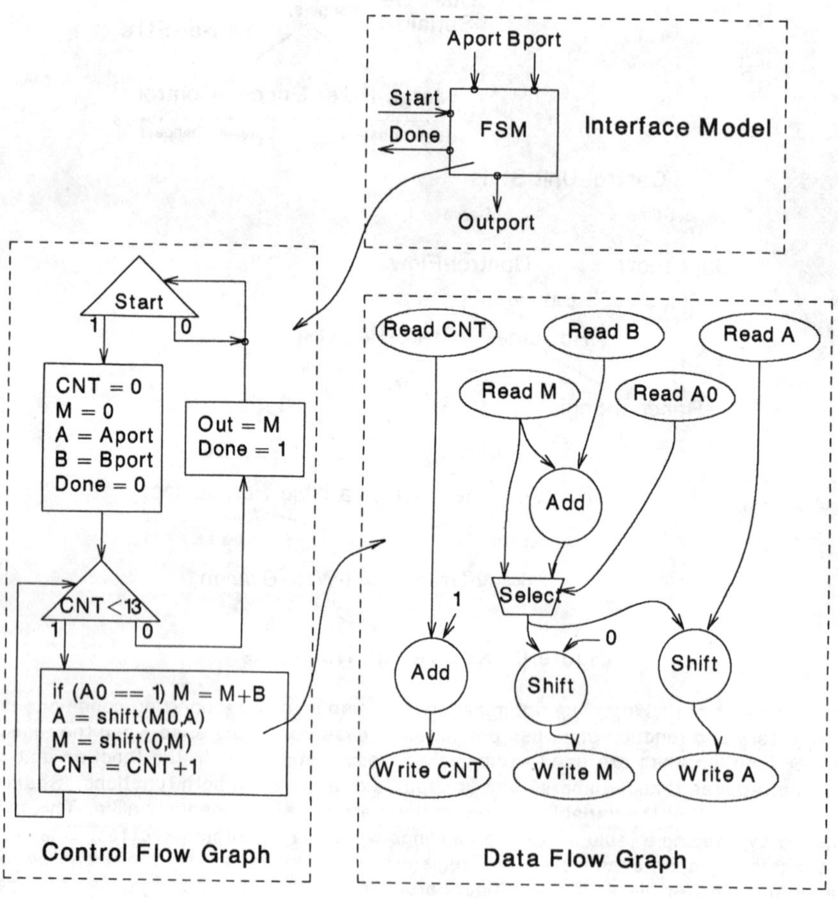

Figure 7. Functional Description and Flow Graph

branching macro states. As a preliminary step the control is assumed to be a non-pipelined FSM and the allocator is given as many functional units as needed to parallelize the data-path. The graph is sliced into states, units are connected into a data-path and the performance is measured. The result is shown in Figure 8 In this case the area includes 4 registers, 2 shift units, 1 incrementer, 1 adder, and the transition based control unit. The time performance is calculated to be n+1 clocks for n-bit operands. For this example the units are given operations times: Mux, Comparator = 20nS, Adder, Shifter = 100nS, PLA = 60nS. The critical path is along the adder which determines the clock cycle to be 240 nS.

Although the time performance is satisfied, the area required by this design is very large. The knowledge network is queried for possible area reduction plans. The control unit is already chosen to be a small style so the data-path strategy network is queried. The resulting refinement is to merge the functional units. The Adder's and the shift and the selector mux are merged into an ALU. This merging caused the A0 control line to be routed to the control unit as the ALU operation has now become condition dependent. Since the value of A0 is available at the beginning of the cycle, this re-routing did not increase the critical path time. The critical path is through the comparator, PLA, Mux, and the ALU. Lastly, a local optimization rule determined that the 'A' shifter can be merged into the 'A' register to produce a shift register (reducing area). This design is depicted in Figure 9. This design has a clock of 220nS but now requires 3n+1 cycles to run. The area requirement is now satisfied but the time performance is too slow.

Once again the knowledge networks are queried, this time the result is a style change in the control unit. Pipeline registers are added to the control to remove the PLA and

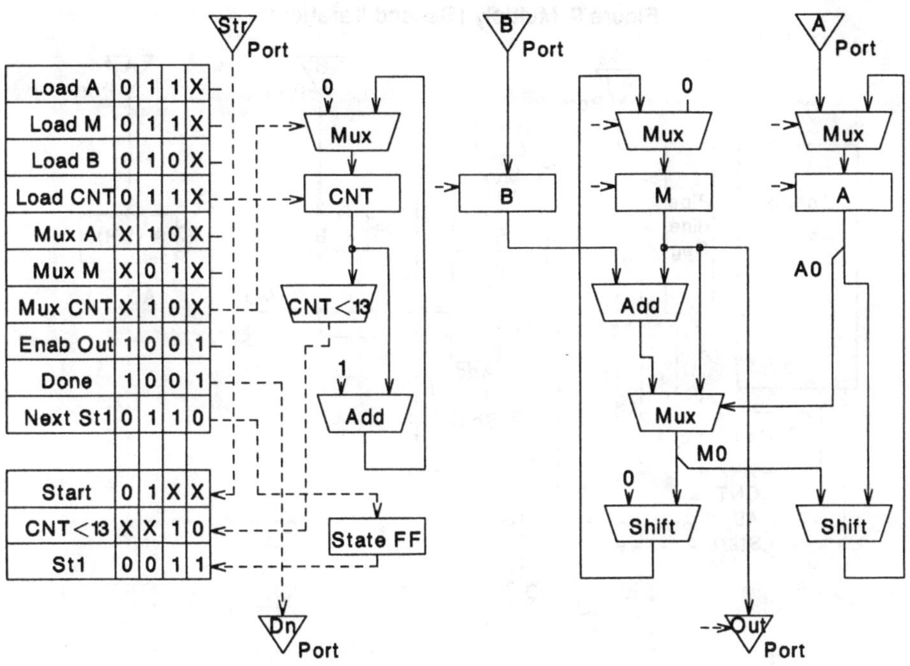

Figure 8. Multiply (First Iteration)

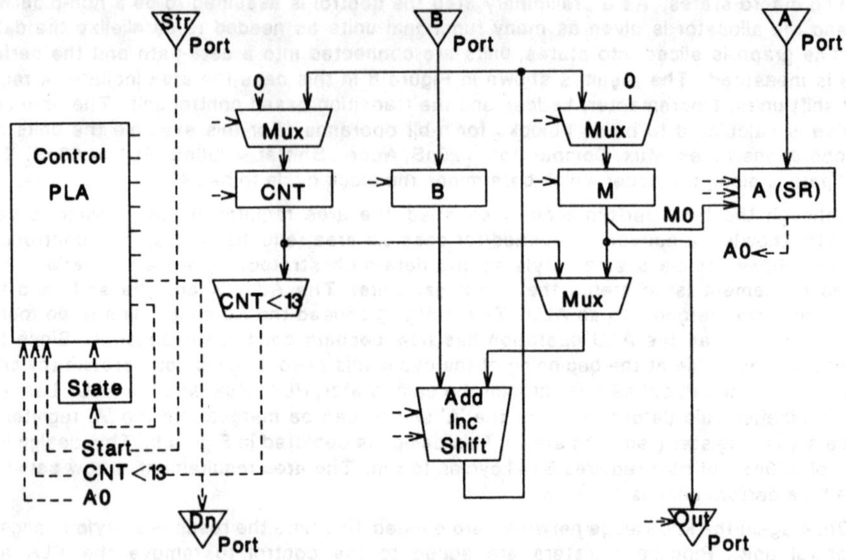

Figure 9. Multiply (Second Iteration)

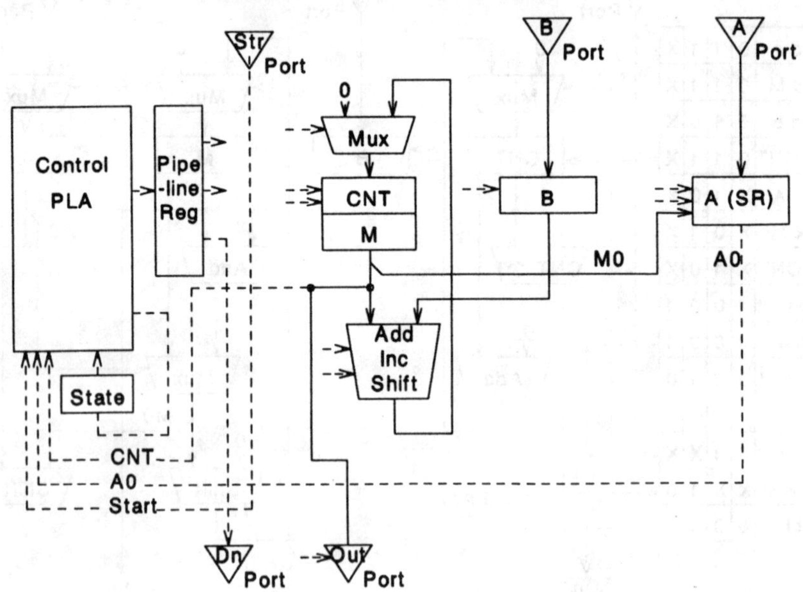

Figure 10. Multiply (Final Design)

Comparator times from the critical path. The clock becomes 140nS and the multiply time is 4n+3 clocks. Optimization in the area caused the registers 'CNT' and 'M' to be merged into a register stack along with the ALU select mux. This also reduced the clock by 10nS to 130nS. Lastly, the comparator can be merged into the PLA, resulting in an area savings. The final design is shown in Figure 10. This trade-off process was shown pictorially in Figure 5. The initial design, second-cut, and final design, are points 1, 2, and 3 respectively.

4.3. Logic Design

Logic Design is the task of translating the behavior of functional blocks (eg. ALU, Adder, Mux etc.) into structures of gates. The functional block behavior is commonly specified in terms of Boolean equations, and functional recurrences of these equations. Normally, gates are functionally represented as 1-bit wide operators of these boolean equations. Creating a functional structure could simply mean instantiating a gate for each instance of the associated operator in the functional specification. There are, however, many functionally equivalent logic circuits so the design task is one of creating a circuit which satisfies the imposed design constraints and is optimized to the imposed goals. In fact we are not nearly so interested in optimality as in meeting the demands of the constraints with an 'acceptable' circuit.

Design Styles There are a very wide variety of design styles available for implementation of logic circuits. Examples are: PLA, ROM, Gate-Array, domino logic, minimum sum-of-product form realization, and finally product and sum form recurrences. Each of these styles satisfies a unique set of design goals in terms of area, delay, power, testability, and design time. For example, PLA's satisfy the goal of high-speed implementation, but are prohibitively expensive in area if the number of minterms in the function to be realized is large. Cascade PLA's (4-level logic) consume less space if the Boolean function is decomposable, but pay for this area advantage by slower operation. ROM's are inefficient if the desired number of outputs is small. If the Boolean function is a recurrence, product-form recurrence realization achieves logarithmic growth of the delay at only polynomial cost in gates. In addition, each style defines a unique set of design components which are used to implement the structural design. These components take advantage of specific features of the implementation technology. For example, PLA implementation in NMOS can use the compact NOR/NOR form, while a domino CMOS technology allows implementation of complex gates.

Style Selection and Strategies The logic design task begins with a behavioral specification and a set of design constraints. These constraints and the domain knowledge of the designer are then used by the planner to propose possible implementation styles and strategies. This choice is further limited by the choice of layout technology. For example, a combinatorial decoder is desired with a time constraint of 15nS. If the technology is low-cost NMOS this imposes a demanding time constraint so the planner would choose a style with a small number of gate levels such as a 2-level PLA. The associated strategy is to expand the Boolean Eq. to Sum of Products form, optimize by term minimization, refine to NOR-NOR form by simple instantiation, estimate the gate delay and complexity and check against the constraints.

Some functional blocks have considerable internal structure which allows functional decomposition or recurrence realization into multi-level logic. An example is an adder replacing the control unit above. These functions would be poorly realized as PLA's even if high speed is desired. The envisioner must therefore, have additional rules to select strategies and styles for these cases. A sample heuristic is:

If (log(# of minterms) $<<$ # of variables)
    then 2-level PLA-type structure
    else Multi-level structure.

For decomposable structures it is infeasible to carry out a sum-of-product expansion and minimization. A better optimization strategy is local symbolic optimization such as cover reduction or pattern substitution to efficient structures in the desired technology.

Refinement Logic design commonly uses two design structure representations: symbolic boolean equations, and instantiated networks of gates. The refinement of symbolic equations allows a much better job by helping to reduce the logical redundancy of the structure.

The first task is to massage the Boolean equations into a form acceptable to the chosen design style (as was done in the PLA example above). Within the constraints of the style, further local trade-offs can be made. For example, the refiner can factor out common terms in expressions, reducing the gate count but increasing the delay. As an example:

$$(x \cdot v + y) \cdot (z + \overline{w}) = x \cdot v \cdot z + x \cdot v \cdot \overline{w} + y \cdot z + y \cdot \overline{w}$$

The first realization requires 4 gates and has delay 3 while the second requires 5 gates (of greater complexity) but only delays 2. Other symbolic techniques are recurrence decomposition and pattern matching. In these cases the trade is between more serial and more parallel gate implementations.

In the second task the design is instantiated by substituting gates for operators in the specification. Care must be taken to locally check that fan-in and fan-out requirements and other style constraints for the gates are not violated. This can be accomplished by style-specific rules which define substitution structures in terms of the components for operators which are not predefined components. If there is no 5-input AND gate for example, a simple structure of existing components could be substituted for that operator. As mentioned before, this kind of substitution can also occur in the optimization phase, where the goal is to minimize some constraint, or make use of advantageous technology. To improve the design system's efficiency, however, the refiner implementation should use the simplest set of rules that achieve reasonable design, leaving the optimizer with the task of improving the design.

Conversely, both of the refinement sub-tasks may simply be fast algorithms for the well-known logic styles. Since refinement of the structure doesn't attempt to optimize the design but merely make the structure amenable to the style goals, we expect such algorithms to be both simple and fast. (This is likely to be true for the well-understood logic design field but less likely for more abstract and less-understood levels such as micro-architecture design.)

Optimization The resulting structure may not be efficient, although it certainly performs the required function. These implementation inefficiencies can be discovered and selectively removed by optimization of the structure. There are two basic optimizations in logic design, one on the symbolic equations, and the other on the designed gate structure [Sang85]. Symbolic reduction of 2-level logic involves removal of redundant symbols and removal of redundant terms (those which are subsumed by others or a combination of others). This amounts to the classic minterm reductions and Karnaugh-map techniques. For multi-level systems reductions are more complex but usually involve local operations to determine term redundancy directly. Global logic minimization has very high algorithmic complexity forcing the use of heuristic and local methods.

Structural optimization replaces local gate structure patterns with more efficient ones (in terms of time, complexity or technology). An example is the replacement of two inverters with a simple connection or removing an inverter and using the complementary output in an ECL implementation. These local optimizations are used successfully in present design systems [Darr84], [HoNe85], [Joyn86].

Evaluation Logic level evaluations include: critical path delay, layout area estimation, power requirements, and cost of testing. Other merit figures include redundancy and wire-ability. In the case of regular implementations (PLA or Standard Cell), fairly accurate estimations of time and area constraints can be made [KuPa86].

4.4. Layout

Layout design involves the translation of abstract structures into physical realizations. Unlike the previous two tasks which refined from behavior to structure, layout refines from a structural to a physical representation. The abstract components of the layout design are rectangular partitions of the chip with defined connections across the boundaries. These partitions are called cells. At the finest granularity cells contain gates, single transistors, or polygons depending on the layout style. Styles with finer granularity components are capable of higher quality layout, but at the cost of increased design time. Since the structure is already defined at the start, the layout task consists of allocation of physical chip space to cells and interconnections.

Design Styles The layout design takes as its input a structure of interconnected components each of which may represent a layout problem at a finer granularity. Because of this, the un-optimized layout cell hierarchy closely corresponds to the functional hierarchy of the structure. To reduce the design time, many functional styles reduce the granularity to logic function blocks that have direct mappings into efficient layouts. Examples are: Distributed Data-path, CMOS gate matrix, Weinberger array, and PLA's. Another style is to reduce the function to a matrix of replicated cells, these cells are then fully compacted to reduce the overall size. This technique is used in memories, associative caches, and multipliers.

There are also several styles for interconnection between the cells. The simplest is to abut the edges of cells along a boundary. This is nearly equivalent to merging the rectilinear cells as the constraints at the boundaries arise from both cells. The cells can be stretched to satisfy the constraints, but this increases the area of the cells. Another technique is to allow channels between the cells for routing the interconnections, but the routing area is otherwise wasted. Finally, certain cells can be redesigned to include interconnect going through the cell itself. This can be very efficient but is computationally the most expensive. Full custom chips commonly include samples of all three interconnection styles. Standard cells are interconnected by a combination of abutment and channel routing. Gate arrays require special techniques dependent on the particular organization of the chip.

Style Selection and Strategies Style selection for layout is more constrained than the behavioral to structure translations because many of the functional styles are specifically designed to layout efficiently. Thus, for many styles the selection is predefined by the structure or the chosen technology. However, the constraints of the layout problem are different from those of the more abstract design levels. The area is still of prime importance, but we also must consider interconnect path lengths, port assignments (ordering and sides), and the desired aspect ratio of the cell. The interconnect path length is important from the non-ideal physics of signal propagation on a chip. The path time delay is related to the path length by both the increased resistance (which limits the current) and the increased distributed capacitance of longer lines. The two-dimensional nature of the chip geometry requires port assignments and aspect ratios as constraints if we wish to allow hierarchical decomposition of the layout.

Although the styles for the lower level structures are commonly predetermined, the layout of compositions of these cells is not. We assume that the basic cells (the leaf-cells of the hierarchy) are strongly constrained (by the predetermined style of the structure). The floor-planner still has the orientation, interconnection, and port assignments of the composed cells to design. A possible strategy at each level of the hierarchy is: Assign ports from exterior constraints, place cells to accommodate the structure, assign ports to the smaller cells, optimize to minimize the path length of critical nets or to reduce the total area. This strategy assumes a top down design, but we must also be able to handle bottom-up constraints of the cells. Another strategy (bottom-up) is to refine the structure in terms of the already design cells to minimize the routing area and path length, then assign the external ports. The overall strategy includes both top-down and bottom-up techniques, repartitioning constraints and redesigning cells to satisfy all of the constraints. The CADRE system [Ackl85] uses a bottom-up technique with similar backtracking constraint partitioning.

Refinement The refinement is split into two tasks. The structure is embedded in a symbolic planar structure and the wires legally connected. The second step is to physically instantiate the cells and interconnection. The second refinement is of primary importance at the bottom level (leaf cell) of the hierarchy to introduce the actual physical constraints. The first task can be implemented by graph algorithms or a set of heuristic rules [Ackl85]. Instantiation is handled by the constraint propagator.

Optimization There are two sorts of optimization procedures, cell compaction and local exchange/modify (combinatorial) compaction. Cell compaction minimizes the area a cell by moving the components and inserting jogs in wires to minimize vacant space. This type of compaction does not modify the topology of the cell. Topological modifications of the layout are usually accomplished by exchange/modify rules. In a hierarchical organization there are two types of topological optimization. The first simply interchanges the ports, orientations, and pairwise placement of the cells to minimize the selected performance parameter. (i.e. the weighted path lengths etc.) This optimization can be greedy or use some annealing schedule [Kirk83], [WoLi86]. The second type of optimization is to merge some members of the design hierarchy and then repartition into cells with better packing capabilities. This approach first locates likely candidates for merging and then removes the local hierarchy constraints to allow application of the first type of optimization. Then the cells are recomposed into the hierarchy to allow further refinement.

Evaluation Evaluation of the floor plan requires simply calculating the performance parameters for the given cell from those of the lower level cells. When the design is first started, however, this information is not yet available so some estimations must be made. The area values are propagated from those calculated in the structural phase, and bounds are calculated from over-constrained and under-constrained versions of the problem [Ackl85].

5. Conclusions

In this paper we have introduced an expert-system design paradigm. The paradigm advocates iterative refinement of the design as opposed to forward design generation. To support this refinement the paradigm describes how to perform closed-loop evaluation of the proposed design and how to propagate the changing constraints to ensure correct design. Closing the loop between evaluation and design modification uses a simple Knob and Gauge approach. The Knobs are design styles and refinement actions applied to the design. The Gauges are the area, delay time, power consumption, and other performance parameters which are conventionally determined by simulation. The paradigm details how the gauges and constraints are used to evaluate trade-offs which select design styles and strategies to produce the refined design.

The paradigm also emphasizes the distinction between Refinement and Optimization. Refinement is the process of creating a structure to implement a desired behavior. Optimization is the process of modifying this structure to improve the efficiency without changing the function. The separation of Refinement and Optimization guarantees a design solution. (First we build the functional structure and then we optimize it). This technique also makes more efficient use of the design time by optimizing only where necessary.

Lastly, we have organized the design knowledge into *Style Networks* and *Strategy Graphs*. The use of these representations simplifies the introduction of new design knowledge and strategies to the design expert system.

6. References

[Ackl85] Ackland et al. "CADRE - A System of Cooperating VLSI Design Experts" *IEEE ICCD Proceedings*, pp. 99-104 June 1985.

[BuMa85] Buric, M. R. and Matheson, T. G. "Silicon Compilation Environments" *Custom Integrated Circuit Conference*, May 20-22, Portland OR, 1985.

[Darr84] Darrenger at al. "LSS: A System for Production Logic Synthesis" *IBM Journal of Research and Development*, vol. 28 n. 5, pp. 537-545, Sept. 1984.

[DeMi84] De Micheli G. "Optimal Encoding of Control Logic" *Proceedings ICCD* Port Chester, NY, 1984.

[HoNe85] Hofmann M. and Newton R. "A Synthesis System for CMOS Domino Logic" *Proceedings 1985 International Symposium on Circuits and Systems*, Kyoto Japan, June 1985.

[Gajs85] Gajski, D. D. "Silicon Compilation" *VLSI Design*, November 1985.

[Joha79] Johannsen, D. "Bristle Blocks: A Silicon Compiler" *Proceedings— 16th Design Automation Conference*, June 1979.

[JoMa84] Johnson, S. C. and Mayor, S. "Silicon Compiler Lets System Makers Design their own VLSI Chips" *Electronic Design*, October 1984.

[JoSi85] Joobbani, R. R. Siewiorek, D. P. "Weaver: A Knowledge Based Routing Expert" *Proceedings— 22nd Design Automation Conference*, June 85.

[Joyn86] Joyner W. H. et al. "Technology Adaptation in Logic Synthesis" *Proceedings— 23rd Design Automation Conference*, Las Vegas 1986.

[Kirk83] Kirkpatrick, S. Gelatt Jr., C. D. Vecchi, M. P. "Optimization by Simulated Annealing," *Science*, vol. 220, no. 4598, May 13, 1983.

[KnPa86] Knapp, D. and Parker, A. "A Design Utility Manager: The ADAM Planning Engine" *Proceedings— 23rd Design Automation Conference*, Las Vegas 1986.

[Kowa84] Kowalski, T. J. "The VLSI Design Automation Assistant: A Knowledge-Based Expert System" *PhD Thesis: Carnegy—Mellon University*, April, 1984.

[KuPa86] Kurdahi F. J. and Parker A. C. "PLEST: A Program foe Area Estimation of VLSI Integrated Circuits" *Proceedings— 23rd Design Automation Conference*, Las Vegas 1986.

[McDe82] McDermott J. "Domain Knowledge and the Design Process" *Proceedings— 18th Design Automation Conference*, 1981.

[OrGa86] Orailoglu A. and Gajski D. D. "Flow Graph Representation" *Proceedings— 23rd Design Automation Conference*, Las Vegas 1986.

[Park84] Parker A. C. "Automated Synthesis of Digital Systems" *IEEE Design and Test of Computers* Nov. 1984 pp. 75-81.

[Sang85] Sangiovanni-Vincentelli A. L. "An Overview of Synthesis Systems" *Proceedings — IEEE 1985 Custom Integrated Circuits Conference*, pp. 221-225.

[WaTh85] Walker, R. A. and Thomas, D. E. "A Model of Design Representation and Synthesis" *Proceedings— 22nd Design Automation Conference*, 1985.

[WoLi86] Wong D. F. and Liu C. L. "A New Algorithm for Floorplanning" *Proceedings— 23rd Design Automation Conference*, Las Vegas 1986.

PART II. SYNTHESIS SYSTEMS

A MIXED-MEDIA APPROACH TO MODULE GENERATOR DESIGN

Hung-fai Stephen Law and Graham Wood

SDA Systems
2461 Mission College Blvd.
Santa Clara, California 95054
(408) 727-7811

ABSTRACT

An interactive procedural design language tightly integrated into a graphical CAD workstation environment and a versatile composition tool for module generation are described. The resultant combination of procedural language, high level module assembly tool and interactive graphics is a powerful and user-friendly vehicle for the design of module generators and parameterized cell generators.

1. INTRODUCTION

Module generators are a class of CAD tools which build some specific functional module of a circuit, such as a shift register or a multiplexor. Essentially, the kernel design of the module is "built in" to the generator. The user of the generator can customize the attributes of the module by specifying values for a set of parameters pertaining to that particular function. Examples of such customization could include the number of bits in a shift register, the driving capacity of a buffer, or the bit pattern in a read-only memory. It is important to note that the degree of control the user has over the output of the generator depends absolutely on what features the designer of the module generator has anticipated and provided. It is for this reason that the success of module generators is often inversely proportional to the distance between the designer of the generator and its user. The key to the profitable use of module generators is to **allow the designers to design the module generators**.

This principle forms the motivation for several recent module generator systems, the so-called "compiler-compilers" or "generator-generators"[1]. The common thread in each of these and earlier[2][3][4] systems is a *procedural design language* in which the kernel design of the module and its various parameterizable attributes may be described.

In this paper, a mixed-media approach to module generation design is proposed. An interactive procedural language called **SKILL** and a general purpose module composition tool are described. Both of these tools are tightly integrated into a graphical CAD workstation environment. The resultant combination is a powerful and user-friendly vehicle for the design of module generators and parameterized cell generators.

1.1. Procedural Design Languages

A procedural design language supporting parameterized cell specifications allows a designer to incorporate a great degree of flexibility into a circuit description. A circuit design expressed in a procedural design language looks like, and has a lot in common with, a computer program. Statements in the program specify circuit elements and relationships between circuit elements, while the control constructs of the language allow selective inclusion or exclusion of features. Parameters can be used to control all the possible variations of the basic module description and they can also be used at the lowest level of description to specify the location or the dimensions of shapes.

Inasmuch as procedural design languages offer a medium capable of supporting the specification of parameterized designs, it would seem that they provide a solution to the problem of how to distribute one designer's expertise in a usable form to a wider community of end users. There is one major problem with procedural design languages, however; namely their **ease-of-use**.

Experience has shown that designing all but the simplest cells in terms of language constructs is a difficult and exceedingly unattractive task. To win designer acceptance, a procedural design language should not be offered as an alternative to the designer's normal circuit design environment. It should be an enhancement, **completely integrated into that environment**.

2. AN INTERACTIVE PROCEDURAL DESIGN ENVIRONMENT

SKILL is a procedural design language. Its significance, however, lies not in its features for describing circuit designs (which it does fully support), but in the way it allows designers to access all the facilities that already are available on its host CAD workstation environment, with which it is tightly coupled. The thesis of this paper is that the use of a procedural design language **in conjunction with** a graphics-based CAD system yields a combination that is more powerful and effective than the sum of all its components. We will illustrate the power of this combination primarily by the use of examples.

2.1. Basic Features of SKILL

The list below highlights the features of **SKILL** that together make it well-suited for use in the design of module generators. Each of the highlighted features is elaborated upon in the remainder of this section.

- Interactive Language
- User Interface Features
- Full Programming Language Capabilities
- Tightly Coupled with Rest of System
- Design Database Access

2.1.1. Interactive Language

SKILL is an interactive language. That is to say, that the user can enter a single **SKILL** statement and observe its effect immediately. If the effect of the **SKILL** statement is to add a shape to a cell design, then that shape is immediately reflected on the pictorial representation of the design shown on the user's graphics screen. This fact in itself is of significant benefit in allowing designers to develop **SKILL** procedures incrementally. For example, the **SKILL** instruction

    rectangle(0:0 25:25)

has the effect of adding a primitive rectangle shape to the design currently being edited (with lower-left and upper-right coordinates (0:0) and (25:25) respectively). If this single instruction is entered interactively, the rectangle is immediately displayed on the graphics screen. Instead of using absolute coordinates for the above rectangle, variable names could be used, or the user can be prompted interactively to specify with the mouse pointing device exactly where on the layout the rectangle should be positioned. This would be performed in **SKILL** with the instruction

```
rectangle(getpoint("please indicate the 1st corner")
        getpoint("and now the 2nd corner"))
```

2.1.2. User Interface Features

SKILL provides a variety of powerful operations supporting interaction between a procedure (the module generator) and the user. The first of these is as introduced above: *getpoint*. *getpoint* prompts the user to indicate a location on the screen by moving the cursor to the spot and clicking one of the mouse buttons. The text with which the user is prompted can be specified as an argument to the function when it is called, as illustrated in the example above. The result returned by *getpoint* is the location of the point indicated by the user, expressed in the coordinate system of the design being viewed in that window.

getresponse is a similar function, except that, in addition to returning the location of the cursor (in both screen coordinates and the coordinate system of the design being viewed in the window at which the cursor is pointing), it also identifies the particular mouse button clicked or the particular keyboard character pressed by the user in response to the prompt. As with *getpoint*, the prompt message may be specified as an argument to the function call, as also may be the prompts associated with each mouse button which are always displayed across the top of the graphics screen.

makeform is used to solicit values of parameters from the user. Its effect is to bring up a form on the screen for the user to fill. A single call to *makeform* is used to prompt the user to provide values for several parameters at once. In the same instruction, a prompt string plus the parameters' purpose, type, initial value, and any constraints on their range may also be specified. For example, the **SKILL** instruction

```
makeform("Please enter values for the parameters indicated"
        (wordlength,"number of bits per word", int, 8, (1 128))
        (memory_size,"number of words in the memory", int, 8, (1))
        (aspect_ratio,"preferred aspect ratio", float, 1, (0.1 10))
        (abort,"abort the generator",Boolean,no))
```

brings up a form which allows the user to alter the values of variables "wordlength", "memory_size", "aspect_ratio" and "abort" within their specified ranges. The format of the form is identical to that which is used throughout the host system, so the user is accustomed to working with this interface. The system checks the value entered by the user for each parameter to ensure that it conforms to the type and range specified in the instruction call. If not, the user is informed of the error and requested to enter another value. On return from **makeform**, the variables have the values either as set by the user or else as initialized in the instruction call.

There are many applications for which pictorial images are more appropriate than words in interacting with the user. **SKILL** provides the capability for the procedure writer to present the user with a pop-up menu containing either pictorial symbols or text, or a mixture of both. The command *popupmenu* brings up a menu, specified by its argument. The menu will previously have been designed using the host system's facilities for user-defined menus. The menu invoked by the instruction call is identical in all respects to the type of menus the user encounters throughout the host system. A typical example of the use of this capability might be an application calling for a parameterized transistor generator. The easiest way for the user to select from the variety of generators provided would be from a menu depicting them all pictorially, as illustrated below.

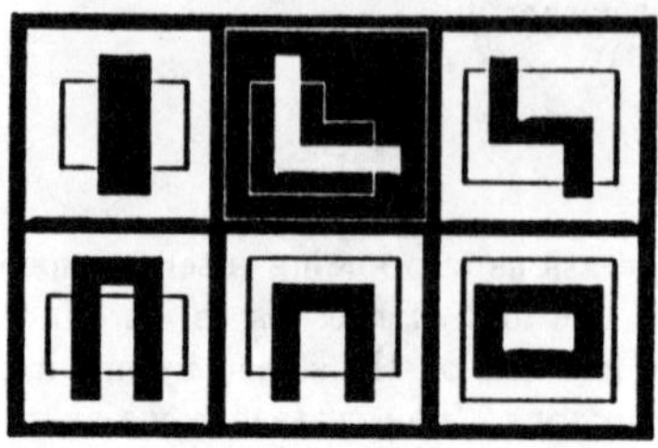

Associated with each menu entry is a string identifying that entry. The string associated with the transistor style selected by the user would be returned as the result of the call to **popupmenu**, allowing the controlling procedure to then select the appropriate specialized generator.

2.1.3. Full Programming Language Capabilities

SKILL is implemented as an extension of IL, a full-functionality, general-purpose programming language, embedded in LISP, but with a more C- or Pascal-like syntax. This means that a full complement of control constructs and data types are available in **SKILL**. As well as the normal conditional and loop control functions found in all languages, IL supports powerful set-oriented control functions, such as *foreach*, which applies a function to each element in a list; or *setof*, which selects from a list all of the elements of the list which satisfy a given condition.

2.1.4. Tightly Coupled with Rest of System

SKILL extends IL by providing access to all the functionality of the host CAD system, so that any operation that can be executed interactively on the host may also also be invoked procedurally from **SKILL**. For example, the instruction

```
graphicseditor("pdvExtract")
```

invokes the host system's physical layout extraction tool. Most of the development effort that has been invested in **SKILL** has gone into providing instructions that tightly integrate the language into the system. Instructions provide direct access to a wide range of system utilities, such as display control primitives (setting the shape of the cursor, for example), user-interface facilities, structure compiler interface, and others. The utilities of the underlying host operating system are also available from within **SKILL**. This allows a designer to call, for example, a program to minimize a set of logic equations from within a **SKILL** procedure, such as a PLA generator. This aspect of the language, namely its interaction with the other features of the rest of the system, is the primary focus of the paper and we will return to examine its implications below.

2.1.5. Design Database Access

SKILL also provides access to the design database. That is, from within a **SKILL** procedure, it is possible to interrogate the design database for information about objects in the database and about the relationships between objects. It is in this realm that the power of the control constructs provided in the underlying IL language can really be exploited. For example, the following fragment of a **SKILL** procedure traverses all the geometric shapes belonging to a logical net (in an extracted physical layout) and prints out the value of the "capacitance" property of each shape:

```
foreach( inst net ->instances
     print(inst ->capacitance))
```

This aspect of the language allows the designer to incorporate design feedback loops right into the module generator itself: to modify an initial cut at a design based on an analysis of its performance.

3. THE STRUCTURE COMPILER

The Structure Compiler is a versatile tool that can be used to build module generators by designers. An *array-structure templates* and a *master cell library* are the only components that a designer has to create to define a module generator. The array-structure template is both a block diagram ("floorplan") and a procedure for generating the module. All dimensional information is extracted from the library cells automatically. The composition tool can be used to build module generators even for modules consisting of multiple arrays and arrays of master cells of different heights and widths. In addition, a procedural mechanism is provided to allow the designer to modify and insert powerful optimization steps inside the module generators. Having graphical and procedural mechanisms in one tool, designers can easily create and modify module generators for structures such as RAMs, ROMs, PLAs, ALUs, shifters and array-multipliers without extensive programming.

3.1. Module Generation with the Structure Compiler

The structure compiler is a tool that automatically generates the final layout of a module from an array-structure template, a library of master cells and an optional *personality* file. The module to be generated may contain multiple blocks of arrays. These arrays can be of two types: homogeneous array and heterogeneous array.

A homogeneous array is an array using only one master cell. Only the number of rows and columns together with the master cell are needed for the generation of a homogeneous array. An array of input buffers to a ROM is an example of homogeneous arrays.

A heterogeneous array is an array of many different master cells. The structure compiler uses a personality file specified by the user to determine which cell is to be placed at different locations. For example, the layout of a ROM in general contains two different cells: one with a transistor and one without. The personality matrix specifies where to insert one cell or the other.

In the next sections, we present how to specify the structure of a module by the array-structure template and how to specify the cell library.

3.2. Structural Specification of a Module

The array-structure template is a block diagram representation of the module to be generated. All generic structural information that is not technology dependent is stored in the array-structure template. Fig. 1 shows a template for a Programmable Logic Array (PLA) structure. Each array-structure template is composed of one or more rectangles called the *array-blocks*. Each array-block signifies a single array of one or more master cells from the user-provided library. The specification of which cell is to be placed at each location is provided in the personality file.

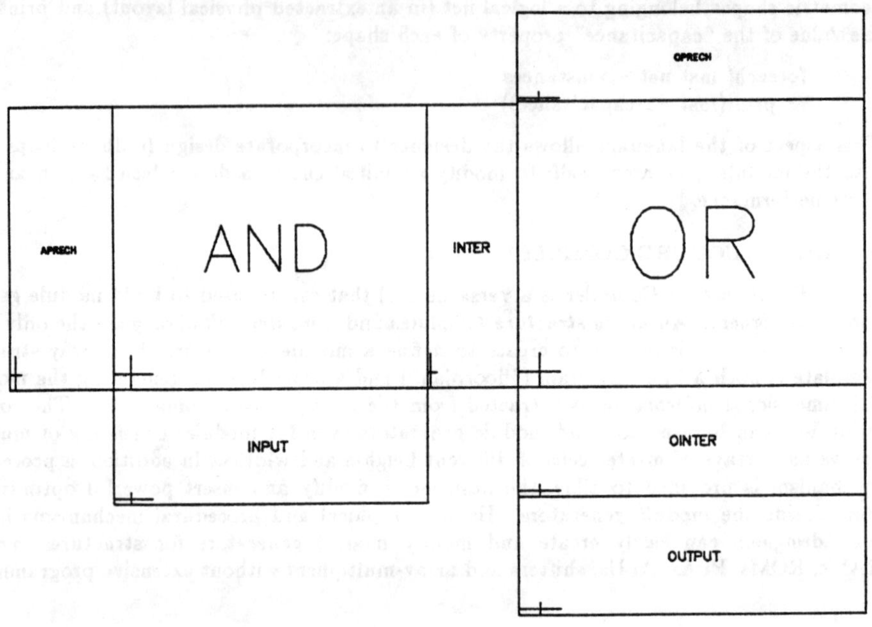

Fig. 1: Array-Structure Template for a PLA

The relative positioning of the blocks of arrays in a module is specified through matching of corners of neighboring array-blocks. The template in Fig. 1 has eight array-blocks each representing a different array with a specific functionality. Each array-block also has a list of properties or parameters, including:

- which section of the personality file to use,
- the type of array (either homogeneous or heterogeneous),
- the mapping of single-character symbols in the personality file to other single-character symbols, (for example, the symbol "0" is mapped into two symbols: "N" followed by "T"),
- the number of columns and rows if needed. If this property is empty the array will be sized automatically by the personality file,
- the ways to translate symbols in the personality matrix to cell names in the library,
- a two dimensional *orientation pattern* to be repeated through out the entire array,
- and a procedure to further manipulate the array.

Fig. 2 shows the property list from the "AND" array-block in Fig. 1. The array-structure template is used as a graphical procedure for the structure compiler to generate the module.

```
name: AND
personality: p0
type: HETEROGENEOUS
set:
map: 1:TN; 0:NT; -:NN;
masters: N:plaTran0 layout;
         T:plaTran1 layout;
orientation pattern: 2  4
                     R180 MY
                     MX   R0
                     MY   R180
                     R0   MX
pin:
procedure: optimizeAnd()
```

Fig. 2: Properties of an Array-Blocks

3.3. Master Cell Library

There is no limitation on what kind of information is in the master cells. The master cells are typically entered using a graphics editor. The only requirement for the cell is a rectangle drawn on a special layer. This rectangle provides the size of the master cell to the structure compiler. Since the dimension information is stored at this level, the structure compiler can assemble arrays that have different spacing for different columns and rows. This is an important feature in order to support heterogeneous arrays. Fig. 3 shows the master cells used by the PLA structure shown in Fig. 1.

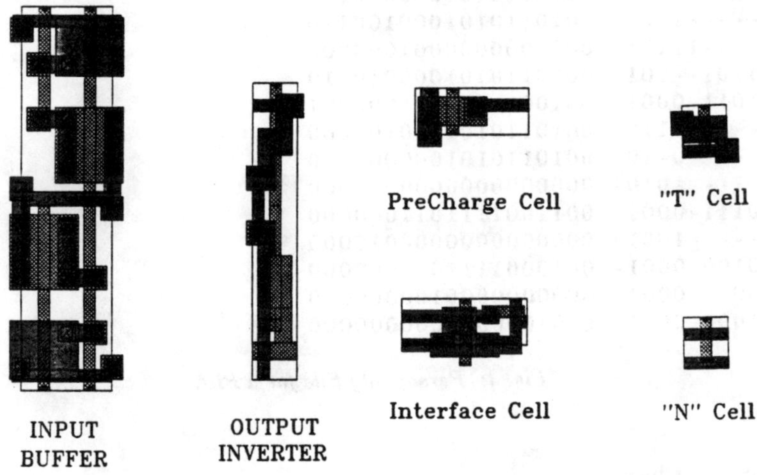

Fig. 3: Master Cells for a PLA

3.4. Personality Matrix File

To generate a module that contains heterogeneous arrays, a personality file may be
provided. A personality file contains connectivity information and is used to assemble the
parts for the overall assembly. This file can be the truth table for a PLA if the module is
a PLA. It can be the character symbolic layout file for a fixed-grid character based symbolic layout system. It can be a relative placement file for the different blocks in a data-
path or array-multiplier. Each character in the personality matrix is a single-character
symbol which can be further mapped to one or more single-character symbols by specifying
the mapping function in the array-block property list. Each of these symbols is then
translated to one of the cells in the library. Fig. 4 shows the content of a personality file
which is the truth table for a PLA.

```
.name alustate
.plane AND OR
----1--10-  0000000000001000000
-----1010-  00101101010001011110
----11-01-  0000000000001000000
00---0001-  00001101010110000000
-00--0001-  00100001010110000000
--011--01-  0000000000001000000
1100-0001-  00001101010000000
0110-0001-  00101101110110000000
1001-0001-  0000111010000000000
1010-0001-  00100111110110000010
0101-0001-  00111101110110000000
-----00001  11000000000000000100
------101-  00101101010000000110
-----1-01-  00101101010001000110
----1-101-  0000000000001000000
1101--001-  00001101010000010110
1011-0001-  00100111110110000001
-----1110-  00101101010101010100
------0-10-  00101101010000001110
------010-  00000000000000010000
0111-0001-  00110011110110000000
-----1001-  00000000000000010000
0100-0001-  00100011110110000000
00-1-0001-  00000000001000000000
1000-0001-  0001000010000000000
```

Fig. 4: Personality File for a PLA

3.5. Array Procedure

In order to allow the user more freedom and power to further optimize the arrays, a
user-provided procedure can be accepted by the structure compiler to manipulate symbols
in an array after it has been generated from a template and a personality file. The procedural language permits the user to inquire about the contents of a certain row and
column location of the personality matrix and can conditionally specify the following
operations:

- insert a column or row,
- replace a symbol at a certain location with another symbol,
- embed a new symbol at a certain location and thus increase the row/column size,
- set new masters, and
- set new orientation patterns.

The array procedure can be used to generate a new personality matrix based on properties provided by the user. Fig. 5 shows a procedure for stripping unused polysilicon from product terms in the "OR" array-block of the PLA structure shown in Fig. 1. The semicolon (i.e. ;) is used to mark the beginning of comments in each line.

```
; Function for stripping unused polysilicon from input lines
; in the AND plane of a PLA.

procedure( optimizeAnd( )
    ; Start at the top of each column of the AND plane,
    ; and work towards the bottom, replacing "no-transistor"
    ; cells with "no-poly" cells, until a "transistor"
    ; cell is encountered.

    prog( (no_transistor no_poly r c stopP theIcon loc)
        no_transistor = "N";
        no_poly = "M";

        for(c 0 columns-1
            r = rows - 1;
            stopP = nil;

            ; Iterate over the current column, replacing cells.

            while( (r >= 0) && !stopP
                loc = xy(c r);
                theIcon = icon(loc);
                if( theIcon == "N" then
                    changecell(loc no_poly orient(loc));
                else
                    stopP = t;
                )
                r--;
            )
        )
    )
)
```

Fig. 5: A Procedure Example

3.6. Steps in Module Synthesis

The structure compiler first generates a list of array-blocks to be processed based upon the array-structure template. Then for every array-block, an array is synthesized according to the properties of the array-block. The user-provided procedure is then executed. After all the arrays are synthesized, the structure compiler then places all the arrays according to how the corners of the array-blocks in the template are matched. For example, in Fig. 1, since the lower left hand corner of the "AND" array-block matches the lower right hand corner of the "APRECH" array-block, the structure compiler will place the synthesized array corresponding to the "AND" array-block so that the lower left hand corner of the array is matched with the lower right hand corner of the array synthesized for the "APRECH" array-block. If the upper corners also matches, the structure compiler will check to make sure that the corresponding synthesized arrays are the same height. This "corner-matching" operation is carried out for all corners of each array-block until all the arrays are placed. Throughout the synthesis process, error messages as well as graphical feedbacks are returned if the master cells or the arrays do not fit or connect properly. Fig. 6 shows the final layout of a PLA using the template in Fig. 1, the library of cells in Fig. 3 and the personality file in Fig. 4.

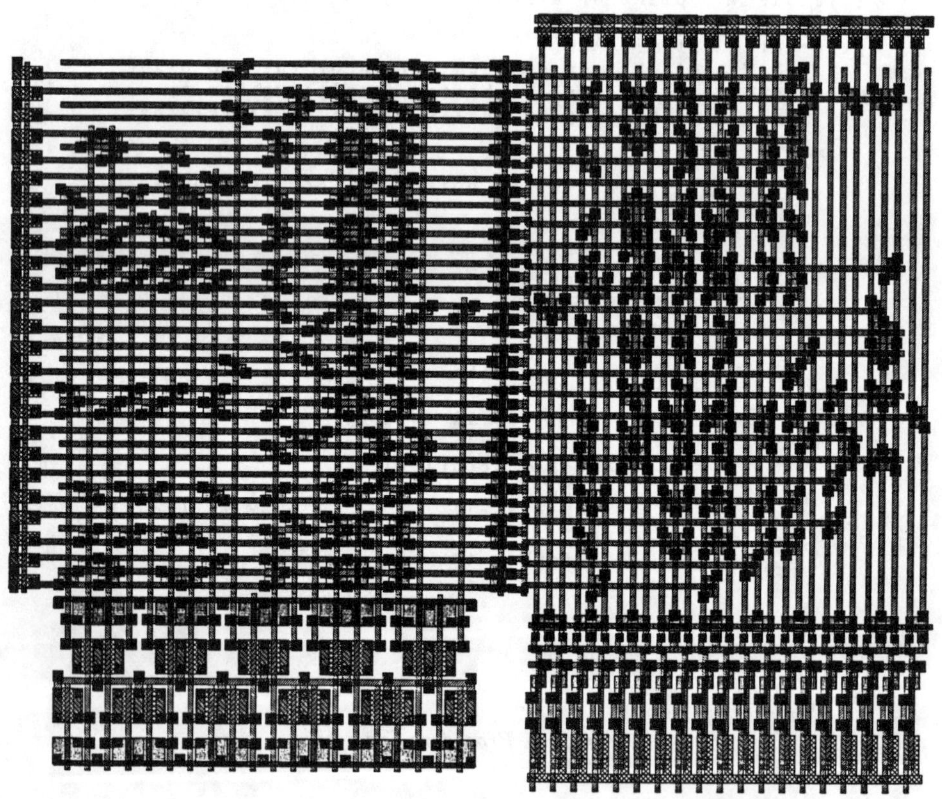

Fig. 6: The generated layout for a PLA

4. Using SKILL and the Structure Compiler to Design Module Generators

We have already expressed our belief that a procedural design language is not the most appropriate medium in which to cast all of the design of a module generator. Here we illustrate how **SKILL** is used in conjunction with the *structure compiler* to facilitate the task of designing sophisticated module generators. We characterize our philosophy as a **mixed-media approach** to module generation, in that the different phases of the task are performed in the expressive medium best suited to that purpose.

The structure compiler, on its own, can be used to assemble a wide variety of structures. The combination of the structure compiler and the **SKILL** language provides a powerful tool for the design of a wide variety of easy-to-use functional module generators. The foundations of the generator are provided by the structure compiler, while **SKILL**, which supports a large body of instructions devoted to interfacing the structure compiler, supplies the flexibility. We will illustrate this with a couple of examples.

The most simple combination utilizes **SKILL** just as a friendly user interface on top of the structure compiler. As noted in the previous section, **SKILL** provides a very powerful primitive instruction, "makeform", for the purpose of interactively soliciting values of parameters from the user. The fact that the system checks the value entered by the user for each parameter to ensure that it conforms to the type and range specified in the instruction call relieves the procedure-writer of the tedious task of validating of the user's data: a task that even experienced programmers are notoriously lax in performing.

On return from the "makeform" instruction, the parameters whose values have been set by the user are usable as variables within the **SKILL** procedure. It is then a simple matter to prime the structure compiler with these values and to invoke it to generate the appropriate module. The following procedure illustrates the essence of the process for a simple shift register example.

```
procedure( shift_reg_gen()
    makeform("Please specify the desired width of the shift register"
             (sh_width, "number of bits wide", int, 16, (1, 256))
             (abort, "abort the shift register generator", Boolean, no))
    if( abort then return())         ; user has set "abort" to TRUE

        ; call procedure to make the shift register
    make_shift_reg("shift_reg layout","shift_cell layout",sh_width)
)

procedure( make_shift_reg(dest, leaf_cell, width)
        ; edit the structure template
    graphedit("shift_reg template")
        ; set #cols in array to shift reg width
    setTemplateField("cells_plane","set",1,width)
        ; set leaf cell
    setTemplateField("cells_plane","master",leaf_cell)

    compilearray(dest,"shift_reg template","none")
)
```

Note that most of the effort of designing the module generator goes into the "true" design effort of designing the leaf cells and their interconnection. The assembly of the module is done by the structure compiler. Only the user interface and the overall control of the generator are programmed in **SKILL**.

By adding "pins" to the leaf cell, the leaf cell also becomes a schematic symbol. The resulting module created by the generator then also serves duty as the schematic of the module. This mechanism relieves the user from the necessity of creating the schematic of the module dynamically and always guarantees the logical representation to be consistent with the physical representation. Timing analysis and simulation can then be run directly on the resulting module if the simulation model for the leaf cell is available.

Note that within the same procedural framework, it is simple to generate different variants of the shift register depending on options specified by the user; static versus dynamic latches, for example. This would be done by altering the leaf cell used in building the shift register array: from a static layout cell to a dynamic layout cell. Again, note that the **makeform** instruction will only allow the value "static" or "dynamic" to be given to the variable "type". The procedure would then look like

```
procedure( shift_reg_gen()
    makeform("Please specify the desired width of the shift register"
        (sh_width, "number of bits wide", int, 16, (1, 256))
        (type, "Type of shift register", string, "static", ("static", "dynamic"))
        (abort, "abort the shift register generator", Boolean, no))
    if( abort then return())                    ; user has set "abort" to TRUE

    if( type === "static" then
      make_shift_reg("shift_reg layout","static_shift_cell layout",sh_width)
    else
      make_shift_reg("shift_reg layout","dynamic_shift_cell layout",sh_width)
    )
)
```

The leaf cell could even be built dynamically from within the procedure, using **SKILL**, if we wanted to vary its attributes depending, for example, on the load it would have to drive. Finally, it is worth noting that the shift register generator can be called by the user interactively as illustrated above, or else it may be called as an integral part of some other generator, which need only call the "make_shift_reg" procedure.

The second module generator example we show, a ROM generator, is somewhat more sophisticated in its use of **SKILL**. As in the preceding example, the language is used to solicit the user's input which it then feeds to the structure compiler. It is also used to customize the ROM row-decoder by post-processing the structure the structure compiler generates for the row-decoder plane. The interaction between generator and user is more sophisticated in this example too: the procedure calculates a variety of possible aspect ratios for the module, depending on the input data, and presents their images graphically, allowing the user to select one of them simply by pointing at it. Finally, the database access features of the language are used to analyze the synthesized circuit and automatically create a symbol for the module, for inclusion in a schematic diagram.

Space constraints permit us to show the top-level procedure in outline only. The procedure used to customize the row-decoder, called by the structure compiler as it is synthesizing that plane, is shown in its entirety.

```
procedure( rom_gen()
    makeform("Please enter values for the parameters indicated"
        (content,"Name of the file containing the coding for the ROM", file )
        (latch_type,"output latch type",string,"static",("static", "dynamic"))
        (abort, "abort the ROM generator", Boolean, no))
  if( abort then return())         ; user has set "abort" to TRUE

    ; Calculate possible aspect ratios based on ROM content and leaf-cell
    ; size. Draws picture of each possibility and asks user to select one.
  aspect_ratio_type = get_aspect_ratio( content, latch_type)

    ; create template and personality file based on aspect ratio selected
  personality_file = create_personality( content, latch_type, aspect_ratio_type)
  template = create_template( content, latch_type, aspect_ratio_type)

    ; call the structure compiler to make the ROM
  compilearray("rom layout", template, personality_file)

    ; create a schematic symbol for the ROM automatically from the layout
  create_symbol("rom layout", "rom symbol");
)


procedure( row_decoder(numrows, numcols)
    ; Decoder initially populated with cells all one type, "zeroSelect".  Want to
    ; selectively replace "zeroSelect" with "oneSelect" to get correct decoder pattern.
    ; Transforms array from:                to:
    ;                       0 0 0           1 0 1
    ;                       0 0 0           0 0 1
    ;                       0 0 0           1 1 0
    ;                       0 0 0           0 1 0
    ;                       0 0 0           1 0 0
    ;                       0 0 0           0 0 0

  addmaster("oneSelect layout", "1")   ; declare leaf cell and its associated icon
  for( col 0 numcols-1  ; from least-significant to most-significant col
    row = 0
    period = 2**col     ; length of consecutive strings of same cell in col
    while( row < numrows
        i = 0
      row = row + period                          ; bypass string of zeros
      while( row < numrows and i < period   ; replace zeros with ones
        changecell(row:col,"1")
        row++
        i++
      )
    )
  )
)
```

Fig. 7 shows the screen of the workstation when the various possible aspect ratios for the ROM have been calculated and drawn on the screen. The user is being prompted to choose one aspect ratio by a call of the **SKILL** instruction **getpoint**.

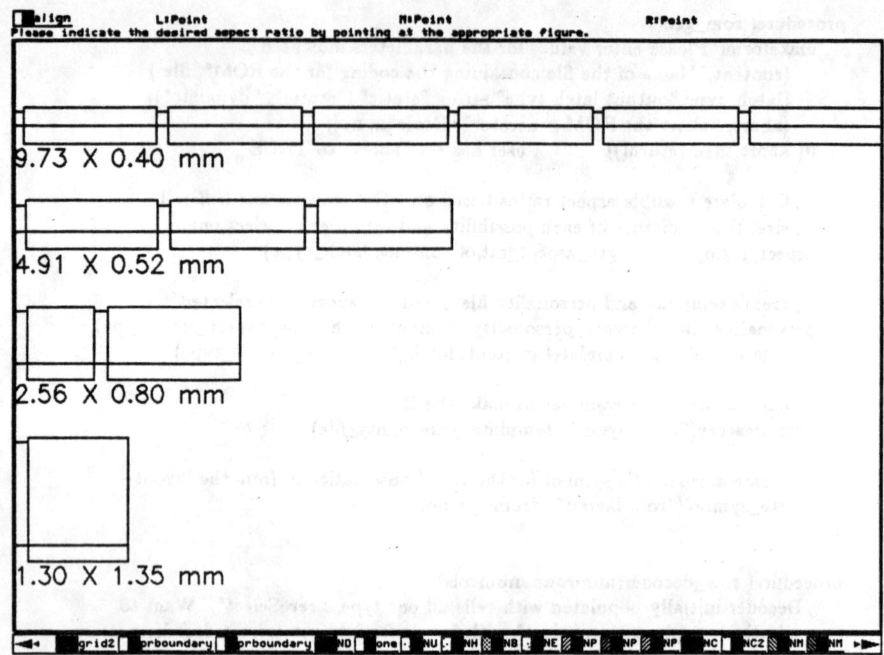

Fig. 7: User Selection of Preferred Aspect Ratio for ROM

5. CONCLUSION

Motivated by the goal of fostering the re-usability of circuit designs, we have incorporated a procedural design language into a full-functionality graphics-based CAD system. In addition to features supporting the expression of parameterized designs, the language supports many operations specifically directed at interaction with the rest of the system and, through the user-interface facilities of the system, with the system user.

We have argued that, if functional module generators are to become a relevant tool in the circuit- or system- designer's arsenal, designers need to retain control of the generators' design and the generators must be friendly to end-users. This means that the design of functional module generators has to be supported unequivocally by CAD systems. We have illustrated how the use of a procedural design language in combination with a structure compiler and other system utilities provides a powerful and supportive environment for the development and use of module generators. It is our belief that only such an integration of language and graphics-based tools can provide the answer to the question of how to make more effective use of the precious expertise of the skilled circuit designer.

REFERENCE

[1]   M.R. Buric and T.G. Matheson, "Silicon Compilation Environments" CICC 85, Portland, Oregon.

[2]   J.P. Gray, I. Buchanan, P.S. Robertson, "Controlling VLSI Complexity Using a High-Level Language for Design Description" ICCD 83, Port Chester, NY.

[3]   J. Batali and A. Hartheimer, *The Design Procedure Language* A.I. Memo 598, M.I.T. Sept 80.

[4]   R. Lipton et al, "ALI: A Procedural Language to Describe VLSI Layouts" DAC 82, Las Vegas, Nevada.

Design of Module Generators and Silicon Compilers

Misha R. Burich

Silicon Design Labs, Inc.

Liberty Corner, N.J. 07938

1. Silicon Compilers and Semiconductor Industry

Silicon compilation is a new branch of integrated-circuit (IC) design automation. There are several reasons for its emergence:

- IC manufacturing technology offers capabilities for integrating highly complex circuits on a single chip: 100,000 to 1 million transistors on a chip are becoming a common occurrence,

- Software technology offers highly sophisticated compiler and data-base products,

- Due to good heuristics, computational algorithms have advanced to the stage of handling complex problems quickly. Placement and routing, logic minimization and synthesis, simulation, fault-grading, and design for testability, are examples of problems that have been aided by good algorithms.

- Work-station technology offers powerful computing capabilities to engineers at ever decreasing cost: 32-bit processors, virtual memory management, tens of mega-bytes of random-access memory, hundreds of mega-bytes of hard-disk memory, floating-point accelerators, and high-resolution graphics are standard features in present work-stations.

Due to these factors, a new type of IC products is gaining popularity. Instead of using standard, off-the-shelf chips, engineers are turning to designing their own application specific integrated circuits (ASIC). Most often, they are integrating many functions into a single chip and combining such devices with standard parts to complete their board, or system, design. The functions they integrate on an ASIC chip would otherwise require many standard parts. The system cost, the power consumption, the cost of the final product, and its maintenance cost become lower with the ASIC parts.

Also, manufacturers of the standard parts are starting to offer many variations of their popular chips. Each variation includes some additional feature, integrated on the same chip. This trend leads to Standard-Application-Specific-Integrated-Circuits. The manufacturers use new methods in design automation, including silicon compilation, to design these SASIC parts.

The relationship between the manufacturers of ASIC devices and their designer customers has been rapidly evolving over the last few years. Gate-Array and Standard-Cell layout styles became the initial dominant technologies. The reasons: the layout design is fully automated, the turn-around time from the design phase to prototypes is short, and the manufacturing reliability is very high.

Figure 1 shows a typical relationship between users and providers of ASIC products. The user takes advantage of work-stations to perform conceptual design in schematic form. The design is verified by simulation, and its test procedure is developed by using test vector generation and fault grading. Physical layout has been traditionally done with gate-arrays and standard cells but a more general Cell-Based technology is emerging.

Cell-based technology relies on libraries of fixed or parameterized function blocks produced by module generators. It is characterized by larger function-blocks mixed together with standard cells where some form of user-assisted floor-planning may be employed.

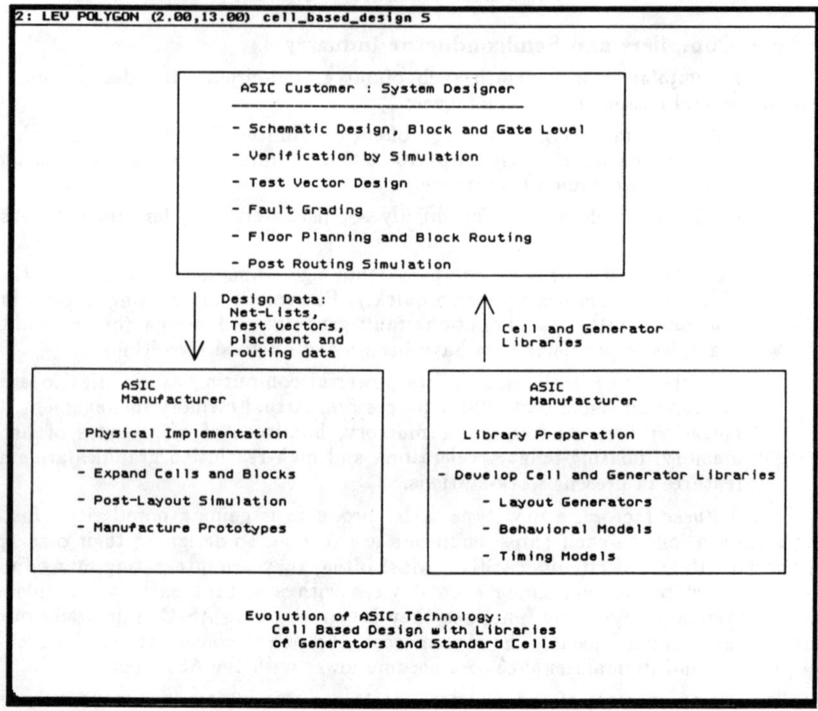

Figure 1.

1.1. The Design Process for Integrated Circuits

In any IC design, including traditional full custom design, designers employ three major levels of abstraction: functional, structural, and geometric. The same levels of abstractions are needed for silicon compilation.

The functional description specifies system behavior, such as instruction set, logic functions, input/output behavior at pins, and timing relationships among signals. The functional description specifies WHAT is to be built.

The structural description specifies the system's architectural components, and their interconnections. The designers partition the whole system into logic sub-systems, data-paths, memory sub-systems, input/output blocks, and sub-system connectivity. The structural description specifies HOW the system is to be built. There is no unique structural description of a system for a given functional description. Indeed, the designer's creativity and skills play the major role in partitioning the system into its components.

The geometric description specifies the physical implementation of the system, such as its floor-plan, the placement of cells, the routing of the blocks, and the layouts of cells. The geometric description also specifies HOW the chip is built, but it is much more specific than the structural description, because it is tied to a specific manufacturing process, such as CMOS, NMOS, bipolar, or GaAs. Again, there is no unique geometric description of the system for a given structural description. Again, the designer's expertise is essential in getting an IC that meets the area and performance criteria.

The IC design process is succinctly described by the Gajski's Y chart in Figure 2. Each axis represents one description of the design: functional, structural, and geometric. The design process a sequence of incremental steps along each of the axes, iterating from functional to structural to geometric, and then back to functional. Each iteration is applied to system's sub-systems, then to their sub-systems, until the smallest sub-systems are reached - gates and transistors. Verifications are performed at each iteration step to insure that the design goals are still satisfied. This is shown as a spiral that reaches the center of the Y chart - the complete design.

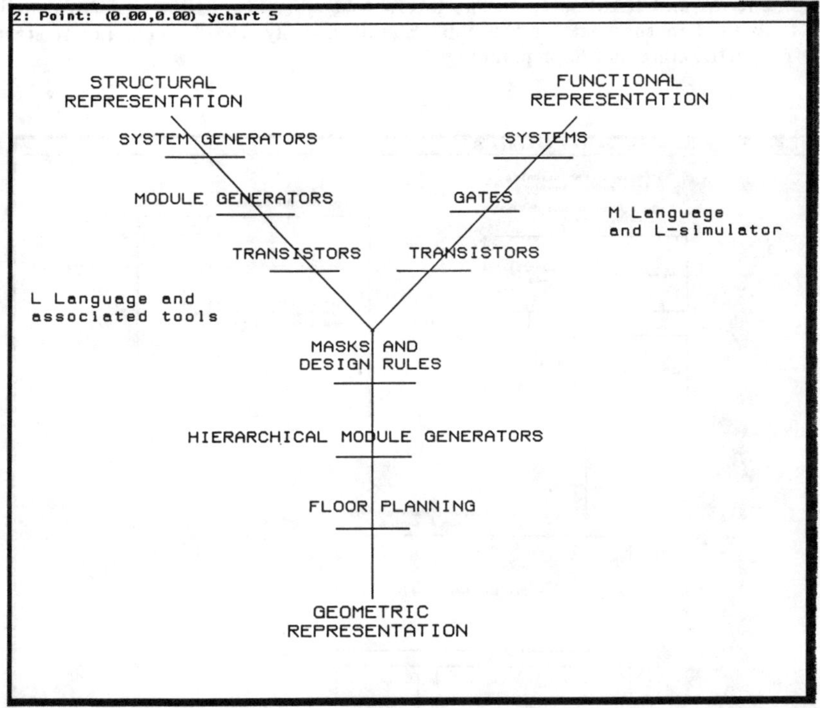

Figure 2.

1.2. Design Environment for Custom Design

A design system for custom ICs has to support all phases of the design process. An alternative view of the Y chart is shown in Figures 3 and 4.

The system designer starts by developing a functional model of the chip. The functional model is also an EXECUTABLE SPECIFICATION. It is executable because it can be simulated. It is a specification because it spells out all details of the system; it is a precise blue-print of the chip behavior. Functional models are developed in a functional description language. Examples include M (Silicon Design Labs), Helix (Silvar-Lisco), Verilog (Gateway Design Automation), and N.2 (Endot).

At this stage the designer develops test-vectors that are applied to the chip. The test will be used throughout the design process to verify the correct behavior of all sub-systems as they are developed.

Figure 3 shows that the designer proceeds by developing a structural description of the chip. This is can be done by interactive graphical editors, or by some structural description language, such as L (Silicon Design Labs), or Model (Lattice Logic). The system might be partioned into a control-logic section, Input/Output section, the data-path, and ROM and RAM section. The interactive schematic editor replaces pencil and paper tools used in the past for this job. Simultaneously, the designer might start the geometric partitioning and floor planning.

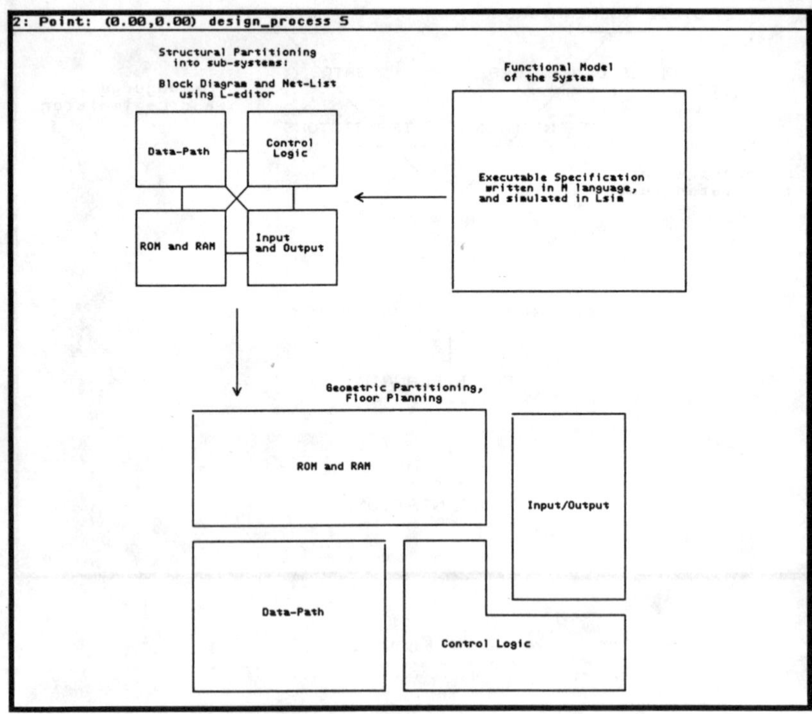

Figure: 3

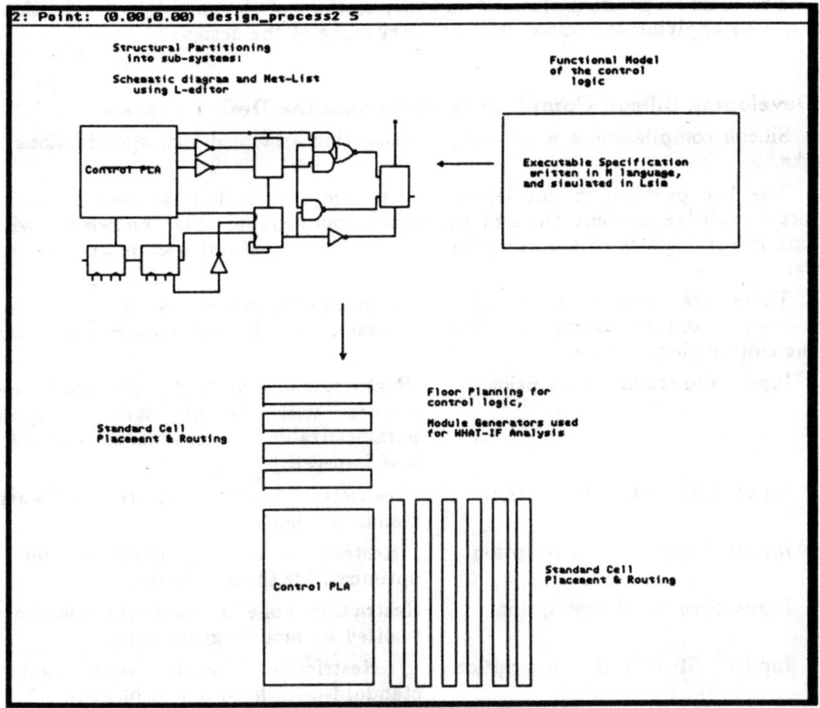

Figure 4

Figure 4 shows the next step, the design of the control-logic sub-system by applying exactly the same steps as before. The functional model of the control logic is developed in a functional description language, and simulated by the simulator. The structural partitioning is described in terms of more primitive building blocks: PLA and gate-level logic. The schematic editor is again used to capture the design intent.

At this point the designer can verify that the detailed control-logic block matches its functional model by using the simulator. The previously developed test-vectors are applied to the whole chip, but the functional model is replaced by the structural model, the net-list.

For geometric description, the module generator for the PLA can be used to provide information about the possible sizes and aspect ratios. The automatic standard-cell placement and routing software is used to evaluate the sizes of other parts of the control-logic.

This whole process can be executed rapidly, if module generators and random-logic placement and routing are available. Therefore, the tools enable the designer to perform many iterations of the WHAT-IF analysis. In many cases custom layouts have to be developed for circuits for which there are no automatic layout tools.

The same steps are performed on other sections of the chip. The verification is carried out by applying the test-vectors at every stage of the design.

2. Developing Silicon Compilers to Automate the Design Process

Silicon compilation is a process of translating system design specifications into IC masks.

The key problem in developing silicon compilers is how to capture and embed expert knowledge in some parts of the silicon compiler, mix that knowledge with algorithms in other parts of the compiler, and provide a natural user interface for the IC designers.

There are several classes of silicon compilers based on the type of input specification, and on assumptions that are made about the architecture and layout style in the chip implementation.

Input: Functional description	Restrictions: none (automatic choice of architecture with layout style composed of parametrizable macro-blocks mixed with standard-cells),
Input: Functional description	Restriction: fixed layout style (gate-arrays or standard-cells),
Input: Functional description	Restriction: fixed architecture, but layout optimized for the architecture,
Input: Structural Description	Restriction: none (all structural components are realized by module generators),
Input: Structural Description	Restriction: layout style (gate-arrays, standard-cells, fixed macro-blocks).

Silicon compilation from functional descriptions to layout with no restrictions on architecture or layout style is most complex. Sophisticated expert knowledge, rules, and algorithms have to be captured in various parts of the silicon compiler. This form of silicon compilation is in its infancy today, and will be evolving for a long time in the future.

Developers of silicon compilers need a suite of design tools. They may have to include knowledge-based system tools, traditional software development tools, and special tools for developing module generators.

Translation from functional models to structural models are most demanding. Design expertise that has to be captured includes architectural knowledge, reasoning about system trade-offs, understanding of technological trade-offs, and an ability to experiment with different options. Today, this translation is done by experienced architects and system designers. The most promising software technology for this component of silicon compilers includes knowledge-based systems, a methodology of capturing and using human expertise.

Translation from structural models to geometric and layout models requires a combination of expert knowledge, good algorithms, and module generators. The expertise that can be captured in this component of silicon compilers includes architecture design, logic design, circuit design, and layout design. The required algorithms include placement and routing, logic verification and simulation, circuit performance verification and simulation, mask layout verification, and fault grading and simulation for test purposes. Traditional software tools are used for this component of a silicon compiler: high level languages, debuggers, and language parsers.

Module generators automate the translation from structural descriptions to geometric descriptions. Expert designers can use the module development tools to capture their knowledge. The generators may be as simple as logic gates in CMOS, NMOS, bipolar, or GaAs technologies. Also, they may be more complex parameterized function blocks, such as data-paths and micro-code sequencers. Finally, they may be very complex function blocks such as microprocessors. No matter what complexity, the tools have to be applied in the same consistent fashion, taking advantage of hierarchy and re-usability of lower level functions. The designers should not have to sacrifice layout density or circuit performance in building generators, and they should be able to build-in mask level process independence. There are several systems for developing module generators: Generator Development Tools (GDT, by Silicon Design Labs), SILC tools (GTE), Genesis (Silicon Compilers, Inc), DPL language (MIT), and SLIC language (Seattle Silicon Technologies).

The developer of a general silicon compiler faces the following steps:

1. Define the input behavioral/functional language. It has to be user-friendly, expressive, and it should not restrict the range of applications. Some graphical input could be useful. In case of interactive input, it is useful to consider Artificial Intelligence (AI) techniques for a Design Assistant/Advisor approach (DAA). The DAA should have access to other parts of the compiler in order to provide viable design alternatives.

2. Develop techniques for mapping functional descriptions to structural descriptions. Knowledge-based systems with captured expert experience are probably the best alternative. The mapping can also be interactive, as in step 1. The mapping should also have access to the estimates of chip size, performance, and power consumption. This is obtained from steps 3 and 4.

3. Develop mechanisms for structural to geometric translation. Here, module generators, floor planning, placement, and routing play the major role. Also develop fast size and performance estimators to be used by procedures in step 2.

4. Develop simulation and verification models to be used in all phases of the compilation. Functional modeling languages and simulators are intended to serve this function.

Fully general silicon compilers will be evolving in the future. So far, the reported compilers always make assumptions about some aspects of the design. As is shown in next sections, the assumptions may be about the architecture, or the layout implementation style. Most promising in the short term are the compilers that start with a structural description of the design. Here, the translation from functional to structural description is left to the experienced designers.

2.1. Compiling Functional Descriptions into Layout

Silicon compilation is a process of translating functional descriptions of systems to their structural descriptions, and then to their geometric descriptions, or mask layouts.

As noted above, there is no unique way to translate functional descriptions to structural descriptions, and no unique way to translate these to geometric descriptions. Therefore every silicon compiler must incorporate some rules, constraints, and expertise in translating one representation to another.

One way to cope with this is to make assumptions about the underlying architecture (structural description), and layout style (geometric description). This simplifies the compilation task, can guarantee correctness by construction, and ease the verification problem.

Silicon Compilation with Fixed Layout Style

PLA design is one of the earliest examples [1]. Functional descriptions can be expressed by finite-state machines. Structural description is assumed to be given by two-level AND-OR logic with registers at some inputs of the AND gates. Geometric description is assumed to be given by NOR-NOR planes with densely packed transistors that share common gate and drain wiring. As a result, automatic compilation from finite-state machine language to mask layouts is possible. In this approach, the PLA minimization algorithms become very important. They include logic minimization and PLA folding algorithms.

Keeping the same PLA layout style, a different functional description language has been reported in [2]. Instead of finite-state machines, the regular expressions were compiled into masks.

A silicon compiler developer could design a compiler of this type with the following steps:

1. Define an input language and write a translator for it. The translator produces an intermediate data representation suitable for required transformations. Recommended tools are yacc, Lex, and the C language, all standard in the UNIX system. Lex can be used to generate a lexical analyzer, and yacc is a parser generator program.

2. Develop algorithms and programs for transforming the intermediate representation to a PLA personality matrix.

3. Develop (or use available) logic minimization algorithms to remove redundant states and logic.

4. Develop (or use available) PLA folding algorithms and programs to minimize the size of the PLA.

5. If testability is needed, develop PLA test strategy and algorithms for adding additional logic to the PLA.

6. Develop a PLA layout generator. Include options for folding, routing over the PLA, and user specified locations for input and output terminals. This is done in a module generator development language, with an aid of the layout editor, design rule checker, and layout compactor. The verification and timing characterization is done with the simulator.

7. Develop a functional model for the PLAs produced by this compiler. This is done in a functional modeling language, so that the user can simulate PLAs as a part of the larger system with the simulator.

8. Develop a program which will automatically invoke all the above programs in the proper sequence. As an example, the designer can use the shell, the standard command interpreter in the UNIX environment. The various parts of the compiler could communicate through temporary files.

A silicon compiler of this type is easy to use if the input language reflects the most frequently needed constructs. An example is a PLA compiler developed by SDL. The PLA is specified by a state-machine language. Logic minimization is performed by ESPRESSO, a UC Berkeley program. PLA folding is done by an SDL program. The layout generator is developed and written in the VLSI design language L.

The following is an example of the input specification program describing the Traffic-Light-Controller problem (Mead, Conway's [3]). The problem consists of designing a device that controls the traffic light at the intersection of a busy highway and a secondary farm road. The farm road will have a green light only if a car is detected on the road, and it will stay on for some limited amount of time, determined by a timer. The timer determines the duration of the yellow light (short) and the green light (long), and it is cleared by the car sensor.

```
/* Clocks */
CLOCKS PHI1 PHI2;     /* clock names to be used by the PLA */
/* Input terminals definitions */
IN INSTATE[2]  instate STATE; /* Two state bits for FSM */
IN CAR                 car;  /* Car sensor input */
IN SEC[2]              sec;  /* timer input: long and short*/


/* Output  terminals definitions */
OUT OUTSTATE[2]        outstate  STATE;
OUT CLRTIME            clrtime;  /* clear timer*/
OUT FARM[2]            farm;
OUT HIWAY[2]           hiway;

/* State assignment */
STATE higreen          "10";
STATE hiyellow         "00";
STATE farmgreen        "11";
STATE farmyellow       "01";

/* highway traffic light is green */
IF (STATE(higreen)) {
      hiway=0x1; farm=0x3; clrtime = 0;
      IF (~ car) NEXT_STATE(higreen);
      IF (sec=="0x") NEXT_STATE(higreen);
      IF (car && sec=="1x") {
            NEXT_STATE(hiyellow); clrtime = 1;
      }
}


/* highway traffic light is yellow */
IF (STATE(hiyellow)) {
      hiway=0x2; farm=0x3; clrtime = 0;
      IF (sec=="x0") NEXT_STATE(hiyellow);
      IF (sec=="x1") {
            NEXT_STATE(farmgreen); clrtime = 1;
      }
}
```

```
        /* farm  traffic light is green */
IF (STATE(farmgreen)) {
        hiway=0x3; farm=0x1; clrtime = 0;
        IF (car && sec==="0x") NEXT_STATE(farmgreen);
        IF (~ car) {
                NEXT_STATE(farmyellow); clrtime = 1;
        }
        IF (sec==="1x") {
                NEXT_STATE(farmyellow); clrtime = 1;
        }
}

        /* farm  traffic light is yellow */
IF (STATE(farmyellow)) {
        hiway=0x3; farm=0x2; clrtime = 0;
        IF (sec==="x0") NEXT_STATE(farmyellow);
        IF (sec==="x1") {
                NEXT_STATE(higreen); clrtime = 1;
        }
}
```

The PLA has five inputs and seven outputs. Two of the inputs are declared to be of the type STATE. These inputs will be connected to the two outputs of the same type, STATE. Other inputs are connected to external signals. Outputs that are not STATE are used to control other functions external to the PLA. The clock signals, PHI1 and PHI2, are used to latch input and output signals. The state-machine behavior is specified by statements of the following type:

```
IF( STATE( state_name ) ) {
        set outputs;
        compute NEXT_STATE;
}
```

The value of the next state is applied to the appropriate inputs on the next clock cycle.

A different regular layout style, named Storage-Logic-Arrays (SLA), was reported in [4]. Once the architecture and layout style was fixed, the authors were able to develop automatic compilation from Register-Transfer-Level (RTL) functional descriptions.

To develop a silicon compiler of this kind the developer would go through similar steps as above. The algorithms would be different, as well as the layout generator, but the principles are the same.

A different target layout style is exemplified by gate-arrays and standard-cells, which have been used extensively in automatic synthesis. The functional description can be a language, such as MODEL described in [5]. The silicon compilation proceeds by translating functional specifications to gate-level interconnections (structural description). Finally, automatic placement and routing software translates this into layout. Simulation and verification programs monitor the success of the compilation.

If the generator development tools contain placement and routing for standard cells, the developer of the silicon compiler would have to pay attention to the higher level steps. The developer would do the following:

1. Develop a functional design language. This might be a special purpose language, such as for signal processing, for communications protocols, or

general purpose RTL. Again, yacc, Lex, and C language are used for this purpose.

2. Develop heuristics and programs for translation from functional to gate level description. Translate these to the net-list format used by a standard-cell place and route package.

3. Develop standard or special cells that are needed as primitive components. Use a module generator language or a layout editor.

4. Develop an interface to the placement and routing package which automatically synthesizes layout.

Silicon Compilation with Fixed Architecture

Architectures based on data-paths controlled by sequencers are popular among designers. The MacPitts silicon compiler uses such an architecture as its target [6]. A design is specified by describing the behavior of the system in a Lisp-like language. The designer controls the architectural resources by specifying the degree of parallelism desired in the circuitry that is output.

Bit-serial architecture was employed in FIRST, the silicon compiler intended for signal-processing applications, and described in [7]. The specification language emulated the equations used by signal processing experts.

A complete parameterizable and programmable core microcomputer was the target of the Plex silicon compiler, described in [8]. The architecture of the Plex microcomputer was fixed, but the user could specify all the major resources of the design. These included the number of registers in the data-path, the size of the data memory, the size of the program memory, the stack depth, the number of interrupt lines, the size of the I/O busses, and the instruction set. The functional specification for the system consisted of the assembly language program that the resulting microcomputer was to execute. The compiler carried out the complete layout generation of the microcomputer, which could be augmented by other logic circuits on its periphery.

The generator development tools are well suited for developing silicon compilers of this type. The fixed architectures are constructed out of well defined but different components. These components are produced by module generators with many parameterized options. For example, the Plex compiler included a ROM generator, a RAM generator, a data-path generator, a PLA generator, random-logic synthesis, and glue generators for binding the pieces together.

The steps in developing such silicon compiler tools are as follows:

1. Define the target architecture, and the specification language for the users. This phase is very important because it sets the tone and range of applications for the silicon compiler.

2. Develop the language parser and the translation to the target structural description. The tools for this phase: yacc, Lex, and C language, all standard in the UNIX environment.

3. Develop module generators in a VLSI design language for all system components. Module generators should be hierarchical and should cover all needed options, both functional and geometric. The top level generator interfaces to the input parser. It calls other generators and asks them to provide needed options. It completes the layout by including input/output pads.

Floor planning, placement, and routing are done hierarchically and procedurally by generators on all levels of hierarchy.

4. Develop the user interface, development aids, and debugging aids. This is done using the C language, and perhaps some graphical utility programs. Develop functional simulation models. These are used in verification of the design.

SDL-2000 is such a silicon compiler for a micro-controller architecture. It is fully compatible with Intel's 80C51 micro-controller. Its purpose is to be used in applications which require a processor, memory, and some user-specified control logic integrated on the same chip. A particular instance of an SDL-2000 processor with some additional I/O glue logic is shown in Figure 5.

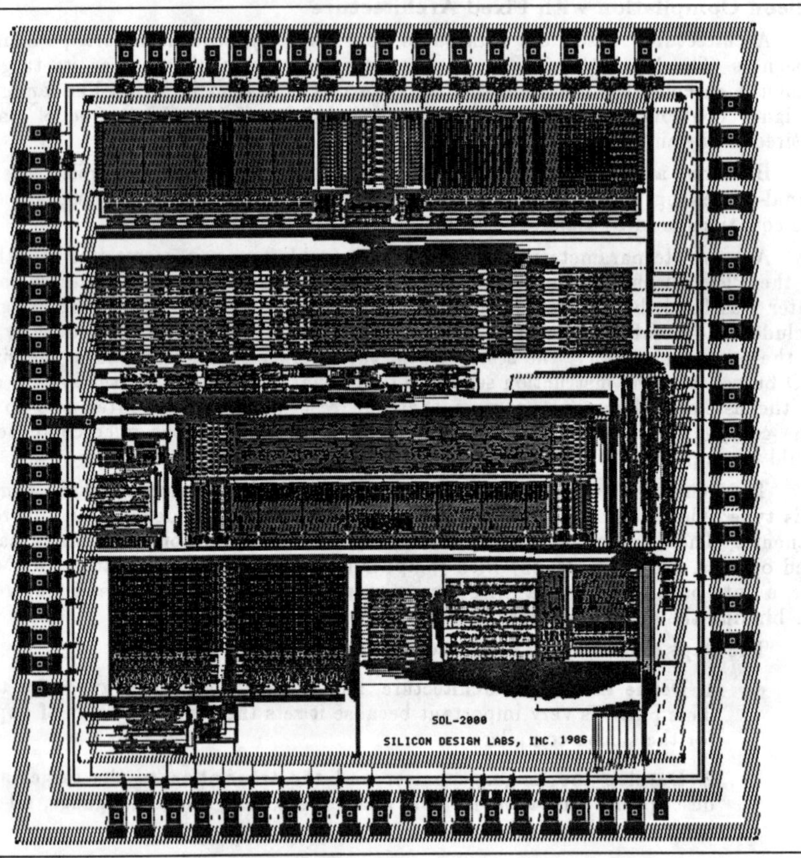

Figure 5.

2.3 Role of Module Generators in Silicon Compilation

Module generators are programs that produce functional, structural, and geometric descriptions from input specifications. They are an integral part of any silicon compiler. They can also be used in full custom designs in a semi-automatic fashion. In order to be useful, the module generators should provide compact layouts and should take advantage of the distinctive features of the target manufacturing technology. They should be functionally parameterizable, and geometrically flexible for easy integration with other modules.

Most common functions provided by generator libraries include arithmetic functions (adders, counters, arithmetic-logic-units (ALUs), shifters, and data-paths), memory storage functions (register banks, RAMs, ROMs), and control-logic functions (PLAs, sequencers, and random-logic synthesizers). They may include complex systems, such as the SDL-2000 microcontroller, shown in Figure 5.

Silicon compilers use generators when the structural description of the system is known. The design has been partitioned into sub-systems and their interconnections. Each of the sub-systems may correspond to a generator from the library, or to random-logic block to be automatically synthesized. The silicon compiler calls the generators from the library, passes parameters to them, and assembles the system by placement and routing procedures.

Input specifications to module generators take many forms, such as property lists, bit patterns, and net-lists. An ALU is an example of a system block that may be specified by by a property list. The list might contain entries for the number of bits, the geometric pitch, the boolean and arithmetic functions to be performed, and the type of output buffering. Tables of bit patterns are used to specify PLAs and ROMs.

In the next example the user fills a form to specify an ALU. The form has spreadsheet capabilities for displaying real-time estimates for size and speed of the block being generated. The user specifies a 4-bit ALU, with shift function, input latch, and output latch, and the tri-state output option.

```
Lform
                        ADDER GENERATOR FORM

              Cell Name: [alu_4bit                    ]

                   View: ↻LAYOUT


          CORE ADDER                 PERFORMANCE ESTIMATE

               Bits: [4]               carry chain delay:      20.00 ns

              Pitch: [69.50]  um       input set-up time:      12.00 ns


             OPTIONS                   total computation time: 32.00 ns

           Subtract: ↻yes             output stage delay:      20.00 ns

 Dynamic Input Latch: ↻yes

Dynamic Output Latch: ↻yes                   TOTAL DELAY:       52.00 ns

         Zero Detect: ↻yes          MAXIMUM FREQUENCY:         19.23 MHz

     Output Tristate: ↻yes

                                          AREA ESTIMATE

   Internal Bus Load: [5     ]  pF
                                               height:        278.00 um

                                               width:         645.00 um

                                          ACTIVE AREA:     179310.00 um[ ]


                                          [reset] [cancel] [ ok ]
```

Once the form is filled, the program constructs a call to the library generator to provide the user with two types of descriptions:

- Simulation Models: Functional model for fast simulation. Transistor or Gate model for accurate timing simulation and fault simulation.

- Geometric Models: Bounding polygons and terminal positions for placement and routing. Complete mask layout for final design verification and manufacturing.

The layout of this ALU is shown in Figure 6.

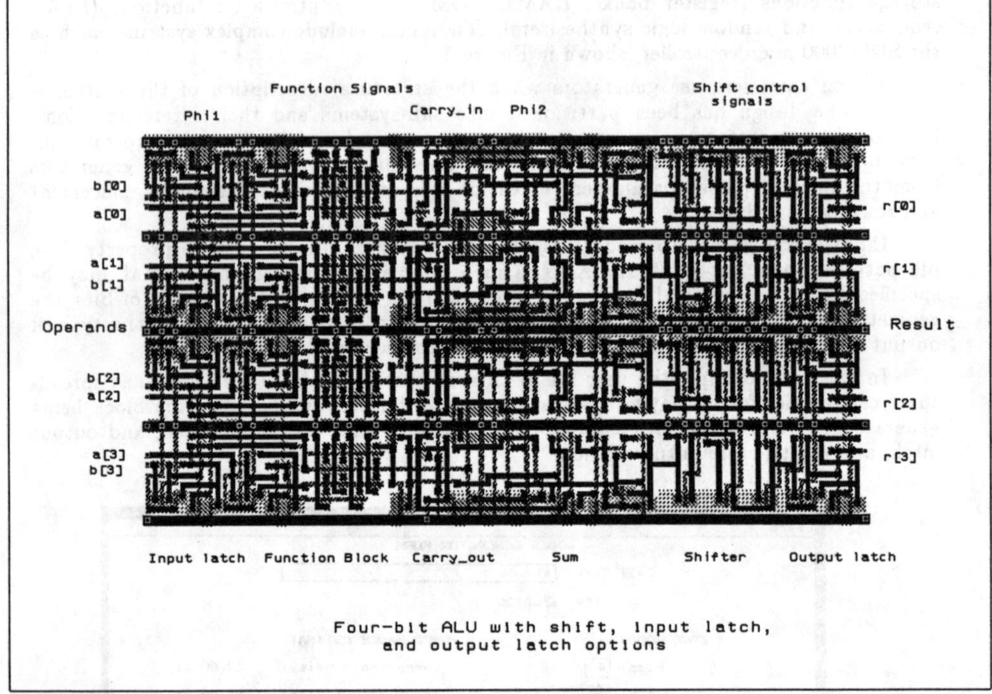

Figure 6.

A standard library of generators has to cover most commonly used system components. These can be divided into arithmetic functions (adder, counter, and ALU), memory functions (RAM, ROM, and latches), and control logic functions (PLA and standard cells with place and route package).

The standard cells in the generator library include inverter, tri-state buffer, nand, nor, static latch, dynamic latch, latched multiplexers, or-and-invert, and-or-invert, exclusive-or, and exclusive-nor gates. The cells are used by the automatic placement and routing program which synthesizes random-logic function blocks. Standard cells are either fixed cells or parameterized generators. If they are parameterized, they might include gates with a variable number of inputs, variable P-channel and N-channel sizes, and variable geometric pitch between power and ground busses. For example, the and-or-invert cell generator can produce 64 different logic gates: one to four AND gates whose outputs are connected to the NOR gate, and each AND gate can have one to four inputs. Figure 7 shows several standard cells produced by the generators.

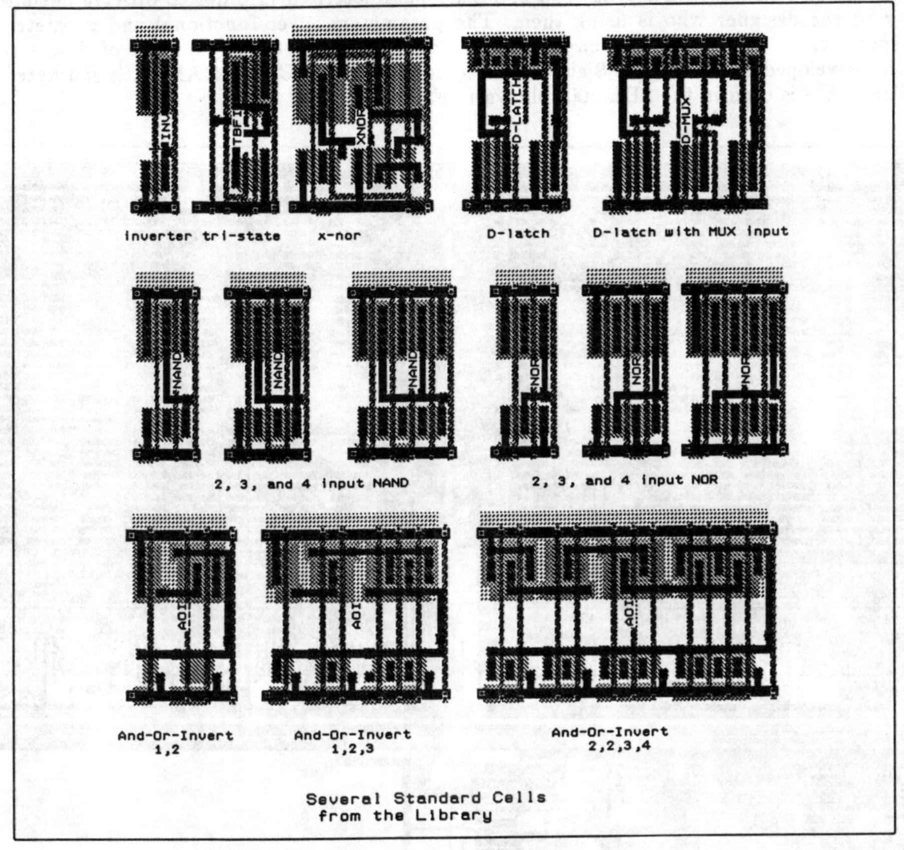

Figure 7.

ALU, adder, shifter, multiplier, and counter generators cover arithmetic functions found in digital designs. The function blocks can have high performance over a range of parameters if an expert designer develops specialized circuits to be used in such a variable environment. Their layout can be very compact if an expert designer develops special leaf-level cells that are optimized for these functions.

Typically, an ALU generator produces a function block whose options include the following: number of bits in the ALU, optional left/right shift function, optional registers at the ALU inputs, optional register at the ALU output, optional tri-state buffer at the ALU output, and programmable geometric pitch between the adjacent bits of the ALU.

Programmable pitch is useful if it is desired to route ALU busses between the bits, thus the number of busses can vary. The ALU performs add, subtract, bit-wise OR, bit-wise XOR, and bit-wise AND functions. If the shifter option is asked for, the ALU can perform a shift in the same clock cycle in which an arithmetic operation is performed.

ROM, register-file, and RAM generators produce memory storage functions, permanent and variable. They have to be highly parameterized in order to provide flexibility to the designer who is using them. The parameters cover functional and geometric aspects of the circuits. Their circuit and layout design reflects the expertise of designers who developed them. Figure 8 shows the layout of a static CMOS RAM with 256 bytes. This RAM is used in the SDL-2000 shown in Figure 5.

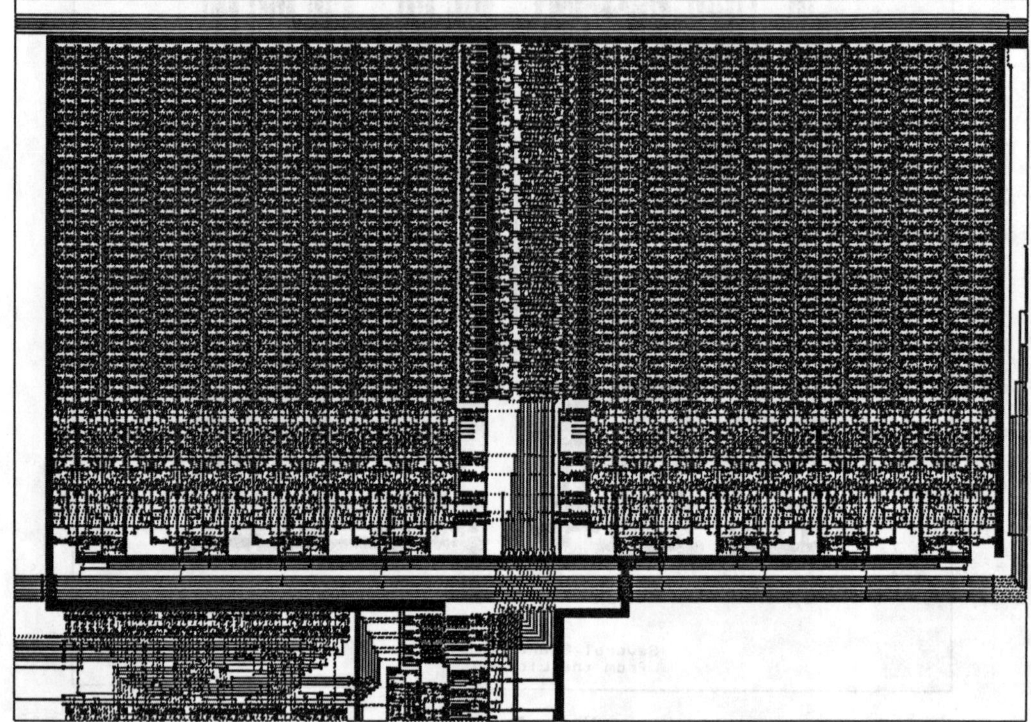

Figure 8.

2.4 Role of Standard Cell Placement and Routing

Standard cells are relatively small circuits, typically simple gates. Their layout is designed such that they can fit together by abutment. Their height is the same, and their terminals appear on the top and bottom. They are placed in rows and the routing is accomplished in one direction in channels between the rows, and in the other direction across the cells.

Their versatility comes from their small size. The placement and routing of standard cells can be accomplished in such a way that different aspect ratios can be obtained for a given circuit. This helps the floor planning process, since standard cell blocks can fit in regions between the rigid macro-blocks.

Figure 9 shows an example of a CRT controller design. The layout consists of two data-paths obtained by module generation and the control logic obtained by standard-cell placement and routing.

Figure 9.

2.5 Design with Libraries of Module Generators

A system designer can use libraries of module generators in an interactive environment consisting of the following components:

420

    a) Schematic capture
    b) Simulation
    c) Floor planning and placement and routing
    d) Library of parameterizable modules

In order to carry out a complete chip design the designer needs the modules in the library to be supported by the following generators:

    a) Icon generator
    b) Functional simulation model
    c) Timing delay model
    d) Power consumption model
    e) Floor-plan model generator
    f) Layout generator
    g) Detailed gate and transistor level simulation model

An icon generator produces a symbolic representation for the module to be used in schematic capture. The designer can quickly identify the function of a module by its icon in the schematic. The icon is parameterized and may change as the parameters change as entered by the designer in the schematic.

For example, the designer may develop a schematic as in Figure 10. Library components are accessed from the schematic editor and parameterized interactively. In this example the designer selected a ROM with 128 words by 8 bits, a RAM with 64 words by 8 bits, an 8-bit ALU, a PLA with 8 inputs and 20 outputs, and a block of random logic. The structure of the design is captured by interconnecting these blocks with busses.

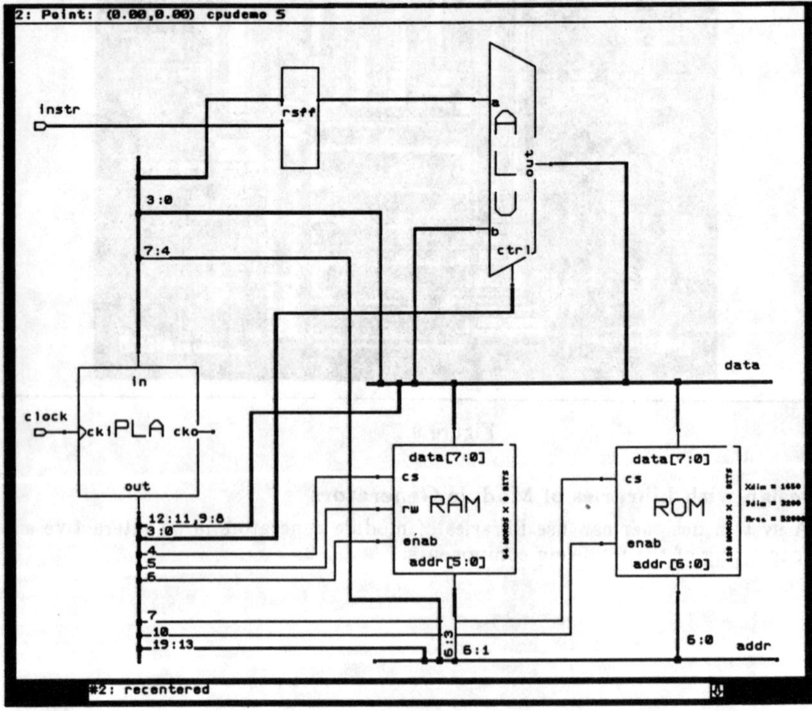

Figure 10.

The next step is to simulate the design. An interactive simulator with mixed-mode capabilities is especially useful for this application.

Mixed-mode capability gives the designer the flexibility of using the same simulator for functional simulation as well as detailed gate and transistor level simulation. In fact, the designer may mix functional blocks with transistor level blocks in the same simulation session.

Functional simulation model exists in the library for every module. These models are used as the schematic is developed. For example, Figure 11 shows an interactive simulation session during the schematic design.

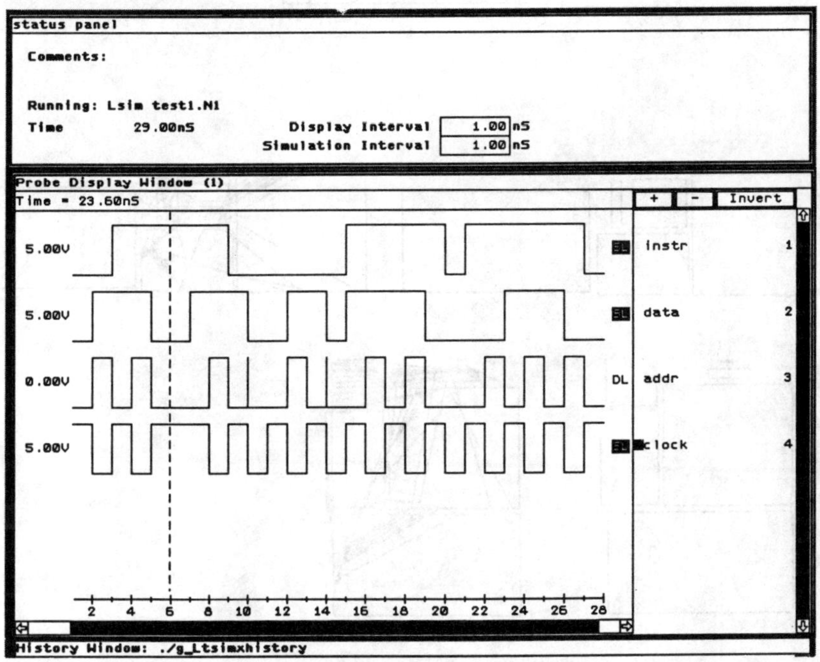

Figure 11.

The next step is to experiment with floor plans. Floor planning is a process of positioning layout blocks for best possible area utilization. The best results are obtained by combining automatic floor-planning methods with interactive designer actions. The parameterized modules from the library contain floor plan model generators which can quickly assist the designer and the automatic tools in experimenting with floor plans. Figure 12 shows several such explorations with the design from the Figure 10. A common method that is used to visually evaluate the floor plans consists of showing direct connections between the signal terminals on the blocks.

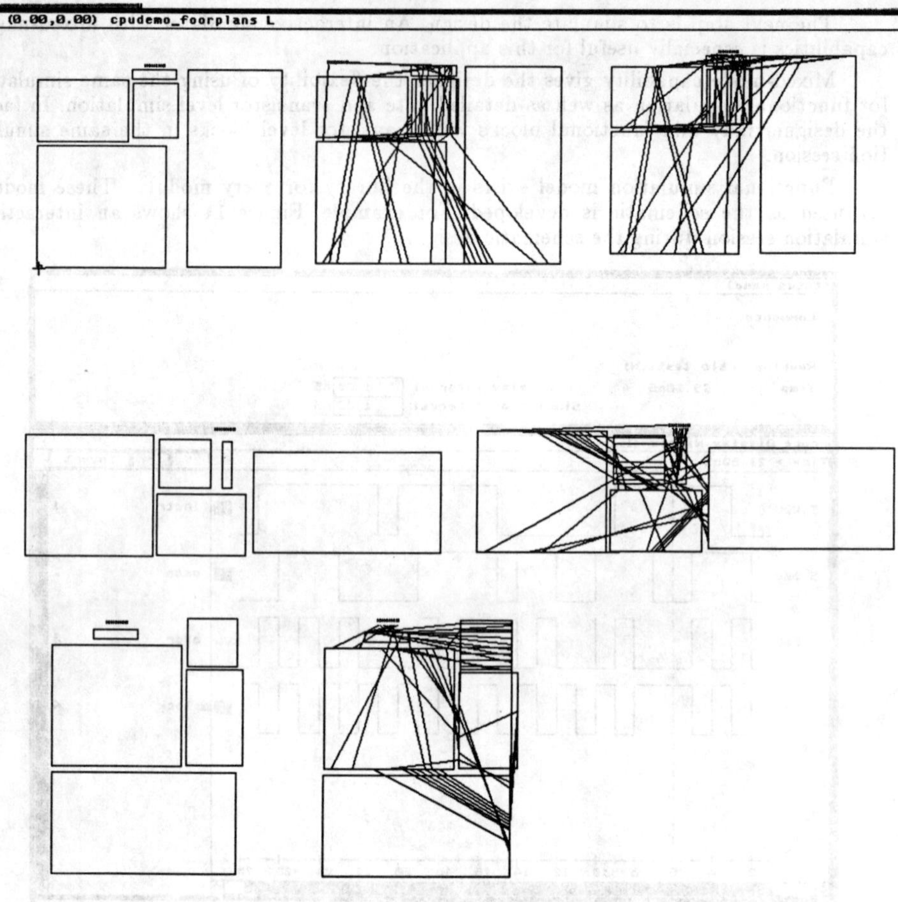

Figure 12.

The next step in the design is to complete detailed placement and routing of the blocks based on the chosen floor plan. The floor planning step and the detailed placement and routing steps are usually applied iteratively until the best solution is found. This is why it is important to use the software tools that guarantee that this iteration will be as smooth as possible.

The library of modules must provide layout generators and power consumption models for the final placement and routing. The power consumption models are used for sizing power and ground distribution lines on the chip.

Figure 13 shows the detailed placement and routing for the design specified in the schematic from Figure 10. Small window shows the detail of the routing channels.

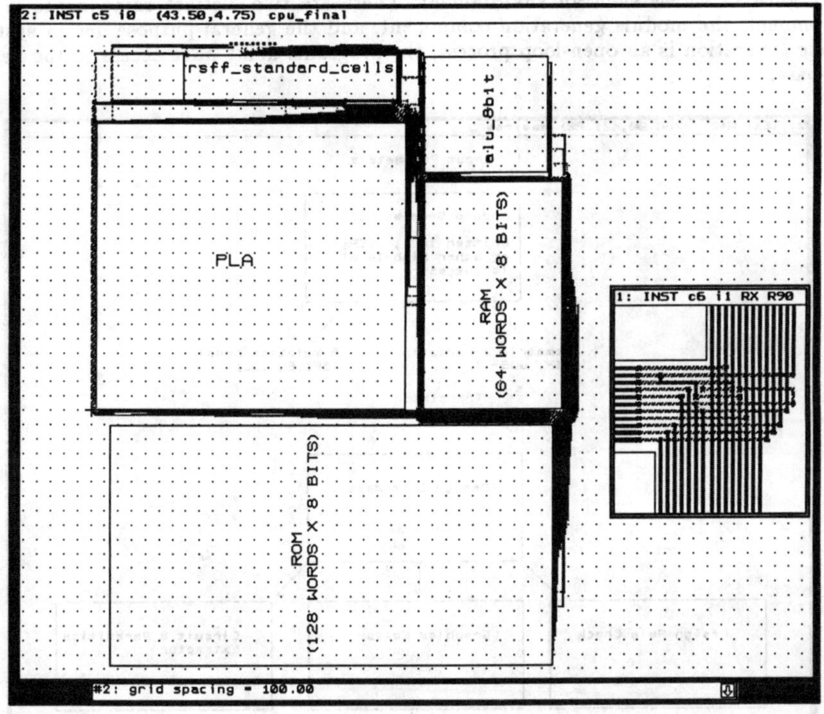

Figure 13.

3. Generator Development Tools and Their Use

The user of silicon compilers views the design process as being top-down. The specification proceeds from functional to structural, and then from structural to geometric. With module generators the geometric translation becomes fairly automatic.

However, in any design process there is a degree of bottom-up reasoning. Module generators offer the designer rather large primitive components for this phase. But what about the designer of the module generators? Here, the primitives might be as small as individual transistors. The tools for developing module generators have to take this into account.

The main challenge designers of generators face is creation of structural and layout generators. The difficulty arises because the nature of layouts is two-dimensional. Each component interacts with neighbors, and has to satisfy manufacturing design rules. The layouts have to be dense, and they should be portable to new design rules.

There have been several approaches to module generators. They are often written using general purpose programming languages. The generators in these cases simply output mask geometries either in geometric data formats, such as GDS-II or CIF, or in symbolic form, such as stick diagrams or character-symbolic representations. Some also output circuit net-lists for simulation purposes.

Figure 14 shows such an environment. There are two distinct components in the environment: the module generation component, and the general purpose design system. Module generation is an open-loop process, the programs don't have access to the design data-base.

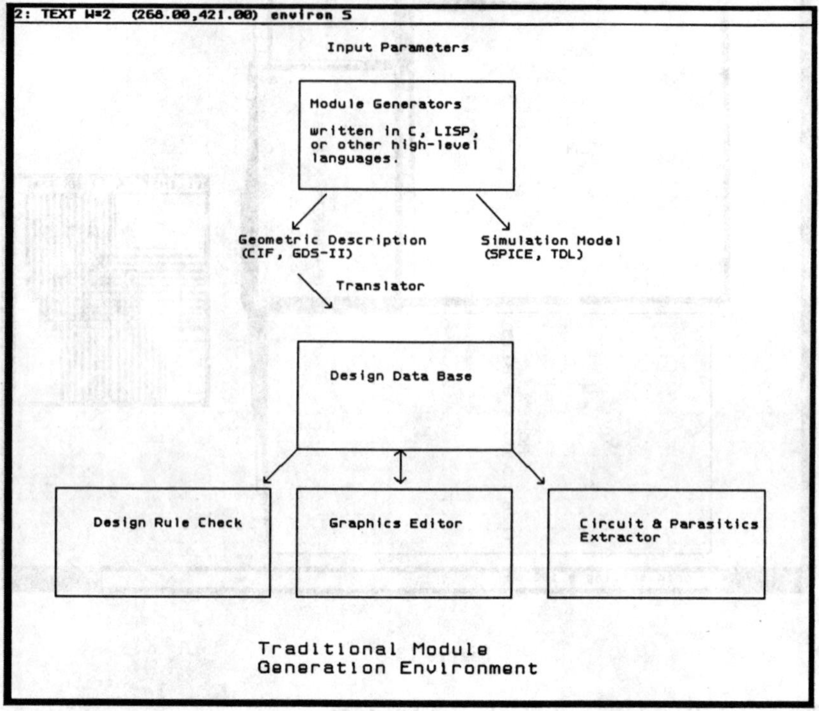

Figure 14.

More powerful environments have been developed that use procedural languages to build module generators. These systems allow the program that produces the layout to be written in the same language as the description of the layout, improving the efficiency of the design capture process. Notable examples from the academic community are the Design Procedure Language (DPL) developed at MIT [9], and ALI and CLAY developed at Princeton University [10].

Silicon Design Labs has developed a design environment called Generator Development Tools. At its heart is a procedural VLSI design language called L. L can capture both the structural and the geometric aspects of a layout simultaneously. This allows a designer to cover two complete branches of the Y chart with one description language (Fig. 2). The third, or functional, branch of the chart is covered by SDL's functional modeling language M, based on C, and the L-Simulator. These capture the behavior of modules ranging from transistors to microcomputers.

3.1. The L database

In contrast to other module generation systems, the L language for module generators, and the L database for design data, are one uniform environment.

During the IC design, the design data has to be easily created, stored, accessed by application programs, and verified against different representations.

The L database (Ldb) is a mechanism for satisfying these functions for all structural and geometric representations of a design. The L language is a procedural mechanism for building the design in the L database. Libraries of L generator programs are stored hierarchically in UNIX file directories. While L generators are general parameterized programs, the Ldb captures a particular design.

Figure 15 shows the interactions among the L program libraries, L compiler, L compiler, L compiler's support programs: placement and routing, and layout compaction, interactive L-editor, and the L database. L compiler accesses L programs in the libraries as requested by procedures within the L programs. It compiles L programs into design data and stores it in Ldb. While compiling, L compiler has access to Ldb for retrieving information needed by L programs. It also has access to utility programs that perform specialized functions such as routing, layout compaction, and placement.

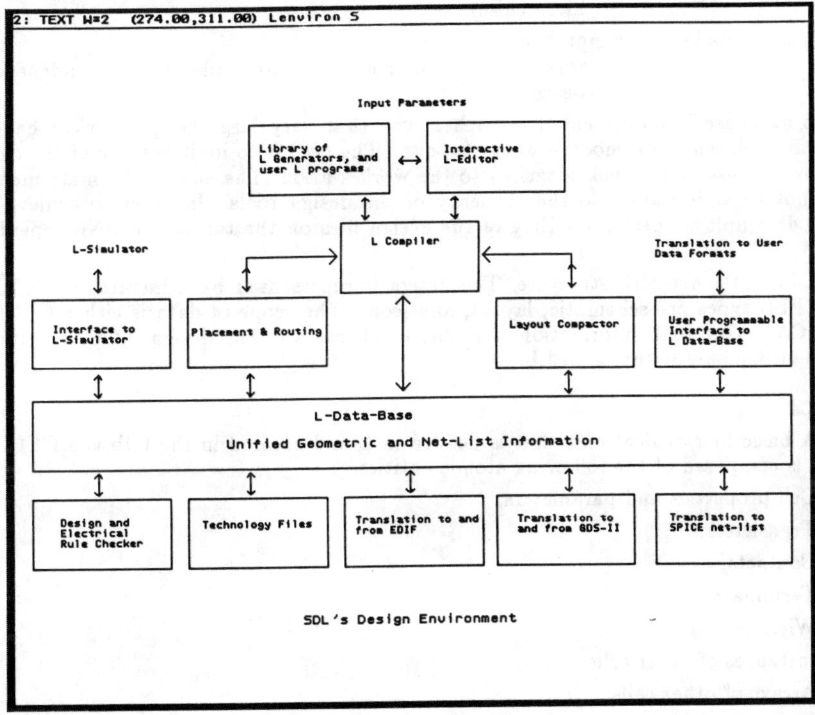

Figure 15.

The main features of L language are:

L is a language   with traditional facilities such as variables, arithmetic and logic expressions, calls to CELL subroutines with arguments, and conditional control structure (IF, ELSE, WHILE, etc). For accessing generator libraries, L supports file input and output and maintenance of directories.

L is a procedural   language for building a design data base (Ldb) for VLSI circuits.

L is technology independent,   because it gets the technology-specific information from technology files, such as definition of physical layers, design rules, transistor types, and contact types.

L describes   parameterized layout generators, circuit and logic schematic diagram generators.

L supports   geometric primitives (polygons, rectangles, lines, etc). On a higher level, it supports electrical circuit primitives and their geometries (transistors, contacts, wires, terminals, circuit hierarchies).

L supports layout compaction,   which is a powerful way to build process independent designs.

L database is implemented in such a way that very large designs can be handled easily and efficiently on modern work-stations. The goal is to limit the size of the design only by the size of the disk attached to the work-station. The size of the main memory must not be a limitation to the efficiency of the design tools. In order to achieve this goal, Ldb employs special handling of the design files on the top of the UNIX operating system.

Ldb has a network structure. The design is represented by a hierarchy of CELLs. The CELL types are schematic, layout, and icon. The scope of data is either GLOBAL or LOCAL. GLOBAL information is visible to all parts of the design. LOCAL information is visible only within a CELL.

CELLs

A basic hierarchical unit that is defined in L and entered in the Ldb is a CELL. A CELL is composed of the following atomic entities:

- Cell properties and parameters,
- Transistors,
- Contacts,
- Terminals,
- Wires ,
- Instances of other cells,
- Arrays of other cells,
- Polygons and Rectangles,
- Properties attached to the atomic entities.

Cell Properties

These include cell type (ICON, SCHEMATIC, and LAYOUT), cell name, bounding polygon, generator parameters which produced the cell, file name of the generator, technology name, etc.

Transistors

A transistor in L and Ldb is a geometric and a circuit object. It is a four-terminal device. In MOS technologies it has a gate, source, drain, channel width and length. Geometrically, it consists of rectangles that define gate, source, drain, implants, well, etc. Every transistor has a name, and its electrical and geometric connection points are well defined and named. These names are used from L programs to access the transistor and its connections.

Contacts

A contact also has a geometric and electrical meaning in L and Ldb. Electrically, it is just a connection point for wires. Geometrically, it consists of rectangles that define contact regions, overlap regions of the layers being connected, and possible implant regions (buried contacts). A contact has a name which is used for accessing the contact from L programs.

Terminals

Terminals are electrical entities without physical dimensions, on a certain layer (metal, polysilicon, diffusion, etc). They are used to export and import signals in and out of cells to the next level of the hierarchy. The types of terminals are IN, OUT, INOUT, VDD, GND, and user defined additional types. The types are used in electrical rule checking. In a schematic cell a terminal may be arrayed for bus connections. A terminal is named so it can be accessed from L programs.

Wires

Wires belong to a physical layer (polysilicon, metal, etc). They may be named, have width, and always connect exactly two vertices. Wires are stretchable such that if one of the connection points moves the wire stays attached. Wires can be attached to terminals, contacts, transistor terminals, instance terminals, array terminals, and other wires. In a schematic cell a wire may be defined to be a bus.

Instances of Cells

An instance of a cell is a virtual copy of the original of the cell, placed in a specific position, and possibly reflected and rotated within the cell in which it is included. There may be many such copies of a cell in a design. If the original cell is changed, all copies will inherit the change.

The only connection points (electrical and geometric) to an instance are the terminal points declared in the original cell. An instance of a cell may be accessed from L programs by its name. In a schematic cell an instance may be arrayed such that its inputs and outputs become connections for busses.

Arrays of Cells

Arrays of cells are similar to instances of cells, in that they contain virtual copies of other cells. However, instead of only one copy of one cell, arrays contain many copies of one or more cells. These copies are arranged in a matrix, and the instances inside the matrix are assumed to be connected to their neighbors.

Polygons and Rectangles

Polygons are closed plane figures with three or more vertices. They may be attached to other objects, in which case they are bound to them. If the object moves its polygon maintains the relative placement. There is also electrical binding such that a polygon and its object share a common net identifier. Rectangles are special cases of polygons.

Properties

Properties are strings that can be declared and attached to objects (such as transistors, instances, terminals, etc.). Properties are used for tagging objects with information that may be used by other L utility programs, or user programs.

There are several built-in types of properties but the user can define additional types. The built-in properties include SIG for naming signals, BUSCONN for specifying bus connections in schematics, VIEWNAME for displaying names of objects in the L-editor, TERMPLACE for specifying physical terminal locations for placement and routing, EQUIV for establishing equivalencies between schematic and layout terminals, SIM-MODE for specifying simulation mode in the L-simulator, and TEXT as a general purpose property.

3.2. Technology Files

Technology data for the L database is obtained from a technology file. A technology file contains information that is specific to a particular manufacturing process. By abstracting technology information to a technology file Generator Development Tools can be used across different technologies. The textual information from the technology files is stored as a special format in the L database for fast access.

Figure 16 shows a few of the technology file entries and their geometric interpretation.

Mask layer types are used to specify manufacturing details and other pseudo-layers for schematic design. Each layer is defined by a name, such as POLY, NDIFF, PDIFF, NWELL, etc. A mask layer can also be a composite of other mask layers. Additional information attached to a layer definition includes minimum feature size, layer resistance, and layer capacitance.

Transistor types are also defined by the user in the technology file. A transistor entry describes the name of transistor type, the general classification of the transistor into p-channel, n-channel, or depletion, the definition of the channel, and layers making up the transistor. Channel length and width are parameters that are instantiated when the transistor is entered into the L database.

Contacts are defined similarly in the technology file. Each contact type has a name and a collection of rectangles. Each contact serves a function of tying two objects that may be defined on different layers. An example of a contact definition in the technology file follows:

The design rules are an essential part of the technology file. They are needed in the L database for the proper operation of the L compiler, routers, compaction, and net-list

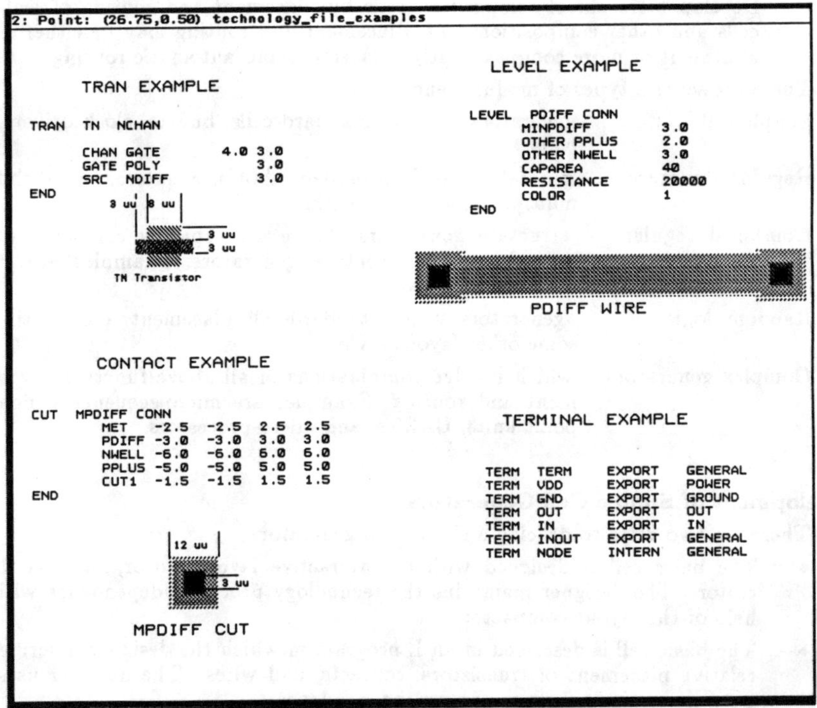

Figure 16.

computation. Some of the design rules are defined within layer definition, transistor definition, and contact definition. Those are minimum feature sizes and overlap and overhang definitions as described above.

The remaining design rules are defined as spacing rules. Since Ldb contains connectivity information, it is possible to perform design rule checking such that the net-list information is used. For example, if two metal wires intersect, it would be a design rule violation if they had different net identifiers. If they have the same net identifier then the intersection is allowed. In more traditional design tools, the intersection would not be considered illegal, but there would have to be net-list extraction followed by net-list comparison against the schematic in order to verify the design.

3.3 Methodology for Developing Module Generators

The development of a module generator is similar to the design of a custom IC. However, there is an additional goal to develop a parameterized and technology portable function block as opposed to a fixed cell layout. These are the steps followed by the designer of a module generator:

- Design the architecture and circuit for the module,
- Develop the floor plan with all of its variations as parameters change,
- Design the generators for the leaf-level cells,

Develop hierarchical composition cells by placement and routing of leaf-level cells and other composition cells. Placement and routing may be either simple abutment or more complex relative placement and automatic routing.

There are several types of module generators:

Simple cell	generators, such as standard-cells, buffers, clock drivers, and pads.
Regular structure	generators, such as adders, ALUs, counters, barrel-shifters, multipliers, and data-paths.
Combined regular	structure generators. These are typically composed out of several regular structure generators. Examples are PLAs, RAMs, and ROMs.
Random logic	generators using standard-cell placement and routing, or some other layout style.
Complex generators	which involve combinations of all above functions by placement and routing. Examples are micro-sequencers, floating-point units, UARTs, and micro-processors.

Development of Simple Cell Generators

There are two ways to develop a simple cell generator:

- The basic cell is designed with an interactive layout editor, such as the L-editor. The designer maintains the technology process independence with the help of the layout compactor.

- The basic cell is described in an L program in which the designer describes the relative placement of transistors, contacts, and wires. The designer uses layout compaction to assure process independence.

The designer may choose the first method when the cell structure does not change as parameters change. The examples include exclusive OR ar NOR circuits, some fixed latches, and signal buffering circuits.

The second method is useful when the circuit structure changes as the parameters change. The simplest example is an AND gate generator if one of the parameters is the number of inputs. As the number of inputs changes so does the number of the transistors and their connectivity.

Figure 7 shows several standard cells from SDL's library. Some cells have been developed by the first method, such as XNOR and latch cells. Others have been developed by using the second method. These are variable input NAND, NOR, and AND-OR-INVERT cells.

Development of Regular Structure Generators

These generators are characterized by being composed of many cells that are same or similar.

All sub-cells are typically connected together by abutment. Memory array in a RAM is a good example. The developer of such generators has to design the hierarchy and a floor plan such that user options can be easily included, such as size of the array and number of bits.

Often, the basic cells that make up the regular structure are best developed by using the interactive editor. These cells are called LEAF cells. The COMPOSITION cells are the ones in which leaf cells are abutted or routed together. The composition cells that are fixed may be edited in the interactive editor. However, if the composition cells are variable and parameterized the designer can use L programs to tile them together.

Figure 17 shows an example of a two-bit adder produced by the adder generator. The generator program is shown below. The generator has the following options:

- Number of bits in the adder
- Zero-detect circuit is optional
- Output tri-state buffer is optional
- Output result latch is optional
- Input operand latch is optional

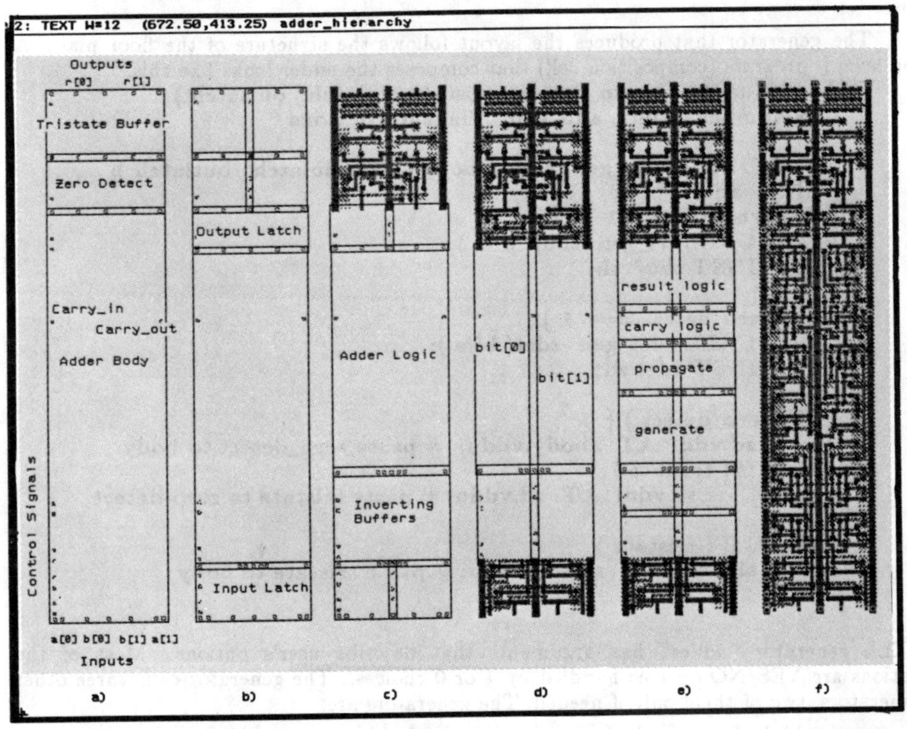

Figure 17.
The optional circuits are attached to the main body of the adder only if needed. The figure shows physical hierarchy of the adder by displaying the same circuit one level deeper going from left to right.

Figure 17 a) shows the view of the adder on the top of the hierarchy. It shows three components: adder body, zero-detect circuit, and tri-state buffer. These circuits are attached because the parameters specified these options. If the parameters did not ask for these options the zero-detect and tri-state buffer would be absent.

The next view in Figure 17 b) shows the circuit one level deeper in the hierarchy. The internal structure of the adder body is shown to consist of the main section, input latch, and the output latch.

Figure 17 c) shows the next level down the hierarchy. It reveals the layout of the zero-detect and tri-state buffer circuits. It also shows that the main section of the adder consists of the inverting buffers and the adder logic.

Figure 17 e) shows that the adder logic contains generate, propagate, carry logic, and result logic circuits. Finally, Figure 17 f) shows the complete layout of the two-bit adder.

It is important to partition the design into such manageable pieces. The layout in Figure 17 f) seemingly has no structure, but the analysis of the floor plan shows otherwise.

The generator that produces the layout follows the structure of the floor plan. A top level L program (composition cell) that composes the adder looks like this:

```
CELL adder( bits , zero_detect , tristate , inlatch , outlatch )
        # number of bits, and 1 or 0 flags for options
{
        CALL add_body_gen  adder_body ( bits , inlatch , outlatch );
        INST adder_body  abody;
        IF( tristate === 1 ) {
                CALL tri_gen  tbuf( bits );
                INST tbuf  tb;
        }
        IF( zero_detect === 1 ) {
                CALL zdet_gen  zdet( bits );
                INST zdet  zd;
        }
        IF( zero_detect ) {
                zd.vddl AT  abody.vddu; # paste zero_detect to body
                IF( tristate )
                        tb.vddl  AT  zd.vddu; # paste tristate to zero-detect
        }
        ELSE IF( tristate )
                tb.vddl  AT  abody.vddu; # paste tristate to body
}
```

CELL generator "adder" has arguments that describe user's options. Most of the options are YES/NO options handled by 1 or 0 choices. The generator calls three other generators, two of them only if needed. The generators are:

- add_body_gen which generates the basic adder,

- tri_gen which generates tri-state buffer. This generator is called only if the tri-state option is asked for.

- zdet_gen which generates zero-detect circuit. This generator is called only if the zero-detect circuit is required.

The decisions are made by IF and ELSE statements. They get all the arguments from the top level generator. These generators create CELLs that get installed in the L-Data-Base by the L compiler at the time the program is run.

Next in the program, INSTances of these cells are created if needed. Finally, placement of the instances is completed by the **AT** statement which ensures relative placement of the cells with respect to each other. The points used for connection are **vddl and vddu** , the lower and upper VDD terminals on the cells.

The **AT** function serves the dual purpose. Firstly, it places the objects by assigning new locations to specified objects. Secondly, it looks for all coincident terminals of the instances being placed, and wires them together.

The generator for the adder body is **add_body_gen**. It is developed using the same methodology. It produces the structure that is visible in Figure 17 b). It constructs the adder body out of the three pieces: input latch, output latch, and the common adder circuit. The input and output latches are optional. The generator looks as follows:

```
CELL add_body_gen( bits , inl , outl )
{
    CALL buffadd  badd( bits );
    INST badd  badd;
    IF( inl === 1 ) {
        CALL inlgen  inlcell( bits );
        INST inlcell inlatch;
        inlatch.vddu AT badd.vddl;
    }
    IF( outl === 1 ) {
        CALL outlgen  outlcell( bits );
        INST outlcell outlatch;
        outlatch.vddl AT badd.vddu;
    }
}
```

The generator calls three other generators:

- buffadd which produces buffered adder circuit,
- inlgen which generates input latch circuit. This generator is called only if the input latch is required.
- outlgen which generates output latch circuit. This generator is also called only if the output latch is required.

The cell obtained by calling these generators are tiled together by using the **AT** statement.

The same procedure is applied to develop other generators in this example. Leaf level cells (the ones with transistors, contacts, and wires) can be developed graphically with the L-editor. Their process independence is maintained by the layout compactor.

3.4. Layout Compaction

The goal of the compaction in the generator development tools is to provide technology portability of layout generators.

Chip layouts can be structured into leaf-level cells and composition cells. In a well structured design, the goal is to contain technology independence to the leaf-level cells, since composition cells combine leaf-level cells by abutment or by automatic routing.

In case the designer uses abutment the compactor enforces the pitch-matching to assure that the cell boundaries connect to each other correctly.

Compaction types include the following:

- compaction of leaf-level cells,
- pitch-matching of composition cells,
- compaction of flattened hierarchy of cells.

The designer can accomplish the technology independence of layout designs by programming the generators in the L language. This is done by describing geometric relationships of objects in terms of the symbolic equations involving design rules. This method is useful, but may require considerable time for verifying many possible combinations of the design rules. Compaction is an alternative method which requires much less engineering time.

A compactor can be used in two ways:

- interactively from an interactive editor such as L-editor,
- procedurally as invoked from L generators.

To compact the cell in an L program the designer can use the compact statement in many different ways. The simplest way is given in the following example:

```
CELL abc()
{
        terminal declarations;
        transistor declarations;
        contact declarations;
        wire declarations;
        COMPACTXY;
}
```

In this L program the cell content is specified first and then at the end of the priogram the compactor is invoked to complete the cell.

Compaction provides design rule portability of a layout design by automatically moving its objects closer together or further apart based on the design rules found in the technology file.

There are four directions of compaction available: X compaction (horizontal), Y compaction (vertical), XY compaction (horizontal followed by vertical), and YX compaction (vertical followed by horizontal).

There are a number of different compaction algorithms. One of these is a virtual-grid method described here. During the compaction of a leaf-level cell a compactor performs the following operations:

- The compactor constructs a list of distinct vertical and horizontal "virtual" grid lines and assigns objects to them. Objects stay together on these lines throughout compaction.

- During the X compaction the compactor compresses the distance among vertical grid lines, so that the objects satisfy design rules against their neighbors to the left.

- During the Y compaction it compresses the distance among horizontal grid lines so that the objects satisfy design rules against their neighbors below.

Figure 18 shows a leaf-level cell, a CMOS inverter. It is shown before compaction, after the X compaction, and finally after the Y compaction.

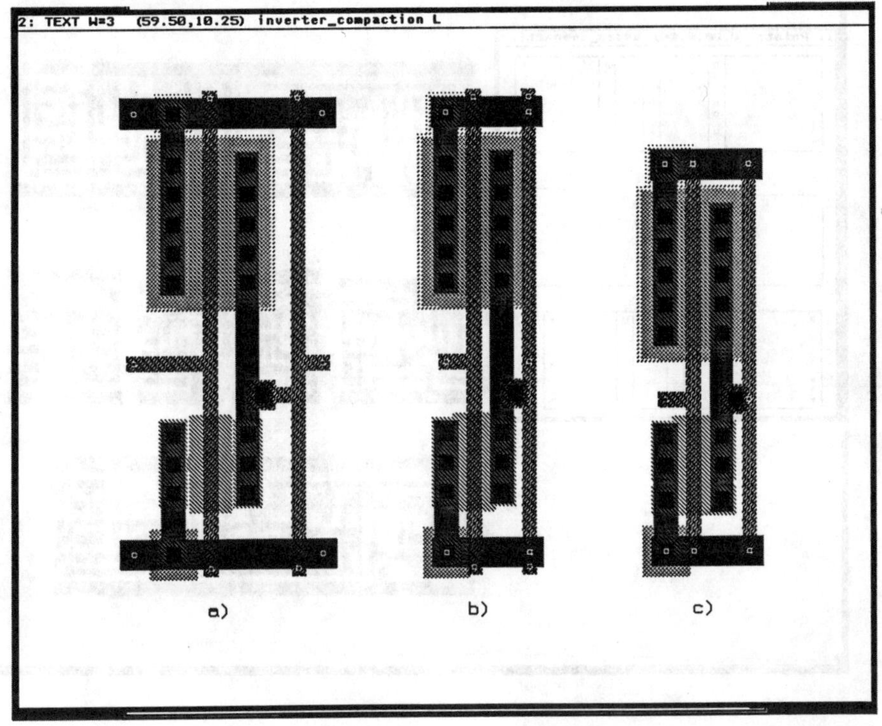

Figure 18.

The compaction with pitch-matching of abuted cells is a process of minimizing the layout area and still keeping the cells well connected. The compactor does pitch-matching by creating a flattened version of the template cell, compacting the flattened version, and then modifying the individual cells to match the result. The flattened version is then thrown away.

Figure 19 shows an adder bit before compaction. The small window shows the floor plan of the adder. Next the figure shows the result of the compaction if it is applied separately on the individual leaf-level cells that the adder uses. It is evident that the individual cells don't match anymore after the separate compaction.

Finally, the figure shows the adder after the compaction with pitch-matching which preserved the abutment of the leaf-level cells.

The complete adder is put together by an L program, by invoking the compactor when. In the L language, the whole procedure may look as follows:

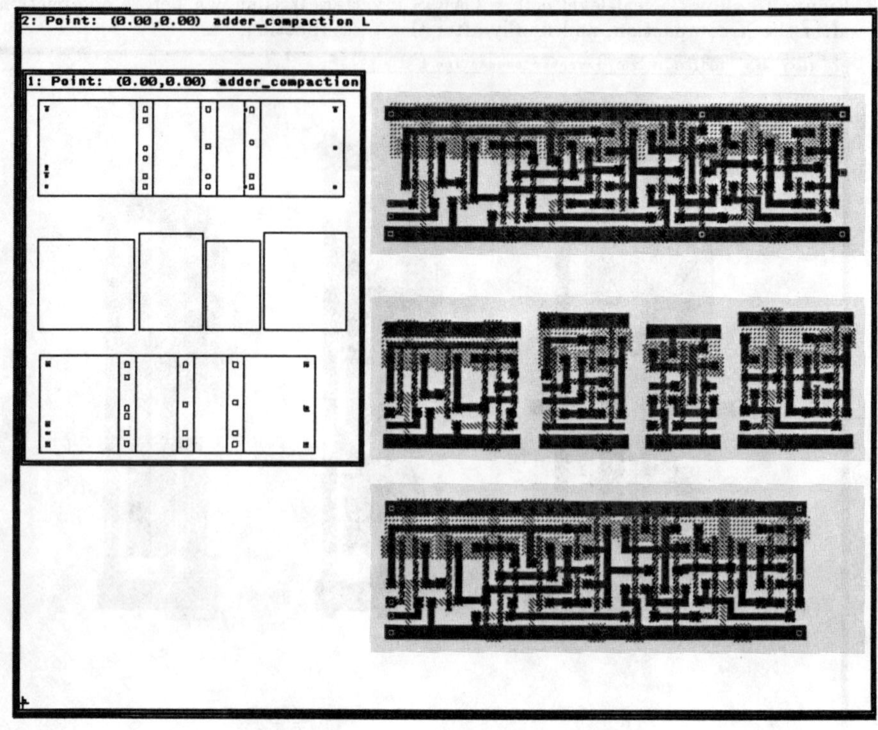

Figure 19.

```
CELL adder_bit()
{
        # instantiate desired cells
    INST generate_cell gen;
    INST propagate_cell prop;
    INST carry_cell carry;
    INST sum_cell sum;

        # abut them together
    prop.vddl AT gen.vddr;
    carry.vddl AT prop.vddr;
    sum.vddl AT carry.vddr;

        # compact
    COMPACTXY;
}
```

Conclusion

Silicon compilation is an emerging computer aided design discipline. Its evolution depends on module generation techniques. Module generation provides the link between structural and physical representation of the systems on a chip.

1. J. L. Hennessy, "SLIM: A Simulation and Implementation Language for VLSI Microcode", *LAMBDA*, April 1981.
2. H. Trickey, J. Ullman, "A Regular Expression Compiler", *IEEE COMPCON, 1982.*
3. Carver Mead, Lynn Conway, "Introduction to VLSI Systems", Addison-Wesley, 1980.
4. Z. Navabi et al., "Storage Logic Array Realization of RTL Descriptions", IEEE CHDL Conference, June 1983.
5. "Design Specification Using the MODEL Language", Lattice Logic, 1982.
6. J. Southard, "MacPitts: An Approach to Silicon Compilation", *IEEE COMPUTER Magazine,* 1983.
7. N. Bergmann, "A Case Study for the First Silicon Compiler", *Proc. Third Caltech Conference on VLSI,* Computer Science Press, 1983.
8. M. R. Buric, T. G. Matheson, C. Christensen, "The Plex Project: VLSI Layouts of Microcomputers Generated by a Computer Program", *IEEE International Conference on Computer-Aided Design,* September 1983, Santa Clara, CA.
9. J. Batali and A. Hartheimer *The Design Procedure Language Manual,* A. I. Memo No. 598, VLSI Memo No.80-31, Massachusetts Institute of Technology, Artifcial Intelligence Laboratory, Sept. 1980.
10. R. J. Lipton, et al., "VLSI Layout as Programming," Trans. ACM, Programming Lang. and Systems, Vol. 5, No.3, pp. 405-421,July 1983.
11. "VLSI Design System", VLSI Technology Inc., 1983.
12. J. Batali, Anne Hartheimer, "The Design Procedure Language", A.I. Memo No. 598, MIT, Sept. 1980.
13. R. Lipton, et al. , "ALI: A Procedural Language to Describe VLSI Layouts", *Proc. Nineteenth Design Automation Conference,* 1982. Princeton University, 1982.

Conclusion

Silicon compilation is an emerging computer-aided design discipline. Its existence depends on module generation techniques. Module generation provides the link between a structural and physical representation of the system layout on a chip.

1. J. Hennessey, "SLIM: A Simulation and Implementation Language for VLSI design," VLMPD, April 1981.

2. R. Tuckey, J. Kuhn, "A Regular Expression Compiler," IEEE CompCon W, 1983.
   ... Mead, Lyon, Conway, "Introduction to VLSI Systems," Addison-Wesley, 1980.

3. ... et al. "Structured Logic Arrays: Basic ... and ... Descriptions," IEEE/SIGDL conference, June 1982.
   ... "Design Specifications from the MODEL ... language," Lattice Logic, 1982.

4. K. Southard, "Macrocells: An Approach to ... Silicon Compilation," IEEE CompCon, Mini/micro, 198...

5. J. Bergmann, "A Code Shuffler for the ... Silicon Compiler," Proc. 2nd ... Conference on LSI Circuit, MIT Science Press, 1982.

6. ... Burr, T.C. Rassmuson, C. Christensen, "The Plex Project / Last Layout of Microcomputer ... Generated by a Computer Program," Proc. R International Conference on Computers in Design, September 1982, Santa Clara, CA.

7. J. Bazille et al. "Another MacDesign Procedure Designers Manual, A," Memo No. 506, VLSI Memo, vol.506, Massachusetts Institute of Technology, Artificial Intelligence Laboratory, Sept. 1980.

8. R. Elson et al. "VLSI Structures Programming," Trans. ... on Programming Languages and Systems, Vol.5, No.3, pp. 326-422, July 1983.

9. VLSI Design Systems, VLSI Technology Inc., 1983.

10. J. Batali, Anne Stavicamp, "The Design Procedure Language," AI Memo No. 564, MIT, September.

11. R. ... et al. "AR: A Procedural Language to Describe VLSI Layout," Proc. ... Design Automation Conference, 1980, Princeton University, 1980.

LAYOUT COMPILATION

RALPH OTTEN
IBM T.J.WATSON RESEARCH CENTER
YORKTOWN HEIGHTS, N.Y., USA

The environment in terms of input, output and data bases is discussed, and the outline of a complete layout design process is proposed. The process is interpreted as an application of the principles of stepwise refinement to the design of complex integrated circuits. Emphasis is on the rationale behind the chosen system decomposition, the accepted restraints, and the methods for cell assembling.

1. STEPWISE REFINEMENT

Stepwise refinement is a technique that has been shown to be effective in the development of computer programs. It was explicitly formulated in a famous paper by Niklaus Wirth [1]. In that paper the design of a structured program was viewed as a sequence of refinement steps. Starting with a clear problem statement, that specifies the relation between the input and the output data, the task is progressively refined, by decomposing it into subtasks, each having an equally clear specification. The sequence of refinement steps terminates when all tasks are specified in a chosen programming language. The constructs of that language should be a direct translation of tasks resulting from the final refinement steps. To be effective, they have to form a small but powerful enough repertoire. This method thus entails a hierarchical structure. (A hierarchy is, either a set of hierarchies, or an atom. In this case each hierarchy represents a task, and each task translatable in a construct is an atom.)

Stepwise refinement can also be viewed as postponing implementation decisions, to avoid committing the program prematurely to a specific implementation. Each decision should leave enough freedom to following stages to satisfy the constraints it created, and at the same time rearrange the available data such that further meaningful decisions are possible in the next step. So, concurrent with the gradual stiffening of the design, the information is progressively organized so that more and more detailed decisions can be derived.

The principles of stepwise refinement obviously apply to any complex design task based on a top-down strategy rather than to a process of combining independently developed subdesigns. Completely specified subdesigns, in general, are difficult to handle, because the flexibility and the information for adapting them to their environment is often not available when they are designed. On the other hand, the application of stepwise refinement in layout design raises a number of questions. Firstly, what information is available in the initial stage of layout design? A difficulty in answering this question is how to separate layout design from the other tasks in a silicon compiler. This will be discussed in section 2, a section devoted to the environment of layout design in a silicon compiler. Another question that immediately arises is what relevant information can be derived at the intermediate stages before fixing the geometrical details in the final stage? This will be the subject of section 3. Section 4 will consider the translation of the results of the last refinement steps

2. LAYOUT DESIGN IN A SILICON COMPILER

In this section the environment of layout design in a silicon compiler is discussed. The various relations of the layout design part with its environment are indicated in figure 2.1. They suggest the division of this section into four subsections. The output of a silicon compiler is the subject of section 2.1. The primitive constructs used to translate the result of the final refinement steps into a layout are discussed. There are some strict constraints on these constructs, the so-called design rules. Their impact on the compiler is outlined in section 2.2. In discussing the information generated by the preceding parts of the silicon compiler, and therefore available to the layout design part as input, the question of the initial data has to be answered. Finally, the interface with other parts of the silicon compiler is considered in section 2.4.

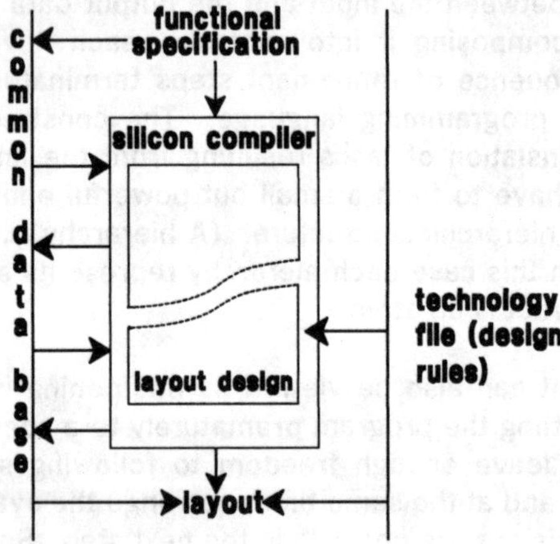

Figure 2.1. The environment of layout design.

2.1 The output: a layout

The ultimate task of a layout design system is to produce a *layout*, a set of data that uniquely and completely specifies the geometry of the circuit. Usually this data is an encoding of patterns, two-block partitions of the plane. The term *mask* will be used for each plane with a pattern, even if that plane is not exactly one of the real masks used in the fabrication. A layout is then translated into a sequence of processes that selectively change the characteristics of the silicon according to those patterns, thus realizing the functional specification available as input to the layout design procedures. Whereas the layout design system has considerable freedom in deriving geometry from functional specifications, the result of that final translation procedure is fixed. Up to forty different patterns are sometimes used in the sequence of selective exposures of the wafer surface. However, many of these are implied by other patterns in the sequence. For present day technologies the geometrical specification of eight to fifteen planes suffices to specify the layout.

From device theory general restrictions on the shapes of the regions can seldom be derived. Lithography techniques, however, do sometimes have their limitations. Quite often only orthogonal artwork is acceptable. This leads to regions that are unions of iso-oriented rectangles. There are examples of circuits that indicate that other layout primitives are more efficient, such as hexagons in some systolic arrays. Rarely is a restriction to rectangles and combinations thereof detrimental, whereas the cell design algorithms and layout data bases profit from such a restraint. The rectangle is therefore accepted as the basic construct in the present system.

The rectangle constraint is also accepted for the compounds of layout primitives that form the atoms, and even the hierarchies in the hierarchy entailed by the method of stepwise refinement. Consequently, each hierarchy will be a rectangle dissection in the final layout, i.e. a rectangle subdivided into nonoverlapping rectangles. The restriction to rectangles might seem rather arbitrary. However, in a truly top-down design it is very difficult to be particular about these shapes, because the shapes of its constituent parts still have to be determined. A good estimate for the shape of the enclosing region is of great value in determining the shape and positions of the constituent parts, and hardly a constraint if these parts have a high degree of flexibility. The earlier estimates become in some sense even self-fulfilling, because the parts mostly can be fit nicely in the environment by using their flexibility. Besides, choosing rectangles as the only constructs in the repertoire simplifies the formulation of design decisions, and lowers the complexity of deriving these decisions, as will be seen later.

2.2 Technological constraints

To improve chances for successful integration of the circuit, and increase yield when the circuit goes into production, patterns are required to satisfy certain rules, the so-called *design rules*. A first classification distinguishes roughly two classes of rules: *numeric rules* quantifying extensions of, and spacings between patterns in a plane and in combinations of planes, and *structural rules*, enforcing and prohibiting certain combinations.

There usually are a large number of numeric rules. Very few, however, are critical in a layout. For example, the spacing between two separate pieces of metal in the same layer is bounded below by different numbers depending on whether or not there are contacts to other layers in one or both of these metal pieces. In a wiring algorithm there rarely is a good reason for trying to use all these different minima. Instead, the maximum of all the rules that might apply is taken as the *pitch* for the metal in that layer. The reason for specifying the different rules for all special cases is mostly for the (manual) optimization of small pieces of layout that are used repetitively, such as memory cells. The numeric rules are almost exclusively specifications of lower bounds. This does not imply that the concerned extensions and spacings can be arbitrarily large. Making them arbitrarily large might impair the functioning of the devices in the circuit and increase delays, and decreases the yield. The rules are formulated as minimum rules only, because it is assumed that the layout design techniques will try to keep the total chip small.

Good algorithms working with these rules produce valid layouts in a wide range of values for these lower bounds. Of course, the algorithms do not produce optimal layouts for all combinations of values in these rules, but they should produce acceptable solutions for all practical value sets. The latter requirement is much more difficult to maintain under changes in structural rules, because these changes often require completely different decisions. The rules usually increase the dependance between different masks. This is particularly problematic if the metal layers are involved. Rules that forbid or enforce certain overlaps between patterns in the metal layer masks and other masks affect the wiring routines which often are generic algorithms solving some cleverly isolated interconnection problem. Introducing structural constraints often invalidate the assumptions made during the isolation.

2.3 The input: the functional hierarchy

Isolating tasks of an integrated approach is dangerous, because of their mutual dependence. Taking this dependance into account by iterations over several design tasks is highly undesirable, because of the time complexities involved and convergence properties. Clearly, since the final result has to be a complete specification of the masks, the later steps are mainly based on layout considerations. And, since the complete functional specification is the input to the compiler the early decisions or refinement steps have to be predominantly based on function and testing arguments. In between, many steps, such as logic decomposition and data path definition, have a significant influence on the final layout, and many functionally almost equivalent decisions may have completely different consequences for the layout and its design process. The boundary between layout design and other tasks of a silicon compiler is therefore quite fuzzy.

The early refinement steps in a silicon compiler will lead to a functional hierarchy, probably reflecting the functional interdependence of the hierarchies (modules). Functional interdependence and connectivity are often highly correlated, and the latter is an important basis for decisions in layout design. The functional hierarchy, therefore, represents useful data for the layout design part. Besides, it is also expedient to keep this hierarchy easily recoverable, as will be argued in section 2.4.

The considerations that lead to a functional hierarchy mostly ignore other important aspects of layout design. For example, in the design of digital systems the isolation and implementation of execution units is often established quite early. The remainder, control and interrupt, is left logically completely specified, but mainly unstructured. A layout with a decomposed, or even partly duplicated control unit, might be more efficient than a layout in which this part has been kept together. Several sections of the control can be placed closer to specific execution units they are heavily connected with. If the connectivity with the rest of the control is relatively low, this might save wiring area. It is also possible that the decomposition goes further than what is useful for layout decisions. The layout design part may, therefore, choose to ignore parts of the decomposition, initially or throughout. Nevertheless it is assumed that, possibly after some clustering around seeds and some pruning, the design data are completely hierarchically structured. That structure is considered part of the initial data for layout design. The hierarchies and atoms are called modules in this context. The formal definition of a module implies such a hierarchy.

A *module* $\mathcal{M}$ is defined to be a collection of modules $\{m_1, m_2,...., m_m\}$ where $m = |\mathcal{M}|$, and an incidence structure $\mathsf{IS} = (\mathcal{M} \cup \{\mathcal{M}\}, \mathcal{P}, \mathcal{N})$. The modules m_i are the *submodules* of $\mathcal{M}$, and *comodules* of each other. $\mathcal{M}$ is their unique *supermodule*. There is exactly one module without a supermodule. This module represents the entire system to be integrated. *Cells* are modules with an empty set of submodules. The others are called *compounds*. The hierarchy can be represented as a rooted tree. The modules are represented by the nodes. The root represents the system. The leaves represent the cells. The internal nodes represent the compounds. Each node representing a submodule is the end of an arc that started in the node representing its supermodule.

With regard to the incidence structure IS the module and its submodules are considered to be subsets of the set of pins $\mathcal{P} = \{p_1, p_2,...., p_p\}$. Also the *.signal nets, forming $\mathcal{N} = \{n_1, n_2,...., n_n\}$, are considered to be subsets of $\mathcal{P}$ (figure 2.2). *Pins* are for the moment merely a mechanism for relating modules and their supermodule with signal nets. The incidence structure can be represented by a bipartite graph $(\{\mathcal{M}\} \cup \mathcal{M} \cup \mathcal{N}, \mathcal{P})$, the *potential graph*.

Cells can be of various types. The type determines the cell's flexibility and how to obtain the layout of a cell. The most rigid cell type is the inset cell. Its config-

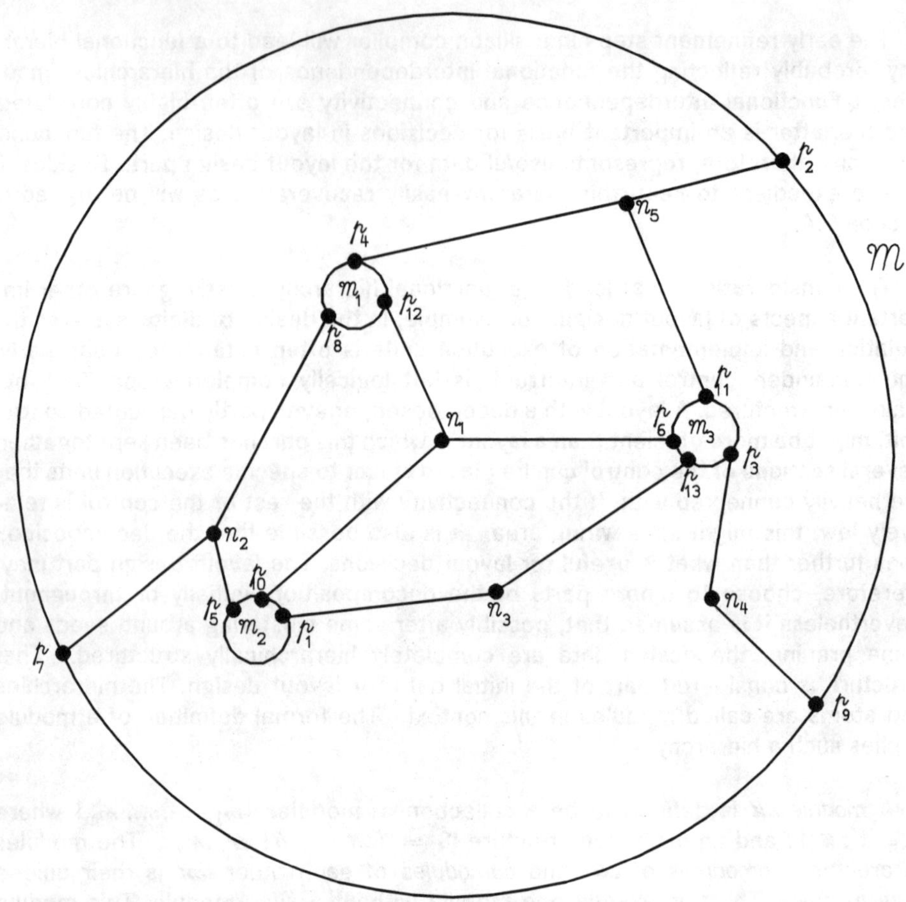

Figure 2.2. Illustration of the concept of incidence structure.

uration and pin positions are fixed and stored in a master or user library or completely a priori implied by the algorithm generating the cell. The layout design system can only assign a location and an orientation to such a cell. Other cells have pointers to algorithms that can determine their layouts. These cells usually have a higher degree of flexibility. The algorithms are such that estimates about the environment can be taken into account. This certainly is true for general purpose cells such as *macros*. These cells have a decomposition of their own into circuits of a particular family. Another, less flexible, example of a general purpose cell is a programmed logic array. Its potential for adapting to its environment is sometimes further diminished by optimizations such as row and column folding that impose stricter constraints on the sequences in which nets enter the cell region. There are also cells, such as cells generated by algorithms that, depending on a few parameters, construct special purpose subcircuits such as arithmetic logic units, rotators and adders. These cells have limited flexibility, such as permitting stretching in one direction. Stretchability is often important in avoiding pitch adjustments in data buses.

2.4 Data integrity

The single most important consideration in designing complex systems is conceptual integrity. An important aspect of this integrity is how to store the data of a design between the various stages. As pointed out in the previous subsection, the design has a hierarchical structure when entering the layout design stage. The modules in that hierarchy may have specific meaning for certain parts of the system. For example, they may have a functional model associated with them. Such a model makes simulation of that module in its environment possible. Extensive circuit simulation will have been performed on the system before the layout is considered. Yet, certain important performance aspects heavily depend on parasitic elements and final device parameters, and these are not known until the layout is determined. Simulation is therefore also important during and after establishing the layout. This requires that the simulation part must be able to find the modules for which a model is known, and assign values to parameters that represent the influence of device realizations and parasitics. It is therefore expedient that the results of the layout design process are stored in a way compatible with the data representation delivered by previous design procedures. In the preceding subsection it was established that that data is hierarchically structured. A hierarchy is mostly represented by an unordered tree. It would be convenient if the layout design procedure could preserve that structure, possibly refined and ordered. Refinement here means that leaves can be replaced by hierarchies of which the root takes the place of the replaced leaf, and subtrees consisting of a module with all its submodules can be replaced by any tree with the same root and the same leaves, but a number of additional internal nodes.

3. FLOORPLAN DESIGN

The question about what meaningful information can be derived at intermediate stages, that is by refinement steps in the layout phase of the design, is addressed in this section. First, some definitions, useful for specifying the flexibility in shape of modules are, introduced. In section 3.2, the need for floorplan design is discussed. A useful restraint in generating floorplan structures is then derived in section 3.3, and some compatible algorithms are indicated in section 3.4. Under the restraint of section 3.3, it is possible to determine the geometry of a floorplan under a variety of constraints imposed on its contour, without violating the shape constraints of the individual modules. The algorithm for this task is presented in section 3.5. Global wiring, that is the assignment of nets to specific regions in a rectangle dissection, is discussed in the last section.

3.1 Shape constraints

For every cell in the hierarchy there is an algorithm that tries to adapt the cell to its estimated environment, while generating its detailed layout. This preliminary environment has to be created on the basis of estimates concerning the area needed by each cell, feasible (rectangular) shapes for it, and the external inter-connection structure. The size and the shape of a cell are constrained by the amount and type of circuitry that has to be accommodated in that cell. It is reasonable to expect one dimension of the enclosing rectangle not to increase if the other dimension is allowed to increase. Constraints satisfying that requirement are called *shape constraints*. The precise definition follows.

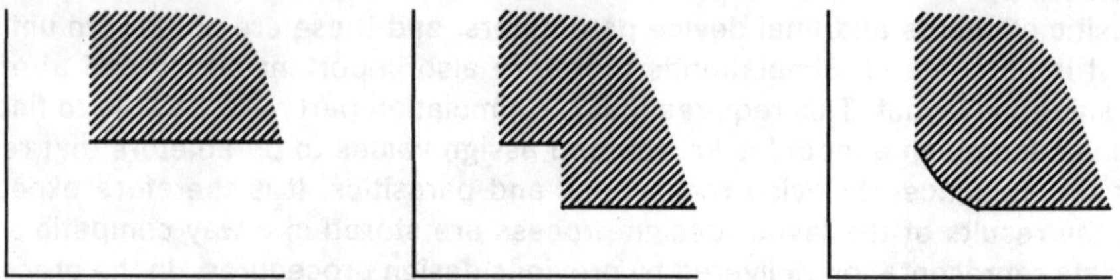

Figure 3.1. Examples of shape constraints. The bounded area is indicated by shading. The examples are an inset cell with a fixed orientation, an inset cell with free orientation, and a cell with a minimum area and minimum dimensions.

A *bounding function* is a right-continuous, non-increasing, positive function of one variable defined for all real values not smaller than a given positive constant.

The *bounded area* of a bounding function f is the set of pairs of real numbers (x,y) such that f(x) is defined and $y \geq f(x)$.

The *inverse* f^{-1} of a bounding function f is a bounding function defining a bounded area with exactly those (y,x) for which (x,y) is in the bounded area of f.

The *shape constraint* of a rectangle is a bounding function of one of its dimensions. The bounded area of that function is the set of all permissible pairs of dimensions. (Some examples are given in figure 3.1.)

Useful shape constraints of compounds in the functional hierarchy can be derived in a straight forward manner from the shape constraints of cells.

3.2 Placement versus floorplan design

Through the shape constraints, the estimation of the rectangle in which the module is going to be realized, is controlled. Guidelines for the position of such a rectangle among all the other rectangles are contained in the functional hierarchy, that gives some indication about which modules belong together functionally, and in the incidence structures associated with the modules. In the context of layout these incidence structures are often called *net lists*.

Utilizing the data (shape constraints, net lists, and functional hierarchy) the cells have to be arranged in a rectangle. This enclosing rectangle is often desired to be as small as possible, sometimes it is constrained in aspect ratio or completely specified. If the cells were fixed objects this would be the classical placement problem. However, in this context the cells are allowed to take any shape not excluded by its shape constraint. This generalization of placement is called *floorplan design*.

Both floorplan design and placement are guided by a number of objectives, not easy to formulate in a single object function. This can be illustrated by the following typical combination of objectives. The first is primarily concerned with the realization of the interconnections. A common figure of merit for it is total wire length, often estimated by summing the perimeters of the rectangles that enclose all module centers connected to the same net. At the same time it is desirable to give the cells rectangular regions in which they can be efficiently allocated. The first objective is of a rather topological nature, working with concepts such as 'close', 'neighbor' and 'connectivity'. The latter is more a geometrical objective. Major concepts for it are 'deformation', 'dead area', 'aspect ratio', and 'wiring space'. To relate the two objectives an additional refinement step, using an intermediate structure capturing much of the data affecting one of the objectives, might be helpful.

3.3 Floorplan topologies [2]

It has already been observed that in the final floorplan the modules will be rectangle dissections in which each submodule is either a rectangle dissection itself, or, as in the case of cells, a rectangle. Creating a preliminary environment for the cells is essentially generating certain aspects of the rectangle dissection, in which each cell is an undivided rectangle. Since the shape of the cells is not yet known in that stage, the geometrical details of the rectangle dissection cannot be determined. Less restrictive aspects of a rectangle dissection are its neighbor relations, i.e. which cells share a particular line segment in the rectangle dissection. The set of neighbor relations is called the *floorplan* of the rectangle dissection. A floorplan is useful information that can be generated at an intermediate stage of the refinement process. Usually, enough freedom is left for the cell assembling procedures after fixing the floorplan, and further decisions concerning the environment of the cells can be derived from it. Therefore, the first task will be designing a floorplan. A reasonable decomposition of that task, certainly in the light of the discussion of section 2.4, is to take one module at a time, starting with the root of the functional hierarchy, and progressing downward such that no module is treated before its supermodule is. This translates the functional hierarchies into nested rectangle dissections.

```
module list:= (root);
while module list is not empty do
    select M from module list;
        if M is compound
            then FLOORPLAN(M)
            else ASSEMBLE(M) end;
        add submodules of M to module list;
        delete M from module list
    end
```

In spite of the constraints accepted so far, manipulating floorplan data is still complex. For example, given a floorplan and the shape constraints of its cells, finding the smallest rectangle dissection is an NP-hard problem. There also is no pseudo-polynomial algorithm for it, since the corresponding decision problem is strongly NP-complete [3]. At this point one may ask whether the class of floorplans for which the previous problem, and hopefully several other problems, can be solved in polynomial time, is still large enough to include an efficient floorplan for all practical cases. To answer that question that class has to be identified.

A concise way of representing the floorplan of a rectangle dissection is by its *polar graph*. This is a plane, directed graph without cycles. There are three bijective relations between elements of this graph and its associated dissection: edges correspond one-to-one with undivided rectangles, vertices with the elements of one set of iso-oriented line segments, and inner faces with the line segments in the other set (figure 3.2). Many floorplans, designed in practice, and all the floorplans of the successful, more special layout styles, have polar graphs that are two-terminal series-parallel digraphs. Such a digraph is one of the following three types:

1. a digraph consisting of two vertices joined by a single arc;
2. a digraph obtained by identifying the sources and identifying the sinks of at least two two-terminal-series-parallel digraphs;
3. a digraph obtained from a sequence of at least two two-terminal-series-parallel digraphs by identifying the source of each (except the first one) with the sink of the preceding one.

It is exactly for the class of floorplans that have such a digraph as polar graph, that the above optimization problem is efficiently solvable.

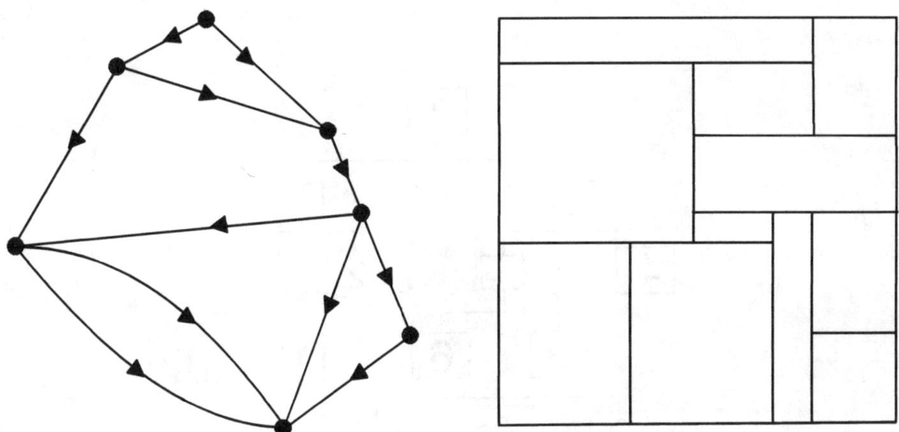

Figure 3.2. A rectangle dissection and one of its polar graphs.

A first observation is that, as any two-terminal series-parallel graph, such a floorplan can be represented by a much easier to handle data structure, namely an ordered tree. By restricting floorplans in this way, the procedure given earlier in this section thus automatically achieves what has been found convenient in the discussion at the end of section 2.4. It orders and refines the given functional hierarchy according to one of the rules there described, and the layout structure can be stored consistently in a natural way.

The tree replacing a two-terminal series-parallel graph is called the *decomposition tree* of that graph. Its leaves correspond with the arcs, and its internal nodes correspond with the two-terminal series-parallel subgraphs of the original graph. Consequently, each leaf represents an undivided rectangle and each internal node represents a rectangle dissection, also with a two-terminal series-parallel graph. The rectangle dissections represented by the endpoints of tree-arcs starting from the same tree-vertex, are placed next to each other in the same order, either from the left to the right, or from the top to the bottom, depending on whether the corresponding two-terminal series-parallel graphs are connected in parallel or in se-

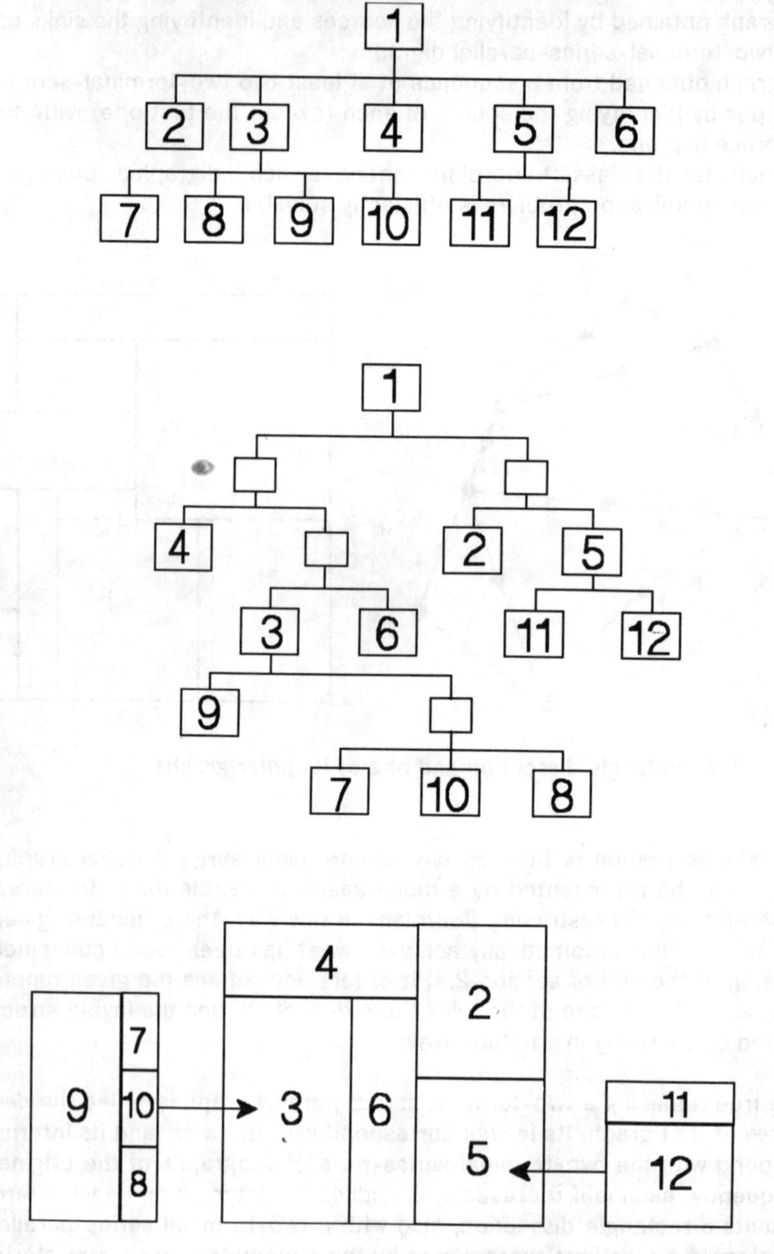

Figure 3.3. A hierarchy, a slicing tree, and the corresponding rectangle dissection.

ries. A rectangle dissection with a two-terminal series-parallel graph as polar graph is a rectangle dissected by a number of parallel lines into smaller rectangles that might be dissected in the perpendicular direction. Such structures are called *slicing structures* (figure 3.3) and the associated tree a *slicing tree*. Each vertex represents a *slice*. Each slice either contains only one cell, or is a juxtaposition of its *child slices*. In the latter case that slice is said to be the *parent slice* of its child slices, and these child slices are the *.sibling slices* of each other. The sibling slices are ordered according to their position in the parent slice (for example, left to right and top to bottom).

When a (preliminary) geometry is associated with the slicing tree, the *longitudinal dimension* of a slice is the dimension it inherits from its parent slice. The other dimension is its *lattitudinal dimension*. The lattitudinal dimension of the common ancestor slice is the length of the side it transmits to its child slices. The other dimension is its longitudinal dimension. The *shape constraint* of a slice is a bounding function of its longitudinal dimension. The bounded area of that function is the set of all permissible pairs of longitudinal and lattitudinal dimensions.

3.4 Implementations

There are several ways of obtaining slicing structures. A well-known method is the min-cut algorithm [5]. If applied in its pure form it leads to binary slicing trees, but one clearly can extend it to produce general slicing trees. The procedure 'mincut' assigns to a set of modules a decomposition type and an ordered two-block partition. Calling the procedure with the set of all modules $\mathcal{M}$ and a given aspect ratio A yields a slicing tree T. Seed selection is fixing the positions of modules in a non-empty subset of $\mathcal{S}$, seed1, relative to the positions of modules in another subset, seed2, disjoint from seed1. The selection can be made on the basis of previous decompositions and the type of decomposition just selected. The partition ($\mathcal{S}_1$, $\mathcal{S}_2$) must be such that $\mathcal{S}_1$ contains seed1, but not seed2, and the difference between the total area of modules in $\mathcal{S}_1$ and the total area of modules in $\mathcal{S}_2$ does not exceed a value depending on the total area of modules in $\mathcal{S}$ and the aspect ratio A. Among the partitions satisfying these requirements one with the minimum (or a small) number of nets common to both blocks is selected.

The mincut method does not use an intermediate structure that captures globally a large part of the topological aspects of the input, as suggested in section 3.2. Each dissection divides the problem into smaller problems, but it is difficult to take into account decisions in one part when handling the other parts.

Methods using an intermediate structure are also known. One such structure is a point configuration in which the topological properties of the input are some-how translated into a closely related geometrical concept, namely distance, and since the configuration will be embedded in a plane, more particular distance in the

```
procedure mincut(𝓜 : set of modules;A : real);
   var q : queue;
       seed1,seed2,𝒮,𝒮₁, 𝒮₂: set of modules;
       typ : 1,2,3;
       exp : 1,-1;
   begin enqu ((𝓜,A),q) ; T := 𝓜;
       while not empty (q) do
          (𝒮,B) :=  dqu(q) ;
          if |𝒮| > 1 then
             if B ≥ 1 then
                C :=  B; typ :=  2; exp :=  1
                else C :=  Bsup -1 ; typ :=  3; exp :=  -1
             end;
             (seed1,seed2) :=  seedselection(𝒮,T,typ);
             (𝒮₁, 𝒮₂) :=  partition(𝒮,C,seed1,seed2);
             T :=  update-tree(T, 𝒮₁, 𝒮, 𝒮₂ );
             enqu ((𝒮₁,(CΣ(𝒮₁)/Σ(𝒮))^exp),q) ;
             enqu ((𝒮₂,(CΣ(𝒮₂)/Σ(𝒮))^exp),q)
          end
       end
   end mincut;
```

two dimensional euclidean space. High connectivity is reflected in relatively short distances. A useful metric in this context is the so-called *dutch metric*. It assigns to each pair of modules a distance equal to

$$d_d(m_1, m_2) = \sqrt{1 - \frac{\sum\{w(n) \mid n \cap m_1 \neq \phi \wedge n \cap m_2 \neq \phi\}}{\sum\{w(n) \mid n \cap m_1 \neq \phi \vee n \cap m_2 \neq \phi\}}}$$

(w is a function assigning weights to the nets.) It has the nice property that the ensueing distance space is always embeddable in a euclidean space, although seldom in the euclidean plane. It is however straightforward to construct a two-dimensional point configuration in which the distances d_p are such that $\Sigma(d_d^2 - d_p^2)$ is minimum. It comes down to determining the eigenvectors of the *schoenberg matrix* with the two largest associated eigenvalues [6]. Also by calculating a partial eigensolution the weighted sum of distances can be minimized, under the constraint that the configuration must have a certain spread [7]. As discussed before, it is important in a top down approach to take the environment into account, and the preferred distance metric in layout design is often minkowski-1, because of the orthogonal artwork required by many lithography techniques. Both eigensolution methods, however, are based on euclidean distances, and cannot take the pin positions on the perimeter into account.

Recently, a probabilistic algorithm became very popular for handling combinatorial optimization problems. This algorithm, called (simulated) annealing [12], can also be used for generating point configurations in the present context [13]. Each state is represented by a pair of permutations over $\mathcal{M}$. To obtain the score of that state, the two permutation are used to give two coordinates to each element in $\mathcal{M}$ by introducing a space between two consecutive elements proportional to the sum of their areas. The perimeter of a box containing all the points representing modules connected by a particular net is used as an estimate for the size of that net. Several refinements (correction for nets connected to many modules, different weights for different directions, weights for nets,etc are possible). To generate a new state a transposition is applied to one of the permutations.

Properties of the final rectangle dissection have to be derived from the point configuration and the shape constraints. The topological considerations should be taken into account by preserving relative positions in the point configuration and keeping modules close together if they are represented by points at a short distance, from each other. The geometrical aspects should be taken care of by keeping track of, for example, deformation implied by the dissections. Shrinking is a method that efficiently achieves these goals for slicing structures. For reasons of clarity its discussion will be limited to designs with exclusively flexible modules. Mostly, shapes close to square are to be preferred for flexible modules, because square shapes require in general less wiring space. Now, assume that a given slice $\mathcal{B}$ with two or more modules is to be sliced in a number of child slices. $\mathcal{B}$ is separated from its sibling slices by lines parallel to one axis, and the slicing must now be tried parallel to the other axis. Supported by figure 3.4, a visualization of a process determining the *shrink factor* of possible *slicing lines* with the given orientation follows.

Think of a module as a square around the point representing it and with its sides parallel to the axes. The sizes of these squares are such that each pair of squares has overlap and the ratio of their areas is equal to the ratio of the area estimates given to the corresponding modules. The squares are simultaneously shrunk leaving their centers fixed and preserving the area ratios. At some point during this shrinking process there will be a line with the given orientation that divides the modules into two blocks without intersecting any of the squares. The amount of shrinking necessary to reach that point is the *shrink factor* to be assigned to the corresponding line. Shrinking is continued and another line with the same orientation and separating other blocks will be intersection free. Again, the corresponding factor will be assigned to that line. The process can be continued until all $|\mathcal{B}|-1$ factors are assigned. The lines with relatively high shrink factors are candidates for slicing lines. The accepted lines partition the set of modules into blocks. Each of these blocks will be treated in the same way. The selection of the

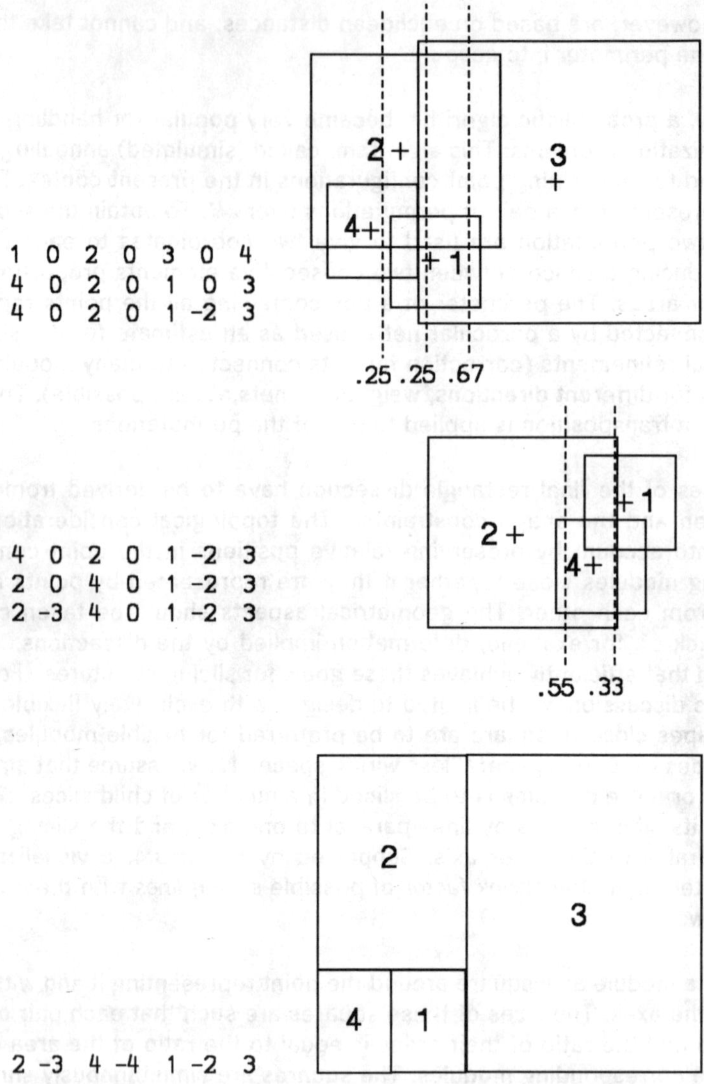

Figure 3.4. The determination of the shrink factors. In this small example the squares around the module centers are drawn in their position before the shrinking process. The amounts of shrinking necessary to have the vertical dashed lines intersection free are given for each line. After accepting the line with highest shrink factor, the three modules at the left are treated in the same way. The consequences for the shorthand tree are given for each step. The final tree and configuration complete the picture.

lines usually takes other aspects into account. One of these, *deformation*, will be discussed after introducing the *shorthand tree*, a data structure that keeps track of the lines that are accepted.

The shorthand tree is useful for storing partial structure trees. It is a one-dimensional array with $2|\mathcal{M}|-1$ integers. The modules are represented as positive integers. After completion of the slicing structure these positive integers will be separated from each other by $|\mathcal{M}|-1$ negative integers representing the slicing lines. The values of these negative integers indicate the level of slicing. Initially all $|\mathcal{M}|-1$ positions for the slicing indicators are filled with zeros. The positive integers are ordered according to the coordinates of the corresponding points on the axis perpendicular to the first set of slicing lines. At the corresponding positions in the array the zeros are replaced by a -2 (0 and -1 are reserved for separating frame modules such as bonding pad macros). In each of the subsequent steps the groups of (two or more) positive integers separated by negative integers are treated in a similar way. They are ordered according to the coordinates not used in the preceding step, and the positions corresponding to the accepted slicing lines are filled with an integer 1 lower than the indicator of the preceding step. The algorithm continues until all zeros in the shorthand tree are replaced by negative integers. The concept is illustrated in figure 3.4, and figure 3.5 shows a point configuration a shorthand tree obtained from that configuration, and a corresponding rectangle dissection. and rectangle dissection.

The preferred shape for flexible modules is a square. It is, however, in general not possible to accommodate the modules of a set $\mathcal{B}$ as squares in a slice with area

$$\Sigma(\mathcal{B}) \;=\; \sum_{m \in \mathcal{B}} \alpha(m),$$

$\alpha(m)$ being the area estimate for module m, to say nothing of such a slice with given dimensions, l_1 and l_2. To assess the deviation from square form of the modules in $\mathcal{B}$ *deformation* defined as follows:

$$\mathrm{def}(\mathcal{B}, l_1, l_2) = \sum_{m \in \mathcal{A}} \left(\frac{\sqrt{\alpha(m)}}{\min(l_1, l_2)} + \frac{\min(l_1, l_2)}{\sqrt{\alpha(m)}} - 2 \right) +$$

$$+\; \mathrm{def}\!\left(\mathcal{B}\backslash\mathcal{A}, \max(l_1, l_2) - \frac{\Sigma(\mathcal{A})}{\min(l_1, l_2)}, \min(l_1, l_2)\right)$$

where $\mathcal{A} = \{m \in \mathcal{B} \mid \alpha(m) > (\min(l_1, l_2))^2\}$, can be used. When $\mathcal{A} = \emptyset$ the deformation, as defined, will be zero. This does not mean that all modules of $\mathcal{B}$ fit into an $l_1 \times l_2$ slice as non-overlapping squares. It only is an indication that the modules are small in comparison with that slice, and a reasonable packing can be expected.

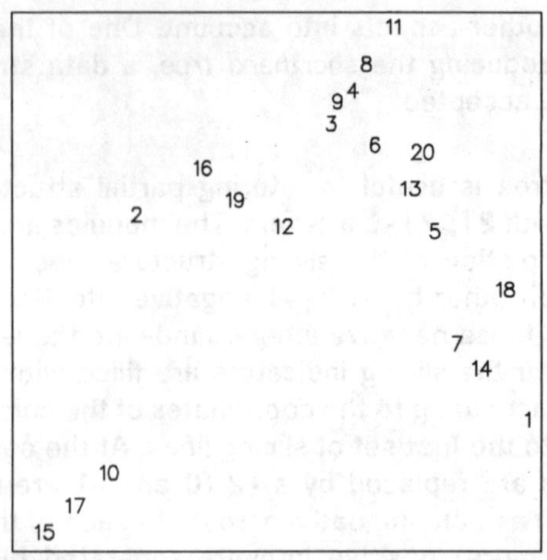

2 –4 16 –5 19 –3 15 –4 17 –4 10 –2 8 –5 4 –5 9 –5 3 –4 11 –5 6 –5 13 –6 20 –3 5 –4 18 –3 7 –5 14 –4 1

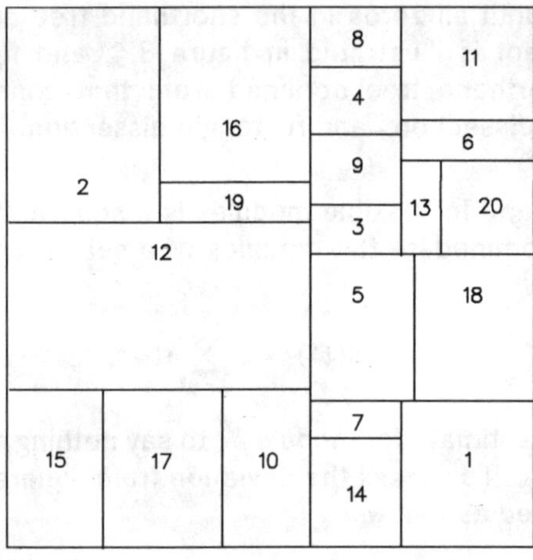

Figure 3.5. A point configuration with a consistent shorthand tree and rectangle dissection.

To use this definition of deformation it must be possible to calculate the dimensions of the slices in a partial structure tree, given that tree, the area estimates of the modules, and the preferred aspect ratio of the supermodule. This is quite easy when all modules are flexible. The calculation starts with deriving the outer dimensions from the total area and the aspect ratio. One of these dimensions is inherited by the child slices of the ancestor. The other dimensions of these child slices are obtained by dividing the sum of the areas of the modules in each of them by the inherited dimension. Proceeding in this way, each slice inheriting the dimension calculated for its father slice, will yield the dimensions of all slices.

```
procedure lot(M: set of modules;T:slicing tree;A : real);
   var q :queue; s,ss : slice;
   begin enqu (ancestor,q);
       dim[ancestor].long := √ A * Σ(M) ;
       dim[ancestor].latt := √ Σ(M)/A ;
       while not empty(q) do
          s :=  dqu(q) ;
          for each child ss of s do
              dim[ss].long :=  dim[s].latt;
              dim[ss].latt :=  dim[s].long * Σ(ss)/ Σ(s) ;
              if not leaf(ss)
              then enqu(ss,q) end;
          end
       end
   end lot;
```

Since the computation is performed top-down, it can also be applied to partial slicing trees. So, for every decision proximity information (shrink factors) and shape information (deformation) is available. For floorplans with a large number of modules, not too different in size, the first slicing lines are mostly determined on the basis of proximity information. Only later, when few modules have to be allocated in a slice, deformation usually dominates the selection. In the final stage, one may even consider to determine the deformation incurred by all possible slicing trees (since the ordering does not affect the deformation one should only generate one tree out of every isomorphism class) that can be formed with the few modules that go into such a low level slice.

Generalization of the process presented here for flexible modules with a given area estimate is not very difficult, but a procedure calculating the dimensions of each slice in a partial slicing tree is needed. Such a procedure will be presented in the next section.

3.5 Floorplan optimization

Given the slicing tree, and the shape constraints of the modules (represented by the leaves of that tree), what is the best rectangle dissection corresponding with that tree? In this section a procedure will be described that produces such a geometry if "best" can be specified in terms of a contour score. A *contour score* c is a function of two variables, defined for a convex subset Γ of the pairs of positive real numbers, which is quasi-concave and monotonously non-decreasing in its two arguments, i.e.

$$\forall_{x0,x1,y0,y1\epsilon\Gamma}[x_1 \geq x_0 \wedge y_1 \geq y_0 \Rightarrow c(x_1,y_1) \geq c(x_0,y_0)]$$

and

$$\forall_{x1,\,x2\epsilon\Gamma}\forall_{0\leq\lambda\leq1}[c(x_1)\leq c(x_2) \Rightarrow c(x_1) \leq c((1-\lambda)x_1 + \lambda x_2)]$$

Area and perimeter are examples of contour scores. Therefore, it is possible, using the procedure of this section, to construct the smallest rectangle dissection with a given slicing structure and given shape constraints for its undivided rectangles. Also, the smallest with a given aspect ratio, or with a lower and upper bound on the aspect ratio can be produced. The complexity of the algorithm given is polynomial. As mentioned in section 3.3, this problem is NP-hard for more general dissections.

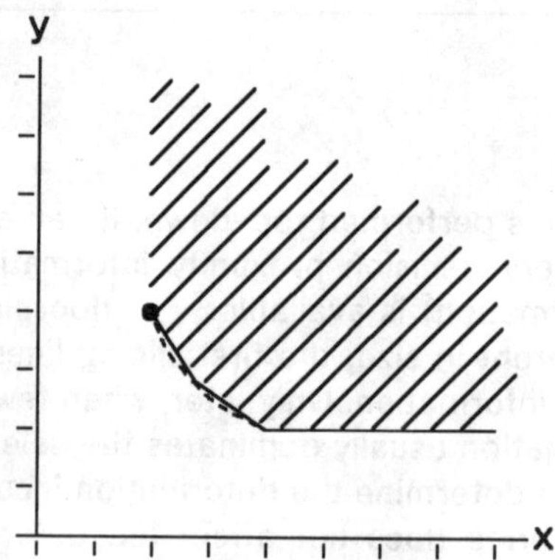

Figure 3.6. Piecewise linear approximation of the shape constraint of a module with given minima on area and dimensions.

Inset cells have piecewise linear shape constraints. Such constraints can be conveniently represented by a sequential list of their breakpoints. This is not the case for flexible cells, and possibly other cell types occurring in practice. Of course, any shape constraint can be approximated by a piecewise linear bounding function with arbitrary accuracy. From the discussion of flexible cells in section 2.3 it is clear that a piecewise linear approximation with three breakpoints suffices considering the limited accuracy of any area estimation for the given examples. (Figure 3.6)

The shape constraint of a compound slice can be derived from the shape constraints of its child slices. In the final configuration these child slices have to have the same longitudinal dimension, which is the lattitudinal dimension of their parent. The inverse of the compound's shape constraint is only defined on the intersection of the intervals on which the shape constraints of its children are defined. Its smallest possible longitudinal dimension for a given feasible lattitudinal dimension x is the sum of the values of the shape constraints of the children at x. So, the shape constraint of a compound is obtained by the addition of the shape constraints of its children in the interval where they are all defined, and determining the inverse of the resulting bounding function. These operations are easy for piecewise linear shape constraints, represented by a list of their breakpoints ordered according to the respective longitudinal dimensions. For each breakpoint of any child of which the first coordinate x is in the mentioned intersection, the shape constraints of all the children have to be evaluated and added. If the result is y, then (y,x) is a breakpoint of the parent's shape constraint. Ordering all these new breakpoints according to the y-values yields a consistent representation for the shape constraint of the corresponding compound slice.

The ability to obtain the shape constraints of a slice by adding the shape constraints of the child slices and inverting the result, enables us to obtain the shape constraint of the enveloping rectangle. The bounded area of that shape constraint is the set of all possible outer dimensions of the total configuration. A contour score always assumes its minimum value over the bounded area at the boundary determined by associated shape constraint. This is a consequence of the monotonicity of shape constraints and contour scores. For piecewise linear shape constraints that minimum will be assumed at at least one of its breakpoints, because of the quasi-concavity of the contour score. So, to find an optimum pair of dimensions for the common ancestor slice the contour score only has to be evaluated at the breakpoints of its shape constraint in the convex set of permissible pairs.

Given the longitudinal dimension of a slice and its shape constraint, its lattitudinal dimension can be found by evaluating its shape constraint for the given longitudinal dimension. After deriving the shape constraint for the common ancestor and determining a dimension pair for which the contour score assumes a minimum, the longitudinal dimensions of its children are known. So, for each of them the lattitudinal dimension, which in turn is the longitudinal dimension of its children, can be calculated. Continuing in this way will finally yield the dimensions of all slices in the configuration. If the shape constraint has a zero right derivative at the point where it has to be evaluated, some slack area will be included, i.e. the slice can be realized in a smaller rectangle without affecting its environment. In

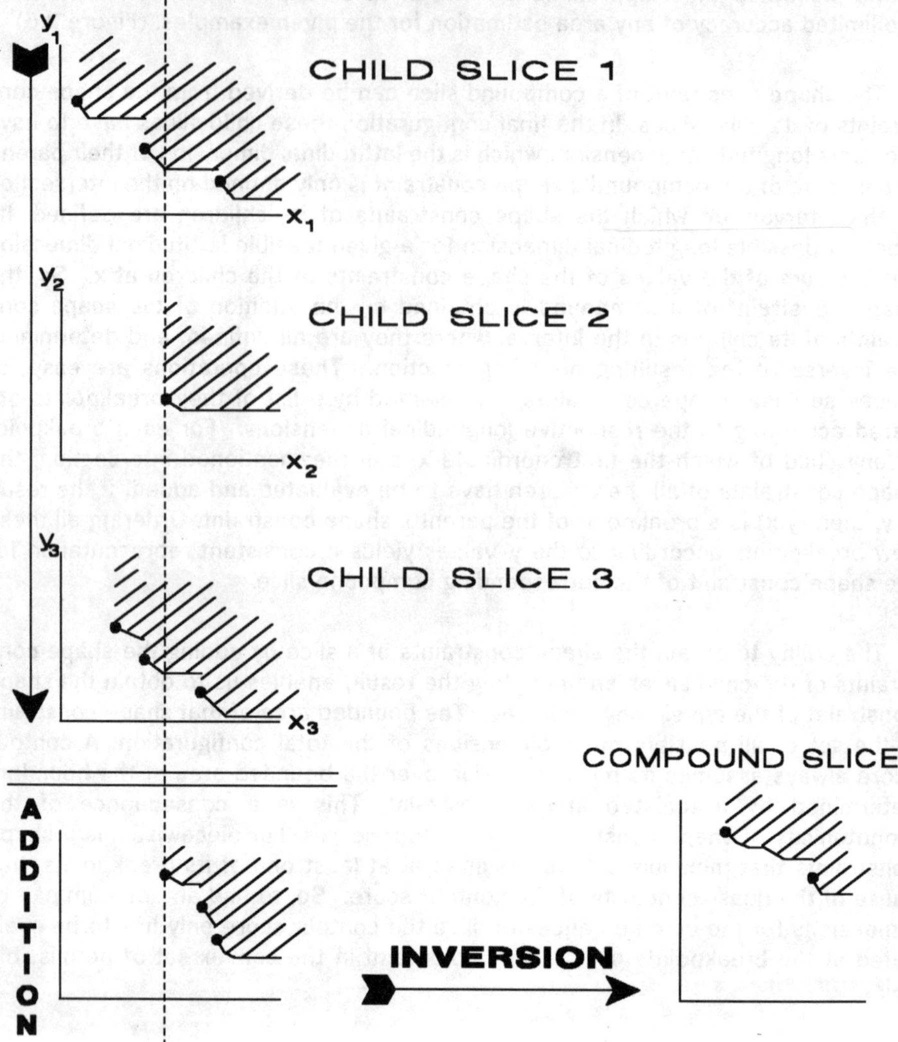

Figure 3.7. Addition and inversion of shape constraints.

order to have the wiring channels connecting to other wiring channels at both ends this slack should be taken up by slices containing only one cell.

The algorithm consists of three parts:

1. Visit the nodes of the slicing tree in depth-first order, and just before returning to the parent determine the shape constraint by adding the shape constraints of its children and inverting the result.

2. Evaluate the contour score for each of the breakpoints of the shape constraint of the common ancestor, and select a dimension pair for which the smallest value of the contour score has been found.

3. Visit the nodes of the slicing tree in depth-first order, and before going to any of its children determine the lattitudinal dimension by evaluating its shape constraint for the inherited longitudinal dimension.

Clearly, when in the first step the same procedure has been applied to all children of a certain slice the shape constraints of these child slices are known and combining them in the way described yields the shape constraint of their parent. The process will end with determining the shape constraint of the common ancestor slice, and then the shape constraints of all slices are known. As explained earlier the contour score will be evaluated at each of its breakpoints. The dimensions associated with the minimum value will become the dimensions of the enveloping rectangle. This means that after completing the second step the longitudinal dimension of the primogenitive (and all the other children) of the common ancestor is known. Together with the shape constraints this is enough information to begin the process of the third step. At the beginning of a visit to a node in the structure tree, representing a certain slice, the lattitudinal dimension of that slice can be determined by evaluating its shape constraint at the value of the dimension that it inherits from its parent slice.

So, completing all three steps yields the dimensions of all slices in an optimum configuration for the given floorplan and cell shape constraints. To determine the position coordinates of the slices from these dimensions and the floorplan is straight forward. Also easy is to determine what orientation the inset cells can have in this optimum configuration.

3.6 Net assignment [8]

The more complex the circuit to be integrated, the more dominant the wiring is in the final layout, as can easily be learnt by examining existing integrated circuits. Though part of the wiring can be realized on top of active devices, particularly when there is more than one metal layer, a considerable portion of the chip is used exclusively for the realization of the incidence structures of the modules. That part of the chip area is called the wiring space. If the cells are realized in iso-oriented rectangular regions, the wiring space can be seen as the union of nonoverlapping rectangles. The selection of the rectangles that together form the wiring space affects the efficiency of the wiring procedures. It constrains the sequence in which

```
procedure fpoptim
    procedure dfs1(s : slice);
        var :  ss : slice; q : queue;
    begin for each ss child of s do
        enqu(ss,q);
            if ss≠leaf then dfs1(ss) end
        end;
        ss := dqu(q);
        F[s] := F[ss];
        while not empty(q) do
            ss := dqu(q);
                F[s] := addition(F[s],F[ss])
            end;
        F[s] := inversion(F[s])
    end dfs1;

    procedure dfs2(s : slice; x : dimension);
        var ss : slice; y : dimension;
    begin y := value(F[s],x);
        dim[s].long := x;
        dim[s].latt := y;
        for each ss child of s do dfs2(ss,y) end;
    end dfs2;

    procedure value(FF : breakpointlist; x : dimension) :  dimension;
        {evaluates FF at x};

    procedure minloc(FF : breakpointlist) : dimension;
            {returns the abscissa of the breakpoint
                with the optimum contour score}

begin dfs1(ancestor);
    chipl := minloc(F[ancestor]);
    dfs2(ancestor,chipl)
end.
```

the wiring can be generated, determines the algorithms that can be used for the wiring, and the number of different algorithms needed to perform that task.

Slicing structures have, again, considerable advantages over general rectangle dissections. Firstly, because they imply a decomposition of the wiring space into the minimum number of rectangles. These rectangles are in a one-to-one correspondence with the slicing lines. To distinguish these undivided rectangles from the ones that correspond with the cells in the functional hierarchy (in the slicing tree both kinds are represented by leaves), they are called *junction cells*. Secondly, feasible sequences for generating the wiring can be easily derived from the slicing tree. A possible rule here is to do the junction cells in a sequence based on the length of the path from the leaf representing the junction cell to the root of the slicing tree, such that the longer this path, the earlier the wiring in that cell has to be generated. And thirdly, all these rectangles can be wired by using the same kind of algorithm, usually called a channel router, though not necessarily in the strictest sense. [2]

Since the wiring consumes quite a high percentage of the total area, it would be useful to have early estimates for this space, so that the sequence of floorplan calls can take these estimates into account when designing the nested floorplans. Several objectives may be important in realizing the interconnections, and many of these are directly related to the size of these nets. This immediately raises two problems. The first one is a consequence of the interpretation of a floorplan as a topology rather than a geometrical configuration. Yet, in order to measure the size of a net, some metric is necessary. The second problem is the need for an ambience in which the wide diversity of objectives can be formulated and optimized.

An often used structure for approaching these problems is the plane graph determined by the rectangle boundaries in the rectangle dissection. Each rectangle corner is a vertex, and the line segments between them are the edges. This graph depends on the geometry of the rectangle dissection, and this geometry is not known in the floorplan design stage. A closely related graph can be defined for slicing structures. For there is a one-to-one correspondence between the junction cells of rectangle dissections with the same slicing structure. Also the relation τ generated by the T-intersections between junction cells is invariant over the rectangle dissections with the same slicing tree [2]. $jc_1 \tau jc_2$ expresses the fact that a junction cell jc_2 intersects with a higher level junction cell jc_1. τ can be stored in a matrix J. This matrix is useful for finding the identity of a T-junction between two junction cells. It has rows for the horizontal junction cells and columns for the vertical junction cells. If $jc_1 \tau jc_2$ then the identifying number of this T-junction is stored in the corresponding matrix position, otherwise there is a 0. If the rows and columns of J are ordered according to the estimated coordinates of the associated junctions cells, the pattern of non-zeros in J then gives a picture of a rectangle dissection having the same floorplan. This ordering of course is not independent of geometry.

The topology of the following graph, the *basic graph*, does not depend on the estimated geometry. τ is its set of vertices, and there is an edge between two vertices if the corresponding T-junctions are on the same junction cell, and there is no T-junction in between. Each edge has a length equal to the difference between the coordinates of the two junction cells it connects. A distance matrix D can be derived from this graph. It is a square matrix with a row and a column for each vertex in τ. Each entry is a number equal to the length of the shortest path between the associated vertices. The lengths of the shortest paths can be computed from the single edge distances by the Yamada-McNaughton algorithm, slightly modified for speed-ups possible because of the properties of slicing structures. For all paths from vertices in one slice to vertices in an abutting sibling slice have to cross vertices on the slicing line that separates them. Using that fact the distances within a compound slice can be computed recursively in a bottom up fashion. Note that the triangle inequality holds for the distance space, thus created.

The above data structures are used in the following net assignment procedure. It proceeds on a net by net basis, in an arbitrary sequence. Each time when a net is routed, extra vertices, representing the modules this net has to connect, are added to the graph. Extra edges connect each new vertex with the T-junctions that surround its module. Pins that can occur only at one side of the module only get the edges that lead to the vertices on that side. The length of the extra edges is used to represent the possibility of duplicating a pin. For a porous module the length of the extra edges is short to make it possible for the net to pass through the module. If this is not allowed then the edges are made long so that passing through the module will be avoided. Also extra vertices are added that represent the pins of the environment. Modules might have multiple pins on their periphery that connect to the same net. In that case a vertex is added to the graph for every pin. The problem to be solved by the net router for each net is to determine a connected subgraph of the extended basic graph. This subgraph has to contain all vertices representing modules to be connected by the net, and the sum of the lengths of its edges has to be small. This subgraph is a steiner tree if that sum is minimum. After the net has been routed the extra vertices and edges are deleted.

This effectively isolates the net routing problem for each net, for an environment in which any steiner tree heuristic can run, is created. The input of the net router is the distance matrix D of the extended basic graph, and the subset of vertices (modules) to be connected. It returns a number of vertex pairs. The edges on a shortest path between vertices of each pair are edges of the connecting subgraph. Each edge corresponds with an interval in the associated junction cell. It is possible that there are several shortest paths. This freedom can be used for optimizing another criterion, like controlling the density in the junction cell.

Any net that passes through a module is split into several nets. Thus every new net has only one pin to a module. To make this possible a new netlist has to be constructed during the net assignment The new nets will inherit the properties and identification from the net that they originated from.

The net assignment is stored as a number of intervals assigned to the junction cells that contain them. If the segment ends at a module vertex then the module is recorded as the terminal. If it ends at a T-junction then the orthogonal junction cell is the terminal. The exact positions of the terminals are determined by the algorithms that generate the modules and route the junction cells. A net in a junction cell is represented by a number of intervals. If those intervals belong to the same net, and are to be connected by the channel cell router then they have an endpoint in common. In this way the net can never become unconnected because of updates in the sizes of the modules afterwards. It can however occur that a loop is created if the sizes of modules change considerably. This happens when two segments in the same junction cell that were originally connected outside the junction cell shift and overlap.

When an interval of a net connects to a module the side on which the pin should preferably be is indicated. This information is used by the algorithm that generates the modules to adapt the positions of the pins to external wiring. The pins are given a position on one of the sides. The net enters the module in the graph via a vertex on its boundary, often one of the vertices on the corners of the module. If the entrance vertex is on one of the sides then that side is chosen. If it is on one of the corners then we choose the side that is facing to the highest level junction cell of the entrance vertex. The only exception is when a two terminal path runs from a module to another module and they are abutting. In this case always the side facing the junction cell that separates them is chosen.

If minimization of the total wire length is the important objective, a steiner tree on the above structure is the best possible. Though slicing yields a considerable saving in computation time, this problem still requires exponential worst-case time (if $P \neq NP$). Therefore, one has to resort to procedures yielding a small connecting subgraph. The edge set of this subgraph is denoted by W in the pseudo programs given at the end of this section. The set of vertices that have to be in that subgraph is denoted by C. Most existing methods for finding steiner trees, exact and heuristic, are based on either a shortest spanning tree algorithm, or a three point steiner tree algorithm. The heuristic, presented here, uses both algorithms in two stages. First it determines a 'topology class' of trees, and then it selects the optimum tree in that class.

The topology class is represented by a binary tree $T(V,B,r)$, where V denotes the set of vertices, B the set of arcs and $r \in V$ the root of the tree. The binary tree itself is a member of the topology class. The other topologies are obtained by contraction of arcs of the tree T. The leafs of the binary tree correspond one to one with the elements of the set C, vertices in the extended graph that have to be connected by a subgraph. The remaining internal vertices of the tree represent candidates for steiner points. In the final result adjacent vertices can represent the same vertex in the basic graph, which means that the arc between them has been contracted.

```
procedure spantree (in: D, C; out: T(V,B,r) );
var p, q, s, S;
begin
   newvertex(r);
   (p,q):⇒(p,q)∈.C×C∧ ∀_{(x,y) | (x,y)∈C×C}[D[p,q] ≤ D[x,y]]
   V :=  {r,p,q}
   S :=  C \ {p,q};
   B :=  {(r,p),(r,q)};

   while not (S = { }) do

      (p,q):⇒(p,q)∈((V∩C)×S)∧ ∀_{(x,y) | (x,y)∈((V∩C)×S)}[D[p,q] ≤ D[x,y]]

      S := S \ {q};
      V := V ∪ {q};
      newvertex (s);
      V := V ∪ {s};
      B:= (B\{(p,pred(p))})∪{(pred(p),s),(s,p),(s,q)}
   end
end spantree;
```

A modification of the nearest neighbor algorithm for generating a shortest spanning tree in a complete graph is used to determine the topology class. The vertices of that graph are the vertices in C. The length of an edge is the length of the path in the extended basic graph as given by the matrix D. The time complexity of that algorithm is $\mathcal{O}(|C|^2)$. Instead of producing the spanning tree the algorithm constructs a binary tree from which the the topology of the spanning tree can be obtained by contracting arcs. The spanning tree is therefore in the topology class represented by the resulting binary tree. The construction, just as the traditional spanning tree algorithm, finds the shortest edge connecting a vertex p already in the tree with a vertex q not yet in the tree. However, instead of simply adding an arc representing this edge, it deletes the arc from pred(p) to p, and incorporates both, p and the new vertex q, in the tree by adding a three point star, the points being pred(p), p and q.

In the second stage the optimum tree in the topology class determined by the binary tree is found in $\mathcal{O}(n^2|C|)$ time, where n is the number of vertices in the basic graph. If there exists a steiner tree with a topology in the determined class then the result will be a steiner tree. In any case the tree that is returned is not longer than the length of the shortest spanning tree, because such a tree is always in the topology class constructed by 'spantree'. The shortest spanning tree cannot be longer than twice the length of the steiner tree.

```
procedure select (in: D[1..n,1..n], T(V,B,r); out:  W ); var s;

  procedure up (t); var i;
    if leaf(t) then
      for i:= 1 to n do t.v[i] :=  D[t,i] end
    else up(left(t)); up(right(t));
      for i:= 1 to n do
      t.v[i] :=  left(t).v[i] + right(t).v[i] end;
      for i:= 1 to n do
        t.v[i] :=  min{s | 1≤j≤n∧s = t.v[j]+D[i,j]} end
    end
  end up;

  procedure down(t,k); var i;
    if not leaf (t) then
      i :⇒ t.v[i]  =  min {t.v[j] | 1 ≤ j ≤ n};
      W :=  W ∪ { (i,k) };
      down(left(t),i);
      down(right(t),i);
    end
  end down;

  W :=  { }
  up(r);
  s :⇒ r.v[s]  =  min {r.v[j] | 1 ≤ j ≤ n}
  down(left(r),s);
  down(right(r),s)
  end select;
```

This stage consists of two depth first visits to the tree. The first one assigns to every vertex t in the binary tree and every vertex i in the basic graph the length of the shortest tree that has a topology in the class of the subtree with t as the root, and contains i and the vertices represented by the leafs in this subtree. From that data the length of an optimum tree in the given class and the identity of the vertex represented by r can be determined. The second depth first traces back how this optimum was constructed, while choosing the steiner points.

After assigning all nets to the junction cells densities in the junction cells can be determined, and on that basis a fairly accurate estimation of the wiring space is possible. This can be used in the floorplan optimization by treating the junction cell as an inset cell with fixed orientation and minimum longitudinal dimension equal to zero. The lattitudinal dimension must be greater than the width estimated on the basis of the cell's density. In addition to obtaining accurate estimates for the shapes of the rectangles, the net assignment also yields information about the location of external nets for the floorplans and cells to be designed later.

4. CELL ASSEMBLERS

The refinement steps described in the previous section determine a topology for every compound of the functional hierarchy. If these topologies are restricted to slicing structures, floorplan design replaces each subtree of which the vertices represent a certain module and all its submodules by another tree of which the root represents the selected module, and the leaves represent its submodules. This is in accordance with one of the rules suggested at the end of section 2.4. The other type of refinement considered there was the replacement of a leaf in the functional hierarchy (a *function cell*) by a tree decomposition. The reasons for not having this decomposition in the initial tree can be quite diverse. For example, the decomposition suitable for the functional design may be be far from optimal for layout design. In that case, such a hierarchy is pruned. Clearly, a data base problem has to be resolved when this happens. It might also happen that the decomposition is suitable for using in the layout program, but more specialized programs are needed than the general floorplanning scheme described. Most often, however, there is no need for further decomposition from the functional design point of view, but flexibility is increased if the layout design procedures use some inherent decomposition.

Algorithms for designing cells, possibly using such a decomposition, are called *cell assemblers*. The task of a cell assembler is to determine the internal layout of its cell (with respect to a reference point in that cell's region) on the basis of a suitable specification and data about its environment. There may be quite a diversity of cell assemblers in a silicon compiler system. The application range of the silicon compiler is highly dependent on the the set of implemented cell assemblers. Whereas most of the decisions during floorplan design are to a high degree technology independent, cell design is dominated by the possibilities and limitations of the target technology. The numeric design rules are stored as numbers of which the value is assigned to certain variables in the cell assembler. The structural rules are to be incorporated in the algorithms, if possible in the form of case statements, so that a variety of rules can be satisfied.

The layout of a slice is obtained by first obtaining the layout of all its child slices except junction cells, and then calling the appropriate assemblers for the junction cells. Visiting the slicing tree in depth first order, and performing the above operations when returning to the parent slice enables the program to determine the 'chip' coordinates (coordinates with respect to a unique point on the chip) of all layout elements in the parent slice before leaving the corresponding vertex. The translation of that result into the rectangles of the various masks is also performed at this point. This translation is straight forward.

In the following two subsections some types of cell assemblers are described. Section 4.1 is devoted to function cells, with special attention to their possible decomposition. Section 4.2 deals with realizing given interconnection lists in a junction cell.

4.1 Function cells

The task of a cell assembler is extremely simple for inset cells of which the internal layout is stored in a master or user library. From the topology determined in the floorplan design process, and the shape constraints of all cells, including the junction cells, an estimate for the rectangle that is going to accommodate that module, can be derived. Reasonably accurate data about the position of the nets to be connected to that cell is generated by the net assignment process. On the basis of that data the assembler has to decide which orientation has to be given to the inset cell, and how it has to be aligned with its sibling slices.

For function cells of which the internal layout is not stored in a library, it may still be implied by the specification. For example, if a cell is going to be realized as a programmed logic array, the specification is either a personality matrix, or a set of boolean expressions. In the former case the assembler does nothing else than performing a straight forward translation and handling the result in the same way as the stored inset cells. If the array is specified by a set of boolean expressions, the assembler must have a so-called pla-generator. The shape of the resulting array is still difficult to control, but the pin positions can be adapted to the results of the net assignment, performed during the floorplan design stage. Some sophisticated pla-generators use techniques such as row and column folding to make the area of the array smaller. This constrains the choice in pin positions considerably, and might lead to a higher area consumption, because of the complex wiring around that array. A pla-generator in a cell assembler should at least be able to take the results of the net assignment into account. [9]

Regular arrays such as programmed logic arrays and memories, have an obvious decomposition into array cells with no or only slight variations in their dimensions. Their positions in the array are heavily constrained, and this decomposition can therefore not be used to manipulate the shape of the array. There also are cells that have a natural or given decomposition that can be used for that purpose. These cells are called *macros*. They are decomposed into circuits that either are selected from a prespecified catalogue, or can be designed with a simple algorithm from a function specification. The reason for keeping such a macro from the floorplan design stage is that the circuits have certain properties that make special layouts very efficient. For example, the catalogue, or the simple algorithm, may have a constraint that gives all cells in the macro the same width and the same positions for general supply pins, such as power supply and clock pins. In that case a pluricell layout style is suitable for the macro. It forces the cells to be distributed over columns, but the number of columns can be chosen freely. Therefore, the aspect ratio of the macro can be influenced. Also the pin distribution around the periphery can be prescribed on the basis of data about the environment. [10]

Decompositions like in macros occur very often, but for special cells that are frequently used, it sometimes is worthwhile to implement a special algorithm producing a highly optimized layout. In word-organized digital systems these special cells often process a number of bit vectors. The layout as a whole may benefit from aligning these cells so that the buses carrying these bit vectors do not have to be matched to the pitch of each individual cell. Also buses that pass over such a cell without making any contact have to be accommodated. These requirements imply a certain kind of flexibility, such as stretchability and variable bus pitches. If possible, such an algorithm must be able to produce these highly optimized cells for several bus dimensions and a range of performance requirements.

4.2 Junction cells

Junction cell assemblers are closely related to channel routers [11], because of the way they are isolated and the moments on which they are called. The junction cell is a rectangular area of which the lattitudinal sides are parts of the longitudinal sides of junction cells that are represented in the slicing tree by vertices closer to the root. When the assembler is called for a certain junction cell, the longitudinal coordinates of the entry points of the nets are known. There are several ways a net may enter the junction cell: from the longitudinal sides of that cell, from a higher metal layer, from the lattitudinal sides, and perhaps in still other ways. The task of the assembler is to realize all the required interconnections in a rectangle with an as small as possible lattitudinal dimension. The longitudinal dimension has to be commensurate with the lattitudinal dimension of the parent slice. Increasing the longitudinal dimension of the channel should therefore be avoided if possible.

References

[1]   N. Wirth: *Program development by step-wise refinement*, Communications of the ACM, Vol 14, 1971, pp. 221-227.

[2]   R.H.J.M. Otten: *Layout structures*, IBM Research Report, RC9657, Thomas J. Watson Research Center, Yorktown Heights, N.Y., 1982. (Preliminary version: Proceedings of IEEE Large Scale Systems Symposium, Virginia Beach, 1982, pp. 349-353.

[3]   L.J. Stockmeyer: *Optimal Orientations of cells in slicing floorplan design*, IBM Research Report, RJ3731, IBM Research Laboratory, San Jose, Cal., 1983.

[4]   R.H.J.M. Otten: *Efficient floorplan optimization*, Proceedings of IEEE International Conference on Computer Design, Port Chester, October 1983, pp. 499-502

[5]   U. Lauther: *A min-cut placement algorithm for general cell assemblies based on a graph representation*, Journal of Digital Systems, Vol. 4, 1980, pp. 21-34.

[6]   R.H.J.M. Otten: *Automatic floorplan design*, IBM Research Report, RC9656, Yorktown Heights, N.Y., 1982, (also: VLSI Technologies in Graphics, H. Fuchs (ed.), IEEE Computer Society, 1983.), Preliminary version in: Proceedings of the 19th Design Automation Conference, Las Vegas, 1982, pp. 261-267.)

[7]   K.M. Hall: *An r-dimensional quadratic placement algorithm*, Management Science, nr. 17, 1970, pp. 219-229.

[8]   L.P.P.P. van Ginneken, R.H.J.M. Otten: *Global routing*, IEEE International Symposium on Circuits and Systems, Kyoto, 1985.

[9]   G. Demicheli, A. Sangiovanni-Vincentelli: *Multiple constrained folding of programmable logic arrays: theory and applications*, IEEE Transactions on Computer-Aided Design, Vol CAD-2, nr.3, 1983, pp. 151-167.

[10]   R.K. Brayton, C.L. Chen, C.T. McMullen, R.H.J.M. Otten, Y.J. Yamour: *Automated implementation of switching functions as dynamic CMOS circuits*, Proceedings of the 1984 Custom integrated Circuits Conference, pp. 346-350.

[11]   M. Burstein: *Channel routing*, Layout Design And Verification, T. Ohtsuki (ed.), chapter 4, pp. 133-168.

[12]   S. Kirkpatrick, C.D. Gelatt Jr., M.P. Vecchi: *Optimization by simulated annealing*, Science, Vol. 220, Nr. 4598, pp. 621-680, May 13, 1983.

[13]   R.H.J.M. Otten, L.P.P.P. van Ginneken: *Floorplan design by simulated annealing*, Proceedings of ICCAD84, Santa Clara, Nov 1984,pp 96-98.

THE SOCRATES LOGIC SYNTHESIS
AND OPTIMIZATION SYSTEM

Aart J. de Geus & David J. Gregory
CALMA, Research Triangle Park
North Carolina, USA

1. INTRODUCTION

A large portion of most designs is combinational logic. Synthesizing this circuitry from functional specifications is a relatively straight-forward process. Optimizing the size and/or the performance of such circuitry in a given technology, however, is considerably more difficult and can consume considerable quantities of expensive design time. For this reason, manual optimization is often not attempted except on the most critical portions of a design. This leads to circuits that are unnecessarily large and slow.

Modern logic synthesis can have both a technical and an economical impact on the circuit design process. Today's logic synthesis and optimization systems often produce circuits that are smaller and faster than those created by human designers. Besides speeding up the circuit design process, these tools make it possible for designers to explore architectural alternatives more thoroughly by providing exact feedback quickly and inexpensively. By exploring a design space more thoroughly, an engineer can make more informed trade-off decisions resulting in higher quality designs.

Automatic synthesis tools can also have a major impact on the cost of designing and fabricating circuits. With the use of these programs optimized logic can be generated in a matter of minutes compared with the days and weeks of work required to do the job manually. Automatic synthesis also guarantees correct results, thus eliminating costly design iterations due to human error. Finally, reducing the size of circuits cuts manufacturing costs, and improving the performance of a design can improve the competitiveness of a final product. For these reasons, automatic synthesis and optimization tools are bound to become an integral part of modern electronic design environment.

This article presents a logic synthesis system called *Socrates* (*S*ynthesis and *O*ptimization of *C*ombinational Logic Using a *R*ule-based and *T*echnology Independant *Ex*pert *S*ystem). The logic synthesis problem has traditionally been approached with boolean and algebraic manipulation and optimization techniques. These techniques can be very powerful, but they fail to account for the circuit characteristics of target technologies. We have developed a rule-based expert system which performs technology specific circuit optimizations and have coupled this program with other programs which perform logic optimization. By combining these two techniques we have achieved results that compete with, and often surpass the results achieved by expert human designers.

2. REQUIREMENTS FOR LOGIC SYNTHESIS AND OPTIMIZATION

Logic synthesis programs are designed to improve engineering productivity by designing combinational circuits automatically. The effectiveness of such programs depends on the quality of the circuits they produce and on their ease of use. A number of factors that affect these characteristics are discussed below:

Verifiable correctness of synthesized circuit. First and foremost, automatically generated circuits should be functionally correct. Although this appears to be a trivial statement, logic synthesis tools have only recently become available, and are still regarded with skepticism by many designers. In addition, modern systems can generate very large circuits which are not easily verified by the designer. For this reason, it is most important that a synthesis system be able to guarantee the correctness of its results.

Quality of synthesized circuits. The quality of a synthesized circuit should be measured against the *constraints* placed on that circuit's performance and fabrication. Circuits are often constrained by the types and characteristics of components available, and by area, delay, power, and testing requirements. Synthesis programs should therefore be capable of generating circuits with competitive area, speed, power, and testability characteristics. However, different constraints are not always compatible. For example, the smallest implementation of a design is rarely the fastest. So, an automatic synthesis program should be able to make tradeoffs between competing constraints. Very often the designer will use goals rather than hard constraints when the "smallest" or the "fastest" circuit is desired.

Technology specific results. There are many circuit technologies available today, and the characteristics of these technologies change often. A synthesis program should be able to work with as many of these technologies as possible, and should also be easily updated to reflect changes in a technology. In order for the resulting circuits to be competitive with human designs, it is not merely enough to implement them correctly in the target technology; it is necessary to *take advantage* of the technology. In the context of logic synthesis, this means that the optimization system should be able to find and use gates that most efficiently implement the circuit in the desired technology.

User friendliness. The tool should be fast, reliable, and easy to use. It should not require expert knowledge to use, nor should it take an inordinately large computer to run. Results should be in a format that is useful to the designer: The system should generate a netlist, but also a schematic, and information on the size, speed and critical paths in the synthesized circuit.

3. Background

Algebraic methods to automatically synthesize and optimize combinational circuitry have been available since the early 1950's[1,2,3]. These methods, and others developed since then[4,5,6,7,8], use logic minimization to reduce the amount of logic in equations which represent the desired functionality of a circuit. Once equations are reduced, they are transformed into circuits by mapping boolean operators to appropriate circuit components.

Since reduced logic normally results in fewer circuit components, logic level techniques are very effective at reducing the size of circuits. Different components can be more or less area efficient, however, and so less logic is not always guaranteed to result in smaller circuits. Other physical characteristics like timing and power consumption vary even more substantially by component and are not accounted for in a strictly boolean representation. For example, in the case of timing, circuit delays depend not only on the number of levels of logic, but also on the relative drive factors and loads of all components on critical paths. Since logic minimization methods do not model these factors, they are at a great disadvantage when optimizing their effects.

Recently, a number of circuit synthesizers have been developed which use circuit models of components[9, 10, 11, 12]. These systems operate by replacing initial circuit components with others that improve overall circuit characteristics. Since the physical characteristics of components are available, actual circuit size and performance can be measured and thus improved. However, because the design representation for these systems is more complex, logic level manipulations which reduce overall complexity are much more difficult to recognize and carry out.

Socrates takes advantage of both the approaches described above by incorporating logic and circuit level manipulation and optimization techniques. Socrates uses one set of programs to reduce the logic of designs, and another set of programs to choose the best components to implement designs.

4. OVERVIEW OF SOCRATES

This section describes the Socrates system. A discussion of Socrates's design representation is followed by a description of its component programs. The Socrates system is illustrated in Figure 1.

Logic level manipulation and circuit level manipulation operate on different aspects of a design. One operates on the logic, and the other operates on the circuit components which implement that logic. To facilitate operations at each level, Socrates uses more than one format for representing designs. Programs performing logic level manipulation use an extended version of *Espresso's*[7] PLA format. Programs performing circuit level manipulations use a net-list format, and a third format is used for entering equations by hand. All three formats share a common constraint specification section and design history section. Translators are also provided for converting designs from one format to another.

A constraint specification allows designers to describe the desired characteristics of their circuits. Designers can specify when signals arrive at inputs, and the drive factor associated with them. Designers can also specify the maximum propagation delays to individual outputs, and the loads that must be driven at those outputs. At present, Socrates does not accept area constraints; it automatically synthesizes the smallest circuit it can under the timing constraints specified.

A design history contains an ordered list of the programs that have run on a design. This list is maintained automatically by all Socrates programs.

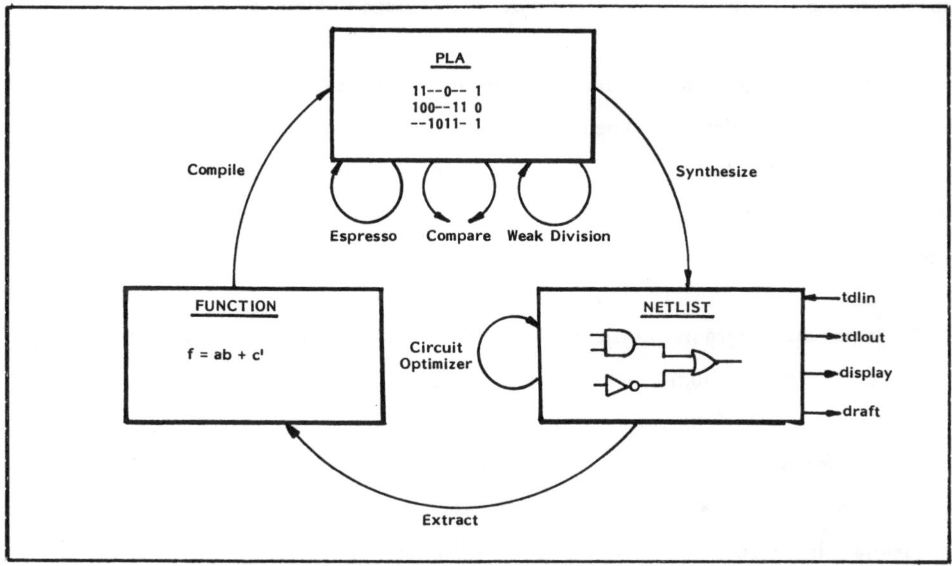

Figure 1:   Socrates System Diagram

Figure 1 shows the three design formats used by Socrates and the various programs that interact with these formats. Each program is described briefly below.

Three programs, *compile*, *synthesize*, and *extract*, translate designs from one format to another. *Compile* converts boolean equations to the two-level *Espresso*[7] PLA format. Multilevel equations are flattened to two-level equations in this step. *Synthesize* converts designs from the PLA format to the netlist format. Generic AND, OR, and NOT gates are used to implement corresponding logic in this step. *Extract* converts a netlist to boolean equations. *Extract* uses a boolean variable to represent each signal in the netlist and writes an equation for each gate.

Two programs, *Espresso*, and *weak division*, perform logic level manipulation on designs. *Espresso* finds a minimal sum of products for each two level function. *Weak division* decomposes two level functions into multiple levels by iteratively dividing out common subexpressions algebraically. *Weak division* is discussed in section 5.

The *circuit optimizer* program manipulates designs at the circuit level. This program improves circuit characteristics by iteratively replacing and rearranging groups of components in the circuit. The *circuit optimizer* is discussed in more detail in section 6.

Compare is a program which takes two designs in the PLA format, and compares them for functional equivalence. When used in conjunction with translators, *compare* can verify designs in any format. This program has proved quite useful for verifying the correctness of design changes made by users, and for debugging portions of the Socrates system.

The remaining programs translate information to and from the netlist format. *Draft* generates a schematic from a netlist. All of the circuits shown in this paper were created using this program. *Tdlin* converts circuits from the TEGAS Design Language (TDL) to the netlist format. *Tdlin* filters out sequential circuit components and stores them so they can be added back later. *Tdlout* converts designs in netlist format back to TDL. Sequential circuit components removed by *tdlin* are automatically reinserted by *tdlout*. Finally, *display* computes and displays circuit statistics including the size and delay of a design.

5. WEAK DIVISION

Weak division is a method originally developed by Brayton and McMullen for algebraically decomposing a set of boolean functions[6]. Weak division decomposes functions by successively dividing the functions by subexpressions that appear more than once. The basic algorithm used is illustrated in figure 2.

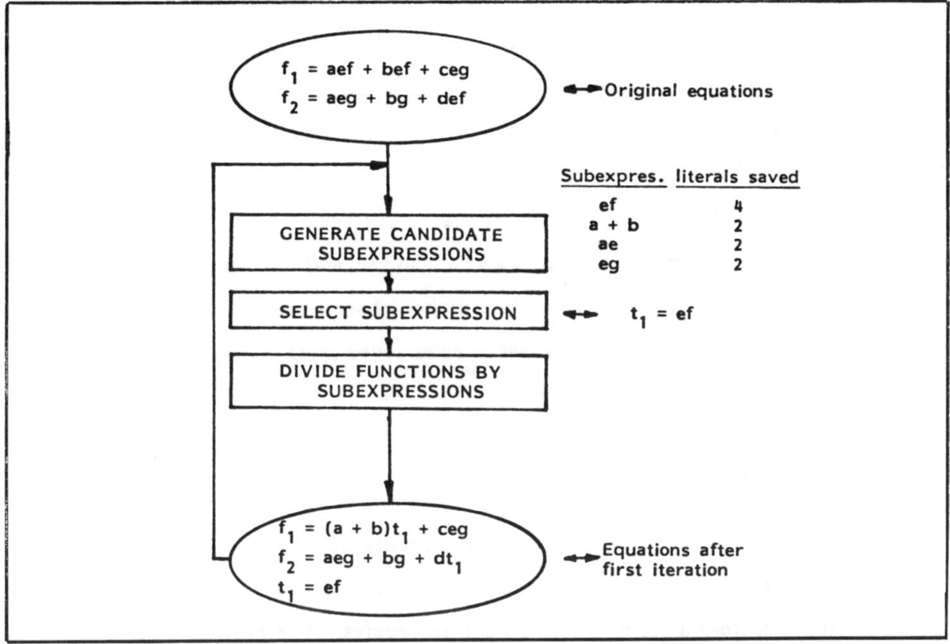

Figure 2:   Weak Division Algorithm

In the first step of the algorithm, a list of candidate subexpressions is generated. The desirability of each candidate is evaluated by a cost function in the second step. The best candidate is then chosen, and divided out of all functions in which it appears. Since the substitution of a subexpression may allow new candidates to be generated in terms of this subexpression, the entire process is repeated. Weak division stops when no more desirable candidates are found.

The cost function has a great effect on the characteristics of the final equations. When optimizing for area, *weak division* uses a cost function which estimates the total number of literals in the set of equations. More complex functions based on levels of logic and predicted circuit delays are used for timing optimization[8].

6. CIRCUIT OPTIMIZATION

The *circuit optimizer* improves measurable circuit characteristics by iteratively replacing and rearranging small portions of the circuit being optimized. A set of transformation rules describes alternative circuit implementations to the system. When a rule is applied, a small configuration of gates in the circuit is replaced by a functionally equivalent but more efficient configuration. A global cost function is used to characterize the quality of the resulting circuit. Transformations are performed with the goal to minimize this cost function. The order in which the transformations are applied to the circuit greatly affects the quality of the final results. The actual sequence of transformations that will be used is determined by a control module which "looks ahead" several rule applications in order to select the most useful transformation. The number of sequences and the length of these sequences is determined by yet another set of rules called *meta-rules*.

6.1. The Cost Function

A cost function is used to characterize the "quality" of the circuit being optimized. During the optimization the cost function is recomputed each time a rule is applied, based on feedback from a set of measurement procedures. The measurement procedures use area and timing values from a user-supplied component library to calculate the size and the speed of the circuits. The size of the circuit is easily obtained by summing the sizes of its constituent components and adding a preset estimate for interconnect wiring. Computing the speed of the circuit is more costly since it involves finding the critical path in the system. To improve the efficiency of the system, both the area and the speed computation are performed incrementally after each transformation to the circuit.

6.2. The Transformation Rules

A transformation rule is a mechanism to replace a portion of a circuit by a functionally equivalent but possibly more desirable circuit portion. Rules are of the form: *if* antecedent, *then* consequent. The antecedent describes the target configuration of gates to be recognized in the circuit being optimized. The consequent describes how to build a replacement configuration. In Socrates rules can be generated automatically from netlists of two gate configurations.

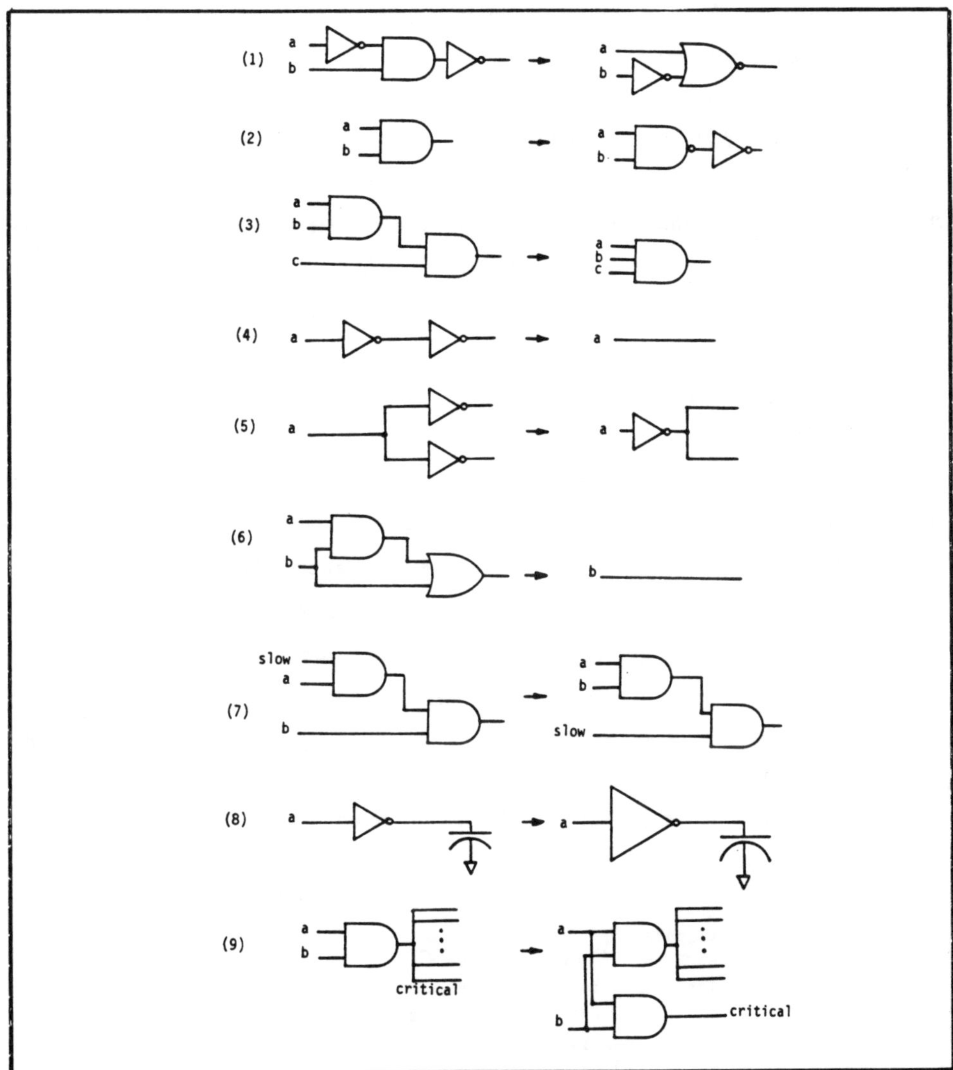

Figure 3: Example rules

Figure 3 shows some example rules. The effect of rules can roughly be classified in three categories: transformations that improve both area and speed, transformations that improve area, generally at the cost of speed, and transformations that improve the speed of the circuit at the cost of its size. Transformations that significantly reduce the number of levels of logic, generally fall in the first category. Rule 4 and 6 in figure 3 are good examples of this case. Rule 3 can improve the area of a circuit but will degrade the timing for the path from c. Rules 8 and 9 are examples in which the timing of the resulting circuit is improved at the cost of either using a larger gate or additional circuitry.

There are currently 163 rules in the *circuit optimizer* program. These rules are grouped and ranked automatically by a program which evaluates their expected effect on a circuit. A rule is formulated to the system by providing two netlists. In order to guarantee that the rules are correct, equations are extracted of each of the netlists and automatically compared. This guarantees the functional correctness of each rule.

6.3. The Search Strategy

The optimization problem can be viewed as a state-space search problem. Each state in the state space is a functionally correct implementation of the desired logic. The edges in the state space graph are possible rule applications. The cost function determines the desirability of each state. Optimization is thus equivalent to finding a path from the initial circuit configuration to the optimal configuration. Socrates approximates a gradient search in traversing the search space, always searching for the transformation that maximally decreases the cost function.

In the first version of the system the rules were ordered in a set of classes. The system then cycled through the rules in each class until no more rules could be applied. The basic algorithm was:

```
select_rule:
for each rule R in some class C
    for each gate G in the circuit
        if the target for R matches at gate G
            apply_rule(R, G)
            go to select_rule
```

The value of C controlled what parameters were being optimized for. For instance, when optimizing for area, the class of *area rules* would have been used.

This simple control strategy proved to be fairly effective for area optimization and fairly poor for timing optimization. One reason for this was that the effect of a *timing rule* is difficult to predict since it is dependant on the way surrounding circuitry is impelemented. Another problem was that this strategy failed to provide for intelligent decisions about where to apply a rule. Figure 4 illustrates a situation in which the same rule can be applied in two different places with dramatically different results. Finally, it appeared that a gradient search could be better approximated by not looking at the effects of the individual rules on the circuit, but by examining the effects of applying a short sequence

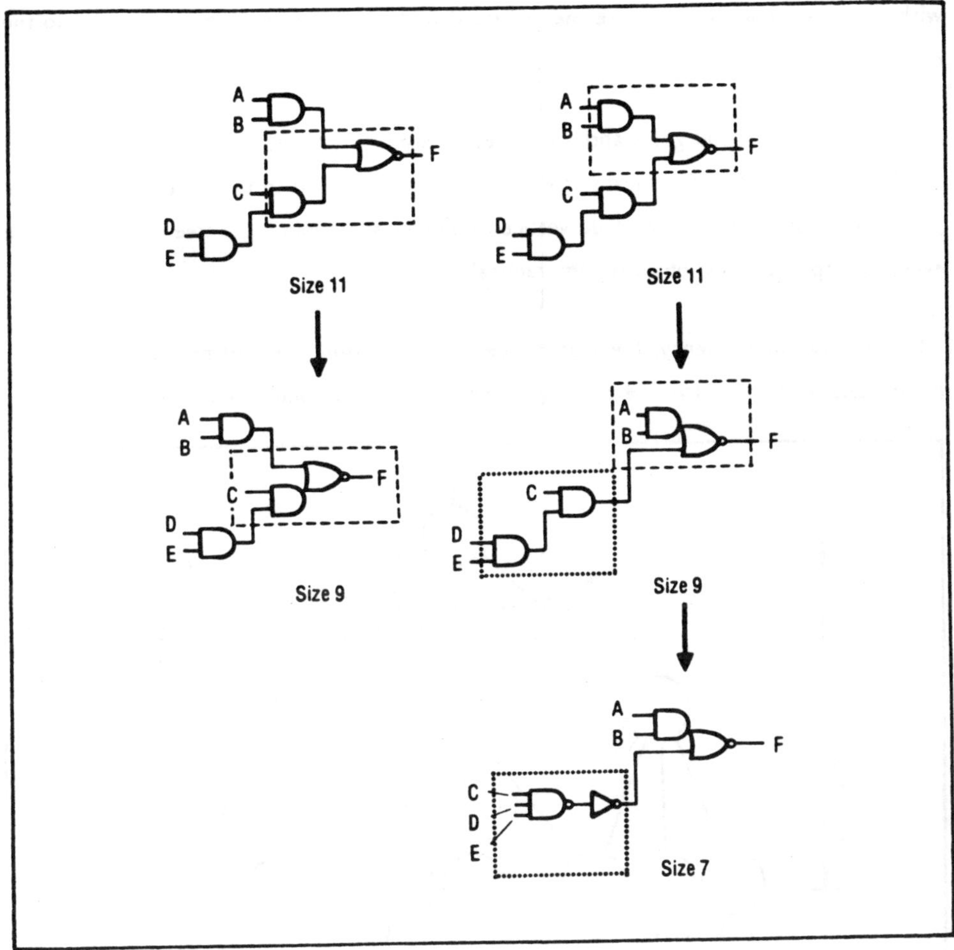

Figure 4: Effect of rule order

of rules. By doing so, the requirement that each individual rule must reduce the cost function could be relaxed, thus reducing the chance to get stuck in a local minimum of the cost function.

6.3.1 Look-Ahead Strategy

For these reasons, a *look-ahead* scheme was implemented. This strategy entails building a portion of the state-space graph, called a *search tree*, explicitly in memory. This tree is then traversed, and the leaves are examined. The rule that leads to the leaf node with the lowest cost function are then selected by the control module and applied to the circuit.

For a circuit with 100 gates and a knowledge base of 50 rules, typical branching factors of 50 were observed. This means that the search tree needed to look ahead two rule applications would contain over 2500 vertices; building and exploring a tree of this size with every rule application is obviously impractical.

For the sake of efficiency, the search tree is pruned using a set of heuristics. The key parameters used to control the pruning of the seach tree are illustrated in Figure 5.

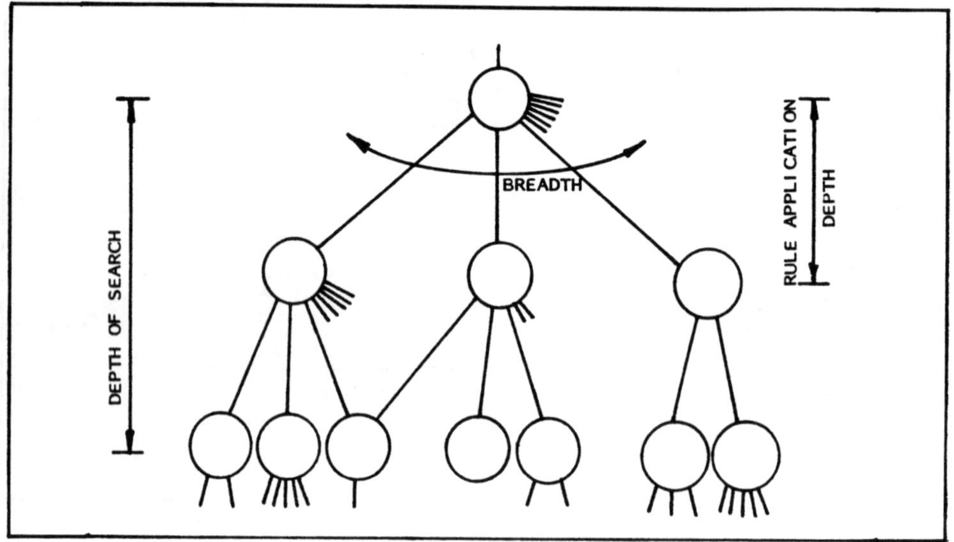

Figure 5: Parameters controlling size of search tree

First, the number of sons of any node that are actually explored is limited to some small number B. This number then corresponds to the maximum *breadth* of the tree. The B sons that the control module choses to explore are the configurations that would be reached by applying the first B applicable rules to the circuit represented by the parent node.

Second, the *depth* of the tree is limited by the parameter D. A depth of three, for example, means that the control module will choose between all reasonable sequences of three rules.

Third, the search tree is restricted to rules that apply to a small area of the circuit. This technique is used to avoid situations where the control module is attempting to decide between two rules that can be applied independently of each other. The size of this local *neighborhood* is determined by the parameter N.

Another parameter that affects the look-ahead strategy is D_{app}, the rule *application depth*. After the search tree is built and explored, rules are chosen to actually apply to the circuit. These rules are a part of the path from the root of the search tree to the lowest-cost node. The length of this path will be D_{App}; for instance if $D_{app}=2$, then two rules will be applied to the circuit each time a tree is build. D_{app}, of course, is always smaller or equal to the depth D.

6.3.2. Metarules

The first look-ahead strategy built and traversed a search tree before applying each rule. The size and the shape of this tree was determined by the parameters described above, but fixed throughout the optimization. A series of experiments fixed the point of diminishing returns for the parameters above at B=3, D=2, and N=D_{app}=1.

Although this simple strategy improved the quality of the results by about 10 percent, it was very inefficient. The optimal amount of look-ahead varies greatly over the course of an optimization; in general, look-ahead is far more useful in later phases of optimization than in earlier ones. Obtaining the full benefit of look-ahead in the last stages of optimization requires fairly large values of B and D.

For these reasons, some way of dynamically varying the search control parameters is needed. A metarule system[13] to handle this problem. This module determines the appropriate values of the control parameters based on the current state of the optimization. The metarule system acts as the supervisor, ensuring effective use of the look-ahead technique. The metarule system also follows the paradigm of a rule-based system. It consists of a series of *metarules*, each of which adjusts a set of search control parameters if and only if a given set of conditions holds.

Using the metarule system reduced the runtime of optimizations by almost 60 percent and modestly improved the quality of the circuits produced, as compared to the old strategy, which used fixed values for the look-ahead parameters. Most importantly, the metarule system has proven invaluable for experimentation with different search strategies and the metarules capture elegantly the experience of the builders of the optimization system.

6.3.3. Optimization for Speed

Being able to satisfy user defined timing constraints is crucial for the success of a logic synthesis and optimization program. Optimization for speed is different from optimization for area in that the cost function computation is more complex and the final results are much more sensitive to the chosen search strategy.

When computing the area of a circuit resulting from a rule application, the replaced gate configuration's area is subtracted from total area, and the new gate configuration's area is added to the total. Computing the timing information is more difficult for two reasons: first, the timing of a new gate configuration is a function not only the gate configurations, but also of the surrounding circuitry, and secondly, changing some gates in the circuit can drastically change the critical paths in the circuit. To address the first point, a timing model is used that approximates actual circuit timing. The models used to calculate the delays are illustrated in Figure 6.

There are two delays associated which each circuit element: The intrinsic delay reflects the time a signal will take to travel through the gate. This value is assumed to be in-

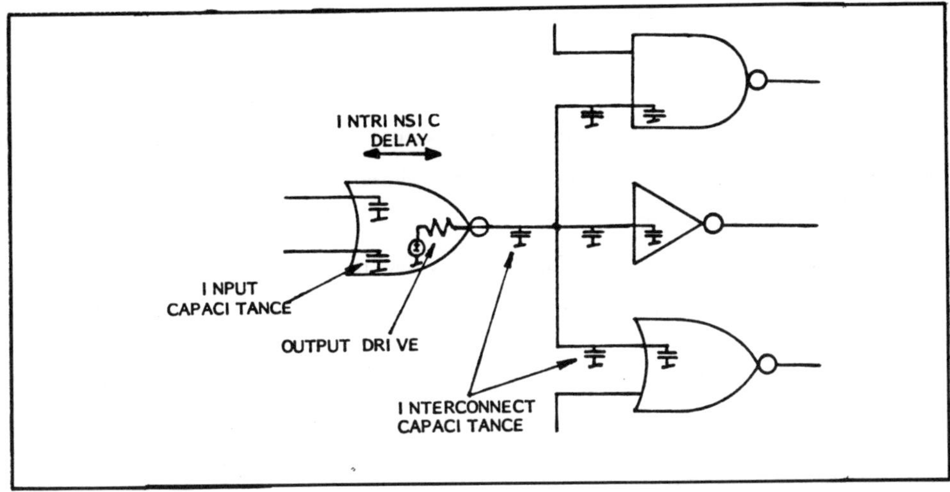

Figure 6: Components of Delay

dependant of any loading and is stored in the technology dependant gate library. The loading delay reflects the delay due to the interconnect lines and the input capacitances of the driven gates. This delay is obtained as follows:

$$Delay_{loading} = Drive\,[C_{interconnect}(fanout) + \sum C_{loads}]$$

A drive factor stored as part of the model data associated with each gate is multiplied with the sum of interconnect and load capacitance. The interconnect capacitance is stored in a table as a function of the fanout and the technology used. This approach allows the user to use statistical interconnect length information if available. The input capacitances are stored as part of the gate models.

After the local delay values have been recomputed, critical paths are reevaluated. Since this is a costly operation entailing a forward and a backward pass through the network[14], the critical path algorithm used in Socrates performs these tasks incrementally, merely recomputing paths in the cone from the new gate to the inputs and in the cone from the gate to the outputs. In this manner the system always has an up-to-date list of the timing slack at each node in the circuit. (Slack is the time that separates a node from being on the critical path; all nodes on the critical path have zero slack.) When Socrates optimizes for speed it concentrates its search to transformations applicable to the critical path.

6.3.4. Changing Technology

The computation of the cost function for a circuit uses the area and timing models for each gate. Different technologies use different values for the models of each gate. These values are supplied to the program by the user in libraries. When competing alternatives are compared during the optimization, the actual area and performance in surrounding circuitry is used as the basis for evaluation. Any new technology is easily described to the system by inserting the appropriate area and timing parameters in a technology library file.

7. DESIGN SCENARIOS

By maintaining distinct programs, Socrates provides a more flexible system than would otherwise be possible. A number of design scenarios are presented below to demonstrate this flexibility and to show how Socrates is used to solve a number of engineering problems. Figure 7 summarizes these scenarios graphically.

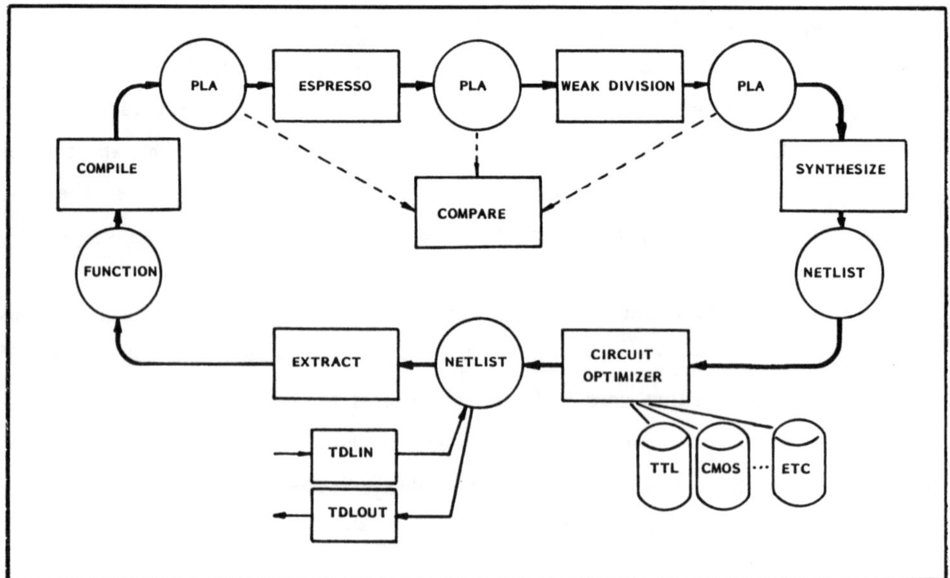

Figure 7:   Example Design Scenarios

Equations to Circuit. In the first scenario, Socrates is used to design a circuit from a set of boolean equations. In the first step of this scenario the user enters equations in the function format to a text file. If there are timing constraints on the circuit, they are entered at this time as well. Figure 8 illustrates how this is done in Socrates: The input and output variables are declared first. Timing constraints are formulated by stating both the signal arrival times and the desired output delays. The function is described by a set of multilevel boolean equations. A don't care set can be described in similar fashion.

```
        # Example input file

        # Variable declarations
          .inputs 11
          .outputs 2
          .inputnames  A B C D E F G H I J K
          .outputnames OUTO OUT1

        # Timing constraints
          .input_arrival    C   2.0
          .input_arrival    F   1.7
          .output_delay OUTO  9.0
          .output_delay OUT1 12.0

        # Description of functions
          .contents function
          OUTO = A B F' + A (B G + C' D') + I J'(A B (C + E));
          OUT1 = A F' I + A I (J + C D) + A C E I + K ;
```

Figure 8: Example Input Description

The *compile* program is run next. *Compile* flags syntax errors and reports them to the user. The equations are now in PLA format. After a syntactically correct file has been created, the logic minimizer *Espresso* is run. If PLA generation tools are available, the designer can create a PLA layout at this point in the procedure. If the goal is to generate a gate level circuit, the programs *weak division*, *synthesize*, and the *circuit optimizer* are run next. To view or analyze the new circuit, the user runs *draft* which creates a schematic of synthesized circuit or *display* which allows the user to interactively inquire on size, speed, critical paths, etc of the circuit. If the circuit meets its constraints, *tdlout* converts the design to TDL format. If the constraints are not met, the user can formulate new constraints. *Weak division* and/or the *circuit optimizer* can then be rerun to create a new circuit. Figure 9 shows the optimized circuit obtained from the input description in figure 8. The system reported that it was unable to meet all the timing constraints and that the output delay for OUT0 was 9.18ns compared to the desired 9.0ns.

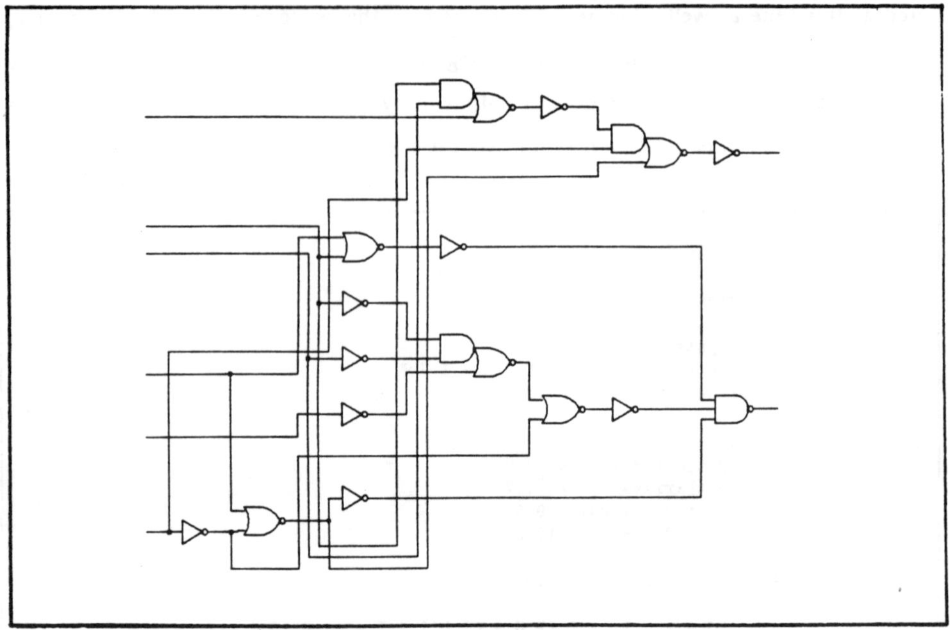

Figure 9: Example Circuit

Optimization of existing circuit. In another scenario, Socrates starts with an existing circuit that needs to be made smaller or faster. The circuit is read in with *tdlin* and a constraint specification section is added. The design can now be massaged at a circuit level with the *circuit optimizer*, or its logic can be reworked first by running *extract, compile, espresso, weak division*, and *synthesize*, and then running the *circuit optimizer*. If the architecture of the original circuit has already been carefully designed and op- timized, a rework by *espresso* and *weak division* may not be necessary or advantageous. However, if the logic of the circuit has not been reduced these programs should be run. The user could also try both alternatives and choose the best circuit. When finished, *tdlout* builds a TDL representation of the circuit and replaces any sequential elements removed by *tdlin*.

Technology translation. To convert a TTL design to CMOS (or visa versa) for example, a design is read in with *tdlin*, and run through the *circuit optimizer* with the appropriate technology library. Any two circuits can be tested for functional equivalence by running them both through *extract* and *compile* and then using the *compare* program.

Independant usage of programs. The modular construction of the Socrates system makes it easy to use its constituent programs independently. For example, PAL or PLA logic can be minimized using *compile, espresso* and *weak division*. Draft can be used to generate a schematic of an existing netlist. Many other scenarios are also possible; some of the experiments described in the next section illustrate this.

8. RESULTS

This section presents the results of running Socrates on a set of benchmark circuits[15] prepared for the 1986 Design Automation Conference. Several experiments were run to demonstrate both the power and the flexibility of the system.

Circuit areas for all of the experiments are quoted in terms of two input gate equivalents, and in terms of the number of actual gates used. Circuit delays are quoted in nanoseconds. Both the sum of the delays to all outputs, and the delay of the slowest output are shown. CPU times are quoted in terms of VAX 11/780 minutes; the VAX 11/780 is a 1 MIPS machine. Except where noted, all circuits were synthesized in General Electric's two micron gate array series, and delays and areas are quoted in terms of this technology. Table 1 shows the gates of this technology used by Socrates, and their two input gate equivalent sizes.

Socrates was run in its fully automatic mode for all experiments. The user's only interaction with the system was through the constraint specification section.

In the first experiment, Socrates was run on each of the benchmark circuits to optimize area. No timing constraints were added. Only *Espresso*, *weak division*, *synthesize*, and the *circuit optimizer* were used. Table 2 summarizes the results of these runs.

Gate	Description	Size
i1x	Standard inverter	0.5
i2x	Power inverter	1.0
nd2	nand(2)	1.0
nd3	nand(3)	1.5
nd4	nand(4)	2.0
nd5	nand(5)	2.5
nr2	nor(2)	1.0
nr3	nor(3)	1.5
nr4	nor(4)	2.0
aoi21	and(2) into nor(2)	1.5
aoi22	two and(2)s into nor(2)	2.0
oai22	two or(2)s into nand(2)	2.0
aoi32	and(3) & and(2) into nor(2)	2.5
aoi222	three and(2) into nor(3)	3.0

Table 1: Technology description

In the second experiment, most of the same circuits were resynthesized for minimum total delay by constraining the delay at each output to one nanosecond. Both *weak division* and the *circuit optimizer* were rerun since both perform timing related optimizations. Table 3 summarizes the results of these runs. On average, these circuits are 53 percent faster and 64 percent larger than those synthesized in experiment one.

Circuit	Size		Delay		
	2input	Gates	Total	Worst	CPU
Alupla	125.5	125	175.2	38.0	38.5
Bw	130.5	125	389.0	20.3	77.6
Duke2	318.0	292	567.7	33.2	355.4
F2	16.0	16	27.2	6.8	2.0
Rd53	29.0	28	54.2	19.4	10.7
Rd73	101.5	100	67.3	33.1	34.9
Sao1	184.5	168	92.2	34.8	106.9
Sao2	154.5	143	106.5	34.6	88.8
Vg2	76.0	73	164.3	31.1	17.3
5xp1	104.5	105	149.2	20.56	35.4
9sym	182.0	131	41.2	41.2	37.4

Table 2: Area Benchmarks

Circuit	Size		Delay		
	2input	Gates	Total	Worst	CPU
Alupla	128.0	128	148.9	33.4	129.2
Bw	185.0	173	25.1	25.1	116.5
F2	16.0	16	25.2	6.3	2.2
Rd53	73.0	62	38.5	15.6	25.4
Rd73	105.5	105	62.2	30.9	85.5
Sao2	191.5	163	75.2	24.1	173.8
Vg2	82.0	84	127.6	21.2	87.3
5xp1	137.5	126	94.9	16.1	132.6
9sym	219.5	173	25.1	25.1	116.5

Table 3: Timing Benchmarks

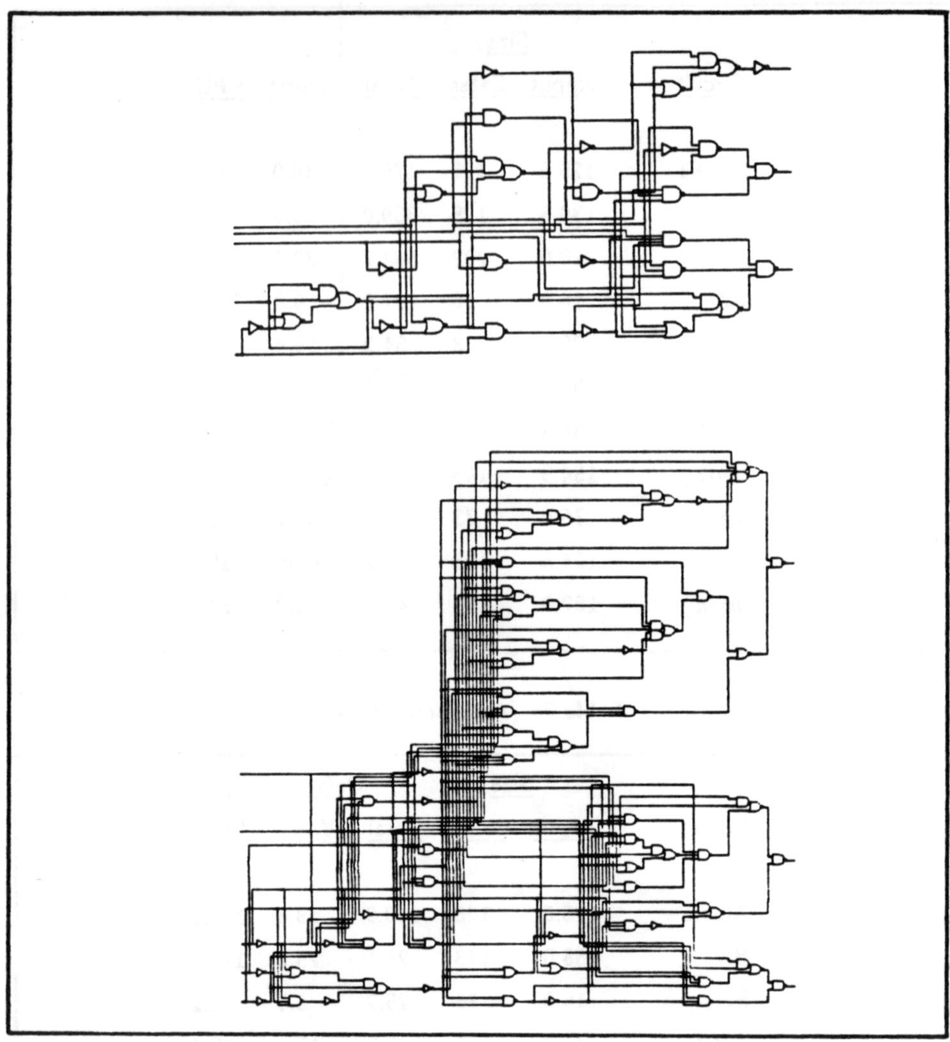

Figure 10: Area Timing Comparison for Example Rd53

Figure 10 shows the two circuits created for example RD53. On the left is the circuit optimized for area, and on the right is the circuit optimized for speed.

In the next experiment example Rd53 was run under increasingly severe timing constraints. The results in Figure 11 show the area/speed trade-offs for this example and demonstrate how Socrates can be used to explore a design space quickly and easily.

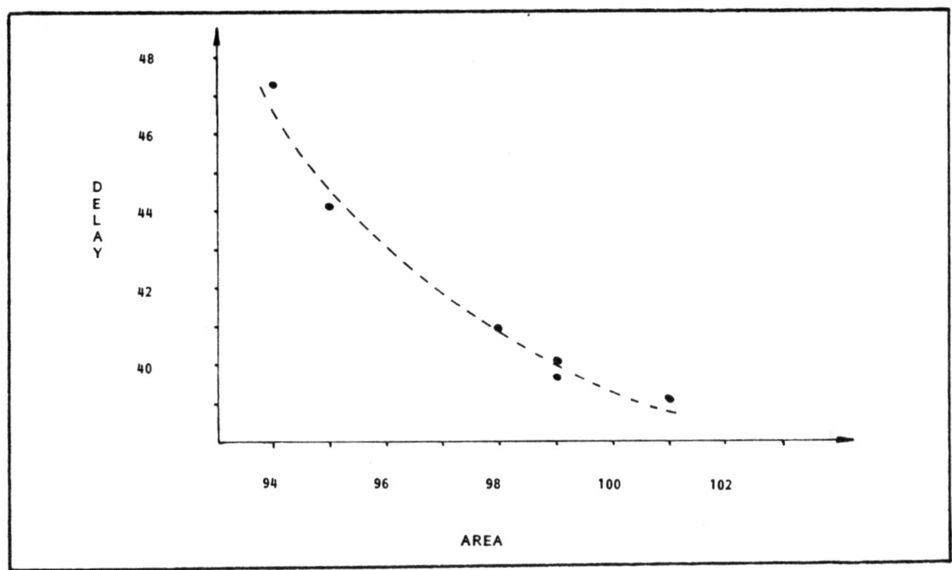

Figure 11: Area Speed Trade-offs for Rd53

To demonstrate Socrates's technology flexibility, a novice user added LSI Logic Inc.'s 5000 series to the system. In addition to different area and timing characteristics, LSI's technology contained five gate types that were previously unknown to Socrates. To allow Socrates to make use of these gates, twenty three new rules were added to the system. It took the new user a total of four man-hours to make all of these changes.

To illustrate how Socrates takes advantage of the gates available in a given technology, example RD53 was synthesized twice. The second time, the model values of the two input NOR gate was changed from small and fast to large and slow. Socrates replaced all of the original NOR gates with NAND gates and inverters. Figure 12 shows the resulting schematics for both cases.

When *espresso* minimizes logic, it creates functions which are prime and irredundant[7]. If a circuit has redundant logic, it may also have undetectable faults. Example C432 from the benchmark circuit set compiled by Brglez and Fujiwara[16], has redundant logic which results in four undetectable faults. C432 was entered into Socrates, and resynthesized using *extract, compile, espresso, weak division, synthesize,* and the *circuit optimizer.* A automatic test vector generation program was able to generate vectors which covered all

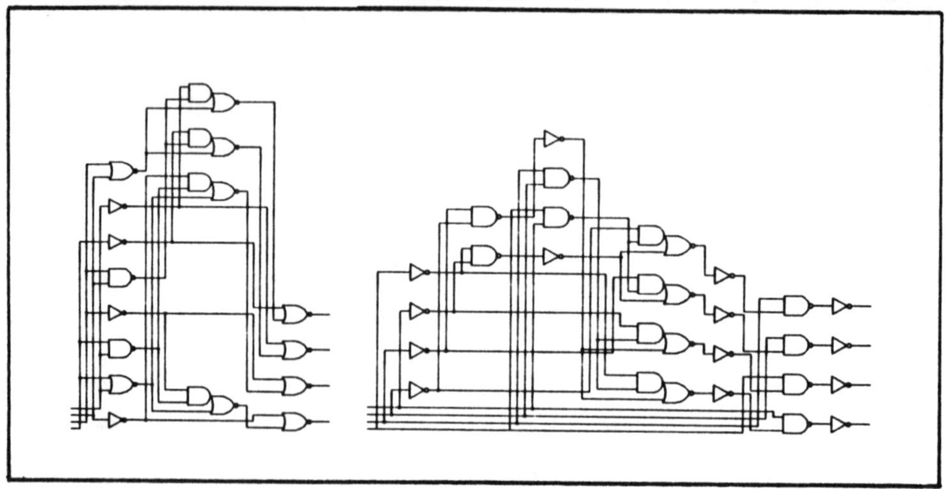

Figure 12:   Example change in technology

of the faults of the resulting circuit.   Resynthesizing C432 with Socrates *eliminated* the untestable faults, reduced the size of the circuit by 2 percent, and increased its speed by 37 percent.

The ultimate test of the quality of an automatic synthesis tool is to compare its circuits with those designed and optimized under the same constraints by experts.   An experiment was performed in which five experienced engineers competed with Socrates to design the smallest circuit for each of three designs[17].   In order to keep the problem complexity at a level manageable for humans, no timing constraints were imposed on the circuits. Socrates won the competition for all three circuits, and beat the best area of the human designs by an average of 13.5%.

Another more recent experiment pitted Socrates against an experienced engineer in the optimization for maximum speed of a standard ALU circuit, TTL part SN54181[18].   Both Socrates and the engineer started with a TTL schematic for the part.   Since the logic of this design was already well optimized, the *circuit optimizer* was run directly without *espresso* or *weak division*.   The total delay of the hand design was 169.6 nanoseconds. The worst delay through the designer's circuit was 31.1 nanoseconds, and the circuit used 100 equivalent two input gates.   Socrates's circuit had a total delay of 170.7 nanoseconds.   The worst delay was 31.8 nanoseconds, and it used 105.5 equivalent two input gates. Overall, the hand designed circuit was faster, but by less than one percent. Socrates used two hours of CPU time to create its circuit; the engineer worked on the design for six weeks.

9. SUMMARY

This paper has presented Socrates, a system of programs which automatically synthesizes combinational circuits from functional specifications. Socrates manipulates the logic descriptions at two distinct levels: On one level algebraic minimization and decomposition techniques reduce the size of the equations and determine their overall structure; on the other level the rule based circuit optimizer implements the circuit in the target technology. Combining these two powerful techniques has resulted in a technology independent system that generates technology specific circuits.

The modular structure of Socrates allows the system to perform a variety of tasks: synthesis of circuits from equations, optimization of existing designs, translation from one technology to another, etc. The circuits generated by Socrates compare favorably in size and speed with those generated by human designers.

10. REFERENCES

1.   W. Quine, "The Problem of Simplifying Truth Functions", *American Math Monthly*, Vol. 59, No. 8, October 1952, pp. 521-531.

2.   R. Mueller, and R. Urbano, "A Topological Method for the Determination of the Minimal Forms of a Bollean Function", *IRE Transactions on Electronics and Computers*, Vol. EC-5, No. 3, September 1956, pp. 126-132.

3.   E. McCluskey, "Minimization of Boolean Functions", *Bell Systems Technical Report*, Vol. 35, No. 5, November 1956, pp. 1417-1444.

4.   S. J. Hong, R. G. Cain and D. L. Ostapko, "MINI: A Heuristic Approach for Logic Minimization", *IBM Journal of Research and Development*, Vol. 18, September 1974, pp. 443-458.

5. T. Sasao, "Multiple-Valued Decomposition of Generalized Boolean Functions and the Complexity of Programmable Logic Arrays", *IEEE Transactions on computers*, September 1981.

6. R.K. Brayton, C. McMullen, "The Decomposition and Factorization of Boolean Expressions", *Proceedings of the International Symposium on Circuits and Systems*, 1982, pp. 49-54.

7. R.K. Brayton, G.D. Hachtel, C.T. McMullen, A. Sangiovanni-Vincentelli, *ESPRESSO-IIC: Logic Minimization Algorithms for VLSI Synthesis*, Kluwer Academic Publishers, Netherlands, 1984.

8. K. Bartlett, G. Hachtel, "Library Specific Optimization of Multilevel Combinational Logic", *Proceedings of the IEEE International Conference on Computer Design*, October 1985.

9. J. Darringer, W. Joyner, L. Bermen, L. Trevillyan, "Logic Synthesis Through Local Transformations", *IBM Journal of Research and Development*, Vol. 25, July 1981, pp. 272-280.

10. D. Gregory, K. Bartlett, A. J. de Geus, "Automatic Generation of Combinatorial Logic from a Functional Specification", *Proceedings of the IEEE International Symposium on Circuits and Systems*, May 1984.

11. T. Uehara, "A Knowledge-Based Logic Design System", *IEEE Design and Test of Computers*, October 1985.

12. K. Enomoto, S. Nakamura, T. Ogihara, S. Murai, "LORES-2: A Logic Reorganization System", *IEEE Design and Test of Computers*, October 1985.

13. W.W. Cohen, K. Bartlett, A.J. de Geus, "Impact of Metarules in a Rule Based Expert System for Gate Level Optimization", *Proceedings of the IEEE International Symposium on Circuits and Systems*, May 1985, pp. 873-876.

14. R. B. Hitchcock, "Timing Verification and the Timing Analysis Program", *Proceedings of the 19th Design Automation Conference*, June 1982, pp. 594-604.

15. A.J. de Geus, "Logic Synthesis and Optimization Benchmarks". The benchmarks are in Logic Interchange Format (LIF) and are available from the author upon receipt of a magnetic tape or floppy diskette.

16. F. Brglez, H. Fujiwara, "Benchmarks for Automatic Test Vector Generation". The benchmarks have been put together for the 1985 International Symposium on Circuits and Systems and are available from the authors upon receipt of a magnetic tape.

17. A.J. de Geus, W.W. Cohen, "A Rule-Based System for Optimizing Combinational Logic", *IEEE Design and Test of Computers*, August 1985, pp. 22-32.

18. National Semiconductor, *Logic DataBook*, National Semiconductor, 1981.

THE SYCO SILICON COMPILER AND ITS ENVIRONMENT

A. JERRAYA - B. COURTOIS

IMAG/TIM3 46, av. Félix Viallet 38031 GRENOBLE Cedex. FRANCE

KEYWORDS/ABSTRACT : silicon compilation / SYCO/ algorithmic description / testable design / electron-beam testing.

The SYCO system is a **silicon compiler** for VLSI ASICs specified by algorithms. The SYCO system starts from an algorithmic description and produces a circuit that realizes the algorithm. It uses a target architecture based on a multiple interpretation level scheme. The SYCO system makes use of a set of specialized compilers to produce the layout of the datapath and of the control section. The whole system is written in Le Lisp. Developments towards the compilation by SYCO of "testable" circuits, and towards electron beam aided debug, are finally given.

I - INTRODUCTION

The SYCO silicon compiler itself is developped by the Computer Architecture Group of the IMAG/TIM3 laboratory in the framework of the SYCOMORE French National Project of CAD for VLSI. The SYCO system is firstly addressed in this paper. Next developments which are in progress towards the compilation of testable circuits and the electron-beam aided debug are addressed.These developments take advantage of works carried out respectively within the EEC supported AIDA and ADVICE projects.

II - INTRODUCTION TO THE SYCO SYSTEM

In this paper, **silicon compilation** will refer to the **automatic synthesis of the layout of an integrated circuit from its behavioral description**. In order to clarify what is meant by an automatic synthesis from a behavioral description, a classification of most of the silicon compilers is firstly quickly given. Next, basic principles for the design of a silicon compiler are reviewed.

II.1 - Previous works
As stated in several new publications [BLA85, KRE85, VLH84..] silicon compilers are popular. The term silicon compiler refers to a program that can produce circuit layout from high level specifications. A lot of software tools can then be considered as silicon compilers.
Several papers [EVA85, GRO83, THO81] establish a good classification of most of the existing tools according to the range of the spectrum (from a behavioral specification of the circuit to the masks) spanned by the compiler.The aim of this section is not to review the existent silicon compilers, but only to place the SYCO system among existing tools.

We will limit our interest to two kinds of silicon compilers. Silicon compilers of the first kind are powerful floor planners, they use for input a functional organization and silicon compilers of the second kind translate an algorithmic description into layout without any interaction with the designer.

SCI (Silicon Compiler Inc) and SST (Seattle Silicon Technology) [EVA85] compilers belong to the first class. They use a menu driven specification table as input. The user has to specify the internal structure of the circuit. Specialized tools are used to generate the layout. The layout blocks are then automatically assembled. Such a system solves the layout problems and makes easier the other design steps : it provides an integrated environment for circuit design and verification. On the other hand the user has a good control of the design process since he controls the circuit partitioning. But such a system still needs expert designers as users. For example, the designer has to give the global layout strategy to minimize the interconnection area wasted by the gobal router.

The Bristle Blocks system [JOH 83] belongs also to this kind of compilers. Bristle Blocks is an experimental compiler designed by Johannsen at the California Institute of Technology; Johannsen is one of the founders of Silicon Compiler Inc (SCI). Bristle blocks solves the global layout problem since it uses a target architecture. It uses also a powerful scheme for the data path, but it restricts the control part to a simple decoder.

The MacPitts system [SIS 83] belongs to the second kind of compilers. It starts from a lisp-like description of a circuit and produces the corresponding layout. This system points out the need of the correspondence between the target architecture and the description language in order to have an efficient translation scheme. Its target architecture consists of a rectangular array of data path units, called organelles, the control part has a Weinberger structure. It seems that the later versions of the Macpitts system (commercialized by MetaLogic Inc) can handle other kinds of control parts. The data path structure wastes a large area for multiplexers, especially in the case of a complex datapath. The SYCO system belongs also to the second kind of silicon compilers. The principles of the SYCO system are detailed below. Table I summarizes the classification. Now, basic principles for the design of a silicon compiler are addressed.

HIERARCHY	INPUT DESCRIPTION		
	ALGORITHMIC ORGANIZATION	FUNCTIONAL ORGANIZATION	
BLOC DESCRIPTION	ALGORITHM	SPECIALIZED DESCRIPTION FSM. DATA PATH	FORMULAS
COMPILER	MACPITTS SYCO SILC	GENESIL FIRST CONCORDE	YORKTOWN

Table I : Silicon compiler classification

II.2 - <u>Basic Principles for the design of a silicon compiler</u>

The classical classification of silicon compilers according to their input description considers only the level of the input description langage (system, algorithm, RTL, boolean expression...) [GAJ 85]. But this classical classification does not take into account the organization of the description. This factor is important from a silicon compilation point of view.

According to the classification given in § II.1, there are two ways to organize a description : the algorithmic organization and the functional organization. With an algorithmic organization, a block of the description describes a part of the algorithm performed by the circuit but not a physical block of the circuit. For example a fetch subroutine in a microprocessor description does not correspond to a physical block in the microprocessor. But with a functional organization, each block of the description corresponds to a physical block of the circuit, such as an operator description block.

In the case of an algorithmic organization, a block is a procedure or a subroutine that contains a set of statements to be executed in a given order. In this case, the description may be independent of the circuit architecture. In the case of a functional organization, a block may be described with a high level langage or with formulas, but the block interface is given. In this case the description gives the global architecture of the circuit. Some compilers allow several specialized languages for the description of different types of blocks and they use specialized compilers. Each kind of organization may lead to facilities and/or complications for the design of a silicon compiler.

Four problems must be addressed :

1) The architectural problems : The architecture of a circuit determines its internal structure, the clocking scheme, and the communication interfaces between the different parts of the circuit. This problem exists mainly if the compiler starts from a description having an algorithmic organization. In this case the compiler must supply itself the architecture of the resulting circuit. The architectural decision spectrum of the compiler must be sufficiently limited firstly to allow an efficient automatic generation of the layout according to the layout restrictions (see the layout problem), but also to have a competitive compilation time.

The compiler can make some low level architectural decisions, such as the decision to fix the number of adders to be implemented in a circuit according to a set of given constraints. But the global architecture is generally fixed and it is called the target architecture.

The target architecture determines also the global layout strategy. It makes the compiler efficient for a given class of IC applications.

A target achitecture may also be used by compilers that start from a description having functional organization (e.g. the first version of the Bristle Blocks compiler [JOH 79]).

2) The layout problems : The VLSI layout designer makes use of a very large amount of layout techniques when designing IC by hand. Most of these techniques are not formalized enough to be automatized and then embedded in a silicon compiler. Some good specialized tools for layout generation exist. These layout generating tools are used to obtain the different blocks of the circuit which are then assembled, using a silicon assembler. The use of a target architecture allows the use of a fixed global layout strategy. In this case the complexity of this problem is brooken down.

If no target architecture is used this problem may become the backbone of the compiler.

3) The description language problem : the input language of the compiler must allow the designer to specify the chip concisely.

If a target architecture is used the input description must also imply efficiently parameters and concerns inherent in the target architecture in order to allow automatic synthesis.

In this case the efficiency of the resulting circuit depends only on its description.

4) The translation problem : this problem may be formulated by the following question : how to describe parameters and concerns of the input language in order to allow an efficient automatic synthesis?

If a target architecture is used, the translation mechanism becomes straightforward if the choice of the description language and the choice of this target architecture is made with a translation scheme in mind. In this case the translation mechanism is precise and easy to understand by the compiler users and the resulting circuits may be influenced by the input description modifications. This way, the user may easily make trade-offs when designing a circuit. For example, it should be possible to make a trade-off between a fast circuit with a large area and a slow circuit with a small area, through modifications of the input description.

On the contrary, trade-offs may also be studied using compilers having a functional description input, but most of the work should be done by the designer, who should be a circuit designer (a better circuit designer than a designer using a compiler having an algorithmic description input). Better trade-offs may be found (compiled designs may be better than hand-made designs), but they result mainly from the designer.

It may be noted that the translation problem may be easily solved for compilers that generate the circuit architecture because of the nature of the input and of the output of the compiler which can be similar : an algorithm, to be executed by a computing architecture (as FORTRAN is easy to compile in order to be executed by the assembler of a Von Neumann machine). This allows quick examination of trade-offs. This is not the case for compilers of the first type.

III - THE SYCO COMPILER PRINCIPLES

Figure 1 gives an example of an interpretation scheme where a large **interpreter** of a high level language is divided into a set of **interpreter levels** which use intermediate languages. This well known technique in the software engineering world, may be applied to the VLSI circuit design. Such an interpretation scheme is used in the INTEL 8085. This recursive scheme is the compiling principle of the SYCO system [JER 86].

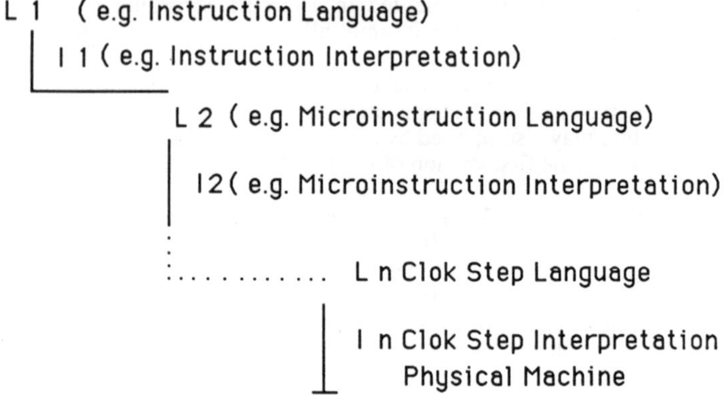

Figure 1 : Interpreter scheme.

This scheme has been proposed in the framework of the CAPRI project [ANC83] in order to fix a methodology for the design of microprocessor-like integrated circuits. The CAPRI project proposed several methodologies for the design of control sections, using one and two levels of interpretation.

A comparison of these methods [OBR82] shows that the best result is obtained with a two interpretation level control section in the case of a redesign of the M6800. More complex designs may need more than two interpretation levels in order to be efficient. The aim of the SYCO compiler is to automatize and generalize this methodology.

The remaining of this section gives the main choices made for the design of the SYCO compiler to solve the four problems stated above.

III.1 - <u>Architecture problem</u>
The circuits generated by the SYCO compiler are regarded as command language **interpreters**. An interpreter is split in a stack of interpreter levels. Each interpreter level breaks the primitives given by the level above into a set of primitives for the interpreter just below. For example an instruction is split into a set of micro-instructions. The bottom **interpreter** executes the elementary operations. It constitutes the data path of the circuit, the other interpreters constitute the control section. Each interpreter level gives to the upper one an execution report (fig. 2). The control interpreters communicate with a control bus that carries control information.

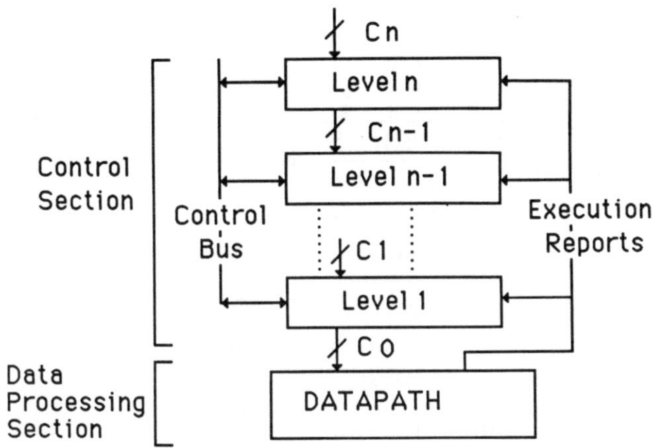

Figure 2 : Architectural Template.

Of course, this architecture restricts the use of the SYCO compiler to sequential machines.

III.2 - <u>Description language problem</u>

The circuit is described in a procedural manner using a common subset of two existing hardware description languages (IRENE [MAR 85], and LDS [LAU 85]). This subset is well adapted to the target architecture as it will be shown throughout this paper. The choice of a subset of existing languages is guided by two factors:

(i)   the users of the compiler can benefit from the environment of these HDLs:existence of simulation tools/verification tools, possibility of compiling a part of a circuit while using other synthesis tools and other methodologies for other parts.

(ii)   the users of these languages have not to learn a new formalism in order to use the SYCO compiler. In the definitive version of the CAD system SYCOMORE, the LDS language will be the only input language.

The user has the possibility of specifying the parallelism and the control, explicitly. The algorithmic description of the circuit is given in a pascal-like syntax, it is constituted by a hierarchy of procedure calls.

III.3 - <u>Translation problem</u>

The translation mechanism is straightforward. The procedure hierarchy is used to define classes of procedures. Each class of procedures describes an interpreter level. The number of slices is implicitly fixed by the user through the description. A user can then influence the resulting layout. In fact the number of levels is a trade-off between the speed and the area of the resulting circuit. This point will be more detailed in the rest of this paper.(cf. XI.3).

III.4 - <u>Layout problem</u>

The circuit layout is composed of a stack of slices; each slice represents the layout of an interpreter level. Several specialized tools are used to produce the layout : a compiler for the generation of the layout slices, a datapath compiler (APOLLON) for the datapath slice layout generation and compilers specialized in the design of control slices based on different architectures. The layout slices are **automatically** designed to be **assembled together.**A block diagram of a typical layout is given in figure 3.

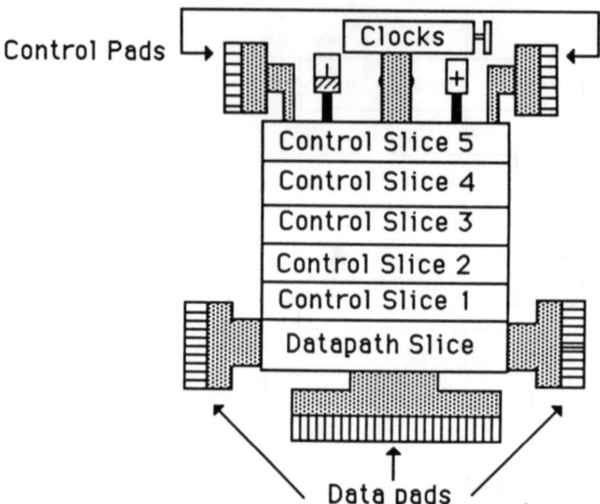

Figure 3 : Typical floor plan

IV - TARGET ARCHITECTURE

Besides the global architecture presented above, the data path compiler and the control section compilers use specific targets for datapath and control slices.

IV.1 - Data path architecture

The datapath slice (fig 4) is connected up with the control section through control command wires (produced by the control section) and execution report wires (produced by the datapath). The control signals are amplified and synchronized by a slice of drivers. The datapath also includes an input/output interface which allows the circuit to read (or write) data from (or to) the external world.

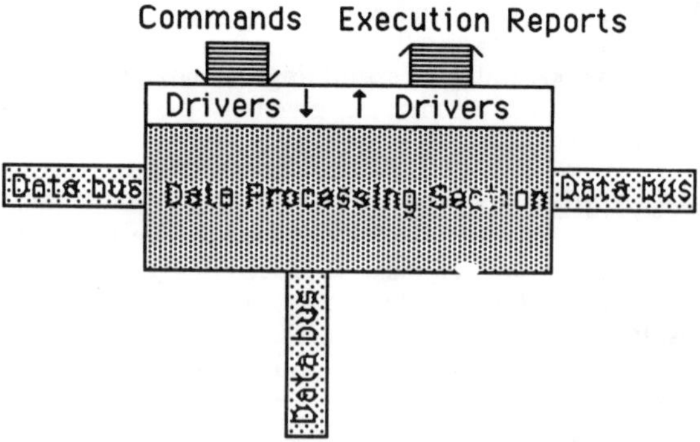

Figure 4 : Data processing section scheme

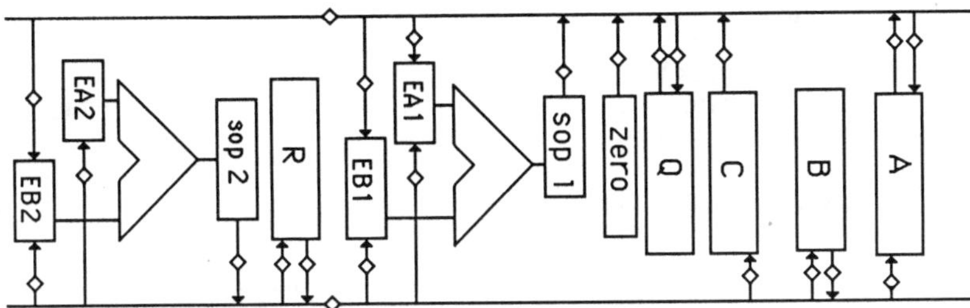

Figure 5 : Typical architecture of a data processing section

The datapath target architecture comes from the MC68000. Similar models are reported in [JOH79] and [MEA80]. It uses two buses which may be cut to define sub-datapaths (fig 5). A datapath of a complex circuit is a set of aligned datapaths running parallel (e.g. data and addresses computations can be performed simultaneously, etc...). The architecture of each sub-datapath is organized around a two bus structure connected with simple or double access registers, simple access constants, I/O buffers, ALUs, simple or double access RAMs, simple access ROMs. Communication between sub-datapaths is made possible by bus switches.

IV.2 - Control section architecture

A control section slice (fig 6) is connected with the neighbor slices: the upper slice gives the current command code, and the slice gives to the lower slice the subcommands to execute. Besides, each slice is connected with two common buses :

- The control bus drives control signals. These signals come either from external control pads or from internal control registers, which allow the different control slices to communicate with one another.

- The execution report bus drives the signals produced by the datapath slice, such as operator reports (ALU flags for example) or register fields used by the control section (the instruction register fields for example).

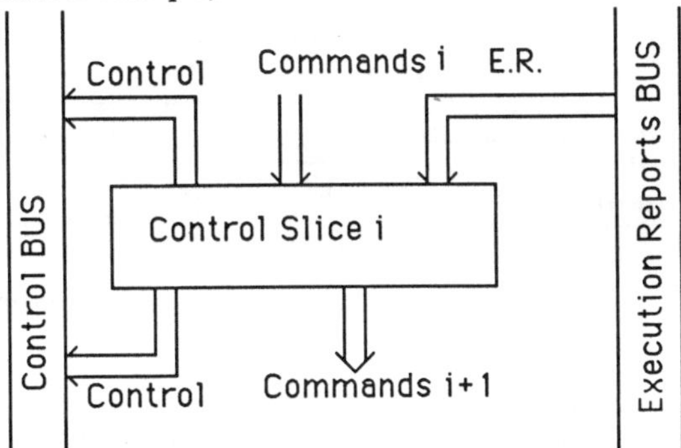

Figure 6 : Control part slice scheme

Several internal organizations may be used for the control section slices. [OBR82] gives seven different ways to organize a control section with a single interpretation level. Three among these models can be easily automatized and fitted into the SYCO compiler. They are the single PLA model, the microprogrammed model, and the time generator model. The choice of an organization for a given control slice section will be made according to the algorithm complexity and to a trade-off between the area and the speed expected for the slice.

IV.2.1 - <u>Single PLA control slice</u>

The control slice is implemented by one PLA (fig 7). The AND matrix decodes the commands given by the upper level, the execution reports, the control bus and the internal sequencing bus. The OR matrix generates the commands for the lower slice and updates the sequencing bus and the control bus.

IV.2.2 - <u>Microprogammed control slices</u>

The control slice description is coded by a microprogram and stored in a ROM. Each micro-instruction records the data necessary to command the lower level, to update the control bus, and to ensure the internal sequencing. The decoding is made by an extra PLA (figure 8). This model is suitable for small complexity control slices.

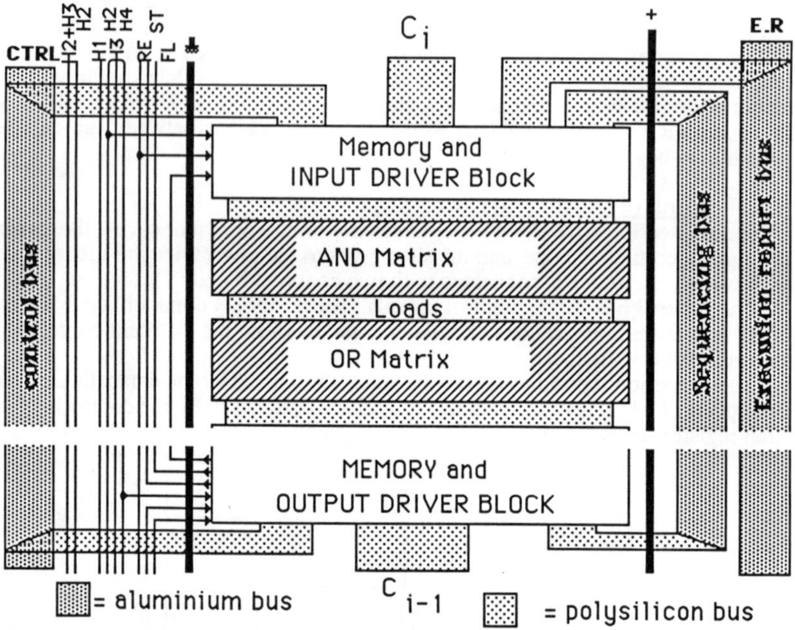

Figure 7 :   Single PLA Control section slice NMOS implementation.

IV.2.3 - <u>Control slice using a timing generator</u>

This approach is based on the fact that each control command is characterized by the time of its activation during the command interpretation. A command decoding PLA (figure 9) generates sub-commands and a set of properties to control the timing generator. The timing generator is a small automaton, which produces the timing for the validation PLA. Sub-commands are generated through the validation of lines coming from the decoding PLA. This technique gives good results for complex control slices if the interpreter level uses a few condition signals (control signals and execution report signals).

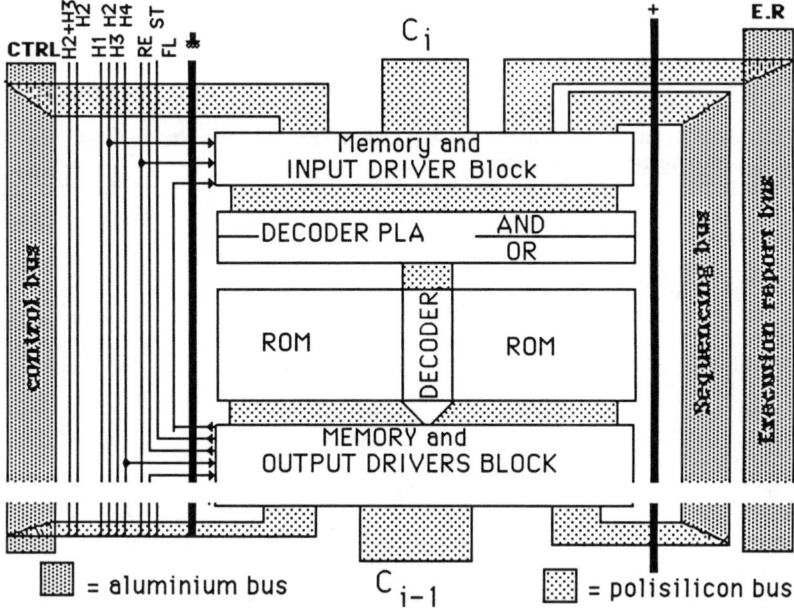

Figure 8 : Microprogrammed control section slice NMOS Implementation

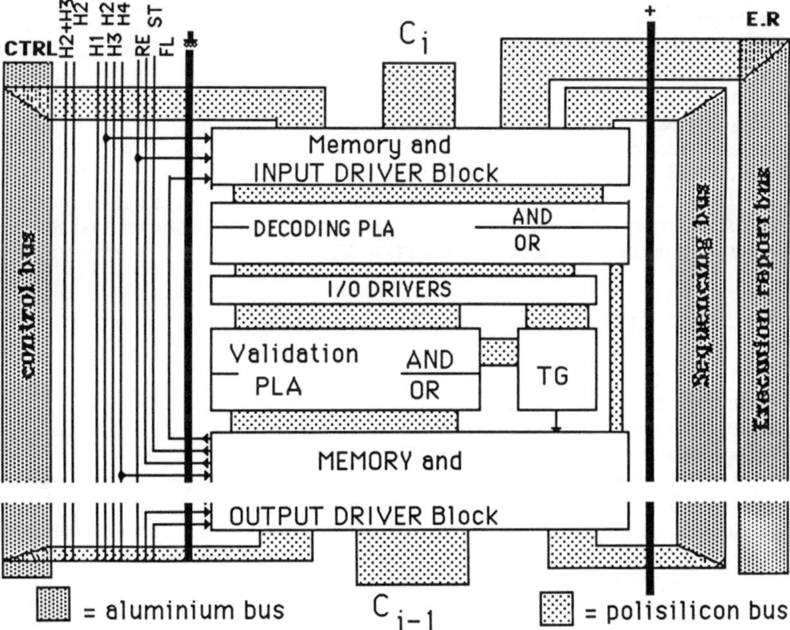

Figure 9 : Time generator control section slice NMOS implementation.

V - LAYOUT STRATEGY

The reason to have a layout strategy is to avoid the need for a placement and routing strategy and then to avoid the induced wasted area for interconnection.
1) The slices must be well adjusted to their environment [ANC83a].
2) The control bus and the execution report bus in the control section must be carefully placed.
These problems can be solved by using sophisticated specialized layout generator tools (for PLAs, ROMs...). These tools must be able to handle dimension and transparency constraints.

V.1 - Datapath layout
The layout is produced by the APOLLON [JAM85] compiler. The layout is organized around a bit slice structure, it is automatically generated. Figure 10 gives a typical layout generated by the APOLLON compiler. APOLLON makes use of a silicon assembler and uses a cell library. Actually, the SYCO system uses the LUBRICK system [SCH 83]. In the near future, it will use the STYX system [JER 85].

V.2 - Control section layout
The present version of the compiler handles only the single PLA architecture for the control slices. It generates a one PLA slice to implement each of the control section interpreters. Figure 7 gives the floor plan of a control section slice layout. The control bus and the execution report bus are placed on the two sides of the PLA. One can note that the PLA has lateral inputs and outputs. It is expected that the next versions of the compiler will be able to handle other internal structures (microprogrammed control sections, and time generator based control sections) to implement control slices.

V.3 - Circuit Layout
Besides the local strategies, for the data path and the control slices, the SYCO SYSTEM uses a global layout strategy. The layout slices are stacked as shown in figure 3. The width of the slices need to be adjusted in order to minimize both routing areas and unused areas.

The data path width fixes the width of the circuit. To adjust the width of the control slices many techniques must be used. in the case of the one PLA control slice, for example, a sophisticated PLA generator like PAOLA [chu 82] is needed. This tool allows the lengthening of the shape of the PLA and the lateral access to the input/output segments, in order to adapt the PLA to its environment. Other techniques based on description, transformations may be used (see XI.3).

VI - THE SYCO COMPILER ALGORITHMS

VI.1 - Circuit description
A description gives:
- The circuit interface: data and control signals.
- The internal variables used in the description :
* Data variables : they may be registers or block of registers. These variables are placed in the datapath.
* Control variables : They are placed in the top slice of the control section. They allow the control section slices to communicate with one another.
- The algorithm description : it is given in a pascal like syntax, with the possibility to express parallelism. The algorithm description is a hierarchy of procedure calls. Figure 11 gives an example of a description.

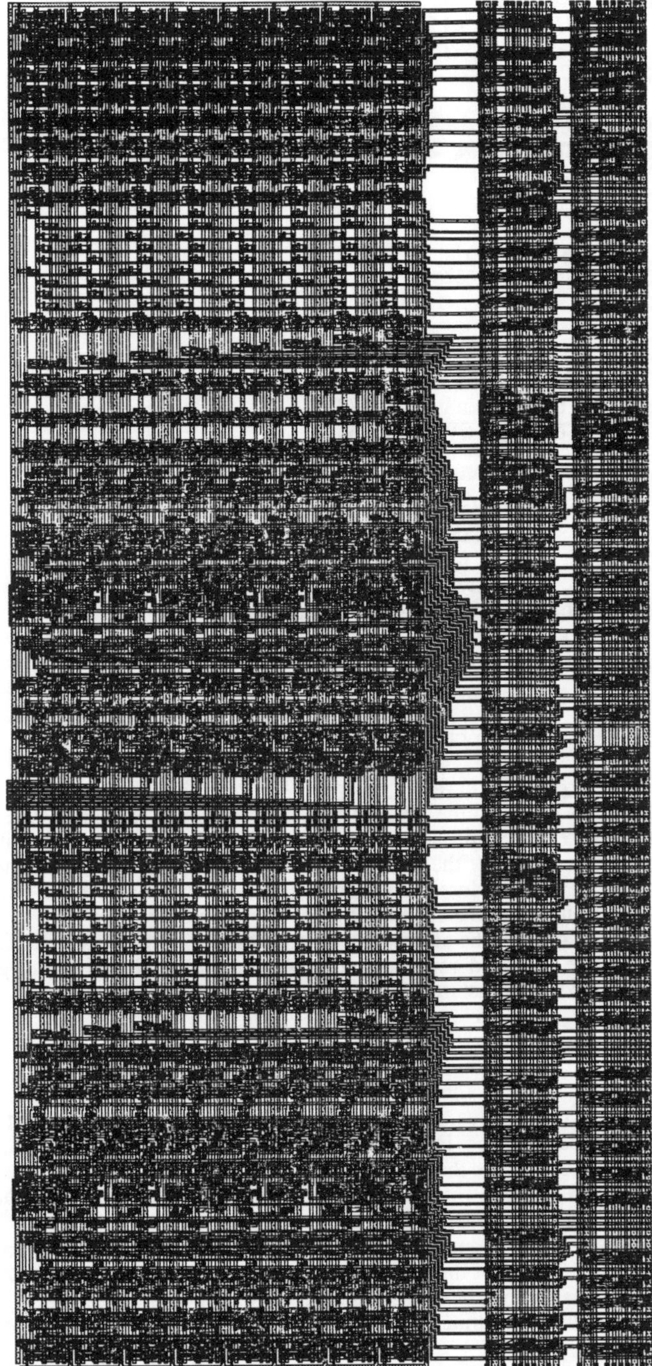

Figure 10 : The Layout of the SYCO-6502 [REIS 86] produced by APOLLON [JAM 85]

510

```
program micro (control in RESTART, out BUSADR <15:0>,
                bi BUSDATA <15:0>)
     reg A<15:0>, B<15:0>, C<15:0>, RI <15:0>, PC <15:0> ;
procedure FETCH ;

          BUSADR <- PC ;
          (RI <- BUSDATA , PC <- PC + 1 ;) ?? instruction to be executed in parallel.
     end
procedure STATEMENT1 ;
          ..........
     end
procedure STATEMENT2 ;
          ..........
     end
..........
begin     %*MAIN PROGRAM *%
$WAIT : if RESTART <> 0 then goto $START
                    else goto $WAIT endif ;
$START : PC <- 0 ;
$EXEC : call FETCH ;
               (case RI<9:8> of
                         1 : call STATEMENT1
                         2 : call STATEMENT2

                         ..........
               endcase ; if RESTART <> 0 then goto $EXEC
                                   else goto $START)
                    endif ;
     end.
```

Figure 11 : a flavour of the input description.

VI.2 -Interpreter level extraction

This step takes for input the algorithmic description of the design. The hierarchy of procedures is used to define classes of procedures. The main procedure forms the first class. The procedures called by the main procedure form the second class, and so on. The system introduces new procedures in order to avoid that one procedure belongs to two different classes, and in order to isolate the elementary primitives.

The last class contains the elementary primitives. An elementary primitive is a set of actions which may be executed simultaneously. An action may be (if A,B,C are registers, and MEM an external data register,and <-, not, + are operators) :
- a transfer as " A <- B ",
- an operation as " A <- B + C " or " A <- not B ",
- an I/O operation (a special transfer) as " A <- MEM " or "MEM <- A",

Each class of procedures describes an interpreter slice. Obviously the last class corresponds to the datapath slice. In fact this class will induce two layout slices, the datapath itself, and a code generator slice that breaks the operative actions in elementary commands.

This step produces the interfaces between the different slices. A slice interface contains :
- The set of procedure names belonging to the class corresponding to the slice.
- The set of the names of the lower slice procedures.
- The signals used: control signals, and execution report signals.

For example, the algorithm given in figure 11 may be executed by three interpreter slices.

Figure 12 gives the initial hierarchy of the procedure calls given by the description (figure 11), and figure 13 gives the resulting hierarchy after the level extraction step. The new procedure "INT1" (in figure 13)b is added in order to allow the main program to access the primitive (PC <- 0) located in the lowest class.

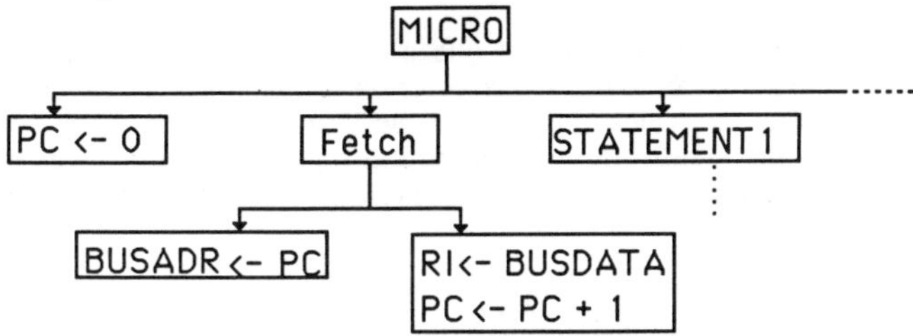

Figure 12 : The initial hierarchy of the procedure call given by the initial description.

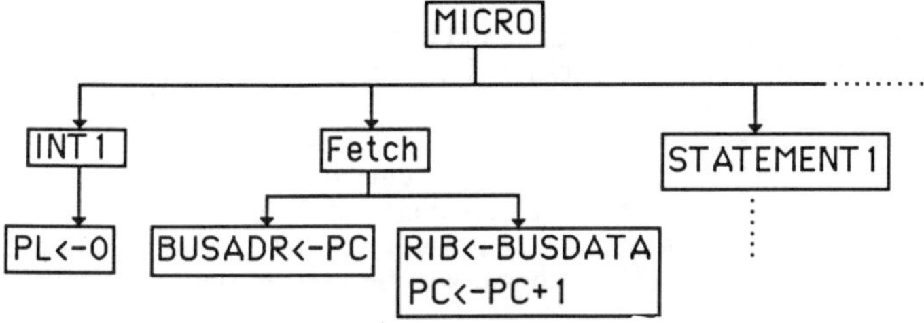

Figure 13 : The classes of procedure extracted from the initial description.

VI.3 - Datapath compiling

This step is performed by the APOLLON datapath compiler. It generates the datapath layout and a complete description of the datapath layout interface from lists of actions such as simple transfers or operations to be executed in parallel. This interface details the connector description of the layout slice, and the correspondence between the symbolic names used and the corresponding connectors. These connectors are used to connect the commands coming from the control unit with the corresponding datapath commands. [JAM85] gives more details about the APOLLON compiler.

VI.4 - Control section compiling

The control section compiling is performed after datapath compiling. It makes use of a control section compiler for each control slice. The order of the slice compiling is bottom-up.

The control section compiler processes in three steps :

- A first step parses the interpreter description and the slice interface. It generates a microcode table that describes the sequencing of the corresponding level interpreter. This step does not depend on the internal organization of the control slice.

- The second step uses this microcode to generate the internal structure of the control slice. The present version produces a single PLA slice, according to the scheme of figure 7. Besides, it produces an assembly description of the slice layout.

- The PLA layout is now generated by using the PAOLA system [CHU82], (monoplane PLAs will be generated in the next version [VAR85]) and the whole slice layout is produced   by running an assembling program, which uses the placement and routing primitives of a silicon assembler.

The control section layout is obtained by simple assembly of the different control slices as shown in fig 7. The control part and the datapath layout are assembled by simple abutment as shown in figure 3, again using a silicon assembler : presently the LUBRICK system [SCH 83], and in the near future STYX [JER 85].

VII - CLOCKING SCHEME

Successively, the basic clock cycle and the slice synchronization are addressed.

VII.1 - Basic clock cycle
The basic machine cycle is a four phase clock (t1, t2, t3, t4), it corresponds to the basic time to execute a register to register transfer on the datapath. To execute one elementary primitive (Figure 12), two cycles are needed : the first one to transfer the operands and the second one to write the result into the destination register. All the slices use this cycle as basic time. A control slice needs one cycle to execute one step (state) of the corresponding algorithm.

VII.2 - Slice synchronization
The classical scheme to synchronize the slices inside such a machine is a sequential one :
- Each cycle, only one slice is active,
- When a slice gives a sub-command to the lower one, it waits until the lower slice ends the sub-command execution before giving another sub-command, or before returning the control to the upper slice.

The present version of the compiler uses this scheme. Another scheme consists in pipe-lining the interpreter execution. In this scheme an active slice returns the control to the upper one, when it generates the last sub-command of the procedure executed. When a slice needs an execution report for the next command generation, it has to wait until all the lower slices finish their execution. This scheme will be used in the next version of the SYCO compiler.

VIII - INTEGRATING THE TEST PROBLEMS

It would be inefficient (even impossible) to generate layout, and next to try to generate test patterns for this layout.

A necessary condition to the development of synthesis/compilation tools is to solve the testing problems inside the synthesis tools : within a CAD system, each tool must take care of the testing problems. It is expected that the SYCO system will produce in the future testable circuits.

VIII-1 Advances in "test pattern generation"
The most advanced steps in test pattern generation are built-in self test and self-checking circuits. Those steps are steps to be considered for the automatic generation of layout.

Built-in self test may be defined as a test method in which either :
- the test patterns are built-in generated,
- the test outputs are built-in analysed,
- or both.

Classical approaches to built-in self-test are the BILBO concept [KON80] and the autonomous test [McC 81]. Those approaches are aimed at the same self-test concept. Those approaches may be used to self-test random logic, or structured logic, like PLAs. Other logic blocks like RAMs may be self-tested using a software algorithm integrated in the circuit : marching algorithms may be used for a built-in test of RAMs [NIC 85-1], according to the general built in test scheme given in figure14.

Self-checking circuits are aimed at the on-line detection of failures occuring during the functionning of the circuit. The general scheme of self-checking circuits is given in figure15.

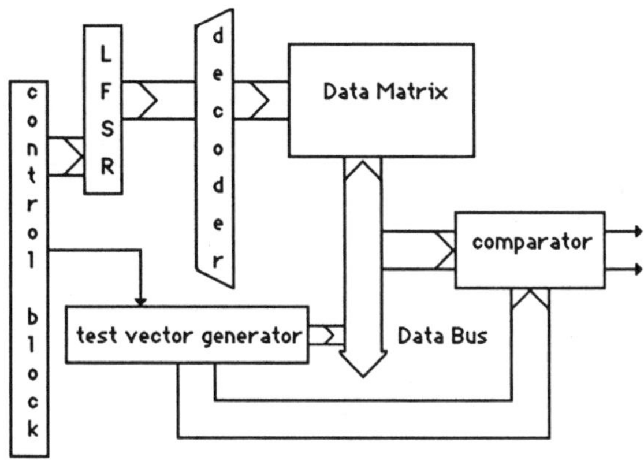

Figure 14 : Built-in test of RAMs

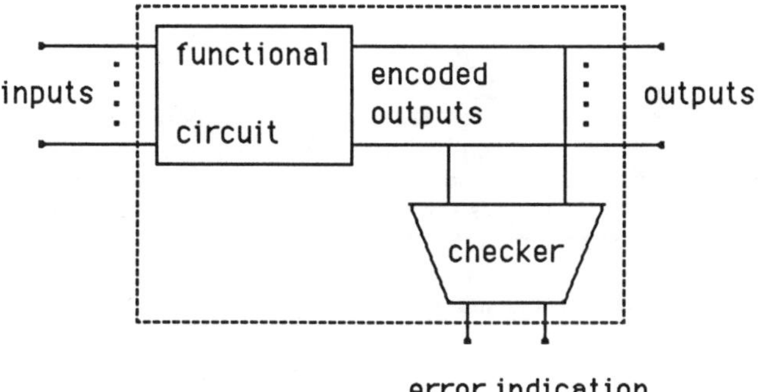

Figure 15 : Self-checking systems

The functional part and the checker must be designed such that the first erroneous output of the circuit must be a non-code word (this is defined as the Totally Self-Checking Goal). The properties of self-checking circuits have been initially defined in 1971 [AND 71]. Then the properties of the functional part have been refined in 1978 [SMI 78] by defining the Strongly Fault Secure (SFS) circuits and the properties of the checker have been refined in 1984 [NIC 84] by defining the Strongly Code Disjoint (SCD) Checkers. It has been demonstrated that a SFS circuit associated to a SCD checker allows to achieve the TSC goal. For several reasons, it has been necessary to define specific checkers [NIC 86], and those allow to define a single design for test scheme unifying built-in test and self-checking properties, in order to achieve both off-line and on-line test. This unified test is represented by figure 16.

In this scheme, the specific checkers have been decomposed in a LFSR part and a checker part. For on-line test, the LFSRs are used to test SFS functional blocks for single and a large number of multiple faults. For on-line test, periodic activation of the test phase during the circuits life ensures the SCD property for any fault (single and multiple) affecting the checkers. It ensures also the complete excercing of functional blocks and for checker, in order to avoid faults latency. Details are given in [NIC 86]. This unified built-in test is probably the "summit" for design for testability.

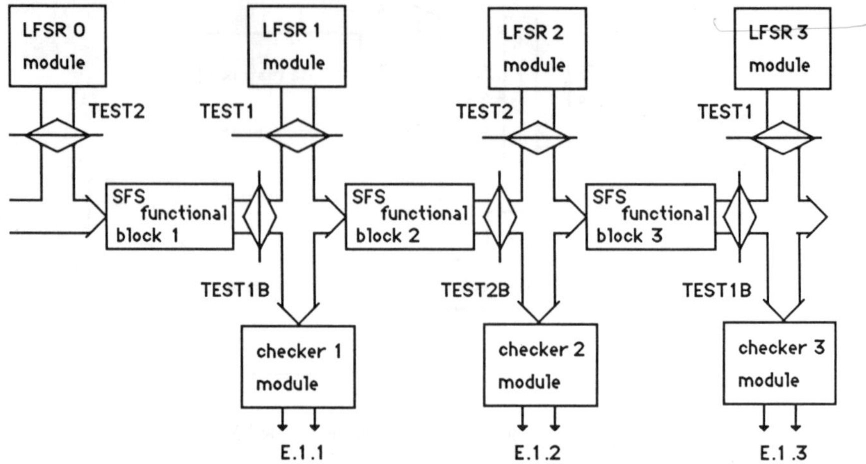

Figure 16 :   Unified BIST approach with decomposition of specific checkers.

VIII-2   Integrating advanced built-in test techniques into silicon compilers.

As previously stated, test facilities must be included in all synthesis tools of a CAD system. If this principle is applied to the SYCO silicon compiler, some basic components must be studied in order to be testable compiled. Slices of the control section make an intensive use of PLAs (this is not specific to the SYCO compiler : PLAs are in fact widely used). A basic requirement is thus to study how to synthetize "testable" PLAs. Further points are to study how to synthetize slices for control sections, control sections and data pathes. Successively, partial results are given concerning PLAs and control/data processing sections.

VIII.2.1 "Testable" PLAs.

Built-in test of PLAs and self-checking PLAs have been already studied by many authors. But, generally, schemes are available for standard PLAs, i.e. for PLAs that are not topologically optimized. Optimized PLAs are nevertheless required. Hence PLAs that are optimized and "testable" are required. But both properties are not always compatible. In the following are given some examples of problems and solutions to get built-in testable optimized PLAs and self-checking optimized PLAs.

Numerous schemes have been given for a built-in test of PLAs. They differ by different parameters, as the number of test vectors (i.e. the testing time), the fault coverage, the area overhead.

Representative schemes are [YAJ 82], [SAL 85] or [FUJ 85]. Built-in self test of PLAs generally require to control inputs and product terms (for test pattern generation) and to analyze outputs.

On the other hand, methods to optimize the area of PLAs are to fold outputs and/or inputs and/or product terms. Folding may be single or multiple. Initially, the technique has been suggested in [GRE 76] and [WOO 79]. The technique has next been extensively developped in [CHU 82] and [MIC 83]. Folding and built-in test are not always compatible. For example, in case of folded inputs, if single folding is used with inputs distributed on each side of the AND matrix, it is necessary to split the control of the inputs into two separate test vector generators. Figure 17 represents such a PLA. But the use of two separate test vector generators implies to use such a scheme only if there is a large benefit due to the folding, because of the overhead due to the test vector generator.

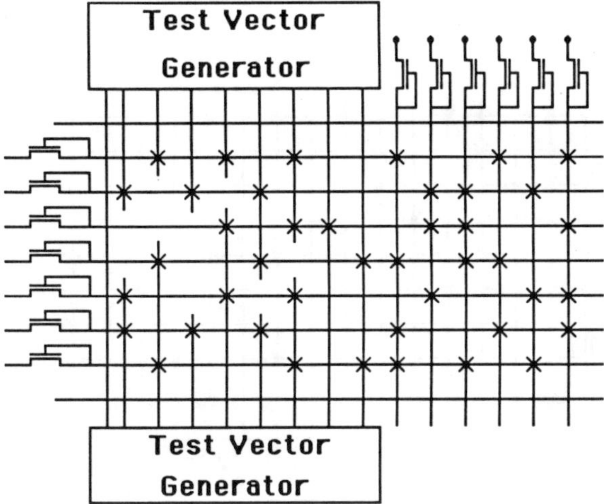

Figure 17 : Single folding of inputs, to be carefully used for built-in test.

Similarly, if product terms are folded, the control of product terms may be splitted, as depicted in figure 18, but problems may occur due to communication between both product terms controlers.

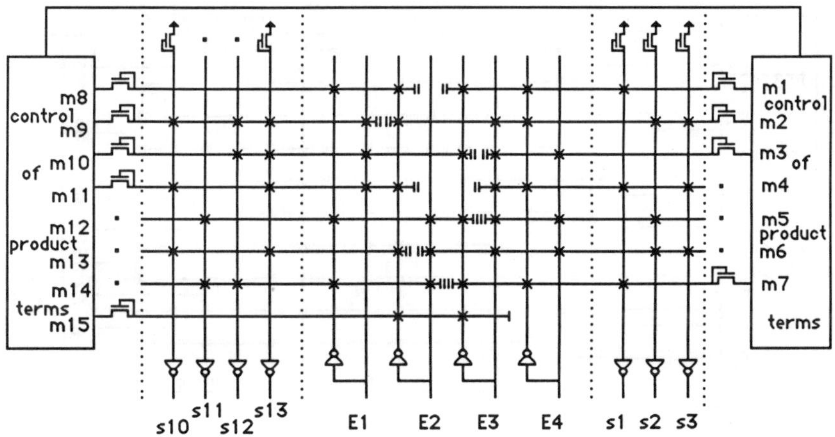

Figure 18 : Single folding of product terms for built-in test.

Built-in test of PLAs requires also to analyse outputs, and this is often made using parity trees on those outputs. Folding may also imply problems which must be carefully studied. In case of single folding of outputs for example, if outputs are taken from both sides of the OR matrix, the parity tree must be splitted into two trees, as depicted in figure 19. These separate trees may imply problems.

[FER 86] details different schemes to be used for folded built-in testable PLAs. One scheme appears to be convenient for both advantages. This scheme is a PLA having folded inputs and folded outputs such that inputs and outputs are parallel to the product terms.

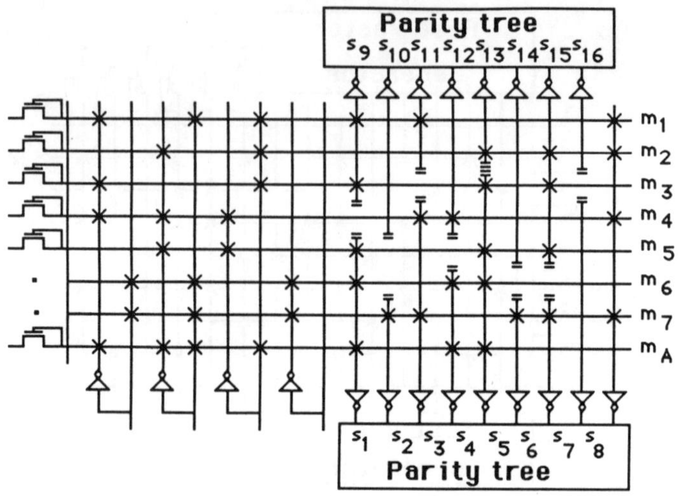

Figure 19 : Single folding of outputs for built-in test.

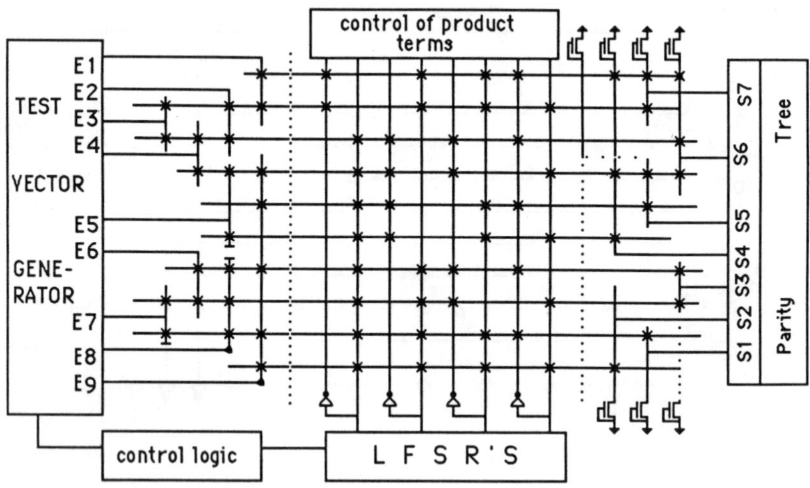

Figure 20 :   PAOLA folded PLA [CHU 82], according to [SAL 85] built-in test

Such PLAs are synthetized by the PAOLA system [CHU 82]. If the built-in self test scheme of   [SAL 85] is used, the control of product terms might be inserted between the AND and the OR matrix, according to figure 20, but the fault coverage should be carefully examined.

Self-checking PLAs require also general features like the on-line control of inputs and/or of outputs and/or of product terms.

In this short summary, only one scheme will be addressed. This scheme has been suggested in 1983 [NIC 83] (the scheme has been later presented in [CHE 85]). The figure 21 represents a self-checking PLA, when the detected errors are single errors. Similarly to what hapens for some built-in self test schemes, if outputs are distributed on both sides of the OR matrix in case of single folding of outputs, it would be necessary to split the control of those outputs into two separate blocks. But in case of multiple folding with the outputs being parallel to the product terms, then the control of the outputs is easy. Folding of the inputs seems difficult since the inputs must be controlled after they have controlled all product terms. This problem is not so crucial if a scheme using a BERGER code encoding of the outputs is used, since in this case, it is sufficient to control the inputs when they feed the PLA [MAK 82]. But this control should be splited in two parts if single folding is used.

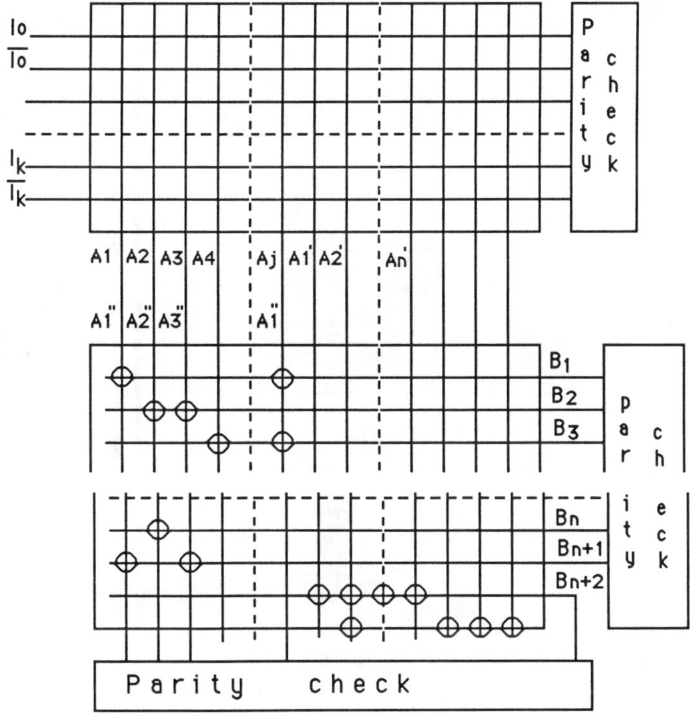

Figure 21 : Self-checking PLA, with respect to single errors.

VIII.2.2 Testable architectures.

Testable folded PLAs should be produced to realize slices of control section for the SYCO silicon compiler. The memory blocks, the sequencing bus, the control bus and execution reports bus will be encoded in order to give "testable" slices.

For example, a self-checking single 7 PLA control slice is depicted in figure 22. In this scheme, the control bus, the execution report bus and the sequencing bus are encoded. The upper slice delivers the Berger code of the commands C_i, and the PLA calculates again the Berger code of the commands it is receiving. Those Berger codes are checked one against the other.

The data processing section architecture will also be designed in order to give "testable" data processing sections. Results obtained on the data processing section of the

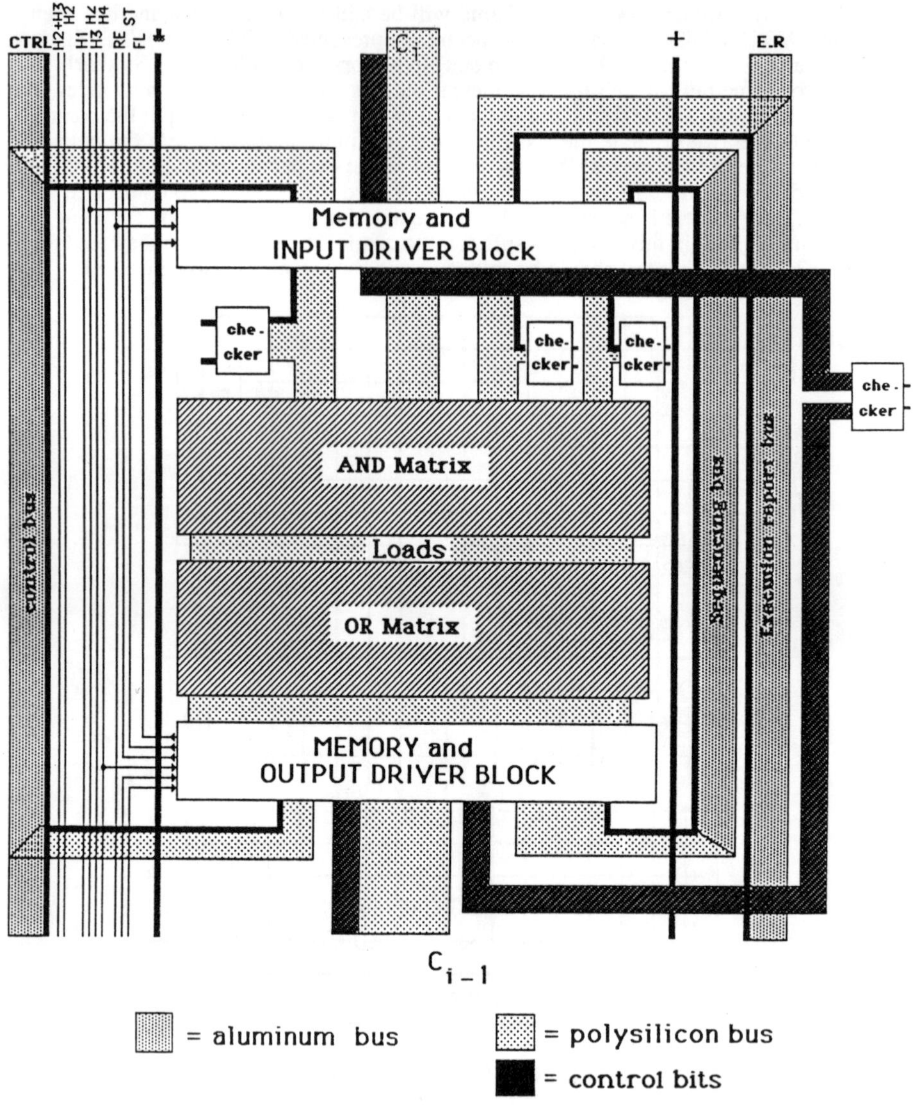

Figure 22 : Self-checking single PLA Control section slice

MC 68000 [NIC 85-2] will be used. This way, the circuits compiled by the SYCO compiler will be testable using built-in test techniques, and will be on-line testable using self-checking architectures.

IX - VALIDATION

Design errors are addressed in this section. It is supposed in this section that circuits are not designed error-free. If circuits can be produced error free using verification techniques or if all circuits could be designed using compilation techniques (supposing the compilation error free and the input description exhaustively simulated), then this section would be of no use. But this assumption will not be made, assuming for example that, even if compilation techniques will be used, it will not be possible to exhaustively validate the input description by simulation, or that implementation problems may occur, leading to discrepancies between simulation and the real behaviour of the fabricated circuit. Moreover the electron-beam validation will be useful in general in the framework of a CAD system, since, even if compilation techniques will be used more and more in the future, some parts will remain "hand" made.

The electron-beam tester is probably the most important tool for the analysis of design errors. But this tool is really efficient if there is a link between the observations of the electron-beam tester and the CAD tools, i.e. if <u>validation</u> facilities are <u>integrated</u> to the CAD. The goal of this section is to briefly report on the work carried out within the Computer Architecture Group. This work is aimed at making easier the work of the designer for the debug of his IC.

In the following, a link between a low level description of cirucits (here after a CALMA GDSII format) and an electron beam tester is detailed. This link would be a basic link between the SYCO compiler (the GDSII format and the STYX format are similar levels of description), but, in the future, there will exist a link between the electron-beam tester and high levels of description, i.e. between the electron beam tester and the variables of the input to the SYCO compiler.

IX.1 <u>Link between a CALMA-type description and the electron beam tester</u>.

There are four successive steps required to link a CALMA type description to the SEM [GUI 85]:
- Calculation of the theoretical coordinates of the observation points from the CALMA name
of the boundaries.
- Transformation of the theoretical coordinates of a point into its practical address within the image issued from the SEM.
- Search for the real location.
- Determination of the logical state of each point from its grey level.

IX.1.1 <u>Calculation of the theoretical coordinates of the observation points</u>

If a CALMA description is used, it is necessary to calculate the theoretical coordinates of the points to be obseved in the CALMA description. This can be made either pointing on a graphic display, or addressing the points by their names. In the future, the SYCO system will take care to give the theoretical location of variables used in the input description.

IX.1.2 <u>Calculation of the practical address of the observation points</u>

At the end of the selection phase, the theoretical coordinates of the observation points are known. "Theoretical" means that these values are issued from the masks description. The problem is then to find the obsevation points within the images obtained with the electron beam tester, i.e. to determine the practical address of these points.

520

In order to achieve a theoretical/practical correspondence, the theoretical coordinates and the practical address of three points, called reference points, are indicated during an initialization step. It is then possible to compute the practical address of each point defined by its theoretical coordinates : a correspondence function that reamains thrue as long as the parameters of the electron beam tester (magnification, stage position, ...) are constant has been determined. At the end of this step, the physical location of the point of interest is known. The problem is that the point of interest is not generally located where it is expected to be, because of errors in determining the physical address, because of optical abrevations, because of linearity problems in the scanning process, etc...

Then a third step is needed aimed at the search for the real location of the point to be observed.

IX.1.3 Search for the real location

The chosen solution is to look for corrections allowing to compensate the distorsions of the electron images, and to consider the shape of the analysed boundary. In order to study the images distortions, the superimposition of a theoretical image and of a practical one has been achieved. This superimposition is an extension of superimposition used for the failure analysis of circuits by comparisons, using the electron beam [BER 86] :

- generation of the CAD masks image in the memory of the image processor

- extraction of the edges of the practical (e-beam given) image of the same area. Up to now, the dimensions and the magnification must be identical for both images.

- display of both outlines, one in black, one in white, on a grey background. It appears that distorsions exist : the practical outline appears dilated in some direction. This is corrected.

- superimposition of the CAD and practical outlines [GUI 86].

A program based on a correlation technique, determines the horizontal and vertical discards between the two outlines. Let two images, S the practical search area, and W the theoretical window be defined as shown in figure 23. The window, that covers KxL pixels, is selected by the operator within the theoretical outline. The limit values of this area, as also the coordinates of the vertices of each boundary are sent to the images processor. In this processor, the window is stored in the form of a set of segments. The total number of segments is kept as also the following informations for each segment : the coordinates of its first point, its number of pixels, and its direction.

The search area is represented by a NxM array of pixels, each pixel assuming one of two grey levels : it is either a black pixel of the background, or a white pixel of the outline. The dimensions of this area depends on the dimensions of the window and on the dislocation between the two images. The images processor calculates the dimensions (K,L)

of the window, and asks for an indice n that allows to determine the Δx and Δy values that are added on both sides of the window (figure 23). The following formulas are applied :

$$\Delta x = K/2^n$$

$$\Delta y = L/2^n, \quad \text{with } 0 \leq n \leq 7$$

Each KxL subimage of S is uniquely defined by the (i, j) coordinates of its lower left corner. The problem, that is to locate the window within the search area, comes to search the subimage of S that is the most similar to W. So the window is compared to each of the $(2\Delta x+1)(2\Delta y+1)$ subimages of S (figure 24). Here are the variation interval of the integers i and j :

$$i \in [-\Delta x, ..., +\Delta x], \quad j \in [-\Delta y, ..., +\Delta y]$$

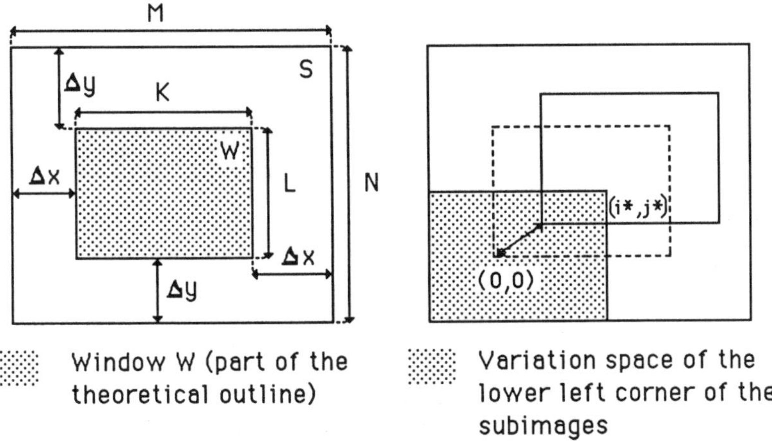

Figure 23 : Search area S (part of
the practical outline)

Figure 24 Subimages of S

A coefficient of correspondence is computed for each subimage. This coefficient is incremented each time that a segment point of W is white in the practical subimage. So the comparison is not made pixel-by-pixel : only the points that belong to the searched outline (i.e. the segments points of W) are analysed. The subimage that gives the greatest coefficient of correspondance is the one that better matches the window. Let (i*,j*) the coordinates of its lower left corner.

The theoretical outline is then shifted according to these optimum values (i* pixels for x, j* pixels for y), and the two outlines are displayed again.

IX.1.4 Determining the logical value of the point.

A correspondance between the grey level and a logical value has finally to be determined. Up to now, fixed thresholds are used, but in the future more sophisticated techniques will be used [BER 85], because of distorsions in grey levels.

IX.2 System configuration

A complete tool is used for experiments. The tool is composed of four devices (figure 25) :

- An home made exerciser which controls the circuit under test [BAU 85]. It allows to apply test vectors, and to store the answers of a circuit of up to 64 pins. To do this, it is linked to the circuit by a cable containing one wire per pin. A board enables to banalize the pins : each pin may be power "1" or "0", an input, an output, an input/output or a clock. The test vectors are loaded either by the operator from a terminal, or by an external system using a RS232 link. They are applied either continuously, or with a break point, or for a given number of cycles.

- A SEM (JEOL 35 C) equipped with a XYZ tilt stage allows to observe a packaged circuit. At present, this stage is manually operated, but its motorization is in progress. The detector is of Everhart/Thornley type, and the source is tungsten or Lab 6.

- An home made image processing system able to produce an image of 512x512 pixels coded on 8 bits [BRE 84]. This image is segmented in order to be processed in parallel by eight

Z8002 processors. The image processor is programmable using an assembler language. The object code is fixed or loaded from a remote computer. Commands are issued

from a terminal and external communications are made using a RS232 connection.
- A VAX 780 remote computer containing the assembler for the images processor and the CAD tools (masks description, simulators, ...)

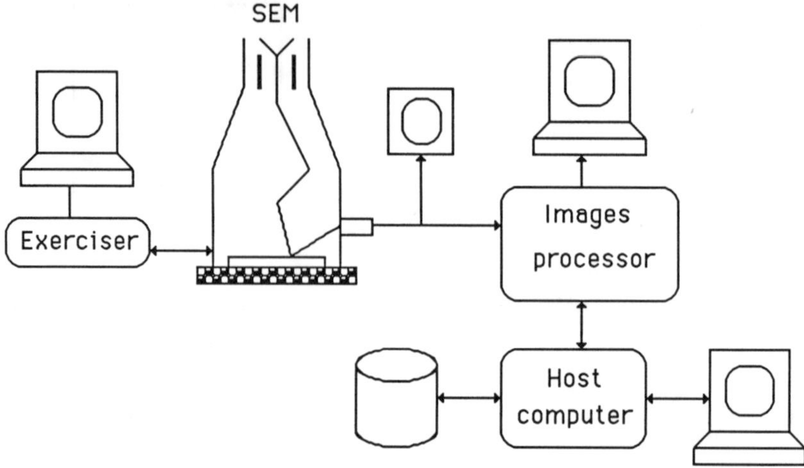

Figure 25 : Structure of the tool

X - CONCLUSION

X-1 Key-points
Several points are expected to be key-points for the success of the SYCO silicon compiler.

1-Complex placement-routing problems are avoided, since the shape ofthe blocks is automatically adjusted to the environnement.

2-The compiler uses target architectures, which have been proved to be efficient.

3-The translation scheme is powerful and speedy because the description language is well suited for the target architecture.

4-Powerful specialized layout generator tools (for PLAs, ROMs...) are used. The present   version of the SYCO compiler uses the PAOLA system for the optimization and the generation of PLAs [CHU 82]. It allows the automatic adjustment of the shape of the slices in order to have the same width, and then to reduce the height of the control section slices. Besides it allows connection mobility ; this way, routing areas are reduced.

X-2 Applications
Some applications of the SYCO compiler have shown the power of this compiling scheme. [REI 86] reports the compilation by SYCO of the 6502 microprocessor. Starting from ten pages of description, the SYCO compiler generates the layout of the datapath and a fine description of the control slices. Figure 26compares the results of the SYCO compiler and the Metalogic compiler (related in [EVA 85]).

The circuit generated by SYCO is smaller, mainly because the SYCO compiler does not flat down the hierarchy of the sequencing. Another example, a master's project, led to the following results : about 50 % of the MC 68000 has been compiled, resulting in a circuit estimated to be 1.37 times larger than the 68000 designed by MOTOROLA using the same technology.

X-3 Further developments
Further developments include the availability of optimization tools and the compilation of "testable" circuits.

COMPILER	TECHNOLOGY	AREA
MACPITTS	NMOS 3μ	7,5 X 6,7 ≈ 50 mm2
SYCO	NMOS 3,5 μ	4,05 X 7,10 ≈ 28.75 mm2

Figure 26 : Comparison of resulting circuits from the 6502 compiling.

The goal of the optimization tools will be to give the possibility to try automatically several architectures starting from the same initial input description, in order to get circuits exhibiting a good trade-off between the speed and the size. The basic parameters on which those tools will not act are the parallelism in the datapath, and the hierarchy of the control section, since the parallelism expressed in the description fixes the datapath complexity and the hierarchy of the procedures calls fixes the number of slices insides the control section.

The optimisation tools will allow to try several solutions starting from a given description by automatically modifying the hierarchy of the control section and by automatically modifying the parallelism inside the datapath.

In order to get a faster circuit the optimisation tools will increase the parallelism of the datapath, and decrease the hierachy of the control section. In the other hand, in order to get a smaller circuit the optimisation tools will reduce the parallelism in order to reduce the datapath size, and will increase the hierarchy in order to decrise the control section size.

The other further development will be made towards the compilation of "testable" circuits.This will be also an important feature, since those developments will integrate the most advanced concepts in the field of built-in self-checking circuits. Those developments are expected to give operational results in 1988. Finally, to link an electron-beam tester with the SYCO silicon compiler will give a precious aid in case of design errors made during either the description of the circuit used as input to the compiler, or if errors exist in the compiler itself (?) or if implementation problems occur (due for example to unexpected variations in the process parameters). Up to now, a link exist between the electron beam and low levels of circuit description, but it is expected to have within some time a link available between the electron beam and high levels circuit descriptions.

REFERENCES

[AND 71]   D.A. ANDERSON
"Design of self-checking digital networks using coding techniques".
Coordinated Science Laboratory, report R/527
University of Illinois, Urbana, September 1971.

[AYR 83]   R.F. AYRES
"Silicon compilation and the art of automatic microchip design
Prentice-Hall", 1983.

[ANC 83]   F. ANCEAU, CAPRI
" A design methodology and a silicon compiler for VLSI circuits specified
by algorithms", Third caltech conference on VLSI.

[ANC 83a]   F. ANCEAU.
"Layout strategies for NMOS-CMOS VLSI", ICCD, Nov. 83 New York.

[BAU 85]   F. BAUDRAND
"Présentation de TESSIE : testeur de circuits intégrés" DAII Report.
IMAG/TIM3, Grenoble, February 1985.

[BER 85]   L. BERGHER
"Analyse de défaillances de circuits VLSI par microscopie électronique à
balayage". Thèse de Docteur-Ingénieur, INPG, June 85.

[BER 86]   L. BERGHER, J. LAURENT, B. COURTOIS, J.P. COLLIN
"Towards automatic failure analysis of complex ICs through E-Beam
testing", will appear in : International Test Conference, Washington USA,
September 86.

[BRE 84]   B.Y. BRETAGNOLLE
"Du traitement d'images dans ses rapports avec l'architecture des
ordinateurs"
INPG Thesis, Grenoble, January 1984.

[BLA 85]   T. BLACKMAN et al.
" The SILC silicon compiler : language and features", 22nd DAC, 1985.

[CHE 85]   C.Y. CHEN, W.K., FUCHS, J.A. ABRAHAM
 "Efficient concurrent Error detection in PLAs and ROMs"
In proc. International Conference on Computer Design.

[CHU 82]   S. CHUQUILLANQUI & T. PEREZ SEGOVIA
" PAOLA : A tool for topological optimization of large PLAs", DAC,
1982.

[COX 84]   W. COX,
"From macroarchitecture to micro-architecture : a design philoşophy",
NATO school, Urbino ITALY, 1984.

[DAR 81]   J.A. DARRINGER et al
"Logic synthesis through local transformations", IBM J. of Res. vol 25 N°
4, July 1981.

[DAV 80]   M. DAVIO & A. THAYSE
" Implementation of algorithms based on automata", Philips J. Res. 35,
1980.

[EVA 85]   S. EVANKZUK
"Results of a silicon compiler design challenge", VLSI design, July 1985.

[FER 86]   A. FERNANDES, B. COURTOIS
"The testable folded PLAs" IMAG/TIM3 research report. To appear.

[FUJ 85]   H. FUJIWARA, R. TRUER et V.K. AGARWAL
"Implementing a built-in self-test PLA design." IEEE, Design & test - April
1985.

[GAJ 85]   D. GAJSKI,
"Silicon Compilation", VLSI System Design, November 1985.

[GRE 76]   D. GREER.
"An Associative Logic Matrix." IEEE, Journal of Solid-State Circuits - vol sc-11, n°5, October 1976, pp.679-691.

[GRO 83]   R. GROSS,
"Silicon compilers : a critical survey", Dep. of computer science, Univ. of north carolina at chapel hill, May 1983.

[GUI 85]   I. GUIGUET
"Liaison d'un outil de description des circuits intégrés à un microscope électronique à balayage" CNAM Thesis, Grenoble, October 1985

[GUI 86]   I. GUIGUET, D. MICOLLET, J. LAURENT, B. COURTOIS
"Electron Beam Observability and Controlability for the Debugging of Integrated Circuits". ESSIRC 86.

[JAM 85]   R. JAMIER, A. JERRAYA, APOLLON
"A datapath compiler", ICCD'85.

[JER 85]   A. JERRAYA, E. ROSIER, F. ROUGEAUX, B. COURTOIS
"A hierarchical symbolic design layout tool" STYX, VLSI'85.

[JER 86]   A. JERRAYA, P. VARINOT, R. JAMIER, B. COURTOIS
"PRINCIPLES of the SYCO COMPILER" 23rd DAC, Las Vegas 1986.

[JOH 79]   D. JOHANNSEN
" Bristle blocks : a silicon compiler" 16th DAC 1979.

[KON 80]   B. KÖNEMANN, J. MUCHA, G. ZWIELHOFF
"Built-in test for complex digital integrated circuits". IEEE journal of solid-state circuits, vol. SC-15, N°3, June 1980.

[KRE 85]   D.E. KREKELBERG et al
" Yet another silicon compiler" 22nd DAC, 1985.

[LAU 85]   D. LAURENT, H. NGUYEN
" LDS : Langage de description de système", Internal report, Sycomore project, June 1985.

[MAR 85]   S. MARINE, E. BOURCIER
" Irene, a language for the description of VLSI digital hardware", IMAG/TIM3 research report N° 534, May 1985.

[MAK 82]   G.P. MAK, J.A. ABRAHAM, and E.S. DAVINDSON
"The design of PLAs with concurrent error detection" in Dig. Pap. 12th Int. FTC Symp., Santa-Monica, CA, June 1982, pp.303-310.

[McC 81]   J. EDWARD, Mc CLUSKEY, S. BOZORGUI-NESBAT
"Design for autonomous test" IEEE trans. on computers, vol. C-30, N°11, November 1981.

[MEA 80]   C. MEAD & L. CONWAY
" Introduction to VLSI systems", Addison-Wesley, 1980.

[NIC 83]   M. NICOLAIDIS, B. COURTOIS
"Design of self-checking system based on analytical fault hypotheses". IMAG, research report RR 353, March 1983.

[NIC 84]   M. NICOLAIDIS, I. JANSCH, B. COURTOIS
"Stongly code disjoint checkers", 14th Fault Tolerant Computing Symposium, Orlando, June 1984. USA.

[NIC 85-1] M. NICOLAIDIS
"An efficient built-in self test scheme for functional test of embedded RAMs3", 15th Fault Tolerant Computing Symposium. Ann Arbor, Michigan, June 1985. USA.

[NIC 85-2] M. NICOLAIDIS
"Evaluation of a self-checking version of the MC 68000 microprocessor", 15th Fault Tolerant Computing Symposium, Ann Arbor, Michigan, June 1985. USA.

[NIC 86]   M. NICOLAIDIS
             "An Unified Built in Self-Test Scheme : UBIST". ESSIRC 86
[OBR 82]   M. OBREBSKA
             "Efficiency and performance of different Design Methodologies for control
             parts of Microprocessors", Microprocessing and microprogramming N° 10
             - 1982   P - 163 -178.
[PAR 85]   N. PARK & A. PARKER
             " Synthesis of optimal clocking schemes", 22nd DAC, 1985.
[REI 86]   R. REIS, A. JERRAYA, R. JAMIER
             " Design of the SYCO 6502 using the SYCO compiler", submitted to the
             ICCD'86
[SAL 85]   J . SALICK, M.R. MERCER et B. UNDERWOOD
             "Built-in self test input generator for programmable logic arrays"
             Proceedings of IEEE International test conference, 1985, pp.115-125
[SCH 83]   J.P. SCHOELLKOPF
             "Lubrick : a silicon assembler and its application to data-path design for
             FISC", VLSI'83.
[SCH 85]   J.P. SCHOELLKOPF
             "SILICIEL : contribution à l'architecture des circuits intégrés et à la
             compilation de silicium", Th. d'état, I.N.P. Grenoble, 1985.
[SHI 83]   S.G. SHIVA
             "Automatic Hardware Synthesis", Proc. IEEE Vol 71, N° 1, Jan. 1983.
[SIS 82]   J.M. SISKIND et al
             " Generating custom high performance VLSI designs from succinct
             algorithmic descriptions", Conference on advanced Res. in VLSI,
             MIT,1982.
[SMI 78]   J.E. SMITH, G. METZE
             "Strongly fault secure logic networks", IEEE Trans. on comp., vol. C 27,
             n°6, June 1978.
[SOU 83]   J. R. SOUTHARD
             "Mac Pitts : An approach to silicon compilation", Computer December
             1983.
[THO 81]   D.E. THOMAS
             " The automatic synthesis of digital system", Proc IEEE, Vol 69, N° 10,
             Oct.1981.
[ULL 84]   J.D. ULLMAN
             " Computationnal aspect of VLSI", Computer science press, 1984.
[VAR 85]   P. VARINOT, S. CHUQUILLANQUI
             " Method of PLA's implementation : the monoplane PLA", ESSIRC'85.
[WER 82]   J. WERNER
             " The silicon compiler : panorama, wishful thinking, or old hat", VLSI
             design sept/oct 1982.
[WOO 79]   R. WOOD
             "A high density programmable logic array chip." IEEE Trans. on
             computers - vol C-28, n° 9, September 1979, pp.602-608.
[YAJ 82]   S.YAJIMA, T. ARAMAKI et H. YASURA
             "The design of autonomously testable PLA."
             Fifth annual IEEE workshop on design for testability, April 1982.

THE M E G A SYSTEM FOR SEMI-CUSTOM DESIGN

U.G. BAITINGER, H.M. LIPP, D.A. MLYNSKI, K. REISS

University of Karlsruhe / Federal Republic of Germany

1. INTRODUCTION

Microelectronics allow to realize even complex system functions with high
reliability at a low cost/performance ratio. Small and medium sized companies with low or even no experience in the field of digital integrated circuit design will more easily switch to microelectronics, if pre-fabricated
semi-custom gate arrays and automated CAD tools are put at their disposal.
Large silicon manufacturers have installed design centers which they offer,
together with their manufacturing processes, to smaller companies. But the
acceptance of this approach is critical, since (small) companies prefer to
keep their (few) designs confidential, and they lack experience with highly
sophisticated interactive CAD tools.

At the University of Karlsruhe/FRG, we therefore developed an alternative:
The decentralized low-cost stand-alone CAD system M E G A ("Modular
Engineering System for Gate Arrays"). Novel algorithms for both the automated logic and physical design directly take into account typical gate
array characteristics, such as low fan-in/fan-out of the logic circuits,
pre-defined complex CMOS logic gates, and geometrically pre-placed basic
gate array cells and wiring channels. Due to the use of these algorithms,
gate array design experience is made available even to the unexperienced
user. Further emphasis has been put on the development of a comfortable
graphic user interface: The automatic design algorithms start at a high-
level flow chart type functional specification.

2. DIGITAL CIRCUIT DESIGN METHODOLOGIES

Each digital system may be split-up into a control part and a process part.
It is common practice to describe the high-level functions of the control
part by a flow-chart of a finite state machine of the Mealy or the Moore
type, respectively. On the other hand, the design of the process part is
less systematic, but more intuitive; its structure is specified by hierarchical block diagrams.

2.1 Simplified Conventional Design Process

The conventional design process of a digital system starts with the functional specification of its control part by a flow-chart and the structural
specification of its process part by block diagrams (Figure 1). This information is communicated by the system designer to a logic design expert.

This designer implements the high-level functional and structural specifications by the low-level logic gate level. Conventionally, this task is supported by an interactive graphic CAD editor. It is typical for interactive design methods, that the result of each design step has to be verified subsequently by simulation.

It is also typical for conventional design methods, that there is a strict separation between logic and physical design. The only communication between user and manufacturer are the logic design data (see figure 1), which usually do not take into account physical and topological constraints. Both the user and the manufacturer tend to keep confidential their system and technology knowledge, respectively.

2.2 Novel M E G A Design System Concept

In the MEGA design system concept, the CAD support is shifted up to the high-level functional and structural specifications of the digital system. The system designer directly puts-in functional flow-charts and structural block diagrams in the graphic form familiar to him (Figure 2).

The flow-chart information is automatically transformed into a specific automaton table format and then directly processed by the logic design algorithms to yield the gate level circuitry. No verification by simulation is required for this automatic generation of design data. The process part design starts with interconnected global blocks. These may be hierarchically refined and detailed step-by-step down to the logic gate level. For this interactive design style, verification by simulation is provided as usual. Both data sets are merged together into a common net list which is communicated to the subsequent placement and routing algorithms. These have been especially developed to fit the pre-defined layout structure of a gate array and its wiring channels.

It is also typical for the MEGA design system concept, that the user/ manufacturer interface is shifted down to detailed mask data level (see figure 2). Thus, the system knowledge is kept confidential, and the design responsibility remains with the user.

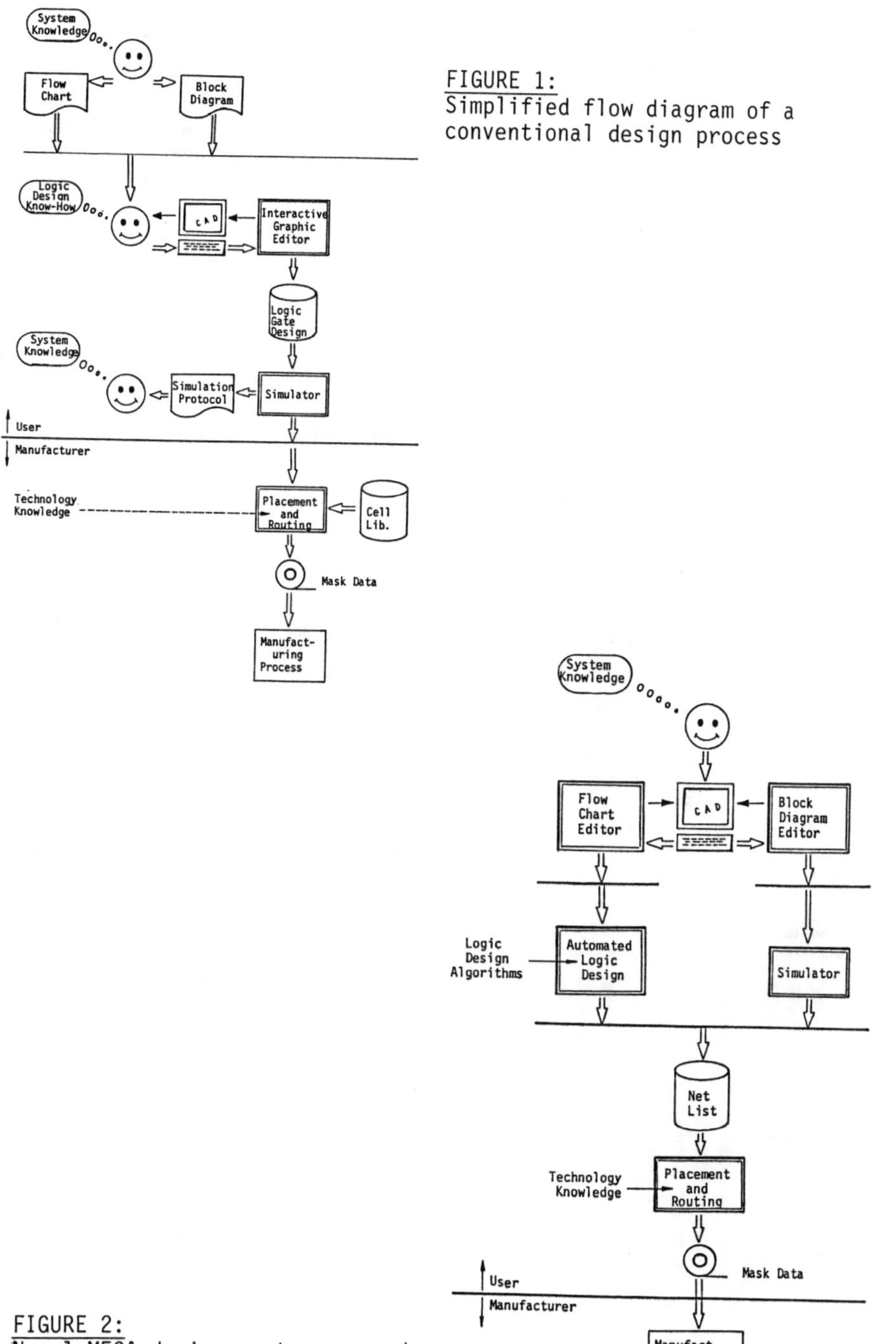

FIGURE 1:
Simplified flow diagram of a
conventional design process

FIGURE 2:
Novel MEGA design system concept

3. THE M E G A USER INTERFACE

3.1 Input to Automated Logic Design

In the MEGA design system, a graphic "Flow-Chart Editor" with extensive on-line checking is at the system designer's disposal. An example flow-chart which has been composed using this editor is represented in figure 3. MEGA offers also a graphic "Block Diagram Editor" to put-in the structure of process parts in a hierarchical manner. As an example, a simple block diagram showing an arithmetic/logic unit is given in figure 4.

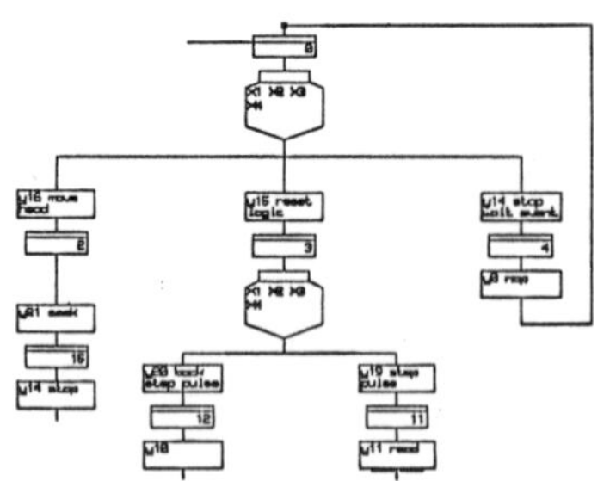

FIG. 3: Flow-chart example
        of the Mealy type

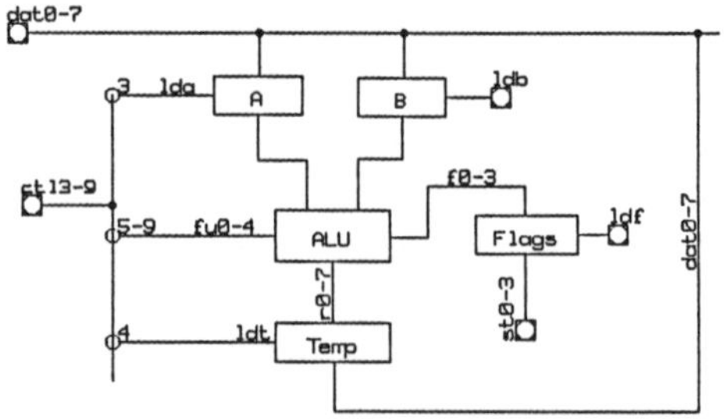

FIG. 4: Block diagram example
        (simplified ALU)

3.2 Output From Automated Physical Design

The essential output of the automated physical design are the mask data of the interconnection levels. In addition to that, the user may plot the result of the "Placement" (Figure 5) as well as that of the "Routing" (Figure 6) as soon as these design steps are completed. But these plots are required for documentation only, not for checking purposes, since the physica design data are generated automatically.

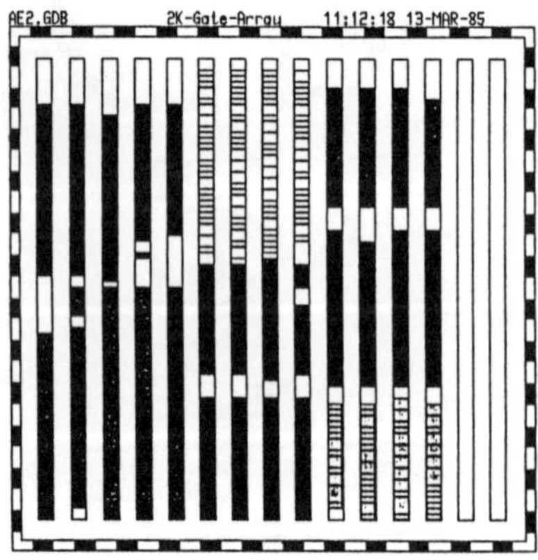

FIG. 5: Placement of gate
        array cells

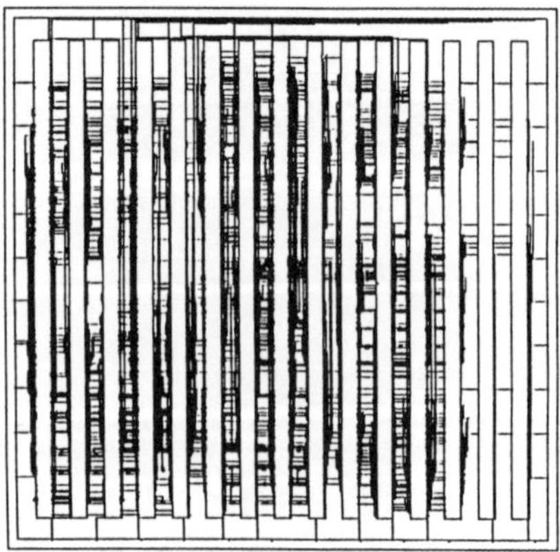

FIG. 6: Routing/wiring of
        gate array cells

4. M E G A   LOGIC DESIGN ALGORITHMS

4.1 State Coding of Sequential Circuits

According to the flip-flop type chosen, a specific automaton table is automatically generated with dual coding for the states of the finite state
machine. The set of state vectors is then partitioned into two subsets,
starting with their high order bits. This procedure creates neighbourhoods
of input and/or output patterns which can then be combined to minimize the
combinational circuits which generate the coded state vectors. This is recursively repeated, until the low order bits of the state vectors are
reached [1].

4.2 Multi-Stage Logic Design Approach

Traditionally, combinational networks are based on the two-stage AND-OR
structure which results from conventional boolean algebra. It is typical for
this approach that there is no upper limit for the number of inputs per
logic gate. Any technological fan-in limitations require a subsequent, non-
optimum expansion into a multi-stage logic circuit.

With the forthcoming of gate arrays, however, their typical constraints and
characteristics should be included into the logic design process from the
very beginning:

- In general, the basic logic gate array cells (NAND, NOR, etc.) have
  only two input lines which implies a multi-stage boolean circuit
  design.

- On a CMOS gate array, pre-defined complex logic cells are available,
  which should be systematically included during the logic design
  process.

Basically, any multi-stage logic network can be described as a sequence of
coding steps. At each stage, the vector of input variables is coded into a
vector of output variables.Figure 7 shows the correspondence between a
circuit "stage" and a coding "step". The design process turns out to be a
multi-step coding process, where one has to find out at each step which
two-input logic function (AND, OR, XOR, EQU) is the most appropriate one.
The overall function to be implemented is partitioned step-by-step into
partial functions of two input variables each. This implies a fan-in = 2
which is the typical value for basic gate array cells. Due to this restriction, computing time and memory space of the algorithm only increase by the
square of the number of input signals [2] .

To ensure the convergence of the multi-step coding process, compatibilities
have to be identified for all pairs of input variables, since their codings
can be made identical. (This is done by the dual investigation of the in-
compatibilities.)

For each pair of input variables there are four possible compatibility configurations:

- <u>1's-Partition</u>: All four partial patterns (00,01,10,11) are compatible, they remain within one compatibility set. This least detailed partition is trivial. The investigated pair of input variables does not influence the value of the output function; both variables can be omitted.

- <u>2's-Partition</u>: The set of partial patterns can be partitioned into two subsets of compatible patterns, either in a symmetric form (two plus two), or in an unsymmetric one (three plus one).This partition can be coded with only o n e binary variable, thus the number of varia- bles is reduced. For an unsymmetric 2's-partition ("vertex partition"), the coding variable results from a vertex function (AND, OR) combining both input variables. In the symmetric case ("symmetric partition"), a symmetric function (XOR, EQU) has to be chosen.

- <u>3's-Partition</u>: The set of partial patterns can be partitioned into three subsets, where two patterns are compatible, whereas the other two are not. To code this partition, t w o binary variables are required. The number of variables is not reduced, but the numer of different patterns, since a "don't care" is generated. This will accel- erate the convergence of the coding process.

- <u>4's-Partition</u>: If all four partial patterns are incompatible, each one corresponds to an own compatibility subset. This trivial partition is the most detailed one. The four partial patterns have to be coded differently, thus no reduction is possible. But a spectral transfor- mation may be applied which possibly reduces the complexity of the residual function to be implemented.

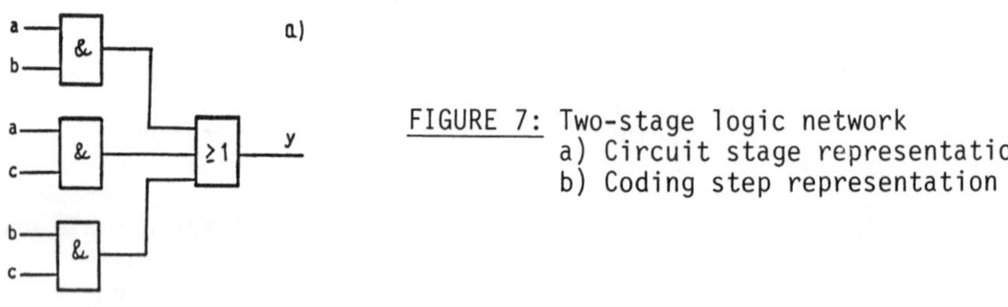

FIGURE 7: Two-stage logic network
a) Circuit stage representation
b) Coding step representation

b)

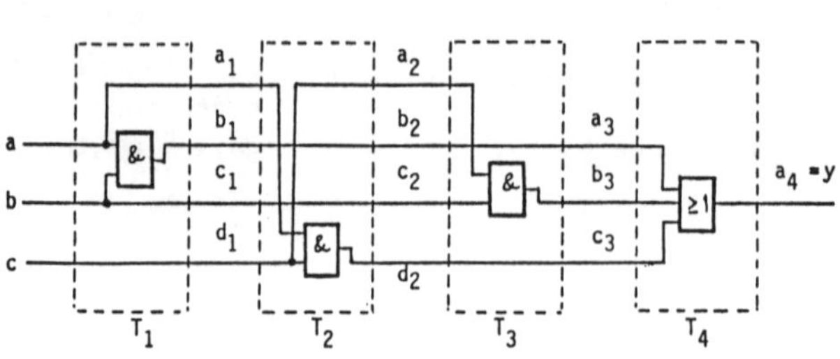

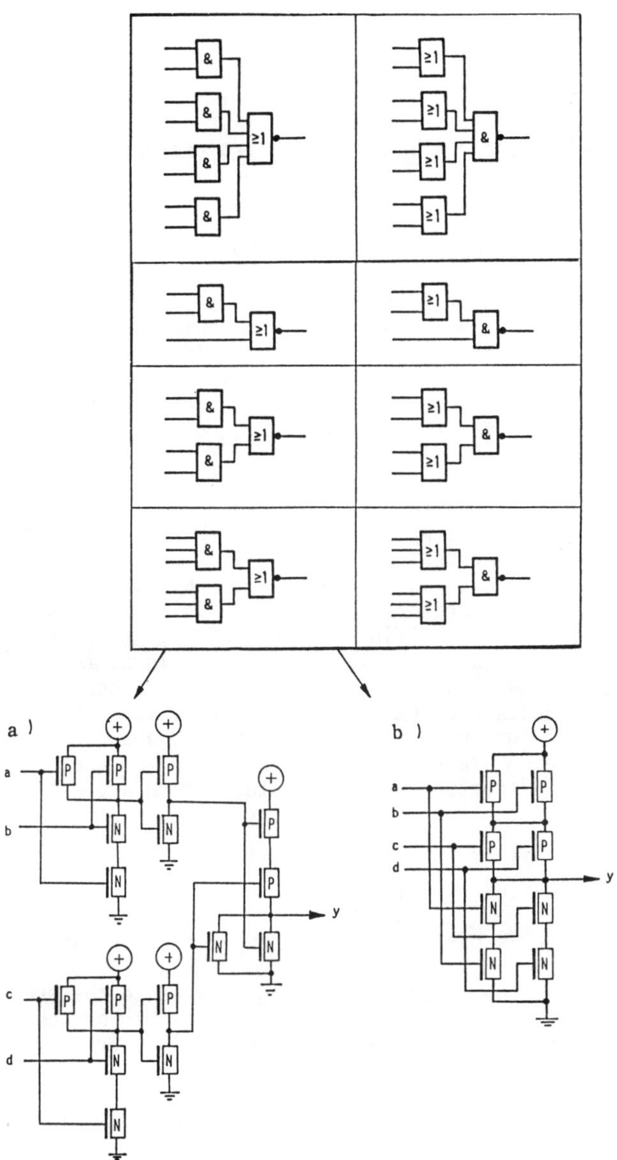

FIGURE 8: CMOS gate array cell library
a) (unrealistic) realization by separate gates
b) realization by a complex CMOS logic gate

4.3 CMOS Complex Gate Cost Function

In the case of CMOS gate arrays, where pre-defined complex one-stage logic gates are available from a design library, the elementary partial functions (AND, OR, XOR, EQU) which result from the multi-stage circuit design steps have to be recombined during the logic design process according to a new cost function.

Traditionally, the cost of a digital circuit is calculated according to the number of elementary logic gates required and by the sum of their input terminals. This cost function is valid for designs using separate logic gates on a printed circuit board. In the case of an integrated CMOS cell library (Figure 8), however, this cost function would correspond to the transistor cicuit shown in figure 8a with each logic gate realized separately. But this is totally unrealistic as can be seen from the real integrated complex logic gate as shown in figure 8b. For this purpose, a new "dynamic" cost function at the transistor level, but in terms of boolean gates, has been defined [2].

At the logic gate level, a CMOS complex gate can basically be considered as a tree structure with an inverting output gate (see figure 8). At the transistor level, only this output gate is a "real" one (fan-out ≥ 1), whereas the internal, non-inverting gates are not physically accessible (see figure 8b). They are, therefore, called "virtual" (fan-out = 1). This implies that all logic gates resulting from the coding steps are to be assigned either to the virtual or to the real level. Thus, the digital circuit under design, which will be composed of multiple stages of elementary logic gates with two inputs each (AND, OR, XOR, EQU), is compared to the cell library during the current logic design process, and not just after its completion. The coding steps are chosen in such a way that preferably CMOS complex logic gate structures are generated. These are embedded between gates with a fan-out > 1 and an inverting output.

5. M E G A PHYSICAL DESIGN ALGORITHMS

The physical design task is partitioned into two closely related subtasks:

- The Placement, i.e. the assignment of logic gates to the pre-defined positions of basic cells on the gate array. This may be considered as an "intra-cell wiring" which is taken from a cell library.

- The Routing, i.e. the wiring between the cells via pre-defined routing channels. This is the application-specific "inter-cell wiring".

5.1 Automatic Placement

The placement is done in two steps. The initial placement ("relative topological placement") yields the initial positions of the logic cells on a continuous plane by a force algorithm [3]. During the final placement ("absolute geometrical placement"), the logic cells are assigned to the fixed positions of the basic cells on the gate array [4].

Depending on the logic design hierarchy, the initial placement is done by a
multi-level hierarchical partitioning using logical groups and macros. It
starts with a force directed placement of logical groups ("floor-planning")
and it is refined successively during the initial placement procedure which
ends at the gate level. Since a user may start without or only some prefixed
positions of pads and I/O-cells, a pad assignment is also done during the
initial placement procedure. Figure 9a shows the result of the initial
placement.

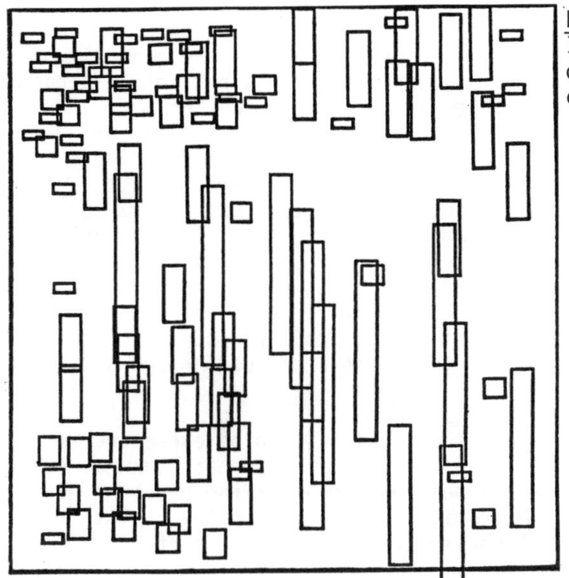

FIGURE 9a:
Initial placement of gate array
cells and macros by a force
directed algorithm

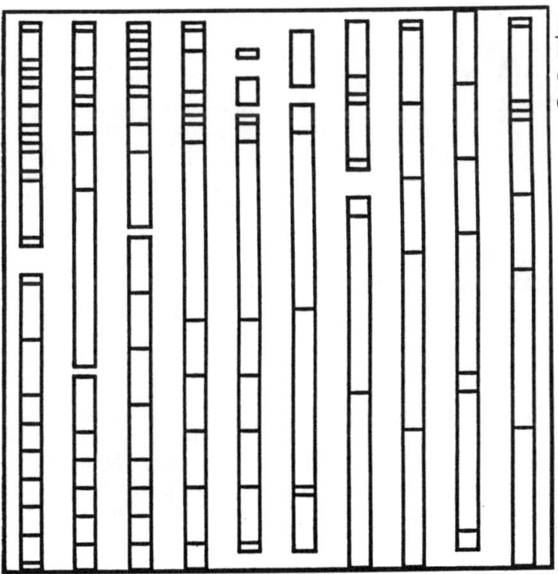

FIGURE 9b:
Final placement, i.e. logic
gate assignment to basic cells
on the gate array

The final placement problem may be viewed as a restricted bin packing problem with optimization features [4]. A logic gate is realized by an integer multiple of basic cells within one gate array column. Given an initial placement, the problem is to assign the logic gates within the columns preserving the positions as good as possible.Based on the initial placement of figure 9a, figure 9b shows the result of the final placement.

5.2 Automatic Routing

The routing task is split into a first step where the nets are globally distributed among the routing channels ("loose routing"), and into a second step where the channels are routed successively ("final routing").

In the case of CMOS gate arrays, there may exist several electrically equivalent terminal pins to connect a signal to a logic gate. Then the loose routing problem can be stated as follows: For each net find a set of subnets such that each subnet is assigned to a single channel, i.e. is given by a set of terminal pins and feed-through pins in a single channel. The subnets are connected by electrically equivalent pins or a feed-through assignment to the net. The given routing capacity for each channel must not be exceeded by the resulting channel density. For this purpose, graph-theoretical methods are used for net partitioning and feed-through assignment [5].

After the loose routing the chosen net terminals have to be connected within pre-defined routing channels. The final routing of each channel can be calculated independently. A new channel routing algorithm has been developed [6]. It combines a graph-theoretical approach with the ability to split constraint loops. The channel routing consists of two parts:

   - detection and splitting of constraint loops
   - embedding of segments

The channel routing problem is completely described by the number of wiring tracks and the terminals at the top- und bottom-side of the channel, respectively. For each net we define a set of horizontal segments. Each terminal of the net is associated with a column, and a segment extends between two successive columns.

The final step in channel routing is to assign all horizontal segments to a minimum number of tracks. For that purpose we use a modified left-edge-algorithm. This algorithm has been tested for several examples. In all cases, we obtained a fully automatic layout [6].

6. CONCLUSIONS

It has been shown that multi-stage gate array circuits which have been directly generated by the MEGA design algorithms need only about 40 per cent of the silicon area of a conventional two-stage solution with subsequent expansion to fulfil the gate array fan-in constraints [2]. A further reduction can be achieved by appropriate automatic state coding of the finite state machine [1].

The logic design algorithms presented here may also be applied to the design of PLA and/or microprogrammed controllers to pre-process their input variables, thus reducing the number of input address lines [7].

The MEGA design system is now running on a VAX 11/750 and on a PCS CADMUS personal computer (QU68000). The acceptance of the graphic high-level MEGA user interface is currently investigated within the scope of an industrial cooperation. The MEGA project will be continued to cover computer aided testing (CAT) of the automatically generated gate array designs.

ACKNOWLEDGEMENT

The MEGA project is financially supported by the German Federal Ministry of Research and Technology (BMFT) in Bonn, the State Ministry of Economics and Technology in Stuttgart/FRG, and by several industrial companies in the Federal Republic of Germany.

REFERENCES

[1] K. H. KAISER, U. G. BAITINGER, "A State Coding Algorithm for Sequential Circuits", Techn. Report, ITIV, University of Karlsruhe (to be published in 1986)

[2] S. DAO-TRONG, "A Design Process for Multi-Stage Switching Circuits Under Gate Array Constraints", Ph.D. Thesis, ITIV, University of Karlsruhe (1984)

[3] G. J. WIPFLER, "A Placement Process for the Bottom-Up Design of VLSI Circuits", Ph.D. Thesis, ITM, University of Karlsruhe (1985)

[4] H. HILLNER, B. X. WEIS, D. A. MLYNSKI, "The Discrete Placement Problem: A Dynamic Programming Approach" (to be published in Proc. ISCAS 1986)

[5] K. WINTER, "Loose Routing Strategies for Gate Arrays", Ph.D. Thesis, ITM, University of Karlsruhe (1986)

[6] F. V. KELLER, D. A. MLYNSKI, "A Graph-Theoretical Channel-Router For Constraint-Loop-Problems" (to be published in Proc. ISCAS 1986)

[7] A. DITZINGER, "Expenditure Reduction for Digital Control Units by Appropriate Pre-Coding of Internal Interfaces", Ph.D. Thesis, ITIV, University of Karlsruhe (1982)

The logic design algorithms presented here may also be applied to the design of PLA and/or microprogrammed controllers to pre-process their input variables, thus reducing the number of input address lines [7].

The MEGA design system is now running on a VAX 11/750 and on a PCS CADMUS personal computer (QU68000). The acceptance of the graphic, high-level VLSI user interface is currently being evaluated within the scope of an improvement computation. The MEGA project will be continued to cover computer aided testing (CAT) of the automatically generated gate-array designs.

ACKNOWLEDGEMENT

The MEGA project is financially supported by the German Federal Ministry of Research and Technology (BMFT) under the State University of Economics and Technology in Stuttgart-FRG, and by several industrial companies in the Federal Republic of German.

REFERENCES

[1] K. H. KAISER, U. G. BAUMGARTNER, "A Silicon Coding Algorithm for Sequential Circuits," Techn. Report, IIIV, University of Karlsruhe. (To be published in 1986.)

[2] S. DAOSTROM, "A Design Process for Multi-Stage Switching Circuits and Gate-Array Constraints," Ph.D. Thesis, IIIV, University of Karlsruhe, 1984.

[3] G. J. MILLER, "A Placement Process for the Better Dimension of VLSI Circuits," Ph.D. Thesis, IIIV, University of Karlsruhe (1985).

[4] H. HILLNER, S. X. WEISS, P. A. MEYER, "The Discrete Placement Problem: A Dynamic Programming Approach." (to be published in Proc. ICCD, 1986)

[5] A. WINTER, "Global Routing Strategies for Gate Arrays," Ph.D. Thesis, IIIV, University of Karlsruhe (1985).

[6] R. KELLER, D. A. MUYISKI, "A Greedy Placement-Channel-Router for Constraint-Shop-Problems," (to be published in Proc. ICCAS, 1986).

[7] A. BLLINGE, "Expenditure Reduction of Digital Control Units by Microprogramming of Logical Interfaces," Ph.D. Thesis, IIIV, University of Karlsruhe (1982).

PART III. SYNTHESIS SYSTEMS FOR DIGITAL SIGNAL PROCESSING

AUTOMATIC GENERATION OF DIGITAL SIGNAL PROCESSING CIRCUITS

S.P. POPE†, J.M. RABAEY‡ and R.W. BRODERSEN

Dept. of Electrical Engineering and Computer Science University of California, Berkeley

I INTRODUCTION

The general problem of automatic generation of integrated circuits from a high level description for an arbitrary application is an exceedingly difficult undertaking. As the level of abstraction increases, the wide variety of possible circuit realizations makes the development of the software quite complex, since it must emulate the decision making capability and originality of a designer. One approach to reduce this design space is to restrict the range of applications and develop a circuit architecture which is optimized for those applications. This approach has been taken for several application areas, which have included audio band signal processing[1,2,3] and image processing [4,5], and is now being extended into robotics and control system applications. The major focus of this paper is the considerations which motivated the target architecture that was chosen for the audio band integrated circuits [1]. The relation between the architecture and a macrocell based layout system is discussed.

1.1 Dedicated Signal Processors

Despite numerous predictions over the past decade that digital signal processing LSI circuits would soon have a huge impact on large segments of the industry, such an effect has not yet materialized. To better understand why this is so, it is necessary to consider how signal processing functions differ from other applications.

A *signal* is a representation of a time-varying physical quantity. Common examples of signals are sounds, vibrations, and electromagnetic waves. A *signal processor* is a device or system which conditions, analyzes, synthesizes, or otherwise modifies or creates signals.

Signal processors may be subdivided into general purpose signal processors and dedicated signal processors. A general-purpose signal processor is one that may, through reconfiguration or reprogramming, perform a variety of fundamentally different algorithms. This paper is concerned with *dedicated* signal processors -- ones that perform a single algorithm. There might be a few variables or parameters that may be modified, but such a system, once created, is pretty much devoted to a single task. Also, the systems studied here are those which work with sampled and digitized (quantized) signals. Digital signal processors such as these have a number of advantages over their analog counterparts: control over dynamic range and other performance aspects, and the ability to be programmed to perform complex tasks.

From an abstract point of view, virtually any computation would satisfy the above definition of signal processing. After all, the inputs and outputs of any computing system can be viewed as time sequences. From a practical point of view, signal processors are distinguished by the following two properties:

(1) Because signals result from ongoing, real-world physical processes, certain types of processing tend to recur in different signal-processing situations. Filtering, modulation, correlation and convolution are all classical signal processing functions.

(2) The signals involved are processed in *real time*. This means that the rate of processing is sufficient to keep up with incoming and outgoing signals. Thus processing could proceed indefinitely without loss of continuity.

---

† S.P. Pope is now with Cyclotomics Corp., Berkeley, CA
‡ J.M. Rabaey is now with IMEC, Leuven, Belgium

1.2 Processing of Sampled Signals

Signal processing algorithms are often presented by means of a signal flow diagram. Fig. 1 is a signal flow diagram for a second-order filter section. Note that the primitive operations involved are multiplication, addition, and delays. All linear filters, however complex, are composed of these same operations.

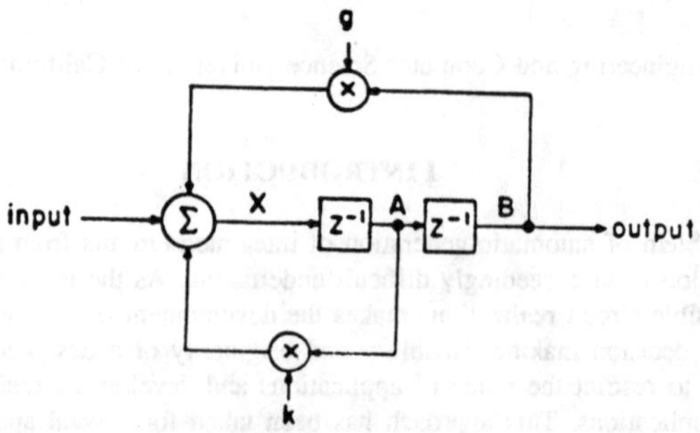

Fig. 1) Flow diagram for a second-order filter

The filter section may be implemented by repeatedly executing the following sequence of operations. The period of repetition is one sample interval. The quantities A and B represent the state (memory) of the filter, while X is a temporary variable. ":=" is the assignment operator.

$$X := input + kA + gB$$
$$output := B$$
$$B := A$$
$$A := X$$

In order for this abstract program to operate properly, the timing of the system must be such that each time the program is repeated the values "input" and "output" refer to new input and output samples respectively. The four-instruction program is executed from top to bottom without branching, and repeats each sample.

The example above is very simple; however, it suggests that all linear digital filters (and possibly other signal processing functions as well) may be implemented by a processor with the following properties:

(1) A simple instruction set
(2) A program which is executed once per sample
(3) A lack of branches in the program flow

This simple model can serve as a basis for the design of reasonably powerful signal processors.

It should be noted that some system functions do not conform to the above model. Consider the Fast Fourier Transform[6], for example. In this case a block of input data (say 128 samples) is processed as a unit, resulting in an algorithm that is difficult to express by a program that repeats at the sample rate. Thus property (2) above would preclude efficient implementation of a Fast Fourier Transformer using the the approach described above.

Many signal processing algorithms make use of a frame. A frame is an interval many samples in length. The concept of a frame is significant because many interesting attributes of a signal, such as second moments and spectral characteristics, vary much less rapidly than the signal waveform itself. This being the case, it is possible to represent the desired attribute as being constant over an interval of considerably greater duration than the sample interval. This sort of data reduction is known as signal analysis and is widely used. The inverse operation, wherein a sampled-data signal is created from lower bandwidth data, is termed signal synthesis.

A typical signal processing system might consist of a front-end in which sample-rate processing is performed, an intermediate level in which frame-rate processing is performed, and additional levels in which processing occurs at increasingly slower data rates. For those levels with a sufficiently slow data rate, conventional computer hardware may be programmed to economically perform the operations required. Other levels may require special-purpose hardware to meet performance goals.

The cell library of Chapter 3 allows a frame rate to be defined and used for the purpose of off-chip I/O. Semi-custom chips designed using the cell library can then serve as special-purpose front-end processors. I/O processing differs qualitatively from the other sorts of processing in a typical system, justifying separate hardware. Also motivating this focus is the perception that the I/O processing is often the major obstacle to a successful hardware realization.

1.3 Architectural Alternatives for Digital Signal Processing

Early digital signal processors were assembled from small- and medium-scale integrated circuits. Since this approach led to rather bulky systems, a considerable degree of specialization is evident in these early designs. Typical examples would be special purpose Fast Fourier Transform (FFT) machines [7], and a computer hard-wired to perform a speech pitch detection algorithm [8]. A considerable emphasis was often placed on fast array multipliers. In fact, much of the research on multiplier implementations was done by those pursuing signal processing applications [9]. Because of the throughput requirements and real-time constraints, pipeline architectures were popular as well.

As the 1980's approached, single-chip digital signal processors started to become available. A few of these were dedicated circuits intended for special applications. An early example would be a speech synthesis chip developed by Texas Instruments for use in a toy known as the "Speak-and-Spell" [10]. Of more interest to signal processing researchers was the availability of general purpose, programmable DSP chips such as those of N.E.C. [11], Bell Laboratories [12], and Texas Instruments [13]. Architecturally, these circuits all followed a standard formula. The instruction set resembles that of a conventional microprocessor. Hardware organization is also conventional, with some pipelining added and a parallel multiplier attached.

Despite heavy promotion, these general purpose processors have failed thus far to achieve widespread use. One probable reason for this is that the general-purpose architecture is seldom a good match for any given application. Such a circuit offers a fixed combination of resources (program memory, data memory, arithmetic capacity) which can not be tailored to suit the designer's needs. The inclusion of a large parallel multiplier results in an expensive part in each case. Although there is a definite need for multiply-accumulate operations in a typical signal processor, it is difficult to make efficient use of the parallel multiplier when programming the circuit to perform a complex system function.

Because of the problems described above, a number of researchers have worked on the problem of designing semi-custom digital signal processing IC's [14-17]. The goal of this work is to combine the cost-effectiveness of a full custom design with the short design times possible using general purpose hardware. The underlying assumption of such work is that similar applications can be efficiently implemented using different configurations of the same hardware.

Semi-custom design approaches have a target architecture to which individual designs must conform. There are several types of target architectures which have been proposed. Three general categories may be identified: (1) processors built around a parallel array multiplier; (2) bit serial processors; and (3) bit-parallel processors without parallel array multipliers.

The parallel multiplier approach is often regarded as the best signal processing solution. This is only partially true. Parallel multipliers by necessity must be constructed with a fixed precision, e.g. a 16 by 16 multiplier. They are therefore best suited for algorithms which make heavy use of multiplies, all of which require about the same precision. Unfortunately, precision requirements in signal processing algorithms vary considerably. Many multiplies require only a low precision coefficient, such as a power of two. The parallel multiplier is in these cases underutilized. Also, there is the architectural problem of getting data to and from the parallel multiplier without bottlenecks occurring. The cost, in terms of busses and registers, is high. In summary, a parallel multiplier is a high cost, high throughput hardware item that fails to see good utilization in actual practice.

The second option, bit serial designs [17], are the opposite end of the architectural spectrum. In an effort to maximally utilize the logic elements of the circuit, bit-level pipelining is used. High clock rates are needed to reap the full advantage of this approach, creating a particularly difficult low-level design problem. An advantage is that bit-serial functional units can be combined in a modular fashion, a plus if the target system is a dedicated processor. One drawback of this approach is that a number of low level functions (such as division) are not compatible with the LSB-first bit-serial format. Due to the hard-wired nature of the systems, control and decision-making functions are not easily implemented. Also, a system-wide wordlength is generally established for all data, in an effort to avoid intractable timing problems. This is a disadvantage, since all data must be represented by words of the same precision.

The third option, bit-parallel processors without parallel multipliers, is the one used for the designs described in this dissertation. An arithmetic unit which can complement, shift and accumulate bit-parallel data is used. This unit may be microprogrammed to multiply signal data by either bit-serial or signed-digit coefficients. (Signed-digit representation provides a low-cost method for fixed coefficient multiplies, as discussed in Section 2.3.3.) Because the arithmetic element is relatively compact, several such processing units may be included on a single IC. Thus the advantage of modularity exhibited by the bit-serial approach is retained here. This results in a better match between algorithm and hardware organization than is possible in designs using large parallel multipliers.

The bit-parallel processors are microprogrammed, resulting in a versatility lacking in hard-wired approaches. Although fully hard-wired architectures might be more efficient in the abstract, programmability is a virtual necessity in a complex system. Control, decision-making and input-output functions are all more readily implemented in a programmable environment.

1.4 Design Methods for Digital Integrated Circuits

Numerous approaches, systems and methodologies have been proposed or are in use for integrated circuit design. One reason for this is that, viewed in the abstract, integrated circuit design is a large process with a high level of complexity and many degrees of freedom. Many methodologies seek to constrain the design problem, to make it less complex while still achieving performance goals.

IC designs are ultimately specified at the geometric level on a number of mask layers. In a *full-custom* design, the designer has complete control over these low-level geometries. Standard cell and gate-array design methods involve specifying the design at a logical level, and creating the mask geometries from a predefined set of primitive elements. Thus standard-cell and gate-array approaches constrain the design problem to make it more tractable.

As a general rule, the more constrained the design problem is made, the more easily a design may be produced. However, this design can not be optimized to the same extent as a full custom design.

Many IC design systems are intended to fill a certain market need. For example, a standard-cell system might be intended to produce circuits which are replacements for systems designed from a standard logic family (for example, the 74-series TTL family). Such systems have the severe drawback that the designer is using logic designs and architectures appropriate for one technology (TTL-MSI) while producing a circuit in a different technology (MOS-LSI).

In addition to the degree of customization, a distinction may be drawn between graphical and procedural methods of specifying IC layouts. In the graphical approach, the designer specifies the mask geometries directly, usually with the aid of a computerized graphics editor. In the procedural approach, the specification is at a higher, functional level; software is used to translate the higher-level specification to the mask level. A typical approach is based on the macrocell design style.

The term *macrocell* has been assigned various meanings in the past. It is now generally agreed that a macrocell is a large block of circuitry containing perhaps several thousand transistors. One typical approach used in assembling the macrocells is to tile (array) smaller rectangular cells in two dimensions. The tiling process is sufficiently flexible that the macrocells can be customized for specific tasks.

One premise of the macrocell technique is that a target application range has been identified. This discussion is concerned solely with digital signal processors. Typical applications would be speech processing, telecommunications and digital audio. However, the macrocell approach may be applied to any group of applications. The more narrow the application range, the more the macrocell-based designs will resemble full-custom designs in terms of architectural efficiency.

Since MOS processes evolve rather rapidly, a desirable property of an IC design system is that it can be applied to a number of different processes. Macrocell based systems can be designed such that only the underlying cell library is technology-dependent.

In summary, the macrocell approach has the following advantages:

(1) Large blocks of circuitry are generated by procedural methods.
(2) The underlying cell library is designed by graphical methods.
(3) Architectures can resemble those of full-custom designs.
(4) The design system may be ported to different technologies.

Of primary importance is the ability to configure the macrocells for a particular application, resulting in a more efficient design. A number of parameters and options may be specified for the macrocells described here. These include data path widths; programming and sizing of control memory, data memory and programmed logic arrays (PLA's); and configuration of the interprocessor communication network. Thus a macrocell is not a fixed block of circuitry, but a class of circuit blocks configured according to explicit rules from the underlying cell library.

Design systems based on standard cells or gate arrays involve considerable supporting software for simulation and layout generation. A similar approach is taken here. The software system consists of an *emulator* and a two-pass *silicon compiler*. The emulator and compiler accept as input a *design file*, which is a text file prepared by the designer. This software system is diagrammed in Fig. 2.

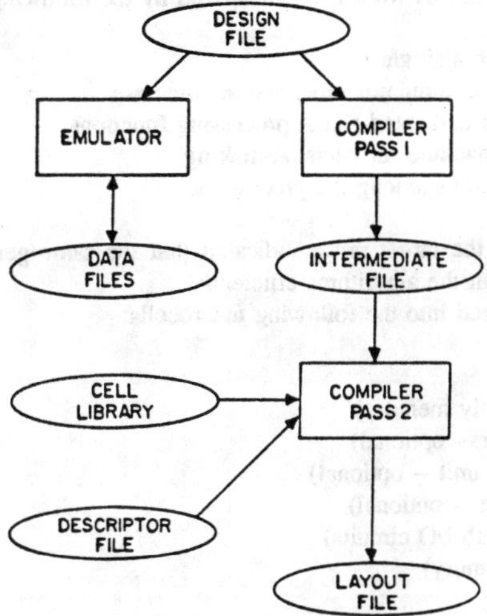

Fig. 2) Macrocell design system software

The design file has a very readable format. Each IC contains several processors operating concurrently. The designer specifies hardware parameters and symbolic microinstructions for each processor. Local constants and variables are declared. Interprocessor and off-chip communications are specified by declaring global variables which are external to the individual processors.

The emulator allows verification of circuit performance prior to fabrication. A full set of debugging commands is available, such as tracing, breakpoints, single-step, and setting or displaying variables. A non-interactive mode allows long simulation runs to be performed.

The first pass of the compiler extracts hardware parameters from the design file. Symbolic microcode is assembled into binary. The resulting hardware description is input to the second pass, which accesses the cell library and assembles the macrocells. Placement and interconnect routines follow. The system is intended to generate automatically a mask-level description of the I.C.

The term *silicon compiler* has been generously applied to specialized IC design systems which generate circuit layouts automatically [18]. In fact, any layout-generating software which does not fit into some other category (graphics editor, router, etc.) is likely to be called a silicon compiler. The system described here satisfies this definition.

A major goal in the design of the software package was process independence. Integrated circuit fabrication processes evolve rapidly. Also, it is dificult for a designer who does not control a captive fabrication facility to ensure continued availability of a given process. For these reasons, a macrocell design system should function for any fabrication process that might become available. This goal was achieved by introducing a small number of parameters which characterize the process for the purposes of the software package. The software may then be used with any process by redesigning the cell library in the new process.

By limiting the focus of the design system to a family of signal processing functions, efficient and specialized architectures may be employed. The range of applications includes: speech processing (vocoders, sub-band and waveform coders, speech recognition), telecommunications (modems, line equalizers, echo cancellers), and digital audio (mixing, equalization, reverberation and other effects). It is estimated that several dozen commercially valuable dedicated IC's will be designed with this system.

The remainder of this section discusses the architecture and hardware organization used in the macrocell system. The next section describes the design file and the software system, followed by example designs using the macrocell system.

II PROCESSOR ARCHITECTURE

The target architecture used in this work is characterized by the following:

(1) Multiple processors on a single IC.
(2) A bit-parallel, single accumulator processor architecture.
(3) Microprogramming of dedicated signal processing functions.
(4) Use of a finite state machine for decision-making.
(5) Bit-serial communications among the processors.

Study of algorithms within the target range indicated that the same general architecture, with suitable parameterization, can perform the algorithms efficiently.

Each processor is organized into the following macrocells:

PC (program counter)
ROM (microcode read-only memory)
SPC (subprogram counter -- optional)
AAU (address arithmetic unit -- optional)
FSM (finite state machine -- optional)
AUIO (arithmetic unit with I/O circuits)
RAM (processor data memory)

A fully configured processor is diagrammed in Fig.3.

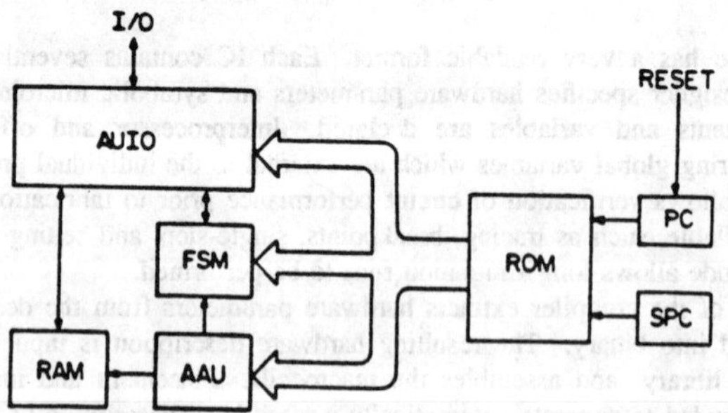

Fig. 3) Macrocells forming a fully configured processor

The processor executes its microprogram (stored in ROM) once per sample interval. The microprogram consist of a "main program", optionally followed by a fixed number of "subprogram" iterations. The SPC macrocell is included only if the subprogram is used. Except for this iteration pattern, there are no branches (conditional or unconditional) in the program execution. The PC, ROM and optional SPC macrocells are collectively known as the control sequencer.

The output of the control sequencer is a sequence of control words. The control word can be subdivided into a number of fields. One of these is the address offset field. It is this field that is used as input to the AAU macrocell. The AAU modifies the address offset to produce the effective address for the processor data memory (RAM macrocell). In some simple cases, no modification is required and the AAU macrocell is not included. In these cases the address offset field of the control word serves directly as the effective address.

Viewed together, the ROM, PC, SPC and AAU present a stream of horizontal control and address words to the AUIO, RAM, and (optional) FSM macrocells.

The AUIO macrocell consists of an arithmetic unit and an I/O interface. The arithmetic unit together with the RAM (data memory) macrocell form the signal data path. This is where all operations on signals are performed. This data path may be microcoded to perform fixed coefficient multiplies, variable coefficient multiplies, add, subtract and accumulate operations, division, and comparisons. In addition, the data memory allows delay operations, introduction of constants and table-lookup.

The final component of the processor assembly is the optional finite-state machine (FSM). This is where logical operations, conditional operations and decision-making are performed. The FSM accepts Boolean inputs from two sources: a comparison in the arithmetic unit; and tests on index registers internal to the AAU. The FSM operates on these inputs and its own internal state. FSM operations are controlled by a field of the control word. Outputs from the FSM are used for two purposes: to control write cycles in the RAM macrocell; and to initiate I/O to an off-chip host.

The processor assembly executes its microprogram once per sample interval. Since primitive operations such as multiplies must be microcoded, it follows that the sample rate must be substantially slower than the processor's clock rate. Otherwise, there would not be time for a significant amount of processing during the sample interval. On the other hand, if the ratio of clock rate to sample rate becomes very large, the architecture exhibits an imbalance wherein the control ROM consumes nearly all of the silicon area, and the data path only a small fraction. This situation is aggravated since the horizontal control words are not densely encoded. Based on the relatively sizes of the library cells, the processor architecture is most efficient if the ratio of clock rate to sample rate is between 50 and 1000. This corresponds to a sample rate range of 5 kHz to 100 kHz, since the cell library is designed to operate at a maximum 5 MHz clock rate.

2.1 Control Sequencer

An important aspect of the macrocell design system is the fact that the individual processors are microprogrammed to perform dedicated tasks. In fact, most of the design file input to the software system consists of microcode. The control sequencers are programmed with this microcode, and therefore control all timing and data-path operations for their processors. In addition, timing strobes generated by the control sequencers can exit the processor assembly to be used for control of interprocessor or off-chip communications.

The control sequencer consists of the ROM macrocell, the PC (program counter) macrocell, and the SPC (subprogram counter) macrocell. The subprogram is optional; if there is no subprogram, the SPC is not included. The outputs of the ROM are known as the control word. The control word is very horizontal: it contains many non-overlapping fields, each of which has an independent function. Rarely are two different functions assigned to the same bit of the control word.

The AUIO macrocell has many control inputs, accounting for perhaps half of the control word. Therefore, provisions are made for defaulting unused signals in this field to ground or Vdd, reducing the typical ROM width.

A few control signals are used to clear and increment the two program counters, PC and SPC. It is required that the various processors on a chip be synchronized, i.e., each processor begins its microprogram on the same clock cycle at the beginning of each sample interval. In order to allow this, one processor is designated as the master processor for timing purposes. Contained in the control word for the master processor is a signal named RESET. This signal is asserted at the end of each sample interval, and serves to reset the PC (program counter) macrocells for all processors on the chip. There are no inputs to the control sequencer aside from this reset input.

There are two possible configurations for the control sequencer: with or without a SPC. The following discussion applies to the case where there is a SPC. The hardware is diagrammed in Fig.4.

The SPC macrocell requires two timing signals JSR and RET as control inputs. The JSR signal is

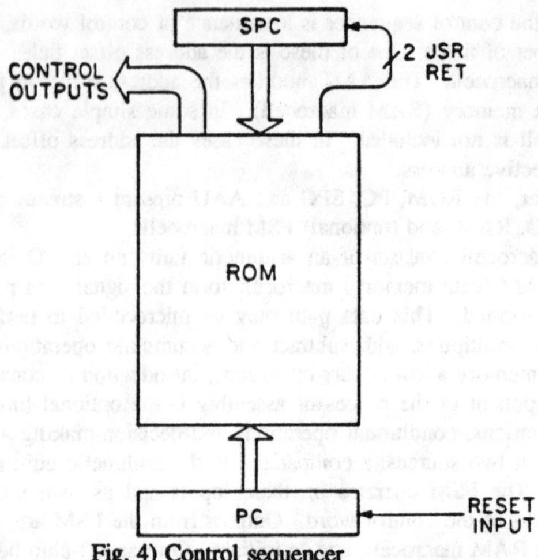

Fig. 4) Control sequencer

asserted at the end of the main program, and at the end of each subprogram iteration. The RET signal is asserted concurrently with the final JSR at the end of the final subprogram iteration. When JSR and RET occur simultaneously, the SPC is cleared and remains cleared until the next JSR. When JSR is asserted but RET is inactive, SPC is loaded with a one and commences counting. Thus, SPC equals zero during the main program and counts from one to an upper limit during each subprogram iteration.

Note that the end of the final iteration of the subprogram need not be coincident with the end of the sample interval. The reason for this is that the total execution times for different processor's microprograms will in general be unequal. Therefore, there may be an idle interval following the end of the microprogram lasting until the end of the sample.

Also note that the subroutine iterations are not explicitly counted. The subroutine iterates the desired number of times only because the RET signal is asserted on the proper cycle.

RESET, JSR, and RET are asserted two cycles before their respective actions are effected. This delay is due to the pipeline register at the output of the control ROM. As an example, suppose that there are 200 clock cycles in a sample interval. The program counter counts from zero to 199. Suppose that there are 60 instructions in the main program, and 40 instructions in each of three subprogram iterations. The following table describes the sequence of events for a complete sample interval:

	PC value	SPC value	Timing signals
main program	0-59	0	
	58	0	JSR
1st subprogram	60-99	1-40	
	98	39	JSR
2nd subprogram	100-139	1-40	
	138	39	JSR
3rd subprogram	140-179	1-40	
	178	39	JSR, RET
idle interval	180-199	0	
	198	0	RESET

It is clear from the above table that the ROM must be programmed so that JSR is asserted whenever PC = 58 or SPC = 39, and that RET must be asserted whenever PC = 178. These signals are readily generated if the split control ROM structure of Fig.4 is employed. In this arrangement, the and-plane (decoder) of the ROM is split, with the upper half addressed by the SPC and the lower half addressed by the PC. The or-plane (core) of the ROM is a single section with bitlines from both halves connected together. This results in a given ROM output being asserted if an instruction in either half of the ROM is programmed to assert the output. Thus, in our example, location 39 in the upper-half ROM and location 178 in the lower-half ROM both assert the JSR signal.

The lower half-ROM in the example only need contain locations 0-59, 178 and 198 (62 words). The upper half-ROM only contains locations 1-40 (40 words). Thus the entire ROM contains 102 words. However, additional locations may be included in the lower-half ROM if they are needed to generate external timing strobes that happen to fall outside the main program, but still do not recur periodically with the subprogram. The dual program counter, split-ROM arrangement makes it possible to generate such strobes, which would not be the case with a more traditional single program counter approach.

Both PC and SPC provide complementary, buffered outputs which drive the and-planes directly. The PC contains an additional output EOS (end-of-sample) which is used for timing purposes in the AAU and in the host interface. The EOS signal is always asserted during the last clock cycle of the sample interval (the cycle following RESET).

2.2 Address Arithmetic Unit

In many traditional computer architectures, address arithmetic and data operations are performed in the same arithmetic element. This can result in congestion and inefficient data flow. In signal processing applications, the types of address arithmetic encountered are less varied and may be categorized and implemented by a special functional unit. This functional unit can then operate in parallel with the arithmetic unit that processes the signal data, resulting in increased throughput. The AAU (address arithmetic unit) macrocell serves this purpose.

The AAU is included only if indexed adressing is used. The alternative is direct addressing, where the address offset field of the control ROM is used as the effective address for the RAM.

The AAU contains one or both of the counters IX and IY. Indexing by either or both of these counters is possible. When indexed addressing is used, the index counter is added to the address offset to give the effective address.

The IX counter counts the subroutine iterations. IX equals -1 during the main program, 0 during the first subroutine iteration, 1 during the second iteration, and so forth.

IX indexing is used if identical operations must be performed on a fixed number of different data sets. The operation is coded into the subprogram of a processor. The subprogram references IX-indexed arrays in the RAM data memory. Indexing by the IX counter allows the subprogram to operate on elements of an array in data memory, accessing a different array element each time the subprogram iterates.

The IY counter has two modes: pointer mode and counter mode. In pointer mode, the counter serves as a register which may be loaded from the data bus of the arithmetic unit. This allows arbitrary address arithmetic to be performed in a moderately efficient fashion. In counter mode, IY is a sample counter with a fixed modulus.

Counter mode indexing by IY allows the main program to operate on elements of an array in data memory, accessing different elements from sample to sample. This permits multiplexed processing of a group of decimated signals.

Pointer mode addressing, on the other hand, is useful for table look-up operations. A good example is a controlled oscillator. A look-up table containing one quadrant of a sinusoid is stored in RAM. The address sequence for this lookup table is computed in the arithmetic unit, to give the desired phase and frequency. The computed address is then transferred to the AAU, where it is used to index into the look-up table.

The disadvantage of pointer-mode addressing is that it no longer preserves the separation of address and signal data paths. It therefore is less efficient than the IX and counter-mode IY indexing modes described above. Fortunately, pointer-mode addressing is needed only in a small number of applications.

Fig.5.a shows a block diagram for a fully configured counter mode AAU, while Fig.5.b shows a fully configured pointer mode AAU. Both types of IY indexing are not available in a single AAU.

One final feature of the AAU, not illustrated in Figs. 5.a and 5.b, is a provision for providing outputs which test the IX and IY counters. Any number of test outputs may be provided. The test outputs

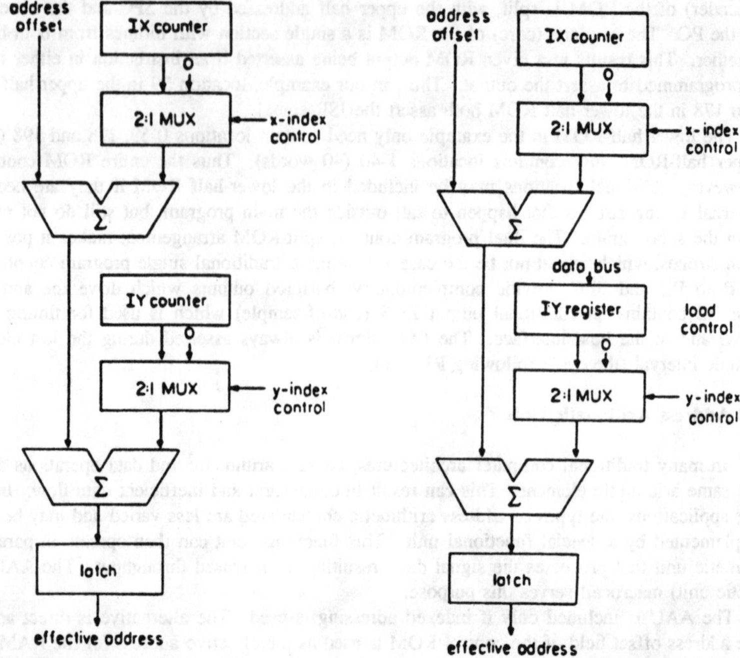

Fig. 5a) AAU in counter mode configuration   5b) AAU in pointer mode

may test either individual bits of a counter, or test for equality between a counter and a constant. These test outputs are used as input to the FSM, and provide a useful hook into the address information. For example, it may be necessary to perform a given operation only during a particular subprogram iteration. By testing IX it is possible to determine whether the current iteration is the one in question, and condition an operation on this information.

2.3 Processor Data Path

2.3.1 Data Path Organization

The data path architecture is of paramount importance in any computational system. As a general rule, the most efficient architectures are highly specific to particular applications. However, the set of target applications for the macrocell system can reasonably be implemented using the same data path architecture. The word length (data path width) may be specified to suit the requirements of the particular application. A bit-slice organization for the arithmetic unit allows this parameterization. The size of the data memory, and the I/O circuits that interface the data path to other processors, may be configured as well.

The general philosophy of the arithmetic unit design was that of simplicity. The number of busses is kept to a minimum, and registers are included only when needed to allow pipelining. This approach results in a compact computational unit whose arithmetic elements can be heavily utilized.

The AUIO macrocell contains the arithmetic unit and the processor I/O section. The processor I/O section is described in Section 2.5. Together with the RAM (data memory) macrocell, the arithmetic unit forms the main processor data path where all signal processing is performed. Fig.6 shows the AUIO and RAM macrocells in block diagram form.

A large number of control signals enters the control section of the arithmetic unit. These signals are decoded, buffered, and transmitted across the bit-slice array.

In addition to control signals, two serial data inputs (COEF1 and COEF2), one serial data output

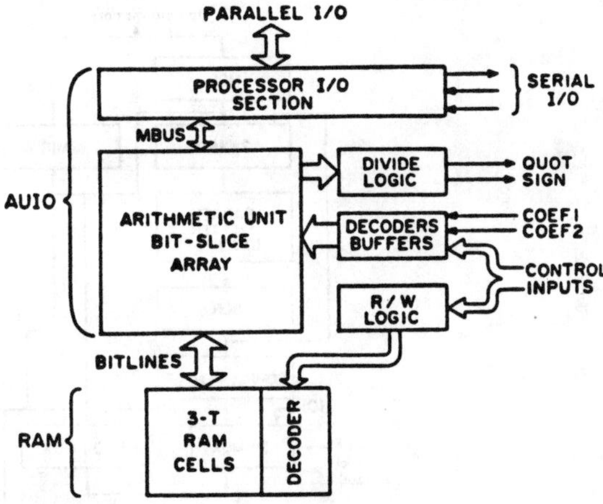

Fig. 6) AUIO and RAM macrocells

(QUOT), and the SIGN output connect to the arithmetic unit control section.

2.3.2 Arithmetic Unit

The processor arithmetic unit (Fig.7) consists of a barrel shifter of depth eight; a complementer; a saturating adder; three master-slave registers (MOR, SOR and ACC); and a transparent latch (MIR). (The value loaded into a master-slave register on one cycle appears at that register's outputs the following cycle, whereas the effect of loading the transparent latch is immediate.) This processor may be microprogrammed for all common signal processing functions. There is a single bus (the "MBUS") through which I/O operations are performed.

All data is in two's complement format.

Every clock cycle, a new set of control signals is transmitted from the control sequencer to the control section of the arithmetic unit. The following describes the effect of these control signals.

The MOR (memory output register) is loaded each cycle with the value on the data memory bitlines. This will be either the result of a read operation, or the inverted write data.

A multiplexer at the input to the barrel shifter selects either the output of the MOR or the output of the SOR.

The barrel shifter itself performs an arithmetic right shift of anywhere from zero to seven bits. Three control signals are decoded to control the amount of shift.

The SOR register is always loaded with the output of the barrel shifter.

A number of controls are provided for the A and B inputs of the saturating adder. The A inputs may be set to zero, or to the true, complement, or absolute value of the SOR register. In addition, the A inputs may be gated (i.e., bitwise and'ed) with COEF1, COEF2 or the inverses of COEF1 or COEF2. This gating allows the microcoding of two's complement multiplies, described in the following section. The B inputs may be set to zero, MOR, or ACC.

The ACC (accumulator) has a single control option, "accumulate if positive", used mainly for microcoding of divisions. When this option is enabled, ACC is loaded with the adder output only if the adder output is non-negative. When the accumulate if positive option is not enabled, ACC is always loaded with the adder outputs.

A single bus (the MBUS) is provided. The MBUS is used for I/O. If an input operation is in progress, the processor I/O section will enable the input data onto the MBUS. At all other times, the ACC will be enabled onto the MBUS.

The transparent latch MIR always takes its inputs from the MBUS. A control is provided to either load or hold the value in the latch.

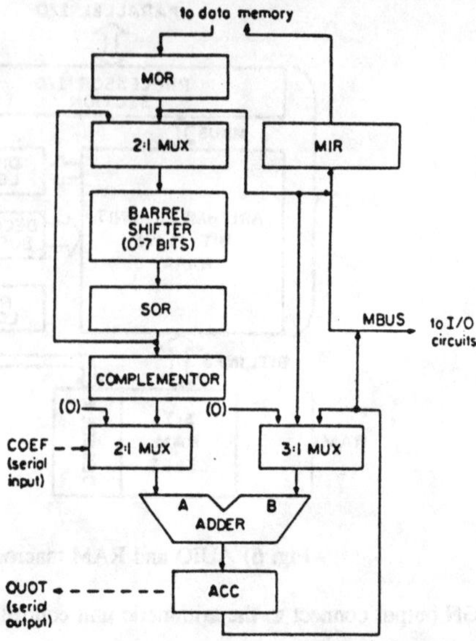

Fig. 7) Arithmetic unit

The COEF1 and COEF2 inputs are used for variable coefficient multiplies and the QUOT output is used for division. The SIGN output is used as input to the FSM macrocell as described in Section 2.4. The two inputs COEF1 and COEF2 are identical in function; providing two such inputs allows transmission of coefficients from two distinct sources (e.g., two procesors) to the same destination.

The pipeline organization of the arithmetic unit allows concurrent memory access, barrel shift and add/accumulate operations. In addition, pipelining around the MBUS prevents bus settling time from impacting throughput. A fast ripple carry adder is used. This allows a flexible bit-slice organization, but gives add times comparable to more complex look-ahead designs.

Other organizations for single-accumulator arithmetic units are possible. One popular arrangement is to put the barrel shifter *after* the accumulator rather than before. In this case the least significant partial products are accumulated first during a mutiply sequence, and the binary weight of the value in the accumulator increases during the course of the multiply. This approach has the advantage that the truncation errors are not as severe, gaining a few bits of precision for the same wordlength. There are however two disadvantages: (1) an additional register is needed to store intermediate results in a multiply-accumulate sequence; (2) an external coefficient must be provided LSB-first. This is incompatible with the MSB-first format which is needed to implement long-division operations (described in the following section).

2.3.3 Multiply and Divide Operations

Essential to any signal processor architecture is the ability to perform multiply-accumulate operations. Two cases are handled separately: multiplying signal data by a variable coefficient external to the processor; and multiplying signal data by a constant.

In the case of variable coefficients, the coefficient is available externally in a bit-serial, MSB first, fractional two's complement format. Each processor arithmetic unit allows two such serial coefficient inputs. In fractional two's complement form, the coefficient k can be expressed as follows:

$$k = -k_{n-1} + \frac{k_{n-2}}{2} + \frac{k_{n-3}}{4} + ...$$

To multiply a signal A by the coefficient k, the following equation is used:

$$kA = -k_{n-1}A + k_{n-2}\frac{A}{2} + k_{n-3}\frac{A}{4} + ...$$

Each term on the right-hand side is a partial product, which is non-zero depending on one of the bits of the coefficient k. To perform these variable-coefficient multiplies, the external coefficient must be multiplexed into the control path for the arithmetic unit. This allows the partial products to be summed sequentially, one per clock cycle.

This method of multiplication is particularly convenient in conjunction with the bit-serial approach to interprocessor communication. Two coefficient inputs (COEF1 and COEF2) are provided for each processor arithmetic unit, along with the control circuitry to enable these coefficients under microprogram control. A coefficient may originate from the same processor that performs the multiply, or it may originate from a different processor and be transmitted over a bit-serial path.

As an example, suppose that one wishes to multiply a variable A in data memory by the 8-bit coefficient k. The following sequence is performed, each line representing one clock cycle:

Read A from RAM into the MOR
Load the SOR from the MOR
If $k_7=1$ (negative) load -SOR into ACC; shift SOR right one bit
If $k_6=1$ add SOR to ACC; shift SOR right one bit
If $k_5=1$ add SOR to ACC; shift SOR right one bit
If $k_4=1$ add SOR to ACC; shift SOR right one bit
If $k_3=1$ add SOR to ACC; shift SOR right one bit
If $k_2=1$ add SOR to ACC; shift SOR right one bit
If $k_1=1$ add SOR to ACC; shift SOR right one bit
If $k_0=1$ add SOR to ACC

If one is performing a sequence of multiplies or multiply/accumulate operations, the 8-bit multiplies can be performed at the rate of one every eight clock cycles.

In the case of a fixed coefficient, it is not necessary to input the coefficient into the processor through the COEF inputs. Instead, a signed-digit representation of the coefficient is embedded in the control stream. This is done by specifying a sequence of shift depths and sign values. Thus, any coefficient g has a signed-digit representation:

$$g = (-1)^{s_0}2^{e_0} + (-1)^{s_1}2^{e_1} + ...$$

Any binary number has a unique such representation with the minimum number of digits. This is known as the canonical signed-digit (CSD) representation [19]. The importance of the CSD representation is that is minimizes the number of clock cycles required to perform a multiply by a fixed coefficient. The multiply is performed by sequencing the barrel shifter and complementer so as to present the desired terms to the accumulator.

Suppose, for example, that a variable A in data memory needs to be multiplied by the constant $g = .10111100$, which may be recoded as

$$2^{-1} + 2^{-2} - 2^{-6}$$

The following sequence performs the multiply:

read A from memory into MOR
load SOR with MOR shifted right one bit
load ACC with SOR; shift SOR right one bit
add SOR to ACC; shift SOR right four bits

add (-SOR) to ACC

The product gA is now in the accumulator.

In both the above cases, a sequence of multiplies may be performed, with the result of each being accumulated. This is an efficient feature of the single-accumulator architecture. The same accumulator is used for adding partial products, and for accumulating the results of several multiplies.

Although not as prevalent as mutiplies in signal processing algorithms, divide operations are also needed in some cases. Usually, a divide is used for normalization.

A common situation requires performing a two-quadrant divide N/D, where $D > |N|$. This will give a quotient between minus one and one. It is desirable to create a two's complement signed quotient. In order to achieve this, first the absolute value of the numerator N is loaded into the accumulator, and the sign bit is saved. The sign bit is immediately outputted as the sign bit of the bit-serial quotient. Then, an unsigned (one quadrant) long division of $|N|/D$ is performed. The resulting quotient bits are exclusive-or'ed with the saved sign bit, giving a two's complement result.

In order to do the long division, the accumulate-if-positive control option is used. This loads the accumulator only if the result of the accumulator operation is not-negative. The inverted sign bit of the result of the accumulator operation is used as the quotient bit. With $|N|$ in ACC, $D/2$ in SOR, and the accumulate-if-positive option enabled, the following operation is repeated:

load ACC - SOR into ACC only if positive ; shift SOR right one bit

Each time this operation is performed, another bit of the quotient is computed, that bit being the sign bit (inverted) of ACC-SOR.

Hardware for performing the absolute value, accumulate-if-positive, sign-bit storage and exclusive-oring of the quotient bit is contained in the arithmetic unit. This allows two-quadrant divides to be performed, with the resulting quotient being in the same bit-serial, two's complement, MSB-first format that is required for coefficients in variable-coefficient multiplies.

2.3.4 Data Memory

The data memory (RAM) macrocell is used for storage of temporary results and state variables of the signal processing algorithm. As such, a given memory location is typically written at least once each sample interval. This generally eliminates the need to explicitly refresh memory locations; thus no hardware refresh is provided.

Read-only and read-write memory locations are intermixed in the RAM. The read-only locations allow introduction of constants into the computation. Note that in traditional architectures, constant introduction is accomplished by the use of immediate operands, requiring a gateway between control and data paths. The fact that the RAM is a macrocell configured for a dedicated application allows a more direct and efficient method of introducing constants.

Read-only locations also allow the implementation of look-up tables, although the current memory size limit of 64 words makes larger look-up tables impractical.

2.4 Finite State Machine

In a traditional computer architecture, the only conditional operations available are conditional branch operations. Usually, condition code flags are set by testing one or more CPU registers. Then, a conditional branch operation may be executed. The branch takes place only if the specified condition is satisfied by the flags.

Another characteristic of the traditional approach is that all decision-making logic is performed in the same ALU that processes numeric data. This is convenient for general-purpose CPU's since the type of decision making is not known in advance, and the hardware cannot be tailored to the application.

A radically different approach to decision-making was taken in the macrocell-based design system. First, a conditional write operation is made available to the programmer, as opposed to the conditional branch. Second, Boolean data is processed in a finite state machine (FSM), separate from and concurrent with the processing of numeric data in the arithmetic unit.

The use of conditional assignments as opposed to conditional branches is not new. A classic result from the theory of programming languages [20] states the the conditional assignment statement is equally as general as the conditional branch, although not necessarily as efficient. Intuitively, a conditional assignment acts as follows:

if (condition is satisfied) then (assign a value to a variable)

while a conditional branch does the following:

if (condition is satisfied) then
    (branch to different stretch of code)

The conditional branch has two major disadvantages if used in signal processors such as those considered here. First, if conditional branches are used the execution time of the program is data-dependent. This is undesirable in a real-time sampled system, since the program must execute in its entirety each fixed-length sample interval. Also, the interprocessor communication (Section 2.7) is dependent upon the programs of the processors being synchronized with one another. Second, the control flow is also data-dependent. This makes it less practical and less efficient to pipeline control sequence generation with data operations (i.e., overlapped fetch/execute).

The conditional write approach has one major drawback. It is much less efficient than the conditional branch if the test selects between two long, substantially different pieces of code. Fortunately, this situation seldom arises in signal processor design.

An important feature of FSM usage in the macrocell approach is that the size and contents of the FSM is tailored to the application. Thus, the state variables of the FSM can be selected to correspond to conditions present in the signal processing algorithm. A field of the horizontal control word is used to modify the state of the FSM; this modification is also made algorithm-specific.

The design file input allows the design of the FSM to be easily specified. This is a very powerful capability, since "FSM instructions" may be specified and then later used in the microprogram for the processor. In effect, the user is able to design part of his instruction set prior to programming the processor.

The FSM is implemented as a PLA (programmed logic array) with an output register for the state, and feedback from the output register to the PLA inputs. Inputs to the FSM come from three sources:

(1) The sign bit of the ACC register in the AUIO macrocell
(2) "test" outputs of the AAU macrocell
(3) A field of the control word, used to modify the FSM.

Outputs of the FSM go to two possible places:

(1) the CC (condition code) output goes to the AUIO macrocell, and must be asserted to enable conditional write instructions
(2) the MOF (middle of frame) and EOF (end of frame) outputs, used to initiate host input and host output operations. These signals go to the Host Interface.

2.5 Input-Output and Communications

2.5.1 Off-chip I/O

In designing any system, the issue of external interface is always important. In designing digital IC's, how the circuit will integrate into larger systems is a matter of concern. Circuits designed with the system described here interface in two distinct ways (Fig.8).

Signal I/O refers to transfers of data at the sample rate over the signal I/O bus. Here, the timing of data transfers is synchronous with the signal processing IC's clock. The transfers are controlled by strobes generated by the signal processing IC. Since the clock rate is considerably higher than the sample rate, transfer of a number of signals is possible each sample interval. These signals are the sampled-data inputs and outputs of the IC.

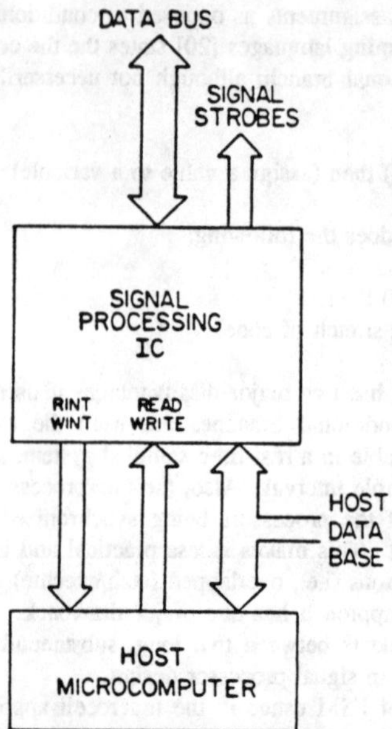

Fig. 8) Signal and host interfaces

Host I/O refers to the transfer of data between the signal processing IC and a host system. The host is typically a microprocessor, with the signal processing IC behaving as a peripheral chip. In general, the host is incapable of sustained data transfers at the signal sample rate, since it has other functions to perform as well. Instead, a slower frame rate is defined. The frame rate is determined by the signal processing IC, which interrupts the host with once-per-frame signals. The host responds to these interrupts by reading and writing data in a buffer (the host interface) contained in the signal processing IC. These data transfers occur asynchronously with the signal processing IC's clock. This simple handshaking arrangement guarantees the validity of the data transfer.

Frame interrupts occur twice per frame. The write interrupt (WINT*) occurs close to the end of frame. The host responds to WINT* by writing into the host input interface. This must be performed before the end of the frame. The new data is available to the signal processor after the beginning of the new frame.

The read interrupt (RINT*) occurs just after the beginning of a frame. The host responds by reading from the host output interface. The data read was collected during the last sample of the previous frame.

The frame discipline described above was designed to support signal analysis and synthesis operations by allowing a frame rate to be defined for the purposes of host I/O. Transfer of sample-rate signals is done on a separate bus to prevent congestion on the host processor's bus. The resulting I/O structure is not completely general, but goes a long way towards integrating the signal processor IC into a system constructed from standard components.

2.5.2 Interprocessor Communication

Internally, the signal processing IC contains one or more processors as shown in Fig. 9. If more than one processor is used the chip designer needs a way of sending signals between processors, as well as on- and off-chip. This is done by providing for serial communication paths among the processors, and between the processors and the host interface. The number and function of these paths can be selected to suit the application. Each path has a source and a destination.

The possible sources are the MBUS of a processor; the QUOT output of a processor; and the host interface. The possible destinations are the MBUS of a processor; the COEF1 or COEF2 inputs of a

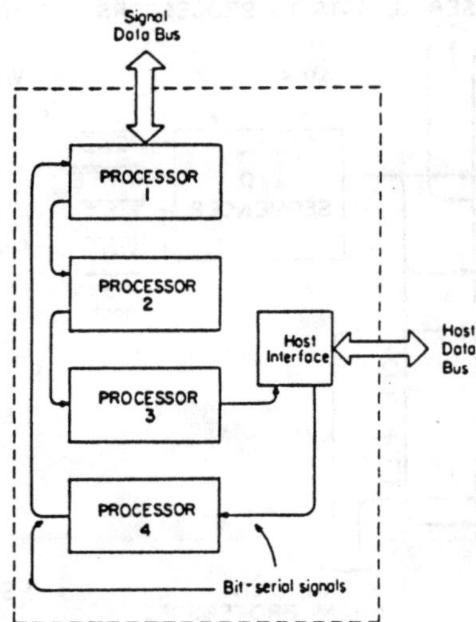

Fig. 9) Interprocessor communication paths

processor; and the host interface.

It is not required that the source and destination be different processors. Although it would be pointless to have the MBUS of a single processor as both source and destination, it is often useful to transmit from the MBUS to a COEF1 or COEF2 input on the same processor. This allows two variables local to the processor to be multiplied together. Similarly, one might wish to transmit from the QUOT output to the MBUS to obtain locally the result of a division.

If the source of a serial communication path is the MBUS, a parallel-serial converter must be provided at the source end since the MBUS data is in bit-parallel form. A separate parallel-serial converter is provided for each communication path. Similarly, a serial-parallel converter is provided at the destination end of a serial communication path if the destination is the MBUS.

The serial-parallel and parallel-serial converters are included in the processor I/O section as needed. A separate option provides for a parallel port. This port is used for signal I/O (discussed above). Only one processor per IC may perform signal IO. The parallel port is also used to provide a pathway from the MBUS to the AAU macrocell if pointer-mode addressing is being used. If neither signal I/O or pointer-mode addressing is being performed, the parallel port hardware is not included in the processor IO section.

Provision is made for including a temporary storage latch in the parallel-serial or serial-parallel converters. This is necessary to decouple the timing of the source and destination processors. Thus, the destination processor may access the data being transmitted over the communication path at any time. The data retrieved will be that which was most recently generated by the source processor (taking into account the delay of clocking the bit-serial data). This allows the microprogram for a processor to perform an input or output operation at any point in time, independently of the other processors.

2.5.3 Host Interface

The Host Interface consists of the I/O sequencer and the HI (host interface) macrocell. This arrangement is shown in Fig. 10a. The HI macrocell consists of FIFO buffers for data being transmitted to or from the host. Thus, there are two major pieces to HI, the host output interface and the host input interface. If only host input (output) is used, the host output (input) interface is not included.

The I/O sequencer is a small finite-state machine which is clocked by the EOS (end of sample) signal. Thus the sequencer never changes states in the middle of a sample. The I/O sequencer controls the reading and writing of data in the FIFO's and determines the sequence of events during the frame. There are four states as shown in Fig. 10b.

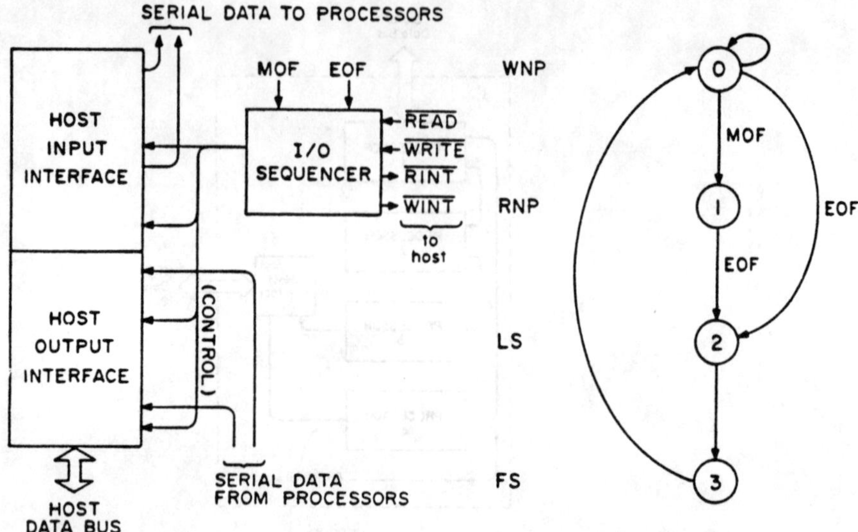

Fig. 10a) Host interface (HI) hardware   10b) State diagram for HI

There are four states:

0	WNP	write not in progress
1	RNP	read not in progress
2	LS	last sample of frame
3	FS	first sample in frame

The signals MOF and EOF (Section 2.4) drive the state sequencer through its sequence. Ordinarily, the frame begins in state 3, progresses through states 0 and 1, and ends in state 2.

The role of the host output interface is to collect the various bit-serial signals coming in from the processors; convert them into parallel form; and store them in a FIFO. Once this is done, the FIFO may be read by the host. The timing is arranged so that the FIFO is loaded with data produced by the processors during the last sample of the frame. (Some of the signals may not arrive until the beginning of the first sample of the following frame due to clocking delays of the serial paths.) At the end of the first sample the RINT* interrupt is asserted and the host responds by reading the data out of the FIFO.

It is assumed that the host has read the complete contents of the FIFO before a nominal two millisecond refresh interval has elapsed. Therefore no refreshing is necessary for the host output interface.

The host input interface is somewhat more complex than the host output interface for two reasons. First, the frame consists of many samples; the host input interface must transmit data to the processors on each sample. The host output interface need only receive data from the processors on the final sample of the frame. Second, the frame may exceed a refresh interval in length. Consequently, the data stored in the host input interface must be refreshed.

To accomodate these added complexities, a double buffered FIFO arangement is used in the host input interface. The MOF flag causes an immediate WINT* interrupt and the sequencer enters state 1. The host proceeds to write data into the "master" section of the double-buffered FIFO. At correct moments near the frame, the data are transfered from the "master" to the "slave" FIFO. The timing is arranged so that the processors receive the new data starting with the first sample of the new frame.

The "master" section of the FIFO is not refreshed; therefore the MOF signal must be asserted less than a refresh interval before the end of the frame, but long enough before that the host is always able to fill the "master" FIFO. The "slave" FIFO is refreshed throughout the frame.

In both the input and output interfaces a split-FIFO arrangement was used. The arrangement for the host output interface is illustrated in Fig.11.

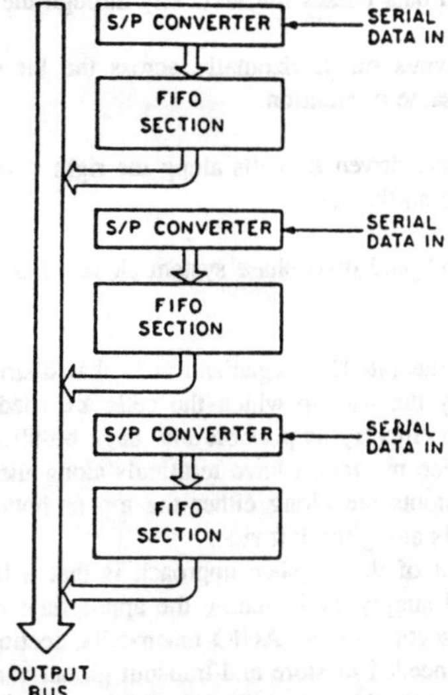

Fig. 11) Split-FIFO arrangement for the host output interface

The idea here is that each serial communication path feeds its own FIFO section. A FIFO section may contain one or more words of storage. Inclusion of more than one word of storage in a single FIFO section is allowed for two reasons: (1) The host interface wordlength may be less that the wordlength of the data coming from the processor, so that more than one word is needed; and (2) an array of data, indexed by the IX or IY counter of the source processor, can be placed in the FIFO section.

Operation of the split FIFO is as follows. For the purposes of writing into the FIFO, the sections operate independently. Each processor sees a separate FIFO which it can clear and write into until it is full. For the purposes of reading out of the FIFO the sections are joined and act as a single, large FIFO. For reading, the FIFO is initialized at the same time the RINT* interrupt is asserted. The host may then read its entire contents one word at a time.

The host input interface is split in an analogous fashion. The split FIFO approach provides an interesting solution to the problem of merging together several unrelated and unsynchronized streams of data and formatting them into a single block.

III CELL LIBRARY

3.1 Characteristics of Library Cells

A major part of the macrocell based design system is the cell library. The library consists of 160 cells, each of which was designed by hand and digitized using the graphics editor *Kic* [21]. This may seem at first a rather large number, but many of the cells are either very simple (containing just a few geometries) or are slightly altered versions of other cells in the library.

The cells are designed specifically for the role of being formed into the macrocells described in the preceding sections. There are several ways in which this requirement affects the design of the library cells.

Many of the macrocells (specifically: AUIO, PC, SPC, AAU, and HI) are organized in a bit-slice fashion. These macrocells have several common characteristics:

(1) A GND (ground) line busses along the left side of the macrocell. This is formed from library cells whose names typically end in the suffix ".gnd".

(2) Each bit of the bit slice has a fixed pitch (for the 3u NMOS cell library, the pitch is 66u).

(3) Generally, polysilicon data busses run vertically through the bit slices.

(4) Aluminum control wires run horizontally across the bit slices. Aluminum wires for power, ground and clocks have the same orientation.

(5) These control wires are driven by cells along the right side of the macrocell. These cells have names typically ending in the suffix ".ctl".

(6) Vdd (power), phi1 and phi2 (two-phase system clocks) bus along the right side of the macrocell, through the ".ctl" cells.

A convenient feature of the bit-slice organization is that distribution of power, ground, clock and control lines is facilitated by the way in which the cells are tiled together. Thus, within limits, it is desirable to collapse as much circuitry as possible into large bit-slice macrocells.

Control inputs to a bit-slice macrocell have terminals along the extreme right-hand side of the macrocell. Data inputs and outputs are along either the top or bottom, connecting to the individual bit slices. There are no terminals along the left side.

Another convenient aspect of the bit-slice approach is that a bit slice array is easily configurable. Word lengths may be varied simply by including the appropriate number of slices. In the case of the HI (host interface) and I/O section of the AUIO macrocells, sections of the bit-slice are stacked vertically to implement circuitry needed to store and transmit global variables. The number and function of the global variables is completely dependent on the application and its design file description. The fact that the AUIO and HI macrocells may be built up from the design file declarations of global variables is one of the more powerful features of the macrocell system.

The two-phase clock (phi1 and phi2) is distributed globally to all macrocells. At this time, off-chip clock drivers are assumed. This clock is symmetrical and non-overlapping. A timing convention for most signals running between macrocells is the following. At the source macrocell, the signal is latched by phi1. At the receiving end the signal is assumed to be valid during phi2. This convention assures that a fraction of the clock cycle is available for signal propagation between macrocells. The exceptions to this convention are: PC and SPC outputs, which feed ROM address inputs; RAM bitlines and control signals which connect to AUIO; and signals going to and from bonding pad circuits.

The cell library is designed to function at the target clock frequency (5 MHz in this case) over all allowable parameterizations of the macrocells. This requires design for "worst case" in many instances. Improved speed/power performance would be achieved if the individual cells could be parameterized (with regards to transistor sizes, for example).

Each cell is a UNIX file [22]. The cells are grouped in a number of UNIX directories, one for each macrocell. Also contained in this directory, and used by the software system, are the files .KIC [23], descriptors , and .techno which are described later.

3.2 Technology Parameters for Macrocells

One goal of the software package is that it be applicable to different semiconductor processes without substantial modification. Clearly, the cell library itself must be designed using the rules and characteristics of the particular process. However, by suitable parameterization of dimensional information, the software package is technology-independent.

A set of technology parameters is provided that describes the width and position of power, ground and clock conductors with respect to the bounding boxes of the macrocells. (Additional technology parameters relating to signal wiring are described in Section 4.3.3.) Dimensions are given below in units of lambda (1.5 micron) for the 3 micron NMOS library.

For all macrocells except RAM and the bonding pad groups, the following parameters apply:

(1) width of Vdd conductor -- 20
(2) width of ground conductor -- 20
(3) width of clock conductors -- 4

(4) spacing from Vdd conductor to phi2 -- 3
(5) spacing from phi2 to phi1 -- 8
(6) spacing from phi1 to edge of bounding box -- 7

For the RAM macrocell, the following are used:

(2) width of ground conductor -- 20
(7) width of Vdd conductor -- 9

For the bonding pad groups, the following apply:

(8) Width of Vdd and ground conductors -- 50
(9) Spacing between Vdd and ground conductors -- 117

IV MACROCELL DESIGN SYSTEM SOFTWARE

4.1 The Design File

A primary goal in creating the format for the design file is that it be easy to prepare and modify. Otherwise, the advantages of automatic layout generation would be partially cancelled. The design file is the main user interface to the system since it is the input to both the emulator and the silicon compiler. In constructing this interface it is helpful to consider the user of the system, and the purpose for which it is used.

Although practical signal processors operate in real-time, much algorithmic research involves non-real-time simulations. There is, however, a gap between that which may be simulated and that which is efficient to implement on a real time architecture. Bridging this gap is a major goal of the system. A designer who knows both the requirements of the algorithm and the capabilities of the hardware is in a position to understand the trade-offs involved and therefore optimize his design.

The design system addresses this situation by providing an interface that allows both simulation and hardware specification. The level of expertise required to use the system is consistent with that expected for a typical signal processing researcher. No prior knowledge of integrated circuit design is necessary.

The design file format may be viewed as a programming language for multiprocessor IC's with a specific architecture. Programming languages generally become easy to use if they involve abstraction of complex concepts. Several examples of abstraction in the design file are: symbolic names for processors, variables, constants, and instructions; a register-transfer description of the data-path; and a simple way of specifying interprocessor and off-chip communications. In each of these cases, much of the hardware complexity is hidden from the user. Yet very little efficiency is sacrificed considering the amount of simplicity gained.

The design file begins by declaring global variables, and specifying which of these are used for off-chip communications over the host or signal interfaces. Globals which are not used for off-chip communications are presumed to be used for interprocessor communications; however, the source and destination processors are not specified explicitly. Instead, this information is implicit in the microcode for the processors which reference these globals.

The following is an example of the global declaration and I/O specification part of a design file.

```
.global
begin

    /* host input globals */
    ks[10]<8>;      /* reflection coefficients */
    pitch<8>;       /* pitch period; zero means unvoiced */
    amplitude<16>;  /* excitation amplitude */
```

562

```
        /* signal output global */
        output<12>;      /* synthetic speech output */

end


.io <8> /* I/O description */
begin
    infile : ks, pitch, amplitude : host_out;
    outfile : output : signal_out;
end
```

(Comments begin and end with the sequences "/*" and "*/".)

The keyword ".global" introduces global declarations. The declaration "ks[10]<8>;" declares an array of ten eight-bit global variables. Arrays of global variables may be used for host input and output, and are indexed by the IY or IX counters of the destination or source processor. Array globals are not permitted for signal I/O or interprocessor communications. The declaration "amplitude<16>" declares a 16-bit scalar (non-array) global variable.

The sequence ".io <8>" introduces the I/O declarations, and indicates that an eight-bit host interface is to be used. (The other option is 16 bits, covering the two most common microprocessor bus widths.) "infile" and "outfile" are UNIX file names to be used for simulations. These are ASCII text files, so that test cases may be easily prepared.

Following these global and I/O specifications is a definition of each of the processors. Each processor section begins by specifying the processor wordlength, and declaring local variables and constants. An optional section specifying a FSM is next, followed by the symbolic microcode for the main program and subprogram, if used.

The processor wordlength must be an even number no larger than 32. Often, this is a critical parameter affecting the dynamic range and noise performance of the IC. Through simulation, the designer may optimize this value.

Local variables and constants all have the same wordlength as the processor itself. The variables may be either scalars or arrays. An array variable is usually used in conjunction with indexing by the IY or IX counters.

The following is an example of the initial section of a processor declaration, showing the local and constant declarations.

```
.processor : filter <18>
begin

    .local              /* local variables for processor */
    begin
    _pitch, _amplitude; /* local copies of pitch, amplitude */
    c[10];              /* lattice filter state variables */
    end

    .constant           /* local constants for processor */
    begin
    ONE = 1;
    FRAMELENGTH = 180;  /* gives frame of 180 samples (22.5 ms) */
    end

    /* ... (fsm declaration and microcode go here) ... */

end  /* end of processor "filter" */
```

The first line above declares an 18-bit processor with symbolic name "filter". The symbolic processor name is useful when running the emulator in debug mode.

A section may be included to define an FSM. This is a powerful device, since symbolic instructions are defined which modify the FSM state. These instructions may then be used in the microcode for the main program and subprogram.

The FSM is described by giving a symbolic name to each state variable, and providing equations which specify the state transitions. These Boolean equations are translated by the software into a truth table format suitable for programming the PLA-based FSM.

One bit of state for the FSM is usually called "cc", for "condition code". It is this bit which must be set (true) if a conditional write operation is to be enabled. Also, one of the FSM inputs is called called "sign", and refers to the sign bit of the accumulator. Thus, recalling that the contents of the accumulator is in two's-complement form, if sign is zero, the accumulator is non-negative.

Simple examples of FSM instructions which may be defined are "SET" and "AND" instructions. The "SET" instruction sets "cc" for non-negative accumulator. The "AND" instruction performs a logical and of "cc" with the condition that the accumulator is non-negative. An FSM declaration defining just these two instructions would be the following:

```
.fsm
begin
SET : cc = !sign;
AND : cc = !sign & cc;
end
```

More complex FSM instruction sets may be defined for specific applications.

If the host I/O is used, one of the processors must be configured with an FSM which defines the MOF and EOF outputs defined in Section 2.5.3.

Following the variable, constant, and FSM declarations is the microcode for the processor. Because of the pipeline nature of the data path, the instruction set is divided into groups, with one instruction per group allowed in a single line of microcode. Each line of microcode corresponds to a single clock cycle.

These are the instruction groups:

(1) Memory group -- reads and writes, which may be indexed or conditional.
(2) The SOR group, which controls the contents of the SOR register.
(3) The accumulator group, which controls the contents of the ACC register.
(4) The "le" (latch enable) instruction, controlling the MIR latch.
(5) The "mbus" group, which enable signals onto the processor's "mbus" (by default, mbus=acc);
(6) The user-defined FSM group.
(7) The "aip" (accumulate-if-positive) instruction
(8) Instructions setting up a correspondence between global variables and the coefficient inputs, quotient outputs, or "mbus"; or assigning the mbus to IY (pointer mode only). (several non-conflicting instructions from this group may be used in the same line.)

Thus, a single line of microcode may contain many of the above instruction groups, as in the following:

```
rx(C[0]), sor:=mor>4, acc:=acc+sor, mbus=pitch, le, SET;
```

There are several thing happening here:

An IX-indexed read from the local array C[] -- group (1);
Assigning MOR, shifted right 4 bits, to SOR -- group (2);
Adding SOR to the accumulator -- group (3)
Enabling the global variable "pitch" onto the mbus -- group (5);
Enabling the MIR latch -- group (4);
Executing the user-defined FSM instruction "SET" -- group (6);

As with any assembler-level programming, familiarity with the processor architecture is needed in order to write efficient code.

In the design file, the main program may be begin with one of the following three types of sequences:

```
.main_pr
.main_pr <8>
.main_pr <*>
```

In the first form, no IY indexing is used by the processor. In the second form, the IY indexing is used in counter mode with modulus (decimation ratio) eight. In the third form, the IY indexing is used in pointer mode.

Similarly, the subprogram might begin as follows:

```
.sub_pr <10>
```

In this case, there would be ten iterations of the subprogram.

Within the main program or subprogram, commas are used to separate the different instructions within a single line of microcode, and a semicolon terminates the line. The entire main program or subprogram is bracketed by the "begin" and "end" keywords.

In summary, the design file contains declarations for global variables and I/O specifications, followed by a block for each processor. The processor block defines local variables and constants; the FSM; and the microcode for the main program and subprogram.

4.2 The Emulator

The emulator allows simulation of the signal processing IC from the design file. This is done in non-real-time. For a typical design file, on a VAX 11-750 or 68000-based Sun Workstation, the ratio of emulator simulation time to real time might be 500 or 1000.

Upon invoking the emulator, the design file is read in and checked for errors. If there are no fatal errors, an interactive mode is entered. A number of debug commands are available, including:

Tracing a variable;
Printing a variable;
Setting a breakpoint at a point in the source program;
Setting the value of a variable;
Running a simulation for a given number of samples;
Aliasing a debug command to a shorter form.

These debug commands considerably reduce the amount of time needed to prepare a design file for a given application. Also, single-stepping through a program is a very good way to familiarize oneself with the architecture of the data path.

Much care was taken to ensure that the emulation exactly matches the hardware itself. Truncation, round-off, and other arithmetic details are emulated precisely.

The ability to simulate the signal processing IC from the design file, rather than from a description of the circuit layout, is critical to the success of the entire macrocell approach. The reason is that signal processors require exceptionally long test sequences to fully verify performance -- perhaps several million clock cycles, or more. For this reason, verification at the circuit level is essentially impossible. Thus the only practical choice for signal processor design is to verify performance from a higher-level description such as the design file. The layout, generated automatically from the design file, is "correct by construction".

4.3 The Silicon Compiler First Pass

The compiler first pass resembles the emulator in its initial phases; in fact, many subroutines which parse the design file are shared between the two programs. After parsing, the first pass generates an intermediate file, a process which is analogous to intermediate code generation in a compiler for a programming language. Besides assembling the microcode into binary, the macrocells must be specified at a hardware level. This involves extracting hardware parameters such as the following:

> Processor word lengths
> Processor data memory sizes
> Constant programming of data memory words
> Configuration of IX and IY counters in the AAU
> Translation of FSM declaration into and-plane and or-plane code
> Program counter modulus
> Subprogram counter modulus
> Programming of microcode in ROM
> Inclusion of parallel-serial and serial-parallel registers in AUIO
> Constructing the host interface from the host-io declarations

4.4 Intermediate File and Macrocell Assembly

The first pass of the compiler outputs an intermediate file. This intermediate file is a hardware description defining the macrocells and how they are interconnected. The macrocells are defined by describing how the library cells are to be tiled (arrayed) to form the macrocells. The interconnect is defined by providing unique names for the macrocells' terminals, and giving a net list.

Several examples of tiling programs exist [24]. Tiling is a useful method for generating a complex block of circuitry from smaller cells. The two most common organizations for a block of circuitry are the bit-slice organization (useful for arithmetic units and register files) and array-type structures such as RAM's, ROM's and PLA's. The second pass of the compiler employs a tiling routine that can generate either of these structures.

Two features (generally lacking in existing programs) were needed for the tiling routine used here. One is that terminal information must be preserved by the tiling process, so that the assembled macrocells could then be used as input to placement and routing routines. The second is that dimensional information must be encapsulated in an easily modifiable form. This makes it possible to use the software with cell libraries designed in different technologies.

To this end, dimensional information defining the sizes of library cells and their terminal locations is contained in a descriptor file. The output of the first pass contains no dimensional information, and is therefore technology independent. The descriptor file contains an entry for each cell in the cell library. A typical entry might look like this:

```
cell ix.ctl 66 88
right eos 66 25
right inc 66 72
```

The keyword "cell" begins the entry for the cell "ix.ctl". The numbers "66 88" are the tesselation dimensions of the cell. The x-dimension is 66 and the y-dimension 88. Tesselation dimensions specify the distance between origins of cells in adjacent rows and columns of an array. Cells are assumed to contain geometries mostly in the first quadrant, with the origin being at or near the lower left corner of the cell.

Note however that the tesselation dimensions are not the same as the dimensions of the cell's bounding box. Thus when cells are tiled, their bounding boxes may overlap, abutt, or be separated by a gap.

The keyword "right" denotes a terminal on the right edge of the cell. Thus the cell "ix.ctl" has two terminals, one of which is named "eos" and has x and y coordinates 66 and 25.

Given the information in the descriptor file, the macrocells are defined in the hardware description by simple listing the cells in each row of the array. This is done row by row, starting at the bottom of the array. The cells in a given row are listed from left to right. The following is an example of a hardware description for a program counter:

```
begin
counter.gnd, counter.0, counter.0, counter.0, pc.ctl;
pc.0, pc.1, pc.1, pc.1, pc.2;
end .
```

To make more complex examples readable, a shorthand notation is allowed for cells within a row. The following description is equivalent:

```
begin
counter(.gnd, .0, .0, .0), pc.ctl;
pc(.0, .1, .1, .1, .2);
end
```

The tiling routine assembles macrocells from these descriptions. The location of terminals relative to the origin of the macrocell is determined. These terminals are given unique names by appending the terminal name in the descriptor file to the macrocell name and the row and column in which the terminal occurs. For example,

```
pr0_pc[1,5]_reset
```

refers to the terminal "reset" in the first row, fifth column of the macrocell pr0_pc (meaning the program counter for processor number zero).

The net-list portion of the hardware description uses these unique terminal names to specify interconnection. A unique node number is assigned to each set of terminals which are interconnected. This allows net-list verification subsequent to routing using a circuit extractor program such as Mextra.

Power, ground and clock connections are not specified in the same way as the signal terminals. Instead, entry points for power, ground and clock connections bear a standard relationship to the bounding box of the macrocell, and are handled separately during placement and routing.

4.5 Floorplan based Processor Assembly

An important step in the layout generation process is the placement and interconnection of the macrocells. The problem has been treated hierarchically, in conformity with the various levels in the chip architecture. General Place and Route (GP&R) optimization was not employed for a number of reasons. The compilation of the individual processors is a restricted problem : the layout generator 'knows' the basic structure of the processor and the nature of the internal interconnections, GP&R techniques ignore this architecture knowledge and start from scratch. Specialized routing constraints are handled by GP&R with great difficulty (e.g. single level routing for certain critical busses, power, ground and clock routing or performance constraints on critical nets). Finally, usage of GP&R at the lowest level can require hours of computer time, which is not compatible with an iterative design process.

A floorplan driven placement and routing tool has therefore been developed. It takes as input the dimensional data, passed by the tiling process, and a set of floorplan files. The floorplan of a structural unit (processor, host-interface, A/D-convertor, ...) contains:

1) the relative placement of the macrocells and the routing channels;

2) the global routing guidance, which includes the assignment of the routing cables (being a set of nets with the same source and destination) to the different routing channels and the determination of the routing strategy for each of the sectors (riverrouting or channelrouting);

3)Power, ground and clock routing information.

A set of alternative floorplans are provided for each unit, so that the floorplan tool (called Flint) can select the optimum strategy for a particular case. Flint first performs the absolute placement for each of the floorplans, based on graph manipulation techniques and an accurate estimation of the required routing space. At the end of this phase, the crossings of nets and routing channel boarders have been fixed, so that the detailed routing phase is a strictly local process. After a selection of the minimal solution, the detailed routing is performed. Both river and channel routing are performed on a gridless base, which allows for denser routing and which fits with the arbitrary placement of terminals on the macrocell boundaries.

4.6 The Functional Front End System

Although the design file language allows the user to describe his algorithm at a symbolic level without knowledge of the underlying hardware, he still has to be aware of the architectural details of the target architecture. Moreover, programming at the microcode level is a labor-intensive and error-prone job, while the register-transfer simulation is too slow to be used at the algorithm development level.

The development of a language that allows for a functional description of signal processing algorithms, is therefore appropriate. This upper level input can then be compiled into the symbolic description of the target architecture using standard software compilation and optimization techniques. The abstractness and the hardware independence of such a description make it perfectly suited as the input language for the extensive functional simulations at the algorithm development level.

A front end system, based on a language optimized for describing signal processing algorithms called 'Silage' is being developed. Silage [25] has been designed as an 'applicative language' in an attempt to describe the signal-flow nature of the signal processing algorithms. An applicative language is a language, whose fundamental operation is function application and that has no variables or assignments. This is equivalent to the signal flow-graph, where nodes represent instances of functions and arcs represents the paths followed by the signals.

The development of the Silage compiler and simulator is currently under way [25]. An important aspect of the compiler design is the optimization of the generated microcode to the pipelined nature of the datapath and the parallelism in the architecture.

V. RESULTS

The automated design system has been used to generate a considerable number of applications, including an audio equalizer, a decision feedback equalizer with timing recovery, a full duplex LPC-vocoder[1], a full duplex 300 baud modem [26], a speech scrambler[27] and a filterbank for speech recognition. The number of parallel operating processors in these applications ranges between one and three, while the LPC-10 vocoder and the audio equalizer also require a host interface.

The layout generator presently uses a 3-4u NMOS library, while a 1.5-3u CMOS library is near completion. The 300 baud modem (4.1mm x 7mm, 20000 transistors, 9.6 kHz cycle) is shown in Figure 12.

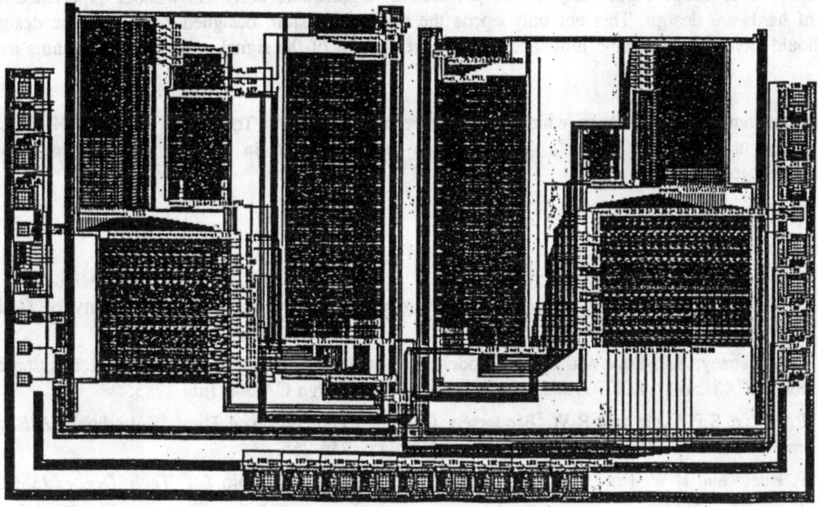

Fig. 12) Plot of the 300 baud modem

The importance of the automatic generation concept is stressed by this observation : the complete 300-baud modem (algorithm development, design file coding and debugging, layout generetion) was generated by a student in less than 2 months [26].

VI. CONCLUSIONS

An integrated automated layout generation system for digital signal processing circuits is presented. The present system, which was considered to be a test vehicle, allows us to draw some important conclusions concerning the target architecture and the software tools needed in the layout generation process. The macrocell based automated design methodology allows for the generation of complex signal processing chips with far less effort than manual design and layout, substantially reducing the probability of errors. The implementation of the complete software system (including the development of the tools) took about one man year. A few successfull applications easily outweigh the development costs of the software and the cell library.

A strong argument in favor of automated layout generation is the technology independency of the software. This minimizes the redesign efforts, when changing or scaling the technology.

The modular and the hierarchical approach allows for a simple adaptation of the software to changes in the target architecture. In fact, the nature of the developed layout tools (tiler, floorplanner) makes them suited for other macrocell based architectures. Future research efforts are being concentrated on the generalization of these software modules.

The modularity, combined with the efficiency of the tools, also supports the iterative nature of the silicon compiler : Due to the almost immediate feedback of dimensional data, the designer can optimize the silicon realization of his application : e.g. decisions at the processor floorplan level or even at the algorithmic level can be based on chip area and power considerations.

The unique input file to both layout generation and simulation assures the consistency between simulation results and hardware realization.

The present target architecture is very flexible and proved its ability to handle most of the applications in the 5kHz-150kHz range. Some weaknesses in the architecture have become apparent and are currently under examination : this includes more support for variable-variable multiplications for e.g. adaptive applications, a more flexible control structure (e.g. more powerful fsm) and extension of the i/o interface (A/D-D/A conversion, serial i/o, digital strobe lines). also under consideration.

Narrowing the range of target architectures allows a high level input description. This makes possible one of the most important aspects of the compilation methodology which is that the design files (or the higher level Silage input files) can be prepared by algorithmic designers without experience in the field of hardware design. This not only opens the world of custom designed chips to these designers, but should have a considerable influence on the development of the signal processing algorithms itself.

Acknowledgements The authors which to thank Peter Ruetz, Mats Torkelson, and Paul Hilfinger for their contributions to the project. This research was supported in part by the Defense Advance Research Projects Agency.

References

[1] S. Pope, *Macrocell Design for Signal Processing*, Ph.D. Dissertation, University of California, Berkeley, CA, December 1984.

[2] J.M. Rabaey, S.P. Pope and R. W. Brodersen, "An Integrated Automatic Layout Generation System for DSP Circuits", *IEEE Trans. on Computer Aided Design* CAD-4, July 1985.

[3] P. Ruetz, S.P. Pope and R.W. Brodersen, Computer Generation of Digital Filterbanks *IEEE Trans. on Computer Aided Design*, CAD-5, April 1986, pp. 256-265.

[4] P. Ruetz and R.W. Brodersen, "A Realtime Image Processing Chip Set" *Tech. Dig. of ISSCC*, Feb. 1986, pp. 148-149.

[5] P. Ruetz, R. Jain and R.W. Brodersen, "Automatic Layout Generation of Real-Time Digital Image Processing Circuits" *Proc. of CICC*, May 1986, pp. 111-115.

[6]   L. R. Rabiner, B. Gold, *Theory and Application of Digital Signal Processing*, Prentice-Hall, Englewood Cliffs N. J., 1975, Ch. 6.

[7]   *Ibid.*, Ch. 10

[8]   N. L. Daggett, "A Computer for Vocoder Pitch Extraction," Tech Note 1966-3, Lincoln Laboratories, Lexington, MA (1966).

[9]   L. R. Rabiner, B. Gold, *op. cit.*, Ch. 8.

[10]   Nelson Morgan, *Talking Chips*, McGraw-Hill, 1984, pp. 75-78.

[11]   Y. Kawakami, et. al., "A Single-Chip Signal Processor for Voiceband Applications", *International Solid State Circuits Conference Digest*, San Francisco, 1980, pp.40-41.

[12]   J. R. Boddie, et. al., "A Digital Signal Processor for Telecommunications Applications", *International Solid State Circuits Conference Digest*, San Francisco, 1980, pp. 44-45.

[13]   S. S. Magar, E. R. Caudel, A. W. Leigh, International Solid State Circuits Conference Digest, San Francisco, 1982, pp. 32-33.

[14]   P. Denyer, D. Renshaw, N. Bergmann, *A Silicon Compiler for VLSI Signal Processors*, Proceedings ESSCIRC 1982, Brussels, pp. 215-218, September 1982.

[15]   J. Rabey IMEC

[16]   J.R. Jasica, S. Noujaim, R. Hartley and M. Hartman, "A Bit-Serial Silicon Compiler" *ICCAD-85*, Nov. 1985, pp. 91-93.

[17]   J. V. Ginderdeuren, H. DeMan, B. De Loore, G. Van Den Audenaerde, "ASIC Filters for HiFi Digital Audio Signal Processing" *Proc. of ICASSP*, April 1986, pp. 1537-1540.

[18]   D. Johansen, "Bristle Blocks: A Silicon Compiler", *Proc., 16th Design Automation Conference*, San Diego, June 1979.

[19]   Hwang, K., *Computer Arithmetic*, Wiley, New York, 1979, p. 149.

[20]   Bohm, C., and Jacobini, G., "Flow-diagrams, Turing Machines, and languages with only two formulation rules", Comm. ACM 9,5, May 1976, pp. 366-371.

[21]   K. H. Keller, A. R. Newton, "KIC2: A Low-Cost Interactive Editor for Integrated Circuit Design", *Digest of Papers, IEEE CompCon 82 Conference*, San Francisco, Feb. 1982, pp. 302-304.

[22]   Ritchie, D. M., Thompson, K., "The UNIX Time-Sharing System", *Bell Systems Technical Journal*, 57(6), pp. 1905-1929, 1978.

[23]   G. C. Billingsley, "Program Reference for KIC", Memorandum No. M83/62, Electronics Research Laboratory, University of California, Berkeley, CA, October 1983.

[24]   R. N. Mayo, J. K. Ousterhout, "Pictures with Parentheses: Combining Graphics and Procedures in a VLSI Layout Tool", *Proc., 20th Design Automation Conference*, June 1983.

[25]   P. Hilfinger, "A High-Level Language and Silicon Compiler for Digital Signal Processing", *Custom Integrated Circuits Conf.* May, 1985, pp.213-216.

[26]   W. Abbott, *Design of a 300 Baud FSK Modem Using Customized Digital Signal Processors*, Memorandum No UCB/ERL M84/93, U.C. Berkeley, August 1984

[27]   M. Torkelson, "Design of a Speech Scrambler", In preparation

CATHEDRAL II : A SYNTHESIS AND MODULE GENERATION SYSTEM FOR MULTIPROCESSOR SYSTEMS ON A CHIP*

H. De Man, J. Rabaey, P. Six

IMEC v.z.w., Kapeldreef 75, B-3030 Heverlee, Belgium

1. INTRODUCTION : GOAL, MOTIVATION AND SUMMARY

In this paper we report about the status of the work on a silicon compiler for automatic synthesis of synchronous multiprocessor system chips starting from a high level behavioral description. The work on this compiler is the subject of ESPRIT project 97 sponsored by the EC and undertaken by IMEC, Philips, Siemens, Bell Telephone Mfg. Co., Silvar Lisco and Ruhr University Bochum. This report is on the work done at IMEC.

A silicon compiler is defined here as a software system supporting chip layout synthesis starting from a behavioral system description.
This work is a continuation of the work on an operational silicon compiler CATHEDRAL-I for bit-serial digital filters which has been reported in [Jai86]. While doing CATHEDRAL-I we have experienced that, in order to be succesful in silicon compilation, it is necessary to start with a careful definition of the target architecture and its associated design strategy.

We feel that "the" silicon compiler simply does not and probably never will exist just as "the" software compiler does not exist since many source and target languages exist, each of which is optimized for a given task. Therefore we believe that efficient silicon compilers will necessarily be strongly tied to a particular target application area for which it is first necessary to define a target architecture. As a result many different silicon compilers will evolve in the future.
A target architecture is defined as the hardware design style into which the behavioral description is to be mapped.
Again, if this is to be succesful, we feel that this design style must be clearly defined and constrained in order to allow for a formal approach. A good compromise must be found between the constrained design style necessary for the compiler to be feasible and the efficiency to realize the target application area.
This requires a very careful study of the architecture and the design methodology in the context of realistic problems in the application area before any tool is developed.

* Work sponsored by the EC under ESPRIT 97 contract.

Based on the result of such study on handcrafted designs then the synthesis formalism can be defined and the implementation of CAD tools and silicon libraries can start. In doing so we can identify target dependent and independent tools. For the latter it is useful to derive the best software development environment to be reused when the next target is to be implemented. <u>We call that the diversification environment.</u>

In CATHEDRAL-II, the target application is a subset of digital signal processing (DSP) algorithms architecturally to be realized by a set of concurrent dedicated bit-parallel processors on a single chip.
Such an architecture is especially suited for so called third generation DSP algorithms involving large blocks of sampled data subject to complex decision making algorithms. It will be shown in this report that such architectures lead very often to much more efficient implementations than on general purpose DSP processors. As shown in Fig. 1.1 we typically address very complex algorithms in the audio and telecom intermediate frequency spectrum. We will show however that the <u>diversification</u> <u>environment</u> also allows applications in the video domain.

Typical applications are : speech synthesis and analysis, modems, digital audio (compact disc signal processing), matrix based processing, ISDN etc...

The content of this paper reflects the above methodology.
First, in section 2 we will discuss the general design methodology on which all is based. It is called "Meet-in-the-middle design strategy". It refers to the fact that we opt for a strict separation between system and silicon design levels. The system designer performs a top down design based on high level language. This design is mapped into an architecture consisting of an interconnection of instances of a <u>minimal</u> <u>set</u> <u>of</u> <u>primitive</u> <u>silicon</u> <u>modules</u> at the level of large arithmetic blocks. These modules are in fact software procedures called by the high level compiler. These procedures have been composed by <u>silicon</u> <u>specialists</u> by using a module generation <u>programming</u> environment. The system designer then can use an interactive floorplanner to compose the chip.
It is as if the system designer meets the silicon designer in the "middle" of the design abstraction levels. This avoids the problem of the "long thin man".

From the design strategy we will define the toolbox necessary to support the design methodology and we will briefly describe the requirements for the individual tools. Basically we will identify two main parts in CATHEDRAL-II : the architectural synthesis part (system designer) and the module generation part (silicon designer). Both parts will have a need for their own verification parts which are based both on simulation and on knowledge based verification, which is made easier by the imposed design constraints.
As stated above, no compilation is possible without a careful

definition of the target architecture. This definition is given in section 3.
We will show, based on a speech processing application, that a general multiprocessor architecture can be defined in an hierarchical fashion.

As modules, basically five so called execution units (EXU) and an address calculation unit (ACU) composed out of functional building blocks (FBB), a controller generator, a switched RAM and I/O generator are sufficient for our purposes. We will also discuss the inter- and intra-processor communication as well as the floorplanning strategy envisioned.

Based on that architecture, in section 4, we discuss the architectural synthesis part. As a high level language we have chosen the SILAGE applicative language developed at UC Berkeley [Hil85]. We will show that the development of the high level compiler involves four steps :

a. partitioning into processors communicating through switched RAM's, FIFO's or MUX.
b. mapping partitioned code into structure defined as EXU's interconnected by a minimal set of dedicated busses;
c. scheduling of the register transfers in terms of the modules in order to generate the microcode which forms the basis for :
d. controller generation.

At this stage of the work we have developed prototype CAD tools for b)c) and d). In our research project we have indeed opted for a strategy to tackle first those problems representing the biggest bottleneck in the design problem. Thereby it was found that the partitioning of an algorithm into parts which can be assigned independently to a single processor is relatively easy to a designer. It is mainly based on an identification of communication bottlenecks caused by an accumulation of sequentially generated data necessary to create new samples into the next processor. This is the place where data storage under the form of switched RAM's can be necessary between separate processors. We feel that this problem can be tackled in the future, based on searching convergence and divergence points in the precedence graph of operations.

In this report we will start from the hypothesis that the partitioning is done and we concentrate on the results obtained in mapping, scheduling and controller generation per single processor executing its share of the SILAGE code.
During the mapping operation an optimal time and area mapping of the SILAGE code into the architecture and this within the troughput requirements of the algorithm is to be found.
Since this is quite strongly dependent on the target architecture we have chosen to do that in a rule based environment. Since this process requires a considerable amount of unification between preprocessed code and the hardware operators and registers, we have selected PROLOG as an

implementation language.

The outcome of this program is the data path of each processor
whereby first all variable communication is done by dedicated
busses.   The data path operators correspond to the ones
available   from   the   module   generators   supporting   the
architecture.
Basically this output is described in a register transfer
language (RTL). Since in our architecture each execution unit
contains a register file at the input and output the RTL can
contain explicitly each operator type, a symbolic opcode as
well as the registers to which a variable is assigned.
The problem of scheduling can now be described as the
technique to find a timing ordering to fire the operations in
such a way that data-precedence is respected, that no resource
allocation conflicts occur and that pipeline constraints in
operators and controllers are respected. It will be shown
that such a problem can be formulated as an integer linear
programming problem (ILP). Since ILP is NP complete we have
to find heuristic solutions to this problem but we will
demonstrate   the   feasibility   of   the   approach with some
examples.

Finally, when scheduling is done, the number of cycles per
processor is known. At this stage the bottleneck processor
can be identified and, using pragmas in the SILAGE
description, one can try to optimize the bottleneck processor
if troughput is not reached. However, more often, one will
have the opposite case where one should again exchange speed
for area. A first mechanism which is present in the system is
a bus merging strategy based on the knapsack algorithm which
is followed by a rescheduling. In the future we will add more
automated optimization to redistribute automatically time for
area.

In section 5 we will describe the status of the work on the
module generation environment(MGE). First we will derive the
requirements for a MGE in a silicon compilation environment.
We will show that a module generator in a silicon compilation
environment has not only to provide a layout view of an
instance of a module but also a functional, timing and test
view respectively for system level simulation, performance
verification and test assembly after the modules have been
placed and routed by the use of a floorplanner.
Furthermore, if we accept the concept of diversification
environment, we also have to face the fact that a module
generator is a programming environment which allows three
activities :
a) Creation : the design of a module (code "writing")
   generator by silicon specialists not used to code writing !
b) Generation : the generation of the views of a module
   instance by a call to the generator.
c) Adaptation : the adaptation to a new technology of a module
   generation library.

In section 5 we will show, how in CATHEDRAL-II we have
developed a prototype MGE which is mainly based on symbolic
layout and compaction of abuttable leaf cells followed by a
procedural, interpretative cell composition including
compaction and routing. This is done in a LISP-PASCAL
environment in which special attention is paid to generate
module generator code based on graphical definitions of
composition procedures.
Verification of the correctness of modules is a special
problem due to their parameterizability. In CATHEDRAL-II we
are working on an expert system based verification system
whereby verification procedures for electrical, logic and
timing verification can be defined in order to perform, as
much as possible, rule based automatic verification at the
electrical, logic and timing level.
This verification system will also provide the non-layout
views for the high level silicon compiler.

Finally in section 6 we will make some conclusions on the
actual status of the work and discuss future extensions.

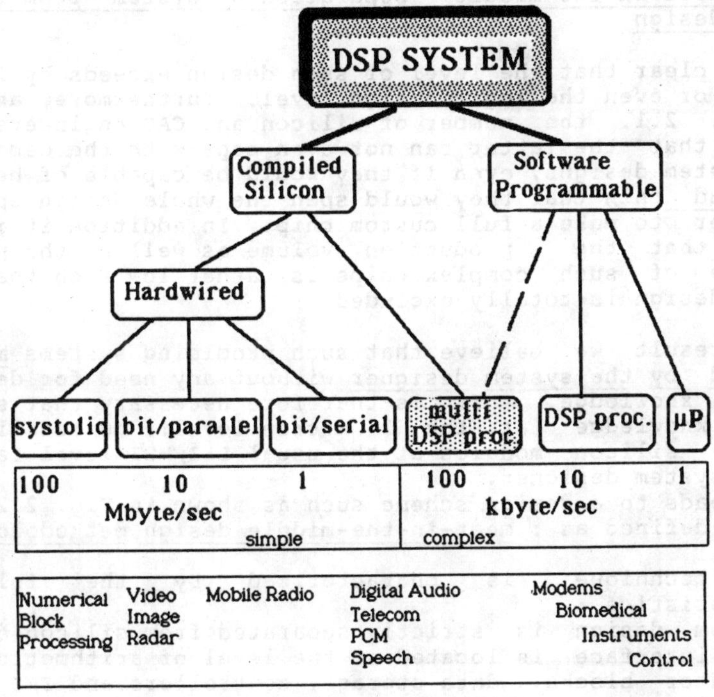

Fig. 1.1 : different implementation styles for DSP
algorithms. CATHEDRAL II aims at the
multiprocessor implementation.

2. THE MEET IN THE MIDDLE DESIGN STRATEGY

It is well known [Tuc85] that by 1990 nearly 50% of all systems implemented in IC's will be ASIC's and nearly 40% are expected to be designed by the end users.
As shown in Fig. 2.1 we can classify VLSI design in a performance complexity plane. Performance is defined as data-rate/max. clock speed and complexity ranges from simple control logic or pure data-flow to implementations of complex decision making algorithms. Today system designer oriented design, truly supported by commercial CAD, in the low-performance/low complexity is supported by standard cell and gate array type of design. The challenge posed by advanced 1.2 micron CMOS is the design of chips containing nearly 300000 devices. This amounts to potentially more than 15 processors of 16 bits each with 256 bytes local memory or near 0,5 million multiplications/sec (0,5 Mega MIPS!) in a data-path high performance realization. This kind of complexity, hard to realize with off the shelf standard DSP processors, opens challenging system applications in the field of telecom, digital audio, video, robotics, speech analysis, image processing, dedicated matrix processing etc...

2.1. : Meet-in-the-middle = separation of system- from silicon design

It is clear that the level of such design exceeds by far the circuit or even the logic design level. Furthermore, as shown in Fig. 2.1, the number of silicon and CAD engineers is so scarce that the latter can not even cope with the demand for such system designs, even if they would be capable of being so long and thin that they would span the whole design spectrum in order to make a full custom chip. In addition it is very likely that the production volume as well as the product lifetime of such complex chips is rather low such that full custom design is totally excluded.

As a result we believe that such promising systems must be designed by the system designer without any need for detailed silicon knowledge. It is therefore necessary that silicon design knowledge (at the 1 micron level!) be localized in reusable silicon modules at the usual LSI/MSI level familiar to the system designer.
This leads to a design scheme such as shown in Fig. 2.2 which we have defined as : meet-in-the-middle design methodology.

This technique is characterized by the following characteristics :
a) System design is strictly separated from silicon design. The interface is located at the level of arithmetic/logic operator blocks, data storage, controllers and I/O units. We will call that in the sequel : functional level. The silicon primitives used at that level are called : modules. Design at the system level consists in translating a system specification into structure which is a netlist of module instances. Placement and routing of layout instances of

   the modules is then the chip design.
b) Silicon modules are reusable just as standard cells are.
   In this way the costly investment in high performance,
   advanced silicon technology design is limited and the cost
   is written off over as many designs as possible.
c) Silicon modules are much more complex than standard cells
   and, in order to save even more in terms of silicon design
   cost, these modules must also be capable to survive a
   number of technology updates. Therefore modules must be
   technology adaptable.
d) Silicon modules, however are not the previledge of a
   particular foundry or CAD vendor. Since the competitive
   edge between system houses will not be in the technology,
   but in the architectural technique and in the
   implementation of it, we expect that module design requires
   a powerful design environment on itself for a local team of
   silicon designers. This may not yet be the case today, but
   we expect this to happen in the future.

Notice that in this design style system designers design as
usual in a top-down fashion down to their usual intermediate
level.   The silicon people design in their usual bottom-up
like fashion, composing the LSI level modules from functional
building blocks (see below) which in turn are composed of
logic leaf cells at transistor level.

It is as if both parties meet each other in the middle of the
design abstraction levels.   In this way scarce talent is
optimally used and the design process corresponds to usual
patterns.

However, due to characteristics b) c) and d) defined above,
some fundamental deviations from classical design at silicon
and system level do occur. They will become clearer after
giving next an outline of the CAD toolbox as we are developing
for CATHEDRAL II.

2.2. : The CAD toolbox for CATHEDRAL II for a
         Meet-in-the-middle design strategy.

Fig. 2.3 shows a CAD toolbox as we are developing it for the
design of multiprocessor implementations of high complexity/
low performance DSP algorithms.
It is clear from the figure that there is indeed in the middle
of the design abstraction a separation between the silicon and
the system designers. The link between them is a "call" to a,
hopefully, limited set of silicon modules which have to be
carefully defined in the target architecture.

The system designer defines his/her system at the behavioral
level in a high level language. In our system, as will be
explained in section 4, we use the SILAGE [2] language which
is especially suited to describe complex DSP algorithms in an
appplicative way e.a. as a set of simultaneous Boolean
equations rather than in a procedural way.
Coupled to this language is a high level simulator to verify

the behavioral correctness of the algorithm.
Based on the troughput requirements and a set of expert design rules, the SILAGE code is first optimized in function of the target architecture. It is subsequently compiled directly into structure i.e. a netlist in terms of the limited set of predefined silicon modules. Clearly most of this synthesis traject is dictated by the target architecture and therefore quite a large part of it is rule based and in our case implemented in a PROLOG program as will be discussed in section 4.

Notice in Fig. 2.3 that we also allow, in principle, the designer to specify the design at intermediate levels, all the way down to structure. However the price to be paid is an increasing amount of "lower" level simulation, a higher redesign risk and a longer time to the market. Perhaps a smaller chip area or a better layout may result. Here the economics of the end product will dictate the choice.

When all calls to the modules are succesful then the chip can be floorplanned in a floorplanner. This can be done either interactively or by auto-place and route techniques.

Notice that in CATHEDRAL II we really aim at a true silicon compiler since we include the synthesis of the algorithm directly into data-paths and control logic. This is in contrast to most commercial systems today which are limited to floorplanning and to the module generation part.

In order for this scheme to work, just as in a software language compiler, succes is intimately linked to a clear definition of the set of modules. Clearly it can not be a set with the same variability as is offered by a TTL or CMOS LSI part catalogue! It would take too long to design it, it would be technologically outdated whenever its design is finished and no compiler would be capable to be powerful enough to map into it.
Neither can we, in view of the evolving technology, design it in a fixed technology but we should be able to adapt it to technology changes.

On the other hand, just as is necessary in traditional PCB design, using LSI components, we need from a module generator a lot more information than just the layout view, which in itself should consist of a Bounding Box View (BBV) and a full Layout View (LV).

In Fig. 2.3 we show that we need a functional view i.e. an RTL level parametrizable function, in case design is done at the structural level since then simulation is necessary. Even if full synthesis is used, still designers, will want to simulate the actual chip.
Furthermore, a timing view is necessary since, although the synthesis compiler can take first order throughput requirements into account, it is only after placement and routing that a full performance check can be done.

In CATHEDRAL II we are following a "bottom-up" hierarchical generation of timing models in a knowledge based program SLOCOP which follows closely the composition procedure of a module as will be discussed in section 5. As the floorplan level the interconnection parasitics are taken into account to check global timing. If unsatisfactory, first buffer sizes are adjusted. If this is not satisfactory, one might consider different placement or, finally, a pragma could be formulated at the SILAGE level to call for higher performance modules or ultimately to increase parallelism in the algorithm.

Finally, although not present in the actual CATHEDRAL II version, there must be a testview with each module. We envision for the future that most modules with a high degree of structure will be C-testable, i.e. dependent on their particular structure a small set of wordlength independent patterns exist which guarantee the module testability. We then want to implement a "test assembly" program at the level of the floorplanner that, for the particular architecture, will generate total testability for the whole chip.
This is a topic for future research but we believe that a silicon compiler without ATPG for the chip is like a hospital building without doctors !

2.3. From fixed standard cells to flexible module generation.

The problem with the above is that the variability of modules is much larger than at the logic standard cell level. In order to have a maximum reusability as well as a technology independence it is not possible to design modules in the standard "CALMA" type way.
Modules must be written as parameterizable procedures or, to be more in vogue with actual object oriented thinking, as classes from which modules are objects. Parameterizability is to be understood in a broad sense ranging from simple wordlength to conditional composition in terms of functional building blocks or size of output buffers etc... Moreover these procedures should also generate the other views needed by the synthesis programs.
In order to reduce the variability, one also has to constrain the design space, based on the target architecture dictated by the application area.
As an example, in CATHEDRAL II the target application area is the high complexity/low performance area needed for DSP applications. The target architecture has the form illustrated in Fig. 2.4. It is oriented towards synchronous sampled systems transforming incoming sampled vector $\underline{x}$ into an outgoing vector $\underline{y}$ potentially sampled at a different rate under control of a sampled control vector $\underline{c}$.

As will be explained in section 3 a detailed study of many applications leads to a multiprocessor structure consisting of a number of dedicated processors communicating through data storage elements which may consist of switched RAMS, FIFO's or simple MUX dependent on the temporal constraints of the

communication bottlenecks in the processors.
Processors are composed of a carefully defined set of so
called execution units (EXU's). These EXU's perform the 4
basic arithmetic/accumulate operations, comparison, scaling,
address calculation and alu. Each processor acts quite
independently from the others and therefore has its own local
controller for which in the next section a special type with
minimum number of NOP's in branching algorithms is proposed.
A supervising controler at the chip level provides
synchronization pulses to each processor and the I/O sections.

We think that the feasibility of our system lies in the fact
that we are restricting ourselves to this carefully chosen set
of EXU's.
In the software world of compilation one can compare this
choice as a choice of the target language for the compiler to
be built.

As is clear from Fig. 2.3 the module level in CATHEDRAL II is
at the level of the EXU's.
The task of the module generator environment in CATHEDRAL II
therefore consists in composing the EXU's from so called
Functional Building Blocks (e.g. adder, comparator,
register...) which in themselves are composed of logic cells.
We can now look at the parts of such a module generator as
shown in Fig. 2.4.

2.4. Anatomy of module generator

Fig. 2.4 shows the anatomy of the module generator as in use
in CATHEDRAL II.
First notice how the different arrow types indicate clearly
the "create" (silicon designer), "generate" (call from silicon
compiler) and "adapt" (to technology rules) functions.

Further notice that the CATHEDRAL II module generation
environment not only provides the layout environment but also
will be equiped with an expert system which is necessary for
the following tasks :
1. Generation of the functional/timing and test models during
   "generate" phase.
2. Verification of the modules during the "create" and "adapt"
   design phase.
3. In the future we would also like to store the design
   history on the accumulated experience.
The design of a module is based on a parameterizable
functional description at the register-transfer level. This
description (and its simulation) can be done by the
HILARICS-LOGMOS system developed at IMEC [Mari86]. This
functional description is the documentation link between the
system and the silicon designer. In case of controller
synthesis it will be shown in section 4 how also finite state
machine description can be an input to the module generator in
order to generate PLA based controller structures. In section
5 it will be described how that is done using the PLASCO
[Bar85] module generator.

The big difference between a module generator and fixed layout is that, because of the parameterizable nature of the description also the layout must be parameterizable and this requires that the relative placement of cells as well as the connectivity needs to be procedural. In CATHEDRAL II, this is done by using a LISP interpretative programming environment whereby as much as possible, the code for procedural composition of a module out of cells is generated from a graphics definition. Only complex mathematical relationships are directly programmed in LISP.

The primitive cells themselves are done mainly by the use of symbolic layout with automated abutment and routing of facilities which can be compaction, called from the LISP procedures. We are now working on an efficient extractor from symbolic layout which provides an input to the, mainly knowledge based verification and model generation part of the module generator.

Symbolic layout provides a good technique to insure technology updatability.

This part contains a timing verifier, a logic and electrical verification and simulation part.

We will now in the next section analyze all these parts in more depth, after a clear definition of the target architecture of CATHEDRAL II.

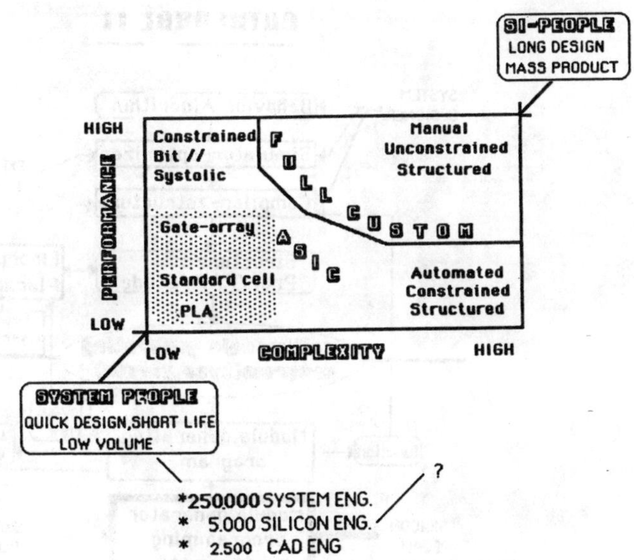

Fig. 2.1 : Evolution of design will be towards ASIC
by the shortage of silicon designers and
the required small design time to product
life ratio.

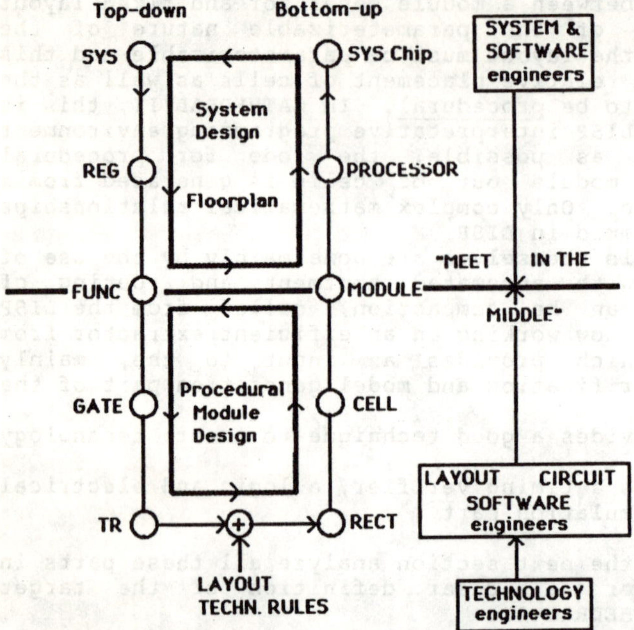

Fig. 2.2 : Meet-in-the-middle design strategy

CATHEDRAL II

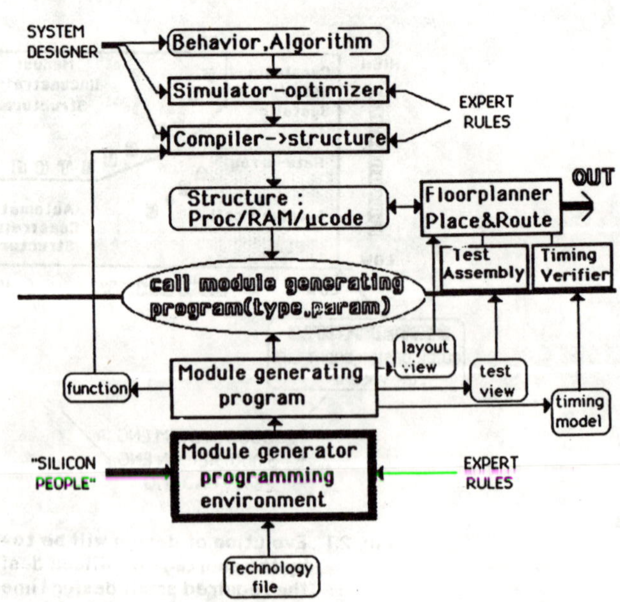

Fig. 2.3 : CATHEDRAL II toolbox.

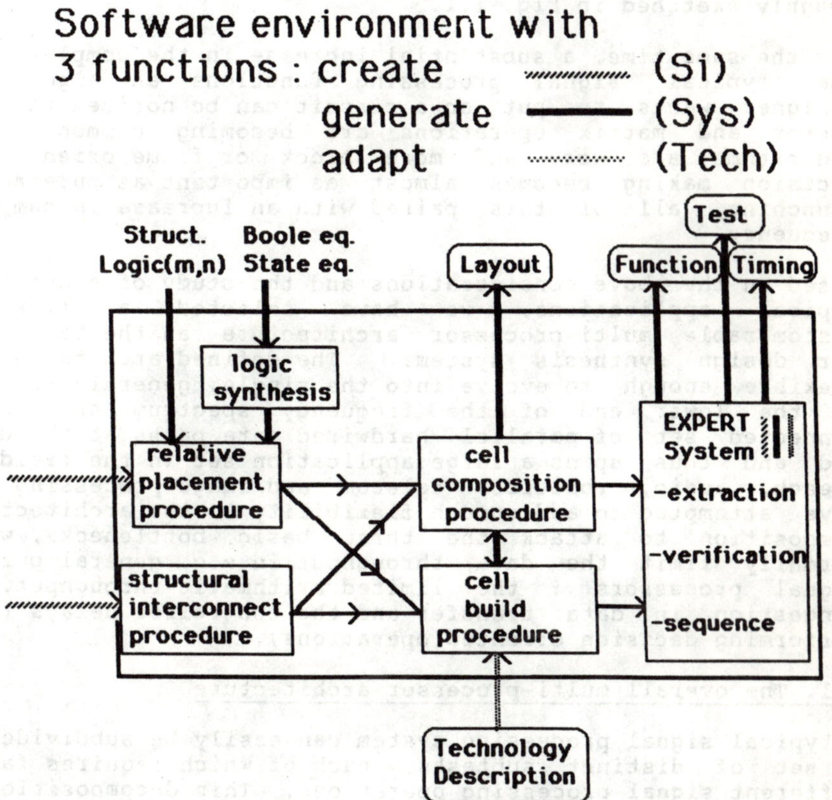

Fig. 2.4 : Anatomy of a module generator

3. THE TARGET ARCHITECTURE

It has already been stated higher that efficient and acceptable design synthesis is only feasible with a restricted target architecture in mind. The definition of such an architecture, even for a (restricted) application field as Digital Signal Processing (DSP), is however not straightforward. In fact, the architectural composition needed for the implementation of a certain algorithm is heavily determined by factors such as the ratio between maximum hardware clocking frequency and the sample frequency, the arithmetic complexity per sample and the regularity or the streamlined nature of the algorithm. Therefore, a large set of totally different architectures have been envisioned as roughly sketched in Fig. 1.1.

At the same time, a substantial increase in the complexity of the typical signal processing functions an algorithmic designer wants to put on a circuit can be noticed [All85]. Vector and matrix operations are becoming common place, algorithms are more and more block or frame oriented and decision making becomes almost as important as pure number crunching, all of this paired with an increase in sampling frequency.

Based on the above considerations and the study of a number of typical applications, we have selected a flexible, customizable multi-processor architecture as the target for our design synthesis system. The defined architecture is flexible enough to evolve into the single, general processor at the lower end of the frequency spectrum and into a connected set of parallel, hardwired data paths at the upper end and thus spans a large application set in the fields of speech, audio, robotics, telecom and image processing. We have attempted to add enough flexibility to the architectural composition to attack the three basic bottlenecks, which normally limit the data throughput in e.g. general purpose signal processors : the limited arithmetic throughput, the congestion in data transfer and the controller delays (when performing decision oriented operations).

3.1. The overall multi-processor architecture

A typical signal processing system can easily be subdivided in a set of distinct subtasks, each of which requires fairly different signal processing operations. This decomposition is reflected in the higher level block-diagram of the system. The proposed architecture is an attempt to map this system decomposition directly into the hardware, by providing a set of parallel operating processors, each of which executes one particular subtask and is optimally tuned to just that one task (Fig. 3.1.). Each of those processors operates relatively independently from its neighbours and communicates with them via a set of transparent communication protocols, exchanging only data which is global between them. Communication with the outside world proceeds over an I/O

frame. This frame can support a large range of I/O protocols,
ranging   from   parallel   to   serial,   from   synchronous   to
asynchronous   and   from   word   oriented   to   buffered   block
oriented.

As stated higher, each of the processors is optimized to
perform one particular part of the algorithm and consists of a
dedicated data path and controller. The data path is composed
of a cluster of strongly connected Execution Units (EXU's),
communicating with each other over a restricted number of
dedicated busses.   In this way, bus contention as occurring in
the case of a single or dual bus architecture is largely
avoided.   The   number   of   selectable   EXU-types   has   been
deliberately   restricted   to   five,   which   are   stored   in   a
parameterized format in the module library.

A multi-branch, microcode based controller structure has been
selected   to   control   the   dataflow   through   the   datapath.   This
structure   is   flexible   and   powerful   enough   to   handle   a   large
span   of   algorithms   in   an   elegant   and   efficient way.   Some of
the   EXU's   (or   their   subblocks)   can   also   have   a   local
controller.   This   helps   to   reduce   the   complexity and the
number   of   instruction   bits   in   the   central   processor-
controller.   Examples of local controllers are the decoder of
the register file and the slave controller of the divider.

It is interesting to see how the hierarchy of the overall
architecture   is   also reflected in the controller hierarchy :
at   the   system   level,   we   have   a   central   controller,   which
governs   the   data   flow   between   the   processors   and to the
outside world.   Each of the processors has its own controller,
determining   the   functional   behavior   of   the processor and at
the   lowest   level,   a   number   of   local   controllers   can be
attached   to   the   EXU's.   This   strict   hierarchical division
allows for a modular approach to the synthesis problem.

A number of the topics raised above will be discussed in more
detail in the sections below.

3.2. The Interprocessor Communication

During the partitioning of the algorithm over the processors,
one of the main goals is to keep the mapping and scheduling of
the   identified   subtasks   as   independent   as   possible.   This
requires that the communication between the processors is
completely   transparent   to   the   internal   operation   and the
timing of those processors.   The type of the protocol used to
achieve this goal is heavily dependent upon the properties of
the   required   transfers : the required transfer capacity, the
ordering of the read and write operations, the synchronicity.
Therefore,   a   set   of   different   communication protocols (and
supporting   hardware)   have   been   provided,   so that for each
particular case an optimal solution can be selected.

The   most   general,   synchronous protocol is presented in Fig.
3.2.   The   communication   hardware   consists of two identical

RAM-units, which are alternatively switched between processors 1 and 2 (and their address computation units). During time frame 1, processor 1 writes its data into RAM1, while processor 2 reads the data from RAM2. This situation is reversed in the next frame. This technique allows for the transmission of large, unordered data blocks at a high rate without any restriction, this at a considerable hardware cost. When the sequence of the generated data is identical to the order in which it should be received, a substantial hardware reduction can be obtained by replacing the switched RAM's by a single FIFO-unit, as shown in Fig. 3.3.

Other simplifications become feasible with decreasing communication rates : in those cases, it is often possible to skew the programs of source and destination processors in such a way that no bus conflicts will occur (Fig. 3.4.). A single bus-structure with connected RAM-buffer will then be sufficient. The last solution however assumes that the absolute timing of the transfers is known at compile time. When this is not the case (e.g. when conditional operations are allowed), a 2-way handshake between the controllers of the communicating processors can be used to synchronize the data transfer.

Other hardware reductions are feasible by e.g. using buffered bit-serial transfers (as in [Pop84]). The communication hardware can also be multiplexed over multiple sources and/or destinations by providing multiplexer and demultiplexer units at the processor sides.

3.3. The Processor Data Path

According to our architectural approach, each of the processors consists of a dedicated data path and a controller. The data path is optimized for (only) the particular tasks it has to perform and is assembled from a set of selected EXU's, interconnected by a restricted number of customized busses. Each of the EXU's contains a register-file (of variable size) at its input side. This local storage helps to resolve the memory bottleneck, which is often a problem in general purpose signal processors.

Studies have shown that following EXU's are sufficient to span most of the intended application space : a parallel multiplier-accumulator for high speed number crunching, a parallel-serial divider/multiplier (takes N cycles for N bit division or multiplication), a shift/add/compare unit for general operations and decision making, a general purpose ALU and an Address Computation Unit (ACU) (address arithmetic, modulo counting, loop counters). All those blocks have been designed in a parameterizable fashion and have been implemented in the module generation environment (see also section 5).
Typical parameters are the wordlength, the size of the register-files, the size of the multiplier array, the type of the adders used (in function of the speed requirements), the

maximum depth of the shift-operations, etc. Not used operators can be removed from the datapath, resulting in a further increase in area-efficiency : e.g. the shift unit of the shift/add/compare EXU can be removed, when no shift operations are required in the task to be performed. In this way, the units can be tailored to the particular requirements of an algorithm.

As an example of such an EXU, we will study the shift/add/compare unit, which is shown in Fig. 3.5. The EXU is constructed of a set of so called 'Functional Building Blocks' (FBB's), which are in this case 2 register files, 2 logarithmic shifters, an adder/subtractor, a comparator unit and some bus-drivers. The parameters are the wordlength, the sizes of the register files, the depth of the shifters, the type of adder unit and the number of bus-drivers. The shifters, adder, comparator and bus-drivers can be removed if not needed. The unit is controlled by a number of control signals (names starting with a C) and a set of status signals (marked with an S), which are routed to the control unit in order to evaluate decisions.

An example of a processor-data path, constructed using the proposed strategy, is shown in Fig. 3.6. This processor is used to compute the amplitude spectrum of a signal, given the complex frequency domain spectrum, and to determine at the same time the maximum amplitude. The real and imaginary values are obtained sequentially from the DFT-processor and are transferred through the I/O buffer and bus1 to the multiplier/accumultor (8x8) where they are squared and added (Ampl = Re*Re + Im*Im). After this, the amplitude values (16 bit) are transferred to the next processor over bus2. At the same time, the comparator unit (stripped Add/Shift/Compare) compares the computed amplitude value with the old maximum, which is updated over bus3. The final maximum is also transmitted to the next processor. Note that the amplitude processor is buffered from the next processor by a switched RAM-structure, as defined in 3.2. The addresses for the data in this RAM are computed concurrently by the Address Computation Unit (7 bit), which also implements the loop counter. This processor with three concurrent units can perform an amplitude computation and a maximum update in 2 cycles (average) !

3.4. The Processor Controller

The main task in the actual implementation of a subsection of an algorithm after an 'optimal' datapath has been selected consists of the design of the associated control section. A single control task can be realized by a large variety of alternative control architectures, ranging from random logic over sequencers to microcoded controllers. The optimality of a structure depends upon the nature of the algorithm to be implemented (repetitivity, number of branches). This can for instance be demonstrated by the case of the general purpose signal processors, the controller of which is optimized for repetitive operations (such as filtering, FFT). This results in a rather poor behavior for decision oriented algorithms. From a synthesis point of view, the availability of optimization and minimization tools for a certain structure is also of crucial importance (logic minimization, state assignment, etc..).

Based on these considerations, we have selected a single, flexible microcode based multi-branch controller as the standard controller structure for our processors. This controller has the flexibility to support heavily decision making oriented as well as regular, repetitive algorithms in an efficient way.

The selected architecture is presented in Fig. 3.7. The core of the controller is identical to the conventional microcode-based controller. It consists of a program-ROM, which stores the horizontal microcode words, and a program counter-register (PC), which is incremented in normal operation mode, but which can be loaded with a jump-address through the multiplexer MUX in jump-mode. In a conventional architecture however, multi-direction branches have to be split into a number of consecutive conditional jumps with the jump-address stored in the program-ROM. In this way, instruction cycles as well as ROM-area are sacrified. The controller architecture of Fig. 3.9 avoids this by storing the jump-addresses (or states) in a seperate memory called ADRMEM. When some type of jump is required, the output of ADRMEM is loaded into the PC. In the case of a data-dependent branch, the status-signals generated in the datapaths have to be interpreted first in a Finite State Machine (FSM) in order to decide on the appropriate next-state address. The updating of the state and the computation of the jump address can proceed simultaneously with the normal arithmetic operations and do not ask for extra cycles.

It should be noted however that three pipeline stages are present in the critical control-path (setting of the status bits, evaluation of the ADRMEM and fetching of the next instruction). In the case of a conditional branch, this results in a required delay of 2 cycles between the test and the actual branch. This could result in a number of NOPs (No Operation), similar to the situation occuring in the RISC-architecture [Pat81]. Most of the time however, the microcode scheduler can introduce independent instructions in between and avoid the loss of those cycles.

The power of the controller can best be illustrated by the following example : a decision feedback equalizer, together with the timing recovery circuit [Tze85] could be executed by this controller in 25 cycles (on a single Add/Shift data path). The decision feedback equalizer only took already 80 cycles on a TMS32010 [Mag84]. The algorithm in question relies heavily on decision making.

Other alternative controller architectures are also under consideration at present. These include the implementation of subroutines, sequencers (instead of the combination PC + ADRMEM), the addition of cache, etc. The most important task here is to identify the application domain of each of those structures in order to help the synthesis process in its selection.

3.5. Floorplan considerations

A number of general chip and processor floorplanning considerations are appropriate at this time, since they will help to determine the area-efficiency of the architecture.

As stated higher, a processor-datapath is composed of a number of EXU's connected by customized busses. Each of those EXU's is composed out of a set of so called FBB's (adders, registers, shifters). The following layout strategy has been applied within the FBB's, which have been developed in a double metal, 3 micron nwell CMOS technology : busses and supply lines are running in metal-2 parallel to the bit-slices, whereas control signals and clocks are routed orthogonally on the metal-1 level. The FBB's normally connect to each other by abutment. When this fails, a number of the metal-2 tracks (normally 5 of them are available per slice) can be used, as shown in Figure 3.8. The remaining tracks can be used to route the inter-EXU bus connections, thus avoiding large routing channels (which are only needed for the routing of the control signals or when not enough spare tracks are available). An example of a floorplan realized in this way is shown in Fig. 3.9.

A general applicable floorplan strategy for the controller structures is difficult to determine, as it depends upon the selected controller architecture, the complexity of the controller and the constraints imposed by the coresponding datapath(s). The floorplan generation is therefore an interactive procedure, in which the aspect ratio's of the controller modules are gradually and iteratively refined in order to obtain an 'optimal' match between datapath and controller.

The placement and routing task at the processor level has to be performed by a floorplanner. The connections will most probably run in channels along the function blocks, as normally too many obstructions already are present on the internal metal-2 tracks. Due to the sometimes very long communication tracks between processors, the capacitive loads can grow reasonably large. Therefore, these transfers have to be latched and consume (at least) one cycle.

3.6. Example

The effectivity and the efficiency of the presented multi-processor architectures have been validated with a number of practical testcases. Applications in the fields of digital audio, telecommunications, speech coding and speech recognition have been studied. The implementation of a high quality pitch extractor for speech [Slu80] will be discussed in more detail.

In this algorithm, the pitch of the speech signal is estimated by studying the matching between the maxima of the frequency domain spectrum of the signal (obtained from a DFT on the windowed signal) and a set of predefined patterns (40 in total). The system can be divided into four subtasks (or processors) as illustrated in Figure 3.10 : in a first step, the amplitude spectrum and the absolute maximum (as threshold) are computed. The first eight maxima (above the threshold) are derived in the second processor and compared with the 40 predefined patterns in processor three. In a last step, the precise value of the pitch period is computed.

The algorithm has been manually and partially automatically (using the emerging synthesis tools as described in section 4) mapped into the defined target architecture. The results of the mapping process are collected in Table 3.1.
The active area needed for the total system equals 37 mm2 in a 3 micron CMOS process. About 20 additional mm2 will be needed for the interconnect. The figures also show that the architecture is flexible enough to handle computation intensive functions (processors 1 and 2, large ratio between data path and control area) as well as decision making oriented tasks (processors 3 and 4).

The exercise gave us important feedback concerning the nature and the quality of the synthesis tools needed. It also shows that synthesis has to be an iterative procedure : in a first step, the synthesis tools generate a sub-optimal solution (e.g. important timing mismatches between the different processors), which can then be gradually refined in the next iteration steps (e.g. load balancing, area-time-power optimization).

This design-exercise also shows the need for an additional data path, which is very general (but slower) in its functionality and can be used in the non-critical paths (e.g. Processor 4). Such a data-path is now being created as a merger of the ALU and the Shift/Add/Compare unit.

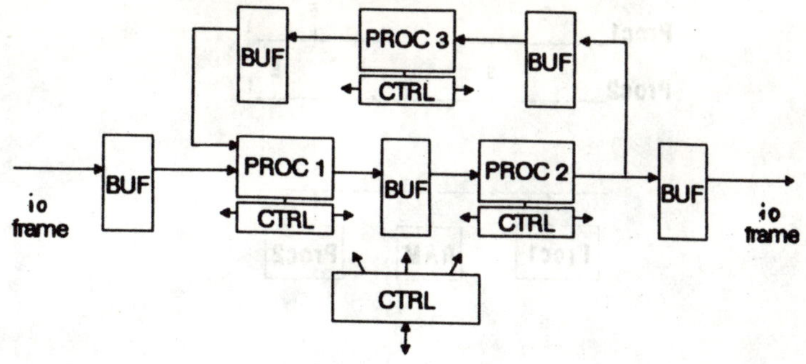

Fig. 3.1. : Composition of multi-processor architecture

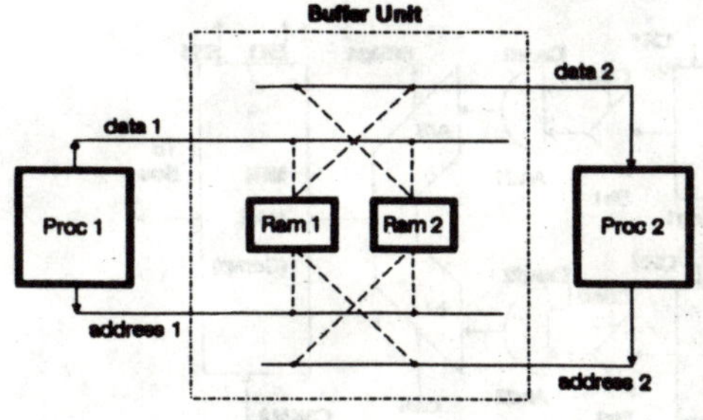

Fig. 3.2. : Interprocessor Communication protocol based on switched RAM's.

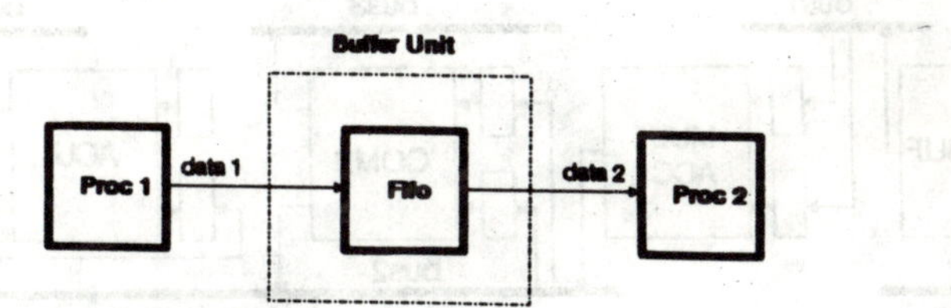

Fig. 3.3. : FIFO based interprocessor communication.

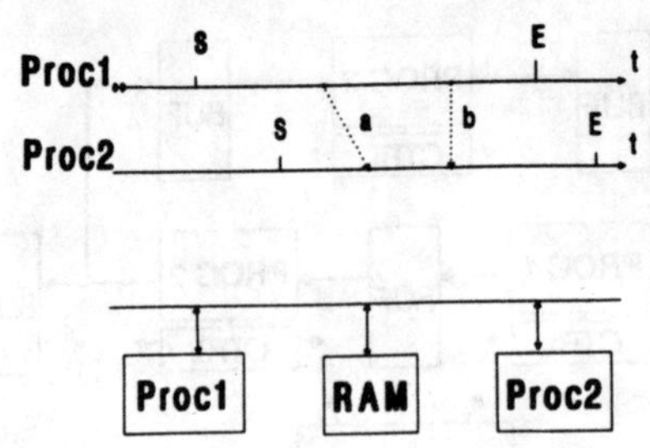

Fig. 3.4. : The application of skew on the programs of two processors makes it possible to use a single bus + RAM buffer protocol.

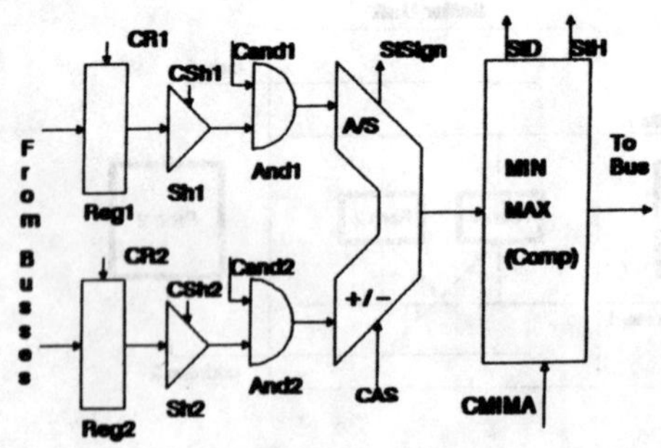

Fig. 3.5. : Shift/Add/Compare EXU.

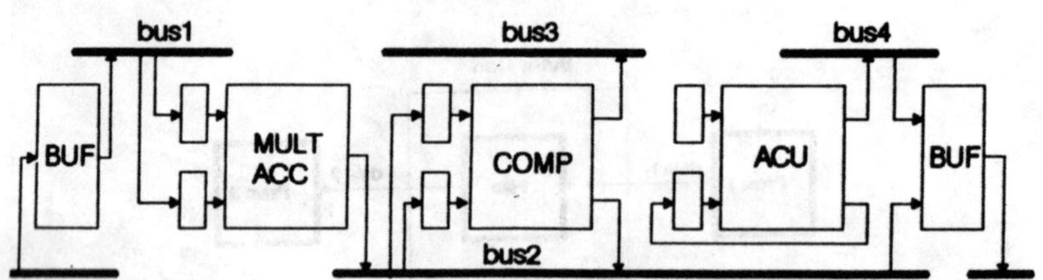

Fig. 3.6. : Example of customized processor data path (processor 1 of the pitch extractor example).

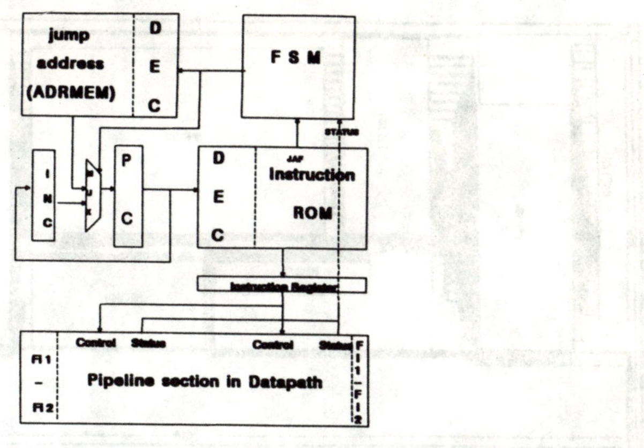

Fig. 3.7. : Multi-branch controller architecture

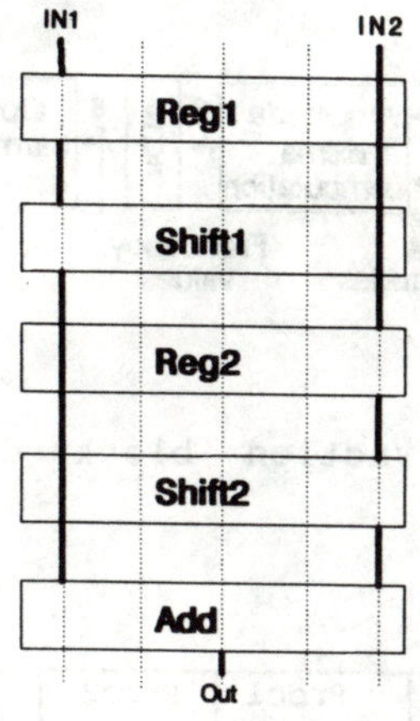

Fig. 3.8. : Inter-FBB and -EXU routing strategy using Metal-2 tracks.

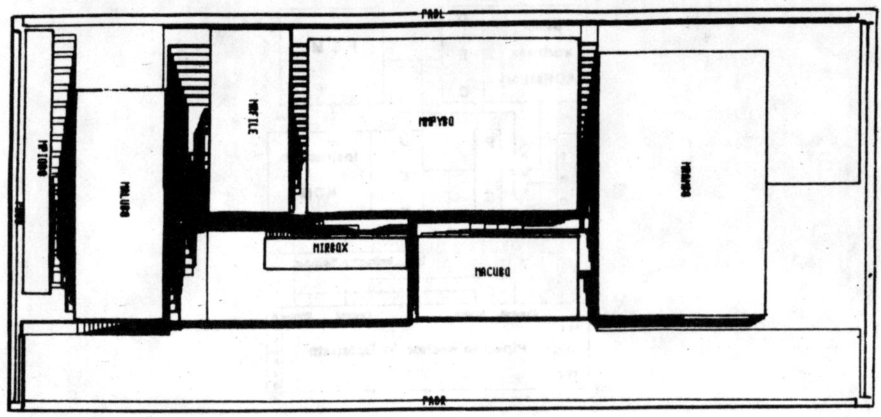

Fig. 3.9. : Processor floorplan using over the EXU-routing.

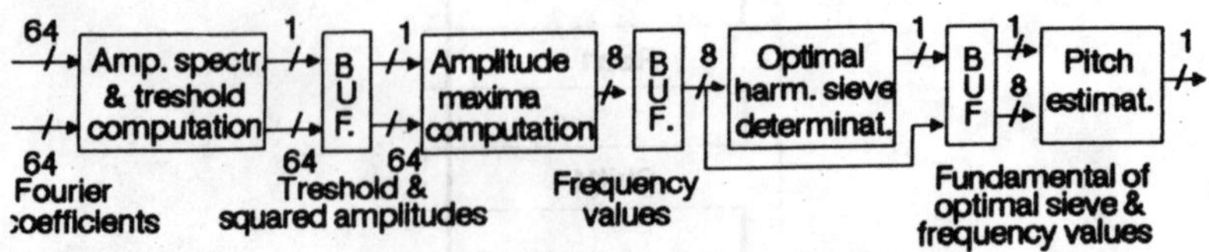

Fig. 3.10.: Pitch extraction block diagram and processor definition.

	Proc1	Proc2	Proc3	Proc4	Total
Data Path (mm2)	3.0	4.3	6.4	6.1	19.8
Control (mm2)	0.6	1.6	2.6	4.1	8.9
Buffer (mm2)	3.3	3.7	1.0	0.5	8.5
Total (mm2)	6.9	9.6	10.0	10.7	37.2
Cycles	1225	661	2854	240	2854

Table 3.1. : Pitch Extractor : results of mapping exercise (for 3 micron nwell CMOS process)

4. THE ARCHITECTURAL SYNTHESIS

4.1. The Design Synthesis Process : General Principles

The automatic synthesis of digital systems from behavioral descriptions has drawn a lot of attention recently. However, most of the techniques, described in literature ([Tho83], [Mar84], [Com85]), adopted a purely top-down methodology : the behavioral model of the system is translated into structure through a sequence of transformations and optimizations. This top-down methodology has the advantage of being very general and flexible, but lacks the ability to incorporate the information about the silicon implementation of the low-level primitives into the synthesis process. Besides, the generality of the approach (and thus the enormous dimensions of the design space to be explored) explains why these techniques failed to produce acceptable designs until now. We attempt to overcome this deficiency by restricting the synthesis process in following ways :

- The synthesis is targetted towards one single (but flexible) architecture, which limits the application range of the compiler but results in far more optimal designs. The architectural limitations result in a considerable pruning of the design search space and make it possible to use dedicated optimization techniques (exploiting the particular properties of the architecture in the form of heuristics). The developed synthesis tools should be flexible enough to accomodate changes or additions in the target architecture. This concept formed the base of the MacPitts [Sis82] and LAGER [Rab85] compilers, which produce acceptible designs but whose architecture is too restricted to span a large application field.
- The architectural primitives are a set of parameterizable modules, which are defined, described and specified in a module generation environment. This environment acts as a 'module knowledge database', which can be queried by the synthesis tools. In this way, design decisions can be based on physical implementation details (speed, area, power). This methodology, which is a mixture of top-down and bottom-up strategies, is the "meet-in-the-middle" strategy of section 2.

The above described ideas are reflected in the overall graph of the CATHEDRAL synthesis system (Fig. 4.1), which consists of three major parts : the architectural synthesis, the module generation and the floorplanning. In this section, we will show how this design philosophy allows for an efficient mapping of a high level behavioral specification of a DSP-algorithm into a customized multi-processor architecture.

An overview of the architectural synthesis process is given in Fig. 4.2. The algorithm to be implemented is described in SILAGE [Hil85], a language which is optimized for the description of DSP algorithms. SILAGE has been designed as an applicative language in an attempt to describe the signal flow nature of signal-processing algorithms. In a first step, the algorithm has to be partitioned over a set of processors. In order to optimize each of those processors for the particular subtask it has to perform, a customized datapath has to be generated (as a composition of the primitive modules or EXU's). This task is performed in the 'datapath synthesis'. The second step in the synthesis consists of the generation of a dedicated controller, which implements the algorithm on the synthesized datapath within the allotted time frame ('microcode scheduling' and 'controller synthesis'). It is interesting to note that the synthesis process can be considered as a stepwise refinement from an applicative (pure behavior, no notion of control, into a procedural (complete definition of the control sequence) representation of the algorithm.

From the design-exercises (cfr. section 3), we have found that the synthesis process has to be considered as an iterative and interactive optimization procedure with the cycle count and the area as main cost functions. Therefore, it is important that the (experienced) system designer can interface with the synthesis process at the different levels. This also implies that he should be able at all those levels to check if his interferences did not cause any disturbances from the requested behavior. Based on those considerations, we decided that the synthesis system should be an open system with multiple entry points (well defined and readable description languages) and verification (simulation) tools at all those levels. Therefore, we have defined a Register Transfer language which can be used to describe the behavior of a processor once the structure is known, a structural specification language and a controller definition language. Furthermore, the "pragma"-concept in SILAGE also allows for user interference with the synthesis process at a high level. User interference at the lower levels should however be avoided as much as possible, while iteration loops should also be kept local as much as possible.

The different tasks defined above will be discussed in detail in the following subsections. The example of the pitch-extractor (cfr. Section 3) for speech will be used as a test vehicle to illustrate the different stages in the compilation process.

4.2. The input Language (SILAGE)

There is no doubt that different application fields require different description or specification languages : a microprocessor designer is definitely using conceptual ideas, different from what e.g. a modem designer is considering. Therefore, we have selected SILAGE [Hil85] as the input description language for the CATHEDRAL II system.

SILAGE allows for the description of signal processing algorithms at a high level. The basic object in SILAGE is a 'signal', which is an infinite vector in time (comparable to a stream) and the basic operation is function application on those signals. In this way, a SILAGE description of an algorithm is equivalent to a signal flow graph, where nodes represent instances of functions and arcs represent the paths followed by the signals. SILAGE allows for the description of equations, decisions, loops, hierarchy, finite word length specifications and quantisation characteristics.

In some cases, the designer might wish to pass some (structural) hints to the compiler in order to guide the synthesis process in a certain direction. In SILAGE a construct named 'pragma' is used to pass this kind of information.

The SILAGE description of the amplitude and threshold computation part of the pitch extractor is presented in Figure 4.3. Notice the use of the function application, the arbitrary ordering of the equations and the use of a pragma to implement the AmplitudeSpectrum function on Processor 1.
The loop construct, used in the definition of the Amplitude Spectrum function, is only used to express repetitivity and does not imply any control or sequencing, as in the case in procedural languages.

A SILAGE simulator is currently under development. Simulation at this level can be used for algorithm development and for code debugging. It was found that a demand driven simulation technique gives the best speed performance.

4.3 The algorithm partitioning

A partitioning of the algorithm into a set of parallel operating processors is the first task in the synthesis process. A processor is defined here as an SIMD machine and consists of one central controller and a set of EXU's. Each processor can be synthesized independently from the other processors, this due to the transparent interprocessor communication protocols.

The partitioning itself can be based on following (often contradictory) observations :

- The identification of a set of algorithmic subtasks with diverging computational requirements (as can be observed from the systems block diagram).
- The existence of a set of cutsets, which can only pass through delay operations and which are selected such that the number of connections crossing the cutsetsis minimal. Function loops with only one delay operator (as e.g. an adaptive predictor) have to be implemented on one processor as a result of this observation.
- Changes in data rates can cause data transfer "bottlenecks", which are a natural separation between processors. This occurs fairly often in block-oriented algorithms, where an algorithmic subtask can only start its execution after a complete block of data (from a previous computation) is

available. This requires complete data-buffering and allows for the introduction of interprocessor 'pipelining'. The inherent capabilities to pipeline the algorithm for blocks of data can be detected from the data-dependency graph, as illustrated in Fig. 4.4. In this directed graph, the vertices represent computations and the edges indicate the required data. Parallel paths with identical operations on the vertices correspond to samples to be processed in the same way. These samples can be processed sequentially on the same processor. A point of convergence (or divergence) of the graph indicates a bottleneck, where data buffering is needed. This technique was used to partitioning the pitch extraction algorithm in four processors, as indicated in Figure 3.10.
- One of the most important factors in the algorithm partitioning is the load balancing, which makes sure that all the hardware is used in an efficient way. At the partitioning time, the load of an algorithmic subtask can only be guessed, based on a rough evaluation of the critical path in the data flow-graph of the algorithm. Therefore, it is clear that the partioning (as well as the whole synthesis process) will be an iterative task with stepwise refinement.

The above observations demonstrate that the processor definition is a non-trivial task, which is guided by area and timing considerations. In most cases however, an educated guess can be made by simple inspection. This is the philosophy we use in this first version of CATHEDRAL, where the processor definition is left to the user by means of pragma statements in the SILAGE description.

4.4 The Data Path Synthesis

The goal of the data path synthesis operation is to map the SILAGE description of a single processor into a minimal structure (in area), which will execute the algorithm within the allotted time frame. The result of the operation is a definition of the structure of the processor (in terms of parameterized modules and their interconnections) and a black box description of the controller, which consists of a unordered list of the operations which have to be performed on this hardware structure. This operation is generally called 'allocation' in literature [Tho83].

The synthesis operation will be heavily influenced by the selected target architecture. Therefore, we have selected a rule based system for most of the process, where the rules are used to describe the basic properties of the architectural components. This technique allows for a flexible definition of the architecture and eases the introduction of architectural changes or additions. However, a number of procedural (algorithm based) tasks can also be identified.

The program flow graph of the processor synthesis is pictured in Figure 4.5. A quick glance at this graph already reveals the basic synthesis guidelines : synthesis is a sequence of

relatively simple design steps. This is the only way to have
sufficiently simple rule bases. Design iterations should be
done locally as much as possible, though overall iterations
cannot be avoided.

The following sequence of design steps, which are currently
being implemented or are in the prototyping phase, can be
identified :
- Intelligent preprocessing of the SILAGE source code.
- Generation of the data path structure of a single processor,
  while also generating the register transfer description of
  the controller.
- Scheduling of the controller register transfer statements
  (mapping the algorithm on the time axis). This task will be
  described in section 4.5.
- Bus allocation and binding.

The implementation of these tasks will now be described in
detail.

4.4.1 The Preprocessor

The SILAGE description is to be parsed, preprocessed and
translated into a set of primitive equations. Preprocessing
includes.
- Expansion of SILAGE function calls and non-primitive
  operations into primitive operations.
- Simplification of the source equations and local
  transformations (finding common sub-expressions, reordering
  of singular variables, removal of constant expressions, loop
  handling, ...).

Preprocessing is not indispensable, but even a small set of
simple rules can save a lot of machine cycles in the final
implementation of the algorithm.

In order to understand the functions of the preprocessor, a
few definitions have to be given first.

Primitive SILAGE equations are equations that only contain
operators and function applications that can directly be
handled by an EXU of the target architecture. Of course this
depends on the available module library.

```
E.g.   s = a + (b / c) - e * e
       max[i] = MAX(a[i], a[i-1])
       IF a=b THEN ... ELSE ...
       FOR i:=1..64 HOLDS ...
       SAVE(thresh[64])
```

Basic equations on the other hand are identities or equations
which can be solved by an EXU in one single machine cycle.

```
E.g.   thresh(i) = thresh(i-1)
       [reg2, alu, a] = a
       s = [reg1, mult, a] * [reg2, mult, b]
```

A singular variable is a variable that actually appears in the
SILAGE description of an algorithm. A compound variable is a
variable that can be looked upon as an intermediate result,
which has no explicit name. The generation of these variables
depends on the kind of mapping rules used.

E. g.   s = a - b + c

In this example, the variables s, a, b and c are singular
variables, whereas the variables 'a-b' and 'b+c' (depending on
the precedence rules of the operators + and -) are compound
variables.

The preprocessing step contains a lot of actions that depend
on the type of EXU's that are available, so a PROLOG knowledge
base will be used. The program is currently under
development.

The output of the preprocessor is an expanded and reordered
list of primitive SILAGE equations, together with an
incomplete list of assignments of variables to registers (in
PROLOG format). This assignment list is fed forward to the
mapping tool and serves as a set of additional constraints to
the next synthesis step. By the way, another way of feeding
this kind of information forward to the mapping tool is by
using pragma's in the SILAGE source text.

E.g.   [reg1, mult, a] states that the variable 'a'
should be used in the first input of a multiplier EXU,
while [_,alu,_] forces the data path synthesis tool to
use an ALU. This information can be important in cases
where an operation can be scheduled on multiple units.

Example : The description of the first processor of the
pitch extractor after the preprocessor is given in
Figure 4.6.

4.4.2 The Data Path Allocation

In order to transform the preprocessed applicative high-level
language description of the ALGORITHM into a customized
processor STRUCTURE, the mapping tool has to assign primitive
SILAGE operations to EXU's and both singular and compound
variables to registers. It must also figure out how the
execution units should be interconnected by parallel busses,
in other words it should assign transfers of SILAGE variables
to busses. Finally it has to decide which parameter sets (as
word lengths, shift ranges, etc.) are to be assigned to the
functional building blocks (FBB's) inside the execution units.

In order to keep the mapping tool flexible with respect to
architectural changes, PROLOG was selected as the
implementation environment because of its excellent pattern
matching capabilities. The basic construct of the program is
a supervising predicate, which executes a READ-ADVICE-RULE

cycle. The READ function supplies a primitive high-level equation to the ADVICE function. ADVICE decides on the format of the particular equation which rules should be tried out in which order.

The simple advice function shown below should be read as :

"If the equation has the format of a subtraction, try rule 2a, if this does not work try rule 2b. If the equation has the format of an addition, try the next rules in the following order: rule 1u, 1v, 1w, 1x. If no rule succeeds, then the equation cannot be solved."

```
adviceEq(_=_-_,    rule2a).
adviceEq(_=_-_,    rule2b):-!.
adviceEq(_=_+_,    rule1u).
adviceEq(_=_+_,    rule1v).
adviceEq(_=_+_,    rule1w).
adviceEq(_=_+_,    rule1x):-!.
```

The RULES try to unify the high-level equation with a low-level equation, of which it knows the basic execution units of the target architecture can deal with. Such low-level equations are for instance the addition, subtraction, division and multiplication of variables, but also comparisons can be handled, allowing for decision making and looping. In most cases the unification can only be done if a number of conditions are satisfied : an addition of two variables is only possible if we succeed in getting those variables in the input registers of the adder execution unit.
Moreover these rules can only succeed if they are consistent with results obtained by rules that were fired earlier.
Intermediate results are stored in the format of an assignment list: for every variable it is recorded in which register of which EXU it is needed.

The conditions which have to be satisfied in order for a unification to succeed, are expressed as a mixed set of new high-level equations, low-level equations and often some identities. This set replaces the original high-level equation in the original set of equations. The READ-ADVICE-RULE cycle continues (and eventually backtracks) until no high-level equations are left. At that moment the equation set contains all the information that is needed to generate both the processor structure and the register transfer description of the processor controller (which can be considered to be one of the parameters of the controller EXU). The assignment list now contains all registers that are used.

Extra conditions which are not consistent with the assignment list might cause backtracking of the ADVICE function. In order to avoid endless backtracking, preprocessing is quite useful.

An example of a simple rule (in natural language) is given below :

"If the equation is of the format VAR = OP1 + OP2, and if
the assignments [reg1, alu, OP1] and [reg2, alu, OP2] are
consistent with the current assignment list, then rule 1v
succeeds, the equivalent set of equations is
```
        [
        VAR = [reg1, alu, OP1] + [reg2, alu, OP2],
        [reg1, alu, OP1] = OP1,
        [reg2, alu, OP2] = OP2,
        OP1=OP1,
        OP2=OP2
        ],
```
you may display the message 'Addition_1'
and finally you may update the assignment list."

For different tasks there are different inference engines
implemented in the program, each of them using its own advice
function and its own rule set. It is important however to
mention that all the rules inside one rule set have the same
global format, so one might consider building a set of
automatic rule compilers doing translation and consistency
check. Anyway, adding a new set of rules to support a new
type of execution unit is a problem that can be handled rather
elegantly. By using this technique, the number of predicates
in the PROLOG database remains limited and controllable.
Moreover, the ordering of the rules in the PROLOG database is
not relevant because of the ADVICE function. There are rule
bases in the program for mapping high-level equations into
low-level equations (as e.g. loop constructs), rules for
mapping conditions into low-level equations, and rules for
interconnecting EXU's.

The result of the mapping operation is a description of the
structure of the processor (in terms of basic components and
interconnections) and a description of the controller as a set
of unordered register transfer operations, which have to be
executed on this structure. Note that at that point, we still
assume that any necessary transfer of the output of an EXU to
any register can be made. This means that dedicated busses for
all variables (both singular or compound variables) are
assumed. After the scheduling of the transfers (cfr. 4.5), it
will be investigated which busses can be merged (4.4.3).

For the example of the Amplitude Computation processor, the
mapping tool produced a processor data path, containing a
multiplier/accumulator and a comparator, as pictured in Figure
3.6. Figure 4.7 contains the generated set of register
transfers, to be executed on the data path. This description
(called ATOMICS) serves as input to the microcode scheduler.

4.4.3 The Bus Assignment

As stated before, the register transfer description of the
controller is generated assigning dedicated busses to all

variables. With this description, the register transfer code is scheduled in time [see section 4.5]. We now know the minimal number of machine cycles that is needed, and the occupation of each bus (Fig. 4.8). It is the intention to merge as many busses as possible in order to minimize the number of busses inside a single processor. Rescheduling is necessary after merging busses. Only when the interconnection structure is decided upon, the micro code can be generated.

A simplified 0-1 Optimization Knapsack algorithm [Pap82] is used to find out what the absolute minimum of busses is, not taking into account that collisions of variables on a bus will increase the number of machine cycles. Even when no collisions take place and therefore the number of cycles would (probably) not increase, it is impossible to find any other combinations of merged busses that would result in a fewer number of busses inside the processor.

The next step in the procedure is dependent upon the cost function. In most cases, the number of busses will be limited to a maximum (number of free metal tracks over the EXU's). The optimization criterion is to find this bus merger, which causes a minimal number of bus conflicts without exceeding the allowed number of busses. Following strategy to realize this is proposed : Generate as many busses as Knapsack indicates. Reorder them, the one with the largest occupation on top. Merge the dedicated busses, generating a minimal number of conflicts by adding the largest unmerged busses to the smallest merged busses with the minimal number of collisions. Proceed with this operation till the maximal number of busses is reached or till no further improvement can be obtained. Other cost functions can also be handled using similar techniques.

At this moment, a rescheduling of the register transfer operations has to be performed. This generates the actual number of cycles needed (after bus merger). Based on this, the designer can decide on extra iteration steps if needed.

> e.g. : for the AmplitudeSpectrum example, it was found that a total of four busses was sufficient to avoid bus contention.

4.4.4 Present Results

The techniques presented above have been implemented and have shown the feasibility to realize design synthesis in an acceptable and yet flexible way. A number of practical test-cases however have shown that the present data path mapping tool performs only local operations and is not able to perform a global optimization with respect to the time and area specifications. Therefore, we are presently studying extensions of the preprocessor to more global optimization, based on critical path analysis and estimation of the mathematical complexity of the algorithm. This will result in a set of pre-assignments for the data path mapper, which will fill in the details (register transfers, bus-assignments, distribution of non-critical operations).

4.5 The Microcode Scheduling Operation

The input to the scheduling problem is a register-transfer (RT) description of the algorithm in a dedicated representation language. The RT's are fundamental operations from a control point of view, since each RT is to be realized in a single machine-cycle. For each transfer, the corresponding RT-statement describes all hardware requirements, as determined by the architecture synthesis tool: source and destination register-elements, the required operation mode of the programmable operators in the data path and the bus-utilization. An example of such an RT-description has been given in Figure 4.7.

In the RT-language, no timing of operations has been imposed. The goal of the scheduling process is the computation of an optimal mapping of the RT-operations to time axis (machine-cycles), i.e. with a globally minimal cycle count. The algorithms are scheduled statically (at compile-time) as opposed to dynamically (at run-time), thus saving complex data-driven control protocols.

Note that besides the RT-input, the scheduler also needs some additional information about the structure of the data path (as generated by the synthesis tool) and the selected controller (the amount of pipelining).

From the scheduled RT's, a symbolic microcode can be derived. This mainly involves : (1) transformation of all RT's, scheduled at the same cycle, into a set of instructions for that cycle which hold under disjoint conditions, (2) computation of conditional jump-addresses for every instruction. This microcode serves as the input to the controller generation environment of subsection 4.6.

In this section, we will consequently discuss the optimization constraints, to be handled by the scheduler, the scheduling techniques used when no repetitive constructs are present, the extension towards loop constructs and some perspectives.

4.5.1 Optimization Constraints

Several timing constraints have to be taken into account during the computation of the optimal schedule. Three basic types of constraints are distinguished (as illustrated in Figure 4.9) :
- Data precedence constraints : An operation writing a variable in a register-element is to be scheduled before any operation reading this variable from the register-element. At the scheduling time, it is assumed that those registers are implemented in register-files with as many storage fields as required by the optimal schedule (unless explicitly defined as e.g. pipeline register). Therefore no

additional optimization constraints are required, to prevent overwriting of still "active" data. An analysis of the variable lifetimes in the optimal schedule will finally allow to dimension all register-files.
- Resource-allocation constraints : different RT's which require the same resources (e.g. source register-files, destination register-files, busses, operators) may not be scheduled in the same machine-cycle. Writing in and reading from the same register-file is allowed in the same cycle.
- Controller pipelining constraints : Conditional operations are executed depending on an evaluation of status signals, produced by operators in the data path. A minimal time difference between the instructions generating the status signals, and the conditional instruction based on these status signals, is required (e.g. 3 cycles for the currently used multibranch controller). This figure is dependent on the data path and controller pipelining scheme. This type of constraint can be modeled as a data precedence constraint, where the depending variable corresponds to the status signal.

The above mentioned constraints can be derived from the RT-description of the algorithm, using a set of data-precedence and resource allocation conflict rules. An efficient data-structure, allowing for fast constraint-detection is required. No constraints have to be taken into account between conditional operations with disjoint condition-fields.

4.5.2 Scheduling Techniques for Non-Repetitive Programs

In a first step, the scheduling technique used for non-repetitive programs will be explained (where non-repetitive means that no loop constructs are allowed). Note that repetitive programs can be transformed towards the non-repetitive form using a technique called 'loop unfolding'. First of all, a mathematical model for the scheduling of non-repetitive programs will be described. In this model, an optimization function is provided and the previously mentioned constraints are described. We will denote an operation by o_i, and the cycle-number in which this operation is scheduled by t_i.

Data precedence constraints can be modeled as weighted edges in a directed loop free graph ("data precedence graph"), in which the vertices represent the operations. The weights are the minimal time differences between the operations. The data precedence constraints are labeled m, for m := 1..D. For any m, the data precedence constraint between operation o_k (m) (to be executed at cycle t_k (m)) and operation o_q (m) (at t_q (m)), where o_k (m) precedes o_q (m) with at least d(m) cycles, can be expressed by following set of inequalities :

$$t_k \; (m) - t_q \; (m) \leqslant = - d(m) \text{ for all } m := 1..D \qquad (4.1)$$

As indicated before, controller pipeline constraints can be

modeled as data precedence constraints, with an appropriate
choice for the d(m)-value.

Constraints to prevent resource allocation conflicts can be
modeled as difference equations. The pairs of potentially
conflicting operations are labeled n, for n := 1..C. A
minimal number of conflict pairs can be found by excluding
those pairs for which there is a directed path with length = 0
in the data precedence graph. The n-th pair is denoted
$(o_i(n), o_j(n))$, the corresponding cycle-numbers are $t_i(n)$ and
$t_j(n)$. For every pair of possibly conflicting operations
$o_j(n)$ and $o_i(n)$, only one node-pair is considered. The
resource allocation restrictions can then be formalized as :

$$t_i(n) \neq t_j(n) \text{ for all } n := 1..C \qquad (4.2)$$

The difference constraints (4.2) are hard to handle by
optimization programs and can be replaced by the set of linear
equalities (4.3), where the additionally introduced variable
d(n) is 0 when $t_i(n) < t_j(n)$ and 1 when $t_i(n) > t_j(n)$.

$$t_j(n) - t_i(n) \leqslant T.d(n)-1 \qquad (4.3)$$
$$t_i(n) - t_j(n) \leqslant T-1-T.d(n)$$
$$d(n) \leqslant 1 \text{ and } d(n) \text{ integer}$$
$$\text{for all } n := 1..C$$

where T is some upper bound for the number of cycles (e.g.
equal to the number of operations). The idea of translating a
difference constraint into a set of inequality constraints is
due to [Pap82].

In order to derive an optimization function, dummy operations
o_{in} and o_{out} are defined as follows :

$$t_{in} \leqslant t_i \text{ for all } o_i \text{ without predecessor} \qquad (4.4).$$

$$t_{out} \geqslant t_j \text{ for all } o_j \text{ without successor}$$

in the data precedence graph.

The optimization target then is : minimize $t_{out} - t_{in}$ (4.5).

(4.1), (4.3) and (4.5) are an instance of integer linear
programming (I.L.P.) problem, i.e. a linear program in which
the solutions (t_i - and d-values) must be integer).

The computation time of algorithms, solving the I.L.P.
problem, is known to grow exponentially with the problem size.
In [Pap82], a general scheduling problem is described and
proved to belong to the class of NP-complete problems. We
expect the above problem to be NP-complete as well. The
NP-complete character is introduced solely by the resource
allocation constraints. When only data precedence constraints
have to be taken into account, polynomial-time leveling
algorithms can be applied to find the optimal schedule.

However, since the programs which have to be scheduled are of a small to medium size (20 to 500 nodes in the precedence graph are typical) and since it is important to approximate the ·minimal execution time as close as possible, a prototype scheduling tool (called ATOMICS) has been implemented based on the I.L.P. approach. Experiments have shown that programs with a moderate number of resource allocation conflict constraints can be scheduled in a reasonable time. As disjoint conditions lead to superfluous constraints between the corresponding conditional operations, the occurrence of many conditional operations in general improves the computational efficiency. It is possible to improve the run-time by introducing heuristic techniques. Heuristics could e.g. be applied to make a guess on whether the difference equation (4.2) would in practice result in an inequality of the type $t_i(n) < t_j(n)$ or rather of the type $t_i(n) > t_j(n)$.

4.5.3 Scheduling Techniques for Repetitive Programs

Due to its applicative nature, program loops do not have any control meaning in the RT-description and can only be considered as a compact notational format (equivalent to the SILAGE loop construct). However, the area-efficiency of the controller can be improved drastically (especially the ROM-area), when loop constructs are allowed in the microcode. Therefore, a FOR-loop construct has been introduced in the RT input-language, which can be considered as a user-pragma, i.e. as a hint to the compiler to exploit the repetitivity of the algorithm to reduce the controller-area. In order to maintain the uniqueness of the variable names (as needed in the applicative context), indexed notations are used. In a first phase, the looping mechanism of the RT-language will be copied directly into the microcode without an attempt towards optimization.

Besides the area gain for the controller blocks, the introduction of FOR-loops also has the advantage that it results in a reduction of the scheduling problem. The loop-construct imposes a limited ordering of the operations and reduces the number of RT-instructions to be scheduled.

The techniques to schedule repetitive programs can be derived by extending the mathematical model for non-repetitive programs described in 4.5.2. We only indicate the most important extensions :

- Schedule times of operations are replaced by so called "potentials" of operations. Operations in successive loop-iterations, represented by the same RT-statement, have the same potential.
- The scheduling problem is now redefined as the computation of an optimal set of potentials for the RT-operations, i.e. so that the overall cycle count is minimized. An optimization function can be derived, as a linear function of the potentials of a set of dummy operations, representing

initial or final operations in the bodies of the respective loops.
- A generalized set of data precedence and resource allocation conflict constraints can be derived. This set also models special requirements resulting from the looping mechanism, e.g. precedence constraints between operations in successive iterations of a loop.

A global I.L.P. can be derived to compute the optimal schedule for the entire program in one shot. Only the N potentials have to be computed where N equals the number of RT-statements. However, in order to master the run time of the scheduler, one can also follow a recursive scheduling algorithm, where the contents of the FOR-loops are scheduled separately using the I.L.P. technique and are considered as fixed blocks at the next higher hierarchy level. This technique, which leads to sub-optimal solutions, has been implemented in the current version of the scheduler.

Besides the microcode definition, ATOMICS also generates a register transfer description of the controller. This can be combined with the processor architecture, as defined by the data path synthesis, to form a simulation model for the complete processor. In this way, it is possible to compare the behavior of the synthesized algorithm with the original specifications, as defined in SILAGE.

The ATOMICS tool has been used to schedule the AmplitudeSpectrum Processor, described by the Register Transfer description of Figure 4.7. This resulted in a total of 326 cycles, where the optimal solution counts only 196 cycles. The difference is caused by the hierarchical loop scheduling technique, which prevents overlaps between the execution of the consecutive loops. It is possible to manipulate the RT-description in such a way that loop overlaps are introduced and that the scheduler also generates the minimal number of cycles (being 196). Techniques to perform this manipulation automatically are currently under investigation.

4.5.4 Perspectives

The presented scheduling tool performs reasonably well on smaller examples, but tends to be too slow for larger programs. The run time becomes important when the scheduler is called iteratively by the synthesis tools, which happens e.g. in the bus scheduling technique described higher. Therefore, we are currently investigating a heuristic driven, graph based scheduling technique, which would replace the I.L.P. routine. Another topic, which is being studied, is the automatic handling of loop overlaps. Finally, we are also considering a new definition of the optimization function, which tries to minimize the number of registers instead of the cycle count and limits the number of cycles to an upper boundary as defined by the specifications.

4.6 The Controller Generation Environment

The result of the scheduling process is a symbolic microcode description of the processor controller. This description is functionally equivalent to a Finite State Machine (FSM) and can be implemented on a variety of controller architectures (as long as the these architectures have the same amount of internal pipelining, since this influences the scheduling process and thus also the obtained microcode). The same microcode can e.g. be implemented on a single FSM or on a ROM and PC based controller. The efficiency of a certain controller depends upon the type of the algorithm under consideration. It is therefore important to be able to compare the implementations of a controller description on a number of available architectures.

This requires a unified environment, as pictured in Figure 4.10. A unique input language is used to describe the requested behavior of the controller. This description tries to capture the FSM-nature of the controller, in terms of inputs, outputs, states and next states. It supports symbolic covers for states and symbolic descriptions of the output behavior. An example of such an input description is given in Figure 4.11. Alfa, Beta, etc. are symbolic notations for the different states of the controller (Note that no coding of the states is implied). Constructs as 'alu : inc' are symbolic output covers (macro's), used to compact the size of the description and to avoid the repetition of identical output patterns. A table driven preprocessor will be used expand this covers into the appropriate output signals (e.g. alu_add = 1,alu_carry = 1, alu_inpa = 1, alu_inpb = 0 for the alu : inc operation).

The body of the Controller Generation Environment (CGE) consists of a number of procedures, which describe how an input description can be mapped into a particular controller architecture. For each architecture, such a procedure has to be provided. Such an assembly procedure normally consists of a number of optimization and minimization steps (state assignment, logic minimization, etc.) and a layout generation phase. The goal of the CGE is to ease the introduction of new architectures and to avoid the implementation of the same functions over and over again. Therefore, the CGE provides a library of common functions as state assignment, logic minimization, multilevel logic synthesis, partitioning. It also supports constructs to assemble regular arrays using symbolic layout techniques.

It is important to notice the similarity between the Controller Generation Environment and the Module Generation Environment, as described in section 5. In both cases, the environment is used for the simple introduction of new regular constructs, which will be used multiple times. The underlying library of support tools is however of a different nature.

At the present time, we are working towards a prototype version of the CGE, using the multi-branch controller of section 3 as a target. The tools used for the generation of this controller are basically the PLA-minimization and generation tools, described in [Bar85].

4.7 Conclusions

The present results, obtained from a number of test-designs, indicate that our synthesis approach, as outlined in Figure 4.2, is a viable strategy, which is capable to produce acceptable designs. The open nature of the system's architecture also allows the experienced designer to perform further optimizations if needed.

This step should however be avoided as much as possible. Therefore, we feel that further research is needed in following areas :
- Extension of the synthesis techniques to include more global optimization techniques with more elaborated cost functions.
- Extended flexibility with respect to architectural changes or additions.
- Algorithm partitioning and load balancing.

These subjects will be the main target of our synthesis research and development efforts in the near future.

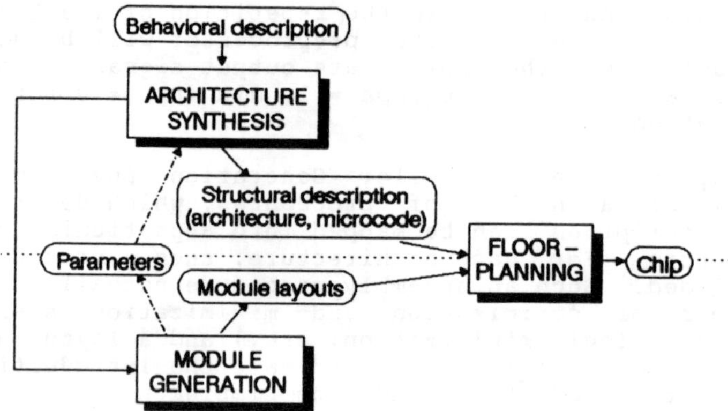

Fig. 4.1. : The basic functions in CATHEDRAL II : Architecture Synthesis, Module Generation and Floorplanning.

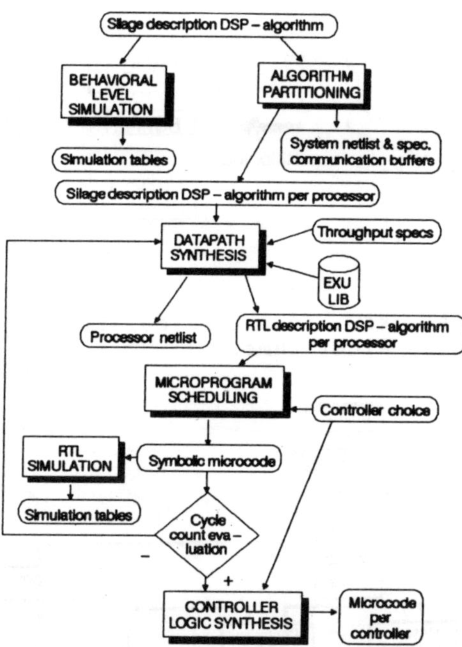

Fig. 4.2. : General Overview of functions, languages and simulators in the CATHEDRAL II synthesis process.

```
/* SILAGE Description of Pitch extractor*/
/* num8 and num16 are defined as 8 and 16 bit fractional
   number representations */

func PitchExtractor (Re[64], Im[64] : num8) Pitch : num8=
begin
   Spectrum = AmplitudeSpectrum(Re, Im);
   ...
end


   func AmplitudeSpectrum(freal, fimag : WORD8[])
                          Ampl : WORD16[]; Thresh : WORD16 =
   begin
      Max[0] = 0;
      Thresh = Max[NrOfPoints];
      (i : 1 .. NrOfPoints)::
      begin
         Max[i] = if (Ampl[i] > Max[i-1]) -> Ampl[i] || Max[i-1] fi;
         Ampl[i] = freal[i-1] * freal[i-1] + fimag[i-1] * fimag[i-1]
      end;
   end;

   pragma "processor (1, AmplitudeSpectrum)";
```

Fig. 4.3. : Silage description of Amplitude and Threshold Computation in the Pitch Extractor.

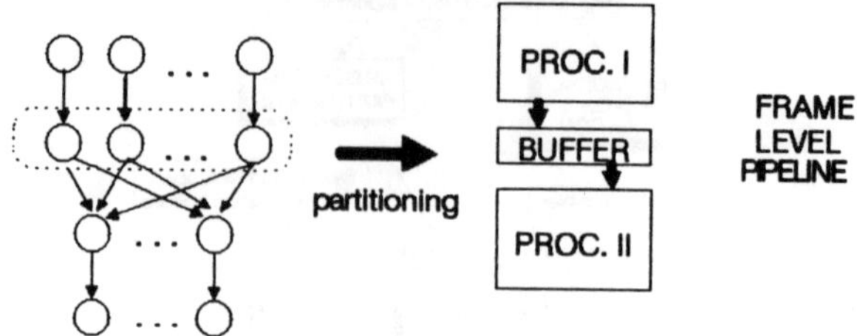

Fig. 4.4. : Bottlenecks   in   the   dataflow   graph   of   block
          oriented algorithms.

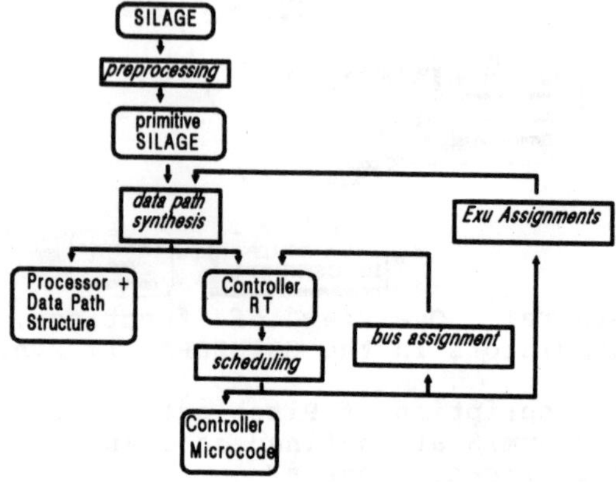

Fig. 4.5. : The Data Path Synthesis : general program flow
          graph.

```
[
    thresh(0) = 0,
    for i:=1..64 holds
    [
        ampl(i) = a(i)*a(i) + b(i)*b(i),
        if thresh(i-1) =< ampl(i)
        then [thresh(i) = ampl(i)      ]
        else [thresh(i) = thresh(i-1)]
    ],
    save(thresh(64))
]
```

together with an initialized assignment list :

```
[_,   mult,  a(i)        ]
[_,   mult,  b(i)        ]
[reg, beram, thresh(64)]
```

Fig. 4.6. : Amplitude Processor after preprocessing.

```
program AmplitudeSpectrum;
{ Note : *Variable denotes the RAM address of the particular
    variable }

begin
    thresh[0]:reg6 <- ZERO:reg5 | bus3 = thresh[0];
    i[0]:reg7 <- ZERO:reg7 | acu = pass;
    *ampl[1]:reg8 <- *ampl[1]:rom | acu = inp;

    for i := 1..64 holds
    begin
            i[i]:reg7 <- i[i-1]:reg7 | acu = modulo;
            a[i]:reg1 <- a[i]:ram1 | bus1 = a[i];
            a[i]:reg2 <- a[i]:ram1 | bus1 = a[i];
            a[i]:reg4 <- a[i]:reg1, a[i]:reg2;
            a[i]:reg3 <- a[i]:reg4;
      b[i]:reg1 <- b[i]:ram1 | bus1 = b[i];
            b[i]:reg2 <- b[i]:ram1 | bus1 = b[i];
            b[i]:reg4 <- b[i]:reg1, b[i]:reg2;
            ampl[i]:reg5 <- b[i]:reg4, a[i]:reg3 |
        mult = accum, bus2 = ampl[i];
            ampl[i]:regr2 <- b[i]:reg4, a[i]:reg3 |
        mult = accum, bus2 = ampl[i];
            ampl[i]:ram2 <- ampl[i]:regr2, *ampl[i]:rega2;
            *ampl[i]:rega2 <- i[i-1]:reg7, *ampl[1]:reg8 |
        acu = add, bus4 = *ampl[i];
            thresh[i]:reg6 <- ampl[i]:reg5, thresh[i-1]:reg6 |
        alu = mima, bus3 = thresh[i];
    end

    thresh[64]:regr2 <- ZERO:reg5, thresh[64]:reg6 |
        alu = mima, bus2 = thresh[64];
    *thresh[64]:rega2 <- *thresh[64]:rom |
        acu = inp, bus4 = *thresh[64];
    thresh[64]:ram2 <- thresh[64]:regr2, *thresh[64]:rega2;
end.
```

Fig. 4.7. : Register Transfer Description of AmplitudeSpectrum

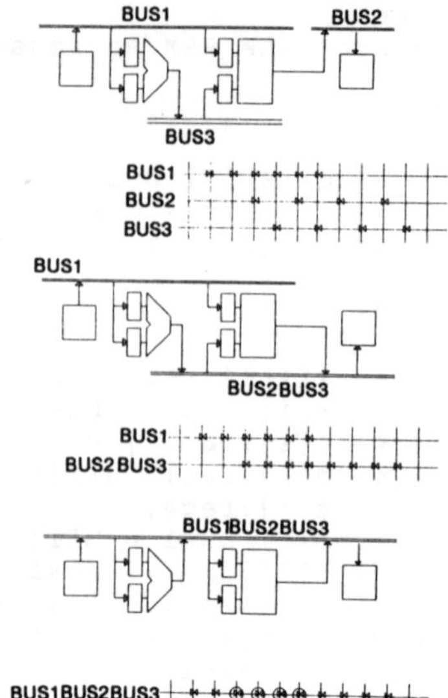

Fig. 4.8. : Bus assignment : occupation and conflict diagram's.

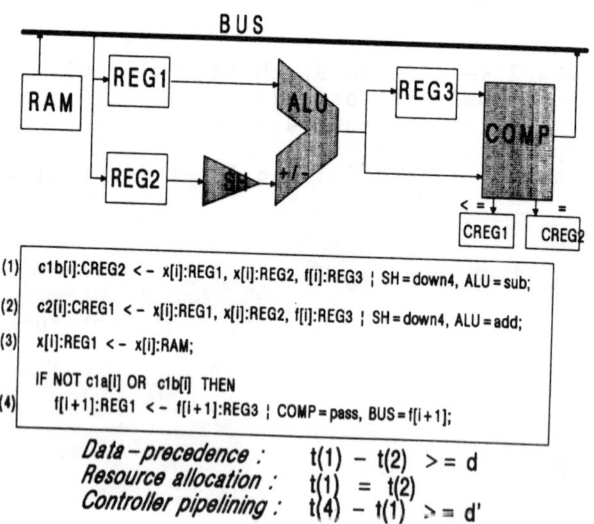

Fig. 4.9. : Constraints in the Microcode Scheduling.

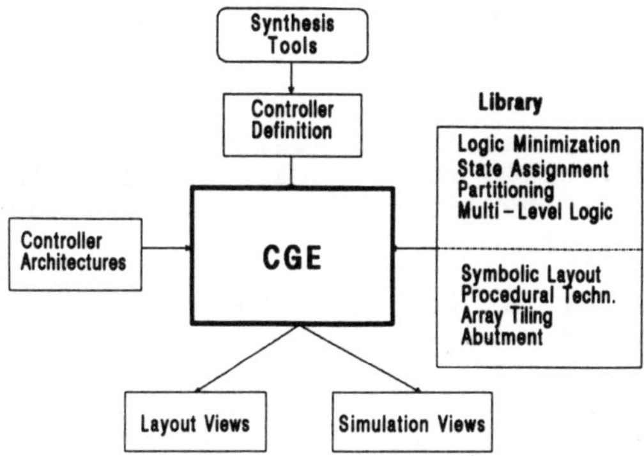

Fig. 4.10 : Controller Generation Environment : block diagram

```
# CONTROL 'pitch_extractor Processor 1';

# LIBRARY 'cathedral.lib'
/* The library contains the macro-models of the control
for the different EXU's */

# FSM
    <Alfa>
                UPD cstart := start_flag;
                JMP IF (cstart) THEN Alfa
                                ELSE Beta;
    <Beta>
                UPD state[1] := alu_sign & acu_overflow;
                CTL alu:inc, acu:modulo,
                    acu_reg1:r[1]w[2];
                JMP Gamma ;
    <Gamma>
                ...
```

Fig. 4.11 : Input Description for CGE.

5.   THE MODULE GENERATOR ENVIRONMENT

5.1. Introduction

In traditional systems the design of a chip is done by composing larger cells out of smaller ones. In this approach each cell is built by interactively placing the smaller cells on specific coordinates.

In modern designs a lot of the blocks have some kind of structure, they are built from smaller cells or groups of cells according to a certain pattern (rows,columns ...). Examples of this type of blocks are adders, multipliers, datapaths etc. It also happens that the same functional block is needed with different complexities, like an 8-, a 12- and a 16 bit adder. In the traditional design style we would create 3 different cells containing respectively 8, 12 and 16 full adder cells. The effort to create and check these functions is repeated over and over again.

A very simple procedure could be written to place and connect N full adder cells. This procedure once written and tested could then generate the netlist or the layout for any type of adder if the designer specifies what value of N he/she needs. Such a procedure is what is called a module generator. A first reason for the introduction of module generators is thus that, once they have been created, they reduce the design time.

A second reason is that, instead of designing at the gate level, system designers decompose their system specification in terms of modules like multipliers, ALU, memory .. which are of MSI and even LSI complexity. The system designers can assemble these modules in nearly the same way as they did in PCB design but they lack the skills to design the internals of these blocks. This is the role of the silicon specialist.

In the CATHEDRAL II system the decomposition is done by the synthesis and optimisation tools, again in terms of the same rather complex blocks. An automatic placement and routing program can assemble the layout of the final chip, but it can not design the internals of eg. an ADD-SHIFT-COMPARE ALU.

Since system designers want to use these functional blocks in a number of applications and combinations, they must be designed in such a way they can be tuned to the application and adapted to their environment. Furthermore if synthesis tools in a silicon compiler try to optimise the global chip area and performance, by changing the parameter values of individual modules, they need fast feedback on the resulting module characteristics. This means that silicon design becomes intermixed with programming since the only way to reach the required flexibility is to make use of parameterized software procedures that generate modules.

5.2. Aspects of module generation

If we want to work with module generators we need to provide facilities for three activities : creation, generation and adaption.

CREATING a module generator means to specify how a module must be constructed. This requires the possibility to define the leaf-cells as well as the composition rules of the assembly.
In contrast to handcrafted designs where cells are fixed and placed at specific coordinates, we must provide powerful constructs and the supporting algorithms to express and implement complex composition rules like : repetitive- and conditional relative placement, automatic fitting and compaction of cells and assemblies as well as interconnection via abutment and routing, to guarantee correct results for the complete set of allowable parameter combinations.

When a system designer needs a module it must be correctly GENERATED for the set of parameter values he/she has specified. The user can be a system designer but in the case of a silicon compiler like CATHEDRAL it can be one of the higher level design tools.

Since the investment in the creation of a module generator is a lot higher than for one specific instance, we want to extend the lifetime of a module generator. This implies that the construction rules should be flexible enough to allow a module generator to automatically ADAPT to design rules of different foundries or a technology upgrade.

5.3. The different views of a module

A lot of people confuse silicon compilation with layout generation. Although the layout of a module is the final goal in any design methodology, a lot more information is needed in the design process. This information can be divided in a number of views, each containing the relevant data for a specific step in the design cycle (eg gate- and transistor views for logic- and circuit level simulation).

In a real silicon compilation environment, including high level synthesis and optimisation, the four views needed are : functional, layout, timing and test (see also Fig. 2.4).

The FUNCTIONAL view must be provided to allow to check the correct operation of assembled processors or chips. It also allows to evaluate architectural trade-offs at the highest level (eg. correct dataflow in parallel- or pipelined processors).

The LAYOUT view must be created as the final description of the module, and will be used to assemble the artwork of the chip.

To check that a module or processor works correctly under the specified timing conditions a TIMING model must be available that accurately describes the input/output delay (or possible setup) relations. Since the decision to split an execution unit into parallel or pipelined units depends on the relative time needed by the different execution units, or to optimise the ordering of instructions, this timing view must be fairly accurate.

The final problem of most IC designs is how to test them. Modules can become rather complex and their repetitive nature does not make them very well suited for automatic test pattern generation. For many of the functions realised as a module generator specific test strategies exist. The TEST view should contain the testpattern to apply the appropriate strategy for the module.

Since the functional- and test views can be provided by the designer in the form of a simple file we will concentrate on the layout and timing views.

5.4. Specifications for a module generation environment

To describe a module we have to design the leaf cells and capture the assembly procedure which consists of four types of information :
- STRUCTURAL :   defines the sub-modules of a module
- CONNECTIVITY : defines how the sub-modules must be interconnected and where are the input and output terminals of a module.
- TOPOLOGICAL : defines the relative placement of the sub-modules.
- CONSTRUCTION : how sub-modules fit together and which of the interconnections are realised by abutment or routing.

The first two are needed in the top down specification of a module and for logic and rough estimate circuit simulation. The last two are required to generate the layout.

Furthermore since the generators must be able to describe a set of modules rather than only one instance, we must be able to define STRUCTURAL PARAMETERS, like the word-length of an adder or the depth of a ROM, as well as CONDITIONAL COMPOSITION PARAMETERS to express the conditional buildup of a module in terms of FBB's (eg the ADD-SHIFT-COMPARE module may be reduced to an ADD-COMPARE if the shift operation is not needed by the algorithm being implemented). The structural and topological data must then be defined as a function of these parameters and constructs like LOOPS and IF .. THEN .. ELSE .. must be provided.

Programs to create modules have been implemented using traditional programming languages like PASCAL and C [Bar85, Bur85, Mat83]. The main disadvantage of these systems is the delay introduced by the compilation and linking steps, from the time the designer specifies the module and the moment he can check the results of certain actions or parameter assignments.

Our idea was to create an <u>interactive</u> environment, where the silicon designer has <u>immediate</u> feedback in graphics form each time a command of <u>the module</u> generator is defined. In this way mistakes are detected exactly where they are made and can be immediately corrected. Furthermore since programming is a discipline traditional silicon designers feel very unconfortable with, we want to hide as many as possible of the programming aspects of creating a module generator for them.

To remain flexible in adapting the input language we decided to implement the command parsing of the first version of the system in LISP while for some of the supporting algorithms, like the compactor and router we use PASCAL and C for reason of efficiency on the VAX or other traditional workstation.

In the rest of the paper we present the main features of MGE, the module generator environment, that is under development at IMEC. In 5.5 we describe CAMELEON our symbolic layout editor and compactor that is used for designing the leaf cells and an efficient algorithm for automatic fitting of cell terminals to allow connection by abutment. In 5.6 an overview of the commands for module composition and the user environment implemented in LISP is given. In section 5.7 we describe DIALOG, a rule based system to check electrical rules, and the SLOCOP program that checks clocking rules and generates accurate hierarchical timing models. Finally some examples are presented in 5.7.

5.5. Leaf cell design and fitting of leaf cells

From the PLASCO project [Bar85] it is our experience that the main problem with procedural description of layout is the large number of relations that is involved. This holds even when a number of small cells has to be fitted to each other like in a PLA. The result is that such a procedural layout description is hard to make, to update and to prove its correctness.

Since the leaf cells of modules are a lot more complex than the PLA cells (an adder contains 20 transistors) we decided to use symbolic layout and compaction techniques for designing the leaf cells of the modules. For ease of use this program can be started as a subprocess from within the module generator environment.

5.5.1 The symbolic layout and compaction program CAMELEON [Rij84]

The aim of symbolic layout [Hsu79, Bal82] is to <u>reduce the</u> <u>layout design time of IC's, without significant loss in area.</u> Symbolic layout [Wes81, Dur85, Vanv85] is a methodology in which point elements, wires and areas are placed in a certain topology, regardless of the design rules. The objects are afterwards rearranged in order to impose the proper rules between objects in the layout. This step is called spacing (or compaction).

Point elements are those objects of which the dimension will not change during a spacing step, e.g. transistors, via's, etc... in a CMOS technology. These point elements are connected by stretchable wires. Areas are used a.o. for building wells.

In principle, layout designers, who have no experience with design rules or the exact construction of point elements, can use CAMELEON to generate correct layouts. During a CAMELEON session, the designer can devote his/her time by searching an optimal topology instead of correcting design rule violations.

In spite of the above mentioned advantages of symbolic layout, it has not been commonly used until now. The complaints were that hand crafted layouts are smaller than those created with symbolic layout. In CAMELEON, this problem is alleviated by allowing more complex point elements, like S-, L- OR U transistors in the layout. These elements can be gathered in rather dense blocks, over which routing is possible. Experiments have shown that layout generated by CAMELEON can compete with manhattan hand layout when these complex primitives are used. An example of a full adder cell using S-shaped transistors is shown in Fig. 5.1.

5.5.2 Overview of the CAMELEON system

The organization of the CAMELEON system, shown in Fig. 5.2. illustrates the basic idea : a layout results from squeezing a connectivity based topology description of a cell (stick diagram created by the LUDIEC editor) in the appropriate area (SPACE) based on technological rules (TDL) and environmental requirements (see further, CONSTRAINTS). Since the technology is coded in a separate file, it is possible to use the CAMELEON system for a wide variety of technologies.

5.5.3 The technology description - TDL

The technology description contains a number of constants (e.g. MIN__MET__WIDTH = 5), the definition of masks and layers, a set of parametrized procedural definitions of the primitive elements in terms of shapes on several layers and the design rules.

Masks concern the production of the integrated circuits, whereas for specifying design rules, it is more easy to work on layers, which are afterwards mapped on (no, one or more) masks.

The defined constants can be used both in the definition of the primitives and in LUDIEC. Normally no literal values should appear in the designs. Stick designs, containing only symbolic constants, generate updated layouts as soon as the constants in TDL change. This is not possible when literals are used. For the same reason, the constants should be used in the expressions defining the coordinates of the geometric elements of the primitives (e.g. the poly-diffusion overlap in a transistor).

The definition of a primitive depends on its type :
- Point elements are described by giving the geometric information: the point element is decomposed into rectangles or polygons on several layers. Each rectangle is described in terms of expressions, involving parameters, formerly evaluated variables and constants and it receives an electrical node. Also terminals are defined where wires can start or end. Each terminal has also an electrical node.
- Wires are described by declaring on which layer they reside, and defining the width as a function of the parameter(s).
- Areas, which can have different rules inside and outside, are defined by specifying the inside and outside layers. These are e.g. used for wells.

The algorithms in the CAMELEON system are adapted to handle non symmetrical primitives, like L-, S- or U- shaped transistors. Fig. 5.3. shows how a description of e.g. an S-shaped transistor might look like and some instances for different combinations of parameters.

In CAMELEON, both spacing- and equipotential spacing (equi) rules can be defined : spacing rules are not applied between segments on the same electrical potential. Equi- rules are always imposed. An example of an equi rule is between polysilicon and diffusion. These layers must never overlap unless a transistor is desired, even if they are at the same electrical potential.

5.5.4 The sticks editor - LUDIEC

The editor is used to add, move, mirror, rotate,... elements to create the topology (relative placement) of a cell. Instances of TDL elements are placed with their instanciated parameters. An other important function of the editor is the detection of electrical connectivity in the stick diagrams.

To provide a means for the designer to guide the SPACE program, CAMELEON supports a number of constraint types that can be imposed on the layout, influencing the final result :
- Upper, lower, fixed : specify that two objects in the layout should be separated at most, at least or exactly the specified distance in the real layout. These constraints are used mostly for abutting cells on each other.
- Grid, half grid : specify that the object should be located at coordinates equal to n & grid or (n + 1/2) & grid. This facility is very useful while designing standard cell libraries.
- Designer : specify that the objects should be separated by exactly the specified distance, regardless of the rules

mentioned in the technology description file. These constraints can be used to simulate conditional rules : the worst case rule appears in the technology description, on all places where less severe rules may be applied, a designer constraint is added. This is not mandatory, but it reduces the area.

CAMELEON guarantees the survival of sticks diagrams between technology versions, as long as the name of the primitives, the function of their parameters and the names of their terminals remain unchanged compared to the older technology. Extra masks and layout rules may be introduced.

Fig. 5.4. shows part of a stick diagram and its textual description, to highlight how the storage format enables us to achieve technology independence of designs produced by CAMELEON.
A unique cross reference number is assigned to each primitive, through which it is later referenced (e.g. 57 for the first met1 wire). The actual parameter of the primitives are stored as expressions, if it is available (the width of PMOS 221 is def_ptr_w, its length is (2*defptrl). The designer is not requested to enter each expression. If the TDL is well defined, the default values of the primitives are specified as expressions.
Connections are stored explicitly, using the cross reference numbers of the primitives and the names of the ports of point elements (wire 57 is connected to 58 and to port cops2 of poly-contact 104). This information allows to restore the connectivity information after a scaling, even though the new physical representation does not.
The constraints are not stored by coordinates, but using the cross reference number of the involved primitives (a horizontal constraint of IO-PITCH is imposed between the metal lines 58 and 61).

The technology conversion is however not "Push the Button". When a new technology is used, old stick diagrams can again be used as starting point for a new optimization step, which takes much less time than a complete redesign (e.g. 1 hour for a full adder). Fig. 5.5. shows the scaling of a 5um CMOS flip-flop to 3um technology.

5.5.5 The compaction module - SPACE

The compaction module, SPACE, is used to squeeze the layout described by the sticks diagram into the smallest possible area, whereby both the design rules and the user imposed constraints should be fulfilled. In the following compaction in the horizontal direction is assumed. The procedure for a vertical compaction is similar.

The compaction module, SPACE, is constraint graph based [Hsu79, Bal82]. The vertices of the constraint graph represent locations (x-coordinates) of objects in the layout. The edges of the graph represent a condition (e.g. a spacing rule) to be fulfilled :

$$x2 - x1 >= w$$

where x1 and x2 are the locations associated with both vertices, and w is the weight associated with the edge of the graph. Fig. 5.6. shows the horizontal constraint graph for a parallel transistor configuration. The constraint graph is built by determining the allowed mutual location of all primitives. Afterwards appropriate locations for each primitive must be calculated. This is done by a critical path algorithm [Bal82]. The critical path (all objects separated by a minimal distance) and possibly the over constrained paths (e.g. when constraints force a contact hole to lay upon the gate of a transistor) are stored and displayed on request, providing the designer with useful information to optimize or to correct his layout.

5.5.6 Automatic fitting of leaf cells

A technique to interconnect cells that is often used in the design of structured modules is abutment. This means that the layout of the cells is done in such a way that the connections are made by just placing them side by side.

In a traditional layout environment abutment of cells can only be achieved by taking into account, while designing the layout of a cell, the requirement to abut to the other cells. This means that for each specific configuration of cells a set of dedicated layouts must be made. If, in a flexible module generating environment, we want to be able change cell parameters (like transistor current drive) under program control, we must be able to fit the layouts of the cells automatically. In our system the designer can express this requirement by the command :

    (ABUT XORY (CELL-LIST)).

The way we solve the abutment of the cells is as follows. Each of the cells is compacted separately by CAMELEON. From the resulting constraint graph of the individual cells we extract what we call a substitute graph (SG) [Sta84]. A node in the SG represents a terminal or a boundary of the cell. For each of the terminals a critical path algorithm is run. The resulting minimum distance requirements (MDR) for the other terminals are translated in an edge in the SG, with a weigth equal to the MDR, between the nodes representing the respective terminals. The SG thus contains the constraints imposed by the internal components of a cell on its terminals. Fig. 5.7. shows the layout and the vertical substitute graph of an invertor.

The substitute graphs of abutting cells are combined by putting constraints between nodes representing terminals to be connected by abutment. Fig. 5.8a and b show three cells and the vertical and horizontal constraint graphs resulting from

the (ABUT X(Y inv1 inv2)inv3) command. The larger graph is solved by the longest path algorithm resulting in constraints to be imposed on the individual cells. The final step is to compact the internals of the individual cells again, but taking into account the constraints imposed from solving the combined substitute graphs.
Fig. 5.8c shows the result of the above ABUT command.

5.6. Cell composition in an interactive environment

In contrast to leafcell design the composition of a module requires a language to express how a module is built. The main features of our module generator environment are the high level composition commands and the interactive graphics feedback provided with the definition of each command. It also allows the user to scroll through the definition and change or delete command lines.

5.6.1 getting started

To start with the design of a new module generator the first thing to do is tell the system we start with a new module. At the same time the parameters for this module are specified. For instance the command

        (EDIT-COMP (RAM N M))

specifies the start of the definition of a block called "RAM" with two parameters N (number of address bits) and M (wordlength).

5.6.2 Specifying structure and topology

In a traditional design environment the structural data is captured via a schematic editor and the topological data is entered on a graphics system like Calma or Applicon. The main difference between a schematic block diagram and a floorplan is that they present the design from a functional versus a layout point of view. Since in essence the same information is contained in both views we have decided to combine these two views in the first version of our system.

Lower level modules can be included in a module by the ADD-COMP command. This command provides a construct to specify the type of component to be added, where it must be located and what transformations (rotate, flip) must be performed on it. The location is given as a grid point or a series of grid points specifying the relative locations of the instances. The command

(ADD-COMP RAMCELL 1..|(* 2 M)| 1..|EXPT 2 (- N 1))| :ROT 90)

will place the component RAMCELL, rotated by 90 degrees on the gridpositions 1 to 2 times M in the x direction and from 1 to 2 to the power N-1 in y, assuming that two columns of memory cells are multiplexed on to one sense amplifier. For N=4 and

M=4 this command would result in the display shown in Fig.5.9a. Fig. 5.9b shows the graphic response after adding the decoder to the right and the multiplexers and sense amplifiers on top of the array. Fig. 5.9c and d show the final topology and layout of the ram after the cells have been fitted and compacted.

If an instance of the component that we want to add WITH THE SPECIFIED SET OF PARAMETERS does exist in the library, its blackbox representation, generated from the compacted layout, is loaded into the system and used for display. If only the definition of the component exists it is executed WITH THE SPECIFIED SET OF PARAMETERS, added to the library, loaded and used. If not even a definition of that component exists the system will give a warning and propose two options : to start the definition of that component first or at least give a "dummy" blackbox it can use to temporary represent the undefined component. If the user chooses to define the missing component an EDIT-COMP command is automatically issued.

5.6.3 Defining connectivity

To provide connection points to the module at a higher level of the hierarchy terminals can be defined. In the module itself the interconnection of the sub-modules is defined by specifying a net and the terminals of the sub-modules and of the module itself that are connected to it. As with the ADD-COMP command facilities to express repetitive connections are provided in the ADD-CON command.

(ADD-CON (CARRY 1..(- N 1)) (COUT 1..(- N 1) 1) (CIN 2..N 1))

would connect all the carry signals in an N bit adder when the cells are placed from left to right. The first loop defines the names of the nets (CARRY 1 to N-1) and the other loops define the pinnames and the component position of the pins connected to those nets (eg pin COUT of the components on gridpoints [1,1] to [N-1,1]). Fig 10 shows the response of the command on the screen for N=6 Note that no assumption is made at this stage of how the connections will be realised. The physical connection will be realised later by the ABUT command or when the router is called to generate the wiring between terminals of the same net.

5.6.4 Picking expressions from grid-points on the screen

In a lot of cases the start or end condition of a loop must be defined as an expression of the parameters, like (N-1) being the last ripple signal. To reduce the amount of typing the system allows to associate an expression with each horizontal and vertical grid line. When the designer later points to a gridpoint the system will automatically place the expression in the x or y part of the command, allowing the graphic definition of parametrized arrays or interconnections (see x-axis in fig 10).

5.6.5 Abutment of cells

In a traditional layout environment abutment of cells can only
be achieved by taking into account, while designing the layout
of a cell, the requirement to abut to the other cells. In our
system the designer can express this requirement by the
command :

> (ABUT XORY (CELL-LIST)).

The procedures described in 5.5.6 will fit the cells and the
(COMPACT XORY) command will then compact the module.

5.6.6 Routing some channels

If the designer chooses not to connect cells by abutment, or
if he finds out that to much area is lost by fitting the
terminals of adjacent cells, the interconnections must be
routed.

Since the cells are placed on relative gridpoints horizontal
and vertical channels are automatically defined. A router can
be called with the command

> (ROUTER DIR XCHAN YCHAN (NET-LIST))

to realise the connections for all the nets included in
NETLIST in a horizontal or vertical channel depending on the
value of DIR and starting at gridpoint (XCHAN,YCHAN).

The routing program is written in C and will automaticaly
decide if all connections can be realised using its river
routing algorithm. If not a two level channel routing
algorithm is used.

5.6.7 Saving the module generator

The final step of the module creation phase is to save the
steps of the interactive session into a procedure that can
later on generate the instances of the module on request of
the system designer. MGE will do this automatically when a
QUIT command is given by the user. Fig. 5.11 shows an 8 bit
ADD-SHIFT-COMPARE module generated by MGE.

5.7. Verification of the modules

One of the difficult problems with module generators is to
prove their correctness over the validity range of the
parameters. The traditional way of doing this is by
simulation. However simulation is a subjective test method
which depends entirely on the patterns defined by the designer
to detect expected problems. It is costly in CPU time and does
not guarantee to capture unexpected problems.

To overcome these drawbacks we have developed DIALOG [DeM85] a set of rule based analysis tools that allows to detect violations against basic design principles and SLOCOP for timing analysis. These tools are based on the LEXTOC language which provides powerful facilities to express rules about MOS circuits. Fig. 5.12 gives an overview of the verification system.

5.7.1 MOS network extraction

To verify the design we first have to extract a model from the layout for the cell on the circuit level. This can be done very easily since the stick diagram contains the devices and the interconnection wires. The internode capacitances can be found by detecting the points where elements cross. Starting from the extracted MOS transistor network the user can perform the following tests :

5.7.2 Decompilation and check of clocking rules

First it is checked if the network belongs to the class of valid circuit configurations, eg. static CMOS using passive multiplexing trees and prescribed registers. All parts of the circuit violating these rules are flagged as errors. The checking is done by partitioning the network into the basic subnetworks that are possible in a given design style. The description of the topology of those subnetworks (static CMOS gates, pass transistor trees with or without bleeder transistors,...) resides in a knowledge base. This knowledge base is written in a LISP-like language called LEXTOC [DeM85]. Therefore, in contrast to e.g. CRYSTAL [Ous84], our system can easily be extended with new types of subnetworks, and can be applied to any definable class of digital MOS circuits (Fig. 5.13).

Next a high level check on the clocking can be done. Starting from the primary clock information the drived clocks are found, latches are detected and separated from combinatorial logic. By assigning the appropriate properties to the nodes in the network illegal combinations can be detected and reported. Examples are :

A DERIVED CLOCK CAN ONLY BE FORMED WITH SIGNALS LATCHED ON THE SAME CLOCK PHASE

A LATCH DATA INPUT MUST NOT BE A PRIMARY CLOCK

Fig. 5.14 shows the response of DIALOG to a clocking rule verification request. The system has identified and highlighted that a loop exists containing only a single clock phase which could cause a race problem.

5.7.3 Electrical rule checking

Starting from the primitive circuits found in the previous step we search for illegal configurations like open, short,

628

odd n- p-MOS combinations etc... For all acceptable circuits, it is checked if they can guarantee correct logic levels. This check is not restricted to static situations but includes dynamic effects like charge redistribution. An example of a possible problem circuit, and the rule involved in the checking procedure is given in Fig. 5.15. It must be noted that to diagnose the occurence of an error on node B caused by charge sharing we have to find the combination of input signals that creates the worst case configuration. This is done by the SETUP-INPUTS procedure. Using these inputs the part of the circuit under investigation is simulated on the circuit level using the SIMMY module [Coc84]. The output waveforms of the simulation are then checked to see if the voltage on node N2 drops below 3.0 volts. If so a message is issued telling the designer charge sharing problems can be expected on node N2. If the designer does not believe this diagnosis a facility can be called to display the conditions creating the problem situation and explain by what successive steps it was detected.

5.7.4 The timing verification program SLOCOP

If the circuit has passed al the general rules we must still check that it does not violate the timing specifications for the module. From the primary clock specifications the edges of the clock inputs for the latches are computed and the maximum allowed delay of the combinatorial blocks can be calculated.

To address the timing characterisation problem, several computer programs have been developed for timing analysis of MOS circuits [Ous84, Jou83]. These programs do not require the user to provide input patterns but pinpoint critical paths. They are mostly based on the assumption that the circuit can be split up into smaller subnetworks. To calculate the delay of those subnetworks, approximative methods (mostly based on RC modeling) are used in these approaches. The RC model is responsible for potentially large errors in the global delay estimation. It also restricts the class of circuits that can be handled.

In contrast, SLOCOP uses local circuit simulation for delay calculation. This method is more time consuming than RC calculations, but it is very reliable, much more accurate, and applicable to any type of subnetwork that can be described in its knowledge base. For the global analysis, graph based algorithms are used since an exhaustive path search is not possible due to the large number of paths. In CRYSTAL, an algorithm is used that calculates the slowest settling time at each node in the circuit. SLOCOP however uses a critical path algorithm which provides the longest signal propagation paths from inputs to outputs, because this information is essential in the timing optimisation phase of the design.

5.7.4.1 Characteristics and algorithmic aspects.

After the partitioning of the transistor network as described in 5.7.2 SLOCOP performs the following sequence of algorithms:

A logic reduction step transforms each subnetwork into an equivalent network of AND, OR, NOT and MUX operators (Fig. 5.16a) [Saa82, Apt82]. The last operator is necessary to model subnetworks which can have a high impedance output state.

Circuits are assumed to be synchronous and clocked with a 2 phase non overlapping clocking scheme. Latches are automatically detected in the equivalent logic network,by checking what multipliers have a high impedance output when the clock is held at zero, and a partitioning in combinational blocks is done. Violations of clocking rules are also reported as explained in 5.7.2. Timing analysis then consists of finding the longest signal paths from latch outputs to latch outputs clocked on the opposite phase.

The SLOCOP method consists of an accurate characterisation of the individual subnetwork delays from a local circuit simulation. The set of input patterns that is needed to fully characterise the subnetwork for all possible signal propagation modes is automatically derived for each type of subnetwork from its logic function. This set consists of a number of dynamic input vectors. The dynamic input vector (DIV) contains one element, that can take one of the four values 0,1,U or D for each of the inputs of a subnetwork. U(p) represents a transition from 0 to 1, D(own) from 1 to 0. We restrict ourselves to DIV's in which only one element has a transition value. This makes it practically feasable to find all DIV's that cause the output to make a transition (Fig. 5.16b).

A signal propagation graph (SPG) is set up. The vertices in the SPG are inputs and outputs of subnetworks described in the knowledge base. Each edge corresponds to one DIV. The tail of the edge is the input that makes a transition, the head of the edge is the subnetwork output (Fig. 5.16c).

A delay value is associated to each edge in the signal propagation graph. The delay calculation is done through an automatic local simulation of the subnetwork at the device level. For this purpose, use is made of SIMMY, a PASCAL simulation module defined as an abstract data type. Input to the module are the transistor network and the DIV corresponding to the edge of the SPG under consideration. To estimate the slope of the input transition, the subnetworks are leveled first. Local simulation is done in topological order. The average of the slope of the output waveforms of the driving subcircuit (which is always simulated first) is taken as the slope of the input waveform. Output of SIMMY is the response waveform at the subnetwork output, with SPICE accuracy. Delay and slope are calculated by interpolation in this piecewise linear waveform. If the delays are measured at the same threshold voltage for each subnetwork, the path delay is the sum of the individual delays of the edges (Fig. 5.17).

An algorithm has been implemented that finds the N longest paths in the SPG. Its time complexitiy is $O(N**2)$, but it is linear in the number of edges and vertices in the graph, which is in turn proportional to the size of the circuit. The result are sequences of edges that form the longest, second longest,... signal paths. These paths can be graphically displayed by highlighting the nodes that make a transition on the schematic. As SLOCOP is capable of handling circuits of up to a few thousand devices, hierarchical backannotation is implemented. This means that several cells in the hierarchy are displayed simultaneously and nets are highlighted at each level (Fig. 5.18).

Each of the path edges corresponds to a local dynamic input vector for a subnetwork. This means that for all subnetwork inputs that do not make a transition, assumptions are made for the logical states of these nodes. Test pattern generation techniques are applied to derive a dynamic input vector at the inputs of the combinational block (primary inputs or latch outputs), which guarantees that those logical states are set. The global DIV has an U(p) or D(own) entry for one of the inputs, and 1 or 0 entries for all other inputs. In some cases the algorithm may find that no global DIV is possible, because of conflicting assumptions imposed by the local DIV's for the subnetworks. In contrast to existing timing verifiers, 'false' paths (signal paths that can never be excited) are automatically excluded in this way. The test pattern allows to check path delays with global simulation.

5.7.4.2 SLOCOP performance and error analysis

Two approximations have to be made when the signal propagation delay through a subnetwork is extracted by simulation :

1. The load of the subnetwork, formed by the inputs of the subnetworks that it drives, is to be modeled by an equivalent capacitor.

2. The waveform used in SLOCOP for the input transition is an average of the linearised output waveforms obtained from the simulations of the driving subnetworks. This may cause errors in two ways :
   - The slope is not exact.
   - Replacing the real waveform by a ramp wave form has a small effect on the output waveform and therefore on the delay measured. This effect is more significant when the waveform is applied to a pass transistor network than to a gate.

The run time of the program in its current version is almost exclusively spent in the subnetwork simulator. Subnetwork partitioning, logic reduction, longest path search and test pattern generation together take only a few percent of the total run time. It is obvious that a tradeoff between accuracy and speed can be obtained by using simulation techniques such as in ELOGIC or CINNAMON [Vid85, Seu85, You84]. This is a direction for further research in SLOCOP.

A variety of circuits have been analysed with SLOCOP. An example is a 10 bit counter in static CMOS. Fig. 5.18 shows the multiwindow hierarchical schematic on which the critical path is highlighted by the program. The deviation between the path delay estimated by SLOCOP and the delay measured from an overall simulation is 4.2% . The critical path is the carry propagation through the half adder cells. Test pattern generation gives us the state of the counter at which the maximal delay occurs.   The circuit contains 240 transistors and was analysed in 157 seconds on a VAX 8600 computer. We want to stress that in the complete process of defining this delay _the designer does not need to define a critical path or specify any simulation pattern._   All these time consuming tasks are automatically handled by the program.

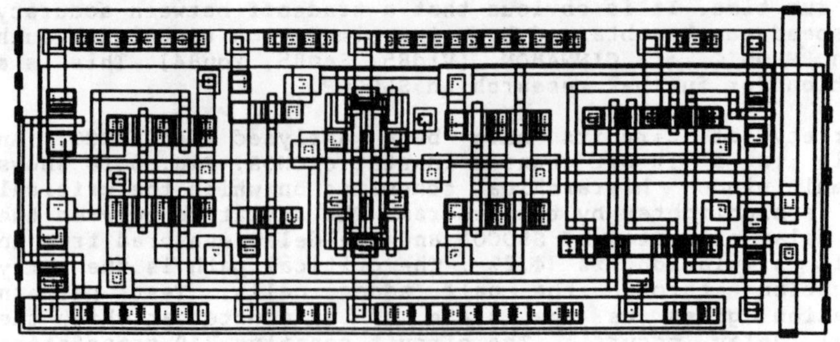

Fig. 5.1. Full adder, containing S- shaped transistors.

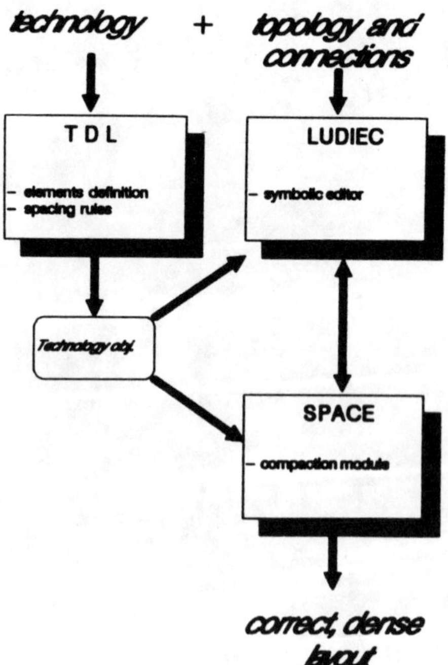

Fig. 5.2. : Overview of the Symbolic Layout and Compaction system CAMELEON.

```
ntr <
  param <
          (Parameter declaration part)
          (name) (minimum) (default)
            w     (difmin)  (cntdffmn)
            l     (polymin) (polymin)  >
  eval <
          (Evaluate some local variables, to simplify)
          (the expressions in the geomel and port - commands)
          (aa := 1/2 + ditrov)
          (bb := w/2 + potrov)
          ...
          (ff := -1/2 + dffdffsp + misaline)
          (gg := 1/2 - 0.5)
        >
  geomel <
          (GEOmetrical - ELectrical description)
          (Define the layout of the element in terms of)
          (rectangles, situated on a certain layer and)
          (having a node name.)
          (layer) (node)        (x1)    (y1)    (x2)    (y2)
          pyga    gate    rect (-bb)   (-1/2)  (bb)    (1/2)
          ndif    source  rect (-w/2)  (-aa)   (w/2)   (-ff)
          ndif    drain   rect (-w/2)  (ff)    (w/2)   (aa)
          gdif    gate    rect (-w/2)  (-ff)   (w/2)   (ff)
          ...
        >
  port <
          (Ports definition)
          (The ports, the locations where wires can connect)
          (to the element, should also be defined. Wires, )
          (which are connected to a certain port, receive the)
          (electrical node, specified below)
          (port-)(wire) (node)           (extent of port)
          (name)                    (x1)   (y1)   (x2)   (y2)
          source ndif   source bott (-w/2) (-gg)  (w/2)  (-gg)
          drain  ndif    drain top  (-w/2) (gg)   (w/2)  (gg)
          gatel  poly    gate left  (-bb)  (-1/2) (-bb)  (1/2)
          gater  poly    gate right (+bb)  (-1/2) (+bb)  (1/2)
        >
      >
```

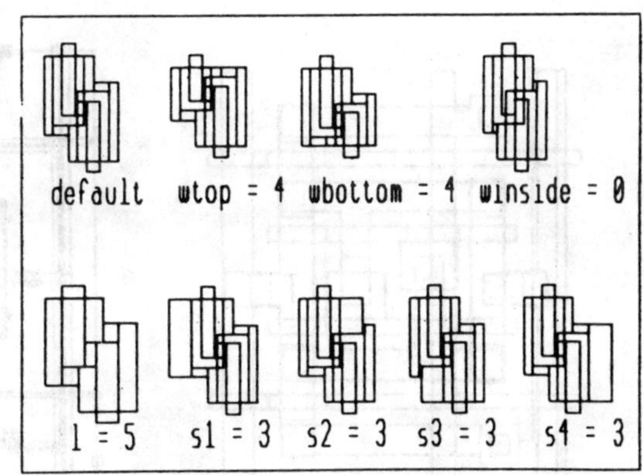

Fig. 5.3 (a) TDL description of a S- shaped transistor.   (b) Some
instances for different combinations of parameters.

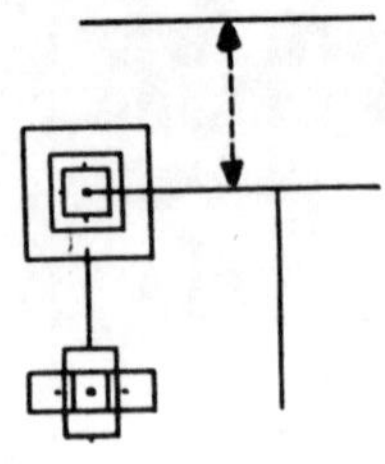

```
....
 add /57 met1 ((def_met1_w)) vdd  #-1 27 53 59 53
 add /58 met1 ((def_met1_w)) vdd  #1  48 53 48 30
 add /61 met1 ((def_met1_w)) rout #-1 27 73 59 73
 add/103 poly ((min_poly_w)) i43  #1  27 47 27 37
 add/104 cops ((def_cont_w),(def_cont_w)) i27   27 53
 add/221 ptr  ((def_ptr_w),((2.000*def_ptr_l))) m270 27 32
....
 connect  /57  /58          48 53
 connect  /57  /104 . cops2 30 53
 connect  /103 /221 . gater 27 37
 connect  /103 /104 . cops8 27 47
....
 add vertical fixed {constraint} (io_pitch)  /57 /61
```

Fig. 5.4. : Part of a stack diagram and its storage format.

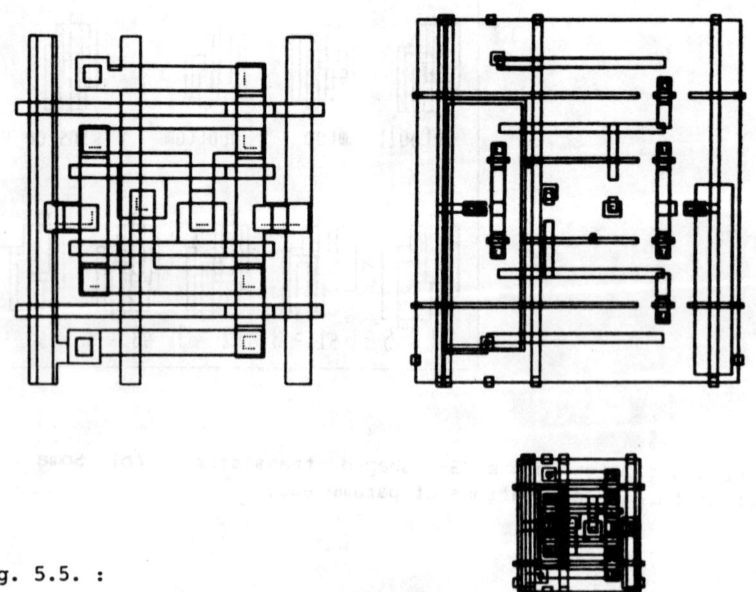

Fig. 5.5. :

Scaling. On the left, a valid layout in a 5 um CMOS. On the right, first a layout in which only the objects are scaled to 3 um CMOS. Finally, the layout obtained after compaction.

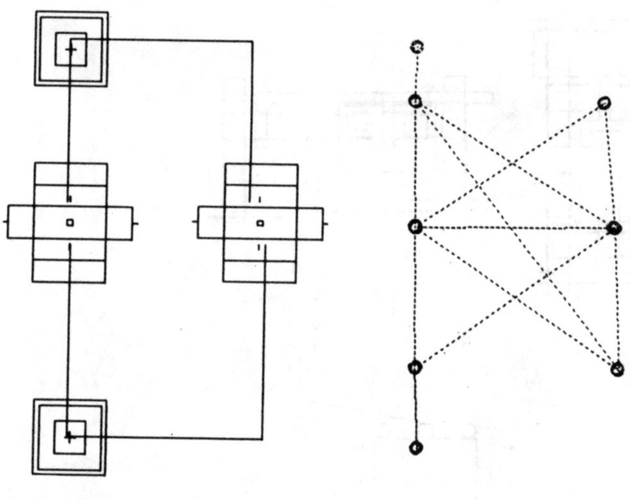

Fig. 5.6. : Stick diagram and horizontal constraint graph

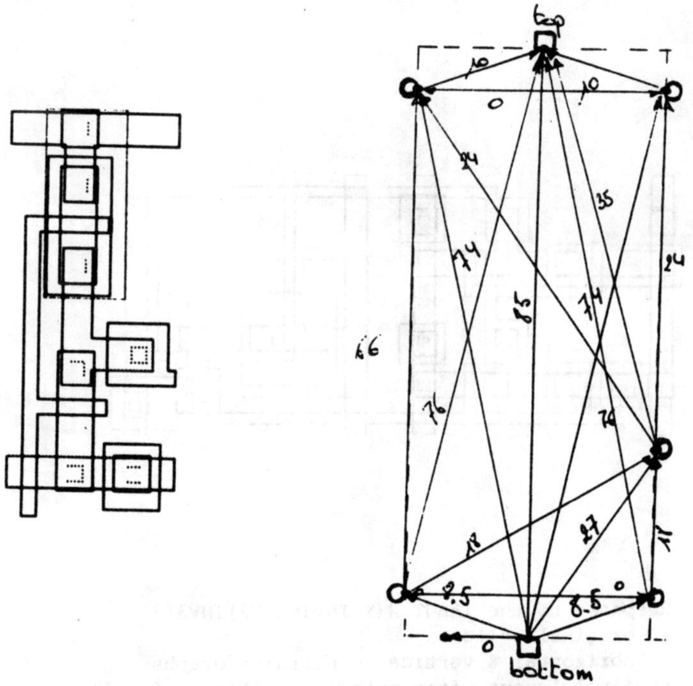

Fig. 5.7. : An invertor cell and its vertical substitute graph.

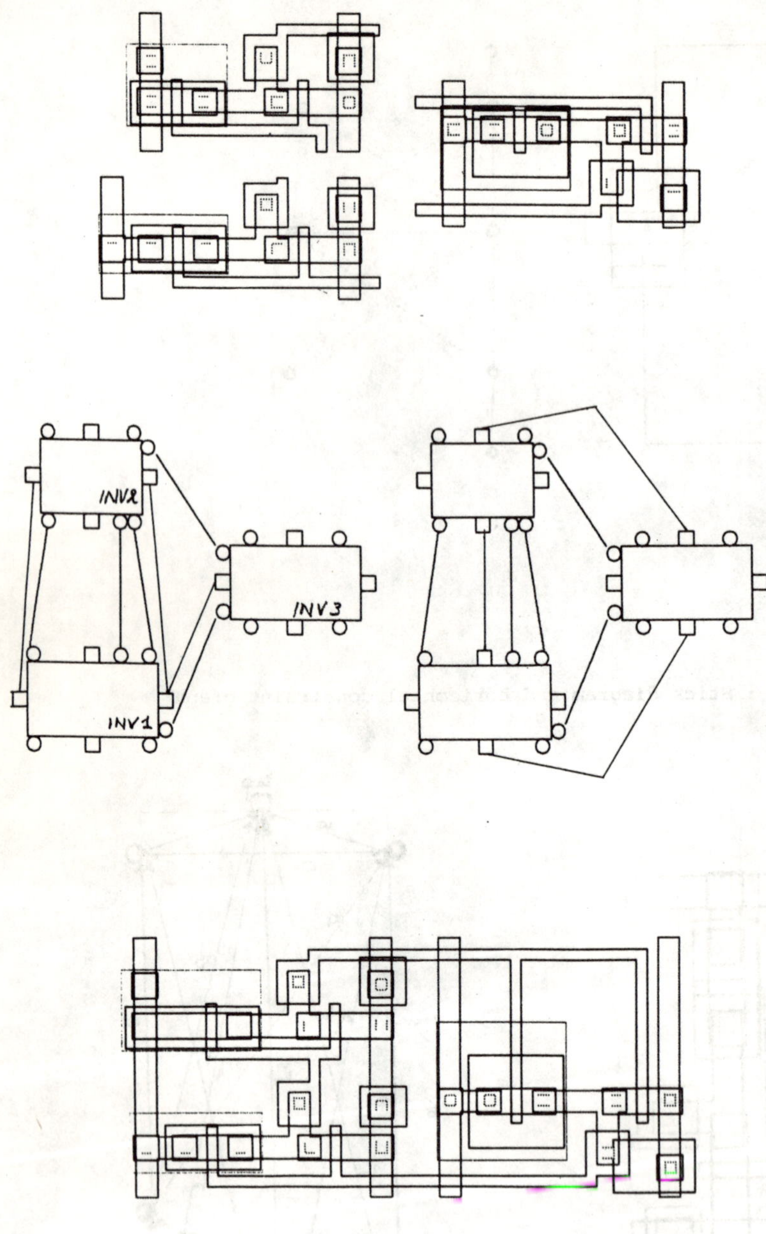

Fig. 5.8. : Response to the (ABUT X(Y INV1 INV2)INV3)
      a. original cells
      b. horizontal & vertical constraint graphs
      c. final layout after automatic fitting of cells

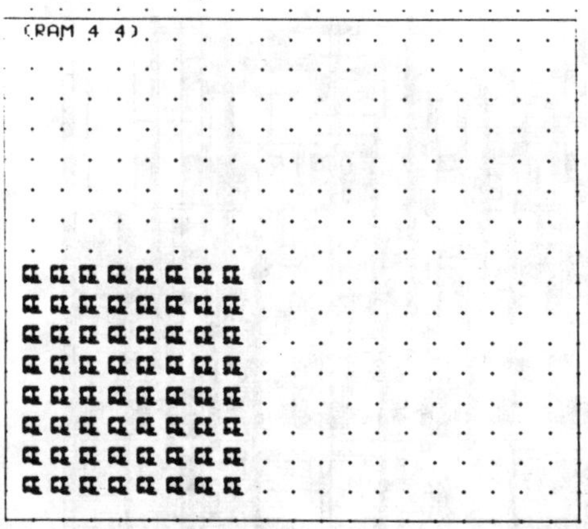

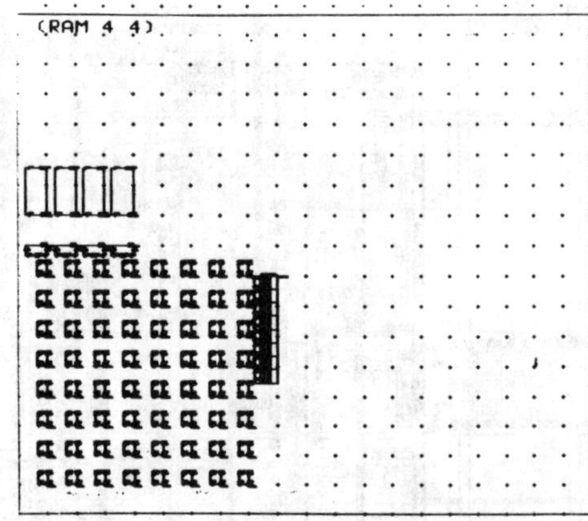

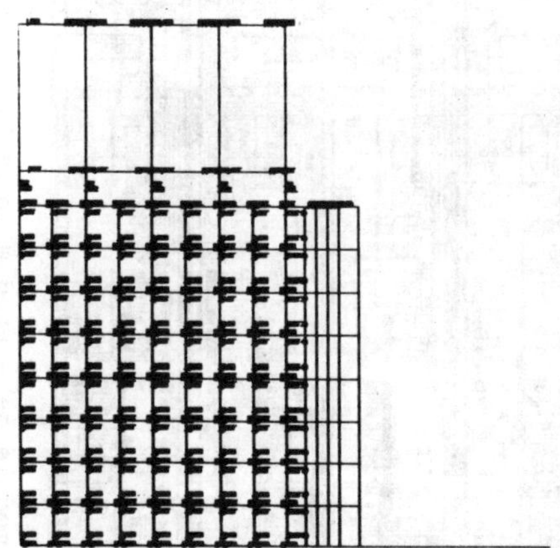

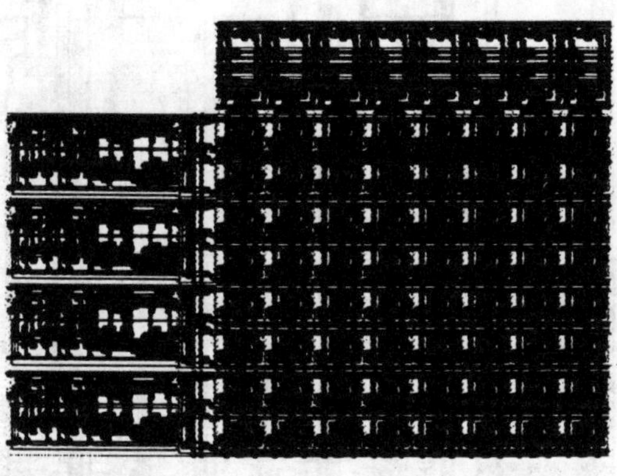

Fig. 5.9. : Interactive design of a RAM module.
a) the array, b) all cells added, c) after fitting cells
d) final layout

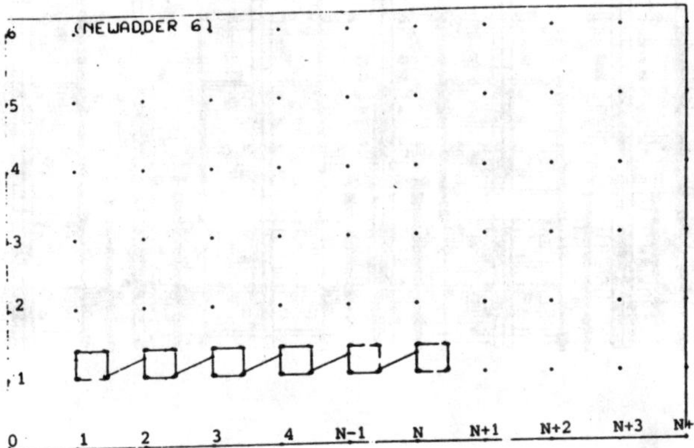

Fig. 5.10 : structure, topology and connections of an adder.

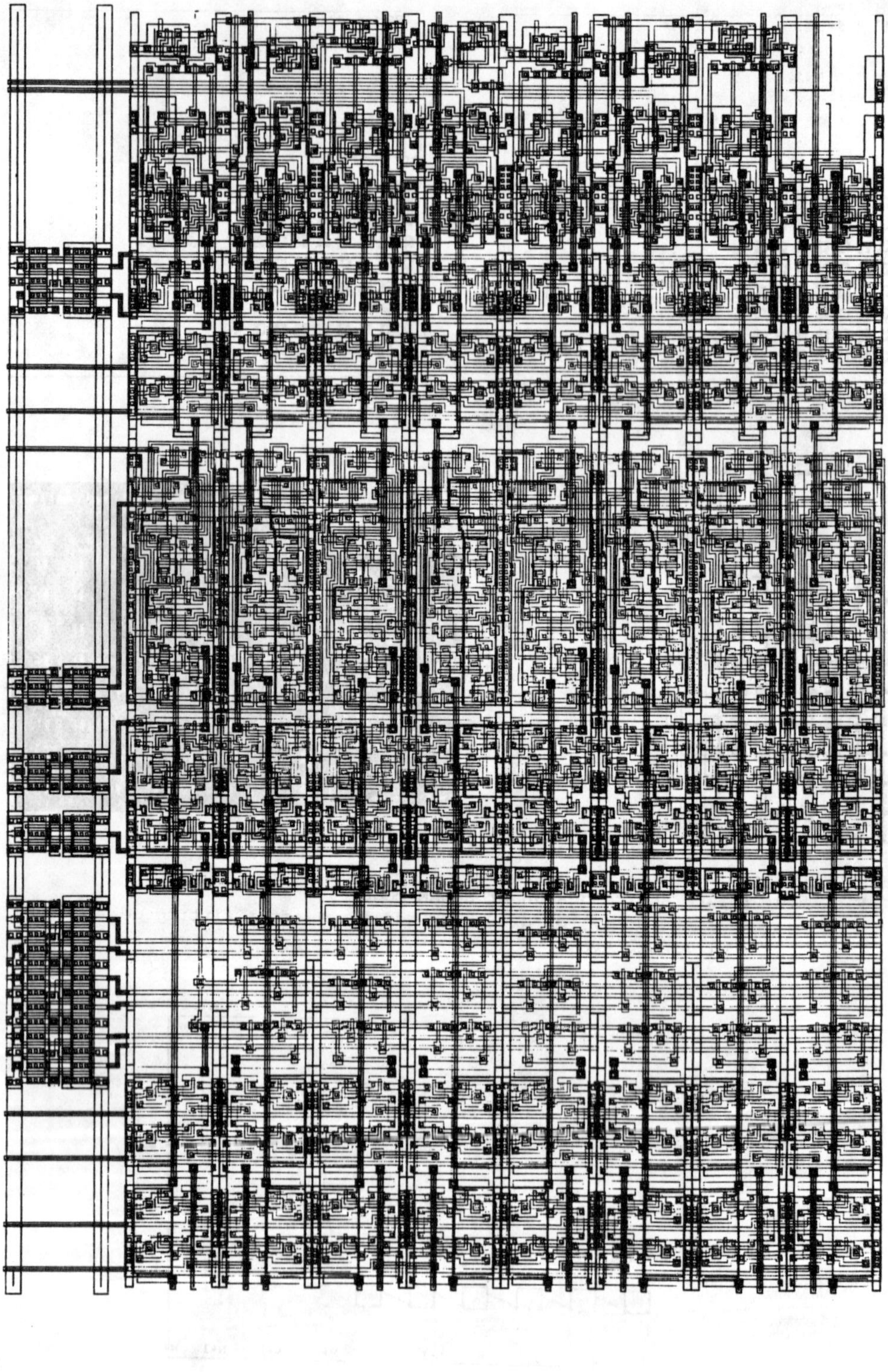

Fig. 5..1. : ADD-SHIFT-COMPARE module generated with MGE.

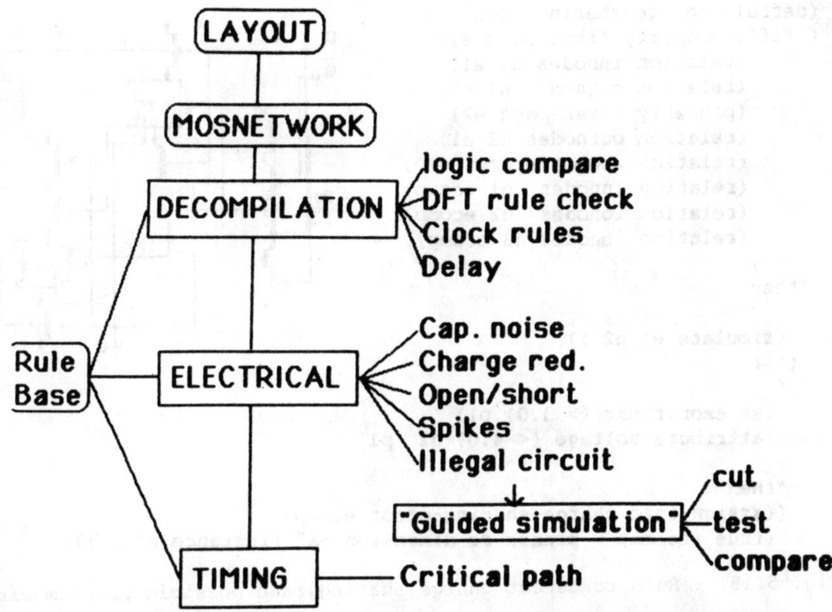

Fig. 5.12. : Overview of the DIALOG expert verification system.

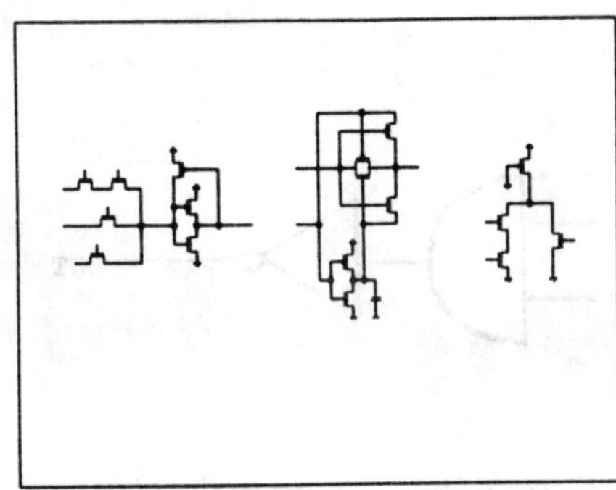

Fig. 5.13. : A sample of the allowed transistor configuration stored in the rule base.

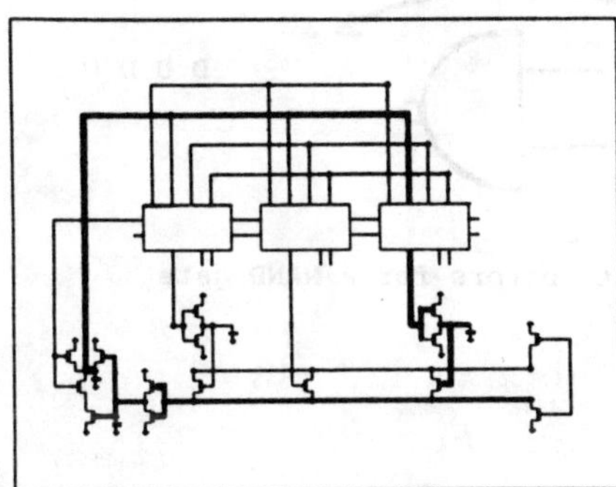

Fig. 5.14. : Simple clock phase loop detected by DIALOG.

```
  (defrule charge-sharing-noral ()
  ( *if ((property fibar-ndyn el)
         (relation innodes nl el)
         (relation outnodes nl e2)
         (property fibar-pdyn e2)
         (relation outnodes n2 el)
         (relation compon ecomp  el )
         (relation innodes  nl ecomp)
         (relation ionodes  n2 ecomp)
         (relation ionodes n3 ecomp)
       )
   *then
    (
     (simulate el n2 1)
    (*if
     (
      (at exor.fibar (> 3.0) pl)
      (attribute voltage (< 4.0) n2  pl)
     )
     *then
     ((assign-info charge-sharing-error ecomp)
      (true (format " please re-dimension ~a" (instance n3))))))))
```

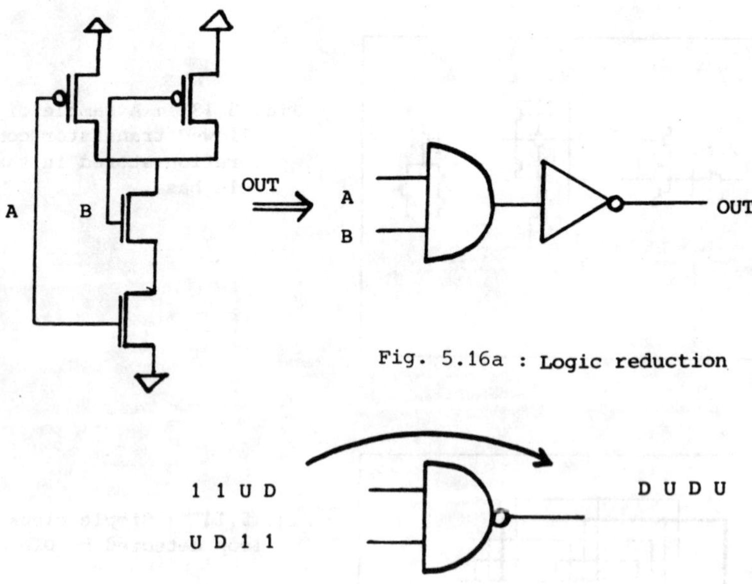

Fig. 5.15. : Rule to detect charge sharing, and possible problem circuit.

Fig. 5.16a : Logic reduction

Fig. 5.16b : Dynamic input vectors for a NAND gate

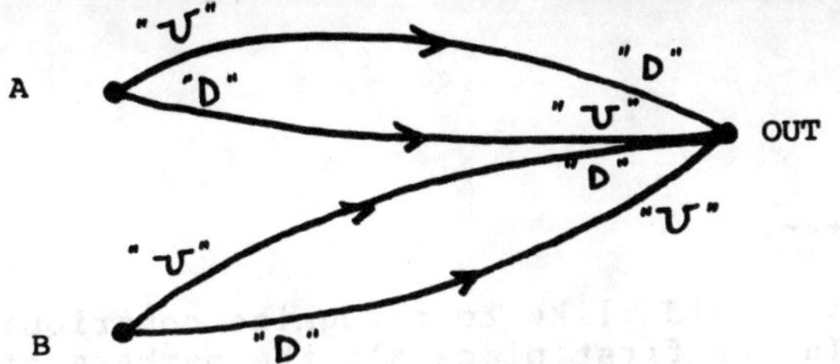

Fig. 5.16c : Corresponding edges in the SPG graph. ("D" = transition 1 -> 0,
"U" = transition 0 -> 1)

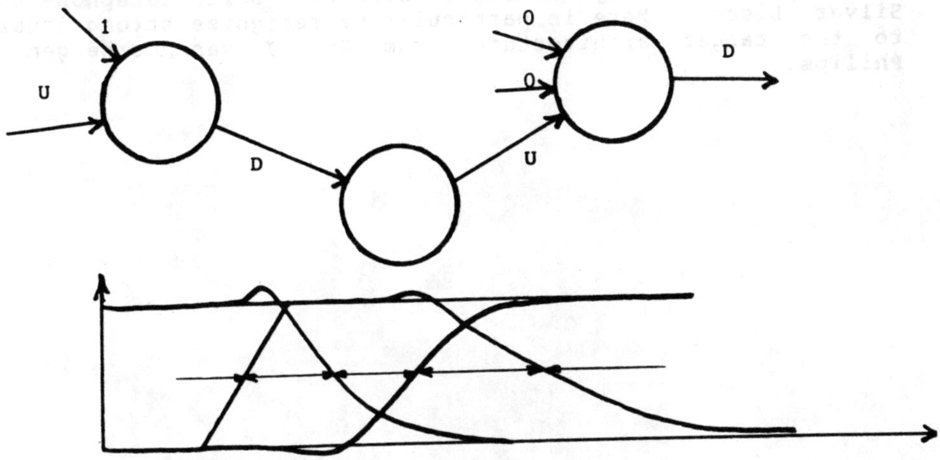

Fig. 5.17 : Path delay = sum of subnetwork delays.

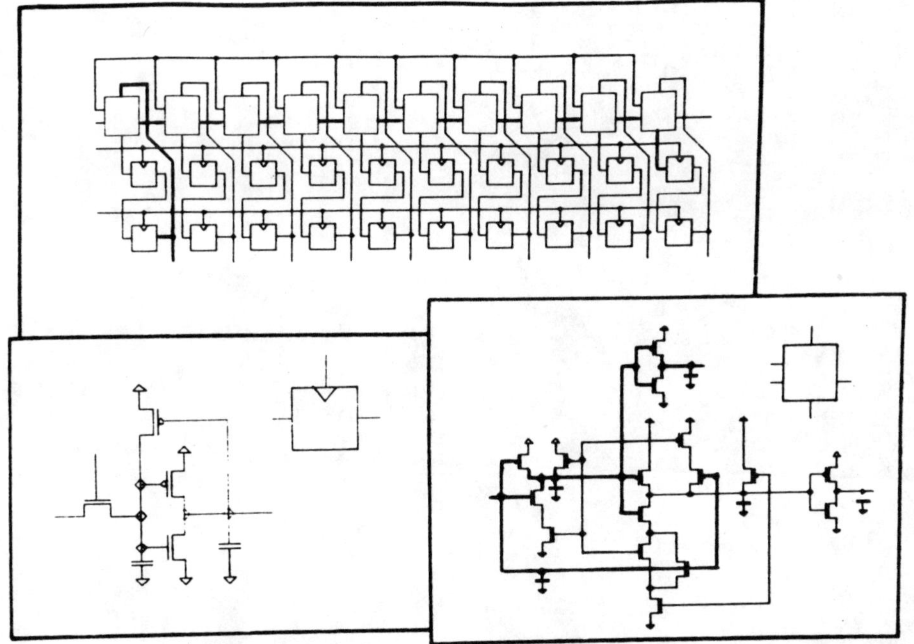

Fig. 5.18. : Hierarchical highlighting of critical delay path by SLOCOP.

Acknowledgement

The authors would like to recognize contributions from many people. In the first place all the members of the IMEC VSDM group and more in particular : K. Croes, L. Rijnders, I. Vandeweerd, M. Pauwels, F. Catthoor, M. Bartholomeus, G. Goossens, J. Vanhoof, E. Vanden Meersch, I. Bolsens.
We also recognize contributions from all our ESPRIT 97 partners, especially Philips, Siemens, Bell Telephone and Silvar Lisco. More in particular we recognize strong inputs to the target architecture from Dr. J. Van Meerbergen of Philips.

REFERENCES

[All85]   J. Allen, "Computer Architecture for Digital Signal Processing", Proc. of the IEEE, Vol. 73, No. 5, pp. 854-873, May 1985.

[Apt82]   Apte R.M., Chang N., Abraham J., "REDUCE, Logic Extraction for NMOS circuits", IEEE Int. Conference on Circuits and Computers, New York, 1982, pp. 324-327.

[Bal82]   M.W. Bales, "Layout rule spacing of symbolic integrated circuit artwork", Memorandum UCB/REL No. M82/72. Electronics Research Labs, University of California, Berkeley, CA 94720.

[Bar85]   M. Bartholomeus, L. Reynders, M. Pauwels and H. De Man : "PLASCO : A Procedural Silicon compiler for PLA Based Systems", Proc. IEEE CICC-conference, Portland, pp. 226-229, May 1985.

[Bur85]   M. Buric et al. : "Silicon Compilation Environments", Proceedings IEEE Custom Integrated Circuits Conference, pp. 208-212, May 1985.

[Coc84]   J. Cockx : "User's manual for SIMMY" KU Leuven, Project MR03KUL report 5-B1-1, Oct. 31, 1984.

[Com85]   R. Composano, "Synthesis Techniques for Digital Systems Design", 22nd Design Automation Conference, Las Vegas, pp. 475-481, 1985.

[DeM85]   H. De Man et al. : "DIALOG : An Expert Debugging System for MOS VLSI Design", IEEE Transactions on Computer Aided Design, Vol. CAD-4, No. 3, pp 303-311, July 1985.

[Dur85]   C. Durward Rogers, J.B. Rosenberg, S.W. Daniel, "MCN's Vertically Integrated Symbolic Design System", 22nd Design Automation Conference, 1985.

[Hil85]   P. Hilfinger, "A High-Level Language and Silicon Compiler for Digital Signal Processing", Proc. IEEE CICC-conf., Portland, pp. 213-216, May 1985

[Hsu79]   M.Y. Hsueh et al. : "Computer Aided Layout of LSI Circuit Building Blocks", Proceedings of the International Symposium on Circuits and Systems Conference, pp. 474-477, July 1979.

[Jai86]   R. JAIN et al. : "Custom Design of a VLSI PLM-FDM Transmultiplexer from System Specifications to Layout using a CAD system", IEEE Journal of Solid-State Circuits, Vol. SC21, No.1, February 1986, pp. 73-85.

[Jou83]   Jouppi, Norman P., "Timing Analysis for NMOS VLSI", IEEE 20th Design Automation Conference, 1983.

[Lau85]   U. Lauther, "An O(NlogN) Algorithm For Boolean Mask Operations", Proc. 22nd Design Automation Conference.

[Mag84]   S. Magar, E. Caudel, A. Leigh, "A MicroComputer with Digital Signal Processing Capabilities", Proc. ISSCC 1982, pp. 32-33, February 1982.

[Mari86] E. Marien et al. : "Manual of LOGMOS V4.2 : a Simulator Covering Register Transfer - Functional - Gate and Switched Level", Available from Silvar-Lisco, Kapeldreef 75, B-3030 heverlee, Belgium.

[Marw84] P. Marwedel, "The MIMOLA Design System : Tools for the Design of Digital Processors", 21st Design Automation Conference, pp. 587-593, 1984.

[Mat83] T.G. Matheson et al. : "Embedding Electrical and Geometrical Constraints in Hierarchical Circuit-layout Generators", Proceedings IEEE International Conference on Computer Aided Design, pp. 3-5, Sept. 1983.

[Mul85] B. Muller et al. : "The Chipgenerator Concept - A New Approach to Full Custom CMOS IC Design", Proceedings of the 11th European Solid-State Circuits Conference, pp. 186-192, Sept. 1985.

[Ous84] Ousterhout, J., "Switch Level Delay Models for Digital MOS VLSI," 21th Design Automation Conference pp 542-548, 1984.

[Pap82] C. papadimitriou, K. Steiglitz, "Combinatorial Optimization, Algorithms and Complexity", pp. 421-424, Prentice Hall 1982.

[Pat81] D. Patterson and C. Sequin, "RISC I : A Reduced Instruction Set VLSI Computer", Proc. 8th international Symposium on Arch., Minneapolis, pp. 443-457, May 1981.

[Pop84] S. Pope, J. Rabaey and R. Brodersen : "Automated Design of SIgnal Processors using Macrocells", in VLSI Signal Processing, pp. 239-251, IEEE Press 1984.

[Rab85] J. Rabaey, S. Pope, R. Brodersen, "An integrated Automated Layout Generation System for DSP Circuits", IEEE Trans. on CAD, Vol. CAD-4, pp. 285-296, July 1985.

[Rij84] L. Rijnders et al. : "CAMELEON Version 1.1, Users Guide", Report nr 5-C1-3 of EEC project MR-03-KUL, available from IMEC.

[Saa82] Saab D., Hajj I., "A Logic Expression Generator for MOS circuits", IEEE Int. Conference on Circuits and Computers, New York, 1982, pp. 328-331.

[Seu85] Seung H. Hwang, Young H. Kim, A.R. Newton, "An Accurate Delay Modeling Technique for Switch-Level Timing Verification". Internal Report Dept. of Electrical and Computer Sciences, Cory Hall, University if california, Berkeley.

[Sis82] J. Siskind, J. Southard and K. Crouch, "Generating Custom High Performance VLSI Designs from Succint Algorithmic Descriptions", Proc. Conf. on Advanced Research in VLSI, Cambridge, pp. 28-40, Jan. 1982.

[Slu80] R. Sluyter, H. Kotmans, A. Van Leeuwaarden, "A Novel Method for Pitch Extraction from Speech and a Hardware Model Applicable to Vocoder Systems", Proc. IEEE ICASSP-conf., pp. 45-48, April 1980.

[Sta84] J.A. Starzyk, "Decomposition approach to a VLSI symbolic layout with mixed constraints", Intern. Symp. on Circuits and Systems ISCAS '84, pp. 457-460.

[Tar72] R. Tarjan, "Depth-first search and linear graph algorithms", SIAM J. Comput., Vol. 1, No. 2, pp. 146-160, June 1972.

[Tho83] D. Thomas, C. Hitchcock, T. Kowalski, T. Rajan, R. Walker, "Automatic Data Path Synthesis", Computer 16(12), pp. 59-70, Dec. 1983.

[Tuc85] B.W. Tucker : "Electronic CAD-CAM - Is it Revolution or Evolution", Proc. 22nd ACM/IEEE DA Conference, pp. 830-834, Las Vegas, Nevada, June 23-26, 1985.

[Tze85] C. Tzeng, "Timing Recovery in Digital Subsriber Loops", Ph.D. Dissertation, Mem. No. UCB/ERL M85/29, April 1985.

[Vand86] I. Vandeweerd, "Module generator environment Tutorial and Users Manual", Report ESPRIT97/IMEC/3.86/2.2.2(1)/1.

[VanV85] S. Van Vierbergen et al., "Symbolic Hierarchical Artwork Generation System", Proc. 22nd Design Automation Conference, pp. 789-793, 1985.

[Vid85] L.M. Vidigal, R.S. Nassif, S.W. Director, "CINNAMON : Coupled Integration and Nodal Analysis of MOS networks." Internal report SRC-CMU Center for Computer-Aided Design. Dept. of Electrical and Computer Engineering, Carnegie-Mellon Univ., Pittsburgh.

[Wes81] N. Weste, "Virtual Grid Symbolic Layout", 18th Design Automation Conference, 1981.

[You84] Young Hwan Kim, J.E. Kleckner, Resve A. Saleh and A.R. Newton, "Electrical-Logic Simulation", Digest 1984 Int. Conf. on CAD, IEEE, November 1984, pp. 7-9.

[Zin82] R. Zinsner et al. : "Technology Independent Symbolic Layout Tools", Proceedings of the IEEE International Conference on Computer-Aided Design, pp. 12-13, Sept. 1982.

[Tho83] D. Thomas, C. Hitchcock, T. Kowalski, P. Nazau, R. Walker, "Automatic Data Path Synthesis," Computer 16(12), pp. 59-70, Dec. 1983.

[Wad85] E.W. Walker: "Electronic CAD-CAM - Is it Revolution or Evolution," Proc. 2nd ACM IEEE DA Conference, pp. 830-834, Las Vegas, Nevada, June 23-26, 1985.

[85] ..., "Timing Recovery via Digital Subscriber ...," Ph.D. Dissertation Mem. No. UCB/ERL No./29, April 1985.

[Van88] Vanderaar, "module generator, environment," Users Manual Report, GENERATE (No.) 46/2.2.1/1/1

[VanV85] S. Van Vleck, et al., "Symbolic Hierarchical Network Generation System," Proc. 22nd Design Automation Conference, pp. 289-293, 1985.

[Wid85] R. Nassar, S.W. Director, "CINNAMON: Coupled Integration and Nodal Analysis of MOS networks," Internal report SRC-CMU Center for Computer-Aided Design, Dept. of Electrical and Computer Engineering, Carnegie-Mellon Univ., Pittsburgh.

[Wea81] N. Weate, "Virtual Grid Symbolic Layout," 18th Design Automation Conference, 1981.

[Kim84] Young Hwan Kim, T.T. Fleckner, Resve A. Salah and A.R. Newton, "Electrical Logic Simulation," Proc. 1984 Int'l Conf. on CAD, IEEE, November 1984, pp. ...

[Zim82] G. Zimmer et al.: "Technology Independent Symbolic Layout Tools," Proceedings of the IEEE International Conference on Computer-Aided Design, pp. 12-13, Sept. 1982.

Index